VARIATIONAL PRINCIPLES AND FREE-BOUNDARY PROBLEMS

VARIATIONAL PRINCIPLES AND FREE-BOUNDARY PROBLEMS

AVNER FRIEDMAN

Department of Mathematics
Northwestern University

A WILEY-INTERSCIENCE PUBLICATION

JOHN WILEY & SONS

New York   **Chichester**   **Brisbane**   **Toronto**   **Singapore**

Library of Congress Cataloging in Publication Data

Friedman, Avner.
  Variational principles and free-boundary problems.

  (Pure and applied mathematics, ISSN 0079-8185)
  "A Wiley-Interscience publication."
  Includes bibliographical references and index.
  1. Boundary-value problems.   2. Variational principles.
I. Title.   II. Series: Pure and applied mathematics
(John Wiley & Sons)

QA379.F74 1982   515.3'5   82-8654
ISBN 0-471-86849-3   AACR2

Printed in the United States of America
10 9 8 7 6 5 4 3 2 1

PREFACE

Important developments in the study of free-boundary problems have been achieved in recent years by introducing variational approach wherever possible. This enables one to conclude without great efforts that a solution to the free-boundary problem exists in some "weak" sense. One can proceed to establish the regularity of the solution and then, hopefully, study the smoothness of the free boundary itself. In fact, within the last five years significant new methods have been developed for analyzing the free boundary and the theory has now reached a certain stage of maturity; its future looks even more exciting. An increasing number of physical and engineering problems are becoming accessible to this growing body of methods.

It is therefore an appropriate time to discuss the main developments in the field in a systematic and self-contained manner. Since some of the main developments were motivated by physical models, we have kept close contact between the general theory and applications to physical examples.

In order to make the book more readable by graduate students and nonspecialists, we have included in the text the detailed statements of the standard theory of elliptic and parabolic operators that is being used. This is done more systematically in the first two chapters.

We have included problems at the end of each section and bibliographical remarks at the end of each chapter.

I would like to thank Luis A. Caffarelli, Hans Wilhelm Alt, and Joel Spruck for several useful conversations, and Ms. Leslie Hubbell for an excellent job of typing the manuscript.

AVNER FRIEDMAN

Evanston, Illinois
June 1982

v

CONTENTS

INTRODUCTION

The Dirichlet problem for the Laplace operator Δ seeks a solution u of $\Delta u = f$ in a given domain Ω satisfying $u = \phi$ on the boundary $\partial\Omega$. Suppose that only a portion S of $\partial\Omega$ is given whereas the remaining portion Γ is not a priori prescribed, and an additional condition is imposed on the unknown part of the boundary, such as $\nabla(u - \phi) = 0$ on Γ (here ϕ is a given function in the entire space). Thus we seek to determine u and Γ satisfying

$$\Delta u = f \text{ in } D, \quad u = \phi \text{ on } \partial D, \quad \nabla(u - \phi) = 0 \text{ on } \Gamma,$$

where D is bounded by the given S and the unknown Γ. This problem is an example of a free-boundary problem.

For a two-dimensional ideal fluid, the density function u satisfies, on the interface Γ between the fluid and the air, the free-boundary conditions

$$u = C_1, \quad |\nabla u| = C_2 \quad (C_1, C_2 \text{ constants}),$$

where either C_1 or C_2 is frequently unknown; Γ is also not prescribed.

Another example of a free-boundary problem occurs when ice and water share a common interface. Here the free-boundary conditions are

$$\theta_i = 0, \quad k_1 \nabla_x \theta_1 \cdot \nabla \Phi_x - k_1 \nabla_x \theta_2 \cdot \nabla_x \Phi = \alpha \Phi_t,$$

where θ_1 is the temperature in the water, θ_2 is the temperature in the ice, α and k_i are positive constants, $\Phi(x, t) = 0$ describes the equation of the free boundary, and θ_i satisfies the parabolic equation

$$\frac{\partial \theta_i}{\partial t} = k_i \Delta \theta_i.$$

There are also free-boundary problems in which the free boundary does not appear explicitly at the outset of the problem. For example, in the case of axisymmetric self-gravitating rotating fluid, the free boundary separating the fluid from the vacuum appears indirectly as the zero set $\{u = 0\}$ of a solution

1

to a nonlinear equation

$$\Delta u + c(u^+)^\beta = f \qquad (c > 0, \beta > 0)$$

(u is a "potential" function depending on the fluid's density ρ and $\{u = 0\} = \partial\{\rho > 0\}$). For gas in a porous medium, the free boundary is the boundary $\partial\{u > 0\}$, where u is the density satisfying the nonlinear degenerate parabolic equation

$$u_t = \Delta u^m \qquad (m > 1).$$

Generally, the initial step in studying a free-boundary problem is to reformulate it in such a way that the free boundary disappears. (There are, however, notable exceptions, especially in the case of one space dimension.) Such reformulations can often be achieved by resorting to variational principles. One of the most celebrated examples is the following: If u minimizes

$$J(v) = \int_\Omega \left[|\nabla v|^2 + 2fv \right]$$

subject to $v = 0$ on $\partial\Omega$, $v \geqslant \phi$ in Ω, then u satisfies (formally, at least)

$$\Delta u = f \qquad \text{in } D \equiv \{u > \phi\},$$

$$u = 0 \qquad \text{on } S \equiv \partial D \cap \partial\Omega,$$

(*)

$$u = \phi \qquad \text{on } \Gamma \equiv \partial D \cap \Omega,$$

$$\nabla(u - \phi) = 0 \qquad \text{on } \Gamma.$$

Thus u is a solution of the free-boundary problem of the type mentioned earlier.

Since the problem of minimizing $J(v)$ has a solution (obtained as a limit of a minimizing sequence), we conclude that there exists a solution u of the free-boundary problem in its variational formulation.

The next two steps after establishing the existence of a solution for the reformulated free-boundary problem are to obtain the best regularity results and then proceed to analyze the free boundary. The latter step often requires much deeper methods.

Thus in the preceding problem (*), the optimal regularity is that u has Lipschitz continuous first derivatives. There are also known sufficient conditions that ensure that the free boundary is smooth, but in general (without such conditions) the free boundary may be quite singular.

In Chapters 1 and 2 we develop the theory of a large class of free-boundary problems called variational inequalities. Chapter 1 deals with the variational approach, existence, uniqueness, and regularity of the minimizer. Chapter 2 is concerned with the study of the free boundary itself.

In Chapter 3 we study a class of variational problems designed for solving problems of jets and cavities of ideal fluids. Whereas in Chapter 1 a typical functional is

$$\int \left[|\, \nabla v\,|^2 + 2 f v\right] \qquad (v \geqslant 0),$$

in Chapter 3 the corresponding functional is

$$\int \left[|\, \nabla v\,|^2 + 2 f I_{\{v>0\}}\right] \qquad (v \geqslant 0),$$

where I_A is the characteristic function of A.

The variational functional in Chapter 4 is of the type

$$-\int\int \frac{\rho(x)\rho(y)}{|x - y|} + \int A(\rho(x)),$$

where ρ is a density function subject to some constraints. Here the free boundary is $\partial\{\rho > 0\}$.

In Chapter 5 we study several free-boundary problems that are not formulated as variational problems; we deal mainly with gas in a porous medium and with the filtration of fluid in a porous dam.

Chapters 3, 4, and 5 are basically independent of each other as far as cross references are concerned. However, they do share some common methods, techniques, and ideas. The material of Chapters 1 and 2 appears in later chapters either directly or indirectly.

There is a large body of literature on time-dependent free-boundary problems in one space dimension. Here the methods are often highly specialized. With a few exceptions, we shall not deal with such problems in the present book.

1

VARIATIONAL INEQUALITIES: EXISTENCE AND REGULARITY

In this chapter we introduce the concept of a variational inequality and establish general existence and uniqueness theorems. Regularity results are proved for some classes of variational inequalities, mainly the obstacle problem, the case of gradient constraint, the biharmonic obstacle problem, and the case of thin obstacles.

We introduce several physical problems to which the existence and regularity theorems are applied, such as (i) the filtration of water in a porous medium, (ii) the elastic–plastic torsion problem, and (iii) the Stefan problem of the melting of a solid.

1. AN EXAMPLE

Let Ω be a bounded domain in R^n. We denote by $W^{m,p}(\Omega)$ the class of functions $u(x)$ in $L^p(\Omega)$ such that all their weak derivatives $D^\alpha u = D_1^{\alpha_1} \cdots D_n^{\alpha_n} u$ of orders $|\alpha| \leq m$ belong to $L^p(\Omega)$; here $\alpha = (\alpha_1, \ldots, \alpha_n)$, $|\alpha| = \alpha_1 + \cdots + \alpha_n$, $D_i = \partial/\partial x_i$, and the weak derivative $D^\alpha u$ is defined by

$$\int_\Omega D^\alpha u \cdot \phi(x)\, dx = (-1)^{|\alpha|} \int u(x) D^\alpha \phi(x)\, dx \qquad \forall \phi \in C_0^\infty(\Omega)$$

[$C_0^\infty(\Omega)$ stands for C^∞ functions with compact support in Ω.]

$W^{m,p}(\Omega)$ is a Banach space with the norm

$$|u|_{m,p} = \left(\sum_{|\alpha| \leq m} \int_\Omega |D^\alpha u|^p\, dx \right)^{1/p}, \qquad |u|_p = |u|_{0,p};$$

4

here $1 \leqslant p < \infty$; for $p = \infty$ we define

$$|u|_{m, \infty} = \sum_{|\alpha| \leqslant m} \|D^\alpha u\|_{L^\infty(\Omega)} \equiv \sum_{|\alpha| \leqslant m} \operatorname{ess\,sup}_\Omega |D^\alpha u|,$$

where $D^\alpha u$ are taken as weak derivatives.

It is well known (see references 94c and 109) that

$$C^\infty(\Omega) \cap W^{m, p}(\Omega) \qquad \text{is dense in } W^{m, p}(\Omega).$$

The closure of $C_0^\infty(\Omega)$ in $W^{m, p}(\Omega)$ is denoted by $W_0^{m, p}(\Omega)$. The notation

$$H^{m, p}(\Omega) = W^{m, p}(\Omega), \quad H^m(\Omega) = W^{m, 2}(\Omega), \quad H_0^m(\Omega) = W_0^{m, 2}(\Omega)$$

is also customary.

Consider the functional

$$(1.1) \qquad\qquad G(u) \equiv \int_\Omega |\nabla u|^2 \, dx - 2 \int_\Omega fu \, dx$$

and the closed convex set in $H^1(\Omega)$,

$$(1.2) \qquad\qquad K = \{u;\, u - g \in H_0^1(\Omega), \quad u \geqslant \phi \text{ a.e.}\},$$

where f is a given function in $L^2(\Omega)$, $\phi(x)$ is a continuous function in $\overline{\Omega}$, and $g \in H^1(\Omega)$. We assume that $g \geqslant \phi$; then K is nonempty.

Consider the problem: Find u such that

$$(1.3) \qquad u - g \in H_0^1(\Omega), \qquad G(u) = \min_{v - g \in H_0^1(\Omega)} G(v).$$

Suppose that u is a solution of this problem. Then

$$G(u + \varepsilon \zeta) \geqslant G(u) \qquad \forall\, \zeta \in C_0^\infty(\Omega), \quad \varepsilon \text{ real,}$$

and we easily deduce that

$$\int_\Omega \nabla u \cdot \nabla \zeta - \int_\Omega f \zeta = 0.$$

Hence, if $u \in H^2(\Omega)$,

$$\int_\Omega (\Delta u + f) \zeta = 0$$

so that

$$(1.4) \qquad\qquad \Delta u + f = 0 \qquad \text{in } \Omega.$$

This equation is satisfied, of course, in the a.e. sense; the condition $u - g \in H_0^1(\Omega)$ is a generalized version of the Dirichlet boundary condition (see references 94c and 109)

$$(1.5) \qquad\qquad u = g \qquad \text{on } \partial\Omega.$$

Thus the solution of the minimization problem (1.3) is also the solution of the Dirichlet problem (1.4), (1.5).

Consider next the variational problem: Find u such that

$$(1.6) \qquad\qquad u \in K, \qquad G(u) = \min_{v \in K} G(v).$$

If u is a solution of this problem, then for any $v \in K$ and $0 < \varepsilon < 1$, $u + \varepsilon(v - u) = (1 - \varepsilon)u + \varepsilon v$ is in K, and therefore

$$G(u + \varepsilon(v - u)) \geqslant G(u).$$

This yields

$$(1.7) \qquad \int \nabla u \cdot \nabla(v - u) \geqslant \int f(v - u) \qquad \forall v \in K; \quad u \in K.$$

If also $u \in H^2(\Omega)$, then we obtain

$$(1.8) \qquad \int (\Delta u + f)(v - u) \geqslant 0 \qquad \forall v \in K; \quad u \in K,$$

and choosing $v = u + \zeta$, $\zeta \geqslant 0$, $\zeta \in C_0^\infty(\Omega)$, we get

$$\Delta u + f \leqslant 0.$$

If u is continuous, then the set

$$(1.9) \qquad\qquad N = \{x \in \Omega; u(x) > \phi(x)\}$$

is open. For any $\zeta \in C_0^\infty(N)$ the function $v = u \pm \varepsilon\zeta$ is in K provided that $|\varepsilon|$ is small enough. We then obtain from (1.8)

$$\Delta u + f = 0 \qquad \text{in } N.$$

Thus we have shown that if u is a solution of (1.6) which belongs to $H^2(\Omega) \cap C(\Omega)$, then

$$(1.10) \qquad \left.\begin{array}{r} \Delta u + f \leqslant 0 \\ u \geqslant \phi \\ (\Delta u + f)(u - \phi) = 0 \end{array}\right\} \qquad \text{a.e. in } \Omega,$$

$$u - g \in H_0^1(\Omega).$$

The problem (1.6) is an example of a *variational inequality*. Any one of the versions (1.7), (1.8), (1.10) is also referred to as a variational inequality.

A more general concept of variational inequality will be given in Section 2, where the functional G will be of more general form and K will be any closed convex set in a Banach space.

The set K is called the *constraint set*, and in the case (1.2) we call ϕ the *obstacle* and (1.6) the *obstacle problem*. In this special case, the set (1.9) is called the *noncoincidence set*, and the set

(1.11) $$\Lambda = \{x \in \Omega; u(x) = \phi(x)\}$$

is called the *coincidence set*; the boundary of the noncoincidence set in Ω,

$$\Gamma = \partial N \cap \Omega,$$

is called the *free boundary*.

It will be proved later that (for suitably smooth f, g, ϕ) the solution u of the obstacle problem is in $C^1(\Omega)$ [in fact, even in $W_{loc}^{2,\infty}(\Omega)$]. Since $u - \phi$ takes its minimum in Ω on the coincidence set, it follows that

(1.12) $$u - \phi = 0, \qquad \nabla(u - \phi) = 0 \text{ on } \Gamma.$$

We may view u as a solution of the Dirichlet problem

$$\Delta u + f = 0 \qquad \text{in } N,$$

(1.13) $$u = g \qquad \text{on } \partial N \cap \partial \Omega,$$

$$u = \phi \qquad \text{on } \partial N \cap \Gamma,$$

with the additional condition

(1.14) $$\nabla u = \nabla \phi \qquad \text{on } \partial N \cap \Gamma$$

compensating for the fact that Γ is not a priori known. This point of view is useful in solving variational inequalities in one space dimension. However, if $n > 1$, then Γ can be quite irregular and we shall therefore not attempt to solve the obstacle problem by the approach of (1.13), (1.14).

Consider the special case $n = 1$, $f = 0$. Then the variational inequality (1.6), (1.2) is to minimize

$$\int_a^b [u'(x)]^2 \, dx$$

subject to $u(a) = u_1$, $u(b) = u_2$, and $u(x) \geqslant \phi(x)$; we shall take $u_1 > \phi(a)$, $u_2 > \phi(b)$.

Suppose that $\phi(x)$ is strictly concave. From (1.13), (1.14) we deduce that the curve $y = u(x)$ [$u(x)$ is the solution] consists of three arcs:

(i)   A line segment l_1 connecting (a, u_1) to a point $(a', \phi(a'))$, tangent to $y = \phi(x)$ at $x = a'$.

(ii) An arc γ: $y = \phi(x)$, $a' < x < b'$.

(iii) A line segment l_2 connecting $(b', \phi(b'))$ to (b, u_2), tangent to $y = \phi(x)$ at $x = b'$.

The free boundary consists of the two points $(a', \phi(a'))$ and $(b', \phi(b'))$. If $\phi''(a') < 0$, then

$$u''(a' - 0) = 0 \neq \phi''(a') = u''(a' + 0);$$

thus $u''(x)$ has a jump discontinuity at $x = a'$.

This example shows that in general $u \notin C^2$; in fact, discontinuities of some second-order derivatives of the solutions of the obstacle problem usually occur across the free boundary.

PROBLEMS

1. Solve the obstacle problem

$$u'' \leqslant 1 - \alpha x \qquad (\alpha \text{ constant}),$$

$$u \geqslant 0,$$

$$(u'' - 1 + \alpha x)u = 0$$

in $0 < x < 1$ under the boundary conditions:
(a) $u(0) = 0$, $u(1) = 1$.
(b) $u(0) = 0$, $u'(1) = 0$.

2. Solve the variational inequality in a ball $B_R = \{|x| < R\}$ in R^n:

$$-\Delta u \geqslant -\mu, \qquad u \geqslant 0, \qquad (-\Delta u + \mu)u = 0,$$

$$u \in H_0^1(B_R),$$

where μ is a real number. (Notice that, for $\mu < 0$, the coincidence set is empty.)

3. Suppose that

$$\left.\begin{array}{r} -\Delta u \geqslant f \\ u \geqslant 0 \\ (\Delta u + f)u = 0 \end{array}\right\} \qquad \text{a.e. in } B_R, \quad B_R \subset R^n$$

and $f \leqslant -\gamma$, $u \leqslant M$, where γ, M are positive constants. Show that if $R^2 > 2nM/\gamma$, then $u(0) = 0$.

[*Hint*: Let $v = u - \gamma |x|^2/2n$. Then $-\Delta v < 0$ in $G = B_R \cap \{u > 0\}$; if $u(0) > 0$, then max v is positive and is attained on ∂B_R.]

4. Consider the problem of minimizing the length $\int_a^b (1 + (u'(x))^2)^{1/2} \, dx$ among all the curves $u = u(x)$ connecting (a, u_1) to (b, u_2) and subject to the constraint $u(x) \geqslant \phi(x)$, where ϕ is strictly concave. Describe the minimizer.

2. GENERAL THEORY OF EXISTENCE AND UNIQUENESS

Let K be a closed strictly convex set in a real Hilbert space H and let y_0 be a point in H. Then there exists a unique point $x_0 \in K$ (called the *projection* of y_0 on K) which is nearest to y_0, that is,

$$\|x_0 - y_0\| \leqslant \|x - y_0\| \qquad \forall x \in K.$$

An equivalent way of writing it is

(*) $$(x_0 - y_0, x - x_0) \geqslant 0 \qquad \forall x \in K.$$

By defining an appropriate H with scalar product

$$(u, v) = \int_\Omega \nabla u \cdot \nabla v$$

one can easily see that problem (*) for x_0 includes (with suitable K) the variational inequalities of Section 1. In this section we shall further extend the notion of problem (*) and obtain an existence and uniqueness for variational inequalities of a very general form, quite sufficient for all the subsequent applications.

Let X be a real reflexive Banach space with dual (conjugate) X', and denote by $\langle , \rangle$ the pairing between X and X'.

A mapping $A : D(A) \to X'$ [with domain $D(A) \subset X$] is called *monotone* if

$$\langle Au - Av, u - v \rangle \geqslant 0 \qquad \forall u, v \text{ in } D(A).$$

When $D(A)$ is convex, A is called *hemicontinuous* if $\forall u, v$ in $D(A)$, the mapping

$$[0, 1] \ni t \to \langle A(tu + (1 - t)v), u - v \rangle$$

is continuous.

For any finite-dimensional subspace M of X, let j denote the injection map $jx = x$ from $M \to X$ and j^* the dual map from $X' \to M'$; that is, if $f \in X'$, then

$j*f$ is the restriction of f to M. If $j*Aj$ is continuous on $M \cap D(A)$ for any such M, then we say that A is continuous on finite-dimensional subspaces of $D(A)$.

Theorem 2.1.   *Suppose that $A : X \to X'$ is monotone and hemicontinuous $[D(A) = X]$. Then for any bounded closed convex subset K of X there exists a $u_0 \in K$ such that*

$$(2.1) \qquad \langle Au_0, v - u_0 \rangle \geqslant 0 \qquad \forall\, v \in K$$

Theorem 2.2.   *Let K be a bounded closed convex subset of X and suppose that $A : D(A) \to X'$ is monotone, $D(A) = K$, and A is continuous on finite-dimensional subspaces of $D(A)$. Then there exists a $u_0 \in K$ such that (2.1) holds.*

Note that in Theorem 2.1 we assume less about the continuity of A and more about the size of the domain of A. The inequality (2.1) is called a *variational inequality*.

The proof of both theorems depends on *Minty's lemma*:

Lemma 2.3.   *Let $A : K \to X'$ be monotone and hemicontinuous. Then u_0 satisfies (2.1) if and only if*

$$(2.2) \qquad \langle Av, v - u_0 \rangle \geqslant 0 \qquad \forall\, v \in K.$$

Here, K is any closed convex set.

Proof.   By monotonicity of A

$$0 \leqslant \langle Av - Au_0, v - u_0 \rangle = \langle Av, v - u_0 \rangle - \langle Au_0, v - u_0 \rangle,$$

so that (2.1) implies (2.2). To prove the converse, note that, for any $w \in K$, $v = tw + (1 - t)u_0 = u_0 + t(w - u_0)$ is in K if $0 < t < 1$. Using (2.2), we get

$$\langle A(u_0 + t(w - u_0)), w - u_0 \rangle \geqslant 0,$$

and taking $t \to 0$, we obtain (2.1) for any $v = w \in K$.

Proof of Theorem 2.2

STEP 1.   Consider first the case $K \subset R^m$, $A : K \to R^m$ and $\langle Au, v \rangle$ replaced by (Au, v), where $(\,,\,)$ denotes the scalar product in R^m. We assume that A is continuous, but we do not require that it is monotone. The variational inequality

$$(2.3) \qquad (Au_0, v - u_0) \geqslant 0 \qquad \forall\, v \in K$$

can be rewritten in the form

$$(u_0, v - u_0) \geqslant (u_0 - Au_0, v - u_0) \qquad \forall v \in K.$$

For any $w \in K$ there exists a unique $u_0 \in K$ such that

$$(u_0, v - u_0) \geqslant (w - Aw, v - u_0) \qquad \forall v \in K,$$

namely, $u_0 = \text{Proj}_K(w - Aw) \ (\equiv Tw)$; that is, u_0 is the nearest element in K to $w - Aw$. The operator $I - A : K \to R^m$ is continuous and $\text{Proj}_K : R^m \to K$ is also continuous, and therefore $T : K \to K$ is continuous. Since K is a bounded closed and convex set in R^m, Brouwer's fixed-point theorem can be applied to conclude that T has a fixed point u_0 in K, that is, $Tu_0 = u_0$; this implies (2.3).

STEP 2.   Now let $K \subset M$, M finite-dimensional Banach space, $A : K \to M'$, A continuous but not necessarily monotone. Then the assertion of the theorem follows from step 1 with slight notational changes. Indeed, introduce a basis $e_1, \ldots, e_m$ in M and a correspondence

$$M \ni u = \sum u_i e_i \leftrightarrow \tilde{u} \in R^m, \qquad \text{where } \tilde{u} = (u_1, \ldots, u_m).$$

We can define $\tilde{A}$ uniquely by

$$(\tilde{A}\tilde{u}, \tilde{v}) = \langle Au, v \rangle;$$

$\tilde{A}$ is continuous. Applying step 1 to $\tilde{A}$, the assertion for A follows.

STEP 3.   For any finite-dimensional subspace $M \subset X$ define $j : M \to X$ by $jx = x$ and let j^* be the dual map from $X' \to M'$. By assumption, the map $j^*Aj : K \cap M \to M'$ is continuous. By step 2 there exists a $y_M \in K \cap M$ solving

$$\langle j^*Ajy_M, z - y_M \rangle \geqslant 0 \qquad \forall z \in K \cap M.$$

Since $\langle j^*Ajy_M, \zeta \rangle = \langle Ay_M, \zeta \rangle \ \forall \ \zeta \in K \cap M$, we get

$$\langle Ay_M, z - y_M \rangle \geqslant 0 \qquad \forall z \in K \cap M,$$

or, by Lemma 2.3,

(2.4) $$\langle Az, z - y_M \rangle \geqslant 0 \qquad \forall z \in K \cap M.$$

For any $v \in K$, set

$$S(v) = \{u \in K; \langle Av, v - u \rangle \geqslant 0\}.$$

Clearly, $S(v)$ is weakly closed subset of K and, since K is bounded and X is reflexive, $S(v)$ is weakly compact. By (2.4), $S(v) \neq \varnothing$. If we show that

$$(2.5) \qquad S(v_1) \cap \cdots \cap S(v_m) \neq \varnothing \qquad \forall\, v_1, \ldots, v_m, \quad 1 \leqslant m < \infty,$$

then it would follow that

$$\bigcap_{v \in K} S(v) \neq \varnothing.$$

But a point u_0 in the last intersection satisfies (2.2), and then also (2.1). Thus it remains to establish (2.5).

Denote by M the linear space spanned by $v_1, \ldots, v_m$. By step 2 there exists a $y_M \in K \cap M$ such that

$$\langle Ay_M, z - y_M \rangle \geqslant 0 \qquad \forall\, z \in K \cap M;$$

hence $\langle Az, z - y_M \rangle \geqslant 0 \; \forall\, z \in K \cap M$ and, in particular,

$$\langle Av_i, v_i - y_M \rangle \geqslant 0 \qquad \text{for } 1 \leqslant i \leqslant m.$$

That means that $y_M \in S(v_i)$ and thus (2.5) holds.

Proof of Theorem 2.1

STEP 1. Let M be any finite-dimensional subspace of X and define j, j^* as before (in step 3). Then

$$(2.6) \qquad \begin{array}{c} j^*Aj \text{ maps bounded sets of } M \\ \text{into bounded sets of } M'. \end{array}$$

Indeed, otherwise there exists a sequence $v_n \in M$, $0 < \|v_n\| \leqslant C$, such that $\|j^*Av_n\| \to \infty$. The monotonicity of A implies that

$$\langle j^*Av_n - j^*Au, v_n - u \rangle \geqslant 0 \qquad \forall\, u \in M.$$

Hence

$$\left\langle y_n - \frac{j^*Au}{\|j^*Av_n\|}, v_n - u \right\rangle \geqslant 0, \qquad \text{where } y_n = \frac{j^*Av_n}{\|j^*Av_n\|}.$$

Since the y_n are elements of M', there exists a subsequence $y_{n'} \to y$, $\|y\| = 1$. We may also suppose that $v_{n'} \to v$. But then $\langle y, v - u \rangle \geqslant 0 \; \forall\, u \in M$, which gives $y = 0$, a contradiction.

STEP 2. If M is as before, then j^*Aj is continuous. Indeed, otherwise there exists a sequence $v_n \in M$, $v_n \to v$, $j^*Av_n \to w$ [we use here (2.6)] and $w \in M'$, such that $w \neq j^*Av$. Since

$$\langle j^*Av_n - j^*Au, v_n - u \rangle \geqslant 0 \qquad \forall u \in M,$$

also $\langle w - j^*Au, v - u \rangle \geqslant 0$ and, by Lemma 2.3,

$$\langle w - j^*Av, v - u \rangle \geqslant 0 \qquad \forall u \in M.$$

Taking $u = v - z$, $z \in M$, we get $w - j^*Av = 0$, a contradiction.

STEP 3. By step 2, A satisfies all the conditions of Theorem 2.2. Thus Theorem 2.1 follows from Theorem 2.2.

Lemma 2.3 implies that the set of all solutions of the variational inequality (2.1) is a closed convex set. We shall now prove uniqueness.
A monotone operator A is said to be *strictly* monotone if

$$\langle Au - Av, u - v \rangle = 0 \qquad \text{implies that } u = v.$$

Theorem 2.4. *If A is strictly monotone, then there exists at most one solution of* (2.1).

Proof. If u_1, u_2 are two solutions, then

$$\langle Au_i, v - u_i \rangle \geqslant 0 \qquad \forall v \in K.$$

Take $v = u_2$ in this relation with $i = 1$ and $v = u_1$ in this relation with $i = 2$. Adding, we obtain

$$\langle Au_1 - Au_2, u_1 - u_2 \rangle \leqslant 0;$$

hence $u_1 = u_2$.
The argument above gives a stability result in case

$$(2.7) \qquad \langle Au - Av, u - v \rangle \geqslant \alpha \|u - v\|^{1+\delta} \qquad (\alpha > 0, \quad \delta > 0).$$

Theorem 2.5. *Let A be as in Theorem 2.1 or 2.2 and let (2.7) hold. If f_1, f_2 are elements of X' and u_i ($i = 1, 2$) is the solution of*

$$(2.8) \qquad \langle Au_i, v - u_i \rangle \geqslant \langle f_i, v - u_i \rangle \qquad \forall v \in K,$$

then

$$(2.9) \qquad \|u_1 - u_2\|_X \leqslant \frac{1}{\alpha} \|f_1 - f_2\|_{X'}.$$

Indeed, substituting $v = u_2$ in (2.8) with $i = 1$ and $v = u_1$ in (2.8) with $i = 2$ and adding, we obtain

$$\langle Au_1 - Au_2, u_1 - u_2 \rangle \leqslant |\langle f_1 - f_2, u_1 - u_2 \rangle|.$$

Using (2.7), the assertion (2.9) follows.

Theorems 2.1 and 2.2 will now be extended to the case where K is unbounded. We shall assume that A is *coercive* in the following sense: There exists a $\phi_0 \in K$ such that

$$(2.10) \qquad \frac{1}{\|v\|} \langle Av - A\phi_0, v - \phi_0 \rangle \to \infty \qquad \text{if } v \in K, \quad \|v\| \to \infty.$$

Theorem 2.6.   *Let K be unbounded closed convex set and let A be as in Theorem 2.1 or 2.2. If A is coercive, then there exists a solution of (2.1).*

Notice that Theorems 2.4 and 2.5 are valid even when K is unbounded.

Proof.   The proof uses an idea of truncation. For any $R > 0$ we introduce the bounded convex set

$$K_R = K \cap \{\|u\| \leqslant R\}$$

and denote by u_R the solution of the variational inequality corresponding to K_R. If

$$(2.11) \qquad\qquad \|u_R\| < R \qquad \text{for some } R > 0,$$

then u_R is a solution of (2.1). Indeed, $\forall\, v \in K$ there exists an $\varepsilon > 0$ sufficiently small so that

$$w = u_R + \varepsilon(v - u_R)$$

belongs to K_R. Hence

$$0 \leqslant \langle Au_R, w - u_R \rangle = \varepsilon \langle Au_R, v - u_R \rangle$$

and (2.1) follows for $u_0 = u_R$ with any $v \in K$. It remains to prove (2.11).

From the condition (2.10) it follows that for any $C > 0$ there is an $R > 0$ such that $R > \|\phi_0\|$ and

$$\langle Av - A\phi_0, v - \phi_0 \rangle \geqslant C\|v - \phi_0\| \qquad \forall\, v \in K, \quad \|v\| \geqslant R.$$

Take $C > \|A\phi_0\|$. Then

$$\langle Av, v - \phi_0 \rangle \geqslant (C - \|A\phi_0\|)\|v - \phi_0\| \geqslant (C - \|A\phi_0\|)(\|v\| - \|\phi_0\|) > 0.$$

Now, if (2.11) is not true, then $\|u_R\| = R$ and upon taking $v = u_R$ in the last inequalities, we get

$$\langle Au_R, u_R - \phi_0 \rangle > 0,$$

a contradiction to the variational inequality satisfied by u_R.

We shall introduce another method to establish existence; it works only in Hilbert spaces, but it has the advantage of simplicity.

Denote the real Hilbert space by V, its dual by V', and the pairing between V and V' by $\langle , \rangle$. Let K be a closed convex set in V and f and element in V'. Let $a(u, v)$ be a bilinear form on $V \times V$ which is bounded, that is,

$$(2.12) \qquad |a(u, v)| \leqslant C\|u\|\|v\|.$$

Assume also that $a(u, v)$ is *coercive*, that is,

$$(2.13) \qquad a(u, u) \geqslant \alpha\|u\|^2 \qquad \forall u \in V \quad (\alpha > 0).$$

Theorem 2.7. *There exists a unique solution u of the variational inequality*

$$(2.14) \qquad u \in K, a(u, v - u) \geqslant \langle f, v - u \rangle \qquad \forall v \in K.$$

Further, the map $f \to u$ is continuous in the sense that

$$(2.15) \qquad \|u_1 - u_2\|_V \leqslant \frac{1}{\alpha}\|f_1 - f_2\|_{V'},$$

where u_i is the solution corresponding to f_i.

This theorem is included in the previous results of this section [if we define A by $\langle Au, v \rangle = a(u, v) - \langle f, v \rangle$; A is a linear operator], but the proof is simpler.

Proof. To prove (2.15) (and uniqueness) take $v = u_2$ in the variational inequality for u_1 and $v = u_1$ in the variational inequality for u_2 and add. We get

$$a(u_1 - u_2, u_1 - u_2) \leqslant \langle f_1 - f_2, u_1 - u_2 \rangle$$

and (2.15) follows.

To prove existence, write $\langle f, v \rangle = (\tilde{f}, v)$, where $(,)$ denotes the scalar product in V and $\tilde{f}$ is some element in V. In case $a(u, v) = (u, v)$, the solution of (2.14) is $u = \text{Proj}_K \tilde{f}$. More generally, if $a(u, v)$ is symmetric, then a scalar product in V is defined by

$$((u, v)) = a(u, v);$$

setting $\langle f, v \rangle = ((\tilde{\tilde{f}}, v))$ the solution u is the projection [in the metric of $((,))$] of $\tilde{f}$ on K.

When $a(u, v)$ is not symmetric, we introduce

$$s(u, v) = \tfrac{1}{2}(a(u, v) + a(v, u)),$$

$$\sigma(u, v) = \tfrac{1}{2}(a(u, v) - a(v, u)),$$

the symmetric and skew-symmetric parts of $a(u, v)$, and set

(2.16) $$a_t(u, v) = s(u, v) + t\sigma(u, v)$$

for $0 \leqslant t \leqslant 1$. For $t = 0$, existence is already established. We now proceed step by step to extend the existence into t-intervals $0 \leqslant t \leqslant t_1$, $t_1 \leqslant t \leqslant t_2$, and so on, until $t = 1$, making use of (2.15) and applying the fixed-point theorem for a contraction mapping.

Suppose that we have already proved existence for all $0 \leqslant t \leqslant t_j$ and set $\tau = t_j$. We rewrite

(2.17) $$a_t(u, v - u) \geqslant \langle f, v - u \rangle$$

in the form

$$a_\tau(u, v - u) \geqslant \langle f, v - u \rangle + (\tau - t)\sigma(u, v - u).$$

Set $F_u(v) = \langle f, v \rangle + (\tau - t)\sigma(u, v)$. This is a bounded linear functional on V and thus can be written as $\langle f_u, v \rangle, f_u \in V'$.

For any $w \in V$, consider the variational inequality

$$a_\tau(z, v - z) \geqslant \langle f_w, v - z \rangle \qquad \forall v \in K; \quad z \in K.$$

It has a unique solution z, which we denote also by Tw, and in view of (2.15),

$$\|Tw_1 - Tw_2\| \leqslant \frac{1}{\alpha} \|f_{w_1} - f_{w_2}\|_{V'} \leqslant C|t - \tau| \|w_1 - w_2\|.$$

Taking $|t - \sigma| \leqslant 1/2C$, we conclude that T is a contraction in V. Hence T has a fixed point u, that is, $Tu = u$, and this implies (2.17).

We conclude this section with a stability result when the convex set K varies. For simplicity we formulate it just in the setting of Theorem 2.7. We shall need the conditions

(2.18) $\qquad \begin{aligned} &K_n \text{ are closed convex sets,} \\ &s - \lim K_n = K = w - \lim K_n. \end{aligned}$

The last condition means that (i) if $x \in K$ then there exist $x_n \in K_n$ with $\|x_n - x\| \to 0$, (ii) the weak limit of any sequence $x_{n'}$ $(x_{n'} \in K_{n'})$ is in K; the condition (ii) is satisfied, for instance, if $K_n \subset K$.

Theorem 2.8. *Let $a(u, v)$ and K be as in Theorem 2.7 and let K_n satisfy (2.18). Let $f_n \in V', f_n \to f$ in V', and denote by u_n the solution of*

$$(2.19) \qquad u_n \in K_n, \quad a(u_n, v - u_n) \geq \langle f_n, v - u_n \rangle \qquad \forall v \in K_n.$$

Then $u_n \to u$ weakly in V.

Proof. If $v \in K_n$,

$$\alpha \|u_n - v\|^2 \leq a(u_n - v, u_n - v) = a(u_n, u_n - v) - a(v, u_n - v)$$

$$\leq \langle f_n, u_n - v \rangle + C\|v\|\|u_n - v\|$$

$$\leq (C\|v\| + \|f_n\|_{V'})\|u_n - v\|.$$

Taking $v = v_n \to v^* \in K$, we deduce that $\|u_n\| \leq C$, with another constant C.

Hence any subsequence of u_n has a weakly convergent subsequence. If we show that $u_n \to w$ weakly implies that w is the unique solution of (2.14), then the assertion of the theorem follows.

Since $u \to a(v, u)$ is continuous,

$$(2.20) \qquad\qquad a(v, u_n) \to a(v, w).$$

By Minty's lemma

$$a(v_n, v_n - u_n) \geq \langle f_n, v_n - u_n \rangle$$

for any $v_n \in K_n$. Take any $v \in K$ and v_n such that $\|v_n - v\| \to 0$. Then

$$|a(v, v - u_n) - a(v_n, v_n - u_n)|$$

$$= |a(v - v_n, v - u_n) + a(v_n, v - v_n)|$$

$$\leq C\|v - v_n\|\|v - u_n\| + C\|v_n\|\|v - v_n\| \leq C\|v - v_n\| \to 0.$$

It follows that

$$a(v, v - u_n) \geq \langle f_n, v_n - u_n \rangle + \varepsilon_n, \qquad \varepsilon_n \to 0.$$

Noting finally that

$$\langle f_n, v_n - u_n \rangle \to \langle f, v - w \rangle$$

and using (2.20), we obtain

$$a(v, v - w) \geqslant \langle f, v - w \rangle \qquad \forall v \in K.$$

Since $w \in K$, w is the (unique) solution of the variational inequality (2.14).

PROBLEMS

1. Suppose that there exists a $\phi_0 \in K$ such that $(1 + \theta)\phi_0 \in K$ for some $\theta > 0$. Show that the condition

    $$\frac{\langle Av, v \rangle}{\|v\|} \to \infty \qquad \text{if } v \in K, \quad \|v\| \to \infty$$

    implies (2.10); A is assumed to be monotone.

2. Let $a(u, v)$ be as in Theorem 2.7. Then there exists a bounded linear operator $A : V \to V'$ such that $a(u, v) = \langle Au, v \rangle \; \forall \; u, v \in V$. For any $f \in V'$, $\langle f, v \rangle = (\wedge f, v) \; \forall \; v \in V$, where $\wedge : V' \to V$, $\| \wedge \| = 1$. Show that if $\|A\| = M$ and $0 < \rho < 2\alpha/M^2$, then there exists a θ, $0 < \theta < 1$, such that

    $$|(u, v) - \rho a(u, v)| \leqslant \theta \|u\| \|v\| \qquad \forall u, v \text{ in } V.$$

    [*Hint*: Write the left-hand side as $(u - \rho \wedge Au, v)$.]

3. Let $F(v)$ be a function from V into $(-\infty, \infty]$. Assume that for any $\lambda > 0$ and for any bounded linear functional L on V, there exists a $\tilde{v}$ in V such that

    (2.21) $(\tilde{v}, \tilde{v}) + \lambda F(\tilde{v}) + L(\tilde{v}) \leqslant (\tilde{v}, v) + \lambda F(v) + L(v) \qquad \forall v \in V.$

    Show that $\tilde{v}$ is unique, and prove that there exists a unique $u \in V$ such that

    (2.22) $a(u, u) + F(u) \leqslant a(u, v) + F(v) \qquad \forall v \in V.$

    [*Hint*: Set $\Psi_u(v) = (u, v) - \rho[a(u, v) + F(v)]$ and prove that for any $u \in V$ there exists a unique $w = Tu$ such that

    $$(w, w) - \Psi_u(w) \leqslant (w, v) - \Psi_u(v) \qquad \forall v \in V,$$

    and that T is a contraction.]

4. Let $F(v)$ be a function from V into $(-\infty, \infty]$, $F \not\equiv +\infty$, F convex and lower semicontinuous in the weak or strong topology (since F is convex, these two are the same). Then there exists a unique $u \in V$ such that

    (2.23) $a(u, v - u) \geqslant F(u) - F(v) \qquad \forall v \in V.$

[*Hint*: To verify the condition of Problem 3 we may replace $\lambda F + L$ by F. Show that (2.21) is equivalent to: $J(v) \equiv \frac{1}{2}\|v\|^2 + F(v)$ has a minimum in V. To prove that a minimum exists, take $F(v_1) < \infty$ and v_0 such that $F(v) - F(v_1) \geq (v_0, v - v_1) \ \forall \ v \in V$. Use this inequality to show that a minimizing sequence has a convergent subsequence.]

5. Show that the result of Problem 4 implies Theorem 2.7.
   [*Hint*: Take $F(v) = -\langle f, v \rangle$ if $v \in K$, $F(v) = \infty$ if $v \notin K$.]

3. $W^{2,\,P}$ REGULARITY FOR THE OBSTACLE PROBLEM

In this section we consider the obstacle problem (introduced in Section 1) for a general second-order elliptic operator. Existence of a solution can be deduced from the general existence theorems of Section 2, but here we are interested mainly in establishing the regularity of the solution. We shall give a new method for proving both existence and regularity at the same time.

We need some general facts from the theory of elliptic equations [94c, 109], which we briefly recall.

Denote by $C^\alpha(\overline{\Omega})$ ($0 < \alpha < 1$, Ω open set in R^n) the space of functions $u(x)$ which are Hölder continuous (exponent α), that is,

$$H_\alpha(u) \equiv \sup_{x,\,y \in \Omega} \frac{|u(x) - u(y)|}{|x - y|^\alpha} < \infty.$$

$C^\alpha(\overline{\Omega})$ is a Banach space with the norm

$$\|u\|_\alpha = \|u\|_0 + H_\alpha(u),$$

where

$$\|u\|_0 = \sup_{x \in \Omega} |u(x)|.$$

Similarly, we say that $u \in C^{m+\alpha}(\overline{\Omega})$ (m positive integer) if

$$\|u\|_{m+\alpha} \equiv \|u\|_m + \sum_{|\beta|=m} H_\alpha(D^\beta u) < \infty,$$

where

$$\|u\|_m = \sum_{|\beta| \leq m} \|D^\beta u\|_0.$$

Consider the operator

$$Au \equiv -\sum_{i,\,j=1}^n a_{ij}(x) \frac{\partial^2 u}{\partial x_i \, \partial x_j} + \sum_{i=1}^n b_i(x) \frac{\partial u}{\partial x_i} + c(x)u.$$

A is said to be *elliptic* in Ω if

$$\sum_{i,\,j=1}^{n} a_{ij}(x)\xi_i\xi_j \geqslant \lambda_x |\xi|^2 \qquad (\lambda_x > 0)$$

for all $x \in \Omega$, $\xi \in R^n$; it is *uniformly elliptic* if $\lambda_x \geqslant \lambda > 0$ for all $x \in \Omega$.

Schauder's Boundary Estimates

Suppose that $\partial\Omega$ is locally in $C^{2+\alpha}$, $f \in C^\alpha(\bar{\Omega})$, $\Phi \in C^{2+\alpha}(\bar{\Omega})$, and

$$\sum \|a_{ij}\|_\alpha + \sum \|b_i\|_\alpha + \|c\|_\alpha \leqslant K,$$

$$\sum a_{ij}(x)\xi_i\xi_j \geqslant \lambda |\xi|^2 \qquad \forall x \in \Omega, \quad \xi \in R^n \quad (\lambda > 0).$$

If $u \in C^2(\bar{\Omega})$ and

$$Au = f \quad \text{in } \Omega,$$

$$u = \Phi \quad \text{in } \partial\Omega,$$

then

$$\|u\|_{2+\alpha} \leqslant C(\|f\|_\alpha + \|u\|_0 + \|\Phi\|_{2+\alpha}),$$

where C is a constant depending only on λ, K, and Ω.

The Schauder interior estimates involve norms $\overline{\|u\|}_{m+\alpha}$ defined as follows:

$$\bar{H}_{j,\alpha}(u) = \sup_{x,\,y \in \Omega} d_{x,y}^{j+\alpha} \frac{|u(x) - u(y)|}{|x-y|^\alpha}, \qquad \bar{H}_\alpha = \bar{H}_{0,\alpha},$$

where $d_x = \operatorname{dist}(x, \partial\Omega)$, $d_{x,y} = \min(d_x, d_y)$,

$$\overline{\|u\|}_\alpha = \|u\|_0 + \bar{H}_\alpha(u),$$

$$\overline{\|u\|}_{m+\alpha} = \sum_{|\beta| \leqslant m} \sup_\Omega |d_x^{|\beta|} D^\beta u(x)| + \sum_{|\beta| = m} \bar{H}_{m,\alpha}(D^\beta u).$$

If $\overline{\|u\|}_{m+\alpha} < \infty$, then we say that u belongs to $\bar{C}^{m+\alpha}(\Omega)$.

Schauder's Interior Estimates

Suppose that Ω is a bounded domain with diameter $\leqslant D$, $f \in \bar{C}^\alpha(\Omega)$, and a_{ij}, b_i, c are measurable functions satisfying

$$\sum \overline{\|a_{ij}\|}_\alpha + \sum \overline{\|b_i\|}_\alpha + \overline{\|c\|}_\alpha \leqslant K,$$

$$\sum a_{ij}(x)\xi_i\xi_j \geqslant \lambda |\xi|^2 \qquad \forall x \in \Omega, \quad \xi \in R^n \quad (\lambda > 0).$$

If $u \in C^2(\Omega) \cap L^\infty(\Omega)$ and

$$Au = f \quad \text{in } \Omega,$$

then

$$\overline{\|u\|}_{2+\alpha} \leqslant C(\overline{\|f\|}_\alpha + \|u\|_0),$$

where C is a constant depending only on λ, K, and D.

If $u \in C^{m+\alpha}(\overline{\Omega}_0)$ for any open set Ω_0 with $\overline{\Omega}_0 \subset \Omega$, then we write: $u \in C^{m+\alpha}(\Omega)$. From the interior Schauder estimates one can deduce that if a_{ij}, b_i, c, and f belong to $C^{m+\alpha}(\Omega)$, then u belongs to $C^{m+2+\alpha}(\Omega)$.

We shall also need the elliptic L^p estimates, where $1 < p < \infty$. Here the assumptions on A, f are weaker:

(3.1) $\qquad |a_{ij}(x) - a_{ij}(y)| \leqslant \omega(|x-y|) \qquad [\omega(t) \to 0 \text{ if } t \to 0],$

(3.2) $\qquad \sum_{i,j=1}^{n} a_{ij}(x)\xi_i\xi_j \geqslant \lambda |\xi|^2 \qquad \forall x \in \Omega, \ \xi \in R^n \ (\lambda > 0),$

(3.3) $\qquad \sum |a_{ij}| + \sum |b_i| + |c| \leqslant K.$

The L^p Estimates

Let $\partial\Omega$ belong to C^2 locally, $f \in L^p(\Omega)$, $\Phi \in W^{2,p}(\Omega)$. If $u \in W^{2,p}(\Omega)$,

$$Au = f \quad \text{in } \Omega,$$

$$u - \Phi \in H_0^1(\Omega),$$

then

$$|u|_{2,p} \leqslant C(|f|_p + |\Phi|_{2,p}),$$

where C is a constant depending only on λ, K, the modulus of continuity ω, and the domain Ω.

Interior L^p elliptic estimates are of the form

$$|u|_{2,p}^G \leqslant C(|f|_p + |u|_p),$$

where G is any compact subdomain of Ω ($|u|_{2,p}^G$ stands for the $W^{2,p}$ norm of u in G) and C is a constant depending only on λ, K, ω, Ω, and G.

We shall sometimes need also local L^p estimates in a subdomain G of Ω for which $\partial G \cap \partial\Omega$ is nonempty. Suppose that G_1 is an open set, $G \subset G_1 \subset \Omega$,

$\partial G \cap \partial \Omega$ is contained in the interior of $\partial G_1 \cap \partial \Omega$, $\partial G \cap \Omega \subset G_1$. If

$$Au = f \quad \text{in } G_1,$$

$$\zeta(u - \Phi) \in H_0^1(G_1)$$

for any $\zeta \in C^\infty(R^n)$, $\zeta = 0$ in a neighborhood of $\partial G_1 \cap \Omega$, then

$$|u|_{2,p}^G \leq C\left(|f|_p^{G_1} + |u|_p^{G_1} + |\Phi|_{2,p}^{G_1}\right).$$

The Strong Maximum Principle

Assume that (3.2), (3.3) hold and that $c(x) \geq 0$. Let u be a function in $H^2(\Omega) \cap C(\Omega)$ satisfying $Au \leq 0$ a.e. in Ω. If u assumes a positive maximum at some point x^0 in Ω, then $u \equiv$ const. in Ω (and then $c = 0$ a.e.).

This result extends also to u, which is not necessarily continuous in Ω; "maximum" of u is replaced by "essential supremum" of u: If ess $\sup_\Omega u$ is positive and coincides with ess $\sup_B u$ for any ball with center x_0 and arbitrarily small radius, then $u = $ const. This implies:

(3.4)
$$\text{If } u \in H^2(\Omega) \cap H_0^1(\Omega), \quad Au \leq 0 \text{ a.e. in } \Omega$$
$$\text{then } u \leq 0 \text{ a.e. in } \Omega.$$

For proof, see references 70 and 43.

The Schauder boundary estimates can be used to solve the Dirichlet problem

(3.5)
$$Au = f \quad \text{in } \Omega,$$
$$u = \Phi \quad \text{on } \partial\Omega,$$

provided that $c \geq 0$, where A, f, Φ, Ω are as in the statement of the estimates. The solution u is in $C^{2+\alpha}(\overline{\Omega})$. Similarly, one can use the interior estimates in order to solve (3.5) with $u \in \overline{C}^{2+\alpha}(\Omega) \cap C(\overline{\Omega})$. Here one requires that A, f are as in the statement of the interior estimates, Φ is continuous on $\partial\Omega$, and for every $x^0 \in \partial\Omega$ there exists a local barrier (see references 74 and 109). A barrier exists whenever there is a ball B such that

$$B \cap \Omega = \varnothing, \quad \partial B \cap \partial\Omega = \{x^0\}$$

(this is called the "outside ball condition").

The L^p estimates can similarly be used to solve the Dirichlet problem (3.5) under conditions on A, f, Φ, Ω as in the statement of these estimates; see references 2 and 109.

Uniqueness for (3.5) follows from the maximum principle stated above.

We shall now recall the main *Sobolev inequalities*. Let

$$\frac{1}{q} = \frac{1}{p} - \frac{1}{n}, \quad \text{where } 1 < p < \infty.$$

If $u \in W^{1,p}(\Omega)$ and $\partial\Omega$ is in C^1, then

$$(3.6) \qquad\qquad |u|_q \leqslant C|u|_{1,p} \qquad \text{if } p < n,$$

$$(3.7) \qquad\qquad \|u\|_\alpha \leqslant C|u|_{1,p} \qquad \text{if } p > n \text{ and } \alpha = 1 - \frac{n}{p},$$

where C is a constant depending only on p, n, and Ω.

From (3.6) it follows that the injection mapping

$$j: W^{1,p}(\Omega) \to L^q(\Omega)$$

is bounded. We also have [94c, 109]

$$(3.8) \quad j: W^{1,p}(\Omega) \to L^r(\Omega) \text{ is compact if } p < n,\, r > 0,\, \frac{1}{r} < \frac{1}{p} - \frac{1}{n}.$$

We denote by $C^{0,1}(\overline{\Omega})$ the space of Lipschitz continuous functions in Ω with norm

$$\|u\|_{0,1} = \|u\|_0 + \sup_{x,\,y \in \Omega} \frac{|u(x) - u(y)|}{|x - y|};$$

this is in fact the space $C^\alpha(\overline{\Omega})$ when we take $\alpha = 1$. Similarly, one defines $C^{m,1}(\overline{\Omega})$ as the space $C^{m+\alpha}(\overline{\Omega})$ with $\alpha = 1$. The space $C^{m,1}(\Omega)$ is defined as $C^{m+\alpha}(\Omega)$ with $\alpha = 1$.

We recall [109] that $u \in C^{m,1}(\overline{\Omega})$ if and only if $u \in W^{m+1,\infty}(\Omega)$. We also recall that if $u \in H^1(\Omega)$, then $u^+ \in H^1(\Omega)$ and a.e.

$$Du^+ = \begin{cases} Du & \text{if } u > 0 \\ 0 & \text{if } u < 0; \end{cases}$$

further, $Du = 0$ a.e. on any set where u is constant.

Let $\phi(x)$ (the obstacle) be a function satisfying

$$(3.9) \qquad\qquad \phi \in C^2(\overline{\Omega})$$

and assume that

$$(3.10) \qquad \begin{array}{l} \text{the coefficients of } A \text{ are in } C^\alpha(\overline{\Omega}), \\ A \text{ is uniformly elliptic in } \Omega,\, c(x) \geqslant 0; \end{array}$$

$$(3.11) \qquad \begin{cases} \partial\Omega \text{ is in } C^{2+\alpha}, \\ f \in C^\alpha(\overline{\Omega}), \qquad g \in C^{2+\alpha}(\overline{\Omega}), \\ \qquad g \geqslant \phi \qquad \text{on } \partial\Omega. \end{cases}$$

Let $\beta_\varepsilon(t)$ $(0 < \varepsilon < 1)$ be C^∞ in t, satisfying

(3.12)
$$\begin{aligned}
&\beta_\varepsilon'(t) \geqslant 0, \\
&\beta_\varepsilon(t) \rightarrow -\infty && \text{if } t < 0, \varepsilon \rightarrow 0, \\
&\beta_\varepsilon(t) \rightarrow 0 && \text{if } t > 0, \varepsilon \rightarrow 0, \\
&\beta_\varepsilon(t) \leqslant C, && \beta_\varepsilon(0) \geqslant -C,
\end{aligned}$$

where C is a constant independent of ε.

Consider the *penalized* problem

(3.13)
$$\begin{aligned}
Au + \beta_\varepsilon(u - \phi) &= f && \text{in } \Omega, \\
u &= g && \text{on } \partial\Omega.
\end{aligned}$$

Lemma 3.1.   *There exists a solution* $u = u_\varepsilon$ *of* (3.13).

Proof.   Set, for any $N > 0$,

$$\beta_{\varepsilon, N}(t) = \max\{\min\{\beta_\varepsilon(t), N\}, -N\}$$

and consider the problem

(3.14)
$$\begin{aligned}
Au + \beta_{\varepsilon, N}(u - \phi) &= f && \text{in } \Omega, \\
u &= g && \text{on } \partial\Omega.
\end{aligned}$$

For each $v \in L^p(\Omega)$ $(1 < p < \infty)$ there exists a unique solution w in $W^{2,p}(\Omega)$ of

(3.15)
$$\begin{aligned}
Aw &= f - \beta_{\varepsilon, N}(v - \phi) && \text{in } \Omega, \\
w - g &\in H_0^1(\Omega),
\end{aligned}$$

and

$$|w|_{2,p} \leqslant R,$$

where R is a constant independent of v. Setting $w = Tv$, we see that T maps the R-ball with center 0 in $L^p(\Omega)$ into itself and T is compact, by (3.8). We can now apply Schauder's fixed-point theorem:

A continuous mapping T from a closed convex set S of a Banach space into a compact subset of S has a fixed point.

It follows that, for some $u = u_{\varepsilon, N}$, $Tu = u$, but that means that u is a solution of (3.14).

Since $u = u_{\varepsilon, N}$ is in $W^{2, P}(\Omega)$, for any $p < \infty$, $\beta_{\varepsilon, N}(u - \phi)$ is Hölder continuous. By a general regularity result for elliptic equations with C^α coefficients, it follows that u is in $C^{2+\alpha}(\Omega)$. We shall now estimate the function

$$\zeta(x) = \beta_{\varepsilon, N}(u - \phi).$$

By definition of β_ε,

$$\zeta(x) \leqslant C, \qquad C \text{ independent of } N, \varepsilon.$$

Consider now the minimum μ of $\zeta(x)$. Suppose that

$$\mu = \zeta(x^0), \qquad \mu \leqslant 0, \quad \mu < \beta_\varepsilon(0).$$

Then $x^0 \notin \partial\Omega$, for otherwise

$$\mu = \zeta(x^0) = \beta_{\varepsilon, N}(g - \phi) \geqslant \beta_{\varepsilon, N}(0) \geqslant \beta_\varepsilon(0).$$

On the other hand, if $x^0 \in \Omega$, then since $\beta_{\varepsilon, N}(t)$ is monotone in t; also, $u - \phi$ takes a minimum at x^0 and the minimum is nonpositive. It follows by the proof of the standard maximum principle that

$$A(u - \phi) \leqslant 0 \qquad \text{at } x^0.$$

But then (3.14) gives

(3.16) $$\zeta(x^0) \geqslant f(x^0) - A\phi(x^0) \geqslant -C.$$

We have thus shown that

(3.17) $|\beta_{\varepsilon, N}(u_{\varepsilon, N} - \phi)| \leqslant C, \qquad C$ independent of ε, N.

We conclude that

$$\|Au_{\varepsilon, N}\|_0 \leqslant C$$

and, by the L^p estimates,

$$|u_{\varepsilon, N}|_{2, p} \leqslant C.$$

Thus if N is large enough, then $u_{\varepsilon, N}$ is a solution of the penalized problem (3.13).

We now take a sequence $\varepsilon = \varepsilon_m \to 0$ such that

$$u_\varepsilon \to u \qquad \text{weakly in } W^{2, P}(\Omega), \quad \forall p < \infty.$$

It follows that

$$u_\varepsilon \to u \qquad \text{uniformly in } \overline{\Omega}.$$

Since $|\beta_\varepsilon(u_\varepsilon - \phi)| \leqslant C$, we deduce that

$$u \geqslant \phi,$$

$$\beta_\varepsilon(u_\varepsilon - \phi) \to 0 \qquad \text{on the set } \{u > \phi\}$$

$$\overline{\lim_{\varepsilon \to 0}} \, \beta_\varepsilon(u_\varepsilon - \phi) \leqslant 0.$$

We conclude that

$$Au = f \qquad \text{a.e. on } \{u > \phi\},$$

$$Au \geqslant f \qquad \text{a.e. in } \Omega.$$

We have thus proved:

Theorem 3.2. *Assume that* (3.9)–(3.11) *hold. Then there exists a solution u of the variational inequality*

$$(3.18) \qquad \left.\begin{array}{r} Au - f \geqslant 0 \\ u \geqslant \phi \\ (Au - f)(u - \phi) = 0 \end{array}\right\} \qquad \text{a.e. in } \Omega,$$

$$u = g \qquad \text{on } \Omega,$$

and $u \in W^{2,p}(\Omega)$ for any $p < \infty$.

Theorem 3.3. *Let u_1, u_2 be solutions in $W^{2,2}(\Omega) \cap C(\overline{\Omega})$ of the variational inequality* (3.18) *corresponding to f_1 and f_2, respectively. If $f_1 \geqslant f_2$, then $u_1 \geqslant u_2$ a.e.*

This comparison theorem implies uniqueness:

Theorem 3.4. *Under the assumptions of Theorem 3.2, the solution of the variational inequality* (3.18) *is unique.*

It follows that the solution u constructed in Theorem 3.2 does not depend on the particular choice of the functions β_ε.

Proof of Theorem 3.3. Suppose the open set G where $u_2 > u_1$ is nonempty. Since $u_2 > u_1 \geqslant 0$ in G,

$$Au_2 = f_2, \qquad Au_1 \geqslant f_1.$$

Consequently, $A(u_2 - u_1) \leqslant 0$ in G. Also, $u_2 - u_1 = 0$ on ∂G. Hence by the maximum principle, $u_2 - u_1 \leqslant 0$ in G, a contradiction.

Theorem 3.2 can be improved by relaxing the conditions on A, f, g, and ϕ. From the point of view of applications, we are interested in particular in weakening the conditions on ϕ. To do this let us recall the concept of mollifiers.

Let $\rho(x)$ be a function in $C^\infty(R^n)$ with support in the unit ball, such that

$$\rho \geqslant 0, \qquad \int \rho \, dx = 1.$$

Set $\rho_\delta(x) = \delta^{-n}\rho(x/\delta)$ for any $\delta > 0$, and consider the mollifier

$$(J_\delta u)(x) = \int_\Omega \rho_\delta(x - y)u(y) \, dy \qquad [u \in L^p(\Omega), \quad 1 < p < \infty].$$

We recall [2, 94c, 109] that $J_\delta u$ is in $C^\infty(R^n)$ and

$$|J_\delta u - u|_{L^p(K)} \to 0 \qquad \text{if } \delta \to 0$$

for any compact subset K of Ω.

A continuous function $\psi(x)$ in an open set $\Omega_0 \subset R^n$ is said to satisfy

$$\frac{\partial^2}{\partial \xi^2}\psi \geqslant 0 \qquad \text{in the sense of distributions, } \mathcal{D}'(\Omega_0),$$

if for any $\zeta \in C_0^\infty(\Omega_0)$, $\zeta \geqslant 0$, there holds

$$\int \psi \frac{\partial^2 \zeta}{\partial \xi^2} \geqslant 0;$$

here $\partial/\partial\xi$ is any directional derivative. Taking $\zeta = \rho_\delta$, we conclude that

$$\frac{\partial^2}{\partial \xi^2}(J_\delta \psi) \geqslant 0 \qquad \text{in the usual sense, in } \Omega,$$

provided that $\overline{\Omega} \subset \Omega_0$ and $\delta < \text{dist}(\Omega, \partial\Omega_0)$.

We now replace the condition (3.9) by

$$(3.19) \qquad \begin{cases} \phi \in C^{0,1}(\Omega_0), \\ \dfrac{\partial^2 \phi}{\partial \xi^2} \geqslant -C \quad \text{in } \mathcal{D}'(\Omega_0), \quad \text{for any direction } \xi, \end{cases}$$

where Ω_0 is a neighborhood of $\overline{\Omega}$.

The last condition means that

$$\frac{\partial^2}{\partial \xi^2}\left(\phi + \tfrac{1}{2}C|x|^2\right) \geqslant 0 \qquad \text{in } \mathcal{D}'(\Omega_0).$$

Setting

(3.20)
$$\phi_\delta = J_\delta\left(\phi + \tfrac{1}{2}C|x|^2\right) - \tfrac{1}{2}C|x|^2,$$

we easily find that

$$|D\phi_\delta| \leqslant C,$$

(3.21)
$$\frac{\partial^2}{\partial \xi^2}\phi_\delta \geqslant -C,$$

$$\phi_\delta(x) \to \phi(x) \qquad \text{uniformly in } \Omega, \text{ as } \delta \to 0,$$

where C is a constant independent of δ.

Theorem 3.5. *Let* (3.10), (3.11), *and* (3.19) *hold. Then the assertion of Theorem 3.2 is valid.*

Proof. We repeat the proof of Theorem 3.2, replacing ϕ by ϕ_δ in the penalized problem and choosing, for instance, $\delta = \varepsilon$. From (3.16) with $\phi = \phi_\delta$ we see that if

(3.22)
$$A\phi_\delta \leqslant C, \qquad C \text{ independent of } \delta,$$

then all the estimates remain valid independently of δ; taking $\delta = \varepsilon \to 0$, we then obtain, as before, a solution of (3.18) in $W^{2,p}(\Omega)$, for any $p < \infty$. Thus it remains to prove (3.22).

Now, for any $x^0 \in \Omega$ we can perform an orthogonal transformation such that at $x = x^0$,

$$\sum a_{ij}(x^0)\frac{\partial^2}{\partial x_i \partial x_j} \qquad \text{becomes } \Delta.$$

Recalling (3.21), the assertion (3.22) follows.

We recall the estimate

$$|\psi|_{L^2(\partial\Omega)} \leqslant C|\psi|_{H^1(\Omega)} \qquad \forall \psi \in C^1(\overline{\Omega})$$

($\partial\Omega$ is in C^1, say). Since $C^1(\overline{\Omega})$ is dense in $H^1(\Omega)$ if $\partial\Omega$ is in C^1 (see references 94c and 109), it follows that any function in $H^1(\Omega)$ has a *trace* on $L^2(\partial\Omega)$: The trace operator is a continuous linear operator from $H^1(\Omega)$ into $L^2(\partial\Omega)$ which coincides with "restriction of ψ to $\partial\Omega$" when $\psi \in C^1(\overline{\Omega})$.

The dual of $H_0^1(\Omega)$ is denoted by $H^{-1}(\Omega)$. Suppose that

(3.23)
$$a_{ij} \in C^{0,1}(\overline{\Omega}), \qquad \text{(3.2) and (3.3) hold,}$$
$$f \in H^{-1}(\Omega), \quad \phi \in H^1(\Omega), \quad g \in H^1(\Omega), \quad \partial\Omega \in C^1$$

and introduce the bilinear form

$$(3.24) \quad a(u,v) = \int_\Omega \left\{ \sum a_{ij} \frac{\partial u}{\partial x_i} \frac{\partial v}{\partial x_j} + \sum \left(b_i + \sum \frac{\partial a_{ij}}{\partial x_j} \right) \frac{\partial u}{\partial x_i} v + cuv \right\}.$$

The variational inequality (3.18) can be transformed into the form

$$(3.25) \qquad a(u, v-u) \geqslant (f, v-u) \qquad \forall v \in K; u \in K,$$

where

$$(3.26) \qquad K = \{ v \in H^1(\Omega); v - g \in H_0^1(\Omega), v \geqslant \phi \text{ a.e.} \}.$$

If $a(u,v)$ is coercive, then Theorem 2.7 gives the existence of a unique solution of (3.25), (3.26). Existence can also be established in case $a(u,v)$ is not coercive, but $c(x) \geqslant 0$; see reference 34.

PROBLEMS

1. Let the assumptions (3.23) hold, and let $a(u,v)$ be coercive and $c \geqslant 0$. Suppose that $f \in L^p(\Omega)$, $\phi \in W^{2,p}(\Omega)$, $\partial\Omega \in C^2$, $g \in W^{2,p}(\Omega)$, $1 < p < \infty$. Prove that the solution u of the variational inequality (3.25), (3.26) belongs to $W^{2,p}(\Omega)$.

   [*Hint*: Multiply the differential equation in (3.14) by $|\beta_{\varepsilon,N}|^{p-2}\beta_{\varepsilon,N}$ and derive

$$\int |\beta_{\varepsilon,N}(u_{\varepsilon,N} - \phi)|^p \leqslant C.]$$

2. Extend Theorem 3.2 to the case where

$$K = \{ v \in H_0^1(\Omega); \phi \leqslant v \leqslant \psi \text{ a.e.} \},$$

   where $\phi, \psi \in C^2(\overline{\Omega})$, $\phi \leqslant v \leqslant \psi$ on $\partial\Omega$. This is a *two-obstacle problem* and the solution u satisfies:
$$Au - f \geqslant 0 \qquad \text{if } u < \psi,$$
$$Au - f \leqslant 0 \qquad \text{if } u > \phi,$$
$$Au - f = 0 \qquad \text{if } \phi < u < \psi.$$

   Extend also theorem 3.5.

[*Hint*: Consider the penalized equation

$$Au + \beta_\varepsilon(u - \phi) + \gamma_\varepsilon(u - \psi) = f,$$

where $\gamma_\varepsilon'(t) \geqslant 0$, $\gamma_\varepsilon(t) \to \infty$ if $t > 0$, $\varepsilon \to 0$, and $\gamma_\varepsilon(t) \to 0$ is $t < 0$, $\varepsilon \to 0$.]

3. Suppose that $a(u, v)$ defined in (3.24) is coercive. Let u_1, u_2 be solutions of (3.25), (3.26) corresponding to f_1, ϕ_1, g_1 and f_2, ϕ_2, g_2, respectively. Show that

$$f_1 \geqslant f_2, \quad \phi_1 \geqslant \phi_2, \quad g_1 \geqslant g_2 \text{ implies that } u_1 \geqslant u_2.$$

[*Hint*: Substitute $v = \max(u_1, u_2)$ into the variational inequality for u_1 and $v = \min(u_1, u_2)$ into the variational inequality for u_2, and conclude that $a(u_1 - u_2, (u_2 - u_1)^+) \geqslant 0$.]

4. Consider the variational inequality

$$\begin{aligned} -u'' + \alpha u &\geqslant f \\ u &\geqslant 0 \qquad \text{a.e. in } R^1, \quad \alpha > 0. \\ (-u'' + \alpha u - f)u &= 0 \end{aligned}$$

Show that if f'' has a finite number N of zeros, then the free boundary consists of at most $N + 1$ points; assume that $u \in W^{2,p}_{\text{loc}}(R^1)$.

5. Let $a(u, v)$ be as in (3.24) and suppose that $a(u, v)$ is coercive in $H_0^1(\Omega)$. We say that u is a local solution of (3.25) with $K = \{v \in H^1(\Omega), v \geqslant \phi \text{ a.e.}\}$ if

$$a(u, \eta(v - u)) \geqslant \int_\Omega f\eta(v - u) \qquad \forall v \in K, \quad \eta \in C_0^\infty(\Omega), \quad \eta \geqslant 0.$$

Prove: If $\phi \in W^{2,p}(\Omega), f \in L^p(\Omega)$, and u is a local solution of (3.25), then $u \in W^{2,p}_{\text{loc}}(\Omega)$.

[*Hint*: For any $\gamma \in C_0^\infty(\Omega)$, $\gamma = 1$ on Ω', $0 \leqslant \gamma \leqslant 1$ in Ω ($\overline{\Omega}' \subset \Omega$), let

$$K_\gamma = \{v \in H_0^1(\Omega), v \geqslant \gamma\phi \text{ a. e. in } \Omega\}.$$

Show that

$$a(\gamma u, v - \gamma u) \geqslant \int_\Omega f\gamma(v - \gamma u)\, dx - \int_\Omega \left[\sum u \frac{\partial}{\partial x_j}\left(a_{ij} \frac{\partial \gamma}{\partial x_i} \right) + 2a_{ij} \frac{\partial u}{\partial x_i} \frac{\partial \gamma}{\partial x_j} \right]$$

$$\times (v - \gamma u)\, dx + \int_\Omega \sum \left(b_i + \sum_j \frac{\partial a_{ij}}{\partial x_i} \right) \frac{\partial \gamma}{\partial x_i} u(v - \gamma u)\, dx \qquad \forall v \in K_\gamma.$$

By Problem 1,

$$|\gamma u|_{2,p} \leqslant C(|u|_{1,p} + |f|_p + |\gamma\phi|_{2,p}).]$$

6. Prove that the solution of the penalized problem is unique.

7. Let Ω be a domain in R^n and let $\Omega_1 = \Omega \cap B_R$, $\Omega_2 = B_R \setminus \Omega$ be nonempty domains, where B_R is a ball of radius R. Suppose that $\partial\Omega \cap B_{R'}$ is in C^1 for some $R' > R$ and set $\Gamma = \partial\Omega \cap B_R$. Suppose that $u_i \in H^1(\Omega_i)$, $u_1 = u_2$ on Γ in the trace sense. Show that

$$u = \begin{cases} u_1 & \text{in } \Omega_1 \\ u_2 & \text{in } \Omega_2 \end{cases}$$

is in $H^1(B_R)$; this result is called the *matching lemma*.
[*Hint*: Take $\Omega = \{x_n > 0\}$ and define $v_2(x', x_n) = u_2(x', -x_n)$, $V = u_1 - v_2$ in Ω_1. There exist $V_m \in C^1(\overline{\Omega}_1)$, $V_m \to V$ in $H^1(\Omega_1)$, $V_m(x',0) \to 0$ in $L^2(\Gamma)$. Consider mollifiers

$$J'_\varepsilon w(x) = \int_{\Omega_1} \rho_\varepsilon(x' - y', x_n - y_n - 2\varepsilon) w(y) \, dy$$

and let

$$\tilde{V}_m = J'_{\varepsilon_m} V_m \qquad \text{in } \Omega_1 \ (\varepsilon_m \to 0),$$

$$Z_m = \begin{cases} v_{1m}(x', x_n) & \text{in } \Omega_1 \\ v_{1m}(x, -x_n) - \tilde{V}_m(x', -x_n) & \text{in } \Omega_2, \end{cases}$$

where $v_{1m} \in C^1(\overline{\Omega}_1)$, $v_{1m} \to u_1$ in $H^1(\Omega_1)$. Show: $Z_m \in H^1(B_R)$, $Z_m \to u$ in $L^2(B_R)$, and (for suitable ε_m) $\{Z_m\}$ is a Cauchy sequence in $H^1(B_R)$.]

4. $W^{2,\infty}$ REGULARITY FOR THE OBSTACLE PROBLEM

We shall assume that

(4.1) $$a_{ij} \text{ belong to } C^{2+\alpha}(\Omega).$$

Theorem 4.1. *Let the assumptions* (3.10), (3.11), (3.19), *and* (4.1) *hold. Then the solution u of the variational inequality* (3.18) *satisfies*

$$u \in W^{2,\infty}_{\text{loc}}(\Omega).$$

Proof. Without loss of generality we may assume that $b_i = c = 0$; otherwise, we replace f by $f - \sum b_i(\partial u/\partial x_i) - cu$, recalling that $u \in W^{2,p}_{\text{loc}}(\Omega)$. We introduce ϕ_δ as in (3.20) and consider the penalized problem:

(4.2)
$$Au + \beta_\varepsilon(u - \phi_\delta) = f \quad \text{in } \Omega,$$

$$u = g \quad \text{on } \partial\Omega.$$

We already know that the solution $u = u_{\varepsilon,\delta}$ satisfies

(4.3)
$$|u|_{2,p} \leqslant C \quad \text{for any } 1 < p < \infty,$$

where C is a constant independent of ε, δ. In the sequel it will be convenient to choose the functions $\beta_\varepsilon(t)$ in such a way that

(4.4)
$$\beta_\varepsilon'(t) > 0, \quad \beta_\varepsilon''(t) \leqslant 0 \quad \text{for all } t.$$

Suppose first that $f = 0$. Note, by elliptic regularity, that $u \in C^{4+\alpha}(\Omega)$. We differentiate the differential equation in (4.2) twice in a direction ξ and get

$$Au_{\xi\xi} + \beta_\varepsilon'(u - \phi_\delta)(u_{\xi\xi} - \phi_{\delta,\xi\xi}) + \beta_\varepsilon''(u - \phi_\delta)(u_\xi - \phi_{\delta,\xi})^2 = I_0,$$

where

$$I_0 = 2\sum \frac{\partial a_{ij}}{\partial\xi} \frac{\partial^3 u}{\partial x_i \partial x_j \partial\xi} + \sum \frac{\partial^2 a_{ij}}{\partial\xi^2} \frac{\partial^2 u}{\partial x_i \partial x_j},$$

Since $\beta_\varepsilon'' \leqslant 0$, we obtain

(4.5)
$$Au_{\xi\xi} + \beta_\varepsilon'(u - \phi_\delta)(u_{\xi\xi} - \phi_{\delta,\xi\xi}) \geqslant I_0.$$

Let Ω_1 be any compact subdomain of Ω and let $\zeta \in C_0^\infty(\Omega)$, $\zeta = 1$ on Ω_1, $\zeta \geqslant 0$. Then

$$A(\zeta u_{\xi\xi}) - \zeta Au_{\xi\xi} = -2\sum a_{ij} \frac{\partial\zeta}{\partial x_i} \frac{\partial^3 u}{\partial x_j \partial\xi^2} - \sum a_{ij} \frac{\partial^2\zeta}{\partial x_i \partial x_j} u_{\xi\xi} \equiv I_1.$$

Substituting $Au_{\xi\xi}$ from (4.5), we get

(4.6)
$$A(\zeta u_{\xi\xi}) + \zeta\beta_\varepsilon'(u - \phi_\delta)(u_{\xi\xi} - \phi_{\delta,\xi\xi}) \geqslant I_1 + \zeta I_0.$$

Each term in $I_1 + \zeta I_0$ of the form $aD^3 u$ can be written in the form $D(aD^2 u) - Da \cdot D^2 u$. Recalling (4.3) and the assumption that the coefficients a_{ij} are in $C^{2+\alpha}$, we see that

(4.7)
$$I_1 + \zeta I_0 = g_0 + \sum_1^n D_i g_i,$$

where

(4.8)
$$\sum_0^n |g_i|_p \leq C, \qquad C \text{ independent of } \varepsilon, \delta.$$

We now need a lemma of Stampacchia [see reference 109, Theorems 8.25, 8.27, and 8.30, and (8.75)].

Lemma 4.2. *Assume that* (3.2), (3.3), *and* $|Da_{ij}| \leq K$ *hold in a bounded domain* Ω *and that* $a(u, v)$ *defined in* (3.24) *is coercive in* $H^1(\Omega)$; *that is,*

$$a(u, u) \geq c_0(|u|_{1,1})^2 \qquad \forall u \in H^1(\Omega) \quad (c_0 > 0).$$

Assume also that for any $x^0 \in \partial\Omega$,

$$\liminf_{R \to 0} R^{-n} \text{meas}\left(B_R(x^0) \setminus \Omega \right) > 0, \qquad B_R(x^0) = \left\{ |x - x^0| < R \right\}.$$

Let $g_1, \ldots, g_n$ *be functions in* $L^p(\Omega)$ *and* g_0 *a function in* $L^{p/2}(\Omega)$ *for some* $p > n$. *Let* ψ *be a continuous function on* $\partial\Omega$. *Then there exists a unique function on* w *in* $C(\overline{\Omega}) \cap H^1(\Omega)$ *such that*

$$Aw = g_0 + \sum_{i=1}^n D_i g_i \quad \text{in } \Omega, \qquad w = \psi \quad \text{on } \partial\Omega;$$

Further,

$$\|w\|_0 \leq C\left(\sum_{i=1}^n |g_i|_p + |g_0|_{p/2} + \max_{\partial\Omega} |\psi| \right),$$

where C *is a constant depending only on* λ, K, Ω.

Without loss of generality we may assume in our case that A is coercive [that is, $a(u, v)$ is coercive]. Indeed, otherwise we simply replace A by a coercive operator $A + k$ ($k > 0$) and add $k\zeta u_{\xi\xi}$ to the right-hand side of (4.6). Thus, by the lemma, there exists a solution w in $C(\overline{\Omega}) \cap H_0^1(\Omega)$ of

$$Aw = I_1 + \zeta I_0, \qquad w \in H_0^1(\Omega),$$

and

(4.9)
$$\|w\|_0 \leq C, \qquad C \text{ independent of } \varepsilon, \delta.$$

Since $I_1 + \zeta I_0$ is in C^α, w is actually in $C^{2+\alpha}(\Omega)$.

The function $V = \zeta u_{\xi\xi} - w$ satisfies

(4.10) $AV + \zeta\beta'_\varepsilon(u - \phi_\delta)(u_{\xi\xi} - \phi_{\delta,\xi\xi}) \geq 0$ in Ω.

We shall now estimate V from below.

Suppose that V takes a negative minimum at some point $x^0 \in \overline{\Omega}$. Since $V = 0$ on $\partial\Omega$, x^0 must belong to Ω. But then

$$AV(x^0) \leq 0,$$

so that by (4.10) and the fact that $\beta'_\varepsilon(t) > 0$,

$$\zeta(x^0)\left(u_{\xi\xi}(x^0) - \phi_{\delta,\xi\xi}(x^0)\right) \geq 0.$$

Hence

$$V(x^0) = \zeta(x^0)u_{\xi\xi}(x^0) - w(x^0) \geq \zeta(x^0)\phi_{\delta,\xi\xi}(x^0) - w(x^0) \geq -C.$$

We have thus proved that $V \geq -C$ and consequently, by (4.9), $\zeta u_{\xi\xi} \geq -C$; that is,

(4.11) $u_{\xi\xi} \geq -C$ in every compact subdomain of Ω.

For any point $y \in \Omega$, we can make an orthogonal transformation of the variable x so that in the new variable

(4.12) $Au = -\Delta u + 2\tilde{b}_i D_i u$ at y.

Since $|\beta_\varepsilon| \leq C$ we deduce from (4.2) that $|Au| \leq C$ and, recalling (4.3), (4.12), we find that

$$|\Delta u| \leq C_0, \qquad C_0 \text{ independent of } \varepsilon, \delta.$$

From this relation and (4.11) it is obvious that $0 \leq u_{\xi\xi} + C \leq C_0 + nC$. Thus $|u_{\xi\xi}| \leq C$, C independent of ε, δ. Since ξ is arbitrary direction, we can express mixed derivatives $u_{x_i x_j}$ at linear combination $u_{\xi\xi} \pm u_{\eta\eta}$ for suitable directions ξ, η and thus obtain $|u_{x_i x_j}| \leq C$ at y. Finally, the estimates $|D^2 u| \leq C$ at y can be taken to be independent of y, if y varies in a compact subdomain of Ω. To complete the proof of the theorem, we recall that $u = u_{\varepsilon,\delta}$ and take $\delta = \varepsilon \to 0$.

We have assumed above that $f \equiv 0$. If $f \not\equiv 0$, then we define u_0 by

$$Au_0 = f \text{ in } \Omega, \qquad u_0 = g \text{ on } \partial\Omega,$$

and work with $\tilde{u} = u - u_0$ and with the obstacle $\phi - u_0$.

We shall next prove $W^{2,\infty}$ in a neighborhood of a boundary point $x^0 \in \partial\Omega$. The proof is local and applies also when $x^0 \in \Omega$ (it is a different proof than that of Theorem 4.1, but it is more complicated).

We assume: There exists a neighborhood N_0 of $x^0 \in \partial\Omega$ such that

$$(4.13) \qquad\qquad a_{ij} \in C^{2+\alpha}(N_0 \cap \overline{\Omega}).$$

Theorem 4.3. *Let the assumptions* (3.10), (3.11), (3.19), *and* (4.13) *be satisfied. Then there exists a neighborhood N of x^0 such that*

$$u \in W^{2,\infty}(N \cap \Omega).$$

Proof. Set

$$\tilde{f} = f - \sum b_i \frac{\partial u}{\partial x_i} - cu,$$

$$A_0 u \equiv -\sum a_{ij} \frac{\partial^2 u}{\partial x_i \partial x_j}.$$

Then the variational inequality for u can be written as a variational inequality with the elliptic operator A_0 and the inhomogeneous term $\tilde{f}$; by Theorem 3.5, $\tilde{f} \in C^{\alpha}(\overline{\Omega})$. Next let

$$A_0 u_0 = \tilde{f} \quad \text{in } \Omega, \qquad u_0 = g \quad \text{on } \partial\Omega,$$

and extend u_0 as a $C^{2+\alpha}$ function into a neighborhood of $\partial\Omega$. Taking $\tilde{u} = u - u_0 + C|x|^2$, $\tilde{\phi} = \phi - u_0 + C|x|^2$, we obtain a variational inequality for $\tilde{u}$ with the obstacle $\tilde{\phi}$ and inhomogeneous term $A_0(C|x|^2)$; notice that $\tilde{\phi}_{\xi\xi} \geqslant 0$ if C is large enough. Thus, without loss of generality, we may assume from the outset that

$$f \in C^{2+\alpha}(N_0 \cap \Omega), \qquad g \in C^{3+\alpha}(N_0 \cap \Omega)$$

$$(4.14) \qquad\qquad b_i \equiv c \equiv 0$$

$$\phi_{\xi\xi} \geqslant 0 \quad \text{in } \mathscr{D}'(\Omega_0), \quad \text{for any direction } \xi,$$

where Ω_0 is a neighborhood of $\overline{\Omega}$.

We shall temporarily assume that

$$x^0 = 0, \qquad \partial\Omega \cap N_0 = \{x_n = 0\} \cap N_0,$$

$$(4.15) \qquad\qquad \Omega \cap N_0 \text{ lies in } \{x_n > 0\},$$

$$\left.\begin{array}{l} a_{in} = 0 \quad \text{if } i < n \\ a_{nn} = 1 \end{array}\right\} \quad \text{on } x_n = 0$$

Set

$$x = (x', x_n), \qquad x' = (x_2, \ldots, x_{n-1}),$$

$$B(r) = \{ |x'| < r \}.$$

$$\Omega(r, \delta_1, \delta_2) = B(r) \times (\delta_1, \delta_2),$$

$$\Omega(r, \delta) = \Omega(r, 0, \delta),$$

$$\partial'\Omega(r, \delta) = \partial\Omega(r, \delta) \setminus B(r) \times \{0\};$$

r and δ are taken sufficiently small. We shall need a lemma similar to Lemma 4.2.

Lemma 4.4. *Let* $v \in C^1(\overline{\Omega(r, \delta)})$, $e(x)$ *measurable*, $0 \leqslant e(x) \leqslant C^*$, *and let*

$$(4.16) \qquad A_0 v + ev = h_0 + \sum_{i=1}^n \frac{\partial h_i}{\partial x_i} \qquad \text{in } \Omega(r, \delta),$$

$$(4.17) \qquad \frac{\partial v}{\partial x_n}\bigg|_{B(r) \times \{0\}} = g, \qquad v\bigg|_{\partial'\Omega(r, \delta)} = \hat{g},$$

where (4.16) is satisfied in an H^1-*sense and* h_i *are* L^p *functions. Set*

$$\gamma = \sum_{i=0}^n |h_i|_{L^p} + |g|_{L^\infty} + |\hat{g}|_{L^\infty},$$

where the norms are taken, say, in $\Omega(2r, \delta)$. *Then*

$$|v|_{L^\infty(\Omega(r, \delta))} \leqslant C,$$

where C *is a constant depending only on* A_0, δ, r, *and* γ.

Proof. Let

$$Av_1 + ev_1 = 0 \qquad \text{in } \Omega(r, \delta),$$

$$(4.18) \qquad \frac{\partial v_1}{\partial x_n} = g \qquad \text{on } B(r) \times \{0\},$$

$$v_1 = \hat{g} \qquad \text{on } \partial'\Omega(r, \delta).$$

One can rewrite this system in the weak formulation

$$(4.19) \qquad a(v_1, \zeta) + \int_{B(r) \times \{0\}} g\zeta = 0 \qquad \begin{array}{l} \forall \, \zeta \in H^1(\Omega(r, \delta)), \\ \zeta = \hat{g} \quad \text{on } \partial'\Omega(r, \delta) \end{array}$$

and establish the existence of a solution by the standard L^2 method. But one can also construct the solution in a different way which will be useful for deriving a uniform estimate on v_1.

Let G_m be mollifiers of v. Since v is in C^1,

$$\hat{g}_m \equiv G_m \rightarrow \hat{g} \qquad \text{uniformly in } \partial'\Omega(r, \delta),$$

$$g_m \equiv \frac{\partial G_m}{\partial x_n} \rightarrow g \qquad \text{uniformly in } B(r) \times \{0\}.$$

If V_m is to be the solution of (4.18) with g_m, $\hat{g}_m$, then $U_m = V_m - G_m$ satisfies

$$AU_m + eU_m = -AG_m - eG_m \equiv F_m \qquad \text{in } \Omega(r, \delta),$$

$$\frac{\partial U_m}{\partial x_n} = 0 \qquad\qquad\qquad\qquad \text{on } B(r) \times \{0\},$$

$$U_m = 0 \qquad\qquad\qquad\qquad \text{on } \partial'\Omega(r, \delta).$$

In order to construct V_m we use the reflection rule

(4.20)
$$\tilde{w}(x', x_n) = \begin{cases} w(x', x_n) & \text{if } x_n \geq 0 \\ w(x', -x_n) & \text{if } x_n < 0 \end{cases}$$

to extend U_m (if existing) into $\tilde{U}_m$, F_m into $\tilde{F}_m$, e into $\tilde{e}$, and the coefficients a_{ij} of A into $\tilde{a}_{ij}$ for all i, j except $i = n$, $j \neq n$ and $i \neq n$, $j = n$; the latter coefficients are extended by

$$a_{ni}(x', x_n) = a_{in}(x', x_n) = -a_{in}(x', -x_n) \qquad \text{if } x_n < 0,$$

and we denote these extended coefficients also by $\tilde{a}_{ij}$, and the extended A by $\tilde{A}$. In view of (4.15), the coefficients $\tilde{a}_{ij}$ are all Lipschitz continuous. It is also clear that

$$\tilde{A}\tilde{U}_m + \tilde{e}\tilde{U}_m = \tilde{F}_m \qquad \text{in } \Omega(r, -\delta, \delta),$$

$$\tilde{U}_m = 0 \qquad \text{on } \partial\Omega(r, -\delta, \delta).$$

This system clearly has a solution in $C^{2+\alpha}$ in $\overline{\Omega(r, -\delta, \delta)} \backslash B(r) \times \{-\delta, \delta\}$, and $V_m = \tilde{U}_m + G_m$ gives a solution of (4.18) with g_m, $\hat{g}_m$.

We next claim that

(4.21)
$$\|U_m\|_{L^\infty(\Omega(r, \delta))} \leq C,$$

where C is independent of m and C^*; recall that $0 < e(x) \leq C^*$. To prove it, let

$$\hat{U} = U_m - \lambda(1 - x_n) \qquad (\lambda > \|g_m\|_{L^\infty}).$$

Then

$$A\hat{U} + e\hat{U} = -e\lambda(1 - x_n) < 0 \qquad \text{in } \Omega(r, \delta),$$

so that $\hat{U}$ cannot take a positive maximum at interior points. It also cannot take a maximum on $x_n = 0$ since

$$\frac{\partial \hat{U}}{\partial x_n} = g_m + \lambda > 0 \qquad \text{on } x_n = 0.$$

It follows that

$$\|\hat{U}\|_{L^\infty} \leq C,$$

where C is independent of m (and C^*). The same holds for U_m. We can now take a subsequence $m \to \infty$ and obtain a solution $v_1 = \lim U_m$ of (4.19). Also,

$$(4.22) \qquad \|v_1\|_{L^\infty(\Omega(r, \delta))} \leq C, \qquad C \text{ independent of } C^*.$$

It remains to estimate $v_2 = v - v_1$. We again reflect and obtain for $\tilde{v}_2$ the equation

$$(4.23) \qquad \tilde{A}\tilde{v}_2 + \tilde{e}v_2 = \tilde{h}_0 + \sum \frac{\partial \tilde{h}_i}{\partial x_i};$$

here $\tilde{h}_j$ for $0 \leq j < n$ are defined by the rule (4.20) but $\tilde{h}_n$ is extended by $\tilde{h}_n(x', -x_n) = -h_n(x', x_n)$. We also have

$$(4.24) \qquad \tilde{v}_2 = 0 \qquad \text{on } \partial\Omega(r, -\delta, \delta).$$

Let v^l be the solution of

$$Av^l = \tilde{h}_l \qquad \text{in } \Omega(2r, -\delta, \delta),$$
$$v^l = 0 \qquad \text{on the boundary.}$$

The existence of v^l follows by the standard existence theory [109]; further, for any $1 < p < \infty$,

$$\frac{\partial}{\partial x_l}\tilde{h}_l - A\left(\frac{\partial v^l}{\partial x_l}\right) = \sum \frac{\partial a_{ij}}{\partial x_l}\frac{\partial^2 v}{\partial x_i \partial x_j} \equiv k_l \in L^p\left(\Omega(\tfrac{3}{2}r, -\delta, \delta)\right)$$

by L^p estimates for v' [which are interior with respect to the corner of the cylinder $\Omega(2r, -\delta, \delta)$, but go up to the boundary of $x_n = \pm\delta$].

Next, let $\hat{v}'$ be the solution of

$$A\hat{v}' = k_l \quad \text{in } \Omega(\tfrac{3}{2}r, -\delta, \delta),$$
$$\hat{v}' = 0 \quad \text{on the boundary.}$$

Then

$$\hat{v}' \in W^{2,p}(\Omega(r, -\delta, \delta)).$$

Finally, let

$$A\hat{v} = \tilde{h}_0 \quad \text{in } \Omega(r, -\delta, \delta),$$
$$\hat{v} = -\sum \left(\frac{\partial v'}{\partial x_l} + \hat{v}' \right) \quad \text{on the boundary.}$$

Then the function

$$V \equiv \sum \left(\frac{\partial v'}{\partial x_l} + \hat{v}' \right) + \hat{v}$$

is a solution of

$$AV = \tilde{h}_0 + \sum_{l=1}^{n} \frac{\partial \tilde{h}_l}{\partial x_l} \quad \text{in } \Omega(r, -\delta, \delta),$$
$$V = 0 \quad \text{on the boundary;}$$

the equation is satisfied in the weak sense.

Take $C \geq \|V\|_{L^\infty}$. Then the function $\tilde{V} = C \pm (V - \tilde{v}_2)$ satisfies

$$A\tilde{V} + \tilde{e}\tilde{V} = (C \pm V)\tilde{e} \geq 0 \quad \text{in } \Omega(r, -\delta, \delta),$$
$$\tilde{V} \geq 0 \quad \text{on the boundary.}$$

We recall [142] that the maximum principle for weak solutions (say in H^1) remains valid if the a_{ij} are in C^2. Thus we may conclude that

$$\tilde{V} \geq 0 \quad \text{in } \Omega(r, -\delta, \delta).$$

It follows that

$$\|\tilde{v}_2\|_{L^\infty} \leq 2C.$$

Combining this with (4.22) and recalling that $v = v_1 + v_2$, the assertion of the lemma follows.

Remark 4.1. Lemma 4.4 extends to the case where the Neumann condition in (4.17) is replaced by

$$(4.25) \qquad v = g \qquad \text{on } B(r) \times \{0\}.$$

It is important to note that this result is not a consequence of Lemma 4.2; the novel point in Lemma 4.4 [and in its version with (4.25)] is that the bound C on $|v|_{L^\infty}$ is independent of the constant C^*; later we shall apply the lemma in cases where we have no a priori bound on C^*.

We recall the penalized problem

$$Au_\varepsilon + \beta_\varepsilon(u_\varepsilon - \phi_\varepsilon) = f \qquad \text{in } \Omega,$$
$$u_\varepsilon = g_\varepsilon \qquad \text{on } \partial\Omega,$$

where $g_\varepsilon = g + C\varepsilon$ is chosen so that $g_\varepsilon \geq \phi_\varepsilon$ on $\partial\Omega$, and the estimate

$$(4.26) \qquad |Au_\varepsilon|_{L^\infty} + |\beta_\varepsilon|_{L^\infty} \leq C, \qquad C \text{ independent of } \varepsilon.$$

Notice that u_ε is in $C^4(N_0 \cap \Omega)$.

Lemma 4.5. *For any small δ, r,*

$$|u_\varepsilon|_{W^{2,\infty}(\Omega(r,\delta))} \leq C,$$

where C is a constant independent of ε.

Proof. Let z^ε be the solution of

$$Az^\varepsilon + \beta_\varepsilon'(u_\varepsilon - \phi_\varepsilon)z^\varepsilon = -\zeta\beta_\varepsilon'(u_\varepsilon - \phi_\varepsilon)A\phi_\varepsilon$$

$$-\zeta\beta_\varepsilon''(u_\varepsilon - \phi_\varepsilon)\sum a_{ij}\frac{\partial(u_\varepsilon - \phi_\varepsilon)}{\partial x_i}\frac{\partial(u_\varepsilon - \phi_\varepsilon)}{\partial x_j}$$

$$(4.27) \qquad\qquad\qquad\qquad\qquad\qquad\qquad\qquad \text{in } \Omega(2r, 2\delta),$$

$$\frac{\partial z^\varepsilon}{\partial x_n} = 0 \qquad \text{on } B(2r) \times \{0\},$$

$$z^\varepsilon = 0 \qquad \text{on } \partial'\Omega(2r, 2\delta),$$

where ζ is a C_0^∞ function in $\Omega(\frac{3}{2}r, -\frac{3}{2}\delta, \frac{3}{2}\delta)$, $\zeta = 1$ in $\Omega(r, -\delta, \delta)$, $0 \leq \zeta \leq 1$, and $\partial \zeta(x', 0)/\partial x_n = 0$. We claim that

(4.28) $$|z^\varepsilon|_{L^\infty(\Omega(2r, 2\delta))} \leqslant C.$$

To prove it, we suppress the index ε and set $\tilde{z} = -Au$. Then

(4.29) $$-\tilde{z} + \beta(u - \phi) = f$$

and

$$\tilde{z} = f \qquad \text{on } x_n = 0,$$

$$\frac{\partial \tilde{z}}{\partial x_n} = -\frac{\partial f}{\partial x_n} \qquad \text{on } x_n = 0 \quad \left[\text{if we take } \beta_\varepsilon(t) = 0 \text{ for } t \geqslant 0\right].$$

Applying $-A$ to (4.29), we get

$$A\tilde{z} + \beta'(u - \phi)\tilde{z} = -Af - \beta'(u - \phi)A\phi$$
$$- \beta''(u - \phi)\sum a_{ij}(u - \phi)_{x_i}(u - \phi)_{x_j}.$$

Multiplying by ζ, we easily compute that

$$A(\zeta\tilde{z}) + \beta'(u - \phi)(\zeta\tilde{z}) = -\zeta Af - \zeta\beta'(u - \phi)A\phi + (A\zeta)\tilde{z}$$
$$-2\sum a_{ij}\zeta_{x_i}\tilde{z}_{x_j} - \zeta\beta''(u - \phi)\sum a_{ij}(u - \phi)_{x_i}(u - \phi)_{x_j}.$$

Comparing with the differential equation for $z = z^\varepsilon$, we find that

$$A(\zeta\tilde{z} - z) + \beta'(u - \phi)(\zeta\tilde{z} - z) = -\zeta Af + (A\zeta)\tilde{z} - 2\sum a_{ij}\zeta_{x_i}\tilde{z}_{xj} \equiv F.$$

Because of (4.26) we can write F in the form

$$h_0 + \sum_{l=1}^{n} \frac{\partial h_l}{\partial x_l}, \qquad \sum |h_j|_{L^p} \leqslant C \quad \text{for any } p < \infty,$$

where C is independent of ε. We can therefore apply Lemma 4.4 [notice that $e(x) = \beta'(u_\varepsilon - \phi_\varepsilon)$ is nonnegative, but we have no a priori bound on e; see Remark 4.1 above]. We conclude that

(4.30) $$|\zeta\tilde{z} - z|_{L^\infty(\Omega(2r, 2\delta))} \leqslant C,$$

where C is independent of ε, and (4.28) thus follows.

We shall now define functions z_{ij}^{ε} for $1 \leqslant i, j \leqslant n$:

(4.31)

$$Az_{ij}^{\varepsilon} + \beta_{\varepsilon}'(u_{\varepsilon} - \phi_{\varepsilon})z_{ij}^{\varepsilon} = \zeta\beta_{\varepsilon}'(u_{\varepsilon} - \phi_{\varepsilon})\phi_{\varepsilon, x_i x_j}$$

$$-\zeta\beta_{\varepsilon}''(u_{\varepsilon} - \phi_{\varepsilon})(u_{\varepsilon} - \phi_{\varepsilon})_{x_i}(u_{\varepsilon} - \phi_{\varepsilon})_{x_j} \quad \text{in } \Omega(2r, 2\delta),$$

$$z_{ij}^{\varepsilon} = 0 \quad \text{on } B(2r) \times \{0\} \text{ if } 1 \leqslant i, j \leqslant n - 1 \text{ or } i = j = n,$$

$$\frac{\partial}{\partial x_n} z_{ij}^{\varepsilon} = 0 \quad \text{on } B(2r) \times \{0\} \text{ if } i = n \text{ or } j = n \quad [(i, j) \neq (n, n)],$$

$$z_{ij}^{\varepsilon} = 0 \quad \text{on } \partial'\Omega(2r, 2\delta).$$

Observe that $u = u_{\varepsilon}$ satisfies

$$u_{\varepsilon, x_i x_j} = g_{x_i x_j} \quad \text{on } B(2r) \times \{0\} \text{ for } 1 \leqslant i, j \leqslant n - 1,$$

$$u_{\varepsilon, x_n x_n} = -f - \sum_{i, j=1}^{n} a_{ij}g_{x_i x_j} \quad \text{on } B(2r) \times \{0\},$$

and

$$\frac{\partial}{\partial x_n} u_{\varepsilon, x_l x_n} = \frac{\partial}{\partial x_l}\left(-f - \sum_{i, j=1}^{n-1} a_{ij}g_{x_i x_j}\right) \quad \text{on } B(2r) \times \{0\} \text{ for } 1 \leqslant l \leqslant n - 1.$$

If we now apply $\partial^2/\partial x_i \partial x_j$ to $Au_{\varepsilon} + \beta_{\varepsilon}(u_{\varepsilon} - \phi) = f$, we obtain an equation similar to that for z_{ij}^{ε} and we can then estimate $\zeta u_{\varepsilon, x_i x_j} - z_{ij}^{\varepsilon}$ in the same way as $\zeta\bar{z} - z^{\varepsilon}$ above. We thus obtain, analogously to (4.30),

(4.32)
$$|\zeta u_{\varepsilon, x_i x_j} - z_{ij}^{\varepsilon}|_{L^{\infty}(\Omega(2r, 2\delta))} \leqslant C,$$

where C is independent of ε. Consequently, if we can show that

(4.33)
$$|z_{ij}^{\varepsilon}|_{L^{\infty}(\Omega(r, \delta))} \leqslant C,$$

then the lemma follows.

Let us consider z_{1n}^{ε}. Choose $C_1 > 0$ such that

(4.34) $\left(C_1 a_{ij}(x) \pm \frac{1}{2}(\delta_{i1}\delta_{jn} + \delta_{j1}\delta_{in})\right)$ is uniformly positive definite.

The function $C_1 z^\varepsilon \pm z^\varepsilon_{1n}$ satisfies

(4.35)

$$A(C_1 z^\varepsilon \pm z^\varepsilon_{1n}) + \beta'_\varepsilon(u_\varepsilon - \phi_\varepsilon)(C_1 z^\varepsilon \pm z^\varepsilon_{1n}) = \zeta\beta'_\varepsilon(u_\varepsilon - \phi_\varepsilon)(-C_1 A\phi_\varepsilon \pm \phi_{\varepsilon, x_1 x_n})$$

$$-\zeta\beta''_\varepsilon(u_\varepsilon - \phi_\varepsilon)\left[C_1 \sum_{i,j=1}^n a_{ij}(u_\varepsilon - \phi_\varepsilon)_{x_i}(u_\varepsilon - \phi_\varepsilon)_{x_j} \pm (u_\varepsilon - \phi_\varepsilon)_{x_1}(u_\varepsilon - \phi_\varepsilon)_{x_n}\right].$$

Since $\phi_{\varepsilon, \xi\xi} \geqslant 0$ for any direction ξ, ϕ_ε is convex and thus $(\phi_{\varepsilon, x_i x_j})$ is positive semidefinite. Recalling (4.34), we then deduce that the first term on the right-hand side of (4.35) is $\geqslant 0$. The second term on the right-hand side is also $\geqslant 0$, by (4.34). Thus

$$A(C_1 z^\varepsilon \pm z^\varepsilon_{1n}) + \beta'_\varepsilon(u_\varepsilon - \phi_\varepsilon)(C_1 z^\varepsilon \pm z^\varepsilon_{1n}) \geqslant 0.$$

Set $W = C_1 z^\varepsilon \pm z^\varepsilon_{1n}$. Then

$$\frac{\partial}{\partial x_n} W = 0 \qquad \text{on } B(2r) \times \{0\},$$

$$W = 0 \qquad \text{on } \partial'\Omega(2r, 2\delta).$$

Applying the maximum principle to W, we get

$$C_1 z^\varepsilon \pm z^\varepsilon_{1n} \geqslant 0 \qquad \text{in } \Omega(2r, 2\delta)$$

and, therefore,

(4.36) $$|z^\varepsilon_{1n}| \leqslant C_1 z^\varepsilon \leqslant C,$$

where (4.28) was used. Similarly, one establishes (4.33) for all the z^ε_{in} and z^ε_{ni}, $i \neq n$.

Finally, since $z^\varepsilon \geqslant 0$ [by (4.36), for instance], we have

$$C_1 z^\varepsilon \pm z^\varepsilon_{ij} \geqslant 0 \qquad \text{on } B(2r) \times \{0\}$$

if $1 \leqslant i, j < n$ or if $i = j = n$. We can therefore proceed as before to establish that

$$|z^\varepsilon_{ij}| \leqslant C_1 z^\varepsilon.$$

This completes the proof of Lemma 4.5.

We can now easily complete the proof of Theorem 4.3. Take for simplicity $x^0 = 0$, $\delta \Omega \cap N_0 = \{x_n = 0\} \cap N_0$, $\Omega \cap N_0 \subset \{x_n > 0\}$, and perform the local diffeomorphism

$$(4.37) \qquad x'_n = \frac{x_n}{\sqrt{a_{nn}}},$$

$$x'_j = x_j - \frac{x_n a_{jn}}{a_{nn}} \qquad \text{if } 1 \leq j \leq n-1.$$

It transforms Au into $A_0 u + A_1 u$, where A_0 is as in (4.16) and $A_1 u$ involves lower-order derivatives of u. The penalized equation can then be written, in $N_0 \cap \Omega$, in the form

$$A_0 u'_\varepsilon + \beta_\varepsilon(u'_\varepsilon - \phi_\varepsilon) = f_1 \qquad [u'_\varepsilon(x') = u_\varepsilon(x)]$$

where $f_1 = f - A_1 u_\varepsilon$. We can reduce this equation into another one with an $f_1 \in C^{2+\alpha}$ (uniformly in ε) and another ϕ, as in the first step in the proof of Theorem 4.3. We now apply Lemma 4.5 and, finally, reformulate the estimate on $u'_\varepsilon(x')$ in terms of $u_\varepsilon(x)$, and take $\varepsilon \to 0$.

PROBLEMS

1. Extend the local $W^{2,p}$ regularity result of Problem 5 of Section 3 to local $W^{2,\infty}$ regularity.

2. Let $\phi_1, \ldots, \phi_m$ be in $C^2(\bar{\Omega})$, Ω a bounded domain. Prove that $\phi = \max\{\phi_1, \ldots, \phi_m\}$ satisfies: $\phi \in C^{0,1}(\bar{\Omega})$ and $\phi_{\xi\xi} \geq -C$ in $\mathcal{D}'(\Omega)$ for any direction ξ (C constant).

3. Set $B_R(x^0) = \{|x - x^0| < R\}$, $B_R = B_R(0)$. Let $v \in C(\bar{B}_R)$, $-\Delta v \geq 0$, in $\mathcal{D}'(B_R)$ (that is, v is superharmonic) with $\mathrm{supp}(-\Delta v) = K$; thus v is harmonic outside K. Suppose that $v \geq 0$ in B_R, $v(0) \leq \lambda$, $\sup_K v \leq \lambda$. Show that for any θ, $0 < \theta < 1$, $v(x) \leq C_\theta \lambda$ in $B_{\theta R}$ (C_θ constant independent of v).

   [*Hint*: Let $\Delta w = 0$ in B_R, $w = v$ on ∂B_R. Then $v - w \geq 0$, $w \geq 0$. By Harnack's inequality,

   $$w(x) \leq \tilde{C}_\theta w(0) \leq \tilde{C}_\theta \lambda \qquad \text{in } B_{\theta R}.$$

   But $v - w \leq v \leq \lambda$ on K and then also in B_R.]

4. Let u be the solution of the variational inequality

   $$u \in K, \qquad \int_\Omega \nabla u \cdot \nabla(v - u) \geq 0 \qquad \forall v \in K,$$

where

$$K = \{v \in H_0^1(\Omega), v \geqslant \phi \text{ in } \Omega\},$$

with $\phi \in C(\overline{\Omega})$, $\phi < 0$ on $\partial\Omega$, Ω a bounded domain in R^n. Suppose that $u(x^0) = \phi(x^0)$ and that there exists a harmonic function h in $B_r(x^0) \subset \Omega$ such that

$$\sup_{B_r(x^0)} |h - \phi| \leqslant \lambda.$$

Then

$$\sup_{B_{r/2}(x^0)} |u - \phi| \leqslant C\lambda, \qquad C \text{ constant.}$$

[*Hint*: $v = u - h + \lambda$ is superharmonic in $B_r(x^0)$ and $v \geqslant 0$, $v(x^0) \leqslant 2\lambda$, $v(x) \leqslant 2\lambda$ on $\text{supp}(-\Delta v)$.]

5. If in Problem 4, $\gamma(r)$ is a modulus of continuity of ϕ and $\gamma_1(r)$ is a modulus of continuity of $\nabla\phi$, both decreasing, then

$$\sup_{B_{r/2}(x^0)} (u - \phi) \leqslant C\gamma(r),$$

$$\sup_{B_{r/2}(x^0)} (u - \phi) \leqslant Cr\gamma_1(r).$$

[*Hint*: Take $h(x) = \phi(x^0)$ or $h(x) = \phi(x^0) + \nabla\phi(x^0) \cdot (x - x^0)$.]

6. Let U be an open bounded set in R^n, $h \in C(\overline{U})$, h harmonic in U. Assume that for any $x^0 \in \partial U$,

$$\sup_{B_\rho(x^0) \cap U} |h(x) - h(x^0)| \leqslant \gamma(\rho) \qquad \forall \rho < \rho_0.$$

Then for any x, x' in U,

$$|h(x) - h(x')| \leqslant \gamma(|x - x'|) \qquad \text{if } |x - x'| < \rho_0.$$

[*Hint*: Apply the maximum principle to $v(y) = h(y + x - x') - h(y)$, where y varies in the domain $U' = \{y \in U, y + x - x' \in U\}$.]

7. Let U, h be as in Problem 6, $g \in C^1(\overline{U})$, and let $\gamma_1(\rho)$ be a modulus of continuity of ∇g, $\gamma_1(\rho) \downarrow$ if $\rho \downarrow$. Show that if for any $x^0 \in \partial U$, $\rho \leqslant \rho_0$ ($\rho_0 > 0$),

$$\sup_{B_\rho(x^0) \cap U} |h(x) - g(x)| \leqslant C_0 \rho \gamma_1(\rho),$$

then $h \in C^1(\overline{U})$ and for any x, x' in U

$$|\nabla h(x) - \nabla h(x')| \leqslant C\gamma_1(2|x - x'|) \qquad (C \text{ constant}).$$

[*Hint*: Let $x \in U$, x^0 the nearest point in ∂U to x, $\rho = |x^0 - x| < \rho_0/2$, $y \in B_{2\rho}(x^0) \cap U$. Then

$$|h(y) - g(x^0) - \nabla g(x^0) \cdot (y - x^0)| \leqslant 2(C_0 + 1)\rho\gamma_1(2\rho).$$

Since $z(y) = h(y) - g(x^0) + \nabla g(x^0) \cdot (y - x^0)$ is harmonic in $B_\rho(x) \subset B_{2\rho}(x^0) \cap U$, $|\nabla z(x)| \leqslant (C/\rho) \sup_{\partial B_\rho(x)} |z| \leqslant C\gamma_1(2|x - x^0|)$. Thus ∇h has continuous extensions to $\overline{U}$ and Problem 6 can be applied.]

8. If u is the solution of the obstacle problem of Problem 4, then:
   (a) If $\phi(x)$ has modulus of continuity $\gamma(\rho)$, then u has modulus of continuity $C_0\gamma(\rho)$.
   (b) If $\nabla_x\phi$ has modulus of continuity $\gamma_1(\rho)$ [$\gamma_1(\rho)\downarrow$ if $\rho\downarrow$], then $\nabla_x u$ has modulus of continuity $C\gamma_1(2\rho)$. The constants C_0, C may depend on the distance δ from $\{\phi > 0\}$ to $\partial\Omega$.

   [*Hint*: Show that in a small Ω-neighborhood of $\partial\Omega$, $u > \phi + \varepsilon_0$ ($\varepsilon_0 > 0$) a.e. Take the standard version of the superharmonic function u defined by

   $$u(x) = \lim_{r \to 0} \frac{1}{|\partial B_r|} \int_{\partial B_r(x)} u \, dS,$$

   where $|\partial B_r| =$ surface area of ∂B_r. Then u is lower semicontinuous and $U = \{u > \phi\}$ is open and contains an Ω-neighborhood of $\partial\Omega$. Apply Problems 6 and 7 with $u = h$.]

9. Let Ω be a bounded domain with $C^{2+\alpha}$ boundary and let $x^0 \in \Omega$. Suppose that h is harmonic in $B_R(x^0) \cap \Omega$ and $h > 0$ in $B_R(x^0) \cap \Omega$, $h = 0$, on $B_R(x^0) \cap \partial\Omega$. Then for any θ, $0 < \theta < 1$,

   $$\sup_{B_{\theta R}(x^0) \cap \Omega} h \leqslant C_\theta h(x^0) \qquad (\text{Harnack's inequality}).$$

   [*Hint*: Express h in terms of Green's function in a domain Ω^* with C^2 boundary, where

   $$B_{\theta_1 R}(x^0) \cap \Omega \subset \Omega^* \subset B_R(x^0) \cap \Omega \qquad (\theta < \theta_1 < 1).]$$

10. (a) Show that if $\partial\Omega \in C^{2+\alpha}$, then the constant C, C_0 in Problem 8 can be taken to be independent of δ; (b) extend the results of Problem 8 to the case where $\phi \leqslant 0$ on $\partial\Omega$.
    [*Hint*: Extend Problem 4, using Problem 9.]

11. The results of Problem 10 extend to variational inequalities with general elliptic operator A, with a_{ij}, b_i, c, f in $C^\alpha(\overline{\Omega})$.

[*Hint*: First approximate ϕ by mollifiers. The corresponding Harnack inequality is given in reference 161e.]

5. THE FILTRATION PROBLEM

A dam made of porous medium (say earth) with parallel vertical walls, situated distance a apart, separates two reservoirs of fluid (say water) at levels $y = H$ and $y = h$. The variable y is the height parameter and the variable x represents the distance from the wall of the higher reservoir. In the stationary case, Darcy's law states that

$$(5.1) \qquad\qquad \mathbf{v} = -k\nabla u$$

where $\mathbf{v}$ is the velocity of the fluid, k is the permeability coefficient, and $u(x, y) = y + p(x, y)$; $y =$ gravity force with normalized units and $p(x, y) =$ inner pressure of the fluid. Notice that we are taking here a two-dimensional dam; this can model a three-dimensional dam whose cross sections with the planes $z = c$ do not vary with c. The function u is called the *piezometric head*.

The law of conservation of mass $\nabla \cdot \mathbf{v} = 0$ then leads to

$$\operatorname{div}(k\nabla u) = 0$$

in the wet part of the dam. We shall assume for simplicity that $k(x, y) \equiv 1$. We then anticipate that the wet part of the dam is given by

$$W = \{(x, y); 0 < x < a, 0 < y < \phi(x)\},$$

where $y = \phi(x)$ is the "free boundary," separating the wet part from the dry part; see Figure 1.1. The function u satisfies:

$$(5.2) \qquad\qquad \Delta u = 0 \qquad \text{in } W,$$

$$(5.3) \qquad\qquad \frac{\partial u}{\partial \nu} = 0 \qquad \text{on } \Gamma: y = \phi(x), \quad 0 < x < a,$$

where ν is the outward normal (that is, the flow at Γ is tangent to Γ),

$$(5.4) \qquad \begin{aligned} u(0, y) &= H \qquad \text{if } 0 < y < H, \\ u(a, y) &= h \qquad \text{if } 0 < y < h, \\ u(a, y) &= y \qquad \text{if } h < y < \phi(x) \end{aligned}$$

[this follows by noting that $p(x, y) = H - y$ inside the higher reservoir,

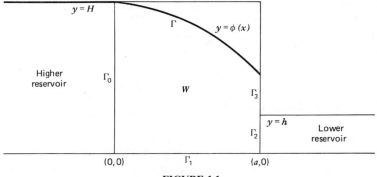

FIGURE 1.1

$p(x, y) = h - y$ inside the lower reservoir, $p(x, y) = 0$, where there is no fluid, and using the fact that $p(x, y)$ is a continuous function]. Also,

$$(5.5) \qquad\qquad u(x, y) = y \qquad \text{on } \Gamma,$$

and finally,

$$(5.6) \qquad\qquad u_y(x, 0) = 0 \qquad (0 < x < a)$$

if the bottom of the dam is impervious.

Suppose that there exists in fact a solution u of (5.2)–(5.6). We claim that the function

$$p(x, y) \equiv u(x, y) - y$$

is then positive in W. Indeed, p is harmonic in W, $p = 0$ on $\Gamma \cup \Gamma_3$, $p > 0$ on $\Gamma_0 \cup \Gamma_2$ and p cannot take negative minimum of Γ_1 since

$$p_y = u_y - 1 = -1 \qquad \text{on } \Gamma_1.$$

Thus, by the maximum principle,

$$(5.7) \qquad\qquad p > 0 \qquad \text{in } W.$$

We shall now transform the problem (5.2)–(5.6) into variational inequality, assuming that $\phi \in C^1$, $u \in C^1(W \cup \Gamma) \cap C(\overline{W})$.

Set

$$w(x, y) = \int_y^{\phi(x)} [u(x, t) - t]\, dt \qquad \text{if } 0 < y \le \phi(x),$$

$$(5.8) \qquad\qquad = 0 \qquad\qquad\qquad\qquad \text{if } \phi(x) < y < H.$$

We compute

(5.9) $w_y(x, y) = -u(x, y) + y,$

$$w_x(x, y) = \int_y^{\phi(x)} u_x(x, t)\, dt \qquad [\text{by (5.5)}],$$

$$w_{xx}(x, y) = \int_y^{\phi(x)} u_{xx}(x, t)\, dt + \phi'(x) u_x(x, \phi(x)).$$

Since

$$-\phi'(x) u_x + u_y = 0 \qquad \text{on } y = \phi(x) \qquad [\text{by (5.3)}]$$

and

$$\int_y^{\phi(x)} u_{xx}(x, t)\, dt = -\int_y^{\phi(x)} u_{yy}(x, t)\, dt = -u_y(x, \phi(y)) + u_y(x, y),$$

we find that

$$w_{xx} = u_y = -w_{yy} + 1 \qquad [\text{by (5.9)}].$$

Thus

(5.10) $\Delta w = 1 \qquad \text{in } W.$

Also, by (5.7),

(5.11) $w > 0 \qquad \text{in } W.$

Let

$$\Omega = \{(x, y); 0 < x < a, 0 < y < H\}.$$

Then clearly

(5.12) $w = 0, \quad \Delta w = 0 < 1 \qquad \text{in } \Omega \setminus \overline{W}.$

One can also check that $\Delta w \leqslant 0$ in the distribution sense, $\mathcal{D}'(\Omega)$. Finally,

(5.13) $w = g \qquad \text{in } \partial\Omega,$

where g is a continuous function on $\partial\Omega$,

$$g(0, y) = \tfrac{1}{2}(H - y)^2,$$

$$g(a, y) = \tfrac{1}{2}(h - y)^2 \qquad \text{if } 0 < y < h,$$

$$g_{xx}(x, 0) = 0 \qquad \text{if } 0 < x < a, \qquad \text{that is,}$$

(5.14) $$g(x, 0) = \frac{H^2}{2}\left(1 - \frac{x}{a}\right) + \frac{h^2}{2}\frac{x}{a},$$

$$g = 0 \qquad \text{elsewhere on } \partial\Omega.$$

Setting

(5.15) $$K = \{v \in H^2(\Omega); v = g \text{ on } \partial\Omega, v \geqslant 0 \text{ in } \Omega\},$$

we find that w is the solution of the variational inequality:

(5.16) $$\int_\Omega \nabla w \cdot \nabla(v - w) \geqslant -\int_\Omega (v - w) \qquad \forall v \in K; \quad w \in K.$$

Conversely, if w is a solution of (5.16) and if the set

$$W \equiv \{(x, y) \in \Omega; w(x, y) > 0\}$$

is given by

$$0 < y < \phi(x), \quad 0 < x < a, \quad \text{and} \quad \phi(x) < H,$$

then the function

$$u(x, y) = y - w_y(x, y)$$

together with ϕ solve the physical problem (5.2)–(5.6) in some generalized sense; if ϕ and w are sufficiently "regular," then u, ϕ solve (5.2)–(5.6) in the usual (classical) sense. We shall now concentrate on the variational inequality (5.16) and establish the existence of a sufficiently regular solution w and free boundary $y = \phi(x)$.

Theorem 5.1. *The variational inequality* (5.16), (5.15) *has a unique solution* w *and* $w \in W^{2, p}(\Omega) \cap W^{2, \infty}_{loc}(\Omega)$.

Proof. The existence of a unique solution w follows from the general theorems of Section 2. In order to prove that $w \in W^{2, p}(\Omega)$, we have to modify the proof of Theorem 3.2 since $\partial\Omega$ is not in $C^{2+\alpha}$; it has four vertices. We claim: There exists a solution of the penalized problem

$$
\begin{aligned}
-\Delta u_\varepsilon + \beta_{\varepsilon, N}(u_\varepsilon) &= -1 \quad && \text{in } \Omega, \\
u_\varepsilon &= g \quad && \text{on } \partial\Omega;
\end{aligned}
$$

(5.17)

u_ε is continuous in $\overline{\Omega}$ (see Problem 2). We can show, as in Section 3, that

$$
|\beta_{\varepsilon, N}| \leqslant C, \qquad |\Delta u_\varepsilon| \leqslant C,
$$

where C is independent of ε, N. By L^p elliptic estimates we deduce that

(5.18)
$$
|u_\varepsilon|_{W^{2, p}(\Omega_0)} \leqslant C,
$$

where $\Omega_0 \subset \Omega$, $\overline{\Omega}_0$ does not contain vertices of $\partial\Omega$. The results of Section 4 can now be applied to deduce that

(5.19)
$$
|u_\varepsilon|_{W^{2, \infty}(\Omega_0)} \leqslant C.
$$

Finally, the $W^{2, p}$ estimates hold also in a neighborhood of each vertex; see Problem 3. Thus

(5.20)
$$
|u_\varepsilon|_{W^{2, p}(\Omega)} \leqslant C.
$$

We can now take $\varepsilon \to 0$ and proceed as in Section 3.

Lemma 5.2. *There holds*:

(5.21)
$$
w_x \leqslant 0, \quad w_y \leqslant 0 \quad \text{in } \Omega.
$$

Proof. Set

$$
W = \{(x, y) \in \Omega; w(x, y) > 0\}.
$$

Since $w \in W^{2, p}(\Omega)$ for any $p < \infty$, the set W is open and w_x is continuous in $\overline{\Omega}$. Notice that $\Delta w_x = 0$ in W. Thus if we can show that w_x in $\overline{W}$ cannot take a positive maximum on ∂W, then it would follow that $w_x \leqslant 0$ in W and thus also in Ω.

Suppose that w_x takes a maximum in $\overline{W}$ at a point x^0. We wish to show that $w_x(x^0) \leqslant 0$.

If $x^0 \in \partial W \cap \Omega$, then $w_x(x^0) = 0$. If $x^0 = (x, 0)$, then $w_x(x^0) = g_x(x^0) < 0$. If $x^0 \in \partial W \cap \{x = 0\}$, then (since $w > 0$ in some Ω-neighborhood of x^0)

$$
(w_x)_x = w_{xx} = 1 - w_{yy} = 1 - g_{yy} = 0 \quad \text{at } x^0
$$

if $x^0 \neq (0, H)$, so that w_x cannot take maximum at this point. The same situation occurs if $x^0 = (a, y)$, $0 < y < h$.

Next, if $x^0 = (a, y)$, $h \leqslant y \leqslant H$, then $w(x^0) = 0$; since, however, $w(x) \geqslant 0$, we get $w_x(x^0) \leqslant 0$. Finally, since $w(x, H) \equiv 0$, also $w_x(x, H) = 0$ and the proof that $w_x \leqslant 0$ is complete.

The proof that $w_y \leqslant 0$ is similar. Here

$$w_y < 0 \qquad\qquad \text{on } \{x = 0\} \cup \{x = a, 0 < y < h\},$$

$$w_y = 0 \qquad\qquad \text{on } \{x = a, h < y < H\},$$

$$w_y \leqslant 0 \qquad\qquad \text{on } \{y = H\} \text{ since } w(x, H) = 0, w \geqslant 0,$$

$$(w_y)_y = 1 - w_{xx} = 1 \qquad \text{on } \partial W \cap \{y = 0\},$$

$$w_y = 0 \qquad\qquad \text{on } \partial W \cap \Omega.$$

From Lemma 5.2 we easily deduce that for any $x \in (0, a)$, there is a number $\phi(x)$ such that

$$(x, y) \in W \qquad \text{if and only if } 0 < y < \phi(x)$$

and $\phi(x)$ is monotone decreasing.

Lemma 5.3. $\phi(x)$ *is continuous.*

Proof. Suppose that $\phi(x_0 + 0) < \phi(x_0 - 0)$ for some $0 < x_0 < a$ and let

$$R = \{(x, y); 0 < x < x_0, \phi(x_0 + 0) < y < \phi(x_0 - 0)\}.$$

The function

$$\tilde{w}(x, y) = w(x, y) - \tfrac{1}{2}(x - x_0)^2$$

is harmonic in R and $\tilde{w} = \tilde{w}_x = 0$ on $x = x_0$. By unique continuation (see Problem 4), $\tilde{w} \equiv 0$ in R, and a contradiction is derived.

Similarly, one proves:

Lemma 5.4. *If $0 < x_1 < x_2 < a$ and $\phi(x_1) < H$, then $\phi(x_2) < \phi(x_1)$.*

Lemma 5.5. $\phi(x) < H$ *if $0 < x < a$.*

Proof. We first show that

$$(5.22) \qquad\qquad\qquad w_y(x, H) = 0.$$

For any $0 < x < x' < a$, we have

$$w_y(x', H) - w_y(x, H)$$

$$= \lim_{\varepsilon \to 0} \frac{1}{\varepsilon} \left[w(x', H) - w(x', H - \varepsilon) - w(x, H) + w(x, H - \varepsilon) \right]$$

$$= \lim_{\varepsilon \to 0} \frac{1}{\varepsilon} \left[w(x, H - \varepsilon) - w(x', H - \varepsilon) \right] \geqslant 0$$

since $w_x \leqslant 0$. Thus $w_y(x, H)$ is monotone increasing in x. Since it vanishes at $x = 0$ and $x = a$, it must vanish identically, and (5.22) follows.

If the assertion of the lemma is not true, then $\phi(x) \equiv H$ for x in some interval $0 < x < x_1$. Since (5.22) holds we can derive a contradiction (by the unique continuation argument) as in the previous two lemmas.

We summarize:

Theorem 5.6. *The free boundary is a curve $y = \phi(x)$, where $\phi(x)$ is continuous and strictly monotone decreasing.*

In Chapter 2 we develop general results on free boundaries for variational inequalities. These results imply for the dam problem that $\phi(x)$ is analytic. We shall also prove that $\phi(x)$ is concave and that $\phi(a - 0) > h$.

PROBLEMS

1. Let $Q_R = \{(x, y); x > 0, y > 0, x^2 + y^2 < R^2\}$, $\Delta u = f$ in Q_R, $u = g$ on $\partial Q_R \setminus \{(0, 0)\}$, where $f \in L^\infty(Q_R)$, g is continuous on ∂Q_R, u continuous in $Q_R \setminus \{(0, 0)\}$, $u \in L^\infty(Q_R)$. Prove that $u(x, y) \to g(0, 0)$ if $(x, y) \to (0, 0)$.
   [*Hint*: Let W be a strong barrier at $(0, 0)$, that is, $\Delta W < -1$ in Q_R, $W \in C(\bar{Q}_R)$, $W > 0$ in $\bar{Q}_R \setminus \{(0, 0)\}$, $W(0, 0) = 0$ (for instance,

   $$W = \gamma \left[\frac{1}{2\alpha^p} - \frac{1}{(x + \alpha)^p + (y + \alpha)^p} \right]$$

   for any $\alpha > 0$ and some $\gamma > 0, p > 0$).

   Show that for any $\varepsilon > 0$, $\eta > 0$, and some $k > 0$,

   $$kW \pm (g(0) - u(0)) + \varepsilon - \eta \log(x^2 + y^2)^{1/2} \geqslant 0 \qquad \text{in } Q_R.]$$

2. Prove that there exists a solution of (5.17).
   [*Hint*: Approximate Ω by smooth Ω_m and g by g_m smooth in $\bar{\Omega}_m$, and denote the corresponding solution of (5.17) by $u_{\varepsilon, m}$. Then $|\beta_{\varepsilon, m}| \leqslant C$,

$|\Delta u_{\varepsilon, m}| \leqslant 1$, $u_{\varepsilon, m} \to u_{\varepsilon}$, and $-\Delta u_{\varepsilon} + \beta_{\varepsilon}(u_{\varepsilon}) = -1$ in Ω. Also, $u_{\varepsilon} = g$ on $\partial\Omega$ except possibly at the vertices.]

3. Let Q_R, u, f, g be as in Problem 1 and suppose that $g \in W^{2, p}(Q_R)$, for some $1 < p < \infty$. Prove that for any $0 < R_0 < R$,

$$|u|_{W^{2, p}(Q_{R_0})} \leqslant C\Big[|g|_{W^{2, p}(Q_R)} + |f|_{L^p(Q_R)}\Big],$$

where C is a constant.

[*Hint*: Suppose that $g = 0$ and define

$$\tilde{u}(x, y) = \begin{cases} u(x, y) & \text{if } y > 0 \\ -u(x, -y) & \text{if } y < 0. \end{cases}$$

Extend $\tilde{f}$ similarly. In the extended domain $\tilde{Q}_R$, $\tilde{u}$ is in H^1 (by the matching lemma, Problem 7 of Section 3). Show that $\Delta\tilde{u} = \tilde{f}$ in $H^1(\tilde{Q}_R)$. The assertion then follows by the L^p estimates in a neighborhood of a smooth portion of the boundary.]

4. Let u be harmonic in a rectangle $R = \{0 < x < a, 0 < y < b\}$, $u \in C^1(\overline{R})$, and $u = u_y = 0$ on $x = 0$. Prove that $u \equiv 0$ in R.

[*Hint*: Extend u by 0 and show that u is harmonic in a neighborhood of $\{x = 0\}$.]

5. Suppose that the permeability coefficient k is not constant, and $k = k(y)$. Replace (5.8) by

$$w(x, y) = \int_y^{\phi(x)} k(t)(u(x, t) - t)\, dt \qquad \text{if } 0 < y < \phi(x)$$

$$= 0 \qquad\qquad\qquad\qquad \text{if } y > \phi(x).$$

Define

$$G(0, y) = \int_y^H k(t)(H - t)\, dt,$$

$$G(a, y) = \int_y^h k(t)(h - t)\, dt \qquad \text{if } y < h,$$

$$G(x, 0) = G(0, 0)\Big(1 - \frac{x}{a}\Big) + G(a, 0)\frac{x}{a} \qquad \text{if } 0 < x < a,$$

$$G = 0 \qquad \text{elsewhere on } \partial\Omega,$$

$$K = \{v \in H^1(\Omega), v = G \text{ on } \partial\Omega, v \geqslant 0\}.$$

Show that w solves the variational inequality

$$\int_\Omega \nabla w \cdot \nabla (v - w) \geq - \int_\Omega (v - w) \qquad \forall v \in K; \quad w \in K.$$

Prove: If $k'(y) \geq 0$, then theorem 5.6 is still valid.

[*Hint*: Show that $w_x \leq 0$ and that $(e^{-h} w)_y \leq 0$, where $e^h = k$.]

6. THE ELASTIC–PLASTIC TORSION PROBLEM: $W^{2,p}$ REGULARITY

We shall consider a variational inequality with gradient constraint: Find u such that

$$u \in K,$$

(6.1)

$$\int_\Omega \nabla u \cdot \nabla (v - u)\, dx \geq \mu \int_\Omega (v - u)\, dx \qquad \forall v \in K,$$

where

(6.2) $$K = \{ v \in H_0^1(\Omega), |\nabla v| \leq 1 \text{ a.e.} \}.$$

Here μ is a given positive number and Ω is a bounded domain in R^n. We first explain how this problem arises in physics.

Let Ω be a bounded simply connected domain in R^2 and let Q be the cylinder

$$\{ (x_1, x_2) \in \Omega, 0 < x_3 < l \}.$$

Q is a rod made up of elastic–perfectly plastic homogeneous material, clamped at the bottom $(x_3 = 0)$ and it is subjected to torsion by an angle α at the top $(x_3 = l)$. It is further assumed that no external forces are acting on the lateral boundary of Q. If α is small enough, then the deformation is small. Assuming that the volume occupied by Q remains unchanged by the torsion, the linear elastic–plastic theory shows that the two nonvanishing components of the stress tensor $\sigma_{x_1 x_3}$ and $\sigma_{x_2 x_3}$ can be written as ∇u, where u (the stress potential) is the solution of (6.1), (6.2) with $\mu = 2\lambda\alpha$, λ the shearing coefficient.

Once $u = u(x_1, x_2)$ is found, the displacement vector (u_1, u_2, u_3) in Q is computed by

$$u_1 = \alpha x_3 x_2, \quad u_2 = \alpha x_3 x_1, \quad u_3 = \alpha \psi(x_1, x_2);$$

where ψ is determined by

$$-x_2 + \frac{\partial \psi}{\partial x_1} = \frac{1}{\mu}(1 + \chi)\frac{\partial u}{\partial x_2},$$

$$x_1 + \frac{\partial \psi}{\partial x_2} = -\frac{1}{\mu}(1 + \chi)\frac{\partial u}{\partial x_1},$$

and $\chi = \chi(x_1, x_2)$ is the solution of

(6.3)
$$\Delta u + \nabla \cdot (\chi \nabla u) = -\mu \qquad \text{in } \Omega,$$

$$\chi = 0 \qquad \text{if } |\nabla u| < 1.$$

Suppose next that Ω is not simply connected and that it has a finite number of "holes," say $\Omega_1, \ldots, \Omega_h$. Thus each Ω_i is a domain with connected boundary $\partial \Omega_i$. Set

$$\Omega^* = \Omega \cup \left(\bigcup_{i=1}^{h} \overline{\Omega}_i \right),$$

and consider the variational inequality

$$u \in K^*,$$

(6.4)
$$\int_\Omega \nabla u \cdot \nabla(v - u)\, dx \geqslant \mu \int_\Omega (v - u)\, dx \qquad \forall\, v \in K^*,$$

where

(6.5) $\qquad K^* = \left\{ v \in H_0^1(\Omega^*), \, |\nabla v| \leqslant 1 \text{ a.e., } \nabla v = 0 \text{ in each } \Omega_i \right\}.$

Here again u has the same physical interpretation as before.

In what follows we shall study (6.1), (6.2) only when Ω is simply connected, and (6.4), (6.5) when Ω has a finite number of holes. Many of the results that we shall obtain for (6.1), (6.2) are valid, however, also if Ω is not simply connected.

We may consider (6.1), (6.2) as a special case of (6.4), (6.5).

By the general results of Section 2, there exists a unique solution of the variational inequality (6.4).

For any sets G_1, G_2, write

$$d(x, G_1) = \text{dist}\,(x, G_1), \qquad d(G_1, G_2) = \text{dist}\,(G_1, G_2).$$

We introduce the functions

(6.6) $$\phi(x) = \max_{0 \leqslant k \leqslant h} \{c_k - d(x, \Omega_k)\} \qquad (x \in R^n),$$

(6.7) $$\psi(x) = \min_{0 \leqslant k \leqslant h} \{c_k + d(x, \Omega_k)\} \qquad (x \in R^n),$$

where $\Omega_0 = R^n \setminus \overline{\Omega}^*$, $c_0 = 0$, and

$$c_k = \text{the restriction of } u \text{ to } \Omega_k,$$

where u is the solution of (6.4). Notice that ϕ, ψ are Lipschitz continuous with coefficient 1. Since $|\nabla u| \leqslant 1$, we easily find that

(6.8) $$|c_k - c_l| \leqslant d(\Omega_k, \Omega_l).$$

Also,

(6.9) $$u = \phi = \psi = c_k \text{ in } \Omega_k, \qquad u = \phi = \psi = 0 \text{ on } \partial\Omega^*.$$

Set

$$J(v) = \tfrac{1}{2} \int_\Omega |\nabla v|^2 - \mu \int_\Omega v.$$

Then the solution u of (6.4) satisfies

$$J(u) = \min_{v \in K^*} J(v).$$

Since $u^+ \in K^*$ and $J(u^+) \leqslant J(u)$ with strict inequality if meas $\{u < 0\} > 0$, it follows that $u \geqslant 0$ in Ω. Thus, in particular, the constants c_k are nonnegative.

Notice also that

(6.10) $$\phi(x) \leqslant u \leqslant \psi(x) \qquad [u \text{ the solution of (6.4)}],$$
(6.11) $$0 \leqslant u \leqslant d(x, \partial\Omega) \qquad [u \text{ the solution of (6.1)}].$$

Thus the solution of (6.1) belongs to

(6.12) $$\tilde{K} = \{v \in H_0^1(\Omega), 0 \leqslant v \leqslant d(x, \partial\Omega) \text{ a.e.}\},$$

and the solution of (6.4) belongs to

(6.13) $$\tilde{K}^* = \{v \in H_0^1(\Omega^*), \phi \leqslant v \leqslant \psi \text{ a.e.}\}.$$

Consider the variational inequalities:

$$u \in \tilde{K},$$

(6.14)

$$\int_\Omega \nabla u \cdot \nabla(v - u) \, dx \geqslant \mu \int_\Omega (v - u) \, dx \qquad \forall v \in \tilde{K},$$

and

$$u \in \tilde{K}^*,$$

(6.15)

$$\int_\Omega \nabla u \cdot \nabla(v - u) \, dx \geqslant \mu \int_\Omega (v - u) \, dx \qquad \forall \, v \in \tilde{K}^*.$$

The first problem is an obstacle problem and the second one is a two-obstacle problem.

Theorem 6.1. (*i*) *The solution of* (6.1) *is also the solution of* (6.14); (*ii*) *the solution of* (6.4) *is also the solution of* (6.15).

Note that in the definition of $\tilde{K}^*$ the constants c_k depend on the solution u of (6.4).

Proof. It suffices to prove (ii). Since $\tilde{K}^* \supset K^*$, it suffices to show that the solution $\tilde{u}$ of (6.15) belongs to K^*.

It will be convenient to work with the auxiliary variational inequality

$$\tilde{u} \in \tilde{K}^*,$$

(6.16)

$$\int_\Omega \nabla \tilde{u} \cdot \nabla(v - \tilde{u}) \, dx + \varepsilon \int_\Omega \tilde{u}(v - \tilde{u}) \, dx \geqslant \mu \int_\Omega (v - \tilde{u}) \, dx \qquad \forall \, v \in \tilde{K}^*,$$

where $\varepsilon > 0$.

Extend $\tilde{u}$ by 0 into $R^n \setminus \Omega^*$. Take any $e \in R^n$ with $\rho = |e|$ small, and consider the functions

$$u^+(x) = \max \{\tilde{u}(x - e) - \rho, \tilde{u}(x)\},$$

$$u^-(x) = \min \{\tilde{u}(x + e) + \rho, \tilde{u}(x)\},$$

and the sets

$$E^+ = \{\tilde{u}(x - e) - \rho > \tilde{u}(x)\},$$

$$E^- = \{\tilde{u}(x + e) + \rho < \tilde{u}(x)\}.$$

Since $\tilde{u}(x + e) + \rho \geqslant \phi(x + e) + \rho \geqslant \phi(x)$ and $\tilde{u}(x - e) - \rho \leqslant \psi(x)$, we have

$$\phi(x) \leqslant u^-(x) \leqslant \tilde{u}(x) \leqslant u^+(x) \leqslant \psi(x) \qquad \text{in } \Omega.$$

Therefore, u^+ and u^- belong to $\tilde{K}^*$. Also,

$$E^+ \subset \Omega, \quad E^- \subset \Omega, \quad E^+ = E^- + e,$$

and a.e.

$$\nabla u^+(x) = \begin{cases} \nabla \tilde{u}(x - e) & \text{in } E^+ \\ \nabla \tilde{u}(x) & \text{in } \Omega \setminus E^+, \end{cases}$$

$$\nabla u^-(x) = \begin{cases} \nabla \tilde{u}(x + e) & \text{in } E^- \\ \nabla \tilde{u}(x) & \text{in } \Omega \setminus E^-. \end{cases}$$

Substituting $v = u^+$ in the variational inequality for $\tilde{u}$, we get

$$\int_{E^+} \nabla \tilde{u}(x) \cdot \nabla(\tilde{u}(x - e) - \tilde{u}(x)) + \varepsilon \int_{E^+} \tilde{u}(x)(\tilde{u}(x - e) - \tilde{u}(x) - \rho)$$

$$\geqslant \mu \int_{E^+} (\tilde{u}(x - e) - \rho - \tilde{u}(x)).$$

Next we substitute $v = u^-$ into the variational inequality for $\tilde{u}$ and, after change of variables $x \to x - e$, obtain

$$\int_{E^+} \nabla \tilde{u}(x - e) \cdot \nabla(\tilde{u}(x) - \tilde{u}(x - e))$$

$$+ \varepsilon \int_{E^+} \tilde{u}(x - e)(\tilde{u}(x) - \tilde{u}(x - e) + \rho)$$

$$\geqslant \mu \int_{E^+} (\tilde{u}(x) + \rho + \tilde{u}(x - e)).$$

Adding the two inequalities, we find that

$$\int_{E^+} |\nabla(\tilde{u}(x) - \tilde{u}(x - e))|^2$$

$$+ \varepsilon \int_{E^+} (\tilde{u}(x) - \tilde{u}(x - e) + \rho)(\tilde{u}(x) - \tilde{u}(x - e)) \geqslant 0.$$

Since

$$\tilde{u}(x) - \tilde{u}(x - e) + \rho < 0 \quad \text{in } E^+,$$

we must have meas $E^+ = 0$. Similarly, meas $E^- = 0$ and thus

(6.17) $$|\tilde{u}(x - e) - \tilde{u}(x)| \leqslant |e|.$$

Recalling that $\tilde{u}$ depends on ε, say $\tilde{u} = \tilde{u}_\varepsilon$, and letting $\varepsilon \to 0$, we obtain the same inequality (6.17) for the solution $\tilde{u}$ of (6.15). Hence $|\nabla \tilde{u}| \leqslant 1$ and $\tilde{u} \in K^*$.

Since the obstacles ϕ, ψ are generally not even in C^1, we cannot deduce further regularity of u by appealing to the results of Sections 3 and 4.

We shall prove regularity by another method, assuming that

(6.18) $\partial \Omega$ is locally Lipschitz and it satisfies the outside ball property.

The last condition means that there is an $R > 0$ such that for any $x^0 \in \partial \Omega$ there is a ball B of radius R satisfying

$$B \cap \Omega = \varnothing, \qquad \bar{B} \cap \partial \Omega = \{x^0\}.$$

Theorem 6.2.   *If* (6.18) *holds, then the solution u of* (6.4) *satisfies*

(6.19) $$\Delta u \in L^p(\Omega) \qquad \text{for all } 1 < p < \infty.$$

Proof.   We construct functions u_ε in K^* which approximate u and which are uniformly regular. More precisely, consider the problem

(6.20) $$-\Delta u_\varepsilon + w_\varepsilon + \left(\frac{|u_\varepsilon - u|}{\varepsilon} \right)^{1/(p-1)} \mathrm{sgn}\,(u_\varepsilon - u) = 0 \qquad \text{in } \Omega,$$

(6.21) $$u_\varepsilon - \phi \in H_0^1(\Omega).$$

We shall prove that for some $w_\varepsilon \in L^p(\Omega)$ there exists a solution u_ε of (6.20), (6.21), and

(6.22) $$|w_\varepsilon|_{L^p(\Omega)} \leqslant C, \qquad C \text{ independent of } \varepsilon.$$

(6.23) $$u_\varepsilon \in K^*.$$

Suppose for the moment that this has been proved. By Minty's lemma

$$\int_\Omega \nabla v \cdot \nabla (v - u) \geqslant \mu \int_\Omega (v - u),$$

and taking $v = u_\varepsilon$, we get

(6.24) $$\int_\Omega \nabla u_\varepsilon \cdot \nabla (u_\varepsilon - u) \geqslant \mu \int_\Omega (u_\varepsilon - u).$$

By (6.22), (6.23), and (6.20), we have that $\Delta u_\varepsilon \in L^p(\Omega)$. Since

$$\int_\Omega \nabla u_\varepsilon \cdot \nabla w = -\int_\Omega \Delta u_\varepsilon \cdot w$$

for any $w \in C_0^\infty(\Omega)$, it follows by approximation that this relation holds for $w \in H_0^1(\Omega)$. Hence we obtain from (6.24)

$$-\int_\Omega \Delta u_\varepsilon \cdot (u_\varepsilon - u) \geqslant \mu(u_\varepsilon - u);$$

substituting $-\Delta u_\varepsilon$ from (6.20), we get

$$-\varepsilon^{-1/(p-1)} \int_\Omega |u_\varepsilon - u|^q - \int_\Omega w_\varepsilon(u_\varepsilon - u) \geqslant \mu \int_\Omega (u_\varepsilon - u),$$

where $q = p/(p - 1)$. Using (6.22), we find that

$$\int_\Omega |u_\varepsilon - u|^q \leqslant C \left(\int_\Omega |u_\varepsilon - u|^q \right)^{1/q} \varepsilon^{1/(p-1)}$$

and thus

(6.25)
$$\left(\int_\Omega |u_\varepsilon - u|^q \right)^{1/q} \leqslant C\varepsilon.$$

We can now further deduce from (6.20) that

(6.26)
$$\int_\Omega |\Delta u_\varepsilon|^p \leqslant C.$$

It follows that $u_\varepsilon \to u$ as $\varepsilon \to 0$ and that

$$\int_\Omega |\Delta u|^p \leqslant C.$$

It remains to construct u_ε, w_ε.

For any $0 < \delta < 1$, let $\beta_\delta(t)$ be a C^1 function satisfying: $\beta_\delta'(t) \geqslant 0$, $\beta_\delta(t) = 1$ if $t > \delta$, $\beta_\delta(t) = -1$ if $t < -\delta$, $|\beta_\delta'(t)| \leqslant C/\delta$, and let γ be any positive number.

We need a comparison lemma.

Lemma 6.3. *Let $\theta(t)$ be a bounded continuous strictly monotone increasing function, $\theta(0) = 0$, and let F_i, ϕ_i ($i = 1, 2$) be $C^0(\overline{\Omega})$ functions, $u_i \in H^1(\Omega) \cap$*

$C(\overline{\Omega})$, Ω *a bounded open set*, $\gamma > 0$, *and*

$$-\Delta u_1 + \gamma\beta_\delta(u_1 - \phi_1) + \theta(u_1 - F_1) \le 0 \qquad \text{in } \Omega,$$

$$-\Delta u_2 + \gamma\beta_\delta(u_2 - \phi_2) + \theta(u_2 - F_2) \ge 0 \qquad \text{in } \Omega.$$

Then

$$(6.27) \quad u_1 - u_2 \le \max\left\{ \sup_{\partial\Omega}(u_1 - u_2),\ \sup_{\Omega}(F_1 - F_2),\ \sup_{\Omega}(\phi_1 - \phi_2) \right\}.$$

Proof. Set $u = u_1 - u_2$, and denote the right-hand side of (6.27) by M. Let $x^0 \in \overline{\Omega}$, $u(x^0) = \max_{\overline{\Omega}} u$. If $u(x^0) > M$, then $x^0 \in \Omega$ and

$$u_1 - \phi_1 > u_2 - \phi_2, \quad u_1 - F_1 > u_2 - F_2 \qquad \text{at } x^0.$$

Therefore,

$$\gamma\beta_\delta(u_1 - \phi_1) \ge \gamma\delta_\delta(u_2 - \phi_2),$$

$$\theta(u_1 - F_1) > \theta(u_2 - F_2)$$

in a neighborhood N of x^0. But then $-\Delta u < 0$ in N, contradicting the maximum principle.

We now take any $\phi_0 \in K^*$, $F \in K^*$, $\phi_0 - \phi \in H_0^1(\Omega)$, $F - \phi \in H_0^1(\Omega)$, and any bounded continuous strictly monotone increasing function $\theta(t)$ with $\theta(0) = 0$. Consider the problem: Find v such that

$$(6.28) \qquad -\Delta v + \gamma\beta_\delta(v - \phi_0) + \theta(v - F) = 0 \qquad \text{in } \Omega,$$

$$(6.29) \qquad\qquad\qquad v - \phi \in H_0^1(\Omega).$$

Notice that the operator A defined by

$$\langle Av, w \rangle = \int_\Omega \left[\nabla v \cdot \nabla w + \gamma\beta_\delta(v - \phi_0)w + \theta(v - F)w \right]$$

is strictly monotone from the convex subset $\phi + H_0^1(\Omega)$ of $H^1(\Omega)$ into the dual of $H^1(\Omega)$. Hence, by Theorems 2.2 and 2.4, there exists a unique solution $v = v_\delta$ of (6.28), (6.29). The function v_δ is in $C^0(\overline{\Omega})$ (see Problem 1).

Let $x^0 \in \partial\Omega_k$ and suppose that the ball $B = \{|x| < R\}$ satisfies: $B \cap \Omega = \varnothing$, $B \cap \partial\Omega_k = \{x^0\}$. For any $w \in K^*$, $w - \phi \in H_0^1(\Omega)$

$$|w(x) - c_k| \le \inf_{y \in \partial\Omega_k} |x - y| \le |x| - R$$

since $w = c_k$ on $\partial\Omega_k$ and $|\nabla w| \leqslant 1$. Setting

$$\delta^+(x) = |x| - R + c_k + \delta,$$

we then have $\delta^+ - w \geqslant \delta$ and, in particular,

$$\theta(\delta^+ - F) \geqslant \theta(\delta) \geqslant 0,$$

$$\gamma\beta_\delta(\delta^+ - \phi_0) \geqslant \gamma\beta_\delta(\delta) = \gamma.$$

Since also

$$-\Delta\delta^+ \geqslant -C \qquad [C \text{ depends on } R \text{ and } \operatorname{diam}(\Omega)],$$

we obtain, after choosing $\gamma \geqslant C$,

$$-\Delta\delta^+ + \gamma\beta_\delta(\delta^+ - \phi_0) + \theta(\delta^+ - F) \geqslant 0.$$

We can now apply Lemma 6.3 to deduce that $v_\delta \leqslant \delta^+$ in Ω. Therefore, if $x \in \Omega$, $x^0 \in \partial\Omega_k$,

$$v_\delta(x) - v_\delta(x^0) \leqslant |x| - |x^0| + \delta \leqslant |x - x^0| + \delta.$$

Similarly [with $\delta^-(x) = -|x| + R + c_k - \delta$],

$$v_\delta(x) - v_\delta(x^0) \geqslant -|x - x^0| - \delta.$$

Thus

(6.30) $\qquad |v_\delta(x) - v_\delta(x^0)| \leqslant |x - x^0| + \delta \qquad (x \in \Omega, \quad x^0 \in \partial\Omega).$

For any point e in R^n with $|e| = \rho$ small, introduce the open set

$$\Omega' = \{x \in \Omega, x + e \in \Omega\}.$$

The function $v_\delta'(x) = v_\delta(x + e)$ satisfies

$$-\Delta v_\delta' + \gamma\beta_\delta(v_\delta' - \phi_0') + \theta(v' - F') = 0 \qquad \text{in } \Omega',$$

where $\phi_0'(x) = \phi_0(x + e)$, $F'(x) = F(x + e)$. We can compare v_δ with v_δ' in Ω', using Lemma 6.3. Recalling (6.30), we have

$$\sup_{\partial\Omega'} |v_\delta - v_\delta'| \leqslant \rho + \delta.$$

Since also $|\nabla\phi_0| \leqslant 1$, $|\nabla F| \leqslant 1$, we obtain, by Lemma 6.3,

(6.31) $\qquad\qquad\qquad \sup_{\Omega'} |v_\delta - v_\delta'| \leqslant \rho + \delta.$

Now take

$$\tilde{\theta}_\varepsilon(t) = \left(\frac{|t|}{\varepsilon}\right)^{1/(p-1)} \operatorname{sgn} t,$$

$$\theta(t) = \begin{cases} \tilde{\theta}_\varepsilon(t) & \text{if } |t| < C \\[2mm] \tilde{\vartheta}_\varepsilon(C) + \dfrac{t-C}{t} & \text{if } t > C \\[2mm] \tilde{\vartheta}_\varepsilon(-C) + \dfrac{t+C}{t} & \text{if } t < -C, \end{cases}$$

where $C > 2 \operatorname{diam}(\Omega)$. In view of (6.30),

$$\theta(v_\delta(x) - F) = \tilde{\theta}_\varepsilon(v_\delta(x) - F) \qquad \text{if } F \in K^*.$$

Choosing $F = u$, we then have for the corresponding v_δ:

$$-\Delta v_\delta + \gamma \beta_\delta(v_\delta - \phi_0) + \tilde{\theta}_\varepsilon(v_\delta - u) = 0.$$

In view of (6.31) we can take a sequence $\delta \to 0$ with $v_\delta \to u_\varepsilon$ uniformly in $\bar{\Omega}$,

$$\gamma \beta_\delta(v_\delta - \phi_0) \to w_\varepsilon \qquad \text{weakly star in } L^\infty(\Omega),$$

and u_ε is then clearly a solution of (6.20). Also,

$$|w_\varepsilon|_{L^\infty(\Omega)} \leqslant C$$

and, by (6.31),

$$|u_\varepsilon(x) - u_\varepsilon(y)| \leqslant |x - y|,$$

which implies that $|\nabla u_\varepsilon| \leqslant 1$. Thus $u_\varepsilon \in K^*$. This completes the proof of Theorem 6.2.

PROBLEMS

1. Show that the solution of (6.28), (6.29) is continuous.
   [*Hint*: For $\Omega_m \uparrow \Omega$, $\partial \Omega_m$ smooth, $\phi_m \to \phi$, ϕ_m smooth, the solutions u_m converge weakly star in $L^\infty_{\mathrm{loc}}(\Omega)$ to u; use barriers.]

2. Compute explicitly the solution of the variational inequality (6.1) in case Ω is a ball $\{|x| < R\}$.

3. Do the same for (6.4) in case Ω is a shell $\{\rho < |x| < R\}$.

4. Take $n = 2$ in Problem 2 and compute χ from (6.3) and the displacement vector of the elastic–plastic torsion problem.

5. Consider the variational inequality with two obstacles (6.14), (6.13). Prove the following local $W^{2,p}$ regularity analogous to Problem 5 of Section 3. Let $\Omega_0 \subset \Omega_1 \subset \Omega$, Ω_i open, $\overline{\Omega}_0 \subset \Omega_1 \cup \Gamma_1$, $\overline{\Omega}_1 \subset \Omega \cup \Gamma_2$, where Γ_i are open subsets of $\partial\Omega$ and $\overline{\Gamma}_1 \subset \Gamma_2$. Suppose that Γ_2 is in $C^{2+\alpha}$ and ϕ, ψ are in $W^{2,p}(\Omega_1)$, $1 < p < \infty$. Then $u \in W^{2,p}(\Omega_0)$.

7. THE ELASTIC–PLASTIC TORSION PROBLEM: $W^{2,\infty}$ REGULARITY

Theorem 7.1. *The solution u of (6.4), (6.5) belongs to $W^{2,\infty}$ in any compact subdomain of Ω, and*

$$(7.1) \qquad |D^2 u(x)| \leq C\left(\mu + \frac{1}{d(x, \partial\Omega)}\right),$$

where C is a universal constant.

Note that no assumptions are made here on the regularity of $\partial\Omega$.

Proof. Consider first the case where

$$(7.2) \qquad |c_i - c_j| < d(\Omega_i, \Omega_j), \qquad \partial\Omega \in C^{2+\alpha}$$

where the c_j are the constant values of u in the domains Ω_j, and $c_0 = 0$. Then the obstacles ϕ, ψ satisfy

$$\phi(x) < \psi(x) \qquad \text{in } \Omega;$$

recall that $\phi(x) = \psi(x) = c_j$ in Ω_j. Introduce ε-mollifiers

$$\tilde{\phi}_\varepsilon(x) = (J_\varepsilon \phi)(x), \quad \tilde{\psi}_\varepsilon(x) = (J_\varepsilon \psi)(x) \qquad (x \in R^n).$$

If $\phi_\varepsilon = \tilde{\phi}_\varepsilon + \delta_\varepsilon$ where $4\varepsilon < \delta_\varepsilon < 5\varepsilon$ then

$$(7.3) \qquad \{x \in \Omega; d(x, \partial\Omega) > 4\varepsilon\} \subset \{x \in \overline{\Omega}; \phi_\varepsilon(x) \leq \psi_\varepsilon(x)\}$$

$$\subset \{x \in \Omega; d(x, \partial\Omega) > \varepsilon\}.$$

Further, by Sard's lemma we can choose δ_ε such that $\partial\{\phi_\varepsilon < \psi_\varepsilon\}$ is in C^∞. It is easily seen that

$$(7.4) \qquad |\nabla\phi_\varepsilon| \leq 1, \quad |\nabla\psi_\varepsilon| \leq 1 \qquad \text{in } R^n.$$

Let $D_\varepsilon = \{x \in \Omega;\ \phi_\varepsilon(x) < \psi_\varepsilon(x)\}$, and denote by u_ε the solution of the variational inequality (6.15), (6.13) with ϕ, ψ, Ω^* replaced by ϕ_ε, ψ_ε, D_ε respectively. Set

$$N_\varepsilon = \{x \in D_\varepsilon;\ \phi_\varepsilon(x) < u_\varepsilon(x) < \psi_\varepsilon(x)\},$$

$$\Lambda_1 = \{x \in D_\varepsilon;\ u_\varepsilon(x) = \phi_\varepsilon(x)\},$$

$$\Lambda_2 = \{x \in D_\varepsilon;\ u_\varepsilon(x) = \psi_\varepsilon(x)\}.$$

Since ϕ_ε, ψ_ε are C^∞ functions, u_ε is in $W^{2,p}(D_\varepsilon)$ for any $p < \infty$ (Section 3, Problem 2), and thus N_ε is an open set and Λ_i are closed sets. Notice that by (7.4),

(7.5) $|\nabla u_\varepsilon| \leq 1$ on ∂D_ε.

For any direction ξ the function $u_{\varepsilon,\xi}$ is harmonic in N_ε and

$$u_{\varepsilon,\xi} = \phi_{\varepsilon,\xi} \leq 1 \qquad \text{on } \partial N_\varepsilon \cap \Lambda_1,$$

$$= \psi_{\varepsilon,\xi} \leq 1 \qquad \text{on } \partial N_\varepsilon \cap \Lambda_2.$$

Recalling also (7.5) and applying the maximum principle, we get $u_{\varepsilon,\xi} \leq 1$ in N_ε. Since ξ is arbitrary, we obtain

(7.6) $|\nabla u_\varepsilon| \leq 1$ in N_ε.

Notice that in D_ε the obstacles ϕ_ε, ψ_ε are never equal. Hence the local behavior of the free boundaries

$$F_i = \partial \Lambda_i \cap D_\varepsilon$$

is the same as in the case of a one-obstacle problem. We shall need here one result on these free boundaries:

Lemma 7.2.

  (i) *If* $y \in N_\varepsilon, y \to x \in F_1$ *then* $\lim \inf (u_\varepsilon - \phi_\varepsilon)_{\xi\xi} \geq 0$.
  (ii) *If* $y \in N_\varepsilon, y \to x \in F_2$ *then* $\lim \inf (\psi_\varepsilon - u_\varepsilon)_{\xi\xi} \geq 0$.

The $\lim \inf$ *are uniformly in* $x \in F_i \cap \Omega'$, *for any closed subset* $\Omega' \subset D_\varepsilon$.

The lemma is contained in Theorem 3.6 of Chapter 2.
We shall use the notation

$$d(x) = d(x, \partial\Omega).$$

Lemma 7.3. *For any direction ξ*

$$(7.7) \qquad \phi_{\varepsilon,\xi\xi}(x) \geq -\frac{1}{d(x)-\varepsilon},$$

$$(7.8) \qquad \psi_{\varepsilon,\xi\xi}(x) \leq \frac{1}{d(x)-\varepsilon}$$

for all $x \in \Omega$ with $d(x) > \varepsilon$.

Proof. Let $x^0 \in \Omega$. Then

$$\phi(x^0) = \max_i \left\{ c_i - d(x^0, \partial\Omega_i) \right\} = c_j - d(x^0, \partial\Omega_j) = c_j - |x^0 - y^0|$$

for some j and for some $y^0 \in \partial\Omega_j$, where $|x^0 - y^0| = d(x^0, \partial\Omega_j) = d(x^0)$. Set

$$\gamma(x) = c_j - |x - x^0|.$$

Then

$$\phi(x) \geq c_j - d(x, \partial\Omega_j) \geq c_j - |x - y^0| = \gamma(x)$$

and $\phi(x^0) = \gamma(x^0)$. It follows that for any unit vector e and for any $0 < h < |x^0 - y^0|$, the pure second differences

$$\Delta^2_{h,e}\phi(x^0) \equiv \frac{\phi(x^0 + he) + \phi(x^0 - he) - 2\phi(x^0)}{h^2}$$

is $\geq \Delta^2_{h,e}\gamma(x^0)$. But by the mean value theorem

$$\Delta^2_{h,e}\gamma(x^0) \geq -\frac{1}{d(\tilde{x})},$$

where $\tilde{x}$ lies in the interval $(x^0 - he, x^0 + he)$. Therefore,

$$\Delta^2_{h,e}\phi(x^0) \geq -\frac{1}{d(x^0) - h}.$$

Since this is true for all $x^0 \in \Omega$ with $d(x^0) > h$, we also have

$$\Delta^2_{h,e}\phi_\varepsilon(x) \geq -\frac{1}{d(x) - h - \varepsilon}$$

if $d(x) > h + \varepsilon$. Taking $h \to 0$, (7.7) follows. The proof of (7.8) is similar.

Lemma 7.4.   *For any direction ξ,*

$$(7.9) \qquad |u_{\varepsilon,\xi\xi}(x)| \leq C\left(\mu + \frac{1}{d(x) - \varepsilon}\right)$$

for a.e. $x \in D_\varepsilon$, C is a universal constant.

Proof.   Since $u_\varepsilon \in W^{2,p}$, we have, for any direction η,

$$(7.10) \qquad u_{\varepsilon,\eta\eta} = \phi_{\varepsilon,\eta\eta} \qquad \text{a.e. on } \Lambda_1.$$

Also, in a D_ε-neighborhood of Λ_1, $u_\varepsilon < \psi_\varepsilon$ and, therefore,

$$(7.11) \qquad -\Delta u_\varepsilon \geq \mu \qquad \text{a.e. on } \Lambda_1.$$

For any a.e. $x \in \Lambda_1$ we then have, by (7.7), (7.10), (7.11),

$$-\frac{1}{d(x) - \varepsilon} \leq \phi_{\varepsilon,\xi\xi}(x) = u_{\varepsilon,\xi\xi}(x) = \Delta u_\varepsilon(x) - \sum_{i=1}^{n-1} u_{\varepsilon,\eta_i\eta_i}(x)$$

$$\leq -\mu - \sum_{i=1}^{n-1} \phi_{\varepsilon,\eta_i\eta_i}(x) \leq -\mu + \frac{n-1}{d(x) - \varepsilon}$$

where ξ and the η_i form an orthogonal system. Thus

$$(7.12) \qquad -\frac{1}{d(x) - \varepsilon} \leq u_{\varepsilon,\xi\xi}(x) \leq -\mu + \frac{n-1}{d(x) - \varepsilon} \qquad \text{a.e. on } \Lambda_1.$$

Similarly,

$$(7.13) \qquad -\mu - \frac{n-1}{d(x) - \varepsilon} \leq u_{\varepsilon,\xi\xi}(x) \leq \frac{1}{d(x) - \varepsilon} \qquad \text{a.e. on } \Lambda_2.$$

In order to estimate $u_{\varepsilon,\xi\xi}$ in a suitable subset of N_ε, we shall apply the maximum principle. But first we need to evaluate this function as we approach the free boundary. To do this, consider first the case where $x \in F_1$. By Lemmas 7.2(i) and 7.3,

$$\liminf_{\substack{y \in N_\varepsilon \\ y \to x}} u_{\varepsilon,\xi\xi}(y) \geq \phi_{\varepsilon,\xi\xi}(x) \geq -\frac{1}{d(x) - \varepsilon}.$$

This is also true for all directions $\eta_i (1 \leq i \leq n - 1)$ which form with ξ an orthogonal system. Thus:

$$\liminf\left(u_{\varepsilon,\eta_i\eta_i}(y) + \frac{1}{d(x) - \varepsilon}\right) \geq 0.$$

Since also

$$\limsup\left[\sum_{i=1}^{n}\left(u_{\varepsilon,\eta_i\eta_i}(y) + \frac{1}{d(x) - \varepsilon}\right)\right] \qquad (\eta_n = \xi)$$

$$= \limsup\left(\Delta u_\varepsilon(y) + \frac{n}{d(x) - \varepsilon}\right) = -\mu + \frac{n}{d(x) - \varepsilon},$$

we get

(7.14)

$$-\frac{1}{d(x) - \varepsilon} \leq \liminf_{\substack{y \in N_\varepsilon \\ y \to x}} u_{\varepsilon,\xi\xi}(y) \leq \limsup_{\substack{y \in N_\varepsilon \\ y \to x}} u_{\varepsilon,\xi\xi}(y) \leq -\mu + \frac{n}{d(x) - \varepsilon}$$

if $x \in F_1$,

and the lim inf, lim sup are uniformly in x. Similarly,

(7.15) $\qquad -\mu - \dfrac{n-1}{d(x) - \varepsilon} \leq \liminf\limits_{\substack{y \in N_\varepsilon \\ y \to x}} u_{\varepsilon,\xi\xi}(y) \leq \limsup\limits_{\substack{y \in N_\varepsilon \\ y \to x}} u_{\varepsilon,\xi\xi}(y) \leq \dfrac{1}{d(x) - \varepsilon}$

if $x \in F_2$.

We shall next establish that

(7.16) $\qquad |u_{\varepsilon,\xi\xi}(x)| \leq C\left(\mu + \dfrac{1}{\rho}\right) \qquad$ if $x \in N_\varepsilon, d(x,\partial D_\varepsilon) = \rho$.

for any fixed and small ρ provided that ε is so small that $\varepsilon < \rho/8$. Notice that

$$\text{dist}\left(B_{2\rho/3}(x), \partial D_\varepsilon\right) > \varepsilon + \rho/12.$$

Let $x^0 \in N_\varepsilon$, $d(x^0, \partial D_\varepsilon) = \rho$ and consider the function $\tilde{u}_\varepsilon(y) = u_\varepsilon(x^0 + \rho y)$ in B_1, where $B_R = \{y; |y| < R\}$. Choose $\sigma \in C_0^\infty(B_{1/2})$ such that $\sigma = 1$ in $B_{1/3}$.

By (7.12), (7.13),

(7.17) $\qquad\qquad\qquad |\Delta u_\varepsilon| \leq C\left(\mu + \dfrac{1}{\rho}\right)$

a.e. in $\Lambda_1 \cup \Lambda_2 \cap \{x; \; |x - x^0| < 2\rho/3\}$. Since also $\Delta u_\varepsilon = -\mu$ in N_ε, (7.17) holds a.e. in $\{|x - x^0| < 2\rho/3\}$. Recalling also (7.6), we obtain

$$|\Delta(\tilde{u}_\varepsilon \sigma)| \leq C\left(\mu + \frac{1}{\rho}\right)\rho^2 \qquad \text{in } B_{2/3}.$$

By elliptic L^p estimates it then follows that

$$|\tilde{u}_\varepsilon \sigma|_{W^{2,p}(B_{1/2})} \leq C\left(\mu + \frac{1}{\rho}\right)\rho^2.$$

for any $p < \infty$ and, in particular,

$$(7.18) \qquad\qquad |\tilde{u}_{\varepsilon,\, y_i y_i}|_{L^p(B_{1/3})} \leq C\left(\mu + \frac{1}{\rho}\right)\rho^2.$$

We next wish to boost this estimate up to $p = \infty$.

For this purpose we introduce a function $\tau \in C_0^\infty(B_{1/3})$, $\tau = 1$ in $B_{1/4}$, and the open set

$$N_{\varepsilon,\rho} = \{y: x^0 + \rho y \in N_\varepsilon\}.$$

In $N_{\varepsilon,\rho}$, $\Delta \tilde{u}_\varepsilon = -\mu\rho^2$. Using (7.18), we find that

$$(7.19) \qquad\qquad \Delta(\tilde{u}_\varepsilon \tau)_{y_i y_i} = \frac{\partial h}{\partial y_i},$$

where

$$(7.20) \qquad\qquad |h|_{L^p(B_{1/3})} \leq C\left(\mu + \frac{1}{\rho}\right)\rho^2.$$

Denote by $V(y)$ the fundamental solution of $-\Delta$, that is,

$$V(y) = -\gamma_n |y|^{2-n} \quad \text{if } n \geq 3 \qquad \left[\gamma_n = \frac{\Gamma(n/2)}{2(n-2)\pi^{n/2}}\right],$$

$$V(y) = -\frac{1}{2\pi} \log \frac{1}{|y|} \quad \text{if } n = 2$$

Then the function

$$g(y) = -\int_{N_{\varepsilon,\rho}} \frac{\partial V(y-z)}{\partial z_i} h(z)\, dz$$

satisfies

$$\Delta g = \frac{\partial h}{\partial y_i} \quad \text{in } N_{\varepsilon, \rho}$$

in the distribution sense; also, in view of (7.20),

$$(7.21) \qquad |g|_{L^\infty(B_{1/3})} \leq C\left(\mu + \frac{1}{\rho}\right)\rho^2.$$

The function $(\tilde{u}_\varepsilon \tau)_{y_i y_i} - g$ is then harmonic in $N_{\varepsilon, \rho} \cap B_{1/4}$ and, by the maximum principle and (7.12)–(7.15), (7.21), we find that

$$|\tilde{u}_{\varepsilon, y_i y_i}(x^0)| \leq C\left(\mu + \frac{1}{\rho}\right)\rho^2;$$

this gives (7.16).

We have thus completed to estimate $u_{\varepsilon, \xi\xi}$ on the boundary of the open set $N_\varepsilon \cap \{d(x, \partial D_\varepsilon) > \rho\}$ by (7.14)–(7.16). Applying the maximum principle, we get

$$|u_{\varepsilon, \xi\xi}| \leq C\left(\mu + \frac{1}{\rho}\right)$$

in this open set. Recalling also (7.12), (7.13), the assertion (7.9) follows.

We can now immediately complete the proof of Theorem 7.1. First we take $\varepsilon \to 0$ and note that $\phi_\varepsilon \to \phi$, $\psi_\varepsilon \to \psi$ uniformly and $u_\varepsilon \to u$ uniformly, u being the solution of (6.4), (6.5); see Problem 4. Next, in order to remove the first restriction in (7.2), we approximate the c_j by θc_j with $0 < \theta < 1$ and then let $\theta \to 1$. In order to remove the second restriction in (7.2) we approximate Ω by domains $\Omega_m \supset \Omega$ with C^∞ boundaries; see Problem 5.

Definition 7.1. Let u be the solution of (6.4), (6.5). The sets

$$E = \{|\nabla u(x)| < 1\} \cap \Omega,$$

$$P = \{|\nabla u(x)| = 1\} \cap \Omega$$

are called, respectively, the *elastic* and *plastic* sets.

PROBLEMS

1. Let

$$P^+ = \{u = \phi\} \cap \Omega,$$

$$P^- = \{u = \psi\} \cap \Omega,$$

$$\tilde{E} = \{\phi < u < \psi\} \cap \Omega.$$

Show that $P = P^+ \cup P^-$, $E = \tilde{E}$.

[*Hint:* If $u(x^0) = \phi(x^0) = c_j - d(x^0, \partial\Omega_j) = c_j - |x^0 - y^0|$, $y^0 \in \partial\Omega_j$, then $u_\xi = \phi_\xi = -1$ along the segment $x^0 y^0$. Conversely, if $u_\xi(x^0) = 1$ and if $\phi(x^0) < u(x^0) < \psi(x^0)$, then apply the maximum principle to u_ξ in the component N_0 of N containing x^0 to derive $u_\xi \equiv 1$ in N_0.]

2. If $x^0 \in P$ and if $u(x^0) = \phi(x^0) = c_j - d(x^0, \partial\Omega_j) = c_j - |x^0 - y^0|$, $y^0 \in \partial\Omega_j$, then the entire segment $x^0 y^0$ is plastic.

[*Hint:* The function $w(x) = u(x) - (c_j - d(x, \partial\Omega_j))$ satisfies $\partial w / \partial l \le 0$ along $x^0 y^0$ (l = the direction from x^0 to y^0) and $w(x^0) = w(y^0) = 0$; hence $w = 0$ along $x^0 y^0$.]

3. Let Ω be a simply connected domain in R^2 with boundary that is piecewise $C^{3+\alpha}$; denote by $V_1, \ldots, V_l$ the vertices of $\partial\Omega$, that is, the endpoints of the $C^{3+\alpha}$ subarcs of $\partial\Omega$. Show that the solution of (6.1), (6.2) is in $W^{2,\infty}(\Omega_0)$ for any subdomain of Ω with $\overline{\Omega}_0 \subset \overline{\Omega} \setminus \{V_1, \ldots, V_l\}$.

[*Hint:* By reference 109, p. 382, $d(x, \partial\Omega)$ is in $C^{3+\alpha}$; use Theorem 7.1 and Problem 1 of Section 4.]

4. Prove that $u = \lim_{\varepsilon \to 0} u_\varepsilon$ is the solution of (6.4), (6.5); $\partial\Omega$ is assumed to be in $C^{2+\alpha}$.

[*Hint:* Consider first test functions $\tilde{v}$, $\phi \le \tilde{v} \le \psi$ with $\tilde{v} = \psi$ in a δ-neighborhood of $\partial\Omega$ and then approximate general v by such functions $\tilde{v}$, noting that $u \in C^1(\overline{\Omega})$ by Problem 2 of Section 3.]

5. If u_m is the solution corresponding to Ω_m and if $\Omega_m \downarrow \Omega$, then $u_m \to u$ a.e.

[*Hint:* If $v \in K^*$ and is properly extended outside Ω, then $J(u) \le \liminf_{m \to \infty} J_m(u_m)$, $J_m(u_m) \le J_m(v) = J(v)$.]

8. PARABOLIC VARIATIONAL INEQUALITIES

Let Ω be a bounded domain in R^n and $Q_T = \Omega \times \{0 < t < T\}$. An operator $u_t + Au$ where

$$Au \equiv - \sum_{i,j=1}^n a_{ij}(x, t) \frac{\partial^2 u}{\partial x_i \partial x_j} + \sum_{i=1}^n b_i(x, t) \frac{\partial u}{\partial x_i} + c(x, t)u$$

with coefficients defined in Q_T is said to be of *parabolic type* at a point (x, t) if

$$(8.1) \qquad \sum_{i,j=1}^{n} a_{ij}(x, t)\xi_i\xi_j \geqslant \lambda |\xi|^2 \qquad \forall \xi \in R^n \qquad (\lambda > 0);$$

it is *uniformly parabolic* in Q_T if (8.1) holds for all (x, t) in Q_T with a constant λ independent of (x, t).

For parabolic operators one can solve the first intitial boundary value problem

$$(8.2) \qquad \begin{aligned} u_t + Au &= f \qquad \text{in } Q_T, \\ u &= g \qquad \text{on } \partial_p Q_T, \end{aligned}$$

where

$$\partial_p Q_T = \partial\Omega \times (0, T) \cup \overline{\Omega} \times \{0\}$$

is the *parabolic boundary* of Q_T. The Schauder estimates stated in Section 3 have their parabolic counterpart. In the definition of the Hölder coefficient one takes

$$H_\alpha(u) = \sup_{(x,t),(x',t')} \frac{|u(x, t) - u(x', t')|}{d^\alpha((x, t), (x', t'))},$$

where

$$d((x, t), (x', t')) = |x - x'| + |t - t'|^{1/2}$$

is the *parabolic distance*. The definition of $\|u\|_\alpha$ for $u = u(x, t)$ defined in Q_T is then analogous to the definition given before for functions $u(x)$ in Ω. We next set

$$\|u\|_{2+\alpha} = \|u\|_\alpha + \|D_x u\|_\alpha + \|D_x^2 u\|_\alpha + \|D_t u\|_\alpha + H^t_{1+\alpha}(D_x u),$$

where

$$H^t_{1+\alpha}(v) = \sup \frac{|v(x, t) - v(x, t')|}{|t - t'|^{(1+\alpha)/2}}.$$

The Schauder boundary estimates then assert that

$$(8.3) \qquad \|u\|_{2+\alpha} \leqslant C(\|f\|_\alpha + \|u\|_0 + \|g\|_{2+\alpha}),$$

where it is assumed that (8.1) holds,

$$(8.4) \qquad \sum \|a_{ij}\|_\alpha + \sum \|b_i\|_\alpha + \|c\|_\alpha \leqslant K,$$

and $\partial\Omega$ is in $C^{2+\alpha}$; the constant C depends only on λ, K, and Q_T.

The interior Schauder estimates are also valid for parabolic operators. For details, see references 94a and 130.

L^p estimates have also been established for parabolic operators. Here one assumes in addition in (8.1) (for all $x \in Q_T$) that a_{ij}, b_i, c are measurable functions satisfying

$$(8.5)$$

$$\sum |a_{ij}(x, t) - a_{ij}(x', t')| \leqslant \omega(|x - x'| + |t - t'|), \qquad \omega(\lambda) \downarrow 0 \text{ if } \lambda \downarrow 0,$$

$$\sum |b_i(x, t)| + |c(x, t)| \leqslant K,$$

and that $\partial\Omega$ is in C^2. Then the solution of (8.2) satisfies

$$(8.6)$$

$$\int_{Q_T} \left[|u|^p + |D_x u|^p + |D_x^2 u|^p + |D_t u|^p \right] dx\, dt$$

$$\leqslant C \int_{Q_T} |f|^p\, dx\, dt + C \int_{Q_T} \left[|g|^p + |D_x g|^p + |D_x^2 g|^p + |D_t g|^p \right] dx\, dt$$

for all $p > 1$, $p \neq \frac{3}{2}$. For proofs, see reference 164 or 88.

For any point $P^0 = (x^0, t^0)$ in Q_T, denote by $C(P^0)$ the set of all points P in Q_T that can be connected to P^0 by a continuous curve $x = x(s)$, $t = t(s)$ $(0 \leqslant s \leqslant s^0)$, contained in Q_T such that $t(s)$ increases in s as the curve is traced from P to P^0.

The Strong Maximum Principle

Assume that $A + \partial/\partial t$ is parabolic in Q_T and $c(x, t) \geqslant 0$. Le u be a function defined in Q_T such that $D_x u$, $D_t u$, $D_x^2 u$ are continuous in Q_T, and

$$Au + \frac{\partial u}{\partial t} \leqslant 0 \qquad \text{in } Q_T.$$

Then, for any point P^0 in Q_T, if

$$u(P^0) = \sup_{C(P^0)} u \geqslant 0,$$

then $u \equiv$ const. in $C(P^0)$.

This result is valid also if $P^0 \in \Omega \times \{T\}$ provided that u is continuous in $Q_T \cup \Omega \times \{T\}$.

Another version of the maximum principle asserts that if u takes a nonnegative maximum in $\overline{Q}_T$ at a boundary point P^0 on $\partial\Omega \times (0, T]$, and if $u < u(P^0)$ in some Q_T-neighborhood of P^0, then

$$\frac{\partial u}{\partial \mu} > 0 \quad \text{at } P^0$$

for any outward spatial direction μ at P^0; it is assumed here that u is C^1 in a $\overline{Q}_T$-neighborhood of P^0 (otherwise, $\partial u/\partial \mu$ is taken as lim inf of finite differences) and there is a ball $B \subset \Omega$ with $P^0 \in \partial B$.

The existence of a unique solution of (8.2) can be established by using either the Schauder estimates or the L^p estimates. The first approach (with the Schauder boundary estimates) requires more smoothness on the data, namely, that (8.4) hold, that $\partial\Omega$ be in $C^{2+\alpha}$, that the right-hand side of (8.3) is finite, and the consistency condition

$$g_t + Ag = f \quad \text{on } \partial\Omega \times \{0\}.$$

The solution then is quite regular; namely, $D_x u$, $D_x^2 u$, and $D_t u$ are Hölder continuous in $\overline{Q}_T$. Using the Schauder interior estimates, one needs only $\partial\Omega$ to satisfy the outside ball property and g to be continuous on $\partial_p Q_T$; then $D_x u$, $D_x^2 u$, $D_t u$ are Hölder continuous in Q_T; see reference 94a.

The second approach requires less smoothness on the data, but the derivatives $D_x u$, $D_x^2 u$, and $D_t u$ are taken only as weak derivatives in L^p.

If $\nabla_x a_{ij}$ exists and is bounded, then we can associate with A the bilinear form

$$(8.7) \qquad a(t; u, v) = \int_\Omega \left(\sum a_{ij} \frac{\partial u}{\partial x_i} \frac{\partial v}{\partial x_j} + \sum \tilde{b}_i \frac{\partial u}{\partial x_i} v + cuv \right) dx,$$

where

$$\tilde{b}_i = b_i + \sum \frac{\partial a_{ij}}{\partial x_j}.$$

We also set

$$(v, w) = \int_\Omega vw \, dx.$$

Then the first initial boundary value problem can be written in the weak form:

$$(8.8) \qquad \begin{aligned} (u_t, v - u) + a(t; u, v - u) &= (f, v - u) \quad \text{for a.a. } t \in (0, T), \\ \forall v \in H^1(Q_T), \qquad v &= g \text{ on } \partial_p Q_T. \end{aligned}$$

The transformation $\tilde{u} = e^{-kt}u$ transforms $u_t + Au = f$ into

$$\tilde{u}_t + (A + k)\tilde{u} = e^{-kt}f$$

and $a(t; u, v)$ into

$$\tilde{a}(t; \tilde{u}, \tilde{v}) = a(t; \tilde{u}, \tilde{v}) + k(\tilde{u}, \tilde{v}).$$

If k is sufficiently large, then $\tilde{a}$ is coercive. Thus without loss of generality we may assume from the start that

$$(8.9) \qquad a(t; u, u) \geqslant \lambda_0 \int_\Omega \left(|\nabla u|^2 + u^2 \right) dx \qquad (\lambda_0 > 0).$$

Since the same transformation can be used also in the variational problems which follow, we may always assume (if $\nabla_x a_{ij} \in L^\infty$) that (8.9) is satisfied and that $c(x, t) \geqslant 0$.

Let K be any closed convex subset of $H^1(Q_T)$. Consider the following problem: Find u satisfying

$$u \in K,$$

(8.10)

$$(u_t, v - u) + a(t; u, v - u) \geqslant (f, v - u) \qquad \text{for a.a. } t \in (0, T), \quad \forall v \in K.$$

This problem is called a *parabolic variational inequality*. We shall study here only one type of variational inequality, which we proceed to describe.

Let $\phi(x, t)$ be a function defined in $\overline{Q}_T$ with uniformly continuous derivatives $D_x\phi$, $D_t\phi$, $D_x^2\phi$, $\phi \leqslant g$ on $\partial_p Q_T$, and consider the convex set

$$(8.11) \qquad K = \left\{ v \in H^1(Q_T), v = g \text{ on } \partial_p Q_T, v \geqslant \phi \text{ a.e.} \right\};$$

here the relation $v = g$ on $\partial_p Q_T$ is taken as usual in the trace sense. The variational inequality (8.10) with K given by (8.11) is called the *parabolic obstacle problem*; ϕ is the *obstacle*.

Uniqueness and stability of solutions of parabolic variational inequalities (8.10) with general K follow easily from the coerciveness condition (8.9), as in Section 2.

In order to prove existence for the obstacle problem, we use the penalty method. Introducing the function $\beta_\varepsilon(t)$ as in (3.12) of Section 3, we consider

$$(8.12) \qquad u_t + Au + \beta_\varepsilon(u - \phi) = f \qquad \text{in } Q_T,$$

$$u = g \qquad \text{on } \partial_p Q_T.$$

We shall assume that

(8.13)   (8.1), (8.4) hold, $c \geq 0$, $\partial\Omega \in C^{2+\alpha}$, and

$$f, g, D_x g, D_x^2 g, D_t g \text{ belong to } C^\alpha(\overline{Q}_T).$$

Lemma 8.1.   *If (8.13) holds, then there exists a solution $u = u_\varepsilon$ of (8.12) and*

(8.14) $$|\beta_\varepsilon(u_\varepsilon - \phi)| \leq C,$$

where C is independent of ε.

The proof uses the same arguments as in Section 3 and is left to the reader.
We can now use the parabolic L^p estimates to deduce that a sequence $u_\varepsilon \to u$ uniformly in $\overline{Q}_T$ and that u is a solution of (8.10); more specifically, u satisfies

(8.15)
$$\left.\begin{array}{r} u_t + Au \geq f \\ u \geq \phi \\ (u_t + Au - f)(u - \phi) = 0 \end{array}\right\} \quad \text{a.e. in } \overline{Q}_T,$$

$$u = g \qquad \text{on } \partial_p Q_T.$$

Uniqueness follows by coerciveness if we assume that

(8.16) $$|D_x a_{ij}| \leq C$$

[so that $a(t; u, v)$ can be defined]. We summarize:

Theorem 8.2.   *If (8.13), (8.16) hold, then there exists a unique solution of the obstacle problem (8.10), (8.11) and*

(8.17) $D_x u, D_x^2 u, D_t u$ *belong to* $L^p(Q_T)$ $\forall\, 1 < p < \infty$.

Note that (8.17) implies that u is Hölder continuous in x and in t, with any exponent $\beta < 1$. We shall now describe another approach to studying the penalized problem which yields better regularity results. For simplicity we assume that

(8.18) a_{ij}, b_i, c, ϕ are independent of t,

$$f_t \text{ is bounded } g = 0.$$

Note that

(8.19) $$\frac{\partial}{\partial t}\beta_\varepsilon(u_\varepsilon - \phi) \cdot \frac{\partial u_\varepsilon}{\partial t} = \beta_\varepsilon'(u_\varepsilon - \phi)\left(\frac{\partial u_\varepsilon}{\partial t}\right)^2 \geq 0 \qquad \text{a.e.}$$

Setting $u_\varepsilon = u$ and differentiating the equation in (8.12) once with respect to t, multiplying by $(\partial u/\partial t)^{2k-1}$ (k positive integer), and integrating over Ω, we obtain, after using (8.19),

$$\int_\Omega \left(\frac{\partial u}{\partial t}\right)^{2k-1} \frac{\partial^2 u}{\partial t^2}\, dx + \int_\Omega A\left(\frac{\partial u}{\partial t}\right)\left(\frac{\partial u}{\partial t}\right)^{2k-1} dx$$

$$(8.20) \qquad\qquad \leqslant \int_\Omega f_t\left(\frac{\partial u}{\partial t}\right)^{2k-1} dx \leqslant C\left[\int_\Omega \left(\frac{\partial u}{\partial t}\right)^{2k} dx\right]^{(2k-1)/(2k)};$$

we have assumed (temporarily) that for a.e. t,

$$(8.21) \qquad\qquad \frac{\partial^2 u}{\partial t^2} \in L^2(\Omega).$$

Using (8.9), we find that the second term on the left-hand side of (8.20) is bounded below by

$$c\int_\Omega \left(\frac{\partial u}{\partial t}\right)^{2k-2}\left[\left|\nabla_x\left(\frac{\partial u}{\partial t}\right)\right|^2 + \left(\frac{\partial u}{\partial t}\right)^2\right] dx \qquad (c > 0).$$

Therefore, the function

$$\Phi(t) = \int_\Omega \left(\frac{\partial u}{\partial t}\right)^{2k} dx$$

satisfies

(8.22)

$$\Phi' + \gamma\Phi \leqslant C\Phi^{(2k-1)/(2k)} \leqslant \frac{\gamma}{2}\Phi + C_1 \qquad \text{for some } \gamma > 0, \quad C > 0, \quad C_1 > 0,$$

which gives [since $\Phi(0) \leqslant C$]

$$(8.23) \qquad\qquad \int_\Omega (\partial u/\partial t)^{2k}\, dx \leqslant C.$$

From this and from Lemma 8.1, we obtain (for $u = u_\varepsilon$)

$$(8.24) \qquad\qquad \int_\Omega |Au|^p\, dx \leqslant C \qquad \forall\, 1 < p < \infty.$$

The assumption (8.21) is not needed if we take finite differences of (8.12) with respect to t and multiply by $(\Delta_h u)^{2k-1}$, where $\Delta_h u$ is the finite-difference quotient of u with respect to t.

We can now appeal to the elliptic L^p estimates and conclude that for a.a. $t \in (0, T)$,

(8.25)
$$\int_\Omega \left[|D_x u|^p + |D_x^2 u|^p + |D_t u|^p \right] dx \leqslant C.$$

Taking $\varepsilon \to 0$, we obtain the same inequality for the solution of the variational inequality. Thus:

Theorem 8.3. *Let* (8.13), (8.16), *and* (8.18) *hold. Then the solution of the variational inequality* (8.15) *satisfies* (8.25).

It follows that, for a.a. t, $D_x u(x, t)$ is Hölder continuous in x with any exponent $\beta < 1$.

The preceding proof of (8.25) is flexible enough to allow also variable coefficients of A and ϕ, g which depend on t [more precisely, if (8.13) holds, if ϕ_t, $D_x \phi$, $D_x^2 \phi$ are in $C(Q_T)$ and if

(8.26)
$$a_{ij}, \nabla_x a_{ij}, b_i, c \text{ and their first } t\text{-derivatives}$$

$$\text{are bounded; } \phi_{tt} \text{ and } g_{tt} \text{ are bounded,}$$

then the assertion (8.25) is valid; for details, see reference 94g].

We next describe briefly the $W^{2, \infty}$ estimate. For simplicity we take $Au = -\Delta u$. We introduce the fundamental solution

$$K(x, t) = (4\pi t)^{-n/2} \exp\left(-\frac{|x|^2}{4t} \right).$$

If $g(x, t)$ is Hölder continuous in Q_T, then (see reference 94a) the function

$$w(x, t) = \int_0^t \int_\Omega K(x - y, t - \tau) g(y, \tau) \, dy \, d\tau$$

is a solution of

(8.27)
$$w_t - \Delta w = g \qquad \text{in } Q_T.$$

Furthermore,

(8.28)
$$|w|_{L^\infty(Q_T)} \leqslant C |g|_{L^p(Q_T)} \qquad \text{if } p > 1 + \frac{n}{2}.$$

Using this fact we can now repeat the proof of Theorem 4.1, replacing Lemma 4.2 by (8.27), (8.28). We obtain

$$u_{\xi\xi}(x, t) \geqslant -C \qquad \text{in any compact subset of } Q_T \, (u = u_\varepsilon).$$

Next, by differentiating the penalized equation once with respect to t, we get

$$(u_t)_t - \Delta u_t + \beta_\varepsilon'(u - \phi)(u_t - \phi_t) = 0 \qquad (\phi = \phi_\varepsilon),$$

and we can proceed as before to derive a bound

$$\pm u_t \geq -C \qquad \text{in compact subsets.}$$

Since also $|\beta_\varepsilon| \leq C$, we have

$$u_t - \Delta u \leq C,$$

and we obtain, as before, $u_{\xi\xi} \leq C$. Thus:

Theorem 8.4. *If in Theorem 8.2, $a_{ij} = \delta_{ij}$, $b_i = 0$, $c = 0$, and if the obstacle ϕ satisfies:*

$$\phi \text{ is Lipschitz in } x \text{ and } t,$$

$\phi_{\xi\xi} \geq -C$ *in $\mathcal{D}'(\tilde{Q}_T)$, for any direction ξ, $\tilde{Q}_T$ a neighborhood of $\overline{Q}_T$, where C is a constant, then the solution u of the variational inequality (8.15) satisfies*

$$D_x u, \; D_x^2 u, \; D_t u \qquad \text{belong to } L_{\text{loc}}^\infty(Q_T).$$

This result extends also to A with variable coefficients.

PROBLEMS

1. Prove Lemma 8.1.

2. Complete the proof of Theorem 8.2.

3. Fill the details in the proof of (8.23).

4. Prove (8.28).

5. Prove a comparison theorem: If u is the solution of (8.10), (8.11) and if $\hat{u}$ is a solution of a similar problem with $\hat{f}, \hat{g}, \hat{\phi}$ then

$$\hat{f} \geq f, \; \hat{g} \geq g, \; \hat{\phi} \geq \phi \qquad \text{implies that } \hat{u} \geq u.$$

6. Suppose that

$$\left.\begin{array}{r} u_t - \Delta u \geq f \\ u \geq 0 \\ (u_t - \Delta u - f)u = 0 \end{array}\right\} \qquad \text{a.e. in } Q_{R,T} = B_R \times (0, T),$$

where $f \leqslant -\gamma$, $u \leqslant M$ (γ, M positive), and $u \in C(\overline{Q}_{R,T})$. Show that if $\theta R^2 > 2nM/\gamma$, $(1 - \theta)T > M/\gamma$ for some $0 < \theta < 1$, then $u(0, T) = 0$.
[*Hint*: See Section 1, Problem 3.]

7. Consider the variational inequality in $R^n \times (0, T)$:

$$(8.29) \quad \begin{array}{l} u \geqslant 0; \quad (u_t - \Delta u)(v - u) \geqslant f(v - u) \qquad \text{a.e. } \forall\, v \geqslant 0, \\ u(x, 0) = u_0(x), \end{array}$$

where $u_0 \geqslant 0$, $u_0 \in L^1$, and f, $f_t \in L^\infty$. Use the penalized problem to deduce the existence of a solution $u_{\varepsilon, R}(x, t)$ in $\{|x| < R, t < T\}$, vanishing on $\{|x| = R\}$ and take $\varepsilon \to 0$, $R \to \infty$ to deduce the existence of a solution u of (8.29) satisfying

$$u_t, D_x u, D_x^2 u \in L^p \qquad \text{in compact subsets of } R^n \times (0, T),$$

$$u(x, t) - \int_{R^n} K(x - y, t)u_0(y)\, dy \to 0 \qquad \text{if } t \to 0,$$

where K is the fundamental solution, and

$$|u_t - \Delta u| \leqslant C \qquad \text{if } u_0 \in L^\infty.$$

8. If in Problem 7, $f \leqslant -\nu$ ($\nu > 0$), then there is $T_0 < \infty$ depending on u_0 and ν such that if $T > T_0$, then $u(x, t) = 0$ if $t > T_0$.
[*Hint*: Compare with $M - \nu t$.]

9. If in Problem 8, u_0 has compact support, then $u(x, t) = 0$ if $|x| > R_0$, for some $R_0 < \infty$.
[*Hint*: Compare with $\mu(R_0 - r)^2$.]

10. If in Problem 8, $u_0(x, t) \to 0$ as $|x| \to \infty$, then for each $t > 0$, $u(x, t) = 0$ if $|x| \geqslant R(t)$, $R(t) < \infty$.
[*Hint*: If $u(x_0, t_0) > 0$, consider

$$v = u - \frac{\nu}{2}\left[(t_0 - t) + \frac{|x - x_0|^2}{2n} \right]$$

in $\{u > 0, |x - x_0| < R, t < t_0\}$; v takes positive maximum on the parabolic boundary.]

11. Let $S(t) = $ support of $t \to u(x, t)$, and assume that f, u_0 are as in Problem 7, $f \leqslant -\nu < 0$, $S(0) = \mathrm{supp}\, u_0$ is bounded, and $u_0 \in L^\infty$. Prove:

$$S(t) \subset S + B\left(C_0\sqrt{t\,|\log t|} \right) \qquad \text{for all small } t,$$

where $B(\rho) = \{|x| < \rho\}$, $C_0 > 0$.

[*Hint*: If $u(x_0, t_0) > 0$, then consider the function $U = u - c|x - x_0|^2$ in $G = \{\text{dist}(x, S(0)) > \varepsilon, t < t_0, u(x, t) > 0\}$. U takes a positive maximum on the lateral boundary of G, at (x^*, t^*). Also,

$$u(x^*, t^*) \leqslant \int_S K(x^* - y, t^*) u_0(y)\, dy + Ct^*.]$$

12. If in Problem 11, $f \geqslant -\nu_0 > 0$ and $u_0 \geqslant \beta > 0$ on $S(0)$, then

$$S(t) \supset S + B\left(c_0\sqrt{t\,|\log t|}\,\right) \qquad \text{for all small } t,$$

for some $c_0 > 0$.
[*Hint*: $u \geqslant \beta \int_S K(x - y, t)\, dy - \nu_0 t.]$

9. THE STEFAN PROBLEM

In this section we describe a problem of melting of a solid and reduce it to a parabolic variational inequality. We shall then derive some specific properties of the solution.

Let G be a bounded domain in R^n whose boundary consists of two connected $C^{2+\alpha}$ hypersurfaces Γ_0 and Γ_1 with Γ_0 lying inside Γ_1 and bounding a simply connected domain G_0 (see Figure 1.2). Let $B_R = \{|x| < R\}$ be a ball containing G and set $\Omega = B_R \setminus G_0$, $Q_T = \Omega \times (0, T)$. Suppose that initially G is filled with water and $R^n \setminus (G \cup G_0)$ is filled with ice at $0°C$. We denote by $\theta = \theta(x, t)$ the water's temperature. We are given the initial temperature

$$(9.1) \qquad \theta(x, 0) = h(x) \qquad \text{on } G \times \{0\}$$

as well as the temperature along the boundary Γ_0 for all t,

$$(9.2) \qquad \theta(x, t) = g(x, t) \qquad \text{on } \Gamma_0 \times [0, \infty).$$

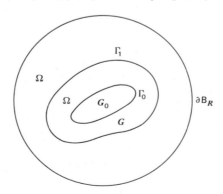

FIGURE 1.2

It is assumed that h and g are positive. Consequently, the ice will begin to melt and the region $N(t)$ occupied by water will grow. The part of $\partial N(t)$ that is adjacent to the ice is called the *free boundary*. Let us suppose that the free boundary is given by an equation $s(x) - t = 0$ (with $s(x) > t$ in the ice region). Then

(9.3)
$$\theta = 0 \qquad \text{(continuity of the temperature)},$$
$$\nabla_x \theta \cdot \nabla_x s = -k \qquad \text{(conservation of energy)}$$

along the free boundary, where k is a positive constant. Finally,

(9.4)
$$\theta_t - \Delta\theta = 0 \qquad \text{in the region } \{\theta > 0\}.$$

The problem of finding a solution θ, s to (9.1)–(9.4) is called the (*classical*) *one-phase Stefan problem*. We shall now transform the problem into a variational inequality for a function u; u is defined by

$$u(x, t) = \int_{s(x)}^{t} \theta(x, \tau)\, d\tau \qquad \text{if } x \in \Omega \setminus G, t > s(x),$$

$$= 0 \qquad \text{if } x \in \Omega \setminus G, t \leqslant s(x),$$

$$= \int_{0}^{t} \theta(x, \tau)\, d\tau \qquad \text{if } x \in G.$$

We compute for $x \in \Omega \setminus G$, $t > s(x)$,

$$u_{x_i}(x, t) = \int_{s(x)}^{t} \theta_{x_i}(x, \tau)\, d\tau - s_{x_i}(x)\theta(x, s(x))$$

$$= \int_{s(x)}^{t} \theta_{x_i}(x, \tau)\, d\tau,$$

$$u_{x_i x_i}(x, t) = \int_{s(x)}^{t} \theta_{x_i x_i}(x, \tau)\, d\tau - s_{x_i}(x)\theta_{x_i}(x, s(x)).$$

Hence

$$\Delta u(x, t) = \int_{s(x)}^{t} \Delta\theta(x, \tau)\, d\tau + k = \int_{s(x)}^{t} \theta_\tau(x, \tau)\, d\tau + k$$

$$= u_t(x, t) + k.$$

Similarly,

$$\Delta u(x, t) = \int_{0}^{t} \theta_\tau(x, \tau)\, d\tau = u_t(x, t) - h \qquad \text{if } x \in G.$$

Setting

(9.5)
$$f(x) = \begin{cases} h(x) & \text{if } x \in G \\ -k & \text{if } x \in \Omega \setminus G, \end{cases}$$

we find that

$$u_t - \Delta u = f \qquad \text{if } u > 0.$$

Also, clearly $u_t - \Delta u = 0 > f$ if $u = 0$. The function u takes on $\partial_p Q_T$ the boundary values ψ, where

(9.6)
$$\psi(x, t) = \begin{cases} \int_0^t g(x, \tau) \, d\tau & \text{if } x \in \Gamma_0, t > 0, \\ 0 & \text{if } t = 0 \text{ or if } |x| = R. \end{cases}$$

Introducing the convex set

(9.7)
$$K = \left\{ v \in H^1(Q_T), v \geq 0 \text{ a.e. in } Q_T, v = \psi \text{ on } \partial_p Q_T \right\},$$

we see that u is a solution of the variational inequality

$$u \in K,$$

(9.8)
$$\int_\Omega u_t(v - u) \, dx + \int_\Omega \nabla u \cdot \nabla(v - u) \, dx \geq \int_\Omega f(v - u) \, dx$$

$$\text{a.e. in } t \in (0, T), \qquad \forall v \in K.$$

We shall now assume that

(9.9)
$$\begin{aligned} g &\in C^{2+\alpha}(\Gamma_0 \times [0, T]), & g &> 0; \\ h &\in C^\alpha(\overline{G}), & h &> 0 \text{ in } G. \end{aligned}$$

From the proofs of Theorems 8.2 and 8.3 we deduce (approximating f by smooth functions) that there exists a unique solution of the variational inequality (9.8) and

$$D_x u, D_x^2 u, D_t u \qquad \text{belong to } L^\infty((0, T); L^p(\Omega)) \quad \text{for any } p < \infty.$$

Notice that since f is not differentiable in x, Theorem 8.4 cannot be directly applied. Nevertheless, we shall prove:

Theorem 9.1. *The solution u of the Stefan problem (9.8) satisfies*

(9.10)
$$0 \leq u_t \leq C, \qquad c < \infty$$

and

(9.11) $D_x u, D_x^2 u$ *belong to* $L_{\text{loc}}^{\infty}(\Omega \setminus G) \times (0, T)$.

Proof.   Let $0 < 3a < \text{dist}(\Gamma_0, \Gamma_1)$ and let $\eta(x)$ be a function in $C_0^{\infty}(R^n)$ such that $0 \leq \eta \leq 1$ and

$$\eta(x) = 1 \qquad \text{if dist}(x, \Gamma_0) < 2a$$

$$= 0 \qquad \text{if dist}(x, \Gamma_0) > 3a.$$

Let f_ε ($\varepsilon > 0$) be a suitable smooth patching of $h(x)$ with $-k$:

$$f_\varepsilon(x) = h(x)\phi_\varepsilon(x) - k(1 - \phi_\varepsilon(x)),$$

where ϕ_ε is ε-mollifier of the characteristic function of $G_\varepsilon \equiv \varepsilon$-neighborhood of G. Finally, let $\beta_\varepsilon(t)$ the standard function used in the penalized problem, with the additional requirement that

$$\beta_\varepsilon(0) = -k, \quad \beta_\varepsilon(t) = 0 \qquad \text{if } t > \varepsilon.$$

Consider the penalized problem, for $u = u_\varepsilon$,

(9.12) $u_t - \Delta u + \beta_\varepsilon(u) = f_\varepsilon$ in Q_T,

$$\begin{aligned} u &= \psi + \varepsilon & &\text{on } \Gamma_0 \times (0, T), \\ u &= \varepsilon\eta & &\text{on } \Omega \times \{0\}, \\ u &= 0 & &\text{on } \partial B_R \times (0, T). \end{aligned}$$

(9.13)

Notice that the initial and boundary conditions agree on $\partial G \times \{0\}$. As usual, $u_\varepsilon \to u$ when $\varepsilon \to 0$, where u is the solution of (9.8). We shall prove that

(9.14) $0 \leq \dfrac{\partial u_\varepsilon}{\partial t} \leq C.$

Differentiating (9.12) once with respect to t, we obtain for $w = \partial u_\varepsilon / \partial t$ the equation

$$w_t - \Delta w + \beta_\varepsilon'(u_\varepsilon)w = 0 \qquad \text{in } Q_T,$$

and

$$\begin{aligned} w &= f_\varepsilon + \varepsilon \Delta\eta - \beta_\varepsilon(u_\varepsilon) & &\text{on } t = 0, \\ &= \psi_t & &\text{on } \Gamma_0 \times (0, T), \\ &= 0 & &\text{on } \partial B_R \times (0, T). \end{aligned}$$

By the maximum principle

$$(9.15) \qquad \min \left\{ \min_{\partial_p Q_t} w, 0 \right\} \leqslant w(x, t) \leqslant \max \left\{ \max_{\partial_p Q_t} w, 0 \right\}.$$

To evaluate the left-hand side, notice that $f_\varepsilon \geqslant f, 0 \leqslant u_\varepsilon(x, 0) \leqslant \varepsilon$ if $\text{dist}(x, \Gamma_0) \leqslant 3a$, and $u_\varepsilon(x, 0) = 0$ if $\text{dist}(x, \Gamma_0) > 3a$. It follows that if ε is small enough,

$$\begin{aligned} w(x, 0) &> 0 & &\text{if } x \in G, \\ w(x, 0) &= f_\varepsilon - \beta_\varepsilon(0) \geqslant k - k = 0 & &\text{if } x \in B_R \setminus G. \end{aligned}$$

It is also easily seen that $w \geqslant 0$ elsewhere on $\partial_p Q_T$. Hence the left-hand side of (9.15) is $\geqslant 0$. One can readily check that the right-hand side of (9.15) is $\leqslant C$, and (9.14) thereby follows. Taking $\varepsilon \to 0$, we obtain (9.10).

For any $t \geqslant 0$, let

$$N(t) = \{x \in \Omega; u(x, t) > 0\};$$

this is an open set. From (9.14) we have that $u_t \geqslant 0$; hence

$$(9.16) \qquad N(t) \subset N(t') \qquad \text{if } t < t'.$$

We claim that

$$(9.17) \qquad \overline{N(0)} \subset N(t) \qquad \text{if } t > 0.$$

Indeed, suppose for contradiction that there is a point $x^0 \in \Gamma_1$ such that $x^0 \in \partial N(t_1)$ for some $t_1 > 0$. In view of (9.16), the segment

$$l_{t_1} = \{x = x^0, 0 < t < t_1\}$$

lies in the boundary of the open set

$$D_{t_1} = \bigcup_{0 < s < t_1} N(s).$$

Since, for a.e. t, $D_x u(x, t)$ is continuous in x, we get

$$(9.18) \qquad D_x u(x, t) = 0 \qquad \text{for a.e. } (x, t) \in l_{t_1}.$$

The function

$$v(x, t) = u(x, t + \delta) - u(x, t), \qquad \delta > 0,$$

is a positive solution of the heat equation in $G \times (0, t_1 - \delta)$ (since $u_t \geqslant 0$) and it vanishes on the boundary segment $l_{t_1-\delta}$. Hence, by the maximum principle,

$$\nabla_x v(x, t) \neq 0 \qquad \text{a.e. on } l_{t_1},$$

contradicting (9.18).

From (9.17) it follows that for $t > \varepsilon_0$ ($\varepsilon_0 > 0$), $f \equiv -k$ in a neighborhood of the free boundary; here ε_0 is any small number. Thus the local regularity result of Theorem 8.4 can be applied in order to deduce (9.11).

The introduction of the ball B_R was just a matter of convenient truncation so that the variational inequality may be considered in a bounded domain. We would therefore like to show that the solution does not depend on R if R is large enough. For this it suffices to show that the free boundary does not intersect $\partial B_R \times (0, T)$ if R is large enough (depending on T). We shall in fact prove:

Theorem 9.2. *There exists a positive constant M such that*

$$(9.19) \qquad N(t) \subset \left\{ |x| < M\sqrt{t+1} \right\} \qquad \text{for } 0 < t < T;$$

M depends on sup g, sup h, G *and* k, *but not on* T.

Proof. The proof is by comparison (see Section 8, Problem 5). We construct a radial solution of the classical Stefan problem in the form

$$\tilde{\theta}(x, t) = f\left(\frac{|x|}{\sqrt{t+1}} \right)$$

with $s(x) - t = 0$ given implicitly by $|x| = M\sqrt{t+1}$ and

$$f(z) = C \int_z^\infty \zeta^{1-n} \exp\left(-\frac{\zeta^2}{4} \right) d\zeta - C'.$$

We choose C, C' so that

$$2CM^{1-n} \exp\left(-\frac{M^2}{4} \right) = k,$$

$$C \int_M^\infty \zeta^{1-n} \exp\left(-\frac{\zeta^2}{4} \right) d\zeta - C' = 0,$$

and then the free-boundary conditions are satisfied. Notice that

$$C = \frac{k}{2} M^{n-1} \exp\left(\frac{M^2}{4} \right) \to \infty \qquad \text{if } M \to \infty,$$

C' remains bounded as $M \to \infty$.

It follows that the data $\tilde{h}, \tilde{g}$ corresponding to $\tilde{\theta}$ (with respect to G) satisfy:

$$\tilde{h} \geqslant h, \qquad \tilde{g} \geqslant g$$

if M is large enough. Choosing $R > M\sqrt{T+1}$, we can compare the variational solution $\tilde{u}$ (corresponding to $\tilde{\theta}$) with u and obtain: $\tilde{u} \geqslant u$, from which the assertion follows.

Remark 9.1. From now on we always take the truncating ball B_R such that $R > M\sqrt{T+1}$. Then the free boundary remains in a compact subset of $B_R \times [0, T]$.

Theorem 9.3. $D_x u$ *is uniformly continuous in* $\{u > 0\} \cap \{\varepsilon < t < T\}$, *for any* $\varepsilon > 0$.

Proof.   It is enough to show that if $(x_m, t_m) \to (x_0, t_0)$, where $u(x_m, t_m) > 0$ and (x_0, t_0) belongs to the free boundary, then $u_{x_i}(x_m, t_m) \to 0$. If the assertion is not true, then suppose for definiteness that

$$(9.20) \qquad u_{x_i}(x_m, t_m) \geqslant \beta > 0.$$

Let $x'_m = x_m + \delta e_i$, $x'_0 = x_0 + \delta e_i$, $\delta > 0$, and e_i the unit vector in the ith direction.

For a.a. t, $D_x u$ is Lipschitz continuous in x with coefficient C independent of t; writing

$$u(x_m, t) - u(x'_m, t) = \int_{x'_m}^{x_m} u_{x_i}(x, t) = \delta u_{x_i}(x_m, t) + \int_{x'_m}^{x_m} \int u_{x_i x_i}(x, t),$$

it follows that

$$|u(x_m, t) - u(x'_m, t) - \delta u_{x_i}(x_m, t)| \leqslant C\delta^2.$$

Since $x_m \in N(t_m)$, $u_{x_i}(x_m, t) \to u_{x_i}(x_m, t_m)$ as $t \to t_m$. Hence

$$|u(x_m, t_m) - u(x'_m, t_m) - \delta u_{x_i}(x_m, t_m)| \leqslant C\delta^2.$$

Taking $m \to \infty$ and using (9.20) and the fact that $u(x'_0, t_0) \geqslant 0$, we get $\delta\beta \leqslant C\delta^2$, which is impossible if $\delta < \beta/C$.

PROBLEMS

1. Suppose that $g(x, t) \geqslant C_0 > 0$ for all $x \in \Gamma_0$, $t > 0$. Show that there exists a small enough $\mu > 0$ and a $T_0 > 0$ such that the solution of the Stefan

problem satisfies $N(t) \supset \{|x| < \mu\sqrt{t}\}$ for all $T_0 < t < T$; T_0 depends on c_0 and G, but is independent of T, h.

2. Show that the free boundary for the Stefan problem with $n = 1$ is given by $x = s(t)$, where $s(t)$ is continuous and strictly monotone increasing in t.

3. Extend the definition of g to $\partial B_R \times (0, T)$ by $g = 0$, and consider the problem: Find bounded measurable $\theta(x, t) \geqslant 0$ such that

$$(9.21) \qquad \iint_{Q_T} (\theta \Delta \zeta + a(\theta)\zeta_t) \, dx \, dt = \int_0^T \int_{\partial\Omega} g \frac{\partial \zeta}{\partial \nu} \, dS \, dt - \int_\Omega a(\theta_0)\zeta(x, 0) \, dx,$$

where ζ is any smooth function in $\overline{Q}_T$, $\zeta = 0$ if $x \in \partial\Omega$ or if $t = T$, and $a(\theta) = a(\theta(x, t))$ is understood to be a measurable function such that

$$a(\theta(x, t)) = \theta(x, t) \qquad \text{if } \theta(x, t) > 0,$$

$$-k \leqslant a(\theta(x, t)) \leqslant 0 \qquad \text{if } \theta(x, t) = 0;$$

finally, $\theta_0(x) = \theta(x, 0)$. It is well known [94b] that this problem has a unique solution which can be obtained by approximating $a(t)$ by $a_m(t)$, $a'_m(t) > 0$. Notice that (9.21) can be written in the distribution sense as

$$(9.22) \qquad -\Delta\theta + \frac{\partial}{\partial t} a(\theta) = 0,$$

$$\theta = g \text{ on } \partial\Omega \times (0, T), \qquad \theta = 0 \text{ on } \Omega \times \{0\}.$$

Set

$$(9.23) \qquad u(x, t) = \int_0^t \theta(x, \tau) \, d\tau \qquad [(x, t) \in Q_T].$$

Prove: θ is the solution of (9.21) if and only if u is the solution of the Stefan problem (9.8).

[*Hint*: If θ solves (9.21), then

$$u_t - \Delta u - f = \gamma(u_t),$$

where $\gamma(u_t) = \theta - a(\theta) \geqslant 0$, so that

$$(9.24) \qquad (u_t - \Delta u)(v - u_t) \geqslant f(v - u_t) \qquad \forall v \geqslant 0, \quad u_t \geqslant 0$$

and conversely (9.24) implies (9.21). Show that if u solves (9.8), then is solves (9.24).]

10. VARIATIONAL INEQUALITIES FOR THE BIHARMONIC OPERATOR

The variational inequality

$$-\Delta u \geqslant 0, \qquad u \geqslant \phi, \quad \Delta u \cdot (u - \phi) = 0$$

in R^2 corresponds to a membrane that must stay above the obstacle ϕ. If instead of a membrane we have a two-dimensional plate, then the variational inequality is

$$\Delta^2 u \geqslant 0, \qquad u \geqslant \phi, \quad \Delta^2 u \cdot (u - \phi) = 0.$$

In order to formulate this problem more precisely, we introduce a bounded domain Ω in R^n with $C^{2+\alpha}$ boundary ($0 < \alpha < 1$). Let $\phi(x)$ be a function in $C^2(\overline{\Omega})$ such that $\phi \leqslant 0$ on $\partial\Omega$, and introduce the convex set

$$(10.1) \qquad K = \left\{ v \in H_0^2(\Omega); v \geqslant \phi \text{ a.e. in } \Omega \right\}.$$

Consider the variational problem: Find u such that

$$(10.2) \qquad u \in K, \qquad \int_\Omega |\Delta u|^2 \, dx = \min_{v \in K} \int |\Delta v|^2 \, dx,$$

or, equivalently,

$$(10.3) \qquad u \in K,$$

$$\int_\Omega \Delta u \cdot \Delta(v - u) \, dx \geqslant 0 \qquad \forall \, v \in K.$$

By Section 2 there exists a unique solution u of this problem. Taking $v = u + \varepsilon \zeta$ [$\varepsilon > 0$, $\zeta \geqslant 0$, $\zeta \in C_0^\infty(\Omega)$] in (10.3), we find that

$$\mu \equiv \Delta^2 u \geqslant 0 \qquad \text{in the sense of distributions.}$$

Hence μ is a measure in K. It follows that $\mu(K) < \infty$ for any compact subset K of Ω; see reference 159.

We shall prove later that $u \in W_{loc}^{2,\infty}(\Omega)$ and that, for $n = 2$, $u \in C^2(\Omega)$.

Lemma 10.1. *There exists a function w satisfying:*

(a) $w = \Delta u$ *a.e. in* Ω.

(b) w *is upper semicontinuous in* Ω.

(c) *For any $x^0 \in \Omega$ and for any sequence of balls $B_\rho(x^0)$ with center x^0 and radii ρ:*

$$\fint_{B_\rho(x^0)} w \, dx \downarrow w(x^0) \qquad \text{if } \rho \downarrow 0.$$

Here we use the notation

(10.4) $$\fint_A w = \frac{1}{|A|} \int_A w, \quad |A| = \text{measure of } A,$$

where A is either a ball or the boundary of a ball.

Proof. Let

(10.5) $$w_\rho(x) = \fint_{B_\rho(x)} \Delta u(y) \, dy.$$

We first show that for any $x^0 \in \Omega$,

(10.6) $$w_\rho(x^0) \quad \text{is decreasing in } \rho.$$

For any $u \in C^\infty$ we can write

$$\Delta u(x^0) = \fint_{S_\rho} (\Delta u) \, dS - \int_{B_\rho} (\Delta^2 u) G_\rho \, dx$$

where $B_\rho = B_\rho(x^0)$, $S_\rho = \partial B_\rho$, and

(10.7) $$G_\rho = \begin{cases} \gamma_n (r^{2-n} - \rho^{2-n}) & \text{if } n \geqslant 3 \ (\gamma_n > 0) \\ \dfrac{1}{2\pi} \log \dfrac{\rho}{r} & \text{if } n = 2 \end{cases}$$

is Green's function for $-\Delta$. Similarly, if $\rho' > \rho$,

$$\Delta u(x^0) = \fint_{S_{\rho'}} (\Delta u) \, dS - \int_{B_{\rho'}} (\Delta^2 u) G_{\rho'} \, dx.$$

Since $G_\rho \leqslant G_{\rho'}$, we get, provided that $\Delta^2 u \geqslant 0$,

$$\fint_{S_\rho} \Delta u \leqslant \fint_{S_{\rho'}} \Delta u.$$

and by integration,

(10.8) $$\fint_{B_\rho} \Delta u \leqslant \fint_{B_{\rho'}} \Delta u.$$

For a general $u \in H^2(\Omega)$ with $\Delta^2 u \geqslant 0$, we can introduce $1/m$-mollifiers

$$V_m = J_{1/m}(\Delta u),$$

where

$$J_\varepsilon(v)(x) = \int j_\varepsilon(x-y)v(y)\,dy, \qquad j_\varepsilon(x) = \varepsilon^{-n} j\left(\frac{x}{\varepsilon}\right),$$

(10.9) $\qquad j(x) = j_0(|x|), \quad j_0 \in C^\infty, j_0(t) = 0 \qquad \text{if } |t| > 1,$

$$j_0(t) \geqslant 0, \qquad \int j_0(|x|)\,dx = 1.$$

Since $\Delta V_m \geqslant 0$, (10.8) holds with Δu replaced by V_m. Taking $m \to \infty$, the inequality (10.8) follows. We conclude that

(10.10) $\qquad\qquad w_\rho(x) \downarrow w(x) \qquad \text{as } \rho \downarrow 0,$

where $w(x)$ is some function.

Since each w_ρ is continuous in x,

(10.11) $\qquad\qquad w(x)$ is upper semicontinuous.

Recalling that $\Delta u \in L^2_{\text{loc}}$, we also have

$$w_\rho(x) \to \Delta u(x) \qquad \text{a.e. in } \Omega.$$

Consequently,

$$w = \Delta u \qquad \text{a.e.,}$$

and together with (10.10), (10.11) the proof of the lemma is complete.

Theorem 10.2. *For any point $x^0 \in \Omega$ that belongs to the support of μ,*

(10.12) $\qquad\qquad w(x^0) \geqslant \Delta\phi(x^0).$

Proof. Extend u into a function in $H^2_{\text{loc}}(R^n)$ and denote by u_ε the ε-mollifier of u. Let $x^0 \in \Omega$. Suppose that for some $\delta > 0$ and a neighborhood W of x^0, the inequality

(10.13) $\qquad\qquad u_\varepsilon(x) - \phi(x) > \delta \qquad \text{for all } x \in W$

holds for all ε sufficiently small.

Let $\eta \in C_0^\infty(W)$, $\eta \geqslant 0$, be such that $\eta = 1$ in a neighborhood W_0 of x^0. Then for any $\zeta \in C_0^\infty(W_0)$, $|\zeta| < \delta/2$, the function

$$v = \eta u_\varepsilon + (1-\eta)u \pm \zeta$$

is in K. Taking this v in (10.3) and then letting $\varepsilon \to 0$, we obtain (since $u_\varepsilon \to u$

in H^2_{loc})

$$\int \Delta u \cdot \Delta \zeta = 0.$$

Thus $\Delta^2 u = 0$ in W_0 and consequently x^0 is not in the support of μ. It follows that if $x^0 \in \text{supp } \mu$, then there exist sequences $x_m \to x^0$ and $\varepsilon_m \to 0$ such that

(10.14)
$$u_{\varepsilon_m}(x_m) - \phi(x_m) \to 0.$$

By Green's formula

(10.15)
$$u_\varepsilon(x_m) = \oint_{S_{\rho, m}} u_\varepsilon \, dS - \int_{B_{\rho, m}} \Delta u_\varepsilon(y) \cdot V(x_m - y) \, dy,$$

where $B_{\rho, m} = \{|y - x_m| < \rho\}$, $S_{\rho, m} = \partial B_{\rho, m}$, and

$$V(z) = G_\rho, \qquad G_\rho \text{ as in (10.7) with } r = |z|,$$

is Green's function for $-\Delta$ in $B_{\rho, m}$. Similarly,

(10.16)
$$\phi_\varepsilon(x_m) = \oint_{S_{\rho, m}} \phi_\varepsilon \, dS - \int_{B_{\rho, m}} \Delta\phi_\varepsilon(y) \cdot V(x_m - y) \, dy.$$

Since $u \geq \phi$, also $u_\varepsilon \geq \phi_\varepsilon$, so that

$$\oint_{S_{\rho, m}} u_\varepsilon \geq \oint_{S_{\rho, m}} \phi_\varepsilon.$$

Using this inequality and (10.14), we obtain, by comparing (10.15) with (10.16), that

(10.17)

$$\liminf_{m \to \infty} \left[\int_{B_{\rho, m}} \Delta u_{\varepsilon_m}(y) \cdot V(x_m - y) \, dy - \int_{B_{\rho, m}} \Delta\phi_{\varepsilon_m}(y) \cdot V(x_m - y) \, dy \right] \geq 0.$$

We can write

(10.18)

$$\int_{B_{\rho, m}} \Delta u_\varepsilon(y) \cdot V(x_m - y) \, dy = \int_{B_{\rho, m}} dy V(x_m - y) \int_{|y-z|<\varepsilon} j_\varepsilon(y - z) \Delta u(z) \, dz$$

$$= \int_{B_{\rho, m}} \int_{|x_m - y - z| < \varepsilon} j_\varepsilon(x_m - y - z) V(z) \, dz \, \Delta u(y) \, dy + \lambda_{\varepsilon, m}$$

$$= \int_{B_{\rho, m}} (J_\varepsilon V)(x_m - y) \Delta u(y) \, dy + \lambda_{\varepsilon, m},$$

where $\lambda_{\varepsilon, m} \to 0$ if $\varepsilon \to 0$ (uniformly in m). A similar relation holds for the second integral in (10.17). Therefore,

$$\liminf_{m \to \infty} \int_{B_{\rho, m}} (J_{\varepsilon_m} V)(w - \Delta\phi)\, dy \geq 0.$$

By the mean value theorem there exist points $x_{m, \rho} \in B_{\rho, m}$ such that

$$w(x_{m, \rho}) - \Delta\phi(x_{m, \rho}) \geq -\delta_m, \qquad \delta_m \to 0 \text{ if } m \to \infty.$$

We may assume that $x_{m, \rho} \to x_\rho$ and then, by the upper semicontinuity of w,

$$w(x_\rho) - \Delta\phi(x_\rho) \geq 0.$$

Taking $\rho \to 0$, $x_\rho \to x^0$, we obtain (again using the upper semicontinuity of w) $w(x^0) - \Delta\phi(x^0) \geq 0$.

Theorem 10.3. Δu is in $L_{\text{loc}}^\infty(\Omega)$.

Proof. Take any point $x^0 \in \Omega$ and denote by B_ρ the ball with center x^0 and radius ρ. Choose R such that $\overline{B}_R \subset \Omega$ and $\zeta \in C_0^\infty(B_R)$, $\zeta = 1$ in $B_{2R/3}$, $0 \leq \zeta \leq 1$ elsewhere. For any $x \in B_{2R/3}$,

$$\Delta u_\varepsilon(x) = \Delta u_\varepsilon(x) \cdot \zeta(x) = -\int_{B_R} V\Delta(\Delta u_\varepsilon \cdot \zeta)\, dy,$$

where $V = G_R, G_\rho$ as in (10.7). Expanding, we get

(10.19)

$$\Delta u_\varepsilon(x) = -\int_{B_{R/2}} V\Delta^2 u_\varepsilon - \int_{B_R \setminus B_{R/2}} V[\Delta^2 u_\varepsilon \cdot \zeta + 2\nabla(\Delta u_\varepsilon) \cdot \nabla\zeta + \Delta u_\varepsilon \cdot \Delta\zeta].$$

Since $u \in H_{\text{loc}}^2(R^n)$,

(10.20) $$\int_\Omega |\Delta u_\varepsilon|^2 \leq C, \qquad C \text{ independent of } \varepsilon.$$

Notice that supp $\nabla\zeta$ is contained in $G_R \equiv B_R \setminus B_{2R/3}$. Hence

$$\int_{B_R \setminus B_{R/2}} V\nabla(\Delta u_\varepsilon) \cdot \nabla\zeta = -\int_{G_R} \Delta u_\varepsilon \cdot \nabla(V\nabla\zeta).$$

Using this and (10.20), we obtain from (10.19),

$$(10.21) \qquad \Delta u_\varepsilon(x) = -\int_{B_{R/2}} V \Delta^2 u_\varepsilon - \int_{B_R \setminus B_{R/2}} V \Delta^2 u_\varepsilon \cdot \zeta + \alpha_\varepsilon(x),$$

where $|\alpha_\varepsilon(x)| \leqslant C$ if $x \in B_{R/2}$.

By integration by parts [analogous to (10.18)] we have

$$(10.22) \qquad \int_{B_{R/2}} V(x-y) \Delta^2 u_\varepsilon(y) \, dy = \int_{B_{R/2}} V_\varepsilon(x-y) \Delta^2 u(y) + \beta_\varepsilon,$$

where $\beta_\varepsilon(x) \to 0$ if $x \in B_{R/2}$, $\varepsilon \to 0$.

Consider now the integral

$$\tilde{V}(x) = \int_{B_{R/2}} V(x-y) \, d\mu(y).$$

It exists in the sense of improper integrals, that is, as

$$\lim_{\delta \to 0} \int_{\substack{\{|x-y|>\delta\} \\ y \in B_{R/2}}} V(x-y) \, d\mu(y), \qquad \text{for a.e. } x.$$

Indeed, this follows from Fubini's theorem since for any $k < \infty$,

$$\int_{y \in B_{R/2}} d\mu(y) \int_{|x|<k} V(x-y) \, dx \leqslant C \int_{B_{R/2}} d\mu(y) < \infty.$$

Notice that $\tilde{V}$ is lower semicontinuous.

Observe next that $V_\varepsilon(z) = V(z)$ if $|z| > \varepsilon$ (since V is harmonic and the mollifier is obtained by taking averages on spheres) and $V_\varepsilon(z) \leqslant V(z)$ if $|z| < \varepsilon$. Therefore,

$$(10.23) \qquad \lim_{\varepsilon \to 0} \int_{B_{R/2}} V_\varepsilon(x-y) \, d\mu(y) \text{ exists and equals } \tilde{V}(x) \text{ a.e.}$$

Analogously to (10.22) we have, for $x \in B_{R/2}$,

$$\int_{B_R \setminus B_{R/2}} V(x-y) \Delta^2 u_\varepsilon(y) \cdot \zeta(y) \, dy$$

$$= \int_{B_R \setminus B_{R/2}} J_\varepsilon(\zeta(y) V(x-y)) \Delta^2 u(y) \, dy + \tilde{\beta}_\varepsilon(x),$$

where $\tilde{\beta}_\varepsilon(x) \to 0$ if $\varepsilon \to 0$. Hence
(10.24)

$$\int_{B_R \setminus B_{R/2}} V(x-y) \Delta^2 u_\varepsilon(y) \cdot \zeta(y) \, dy \to \int_{B_R \setminus B_{R/2}} V(x-y) \Delta^2 u(y) \cdot \zeta(y) \, dy$$

$$\text{if } x \notin B_{R/2}, \varepsilon \to 0.$$

We can write

$$\Delta u_\varepsilon(x) = \int u(z)\,\Delta j_\varepsilon(x-z)\,dz = \int \Delta u(z)\cdot j_\varepsilon(x-z)\,dz$$

$$= \int w(z)j_\varepsilon(x-z)\,dz = \int_0^\infty \int \lambda_\varepsilon(\rho)w(\rho,\theta)\,dS_\theta\,d\rho$$

where $(\rho,\theta) = (\rho,\theta_1,\ldots,\theta_{n-1})$ are the spherical coordinates about x and $\lambda_\varepsilon(\rho)$ is a smooth nonnegative function. Since

$$\frac{1}{\omega_n}\int w(\rho,\theta)\,dS_\theta \downarrow w(x) \qquad \text{as } \rho \downarrow 0,$$

where ω_n is the area of the unit sphere, the mean value theorem gives

$$\Delta u_\varepsilon(x) \to w(x) \qquad \text{as } \varepsilon \to 0.$$

Combining this with (10.22), (10.23), (10.24), we deduce from (10.21) upon taking $\varepsilon \to 0$, that

(10.25)

$$w(x) = -\tilde{V}(x) - \int_{B_R\setminus B_{R/2}} \zeta(y)V(x-y)\,\Delta^2 u(y)\,dy + \delta(x),\, |\delta(x)| \leqslant C$$

$$\text{for } x \in B_{R/2}.$$

We shall need the following (Evans) maximum principle:

Lemma 10.4. *Let Z be a superharmonic function in R^n; that is, Z is lower semicontinuous in R^n and $\nu = -\Delta Z$ is a measure. suppose that $Z(x) \to 0$ if $|x| \to \infty$ and set $S = \text{supp}\,\nu$. If $Z \leqslant M$ on S, where $M \geqslant 0$, then $Z \leqslant M$ in R^n.*

For proof, see reference 66 or 132.

We apply this result to $Z = \tilde{V} - \tilde{V}(\infty)$; the measure ν is the restriction of μ to $B_{R/2}$. By Theorem 10.2

$$w(x) \geqslant \Delta\phi(x) \qquad \text{on supp } \nu.$$

Since the integral on the right-hand side of (10.25) is $\geqslant 0$, we see that

$$\tilde{V}(x) \leqslant -w(x) + \delta(x) \leqslant C \qquad \text{on supp } \nu,$$

and Lemma 10.4 gives

$$\tilde{V}(x) \leqslant C \qquad \text{in } R^n.$$

Observing finally that the integral in (10.25) is bounded in $B_{R/3}$, we conclude that $|w| \leqslant C$ in $B_{R/3}$, and the proof is complete.

Theorem 10.5. $u \in W_{\text{loc}}^{2,\infty}(\Omega)$.

Proof. We can write

$$u_\varepsilon(x) = \int_{B_R} W(x-y)\, \Delta^2(\zeta u_\varepsilon)(y)\, dy, \qquad x \in B_{R/2},$$

where W is the fundamental solution of Δ^2:

$$
\begin{aligned}
W(x) &= \gamma_n |x|^{4-n} && \text{if } n = 3 \text{ or } n \geqslant 5 \\
&= \gamma_n \log \frac{1}{|x|} && \text{if } n = 4 \\
&= \gamma_n |x|^2 (\log|x| - 1) && \text{if } n = 2,
\end{aligned}
$$

where γ_n are constants, chosen such that

$$\Delta^2 W = \delta \qquad (\delta = \text{Dirac measure}).$$

Expanding $\Delta^2(\zeta u_\varepsilon) = \zeta \Delta^2 u_\varepsilon + \cdots$ and performing integration by parts, we find, after making use of the fact $|\Delta u_\varepsilon| \leqslant C$ (which follows from Theorem 4.3), that

$$(10.26) \qquad u_\varepsilon(x) = \int_{B_{2R/3}} W(x-y)\zeta(y)\,\Delta^2 u_\varepsilon(y)\, dy + \beta_\varepsilon(x),$$

where $|D^\alpha \beta_\varepsilon(x)| \leqslant C$ in $B_{R/2}$ for any derivative D^α.

One can directly verify that

$$(10.27) \qquad \left(\frac{\partial^2}{\partial x_j^2} - \frac{1}{2}\Delta \right) W \geqslant -c \qquad (c \text{ positive constant})$$

and

$$(10.28) \qquad \left(\frac{\partial^2}{\partial x_j^2} - \frac{1}{2}\Delta \right) W = \gamma_2 \frac{2x_j^2}{|x|^2} \qquad \text{if } n = 2.$$

Applying $\partial^2/\partial x_j^2 - \Delta/2$ to both sides of (10.26) and using (10.27) and the fact that $\zeta \Delta^2 u_\varepsilon \geqslant 0$, we obtain

$$\left(\frac{\partial^2}{\partial x_j^2} - \frac{1}{2}\Delta \right) u_\varepsilon \geqslant -c\int_{B_{2R/3}} \zeta(y)\,\Delta^2 u_\varepsilon(y)\, dy - C.$$

Since the last integral is equal to

$$\int_{B_{2R/3}} \zeta_\varepsilon(y) \, \Delta^2 u \, dy - \tilde{C}, \qquad |\tilde{C}| \leqslant C \text{ if } x \in B_{R/2},$$

with another C, we conclude that

$$\left(\frac{\partial^2}{\partial x_j^2} - \frac{1}{2}\Delta \right) u_\varepsilon \geqslant -C \qquad \text{in } B_{R/2}.$$

Since, by Theorem 10.3, Δu is locally bounded, the same is true of Δu_ε, and consequently

$$\frac{\partial^2 u_\varepsilon}{\partial x_j^2} \geqslant -C \qquad \text{in } B_{R/2}.$$

But then also

$$\frac{\partial^2 u_\varepsilon}{\partial x_j^2} = \Delta u_\varepsilon - \sum_{i \neq j} \frac{\partial^2 u_\varepsilon}{\partial x_i^2} \leqslant \Delta u_\varepsilon + (n-1)C,$$

and we obtain

$$\left| \frac{\partial^2 u_\varepsilon}{\partial x_j^2} \right| \leqslant C \qquad \text{in } B_{R/2}.$$

Taking $\varepsilon \to 0$ and noting that x_j can be in any direction, the assertion of the theorem follows.

In the next theorem we restrict the dimension n to be equal to 2.

Theorem 10.6.   *If $n = 2$, then $u \in C^2(\Omega)$.*

Proof.   By Green's formula, if $\overline{B_r(x^0)} \subset \Omega$,

$$\Delta u_\varepsilon(x^0) = -\int_{\partial B_r(x^0)} \Delta u_\varepsilon \cdot \frac{\partial G}{\partial \rho} \, dS - \int_{B_r(x^0)} G \, \Delta^2 u_\varepsilon \, dx,$$

where $G = (1/2\pi)\log r/\rho$ is Green's function with pole at x^0 and $\rho = |x - x^0|$. Since $|\Delta u_\varepsilon| \leqslant C$ in compact subsets of Ω,

$$\int_{B_r(x^0)} \log \frac{1}{|x^0 - y|} \Delta^2 u_\varepsilon(y) \, dy \leqslant C.$$

It follows that the measure $\mu_\varepsilon = \Delta^2 u_\varepsilon$ satisfies

$$(10.29) \qquad \mu_\varepsilon\left(B_r(x^0)\right) \leqslant \frac{C}{\log(1/r)} \qquad \text{if } B_r(x^0) \subset K, \bar{K} \subset \Omega,$$

where C depends on K.

Applying the operator

$$\square = \frac{\partial^2}{\partial x_1^2} - \frac{\partial^2}{\partial x_2^2}$$

to both sides of (10.26), we get

$$\square u_\varepsilon(x) = \int_{B_{2R/3}} F(x, y)\, d\mu_\varepsilon(y) + \gamma_\varepsilon(x),$$

where $\gamma_\varepsilon(x)$ is continuous in $x \in B_{R/2}$ uniformly with respect to ε and, by (10.28), $F(x, y)$ is a bounded function, continuous in (x, y) if $x \neq y$. Since, by (10.29),

$$\mu_\varepsilon\left(B_r(y)\right) \to 0 \text{ if } r \to 0, \quad \text{uniformly with respect to } y,\ \varepsilon,$$

a standard argument in potential theory shows that

$$(10.30) \qquad\qquad \square u_\varepsilon(x) \text{ is uniformly continuous in } x,$$

when x is restricted to any compact K subset of Ω, uniformly in ε, for $0 < \varepsilon \leqslant \varepsilon_0(K)$.

It follows that for a sequence $\varepsilon \to 0$,

$$(10.31) \qquad\qquad \square u_\varepsilon \to \square u \text{ uniformly in } x \in K,$$

where $\square u$ is a version of the distribution derivative $\square u$. Thus there is a version of $u_{x_1 x_1} - u_{x_2 x_2}$ that is continuous in Ω. By change of variables

$$x_1 \to \frac{x_1 + x_2}{\sqrt{2}}, \qquad x_2 \to \frac{x_1 - x_2}{\sqrt{2}},$$

we find that also $u_{x_1 x_2}$ has a continuous version in Ω.

Let us explain this last fact more carefully. For a sequence $\varepsilon \to 0$,

$$u_{\varepsilon, x} \to u_x, \quad u_{\varepsilon, y} \to u_y, \quad u_{\varepsilon, xy} \to U$$

uniformly in compact subsets. Hence $(u_x)_y$ and $(u_y)_x$ exist and coincide with U. We set, of course, $U = u_{xy}$. Thus

$$(10.32) \qquad (u_x)_y = (u_y)_x = u_{xy} \qquad \text{exists and is continuous.}$$

We next prove:

Lemma 10.7. *w is continuous in Ω.*

Proof. Denote by S the support of $\mu = \Delta^2 u$. By a continuity theorem for subharmonic functions [66], if w restricted to S is continuous, then w is continuous in Ω. Thus it suffices to show that

(10.33)   if $P_0 = (x_0, y_0) \in S$,   then $w|_S$ is continuous at P_0.

Let $P_m = (x_m, y_m) \in S$, $P_m \to P_0$ be such that if α_m is the angle between $\overline{P_0 P_m}$ and the y-axis, then

(10.34) $\alpha_m \to 0$ if $m \to \infty$,

$$| P_{m+1} - P_0 | < \tfrac{1}{5} | P_m - P_0 | .$$

We shall prove that

(10.35) $w(P_m) \to w(P_0).$

Take for simplicity $(x_0, y_0) = (0,0)$ and introduce the squares

$$R_m = \{ 0 < y < y_m, |x| < \tfrac{1}{2} y_m \}.$$

We can write

(10.36) $w = \Delta u = u_{xx} - u_{yy} + 2(u_{yy} - \phi_{yy}) + 2\phi_{yy}$ a.e.

Since $u - \phi = 0$, $\nabla(u - \phi) = 0$ at P_m,

(10.37) $\displaystyle\iint\limits_{R_m} (u_{yy} - \phi_{yy})\, dx\, dy$

$$= \int_{-(1/2)y_m}^{(1/2)y_m} (u_y - \phi_y)(x, y_m)\, dx - \int_{-(1/2)y_m}^{(1/2)y_m} (u_y - \phi_y)(x,0)\, dx$$

$$= \int_{-(1/2)y_m}^{(1/2)y_m} \int_{x_m}^{x} (u_{xy} - \phi_{xy})(\xi, y_m)\, d\xi\, dx$$

$$- \int_{-(1/2)y_m}^{(1/2)y_m} \int_{0}^{x} (u_{xy} - \phi_{xy})(\xi, 0)\, d\xi\, dx.$$

From the relations (with $\tilde{u} = u - \phi$)

$$0 = \tilde{u}_x(x_m, y_m) - \tilde{u}_x(0,0)$$

$$= [\tilde{u}_x(x_m, y_m) - \tilde{u}_x(x_m, 0)] + [\tilde{u}_x(x_m, 0) - \tilde{u}_x(0,0)]$$

$$= \int_0^{y_m} \tilde{u}_{xy}(x_m, y)\, dy + \int_0^{x_m} \tilde{u}_{xx}(x, 0)\, dx,$$

we obtain

$$\left(\tilde{u}_{xy}(0,0) + \delta_m\right)y_m + O(|x_m|) = 0 \qquad (\delta_m \to 0)$$

since $\tilde{u}_{xy}$ is continuous and $|\tilde{u}_{xx}| \leq C$. Using (10.34) we see that $\tilde{u}_{xy}(0,0) = 0$. We can therefore deduce from (10.37) that

$$(10.38) \qquad \frac{1}{|R_m|} \iint_{R_m} (u_{yy} - \phi_{yy})\, dx\, dy = \tfrac{1}{2}\sigma_m, \qquad \sigma_m \to 0 \text{ if } m \to \infty,$$

where $|R_m| = $ area of R_m.

Let U be a continuous version of the distribution derivative $u_{xx} - u_{yy}$ and introduce the continuous function

$$g = U + 2\phi_{yy}.$$

Then (10.38) gives

$$(10.39) \qquad \frac{1}{|R_m|} \iint_{R_m} (w - g)\, dx\, dy = \sigma_m, \qquad \sigma_m \to 0 \text{ if } m \to \infty.$$

Let B_m be the ball with center P_0 and radius y_m and let $P \in B_m$. Denote by $B_\rho(P)$ the ball with center P and radius ρ. Clearly,

$$R_m \subset B_{4y_m}(P).$$

Let M be an upper bound on $w - g$ in a compact subset of Ω which contains P_0. By (10.39),

$$\fint_{B_{4y_m}(P)} (w - g) \leq \lambda M + (1 - \lambda)\sigma_m \qquad (0 < \lambda < 1)$$

where λ is independent of m, provided that m is sufficiently large. Since w is subharmonic and g is continuous, the left-hand side is

$$\geq w(P) - g(P) + \eta_m, \qquad \eta_m \to 0 \text{ if } m \to \infty.$$

Hence

$$w(P) - g(P) \leq \lambda M + (1 - \lambda)\sigma_m + \eta_m \leq \lambda' M$$

for any $\lambda < \lambda' < 1$, provided that m is sufficiently large.

Thus

$$w - g \leq \lambda' M \qquad \text{if } P \in B_m, m \geq m_0.$$

Repeating the process, we obtain

$$w - g \leq (\lambda')^k M \qquad \text{in } B_{m_k},$$

and then

$$\frac{1}{|B_{m_k}|} \int_{B_{m_k}} (w - g) \leq (\lambda')^k M \to 0 \qquad \text{if } k \to \infty.$$

Since w is subharmonic and g is continuous, the left-hand side is $\geq w(P_0) - g(\tilde{P}_k)$, $\tilde{P}_k \in B_{m_k}$. It follows that

(10.40) $$w(P_0) \leq g(P_0).$$

For any small $h > 0$, consider the rectangle

$$T_h = \{x_m < x < x_m + h^2, y_m < y < y_m + h\}.$$

We can write

$$\iint_{T_h} (y_m + h - y)(u - \phi)_{yy} \, dx \, dy$$

$$= \int_0^{h^2} (u - \phi)(x_m + x, y_m + h) \, dx$$

$$- \int_0^{h^2} (u - \phi)(x_m + x, y_m) \, dx - \int_0^{h^2} h(u - \phi)_y(x_m + x, y_m) \, dx$$

$$\equiv J_1 - J_2 - J_3.$$

Since

$$|(u - \phi)(x_m + x, y_m)| \leq C|x|^2,$$

$$|(u - \phi)_y(x_m + x, y_m)| \leq C|x|,$$

we get

$$|J_2| \leq Ch^6, \qquad |J_3| \leq Ch^5.$$

Noting also that $J_1 \geq 0$, it follows that

$$\operatorname*{ess\,sup}_{T_h} \left[(y_m + h - y)(u - \phi)_{yy} \right] \geq -Ch^2.$$

Therefore, there exists a point $Q_h \in T_h$ such that

$$w(Q_h) - g(Q_h) \geq -Ch.$$

Taking $h \to 0$ and using the upper semicontinuity of w, we find that

$$w(P_m) \geq \lim \sup w(Q_h) \geq \lim g(Q_h) = g(P_m).$$

Thus

(10.41) $\qquad\qquad w(P_m) \geq g(P_m) \qquad$ for any P_m.

The proof is valid also for P_0, so that $w(P_0) \geq g(P_0)$. Recalling (10.40) it follows that

$$w(P_0) = g(P_0),$$

and (10.41) yields

$$\lim_{m \to \infty} \inf w(P_m) \geq g(P_0) = w(P_0).$$

Since w is also upper semicontinuous, (10.35) follows.

To complete the proof of (10.33), suppose for contradiction that $P_m \in S$, $\tilde{P}_m \in S$, $P_m \to P_0$, $\tilde{P}_m \to P_0$ and

$$w(P_m) \to A, \quad w(\tilde{P}_m) \to \tilde{A} \qquad \text{with } A \neq \tilde{A}.$$

By extracting a subsequence we may assume that (10.34) hold. Then $A = \lim w(P_m) = w(P_0)$. Similarly, $\tilde{A} = w(P_0)$ and we get a contradiction.

We have proved so far that $u_{xx} + u_{yy}$, $u_{xx} - u_{yy}$ have continuous versions. Hence also u_{xx} has a continuous version, say U_0. We can write, for any $(x, y) \in \Omega$, $|x - x_0|$ small,

$$u_x(x, y) - u_x(x_0, y) = \int_{x_0}^{x} U_0(\xi, y) \, d\xi$$

and thus conclude that $u_{xx}(x, y)$ exists and coincides with U_0. Similarly, u_{yy} exists and is continuous. Recalling also (10.32), the proof of the theorem is complete.

Remark 10.1. The results of this section are valid also for the variational inequality (10.2), (10.1), in which $H_0^2(\Omega)$ is replaced by $H^2(\Omega) \cap H_0^1(\Omega)$.

The variational inequality (10.2), (10.1) is called the *obstacle problem*. Other interesting variational inequalities arise with constraints on the second derivatives of u; see Section 12.

PROBLEMS

1. Consider the penalized problem

$$\Delta^2 u_\varepsilon + \beta_\varepsilon(u_\varepsilon - \phi) = 0 \text{ in } \Omega, \qquad u_\varepsilon \in H_0^2(\Omega),\cdot$$

where $\beta_\varepsilon(t) = \gamma'_\varepsilon(t)$, $\gamma_\varepsilon(t) = t^2/\varepsilon$ if $t < 0$, $\gamma_\varepsilon(t) = 0$ if $t > 0$. Show that this problem has a solution u_ε obtained as a minimizer of

$$\int_\Omega \left[|\Delta v|^2 + \gamma_\varepsilon(v - \phi) \right] dx, \qquad v \in H_0^2(\Omega).$$

Show also that

$$\int_\Omega |\Delta u_\varepsilon|^2 \leqslant C, \qquad \int_\Omega \gamma_\varepsilon(u_\varepsilon - \phi) \leqslant C$$

and $u_\varepsilon \to u$ weakly in $H_0^2(\Omega)$, where u is the solution of (10.2), (10.1).

2. Let $\Omega = \{|x| < 1\} \subset R^2$, $\phi = \varepsilon - r^2$, $r = |x|$, $0 < \varepsilon < 1$. Prove that the corresponding solution of (10.2), (10.1) is a function $u = u(r)$, and that:
    (a) $\{u > \phi\}$ coincides with $\{r > \delta\}$ for some $0 < \delta < 1$.
    (b) $\gamma \equiv \Delta u(1)$ is $\geqslant 0$ and

$$\Delta u = -\frac{4 + \gamma}{\log(1/\delta)} \log\frac{1}{r} + \gamma \qquad \text{if } r > \delta.$$

    (c) $\delta = \delta_\varepsilon \to 0$ if $\varepsilon \to 0$ and, consequently, the functions Δu do not have a modulus of continuity that is uniform with respect to ε.

3. Consider the variational inequality

$$u \in K, \qquad \int_\Omega \Delta u \cdot \Delta(v - u) \geqslant \int_\Omega f(v - u) \qquad \forall v \in K,$$

where $\partial\Omega$ is in C^1 and

$$K = \left\{ v \in H_0^1(\Omega) \cap H^2(\Omega), \alpha \leqslant \Delta v \leqslant \beta \text{ a.e.} \right\}$$

and $\alpha < 0 < \beta$. Show that the solution is given by $u = \tau(F)$, where $\Delta F = f$ in Ω, $F \in H_0^1(\Omega)$ and $\tau(t) = t$ if $\alpha \leqslant t \leqslant \beta$, $\tau(t) = \alpha$ if $t < \alpha$, $\tau(t) = \beta$ if $t > \beta$. It follows that $\Delta u \in W_0^{1,p}(\Omega)$ for any $2 \leqslant p < \infty$ (but $u \notin C^3$ in general).

4. Introduce the second finite-difference quotients

$$\Delta_{ih}v(x) = \frac{v(x + he_i) - 2v(x) + v(x - he_i)}{h^2}.$$

Let $u \in H^2(\Omega)$, $u \geqslant \phi$, $\phi \in C^2(\Omega)$, $\psi \in C_0^2(\Omega)$, $0 \leqslant \psi \leqslant 1$. Show that if ϕ is convex, then

$$u + \varepsilon\psi\Delta_{ih}u \geqslant \phi \qquad \text{if } \varepsilon, h \text{ are small enough.}$$

5. Let $g(x) = 1 - \frac{1}{2}\exp[r\Sigma_{i=1}^{n}(x_i - x_i^0)]$, where $x^0 = (x_1^0, \ldots, x_n^0)$ is a point in Ω. Show that if $\phi \in C^2(\Omega)$, then there exist sufficiently large numbers r, a and a sufficiently small neighborhood U of x^0 such that $g \geq \frac{1}{2}$ and $(\phi - a)g$ is convex in U.

6. Prove that the solution of (10.3), (10.1) is in $H_{\mathrm{loc}}^3(\Omega)$.
    [*Hint*: Take $v = u + \varepsilon g^{-1}\zeta^2 \Delta_{ih}[(u - a)g]$ in (10.3), $\zeta \in C_0^2(\Omega)$, $0 \leq \zeta \leq 1$.]

11. THIN OBSTACLES

Let Ω be a bounded domain in R^n with C^3 boundary and let S_0 be a C^3 hypersurface in R^n such that

(11.1) S_0 divides Ω into two domains, Ω_+ and Ω_-.   (See Figure 1.3.)

We shall use the notation

$$S = S_0 \cap \Omega,$$

so that $\partial S \subset \partial \Omega$.

Let ϕ be a continuous function defined on $\bar{S}$, C^2 on S, and

(11.2) $\phi < 0$ on ∂S, $\max_{\bar{S}} \phi > 0$,

and introduce the convex set

(11.3) $K = \{v \in H_0^1(\Omega), v \geq \phi \text{ a.e. on } S\}.$

We also introduce a bilinear form

$$a(u, v) = \int_\Omega \sum_{i, j=1}^n a_{ij}(x)\frac{\partial u}{\partial x_i}\frac{\partial v}{\partial x_j}dx,$$

where

(11.4) $\sum a_{ij}(x)\xi_i\xi_j \geq \kappa|\xi|^2$ $\forall \xi \in R^n,$ $x \in \Omega$ $(\kappa > 0),$

$$\sum \|a_{ij}\|_{1+\alpha} \leq K.$$

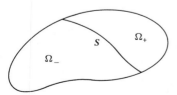

FIGURE 1.3

Consider the variational inequality:

(11.5) $u \in K$,

$$a(u, v - u) \geqslant 0 \qquad \forall \, v \in K.$$

We call it the *interior thin-obstacle problem*; ϕ is the *thin obstacle*.

Before we begin with the study of this problem, we prove a lemma that applies to general elliptic operators:

$$Au \equiv - \sum_{i, j=1}^{n} a_{ij}(x) \frac{\partial^2 u}{\partial x_i \, \partial x_j} + \sum_{i=1}^{n} b_i(x) \frac{\partial u}{\partial x_i} + c(x) u$$

provided that

(11.6) $\sum a_{ij}(x) \xi_i \xi_j \geqslant \kappa |\xi|^2 \qquad (\kappa > 0),$

$$\sum \|a_{ij}\|_1 + \sum \|b_i\|_1 + \|c\|_1 \leqslant K \qquad (c \geqslant 0).$$

Lemma 11.1. *Suppose that $Au = f$ in Ω, Ω a bounded domain, and $u \in C^3(\Omega) \cap C^1(\bar{\Omega})$,*

(11.7) $\| f \|_0 \leqslant M, \qquad \| \nabla f \|_0 \leqslant M \qquad (M \geqslant 1).$

If

$$|u| \leqslant M, \quad |\nabla u| \leqslant M \quad \text{on } \partial\Omega,$$

then

$$|u| \leqslant CM, \quad |\nabla u| \leqslant CM \qquad \text{in } \Omega$$

where C is a constant depending only on κ, K, and Ω.

Proof. Suppose that $\Omega \subset \{x_1 > h\}$ and take

$$v = 1 - e^{-\lambda(x_1 - h + 1)} \qquad (\lambda > 0).$$

If λ is sufficiently large then $v > 0$, $Av > 0$ in $\bar{\Omega}$. Applying the maximum principle to $Cv \pm u$, we obtain $|u| \leqslant CM$.

To estimate $|\nabla u|$, we use the Bernstein trick of applying the maximum principle to the function

$$w = u_\lambda u_\lambda + \gamma u^2 \qquad \text{(with suitable } \gamma > 0)$$

where $v_\lambda = \partial v / \partial x_\lambda$ and the summation convention is used. We compute

$$w_i = 2 u_\lambda u_{\lambda i} + 2 \gamma u u_i,$$

$$w_{ij} = 2 u_{\lambda i} u_{\lambda j} + 2 u_\lambda u_{\lambda ij} + 2 \gamma u_i u_j + 2 \gamma u u_{ij},$$

so that

$$Aw = -2a_{ij}u_{\lambda i}u_{\lambda j} - 2u_\lambda a_{ij}u_{\lambda ij} - 2\gamma a_{ij}u_i u_j - 2\gamma u a_{ij}u_{ij}$$

$$+ 2b_i u_\lambda u_{\lambda i} + 2\gamma u b_i u_i + cu_\lambda u_\lambda + \gamma cu^2.$$

Noting that

$$(Au)_\lambda = Au_\lambda - a_{ij,\lambda}u_{ij} + b_{i,\lambda}u_i + c_\lambda u,$$

we get

$$(11.8) \qquad Aw = 2\gamma Au + 2u_\lambda (Au)_\lambda + 2u_\lambda (a_{ij,\lambda}u_{ij} - b_{i,\lambda}u_i - c_\lambda u)$$

$$- 2a_{ij}u_{\lambda i}u_{\lambda j} - 2\gamma a_{ij}u_i u_j - \gamma cu^2 - cu_\lambda u_\lambda.$$

Using the assumptions (11.6), (11.7) we find that

$$Aw \leq C|\nabla u|\left[|\nabla^2 u| + |\nabla u| + M\right] + \gamma CM - 2\kappa u_{\lambda i}u_{\lambda i} - 2\gamma\kappa|\nabla u|^2$$

$$\leq -\kappa u_{\lambda i}u_{\lambda i} + C|\nabla u|^2 + CM^2 + \gamma CM - 2\gamma\kappa|\nabla u|^2,$$

where C is a constant depending on κ, K but not on γ. Choosing γ sufficiently large, we obtain the inequality $Aw \leq CM^2$ with another C. We can now apply the maximum principle to $CM^2 v - w$ and derive the estimate $w \leq CM^2$ in Ω; hence $|\nabla u| \leq CM$.

We shall also need the following result [161f] (see also reference 109).

Lemma 11.2. Set $d(x) = \text{dist}(x, \partial\Omega)$ for $x \in \overline{\Omega}$, and let m be an integer ≥ 2, $0 < \alpha < 1$. If $\partial\Omega$ is in $C^m (C^{m+\alpha})$, then there exists an Ω-neighborhood N of $\partial\Omega$ such that $d(x)$ is in $C^m(\overline{N}) [C^{m+\alpha}(\overline{N})]$.

We return to the variational inequality (11.5), and suppose that u is a solution. By choosing $v = u + \varepsilon\zeta$ with $\zeta \in C_0^\infty(\Omega)$, $\zeta \geq 0$ on S we find that u is a weak solution of

$$(11.9) \qquad\qquad -\sum \frac{\partial}{\partial x_i}\left(a_{ij}\frac{\partial u}{\partial x_j}\right) \geq 0 \qquad \text{in } \Omega;$$

Also,

$$(11.10) \qquad\qquad \sum \frac{\partial}{\partial x_i}\left(a_{ij}\frac{\partial u}{\partial x_j}\right) = 0 \qquad \text{in } \Omega \setminus S.$$

If, further, u belongs to $C^1(\overline{\Omega}_+)$ and to $C^1(\overline{\Omega}_-)$, then

$$\frac{\partial u}{\partial \mu^+} + \frac{\partial u}{\partial \mu^-} \geqslant 0 \quad \text{on } S;$$

here $\mu^\pm$ is the outward conormal to $\Omega_\pm$ at S; that is, $\mu^\pm$ is in the direction of the vector

$$\left(a_{ij}(x) \cos\left(x, \nu_j^\pm \right) \right),$$

where $\nu_j^\pm$ are the component of the outward normal $\nu^\pm$ to $\Omega_\pm$ at S. If $u(x) > \phi(x)$, we actually get equality; thus

$$(11.11) \qquad \left. \begin{array}{c} u \geqslant \phi \\[4pt] \dfrac{\partial u}{\partial \mu^+} + \dfrac{\partial u}{\partial \mu^-} \geqslant 0 \\[8pt] (u - \phi)\left(\dfrac{\partial u}{\partial \mu^+} + \dfrac{\partial u}{\partial \mu^-} \right) = 0 \end{array} \right\} \quad \text{on } S.$$

Theorem 11.3. *There exists a unique solution u of the variational inequality* (11.5) *and* $u \in C^{0,1}(\overline{\Omega})$.

Proof. Existence and uniqueness follow from Section 2. The set

$$S_* = \{ x \in S, \phi(x) \geqslant 0 \}$$

is a compact subset of Ω.

Let $x^i = f^i(s)$ be a local parametrization of S. Then the hypersurfaces

$$S^\delta: x^i = f^i(s) + \nu^i(s)\delta \qquad (|\delta| \text{ small}),$$

where $(\nu^1(s), \ldots, \nu^n(s))$ is the vector $\nu = \nu^-$ at $f(s)$, are parallel to S at distance δ and they lie in Ω_+ if $\delta > 0$ and in Ω_- if $\delta < 0$, provided that $f(s)$ is restricted to a small neighborhood of S_*.

We extend ϕ as a C^2 function in $\overline{\Omega}$ such that, for some sufficiently small $\delta_0 > 0$,

$$\phi(f(s) + \nu(s)\delta) = \phi(f(s))$$

if $|\delta| < \delta_0$ and dist$(f(s), S_*) < \delta_0$, and $\phi < 0$ on $\partial\Omega$.

Let $d^+(x)$ be a function in $C^2(\overline{\Omega}_+)$ which coincides with dist(x, S) if dist$(x, S_*) < \delta_0$ and is $\geqslant \delta_0/2$ elsewhere in Ω_+. Define $d^-(x)$ in a similar manner with respect to Ω_-. Set $d(x) = d^\pm(x)$ in $\overline{\Omega}_\pm$.

For any small $\varepsilon > 0$ we now consider a "thick obstacle"

$$(11.12) \qquad \phi_\varepsilon(x) = \phi(x) - \frac{d^2(x)}{\varepsilon}.$$

This obstacle is in $C^2(\overline{\Omega})$, is positive in the interior of S_*, and is strictly negative in a neighborhood of $\partial\Omega$ and on $S \setminus S_*$.

Consider the variational inequality

(11.13) $$u_\varepsilon \in K_\varepsilon,$$

$$a(u_\varepsilon, v - u_\varepsilon) \geq 0 \qquad \forall v \in K_\varepsilon,$$

where

(11.14) $$K_\varepsilon = \{v \in H_0^1(\Omega), v \geq \phi_\varepsilon \text{ in } \Omega\};$$

and set

$$Av \equiv -\sum \frac{\partial}{\partial x_i}\left(a_{ij} \frac{\partial v}{\partial x_j}\right).$$

If u_ε is the solution of (11.13), then

$$Au_\varepsilon \geq 0 \qquad \text{a.e. in } \Omega.$$

Since also $u_\varepsilon \in H_0^1(\Omega)$, the maximum principle gives

(11.15) $$u_\varepsilon > 0 \qquad \text{in } \Omega.$$

Hence the coincidence set

$$\Lambda_\varepsilon = \{x \in \Omega; u_\varepsilon = \phi_\varepsilon\}$$

must be contained in the set $\{\phi_\varepsilon > 0\}$. It follows that

(11.16) $$\Lambda_\varepsilon \subset \{\phi(x) > 0, d^2(x) < \varepsilon\phi(x)\},$$

$$\Lambda_\varepsilon \cap S \subset S_*.$$

Let w be the solution of

$$Aw = 0 \qquad \text{in } \Omega,$$

$$w = \phi \qquad \text{on } S,$$

$$w = 0 \qquad \text{on } \partial\Omega \setminus S.$$

By comparison, $u_\varepsilon \geq w$ in Ω_+. Hence if $x \in \Lambda_\varepsilon$, then

$$w(x) \leq \phi(x) - \frac{d^2(x)}{\varepsilon}.$$

Since w is Lipschitz continuous in some $\overline{\Omega}_+$-neighborhood of S_*,

$$w(x) \geqslant w(x^0) - C|x - x^0|,$$

where x^0 is the nearest point to x on S [that is, $|x - x^0| = d(x)$; note that $|x - x^0|$ is $\leqslant \mathrm{const}\sqrt{\varepsilon}$ by (11.16)]. Since, by definition of $\phi(x)$ in $\Omega \setminus S$

$$\phi(x) = \phi(x^0) \qquad \text{if } |x - x^0| \text{ is small enough,}$$

we get

$$-Cd(x) \leqslant -\frac{d^2(x)}{\varepsilon},$$

so that $d(x) \leqslant C\varepsilon$. Thus we have improved (11.16):

(11.17) $\Lambda_\varepsilon \cap \Omega$ lies in a $C\varepsilon$-neighborhood of S_*.

The function u_ε satisfies

$$u_\varepsilon = 0 \qquad \text{in } \Omega \setminus \Lambda_\varepsilon.$$

By (11.17)

$$|u_\varepsilon| = |\phi_\varepsilon| \leqslant C \qquad \text{on } \Lambda_\varepsilon,$$

$$|\nabla u_\varepsilon| = |\nabla \phi_\varepsilon| \leqslant C \qquad \text{on } \Lambda_\varepsilon.$$

Applying Lemma 11.1, we obtain

(11.18) $|u_\varepsilon| + |\nabla u_\varepsilon| \leqslant C \qquad \text{in } \Omega \setminus \Lambda_\varepsilon.$

Now take a sequence $u_\varepsilon \to \tilde{u}$ uniformly in Ω, such that $\nabla u_\varepsilon \to \nabla \tilde{u}$ weakly star in $L^\infty(\Omega)$. By Minty's lemma

$$a(v, v - u_\varepsilon) \geqslant 0$$

if $v \in H_0^1(\Omega)$, $v \geqslant \phi_\varepsilon$; this holds in particular if $v \in C_0^\infty(\Omega)$, $v > \phi$ on S, provided that ε is small enough. It follows that for such v

$$a(v, v - \tilde{u}) \geqslant 0.$$

By approximation, this inequality holds also for any $v \in H_0^1(\Omega)$, $v \geqslant \phi$ on S. Thus $\tilde{u}$ coincides with u and, from (11.18),

$$|\nabla u| \leqslant C.$$

This completes the proof of the theorem.

Consider now the case where the hypersurface S satisfies

(11.19) $$S \subset \partial\Omega, \qquad \overline{S} \neq \Omega$$

and suppose that $\phi < 0$ on ∂S. Let

(11.20) $$K = \{v \in H^1(\Omega), v = 0 \text{ on } \partial\Omega \setminus S, v \geq \phi \text{ on } S\}.$$

The corresponding variational inequality (11.5) is called the *boundary thin-obstacle problem*, or the *Signorini problem*. In this case we formally have

(11.21) $$\left.\begin{array}{r} u - \phi \geq 0 \\[4pt] \dfrac{\partial u}{\partial \mu} \geq 0 \\[6pt] (u - \phi)\dfrac{\partial u}{\partial \mu} = 0 \end{array}\right\} \quad \text{on } S$$

where μ is the outward conormal. Theorem 11.3 extends to the present case. In fact, since

$$\int_\Omega |v|^2 \leq C \int_\Omega |\nabla v|^2$$

for any $v \in C^1(\Omega)$, $v = 0$ on $\partial\Omega \setminus S$ (see Problem 4), $a(u, v)$ is coercive and thus the Signorini problem has a unique solution u. The proof that $u \in C^{0,1}(\overline{\Omega})$ is similar to the proof in Theorem 11.3.

We return to the interior thin-obstacle problem and prove additional regularity of the solution. Consider first the special case of

(11.22) $$\int_\Omega \nabla u \cdot \nabla(v - u) \geq 0 \qquad \forall v \in K; \quad u \in K$$

and

(11.23) $$S_0 \text{ is a hyperplane.}$$

We set $x = (x_1, \ldots, x_m)$, where $m = n - 1$, $y = x_n$, and $X = (x, y)$, and take $S_0 = \{y = 0\}$. Recall that the coincidence set

$$\Lambda = \{(x, 0); u(x, 0) = \phi(x)\}$$

is a closed subset of Ω.

We shall often identify the points x with $(x, 0)$.

Theorem 11.4. *The solution u is in $C^{1+\alpha}(\overline{\Omega}_+)$ and in $C^{1+\alpha}(\overline{\Omega}_-)$, for some $0 < \alpha < 1$.*

The proof will be based on several lemmas. We begin with the penalized problem

$$(11.24) \qquad -\Delta u_{\varepsilon, \delta} + \beta_\delta(u_{\varepsilon, \delta} - \phi_\varepsilon) = 0 \qquad \text{in } \Omega,$$

$$u_{\varepsilon, \delta} = 0 \qquad \text{on } \partial\Omega,$$

where $\beta_\delta(t) < 0$, $\beta_\delta'(t) > 0$, $\beta_\delta''(t) < 0$, $\beta_\delta(t) \to 0$ if $t \geqslant 0$, $\delta \downarrow 0$, $\beta_\delta'(0) \to 0$ if $\delta \downarrow 0$, and $\beta_\delta(t) \to -\infty$ if $t < 0$, $\delta \downarrow 0$.
Then

$$-\Delta u_{\varepsilon, \delta} \geqslant 0$$

and therefore $u_{\varepsilon, \delta} \geqslant w$ (w as in the proof of Theorem 11.3). We can proceed as before to show that

$$(11.25) \qquad |\nabla u_{\varepsilon, \delta}| \leqslant C.$$

Since $u_{\varepsilon, \delta} > \phi_\varepsilon$ in an Ω-neighborhood of $\partial\Omega$ that is independent of ε, δ, $\Delta u_{\varepsilon, \delta} = 0$ in this neighborhood and, consequently,

$$(11.26) \qquad |D^2 u_{\varepsilon, \delta}| \leqslant C \text{ in an } \Omega\text{-neighborhood of } \partial\Omega.$$

Differentiating the penalized equation twice with respect to any tangential direction τ in the x-space, we get

$$-\Delta(D_{\tau\tau} u_{\varepsilon, \delta}) + \beta'(u_{\varepsilon, \delta} - \phi_\varepsilon) D_{\tau\tau}(u_{\varepsilon, \delta} - \phi_\varepsilon) \geqslant 0.$$

If $D_{\tau\tau} u_{\varepsilon, \delta}$ takes negative minimum at a point $X^0 \in \Omega$, then we deduce that

$$D_{\tau\tau} u_{\varepsilon, \delta} \geqslant D_{\tau\tau} \phi_\varepsilon \geqslant -C \qquad \text{at } X^0.$$

Recalling also (11.26), we conclude that

$$(11.27) \qquad D_{\tau\tau} u_{\varepsilon, \delta} \geqslant -C \qquad \text{in } \Omega.$$

Taking $\delta \to 0$ and then $\varepsilon \to 0$, we obtain:

Lemma 11.5. $u_{\tau\tau} \geqslant -C$, $u_{yy} \leqslant C$ in $\Omega \setminus \Lambda$.

The next lemma is a comparison result.

Lemma 11.6. Let $\sup_S \phi_{\tau\tau} \leqslant M$ for all τ. If $A \gg M$, $(x_0, 0) \notin \Lambda$ and

$$(11.28) \quad h_{x_0}(X) = \phi(x_0) + \nabla\phi(x_0) \cdot (x - x_0) + A(|x - x_0|^2 - my^2),$$

then for any open set Q such that

$$(x_0, 0) \in Q \subset \Omega,$$

there holds:

(11.29) $$\sup_{\partial Q} (u - h_{x_0}) > 0.$$

Proof. At $(x_0, 0)$ $u \geqslant h_{x_0}$ [since $u(x_0, 0) > \phi(x_0)$]. Noticing that $u - h_{x_0}$ is harmonic in $Q \setminus \Lambda$, the maximum principle gives

$$\sup_{\partial(Q \setminus \Lambda)} (u - h_{x_0}) \geqslant 0.$$

Since however $u - h_{x_0} \leqslant 0$ on Λ if $A \gg M$, the assertion follows.

Lemma 11.5 implies that $u_y - Cy$ is a bounded monotone function, for $y > 0$ and for $y < 0$, and therefore the limits

$$\sigma_1(x) = \lim_{y \downarrow 0} u_y(x, y),$$

$$\sigma_2(x) = -\lim_{y \uparrow 0} u_y(x, y).$$

exist. We shall temporarily assume that

(11.30) Ω is symmetric with respect to $\{y = 0\}$.

Then $u(x, y) = u(x, -y)$ and $\sigma_1(x) = \sigma_2(x)$; set $\sigma(x) = \sigma_1(x)$. In view of (11.11),

(11.31) $\sigma(x) \leqslant 0$ if $(x, 0) \in \Lambda$, i.e., if $u(x, 0) = \phi(x)$,
(11.32) $\sigma(x) = 0$ if $(x, 0) \in S \setminus \Lambda \equiv N$, i.e., if $u(x, 0) > \phi(x)$.

We are interested in proving Hölder continuity of σ. We begin with continuity in measure.

Lemma 11.7.   *Let $(x_0, 0) \in N$. There exist positive constants $C, \overline{C}$ such that for any small $\gamma > 0$, there exists a ball $B_{C\gamma}(\bar{x})$ contained in $B_{\overline{C}\gamma}(x_0) \cap \lambda_\gamma$, where $\lambda_\gamma = \{x \in S; \sigma(x) > -\gamma\}$.*

Proof.   Take a cylinder

$$Q = B_{C_1\gamma}(x_0) \times (-C_2\gamma, C_2\gamma) \qquad \text{with } C_2 \ll C_1$$

and apply Lemma 11.6. We distinguish between two cases:

(i) $\sup_{\partial Q}(u - h_{x_0})$ is attained at a point (x_1, ε) on the lateral boundary of the cylinder.

(ii) $\sup_{\partial Q}(u - h_{x_0})$ is attained on one of the bases of the cylinder, say at $(x_1, C_2\gamma)$.

In case (i),

$$u(x_1, \varepsilon) \geq h_{x_0}(x_1, \varepsilon) = \phi(x_0) + \nabla\phi(x_0) \cdot (x_1 - x_0)$$

$$+ A(|x_1 - x_0|^2 - m\varepsilon^2) \geq \phi(x_1) + C_3\gamma^2$$

since $\varepsilon \leq C_2\gamma \ll C_1\gamma = |x_1 - x_0|$. Also, if

(11.33) $|x_2 - x_1| \leq C_4\gamma, \qquad (x_2 - x_1) \cdot \nabla(u - \phi)(x_1, \varepsilon) \geq 0$

then (since $u_{\tau\tau} \geq -C$)

$$(u - \phi)(x_2, \varepsilon) - (u - \phi)(x_1, \varepsilon) = (x_2 - x_1) \cdot \nabla(u - \phi)(x_1, \varepsilon)$$

$$+ \int\int (u - \phi)_{\tau\tau} > -C_3\gamma^2$$

provided that C_4 is small enough. It follows that

$$(u - \phi)(x_2, \varepsilon) > 0.$$

We claim that for x_2 as in (11.33), $\sigma(x_2) > -\gamma$. Indeed, if $\sigma(x_2) \leq -\gamma$, then $(u - \phi)(x_2, 0) = 0$ [by (11.32)] and then

$$(u - \phi)(x_2, \varepsilon) = (u - \phi)(x_2, \varepsilon) - (u - \phi)(x_2, 0)$$

$$= \varepsilon\sigma(x_2) + \int\int u_{yy} \leq -\gamma\varepsilon + C\varepsilon^2 < 0$$

if C_2 is small enough (since $\varepsilon \leq C_2\gamma$).

The set of points x_2 satisfying (11.33) certainly contains a ball $B_{C\gamma}(\bar{x})$. Since $\sigma(x_2) > -\gamma$ in this ball, the proof is complete.

In case (ii) we write

$$u(x_1, C_2\gamma) \geq h_{x_0}(x_1, C_2\gamma)$$

$$\geq \phi(x_0) + \nabla\phi(x_0) \cdot (x_1 - x_0) + A(|x_1 - x_0|^2 - m(C_2\gamma)^2)$$

$$\geq \phi(x_1) - C_5(C_2\gamma)^2,$$

and, as before, if

(11.34) $|x_2 - x_1| \leqslant C_6(C_2\gamma)^2,$ $(x_2 - x_1) \cdot \nabla(u - \phi)(x_1, C_2\gamma) \geqslant 0$

then

$$u(x_2, C_2\gamma) - \phi(x_2) \geqslant -C_7(C_2\gamma)^2.$$

Since, on the other hand,

$$u(x_2, C_2\gamma) - \phi(x_2) = \sigma(x_2)C_2\gamma + \iint u_{yy} \leqslant \sigma(x_2)C_2\gamma + C(C_2\gamma)^2,$$

we conclude that $\sigma(x_2) > -\gamma$ if C_2 is small enough. Finally, the set of points x_2 satisfying (11.34) certainly contains a ball $B_{C\gamma}(\bar{x})$.

In the next lemma we shall use the following simple fact:

(11.35)
> If w is nonnegative harmonic function in
> $B_1(\bar{x}) \times [0, 1]$, $w \geqslant 1$ on $B_\delta \times \{0\}$ where $\delta > 0$,
> $B_\delta \subset B_1(\bar{x})$, then $w \geqslant \varepsilon(\delta) > 0$ in $B_{1/2}(\bar{x}) \times (\frac{1}{4}, \frac{3}{4})$.

Lemma 11.8.   *If $(x_0, 0) \in N$, then there exists a number α, $0 < \alpha < 1$, such that*

$$\sigma(x) \geqslant -C|x - x_0|^\alpha$$

for some positive constant C.

Proof.   We shall prove inductively that

(11.36) $u_y \geqslant -\beta^k$ in $B_{\gamma^k}(x_0) \times [0, \gamma^k],$

where $0 < \gamma \ll \beta < 1$. By scaling u we may assume that (11.36) holds for $k = 0$. To proceed from k to $k + 1$, consider for small $\mu > 0$ the function

(11.37) $$w = \frac{u_y + \beta^k}{-\mu\gamma^k + \beta^k}.$$

This function is harmonic in $B_{\gamma^k}(x_0) \times (0, \gamma^k)$. By Lemma 11.7 with γ replaced by $\mu\gamma^k$, we have that

$$w \geqslant 1   \text{in } B_{\delta\gamma^k}(\bar{x}) \times \{0\};$$

here μ has to be chosen sufficiently small [but independently of k, γ, and

$\delta = \delta(\mu)$ is also independent of k, γ]. Using (11.35) (after appropriate scaling), we deduce that

$$\frac{u_y + \beta^k}{-\mu\gamma^k + \beta^k} \geq c > 0 \quad \text{in } B_{\gamma^k/2}(x_0) \times \left[\tfrac{1}{4}\gamma^k, \tfrac{3}{4}\gamma^k\right].$$

Since $\gamma \ll \beta$ we obtain

$$u_y \geq -\beta^k + \frac{\lambda}{2}\beta^k$$

for some small $\lambda > 0$, λ independent of β, γ, k.
   If $0 < y < \gamma^k/4$, then

$$u_y(x, y) = u_y\left(x, \tfrac{1}{4}\gamma^k\right) - u_{yy}(x, \tilde{y})\left(y < \tilde{y} < \tfrac{1}{4}\gamma^k\right)$$

$$\geq -\beta^k + \frac{\lambda}{2}\beta^k - C\gamma^k \geq -\left(1 - \frac{\lambda}{4}\right)\beta^k.$$

Choosing $\gamma < 1/2$, $\beta > 1 - \lambda/2$, the proof for $k + 1$ follows. Finally, the assertion of the lemma follows immediately from the inequalities (11.36).

Completion of the Proof of Theorem 11.4.   Since $\sigma = 0$ on N, Lemma 11.8 gives

$$\sigma(x) - \sigma(x_0) = \sigma(x) \geq -C|x - x_0|^\alpha$$

if $x_0 \in N$, $x \in S$. Since also $\sigma(x) \leq 0$, we get

$$(11.38) \qquad |\sigma(x) - \sigma(x_0)| \leq C|x - x_0|^\alpha, \qquad x_0 \in N, \quad x \in S.$$

Now, in int Λ, $u = \phi$ and u is then regular on either side Ω_+ or Ω_- (up to the boundary). If x_1 and x_2 belong to Λ and

$$|x_1 - x_2| < \min\left\{d(x_1, N), d(x_2, N)\right\}^2 \equiv d^2$$

then, by interior regularity,

$$|\sigma(x_1) - \sigma(x_2)| \leq \frac{C|x_1 - x_2|}{d} \leq C|x_1 - x_2|^{1/2}.$$

If $|x_1 - x_2| > d^2$, then let x_0 be a point in N whose distance to either x_1 or x_2 is $\leq |x_1 - x_2|^{1/2}$. Applying (11.38) with $x = x_1$, $x = x_2$, we get

$$|\sigma(x_1) - \sigma(x_2)| \leq C|x - x_0|^{\alpha/2}.$$

We have thus proved that $\sigma(x)$ is Hölder continuous in some Λ-neighborhood of $\partial\Lambda$. Since $\sigma(x) = 0$ in N, it follows that $\sigma(x)$ is Hölder continuous, say with exponent α, in some neighborhood of $\partial\Lambda$. But then, by a standard result in potential theory, $u \in C^{1+\alpha}$ up to the boundary, from either side of S; see Problem 4.

We have completed the proof of Theorem 11.4 under the assumption (11.30). In the general case we take a domain $\Omega_0 \subset \Omega$ symmetric with respect to $\{y = 0\}$ such that $\Omega_0 \cap \{y = 0\}$ contains $S^* = \{x \in S, \phi(x) \geqslant 0\}$, and let

$$\Delta V = 0 \qquad \text{in } \Omega_0,$$

$$V = u \qquad \text{on } \partial\Omega_0.$$

Then $\tilde{u} = u - V$ is a solution of the thin obstacle problem in Ω_0, with obstacle $\tilde{\phi}(x) = \phi(x) - V(x,0)$, and $\tilde{u} = 0$ on $\partial\Omega_0$. We can now apply Theorem 11.4 to $\tilde{u}$ and thereby obtain the assertion for u.

We shall next generalize Theorem 11.4 to the general variational inequality (11.5), assuming that

$$\sum a_{ij}(x)\xi_i\xi_j \geqslant \kappa|\xi|^2 \qquad \forall \xi \in R^n, \quad x \in \Omega \quad (\kappa > 0),$$

(11.39)

$$a_{ij} \in C^2(\overline{\Omega}), \qquad \phi \in C^4.$$

We shall also assume that

$$S = \{y = 0\},$$

(11.40)

$$a_{in}(x,0) = 0 \qquad \text{for } 1 \leqslant i \leqslant n - 1.$$

Here we use the notation $X = (x, y)$, $x = (x_1, \ldots, x_{n-1})$, as before.

Theorem 11.9. *If (11.39), (11.40) hold, then the solution u of the variational inequality (11.5) is in $C^{1+\alpha}(\overline{\Omega}_+)$ and in $C^{1+\alpha}(\overline{\Omega}_-)$ for some $0 < \alpha < 1$.*

Remark 11.1. The conditions in (11.40) are not very restrictive. Indeed, suppose that (11.40) is not satisfied and let X^0 be a point of S. Then there is a diffeomorphism of a neighborhood V of X^0 onto a domain V' such that $V \cap S$ is mapped onto a planar region in $\{y' = 0\}$ [where (x', y') are the new coordinates] and the new coefficients a'_{ij} satisfy

$$a'_{in}(x',0) = 0, \qquad 1 \leqslant i \leqslant n - 1;$$

see Section 4 for the construction of such a diffeomorphism. The new operator will have first-order terms, but this will not affect the proof of Theorem 11.9.

Thus if for some small neighborhood V of X^0

$$(11.41) \qquad\qquad u > \phi \qquad \text{on } \partial V \cap S,$$

then we can apply Theorem 11.9 in V.

We begin with a generalization of the first part of Lemma 11.5.

Lemma 11.10. *For any small $\delta > 0$, there is a constant $C > 0$ such that if $\rho_0 = \text{dist}((x_0, y_0), \Lambda) > 0$, then*

$$(11.42) \qquad\qquad u_{\tau\tau}(x_0, y_0) \geqslant -\frac{C}{\rho_0^\delta}$$

for any tangential direction τ.

Proof. Applying D_τ and $D_{\tau\tau}$ to the penalized equation in (11.24) (with Δ replaced by A), we obtain, after setting $u = u_{\varepsilon,\delta}$, $\phi = \phi_\varepsilon$, $\beta = \beta_\delta$,

$$(11.43) \qquad Au_\tau + \beta'(u - \phi)(u_\tau - \phi_\tau) = \sum D_i(D_\tau a_{ij} \cdot D_j u),$$

(11.44)

$$\zeta Au_{\tau\tau} + \zeta\beta'(u - \phi)(u_{\tau\tau} - \phi_{\tau\tau}) \geqslant \zeta\sum D_i\big[(2D_\tau a_{ij} \cdot D_{\tau j}u) + D_{\tau\tau}a_{ij} \cdot D_j u\big]$$

$$\equiv \zeta J,$$

where ζ is any nonnegative function.

Note that the proofs of (11.25), (11.26) remain unchanged in the present case of the operator A. Multiplying (11.43) by $\zeta(u_\tau - \phi_\tau)$ [$\zeta \geqslant 0$, $\zeta \in H_0^1(\Omega)$] and integrating over Ω, we get, after integration by parts,

(11.45)

$$\kappa \int \zeta |\nabla(u_\tau - \phi_\tau)|^2 + \int \zeta A\phi_\tau(u_\tau - \phi_\tau) \leqslant C\int |\nabla\zeta| \|u_\tau - \phi_\tau\| |\nabla(u_\tau - \phi_\tau)|$$

$$+ \int \sum \zeta D_i[D_\tau a_{ij} \cdot D_j u](u_\tau - \phi_\tau).$$

Integrating by parts in the last integral and using (11.25), (11.26), we find that

$$(11.46) \qquad\qquad \int |\nabla u_\tau|^2 < C,$$

where C is a constant independent of δ, ε.

Let

$$v = \min\{u_{\tau\tau} - \phi_{\tau\tau} + \lambda, 0\}.$$

By (11.26) we can choose $\lambda > 0$ sufficiently large so that $v = 0$ in a neighborhood of $\partial\Omega$.

Let k be an odd positive integer and take $\zeta = -v^k$ in (11.44); then

(11.47) $$-\int v^k A(u_{\tau\tau} - \phi_{\tau\tau}) - \int v^k A\phi_{\tau\tau} \geq -\int v^k J.$$

The first term on the left is

$$\geq c\int |\nabla v^{(k+1)/2}|^2 \qquad (c > 0).$$

On the right-hand side we get terms of the form

$$L_1 = \int v^k D_i(aD_{\tau j}u), \qquad L_2 = \int v^k D_i(bD_j u).$$

By integration by parts,

$$L_1 = -k\int v^{k-1}v_i aD_{\tau j}u = -k\int v^{(k-1)/2}v_i v^{(k-1)/2}aD_{\tau j}u,$$

so that

$$|L_1| \leq C\left(\int |\nabla v^{(k+1)/2}|^2\right)^{1/2}\left(\int v^{k-1}|D_j u_\tau|^2\right)^{1/2}.$$

Also, by integration by parts and (11.25),

$$|L_2| \leq C\int |D_i v^k|.$$

Thus (11.47) gives

(11.48) $$\int |\nabla v^{(k+1)/2}|^2 \leq C\int |\nabla v^k| + C\int v^{k-1}|\nabla u_\tau|^2.$$

To estimate the last integral we take $\zeta = v^{k-1}$ in (11.45). We then obtain after integrating by parts the last term on the right-hand side and using (11.25), the bound

$$C\int |\nabla v^{k-1}| + C\int |\nabla v^{k-1}|\,|\nabla u_\tau| + C\int v^{k-1}|\nabla u_\tau|.$$

It follows that

(11.49) $$\int |\nabla v^{(k+1)/2}|^2 \leqslant C\int |\nabla v^{k-1}| + C\int |\nabla v^k|$$

$$+ C\int v^{k-1}|\nabla u_\tau| + C\int |\nabla v^{k-1}|\,\|\nabla u_\tau|.$$

Taking $k = 1$ and using (11.46), we deduce that

(11.50) $$\int |\nabla v|^2 \leqslant C; \quad \text{hence also } \int v^2 \leqslant C.$$

Next we refine (11.39) for $k \geqslant 3$, using

$$\tfrac{1}{2}\int |\nabla v^k| \leqslant \int |\nabla v^{(k+1)/2}|\, v^{(k-1)/2}$$

$$\leqslant \eta\int |\nabla v^{(k+1)/2}|^2 + \frac{1}{\eta}\int v^{k-1}, \quad \eta \text{ small},$$

$$\int v^{k-1}|\nabla u_\tau| = \int v^{(k+1)/2} v^{(k-3)/2}|\nabla u_\tau|$$

$$\leqslant \eta\int v^{k+1} + \frac{1}{\eta}\int v^{k-3}|\nabla u_\tau|^2,$$

$$\int |\nabla v^{k-1}|\,\|\nabla u_\tau| \leqslant \int |\nabla v^{(k+1)/2}|\, v^{(k-3)/2}|\nabla u_\tau|$$

$$\leqslant \eta\int |\nabla v^{(k+1)/2}|^2 + \frac{1}{\eta}\int v^{k-3}|\nabla u_\tau|^2.$$

The last term can be estimated in the same way as the last term in (11.48); we get a bound for it of the form

$$C\int |\nabla v^{k-3}| + C\int |\nabla v^{k-3}|\,\|\nabla u_\tau| + C\int v^{k-3}|\nabla u_\tau|.$$

Putting the estimates above together in (11.49), we find that, for $k \geqslant 3$,

(11.51) $$\int |\nabla v^{(k+1)/2}|^2 \leqslant C\int |\nabla v^{k-1}| + C\int |\nabla v^{k-3}|$$

$$+ C\int v^{k-3}|\nabla u_\tau| + C\int |\nabla v^{k-3}|\,\|\nabla u_\tau|.$$

Taking $k = 3$, we obtain, after using (11.46) and (11.50),

$$(11.52) \qquad \int |\nabla v^2|^2 \leqslant C; \qquad \text{hence also } \int v^4 \leqslant C.$$

We can proceed now to refine (11.51) for $k \geqslant 5$, then substitute $k = 5$ to deduce that

$$\int |\nabla v^3|^2 \leqslant C, \qquad \int v^6 \leqslant C,$$

and so on. In conclusion, $v^k \in W^{1,2}$ for all k and, in particular,

$$(11.53) \qquad u_{\tau\tau}^- \in L^p \qquad \text{for all } 1 < p < \infty.$$

Since the estimates above are independent of δ, ε, they hold true also for $u = \lim u_{\delta, \varepsilon}$. Hence (11.53) is valid for the solution u of the variational inequality.

To complete the proof of the lemma, we represent $u_{\tau\tau}(X_0)$ in $B_r(X_0)$ [where $X_0 = (x_0, y_0)$, $\frac{1}{4}\rho_0 < r < \frac{1}{2}\rho_0$] by Green's function G_r for A (in $B_r(X_0)$):

$$(11.54) \qquad D_{\tau\tau} u(X_0) = -\int_{\partial B_r(X_0)} \frac{\partial G_r}{\partial \nu} D_{\tau\tau} u + \int_{B_r(X_0)} G_r A(D_{\tau\tau} u).$$

Notice that in $B_{\rho_0/2}(X_0)$,

$$(11.55) \qquad A(D_{\tau\tau} u) = \sum D_i \gamma_i, \qquad \text{where } |\gamma_i| \leqslant \frac{C}{\rho_0}$$

by interior elliptic estimates, since u satisfies the elliptic equation $Au = 0$ in $B_{\rho_0}(X_0)$ and $|Du| \leqslant C$.

If we integrate both sides of (11.54) with respect to r, $\frac{1}{4}\rho_0 < r < \frac{1}{2}\rho_0$, we obtain

$$(11.56) \quad D_{\tau\tau} u(X_0) = \int_{B_{\rho_0/2}} m(X) D_{\tau\tau} u(X) + \int_{B_{\rho_0/2}} M(X) A(D_{\tau\tau} u),$$

where (see Problem 5)

$$(11.57) \qquad 0 \leqslant m(X) \leqslant \frac{C}{\rho_0^{n-1}}, \qquad |\nabla M| L^1(B_{\rho_0/2}) \leqslant C\rho_0.$$

Using (11.53), (11.55), and (11.57), we obtain from (11.56)

$$D_{\tau\tau} u(X_0) \geqslant -C - C\rho_0^{(n-1)(1/q-1)} |(D_{\tau\tau} u)^-|_{L^p},$$

where $1/p + 1/q = 1$. Since p can be taken arbitrarily large, (11.42) follows.

Since, by (11.40), (11.25), and interior estimates,

$$|a_{in}D_{in}u| \leqslant C \qquad \text{if } 1 \leqslant i \leqslant n-1,$$

we obtain from Lemma 11.10 that

$$(11.58) \qquad u_{yy} \leqslant \frac{C}{|y|^{\delta}} \qquad \text{for any small } \delta > 0.$$

It follows that $u_y - Cy^{1-\delta}$ is a bounded monotone function for $y > 0$ or $y < 0$, and thus the limits $u_y(x, 0 \pm)$ exist.

Now take a domain Ω_0 symmetric with respect to $\{y = 0\}$ such that $\Omega_0 \cap \{y = 0\}$ contains the set S_* (where $\phi \geqslant 0$), and let

$$u^*(x, y) = u(x, y) + u(x, -y).$$

This function is symmetric in y, and $u^*(x, 0) > 2\phi(x)$ near $\partial\Omega_0 \cap S$. We have

$$\sum \frac{\partial}{\partial x_i}\left[(a_{ij}(x, y) + a_{ij}(x, -y)) \frac{\partial u^*(x, y)}{\partial x_j} \right] = f \qquad \text{in } \Omega_0$$

where

$$f = \sum \frac{\partial}{\partial x_i}\left[(a_{ij}(x, y) - a_{ij}(x, -y)) \frac{\partial u(x, -y)}{\partial x_j} \right]$$

$$- \sum \frac{\partial}{\partial x_i}\left[(a_{ij}(x, y) - a_{ij}(x, -y)) \frac{\partial u(x, y)}{\partial x_j} \right].$$

It follows that f is a bounded function.

Define $\phi^*(x) = 2\phi(x)$ and $\sigma^*(x) = u_y^*(x, +0)$. We now proceed to establish Lemma 11.6 for u^* (with obstacle ϕ^*) taking

$$h_{x_0} = \phi^*(X_0) + \nabla\phi^*(X_0) \cdot (x - x_0) + A_0(|x - x_0|^2 - Ny^2);$$

N is sufficiently large so that $Ah_{x_0} \geqslant 0$ and $A_0 \gg \sup|D^2\phi|$. Since

$$u_{yy}^* \leqslant \frac{C}{|y|^{\delta}}$$

(instead of $u_{yy} \leqslant C$), the proof of Lemma 11.7 needs to be slightly modified; the assertion now is that

$$B_{C\gamma}(\bar{x}) \subset B_{\bar{C}\gamma}(x_0) \cap \lambda_{\gamma^\theta} \qquad \text{for any } 0 < \theta < 1.$$

Proceeding to extend Lemma 11.8 we construct a function w_0 such that

$$A(w_0 + u_y) = 0 \quad \text{in } K,$$

$$w_0 = 0 \quad \text{on } \partial K$$

where K is the cylinder $B_{\gamma^k}(x_0) \times (0, \gamma^k)$. Since $w_0 \in C^\alpha$, we find (after scaling) that

$$|w_0| \leqslant C\gamma^{k\alpha}.$$

We now replace w in (11.37) by

$$w = \frac{u_y + \beta^k + w_0 + C\gamma^{k\alpha}}{-\mu\gamma^{k\theta} + \beta^k}$$

for any $0 < \theta < 1$. Since $Aw = 0$, we can proceed as before to establish (11.36) by induction. Having thus extended Lemma 11.8, we have that $\sigma^*(x) \in C^\alpha$ in a neighborhood V of $\partial\Lambda$.

The functions

$$u_1(x, y) = u(x, y), \qquad u_2(x, y) = u(x, -y)$$

satisfy, in Ω_+,

$$A_i u_i = 0 \quad (A_i \text{ elliptic}),$$

and

$$u_1 - u_2 = 0, \quad \frac{\partial u_1}{\partial y} - \frac{\partial u_2}{\partial y} = 2\sigma^* \in C^\alpha \quad \text{in } V.$$

By a standard regularity result for elliptic systems [3b], it follows that $u_i \in C^{1+\alpha}$ in $\Omega^+ \cup V$. This yields the assertion of Theorem 11.9.

Remark 11.2. Theorem 11.9 extends with obvious modifications to the Signorini problem.

Remark 11.3. The function $u = \mathrm{Re}\{z^{3/2}\} = r^{3/2}\cos(3\theta)/2 (-\pi \leqslant \vartheta \leqslant \pi)$ is a solution of an interior obstacle problem with $\varphi = 0$ on the x-axis; it does not belong to $C^{1+\alpha}$ in $\{y \geqslant 0\}$ for any $\alpha > 1/2$.

PROBLEMS

1. Consider the Signorini problem

$$u \in K, \quad \int_\Omega \nabla u \cdot \nabla(v - u)\, dx \geqslant \int_\Omega f(v - u)\, dx \quad \forall v \in K,$$

where

$$K = \{v \in H^1(\Omega), v \geqslant 0 \text{ on } \partial\Omega\}.$$

Prove: If there exists a solution, then $\int_\Omega f\, dx \leqslant 0$. (The condition $\int_\Omega f\, dx \leqslant 0$ is sufficient for existence; see reference 141.)

2. Solve explicitly the solution of Problem 1 in the case $\Omega = \{|x| < 1\}$, $f \equiv \text{const.}$

3. Solve: $u'' = \gamma \quad$ if $-1 < x < 1$,
    $u(-1) = 0$,
    $u(1) \geqslant 0, \quad u'(1) \geqslant 0, \quad (uu')(1) = 0.$

4. If $\Delta u = 0$ in $D_R = \{(x, y); y > 0, |x|^2 + y^2 < R^2\}$ and $u_y = \sigma$ on $y = 0$, $\sigma \in C^\alpha$, then $u \in C^{1+\alpha}(D_{R_0})$ for any $R_0 < R$.
    [*Hint*: Represent u by Neumann's function in the half-space $y > 0$.]

5. Prove (11.57). [*Hint*: Introduce $X' = X_0 + \rho_0 X$.]

6. Consider the Signorini problem

    $$\int_\Omega \nabla u \cdot \nabla(v - u) \geqslant -\int_\Omega f(v - u) \qquad \forall v \in K; u \in K,$$

    where K is defined by (11.20) with $\phi \equiv 0$, and $f \in C^\alpha(\overline{\Omega})$ and (11.19) holds. Set

    $$\tilde{f}(\xi) = -\int_\Omega f(x) \frac{\partial}{\partial \nu_\xi} G(x, \xi)\, dx \qquad (\xi \in \partial\Omega).$$

    where $G(x, \xi)$ is Green's function for $-\Delta$ in Ω. Prove: If $\xi_0 \in \text{int } S$ and $\tilde{f}(\xi_0) < 0$, then $u(\xi_0) > 0$.

7. Consider the thin-obstacle problem under the assumption of Theorem 11.4, and assume that $n = 2$ and that $\phi(x)$ is analytic. Prove that Λ consists of a finite number of intervals.
    [*Hint*: Let v be the harmonic conjugate of u in $\text{Im } z > 0$ ($z = x + iy$), $F(z) = u(x, y) + iv(x, y)$, $F'(z) = U + iV$, $\Phi(z) = $ holomorphic extension of $\phi(x)$. Then

    $$(F'(z) - \Phi'(z))^2 = (U + iV - \Phi')^2$$

    is holomorphic in $\text{Im } z > 0$ and, on $y = 0$, it is equal to

    $$(U - \Phi'(x))^2 - V^2 + 2i(U - \Phi'(x))V = \text{real.}$$

Hence $g = (U - \Phi')^2 - V^2$ has analytic extension to $\{y \leqslant 0\}$. Next, $g \neq 0$ since, on Λ, $g = -V^2$ and $\int V(x, 0) \, dx \neq 0$. If $g(x) < 0$, then $-V(x) > 0$ and $x \in \Lambda$; if $g(x) > 0$, then $x \in N$; if x is a free boundary point, then $g(x) = 0$.]

8. Consider the thin-obstacle problem for a plate:

$$\int_\Omega \Delta u \cdot \Delta(v - u) \geqslant 0 \qquad \forall \, v \in K; \quad u \in K$$

where Ω is a circle $\{x^2 + y^2 < 1\}$ and $K = \{v \in H_0^2(\Omega), v(x, 0) \geqslant \phi(x)\}$. Assume that $\phi \in C^2$ and $\phi(\pm 1) < 0$. Prove that $u \in W^{3, p}(\Omega) \; \forall \; p < 2$ [and thus $u \in W^{2, r}(\Omega) \; \forall \; r < \infty$].

[*Hint*: For any $-1 < x_0 < 1$, there is a neighborhood U of $(x_0, 0)$ such that if $u \in K$, then

$$v_\varepsilon = u + \varepsilon g^{-1} \zeta^2(x) \Delta_h^2[(a + v)g] \qquad \text{is in } K$$

for any small $\varepsilon > 0$, where $\zeta \in C_0^\infty$, $\zeta = 0$ if $(x, 0) \notin U$, $0 \leqslant \zeta \leqslant 1$, a is some real number and

$$g(x) = 1 - \tfrac{1}{2}(e^{rx} - e^{rx_0}) \qquad \text{for some } r > 0;$$

see Problems 4 to 6 of Section 10. Substitute v_ε into the variational inequality and deduce that $u_x \in H^2(\Omega)$. Next, by symmetry, $(\partial/\partial v) \Delta u$ keeps a sign on $y = 0$ (it is taken in the distribution sense). Since Δu is harmonic in Ω_+, it follows that

$$\int_{y=0} \left| \frac{\partial}{\partial y} \Delta u \right| < \infty.$$

Represent $\partial(\Delta u)/\partial y$ by Green's function in Ω_+ and deduce that it is in $L^p(\Omega_+)$ for any $p < 2$.]

12. BIBLIOGRAPHICAL REMARKS

There are two recent books on variational inequalities; by Baiocchi and Capelo [20] and by Kinderlehrer and Stampacchia [126b]. There is also a brief survey by Kinderlehrer [122c]. A systematic existence theory for variational inequalities was first developed by Lions and Stampacchia [141]. The treatment in Section 2 is based on Stampacchia [167a]; the method outlined in problems (1.1)–(1.5) is taken from reference 141. $W^{2, p}$ estimates for the obstacle problem were first obtained by Lewy and Stampacchia [138a–c]; the penalty method was used by Brezis and Stampacchia [53a]; see also Brezis [45a] (from which Problem 5 of Section 3 is taken), Gerhardt [104a], and Lions [140]. Similar results for the Neumann boundary condition were developed by Murthy and Stampacchia [149].

The interior $W^{2,\infty}$ regularity (Theorem 4.1) is due to Brezis and Kinderlehrer [49]; a special case was earlier obtained by Frehse [93a]. The boundary $W^{2,\infty}$ regularity (Theorem 4.3) was established by Jensen [116c]. Problems 3 to 11 of Section 4 are based on the work of Caffarelli and Kinderlehrer [62]; see also reference 58i. Frehse and Mosco [184] and Frehse [93e] proved Hölder continuity for solutions of the obstacle problem with irregular obstacle.

$W^{2,p}$ regularity for two obstacles was derived by Brezis [45a], and interior $W^{2,\infty}$ regularity by Caffarelli and Kinderlehrer [62] and Chipot [71a].

The treatment of the filtration problem in Section 5 is due to Baiocchi [19a]; related problems are described in Bear [26].

The physical problem of Section 6 is described by Lanchon [131], where the variational principle is also stated; see also Duvaut and Lions [81]. Brezis [45b] solved the problem (6.3). Theorem 6.1 is due to Brezis and Sibony [52]. $W^{2,p}$ regularity (Theorem 6.2) was established by Brezis and Stampacchia [53a] for Ω convex; the present version is due to Gerhardt [104b]. Local $W^{2,p}$ regularity for general gradient–constraint variational inequalities was recently obtained by Jensen [116d].

Theorem 7.1 ($W^{2,\infty}$ regularity) is due to Caffarelli and Riviere [63e]. Another method based on a penalized problem $-\Delta u + \beta_\varepsilon(|\nabla u|^2 - 1) = -\mu$ was more recently given by Evans [85b] in the case of simply connected domains and extended by Wiegner [181]; see also Ishii and Koike [191]. The simple but useful result of Problem 2 of Section 7 was first noticed by Ting [172b].

$W^{2,p}$ regularity for parabolic variational inequalities with one or two obstacles were established by Brezis [45a] and Friedman [94d, e, g]. The estimates on the support of solutions given in Problems 8 to 12 of Section 8 are due to Brezis and Friedman [48]; Brezis [45a] studied variational inequalities (8.10) with general closed convexed sets K. The method of proof outlined in Problems 10 and 11 of Section 8 is taken from Evans and Knerr [87a].

Duvaut [80] has transformed the one-phase Stefan problem into a variational inequality. Theorems 9.1 and 9.2 and Problem 9.3 are taken from Friedman and Kinderlehrer [97]. For the two-phase Stefan problem, see Friedman [94a, b, h] and Rubinstein [156] and the references given there. A one-phase Stefan problem with supercooled water was studied by Friedman [94f], Jensen [116b], and van Moerbeke [144a, b].

The material of Section 10 is based on the paper of Caffarelli and Friedman [58d]. Frehse [93b, c] had earlier proved $W^{2,\infty}$ and $W^{3,2}$ regularity. Caffarelli, Friedman, and Torelli [60b] proved $W^{2,\infty}$ regularity for the corresponding two-obstacle problem, and gave a counterexample showing that the solution is, in general, not in C^2 if $n \geqslant 2$.

Problem 3 of Section 10 is based on a paper by Brezis and Stampacchia [53d]. They considered also the variational inequality with a set K given by

$K_{\alpha, \beta} = \{v \in H_0^2(\Omega), \ \alpha \leqslant \Delta v \leqslant \beta\}$. Caffarelli, Friedman, and Torelli [60a] considered this variational inequality with $-\alpha = \beta = \varepsilon > 0$ and proved that, for Ω a rectangle or a regular triangle, the solution $u = u_\varepsilon$ satisfies $(1/\varepsilon)\Delta u_\varepsilon \to U$ as $\varepsilon \to 0$, where U solves the obstacle problem for $-\Delta$ with obstacle $d^2(x) = (\text{dist}(x, \partial\Omega))^2$.

Evans and Knerr [87b] studied elastic–plastic problems for plates; the constraint is a differential inequality involving second derivatives of u. They derived $W^{3,2}$ regularity.

Problems 4 to 6 of Section 10 are based on Frehse [93b].

Section 11 is based on Caffarelli [56d]. More recently Kinderlehrer [122d] has obtained similar results using a different penalization. Lewy [137a–c] has studied the Lipschitz continuity of the solution in the case of two dimensions and showed that the coincidence set consists of a finite number of intervals in case of an analytic obstacle (this is the result given in Section 11, Problem 7); related results were obtained by Athanasopoulos [16]. Frehse [93d] has proved the continuity of the first derivatives in any number of dimensions; see reference 93d also for related results for thin obstacles for minimal surfaces. The result of Problem 8 of Section 11 is due to Caffarelli and Friedman. Recently Schild [188] established $W^{2,\infty}$ regularity in n-dimensions for the thin obstacle problem for Δ^2, and C^2 regularity in case $n = 2$.

The Signorini problem for systems models deformations in linear elasticity. It was first treated by Fichera [89] who established the existence of solutions in H^1. Recently, Kinderlehrer [122e, f] proved that the solutions belong to H^2 and derived estimates on the coincidence set; for $n = 2$ the solution was shown to be in $C^{1,\alpha}$.

In a recent monograph [192] Elliott and Ockendon survey a large number of free boundary problems which arise in physics and engineering.

2

VARIATIONAL INEQUALITIES: ANALYSIS OF THE FREE BOUNDARY

In this chapter we study the regularity of the free boundary and its shape. It is shown that if the noncoincidence set has positive density at a point x_0 of the free boundary, then the free boundary is smooth in a neighborhood of this point. Some theorems are established which can be used to show (under suitable conditions) that the noncoincidence set does indeed have a positive density at x_0.

For those variational inequalities studied in Chapter 1, which arise in physical problems, we establish in this chapter the smoothness of the free boundary. We also study the shape of the boundary. The chapter includes stability results for the free boundaries and examples of variational inequalities whose free boundaries have singularities.

1. THE HODOGRAPH–LEGENDRE TRANSFORMATION

In this section we show that if the free boundary is C^1, then it is in fact arbitrarily smooth provided that the coefficients of the equation and the obstacle are sufficiently smooth. Consider for example the variational inequality

$$(1.1) \qquad \Delta u \geqslant f, \qquad u \geqslant 0, \qquad (\Delta u - f)u = 0$$

in a domain G of R^n and denote by Ω the noncoincidence set $\{u > 0\}$ and by Γ

the free boundary. Thus

$$(1.2) \qquad\qquad \Delta u = f \qquad \text{in } \Omega,$$

$$(1.3) \qquad\qquad u = 0, \qquad \nabla u = 0 \qquad \text{on } \Gamma.$$

We shall require that

$$(1.4) \qquad\qquad \Gamma \text{ is in } C^1,$$

$$(1.5) \qquad\qquad u \in C^2 \qquad \text{in } \Omega \cup \Gamma,$$

where the last condition means that $D^2 u$ is continuous in Ω and has continuous extension into Γ.

In the theorem to follow we do not require that the u, Γ, Ω come necessarily from the variational inequality (1.1); Ω is any domain in R^n and Γ is an open portion of $\partial\Omega$. We are interested in proving regularity of Γ in a neighborhood of any given point $x^0 \in \Gamma$.

Theorem 1.1. *Assume that* (1.2)–(1.5) *hold and that* $x^0 \in \Gamma$. *Then*:

    (i) *If* $f(x^0) \neq 0$ *and* f *is in* C^1 *in a neighborhood of* x^0, *then* Γ *is* $C^{1+\alpha}$ *for any* $0 < \alpha < 1$, *in some neighborhood of* x^0.

    (ii) *If further* $f \in C^{m+\beta}$ *(m integer $\geqslant 1$, $0 < \beta < 1$) in a neighborhood of* x^0, *then* Γ *is in* $C^{m+1+\beta}$ *in some neighborhood of* x^0.

    (iii) *If* f *is also analytic in a neighborhood of* x^0, *then* Γ *is analytic in some neighborhood of* x^0.

Proof. Take for definiteness $f(x^0) > 0$. We shall perform a local diffeomorphism $x \to y$ and a transformation from $u(x)$ into another function $v(y)$ in such a way that Γ is transformed locally into $y_0 = 0$ and v satisfies an elliptic equation in $\{y_1 < 0\}$ with zero Dirichlet data on $y_1 = 0$. Using standard boundary regularity results for elliptic equations, the assertions of the theorem will follow.

Set $u_i = \partial u / \partial x_i$, $u_{ij} = \partial^2 u / \partial x_i \partial x_j$.

For simplicity we take $x^0 = 0$ and assume that the inward normal to Γ at x^0 is $(1, 0, \ldots, 0)$. We extend u_1 as a C^1 function into a full neighborhood of 0.

The change of variables is given by

$$(1.6) \qquad\qquad y_\alpha = x_\alpha \quad (2 \leqslant \alpha \leqslant n), \qquad y_1 = -u_1$$

and the new function $v(y)$ is defined as

$$(1.7) \qquad\qquad v(y) = x_1 y_1 + u(x).$$

Then

$$u_i = 0 \qquad \text{on } \Gamma,$$

$$u_{i\alpha}(0) = 0 \qquad (1 \le i \le n, \quad 2 \le \alpha \le n),$$

and (1.2) gives

$$u_{11}(0) = f(0) > 0.$$

It follows that $\det(\partial y / \partial x) \ne 0$ at 0, so that (1.6) is indeed a local diffeomorphism. It maps a neighborhood of 0 in Ω onto an open set W in $\{y_1 < 0\}$ and a neighborhood of 0 in Γ onto an open set S in $y_1 = 0$; $0 \in S$.
    Notice that

$$(1.8) \qquad dv(y) = x_1 \, dy_1 + y_1 \, dx_1 + du$$

$$= x_1 \, dy_1 - u_1 \, dx_1 + \sum_{j=1}^{n} u_j \, dx_j$$

$$= x_1 \, dy_1 + \sum_{\alpha=2}^{n} u_\alpha \, dx_\alpha = x_1 \, dy_1 + \sum_{\alpha=2}^{n} u_\alpha \, dy_\alpha,$$

so that

$$(1.9) \qquad v_1 = x_1, \qquad v_\alpha = u_\alpha \quad (2 \le \alpha \le n),$$

where $v_j = \partial v / \partial y_j$. Since S is given by $y_1 = 0$ with arbitrary y_α, $\Sigma_{\alpha=2}^{n}(y_\alpha)^2$ small, the corresponding points in the x_i coordinates are

$$x_1 = v_1(0, y_2, \ldots, y_n), \qquad x_\alpha = y_\alpha \quad (2 \le \alpha \le n).$$

Thus

$$(1.10) \qquad x_1 = \frac{\partial v}{\partial y_1}(0, x_2, \ldots, x_n)$$

is a local parametrization of Γ near 0, and it remains to establish the smoothness of v in $W \cup S$.

In order to find a differential equation for v, we compute

$$\left(\frac{\partial y}{\partial x}\right) = \begin{pmatrix} -u_{11} & -u_{12} & \cdots & -u_{1n} \\ 0 & 1 & \cdots & 0 \\ \cdots & & & \\ 0 & 0 & \cdots & 1 \end{pmatrix}.$$

On the other hand, since $x_1 = v_1$, $x_\alpha = y_\alpha$ $(2 \leqslant \alpha \leqslant n)$, we have

$$\left(\frac{\partial x}{\partial y}\right) = \begin{pmatrix} v_{11} & v_{12} & \cdots & v_{1n} \\ 0 & 1 & \cdots & 0 \\ \cdots & & & \\ 0 & 0 & \cdots & 1 \end{pmatrix}.$$

Since $(\partial y/\partial x) = (\partial x/\partial y)^{-1}$, we obtain

(1.11) $\qquad u_{11} = -\dfrac{1}{v_{11}}$ and $u_{1\alpha} = \dfrac{v_{1\alpha}}{v_{11}}$ $\quad (2 \leqslant \alpha \leqslant n).$

Then also

(1.12) $\qquad u_{\alpha\beta} = \dfrac{\partial v_\alpha}{\partial x_\beta} = \sum_{i=1}^{n} \dfrac{\partial v_\alpha}{\partial y_i}\dfrac{\partial y_i}{\partial x_\beta} = -v_{\alpha 1}u_{1\beta} + v_{\alpha\beta}$

$$= v_{\alpha\beta} - \frac{v_{1\alpha}v_{1\beta}}{v_{11}} \qquad (2 \leqslant \alpha, \beta \leqslant n),$$

and (1.2) gives

(1.13) $\qquad -\dfrac{1}{v_{11}} - \dfrac{1}{v_{11}} \sum_{\alpha=2}^{n} v_{1\alpha}^2 + \sum_{\alpha=2}^{n} v_{\alpha\alpha} - f(v_1, y') = 0 \qquad$ in W

where $y' = (y_2, \ldots, y_n)$. From (1.3) we get

(1.14) $\qquad\qquad\qquad v(0, y') = 0 \qquad$ on $S.$

The equation (1.13) has the form

$$\Phi(y, v, v_1, \ldots, v_n, v_{11}, \ldots, v_{1n}, \ldots, v_{n1}, \ldots, v_{nn}) = 0$$

or, more, briefly,

$$\Phi(y, v, Dv, D^2v) = 0.$$

Such an equation is said to be *elliptic* at y^0 with respect to a specific solution $v(y)$ if the matrix

$$(1.15) \qquad \left(\frac{\partial \Phi}{\partial v_{ij}} \right) \qquad \text{is positive definite}$$

when evaluated for $v = v(y)$ at $y = y^0$. In our case $\Phi = 0$ is elliptic at $y = 0$ and in fact the matrix (1.15) is the diagonal matrix (λ_{ij}) with $\lambda_{11} = -1/v_{11}(0) > 0$, $\lambda_{\alpha\alpha} = 1$ $(2 \leqslant \alpha \leqslant n)$.

We can now apply local regularity results for the elliptic problem (1.13), (1.14): (i) follows from Theorem 11.1′ in reference 3a, (ii) follows from Theorem 11.1 of reference 3a, and (iii) follows from Section 6.7 of reference 145 or from reference [94c].

Theorem 1.1 can be generalized to solutions of

$$(1.16) \quad F(x, u, Du, D^2u) = 0 \text{ in } \Omega, \qquad F \in C^1 \text{ in all its variables.}$$

Theorem 1.2. *Let* u, Γ *satisfy* (1.3)–(1.5), (1.16) *and assume that F is elliptic with respect to* $u(x)$ *at some point* $x^0 \in \Gamma$ *and that*

$$F(x^0, 0, 0, D^2u(x^0)) \neq 0.$$

Then:

    (i) Γ *is* $C^{1+\alpha}$ *for any* $0 < \alpha < 1$ *in a neighborhood of* x^0.
    (ii)   *If* $F \in C^{m+\beta}$ $(m \geqslant 1, \ 0 < \beta < 1)$ *in all its arguments, then* $\Gamma \in C^{m+1+\beta}$ *in some neighborhood of* x^0.
    (iii) *If F is analytic in all its arguments in a neighborhood of* x^0, *then* Γ *is analytic in some neighborhood of* x^0.

Proof. We proceed as in the preceding proof. The only point that needs to be checked is that if Φ is defined by

$$F(x, y, Du, D^2u) = \Phi(y, v, Dv, D^2v),$$

then Φ is elliptic at $y = 0$. From (1.11), (1.12) we find that, at $y = 0$,

$$\frac{\partial u_{\alpha\beta}}{\partial v_{\alpha\beta}} = 1, \qquad \frac{\partial u_{1\alpha}}{\partial v_{1\alpha}} = \frac{1}{v_{11}}, \qquad \frac{\partial u_{11}}{\partial v_{11}} = \frac{1}{v_{11}^2}.$$

and all the other partial derivatives $\partial u_{ij}/\partial v_{km}$ vanish. Therefore,

$$\sum \frac{\partial \Phi}{\partial v_{km}} \xi_k \xi_m = \sum_{k,m} \sum_{i,j} \frac{\partial F}{\partial u_{ij}} \frac{\partial u_{ij}}{\partial v_{km}} \xi_k \xi_m = \sum_{i,j} \frac{\partial F}{\partial u_{ij}} \bar{\xi}_i \bar{\xi}_j$$

where $\bar{\xi}_1 = \xi_1/v_{11}(0)$, $\bar{\xi}_\alpha = \xi_\alpha$ $(2 \leqslant \alpha \leqslant n)$, and the ellipticity of Φ follows from the ellipticity of F.

The transformations (1.6), (1.7) can be generalized to

(1.17) $y_\alpha = x_\alpha$ $(k + 1 \leqslant \alpha \leqslant n)$, $y_\beta = -u_\beta$ $(1 \leqslant \beta \leqslant k)$,

(1.18) $$v = \sum_{\alpha=1}^{k} x_\alpha y_\alpha + u,$$

and again

(1.19)   if $F(x, y, Du, D^2u) = \Phi(y, v, Dv, D^2v)$, and
          if F is elliptic, with respect to u at 0,
          then Φ is elliptic with respect to v at 0.

The transformation

(1.20) $$y_i = -u_{x_i}   (1 \leqslant i \leqslant n)$$

is called the *hodograph transformation*; (1.17) is called the partial hodograph transformation. The transformation given by (1.20) and

$$v = \sum_{i=1}^{n} x_i y_i + u$$

is called the *Legendre transformation*. We shall refer to (1.6), (1.7) and more generally to (1.17), (1.18) as the *hodograph–Legendre* transformation.

Theorem 1.1 can be used in the obstacle problem provided we already know that the free boundary is C^1 and that the solution u is C^2 up to the free boundary.

We shall next establish an analogue of Theorem 1.2 for parabolic equations. For simplicity we shall deal only with the C^∞ case.

We assume: Ω is an open set in the space R^{n+1} of variables $(x, t) = (x_1, x_2, \ldots, x_n, t)$, Γ is an open portion of $\partial\Omega$, and

(1.21)

$$0 \in \Gamma,$$
$$\Gamma \text{ is in } C^1,$$
$$\Gamma \text{ is not tangent to } t = 0 \text{ at the origin,}$$
$$u \text{ and } D_x u \text{ belong to } C^1(\Omega \cup \Gamma),$$
$$u = 0, \text{ grad}_x u = 0 \quad \text{on } \Gamma,$$
$$u_t - F(x, t, u, D_x u, D_x^2 u) = 0 \quad \text{in } \Omega.$$

Theorem 1.3. *If* (1.21) *holds and if F is elliptic with respect to $u(x, t)$ at 0, $F \in C^\infty$ in all its arguments and $F(0, 0, 0, 0, D^2 u(0)) \neq 0$, then Γ is C^∞ in some neighborhood of 0.*

Proof. We may suppose that Γ is given locally by

$$x_1 = \psi(x_2, \ldots, x_n, t)$$

where $\psi \in C^1$ and $x_1 < \psi$ in Ω; further,

$$\psi_{x_i}(0) = 0, \qquad u_{x_i x_j}(0) = 0 \text{ except for } u_{x_1 x_1}(0) > 0.$$

We extend u_1 as a C^1 function into a full neighborhood of 0 and consider the transformation

$$y = (-u_1, x_2, \ldots, x_n), \qquad s = t, \quad v = x_1 y_1 + u.$$

Then v satisfies

(1.22) $$v_t - \Phi(x, t, v, Dv, D^2 v) = 0$$

where Φ is elliptic; also, $v = 0$ on $y_1 = 0$. We now apply boundary regularity results for the nonlinear parabolic equation (1.22); for further study, see reference 123a.

For other free-boundary problems, one may apply related transformations. We shall consider here one important case:

(1.23) $$F(x, u, Du, D^2 u) = 0 \quad \text{in } \Omega,$$

(1.24) $$u = 0, \qquad g(x, \text{grad } u) = 0 \text{ on } \Gamma.$$

Theorem 1.4. *Assume that* (1.4), (1.5), (1.23), (1.24) *hold, that $g(x, p_1, \ldots, p_n)$*

is in C^2,

$$\frac{\partial g}{\partial p_n}(0, \operatorname{grad} u(0)) \neq 0, \qquad u_n(0) \neq 0,$$

and that F is elliptic with respect to u at 0. *Then*:

(i) $\Gamma \in C^{2+\alpha}$ *for any* $0 < \alpha < 1$ *in some neighborhood of* 0.

(ii) *If* $F \in C^{m+\beta}$, $g \in C^{m+1+\beta}$ $(m \geqslant 1, 0 < \beta < 1)$ *in all the arguments, then* $\Gamma \in C^{m+2+\beta}$ *in some neighborhood of* 0.

(iii) *If F, g are analytic in all the arguments in a neighborhood of* 0, *then* Γ *is analytic in some neighborhood of* 0.

Proof.   Suppose that the outward normal to Γ is in the direction of the positive x_n-axis and extend u as a C^2 function into a full neighborhood of 0. We suppose that $u_n(0) > 0$ and make the local diffeomorphism

$$(1.25) \qquad\qquad y = (x_1, \ldots, x_{n-1}, u(x));$$

the function $v(y)$ is now chosen to be

$$(1.26) \qquad\qquad v(y) = x_n.$$

Then Γ is transformed into $y_n = 0$ and

$$\left(\frac{\partial x}{\partial y}\right) = \left(\frac{\partial y}{\partial x}\right)^{-1} = \left(\begin{array}{ccc|c} I & & & 0 \\ \hline u_1 & \cdots & u_{n-1} & u_n \end{array}\right)^{-1}$$

$$= \left(\begin{array}{ccc|c} I & & & 0 \\ \hline -\dfrac{u_1}{u_n} & \cdots & -\dfrac{u_{n-1}}{u_n} & \dfrac{1}{u_n} \end{array}\right).$$

Consequently,

$$v_\alpha = \frac{\partial x_n}{\partial y_\alpha} = -\frac{u_\alpha}{u_n} \quad (1 \leqslant \alpha < n),$$

$$v_n = \frac{\partial x_n}{\partial y_n} = \frac{1}{u_n},$$

$$\frac{\partial y_n}{\partial x_n} = u_n = \frac{1}{v_n}, \qquad \frac{\partial y_n}{\partial x_\alpha} = u_\alpha = -\frac{v_\alpha}{v_n} \quad (1 \leqslant \alpha < n).$$

It follows that

$$u_{nn} = -\frac{v_{nn}}{v_n^2}\frac{\partial y_n}{\partial x_n} = -\frac{v_{nn}}{v_n^3},$$

$$u_{n\alpha} = -\frac{v_{n\alpha}}{v_n^2} - \frac{v_{nn}}{v_n^2}\frac{\partial y_n}{\partial x_\alpha} = -\frac{v_{n\alpha}}{v_n^2} + \frac{v_\alpha v_{nn}}{v_n^3} \quad (1 \leq \alpha < n),$$

$$u_{\alpha\beta} = -\frac{v_{\alpha\beta}}{v_n} + \frac{v_{\alpha n}}{v_n^2}v_\beta + \frac{v_{\beta n}}{v_n^2}v_\alpha - v_\alpha v_\beta \frac{v_{nn}}{v_n^3} \quad (1 \leq \alpha, \beta < n).$$

Using these formulas one computes, for instance, that

$$(1.27) \qquad \Delta u = -\frac{1}{v_n}\sum_{\alpha=1}^{n-1}v_{\alpha\alpha} + \frac{2}{v_n^2}\sum_{\alpha=1}^{n-1}v_\alpha v_{\alpha n} - \frac{v_{nn}}{v_n^3}\left(1 + \sum_{\alpha=1}^{n-1}v_\alpha^2\right),$$

and more generally

$$(1.28) \qquad 0 = F(x, u, Du, D^2u) \equiv \Phi(y, v, Dv, D^2v),$$

and Φ is elliptic at 0.

The function v also satisfies the boundary condition

$$g\left(y_1,\ldots,y_{n-1}, v, -\frac{v_1}{v_n},\ldots, -\frac{v_{n-1}}{v_n}, \frac{1}{v_n}\right) = 0 \qquad \text{on } y_n = 0,$$

and since $\partial g/\partial p_n \neq 0$, we can write this condition in the form

$$v_n = h(y_1,\ldots,y_{n-1}, v, v_1,\ldots,v_{n-1})$$

with h being as smooth as g. This boundary condition is of Neumann type and the boundary regularity results quoted above hold for this condition.

Note finally that the boundary Γ near 0 is represented by $x_n = v(x_1,\ldots,x_{n-1},0)$; thus Γ has the same smoothness as v.

Remark 1.1. Suppose that $n = 2$ and

$$(1.29) \qquad \begin{array}{ll} \Delta u = 0 & \text{in } \Omega, \\ \dfrac{\partial u}{\partial \nu} = 0, & |\operatorname{grad} u| = 1 \text{ on } \Gamma, \quad 0 \in \Gamma, \end{array}$$

where ν is the normal, and u, Γ satisfy (1.4), (1.5). Then the harmonic conjugate v, with $v(0) = 0$, satisfies

$$\Delta v = 0 \qquad \text{in } \Omega,$$

$$v = 0, \qquad \frac{\partial v}{\partial \nu} = 1 \text{ on } \Gamma,$$

and $v \in C^2 (\Omega \cup \Gamma)$. Applying Theorem 1.4, we conclude that Γ is analytic. In $n > 2$, this conclusion is no longer true. In fact, the function $u(x) = x_1$ satisfies (1.29) with Γ any C^1 hypersurface of the form $x_n = \psi(x_2, \ldots, x_{n-1})$.

PROBLEMS

1. Prove (1.19); here

$$v_\alpha = x_\alpha, \qquad u_\alpha = -y_\alpha \quad (\alpha \leqslant k),$$

$$v_\beta = u_\beta \qquad (\beta > k),$$

$$dv = \sum_{\alpha \leqslant k} x_\alpha \, dy_\alpha + \sum_{\beta > k} u_\beta, \, dy_\beta,$$

$$\left(\frac{\partial x}{\partial y} \right) = \begin{pmatrix} \overset{k}{v_{\alpha\beta}} & \overset{n-k}{v_{\alpha\gamma}} \\ \hline 0 & I \end{pmatrix} = \left(\begin{array}{c|c} A & B \\ \hline 0 & I \end{array} \right),$$

$$\left(\frac{\partial y}{\partial x} \right) = \left(\begin{array}{c|c} A^{-1} & -A^{-1}B \\ \hline 0 & I \end{array} \right),$$

$$-u_{\alpha j} = \text{element } (\alpha, j) \text{ in } (\partial y / \partial x) \equiv y_j^\alpha \quad (\alpha \leqslant k),$$

$$u_{r\beta} = -y_r^\beta = \sum_{\alpha \leqslant k} v_{r\alpha} y_\beta^\alpha \quad (\beta \leqslant k < r),$$

$$u_{rs} = v_{rs} + \sum_{\alpha \leqslant k} v_{r\alpha} y_s^\alpha \quad (k < r, s).$$

2. Let the assumptions of Theorem 1.1 hold with $x^0 = 0$ and $\Gamma : x_1 = \psi(x_2, \ldots, x_n)$. Can Theorem 1.1 be proved by using the transformation

$$y_\alpha = x_\alpha \quad (2 \leqslant \alpha \leqslant n), \qquad y_1 = x_1 - \psi(x_2, \ldots, x_n)$$

(which maps Γ into $y_1 = 0$) with $v(y) = u(x)$?

3. For the Stefan problem in one dimension

$$\theta_t - \theta_{xx} = 0 \text{ if } x < s(t),$$

$$\theta = 0, \qquad \theta_x = -s'(t) \text{ if } x = s(t),$$

prove that if $s \in C^1$ and $\theta \in C^2$ up to the free boundary, then $s \in C^\infty$.
[*Hint*: Use the transformation $y = x - s(t)$ or $y = x/s(t)$ and set $v(y, t) = u(x, t)$.]

2. REGULARITY IN TWO DIMENSIONS

We define

$$D_z = \frac{\partial}{\partial z} = \frac{1}{2}\left(\frac{\partial}{\partial x_1} - i\frac{\partial}{\partial x_2}\right),$$

$$D_{\bar{z}} = \frac{\partial}{\partial \bar{z}} = \frac{1}{2}\left(\frac{\partial}{\partial x_1} + i\frac{\partial}{\partial x_2}\right)$$

and use the notation $w(x_1, x_2) = w(z)$, where $z = x_1 + ix_2$. As usual, $B_R = \{|z| < R\}$, $B_R(z_0) = \{|z - z_0| < R\}$.
We shall need Green's formula in complex form. Let $w = u + iv$. Then for any bounded open set E with piecewise C^1 boundary ∂E and for any $v \in H^{1,s}$ $(2 < s < \infty)$,

$$(2.1) \qquad \int_E v_{\bar{z}} \, dx_1 \, dx_2 = -\frac{i}{2}\int_{\partial E} v \, dz;$$

notice that $v \in C^\lambda(E)$, $\lambda = 1 - 2/s > 0$.
Let z_0, z_1 be two points in E, $z_0 \neq z_1$, and define

$$E_\varepsilon = E \cap \{|z - z_0| > \varepsilon, |z - z_1| > \varepsilon\}, \qquad \varepsilon > 0.$$

Consider the function

$$v(z) = \frac{w(z) - w(z_0)}{(z - z_0)^m (z - z_1)}, \qquad m \text{ integer} \geq 1,$$

and assume that

$$w \in H^{1,s}(E), \qquad s > 2,$$

$$(2.2) \qquad |w(z) - w(z_0)| \leq C|z - z_0|^{m-1+\alpha},$$

$$|w_{\bar{z}}(z)| \leq C|z - z_0|^{m-2+\alpha} \qquad (0 < \alpha < 1).$$

Applying (2.1) with this v and with E replaced by E_ε, and then taking $\varepsilon \to 0$, we obtain Green's formula:

(2.3)

$$\frac{w(z_1) - w(z_0)}{(z_1 - z_0)^m} = -\frac{1}{\pi} \int_E \frac{1}{(z - z_0)^m} \frac{1}{z - z_1} w_{\bar{z}}(z) \, dx_1 \, dx_2$$

$$+ \frac{1}{2\pi i} \int_{\partial E} \frac{1}{(z - z_0)^m} \frac{1}{z - z_1} w(z) \, dz, \qquad w \text{ as in (2.2).}$$

In this section we derive regularity results on the free boundary for two-dimensional problems. Unlike the results of Section 1, which presuppose a C^1 free boundary, here we only assume that the free boundary is a Jordan arc.

Theorem 2.1. *Let Ω be a simply connected domain in R^2 whose boundary $\partial\Omega$ is a Jordan curve, and let Γ be a Jordan arc, $\Gamma \subset \partial\Omega \cap B_R$ (for some $R > 0$). Let u be a function in $C^1(\Omega \cup \Gamma)$ and let f, ϕ be analytic functions in B_R. If*

(2.4)
$$\Delta u = f \quad \text{in } \Omega \cap B_R,$$

(2.5)
$$u = \phi, \quad \nabla u = \nabla\phi \quad \text{on } \Gamma,$$

(2.6)
$$f - \Delta\phi \neq 0 \quad \text{in } B_R,$$

then Γ admits an analytic parametrization.

We shall prove the existence of a local analytic representation; this actually implies global analytic representation [27; p. 376].

Proof. The function

(2.7)
$$\tilde{f}(z) = \frac{1}{2\pi} \int_{B_R} \log \frac{1}{|z - t|} f(t) \, dt_1 \, dt_2$$

is real analytic since $\Delta\tilde{f} = -f$. Therefore, it suffices to prove the theorem in case $f = 0$, for otherwise we consider $u - \tilde{f}$ instead of u.

We establish an analytic parametrization in a neighborhood of a point $z^0 \in \Gamma$, and for simplicity take $z^0 = 0$. Set

$$\nabla^* u = u_{x_1} - i u_{x_2} = 2u_z,$$

$$\nabla^* \phi = \phi_{x_1} - i \phi_{x_2} = 2\phi_z.$$

One can immediately check that $2D_{\bar{z}} \nabla^* u = \Delta u = 0$ in Ω that is, $\nabla^* u$ is holomorphic in Ω.

The function $\nabla^*\phi(x_1, x_2)$ has power series expansion

$$\sum a_{\lambda\mu} x_1^\lambda x_2^\mu \qquad (|x_1| < 2\delta, |x_2| < 2\delta)$$

for some $\delta > 0$. Writing

$$x_1 = \frac{z + \bar{z}}{2}, \qquad x_2 = \frac{z - \bar{z}}{2i}$$

we get an expansion

$$\nabla^*\phi(z, \bar{z}) = \sum b_{\lambda\mu} z^\lambda \bar{z}^\mu,$$

where the $b_{\lambda\mu}$ are now complex coefficients, and $\nabla^*\phi(z, \bar{z}) \equiv \nabla^*\phi(x_1, x_2)$. This series is convergent for $|z| < \delta$, $|\bar{z}| < \delta$, and we can extend it into a holomorphic function of two variables, z and ζ, by

$$\nabla^*\phi(z, \zeta) \equiv \sum b_{\lambda\mu} z^\lambda \zeta^\mu.$$

Consider the equation, for ζ,

(2.8) $$\nabla^*\phi(z, \zeta) = \nabla^*u(z).$$

By assumption

$$D_{\bar{z}}\nabla^*\phi(z, \bar{z}) \equiv D_{\bar{z}}\nabla^*\phi(x_1, x_2) = \tfrac{1}{2}\Delta\phi \neq 0 \qquad \text{in } B_R$$

and, by (2.5), (2.8) does hold if $z \in \Gamma$, $\zeta = \bar{z}$. We can therefore use the implicit function theorem. It implies that there exists a unique solution

$$\zeta = \zeta(z, \nabla^*u(z)) \equiv \zeta^*(z)$$

of (2.8) in a neighborhood of $z = 0$, $\zeta = 0$, and that $\zeta(z, w)$ is holomorphic in (z, w) in a neighborhood of $(0, 0)$. Recalling that

(2.9) $$\zeta^*(z) = \bar{z} \qquad \text{if } z \in \Gamma,$$

we see that the conjugate points of Γ are boundary values of a holomorphic function.

In order to get analytic parametrization we introduce a conformal mapping $t \to g(t)$ from the half-disc $G = \{|t| < 1, \text{Im } t > 0\}$ onto Ω. Since Ω is simply connected and Γ is a Jordan arc, there exists a g that maps the real interval

$$I = \{t_1 + 0i; -1 < t_1 < 1\}$$

onto Γ and g is continuous and $1 - 1$ from $G \cup I$ onto $\Omega \cup \Gamma$; see reference 27, p. 369. Thus g gives a continuous parametrization of Γ, and we shall prove that this parametrization is analytic. For simplicity, let $g(0) = 0$.

Consider the function

(2.10)
$$\Phi(t) = \begin{cases} g(t) & \text{if Im } t \geqslant 0, \\ \zeta^*(\overline{g(\bar{t})}) & \text{if Im } t < 0. \end{cases}$$

This function is holomorphic if $|t| \leqslant \varepsilon$, Im $t \neq 0$ [where ε is sufficiently small so that $g(t)$ is in the domain where ζ^* is holomorphic]. Since $\Phi(t)$ is also continuous along Im $t = 0$, the theorem of Morera can be applied to deduce that $\Phi(t)$ is holomorphic in $\{|t| < \varepsilon\}$. Thus, in particular, $g(t)$ is analytic for t real, $|t| < \varepsilon$.

In the remainder of this section we shall relax the conditions of analyticity, replacing them by $C^{m+\lambda}$ smoothness conditions. It will be convenient to use complex notation also for the elliptic equation. Thus we shall work with an equation

(2.11)
$$w_{\bar{z}} = k \qquad \text{in } \Omega \cap B_R$$

and assume that

(2.12)
$$w = 0 \qquad \text{on } \Gamma,$$

(2.13)
$$k \neq 0 \qquad \text{in } B_R.$$

In the applications to (2.4)–(2.6) we take

(2.14)
$$w = u_z - \phi_z$$

and then (2.4)–(2.6) yield (2.11)–(2.13) with $4k = f - \Delta\phi$.

Theorem 2.2. *Let* Ω, Γ *be as in Theorem 2.1 and let* $g(t)$ *be the conformal mapping* $G \to \Omega$ *(continuous and* $1 - 1$ *from* $G \cup I \to \Omega \cup \Gamma$, *as before). Assume that* $w \in H^{1, \infty}(\Omega)$, $k \in C^\lambda(B_R)$ *for some* $0 < \lambda < 1$, *and that* (2.11)–(2.13) *hold. Then, for any* $0 < \mu < \lambda$ *and* $0 < \rho < 1$, $g(t) \in C^{1+\mu}$ *in* $\{t$ *real,* $-\rho < t < \rho\}$.

Proof. Extend k as a C^λ function in R^2. Let

$$A(z) = -\frac{1}{\pi} \int_{B_{R'}} \frac{k(\zeta)}{\zeta - z} d\zeta_1 \, d\zeta_2 \qquad (R' > R),$$

and for some fixed $z_0 \in \Gamma$ ($|z_0| < 1$), let

$$w^*(z) = A(z) - A(z_0) - A_z(z_0)(z - z_0).$$

Then $w^* \in C^{1+\lambda}(B_R)$ and

(2.15) $$w^*_{\bar{z}} = k, \qquad w^*(z_0) = w^*_z(z_0) = 0.$$

Writing

(2.16) $$w^*(z) = k(z_0)(\bar{z} - \bar{z}_0) + R(z, z_0) \qquad (z \in B_R),$$

we find that

(2.17) $$|z - z_0|^{-1} |R(z, z_0)| + |R_z(z, z_0)| + |R_{\bar{z}}(z, z_0)| \leq C |z - z_0|^\lambda.$$

Notice [by (2.13), (2.15)] that for any small $\varepsilon > 0$,

(2.18) $$\left| \frac{w^*_z}{w^*_{\bar{z}}} \right| < \varepsilon < 1 \qquad \text{in } B_r(z_0)$$

if r is small enough. The Jacobian of the mapping $z \rightarrow w^*(z)$ is

$$\tfrac{1}{2}\left(|w^*_z|^2 - |w^*_{\bar{z}}|^2 \right) < 0 \qquad \text{in } B_r(z_0).$$

Therefore, this mapping is a $C^{1+\lambda}$ diffeomorphism from $B_r(z_0)$ into a neighborhood of 0; we denote its inverse by w^{*-1}.

The function

$$h(z) = w^*(z) - w(z)$$

is holomorphic in $\Omega \cap B_R$, continuous in $\Omega \cup \Gamma$ and, by (2.12),

(2.19) $$h(z) = w^*(z) \qquad \text{on } \Gamma.$$

Define in B_δ

(2.20) $$\Phi(t) = \begin{cases} g(t) & \text{if Im } t \geq 0, \\ w^{*-1}(h(g(\bar{t}))) & \text{if Im } t < 0, \end{cases}$$

where δ is chosen small enough so that $g(B_\delta \cap \{\text{Im } t \geq 0\}) \subset B_r(z_0)$; we assume for simplicity that $g(0) = z_0$.

Note that

$$\int_G | \nabla g |^2 \, dt_1 \, dt_2 = \text{area of } \Omega < \infty,$$

so that $g \in H^1(G)$. Since h is Lipschitz, it follows that Φ restricted to Im $t > 0$ (< 0) is in H^1. Applying the matching lemma, we conclude that

$$\Phi \in H^1(B_\delta).$$

The following chain rule formulas can be directly established for a function $G(z)$ with $z = k(t)$:

$$D_t G(k(t)) = D_z G(k(t)) D_t k(t) + D_{\bar{z}} G(k(t)) D_t \overline{k(t)},$$

$$D_{\bar{t}} G(k(t)) = D_z G(k(t)) D_{\bar{t}} k(t) + D_{\bar{z}} G(k(t)) D_{\bar{t}} \overline{k(t)}.$$

We also have

$$\overline{D_t k(t)} = D_{\bar{t}} \overline{k(t)}, \qquad D_t \overline{k(t)} = \overline{D_{\bar{t}} k(t)}.$$

If $G(z)$ has inverse $G^{-1}(z)$, then by applying $D_{\bar{z}}$ to $G^{-1}(G(z)) = z$, we get

$$\left| \frac{D_z G^{-1}}{D_{\bar{z}} G^{-1}} \right| = \left| \frac{D_z G}{D_{\bar{z}} G} \right|.$$

Using these rules, we deduce from (2.20) that

$$\frac{| \Phi_{\bar{t}} |}{| \Phi_t |} = \frac{| w_z^* |}{| w_{\bar{z}}^* |} < \varepsilon < 1 \qquad \text{by (2.18), if } | t | < \delta, \quad \text{Im } t < 0.$$

Since also $\Phi_{\bar{t}} = g_{\bar{t}} = 0$ if Im $t > 0$, the function $\Phi(t)$ satisfies

$$\frac{| \Phi_{\bar{t}} |}{| \Phi_t |} < \varepsilon < 1 \qquad \text{in } B_\delta.$$

A function Φ in $H^1(B_\delta)$ satisfying such an inequality is called ε-*quasiconformal*; by reference 40, p. 269.

$$\Phi \text{ belongs to } H^{1,s}(B_{\delta/2}), \qquad \text{where } s = s(\varepsilon) \to \infty \text{ if } \varepsilon \to 0.$$

It follows that g belongs to $H^{1,s}$ in $B_{\delta/2} \cap \{ \text{Im } t > 0 \}$. If instead of $t_0 = 0$ we take any other point $t_0 \in (-1, 1)$, we conclude that for any small $\delta > 0$,

$$(2.21) \qquad g \in H^{1,s}(G \cap B_{1-\delta}) \qquad \text{for any } 1 < s < \infty.$$

To obtain more regularity for g, we shall use another extension of $g(t)$ into $\operatorname{Im} t < 0$. First define $g^*(t)$ in G by

$$(2.22) \qquad h(g(t)) = k(z_0)(g^*(t) - \bar{z}_0) + R(g(t), \bar{g}(t)),$$

where $R(z, \bar{z})$ is defined by [see (2.16)]

$$w^*(z) = k(z_0)(\bar{z} - \bar{z}_0) + R(z, \bar{z}).$$

It follows from (2.19) that

$$g^*(t) = \overline{g(t)} \qquad \text{if } \operatorname{Im} t = 0.$$

Applying $D_{\bar{t}}$ to (2.22) we obtain, since $h(g(t))$ is holomorphic,

$$g_{\bar{t}}^*(t) = -\frac{1}{k(z_0)} D_{\bar{t}} R(g(t), \overline{g(t)}),$$

so that, by (2.17),

$$(2.23) \qquad |g_{\bar{t}}^*(t)| \leqslant C |g'(t)| |g(t) - z_0|^{\lambda}, \qquad z_0 = g(0).$$

Now, by (2.21), $g \in H^{1,s}$ and thus $g \in C^{1-2/s}$, $g' \in L^s$. It follows that for any $\delta > 0$,

$$(2.24) \qquad |t|^{-\sigma} |g_{\bar{t}}^*| \in L^s(G \cap B_{1-\delta}), \qquad \sigma = \lambda\left(1 - \frac{2}{s}\right).$$

Note also that $g_{\bar{t}}^* \in L^s(G)$. Thus the function

$$(2.25) \qquad \Psi(t) = \begin{cases} g(t) & \operatorname{Im} t \geqslant 0 \\ \overline{g^*(\bar{t})} & \operatorname{Im} t < 0 \end{cases}$$

is in $H^{1,s}(B_{1-\delta})$, by the matching lemma.

We shall need the following fact:

Lemma 2.3. *Suppose that $v(t) \in H^{1,s}(B)$, where $B = B_1$ and*

$$|t|^{-\sigma} v_{\bar{t}} \in L^s(B),$$

where $\sigma > 0$, $s > 2$ and $\sigma - (2/s) \geqslant \tau$, $0 < \tau < 1$. Then there exists a complex number c such that

$$\left| \frac{v(t) - v(0)}{t} - c \right| \leqslant C\Lambda |t|^\tau \qquad \text{if } t \in B_\rho, \quad 0 < \rho < 1,$$

$$|c| \leqslant C\Lambda,$$

where

$$\Lambda = \| |t|^{-\sigma} v_{\bar{t}} \|_{L^s(B)} + \|v\|_{L^\infty(B)}$$

and C depends only on ρ, s, τ.

For proof, see Problem 1.
Applying the lemma to Ψ, we obtain

$$\left| \frac{\Psi(t) - \Psi(0)}{t} - c \right| \leqslant \text{const.} |t|^\tau$$

if $|t| < \rho < 1$; this holds in particular for $g(t)$. Since $t_0 = 0$ can be replaced by any $t_0 \in (-1, 1)$, we obtain, for any $0 < R < 1$,

$$(2.26) \qquad \left| \frac{g(t) - g(t_0)}{t - t_0} - c(t_0) \right| \leqslant C_R |t - t_0|^\tau, \qquad t_0 \in (-R, R), \quad |t| < R,$$

where $|c(t_0)| \leqslant C_R'$, and C_R, C_R' are constants depending on R, s, τ, and the $H^{1,s}$ norm of g. It follows that $g'(t_0)$ exists and $g'(t_0) = c(t_0)$. Further, interchanging t and t_0 in (2.26), we find that

$$|c(t) - c(t_0)| \leqslant 2C_R |t - t_0|^\tau,$$

so that $g \in C^{1+\tau}$ on $(-R, R)$.

We shall next extend Theorem 2.2 to higher differentiability of the parametric representation $g(t)$. We use the notation of Theorem 2.1.

Theorem 2.4. *Let Ω, Γ be as in Theorem 2.1, and let $u \in C^{1,1}(\Omega \cup \Gamma)$, $f \in C^{m+\alpha}(B_R)$, $\phi \in C^{m+2+\alpha}(B_R)$, where $m \geqslant 0$, $0 < \alpha < 1$. If u satisfies (2.4)–(2.6), then the parametric representation $z = g(t)$ of Γ is in $C^{m+1+\alpha}$.*

Proof.   The proof is by induction. The inductive assumption is that for any $-1 < t_0 < 1$, $t \in G \cap B_R$,

$$g(t) = \sum_{i=0}^{m} c_i(t_0)(t - t_0)^i + O(1)(t - t_0)^{m+\alpha},$$

where $c_i(t_0)$ are functions of t_0 and

$$\sum_{i=0}^{m} |c_i(t_0)| + |O(1)| \leq C_R \qquad \text{if } -R < t_0 < R \quad (R < 1),$$

where C_R depends on ϕ, f and R. The case $m = 0$ was established in Theorem 2.2. Without loss of generality we may take $f = 0$ [otherwise, consider $u - \tilde{f}, \tilde{f}$ as in (2.7)].

Suppose for simplicity that $g(t_0) = 0$. We can write

$$\nabla^* \phi(x_1, x_2) = P_{m+1}(z, \bar{z}) + \Lambda(x_1, x_2),$$

where

$$P_{m+1}(z, \zeta) = \sum_{i+j \leq m+1} \frac{1}{i! j!} \left(D_z^i D_{\bar{z}}^j \nabla^* \phi(0) \right) z^i \zeta^j,$$

$$|\Lambda(x_1, x_2)| \leq C |z|^{m+1+\alpha}.$$

Set $\Lambda(z) = \Lambda(x_1, x_2)$ and consider the equation for ζ:

(2.27) $$P_{m+1}(z, \zeta) + \Lambda(z) = \nabla^* u(z).$$

Observe that, by (2.6) with $f = 0$,

$$P_{\bar{z}} P_{m+1}(z, \bar{z}) = D_{\bar{z}} \nabla^* \phi(z) + O(|z|^{m+\alpha})$$

$$= \tfrac{1}{2} \Delta \phi(z) + O(|z|^{m+\alpha}) \neq 0$$

if $|z|$ is small enough, and, by (2.5),

$$P_{m+1}(z, \zeta) + \Lambda(z) = \nabla^* u(z) \qquad \text{if } z \in \Gamma, \zeta = \bar{z}.$$

Hence by the implicit function theorem, we can solve (2.27) uniquely. The solution $\zeta^*(z)$ has the form

$$\zeta^*(z) = \zeta(z, \nabla^* u(z) - \Lambda(z)),$$

where $\zeta(z, w)$ is holomorphic in z, w. Since ∇^*u is Lipschitz continuous and since Λ is Lipschitz continuous, also $\zeta^*(z)$ is Lipschitz continuous in a neighborhood of $z = 0$.

Taking $\zeta = \zeta^*(z)$ in (2.27) and applying $D_{\bar{z}}$, we get, since $\Delta u = 0$,

$$D_{\zeta} P_{m+1}(z, \zeta^*(z)) \zeta_{\bar{z}}^* + D_{\bar{z}} \Lambda(z) = 0 \text{ in } \Omega.$$

But the first term on the left is $\neq 0$; hence

$$|\zeta_{\bar{z}}^*(z)| \leq C |z|^{m+\alpha} \text{ in } \Omega.$$

We now extend $g(t)$ into $\text{Im } t < 0$ by

$$\Psi(t) = \begin{cases} g(t) & \text{if } \text{Im } t \geq 0, \\ \zeta^*(g(\bar{t})) & \text{if } \text{Im } t < 0 \end{cases}$$

and let

$$\tilde{\Psi}(t) = \Psi(t) - \sum_{i=0}^{m} c_i(t_0)(t - t_0)^i.$$

Then for t with $\text{Im } t \neq 0$,

$$(2.28) \qquad |\tilde{\Psi}_{\bar{t}}| \leq C |t - t_0|^{m+\alpha}.$$

We can therefore apply Green's formula (2.3):

$$\frac{\tilde{\Psi}(t) - \tilde{\Psi}(t_0)}{(t - t_0)^{m+1}} = -\frac{1}{\pi} \int_{|z|<1} \frac{1}{(z - t_0)^{m+1}} \frac{1}{z - t} \tilde{\Psi}_{\bar{z}}(z) \, dx_1 \, dx_2$$

$$+ \frac{1}{2\pi i} \int_{|z|=1} \frac{1}{(z - t_0)^{m+1}} \frac{1}{z - t} \tilde{\Psi}(z) \, dz.$$

Defining $c_{m+1}(t_0)$ to be the value of the right-hand side when $t \to t_0$, we can now prove (see Problems 1 and 2) that $c_{m+1}(t_0)$ exists and

$$\left| \frac{\tilde{\Psi}(t) - \tilde{\Psi}(t_0)}{(t - t_0)^{m+1}} - c_{m+1}(t_0) \right|$$

$$(2.29) \qquad \leq \frac{1}{\pi} \left| \int_{|z|<1} \frac{t - t_0}{(z - t_0)^{m+2}(z - t)} \tilde{\Psi}_{\bar{z}}(z) \, dx_1 \, dx_2 \right|$$

$$+ \frac{1}{2\pi} \left| \int_{|z|=1} \frac{t - t_0}{(z - t_0)^{m+2}(z - t)} \tilde{\Psi}(z) \, dz \right|$$

and the right-hand side can be estimated by const. $|t - t_0|^\alpha$, using (2.28).

We have thus derived the expansion

$$g(t) = \sum_{i=0}^{m+1} c_i(t_0)(t - t_0)^i + O(1)(|t - t_0|^{m+1+\alpha}),$$

with

$$\sum_{i=0}^{m+1} |c_i(t_0)| + |O(1)| \leq C \qquad (-R < t_0 < R)$$

for $t \in G \cap B_R$, for any $0 < R < 1$. The assertion that $g \in C^{m+1+\alpha}$ now follows by standard arguments (see Problem 3).

We finally wish to prove, under the conditions of Theorem 2.4 with $m \geq 1$, that the parametric representation $z = g(t)$ of Γ is nondegenerate:

Theorem 2.5. *Let the conditions of Theorem 2.4 with $m = 1$ hold. Then, for any $-1 < t_0 < 1$, there holds*

$$(2.30) \qquad \lim_{t \to t_0} \frac{g(t) - g(t_0)}{(t - t_0)^n} \neq 0 \qquad \text{for either } n = 1 \text{ or } n = 2.$$

Proof. Consider the function $g^*(t)$ defined in (2.22). From (2.23) we have (since now we can take $\lambda = 1$)

$$|g_t^*(t)| \leq C |g'(t)| \, \|g(t) - z_0|, \qquad C \text{ constant.}$$

Since

$$|D_{\bar{t}} \, \overline{g^*(\bar{t})}| = |D_{\bar{z}} \, g^*(z)|_{z = \bar{t}}$$

and since $D_{\bar{t}} g(t) = 0$, it follows that the function $\Psi(t)$ defined in (2.25) satisfies

$$|\Psi_{\bar{z}}(z)| \leq q(z) \sup_{|t| = |z|} |\Psi(t) - \Psi(0)|,$$

where

$$q(z) = \begin{cases} 0 & \text{in } \{\operatorname{Im} z > 0\} \cap B_r \\ C |g'(\bar{z})| & \text{in } \{\operatorname{Im} z < 0\} \cap B_r \end{cases}$$

for some $r > 0$. Recall also that $\Psi \in H^{1,s}$ for any $s > 2$ (r depends on s).

We shall need the following lemma.

Lemma 2.6. *Suppose*

$$w \in H^{1,s}(B) \qquad \text{for some } s > 2 \quad (B = B_1),$$

(2.31)
$$|w_{\bar{z}}(z)| \leq q(z) \sup_{|t| = |z|} |w(t) - w(0)|, \qquad z \in B,$$

$$q \in L^s(B).$$

If $w \not\equiv 0$ in some neighborhood of $z = 0$, then there exists an integer $n \geq 1$ such that

$$\lim_{z \to 0} \frac{w(z) - w(0)}{z^n} = c \neq 0.$$

Assuming that the lemma is true, we can apply it to $\Psi(rz)$ and deduce that (2.30) holds for some integer $n \geq 1$. If $n \geq 3$, then $g(t) \sim g(t_0) + c(t - t_0)^n$ ($c \neq 0$) cannot be a $1 - 1$ mapping from G into Ω; hence $n \leq 2$ and the theorem follows.

Proof of Lemma 2.6. We first claim that if $W \in H^{1,s}(B_R)$ for some $s > 2$, $0 < R \leq 1$, and

$$|z|^{-n-\sigma} W_{\bar{z}} \in L^s(B_R),$$

$$|z|^{-n}(W(z) - W(0)) \in L^\infty(R_R),$$

where $\sigma \geq 0$, n is an integer ≥ 0, then

(2.32)
$$\left| \frac{W(z) - W(0)}{z^n} - c \right| \leq C|z|^\tau, \qquad |z| \leq \rho < R,$$

where $\lambda = 1 - 2/s$, $\tau = \min(\sigma + \lambda, 1)$,

(2.33) $\quad c = -\frac{1}{\pi} \int_{B_R} t^{-n-1} W t \, dt_1 \, dt_2 + \frac{1}{2\pi i} \int_{B_R} t^{-n-1}(W(t) - W(0)) \, dt,$

$$C = c_s \| t^{-n-\sigma} W_{\bar{t}} \|_{L^s(B_R)} + R^{1-n-\tau} \| W - W(0) \|_{L^\infty(\partial B_R)} (R - \rho)^{-1}$$

and c_s depends only on s; for proof, see Problem 4.

We next wish to show that if for the function w in the lemma

(2.34)
$$\lim_{|z| \to 0} |z|^{-n}(w(z) - w(0)) = 0, \qquad n \text{ integer} \geq 0,$$

then for each $0 < R \leqslant 1$ and $0 < \delta < 1$,

(2.35)

$$|z|^{-n-1}|w(z) - w(0)| \leqslant C \|t^{-n}(w(t) - w(0))\|_{L^\infty(B_R)}, \qquad z \in B_{(1-\delta)R},$$

where C depends only on s, δ, and the $L^s(B_R)$ norm of q.

The proof is by iteration. Take for simplicity $w(0) = 0$ and suppose that

(2.36) $\qquad |w(z)| \leqslant A |z|^{n+\sigma} \qquad$ for $|z| = \rho \leqslant R, \quad A > 0, \quad \sigma \geqslant 0.$

In view of the assumptions on w, the conditions for (2.32) are then satisfied with $W = w$. Hence

(2.37) $\qquad \rho^{-n-\tau}|w(z)| \leqslant c_s A \|q\|_{L^s(B_R)} + R^{1-n-\tau}\|w\|_{L^\infty(\partial B_R)}(R - \rho)^{-1}$

where $\tau = \min(\sigma + \lambda, 1)$.

If $\tau = \sigma + \lambda$, then using $R^{-n-\sigma}|w| \leqslant A$ on ∂B_R, we get

(2.38) $\qquad \rho^{-n-\sigma-\lambda}|w(z)| \leqslant \left\{ c_s \|q\|_{L^s(B_R)} + R^{1-\lambda}(R - \rho)^{-1} \right\} A,$

whereas if $\tau = 1$, then we similarly get

(2.39) $\qquad \rho^{-n-1}|w(z)| \leqslant \left\{ c_s \|q\|_{L^s(B_R)} + R^{\sigma}(R - \rho)^{-1} \right\} A.$

Since (2.36) certainly holds when $\sigma = 0$ [by (2.34)] we deduce (2.38) with $\sigma = 0$. We now proceed by iteration. Let m be a positive integer such that

$$m + 1 < \frac{1}{\lambda} \leqslant m,$$

and choose small $\eta > 0$ and $\sigma = (j - 1)\lambda$ for $j = 1, 2, \ldots, m$. From (2.38) we deduce, for each j,

(2.40) $\quad \|\rho^{-n-j\lambda}w\|_{L^\infty(B_{(1-\eta)R})} \leqslant \left\{ C + \eta^{-1}R^{-\lambda} \right\} \|\rho^{-n-(j-1)\lambda}w\|_{L^\infty(B_R)},$

and from (2.39) with $\sigma = (m - 1)\lambda$, we deduce that

(2.41) $\quad \|\rho^{-n-1}w\|_{L^\infty(B_{(1-\eta)R})} \leqslant \left\{ C + \eta^{-1}R^{-1+(m-1)\lambda} \right\} \|\rho^{-n-(m-1)\lambda}w\|_{L^\infty(B_R)}.$

Applying (2.40) with $R = R_{j-1} = R_0(1 - \eta)^{j-1}$ for $j = 1, 2, \ldots, m$ and then applying (2.41) with $R = R_m = R_0(1 - \eta)^m$, then multiplying together all the left-hand sides and all the right-hand sides of these inequalities, we obtain the assertion (2.35) with $1 - \delta = (1 - \eta)^m$.

For $n = 0$ (2.34) holds. Hence (2.35) gives

$$|z|^{-n-1}|w(z)| \leqslant C$$

and recalling (2.31), we see that (2.32) can be applied to $W = w$ to deduce that

(2.42) $$c \equiv \lim_{z \to 0} \frac{w(z) - w(0)}{z^n} \text{ exists.}$$

If $c = 0$, we repeat this procedure with $n = 1$. We deduce that (2.42) holds with $n = 1$. Again, if $c = 0$, we can deduce (2.42) with $n = 2$, and so on. Thus to complete the proof of the lemma it remains to show:

If w is as in the lemma, then

(2.43) $$\lim_{z \to 0} |z|^{-n} w(z) = 0 \text{ cannot hold}$$

for all $n \geqslant 0$.

The proof is outlined in Problem 5.

PROBLEMS

1. Prove Lemma 2.3.

   [*Hint*: By (2.3),

   $$\frac{v(t) - v(0)}{t} = -\frac{1}{\pi}\int_B \frac{1}{z(z-t)} v_{\bar{z}}(z)\, dx_1\, dx_2 + \frac{1}{2\pi i}\int_{\partial B}\frac{v(z)}{z(z-t)}\, dz.$$

   Define c by the right-hand side when $t = 0$:

   $$c = -\frac{1}{\pi}\int_B z^{-2} v_{\bar{z}}\, dx_1\, dx_2 + \frac{1}{2\pi i}\int_{\partial B} z^{-2} v\, dz.$$

   Show that c is finite and that

   $$\frac{v(t) - v(0)}{t} - c = -\frac{t}{\pi}\int_B \frac{1}{z^2(z-t)} v_{\bar{z}}\, dx_1\, dx_2 + \frac{1}{2\pi i}\int_{\partial B}\frac{v}{z^2(z-t)}\, dz.$$

   The first integral is estimated by

   $$\| |z|^{-\sigma} v_{\bar{z}} \|_{L^s}\left[\int |z|^{(\sigma-2)s'}|z-t|^{-s'}\, dx_1\, dx_2\right]^{1/s'},$$

where $1/s + 1/s' = 1$. If $\sigma < 2$, break the domain of integration into three parts:

$$A_0 = \{|z - t| < \tfrac{1}{2}|t|\}, \qquad \{|z| < \tfrac{3}{2}|t|\} \setminus A_0, \qquad \{|z| > \tfrac{3}{2}|t|\} \setminus A_0.]$$

2.   Show that the right-hand side of (2.29) is bounded by const. $|t - t_0|^{\alpha}$.

3.   Let $g(t)$ be defined for $0 < t < 1$ and assume that for any $0 < s < 1$,

$$g(t) = \sum_{i=0}^{n} c_i(s)(t - s)^i + 0(1)(|t - s|^{n+\alpha}) \qquad (0 < \alpha < 1)$$

where $|c_i(s)| \leqslant C$, $|0(1)| \leqslant C$, C constant. Prove that $g \in C^{n+\alpha}$.
[*Hint*: Take $n = 2$. Clearly, $c_0(s) = g(s)$, $c_1(s) = g'(s)$ and developing

$$g(a + h) - g(a) = hc_1(a) + O(h^2),$$

$$g(a) - g(a + h) = -hc_1(a + h) + O(h^2),$$

it follows that $c_1 \in C^1$. Developing about $a - h$

$$g(a + h) + g(a - h) - 2g(a) = 4h^2 c_2(a - h) + O(h^{2+\alpha}),$$

then replacing h by $-h$ and comparing, yields $c_2 \in C^{\alpha}$. The limit of the second finite-difference quotients $\overline{D}^2 g = \lim \Delta_h^2 g(t)$ is $c_2(t)$. From

$$g(a + h) - g(a) = hc_1(a) + h^2 c_2(a) + O(h^{2+\alpha}),$$

$$g(a + 2h) - g(a + h) = hc_1(a + h) + h^2 c_2(a + h) + O(h^{2+\alpha}),$$

it follows that

$$\Delta_h^2 g(a + h) = \frac{c_1(a + h) - c_1(a)}{h} + o(1)$$

and $c_1'(a)$ exists and equals $c_2(a)$.]

4.   Prove (2.32).
[*Hint*: Express $(W(z) - W(0))z^{-n}$ by Green's formula and adapt the procedure in Problem 2.]

5.   Prove (2.43).
[*Hint*: Express $z^{-n} w(z)$ by Green's formula in B_R and deduce for $f(\rho) = f_n(\rho) = \sup_{|z|=\rho} |z^{-n} w(z)|$,

$$F(\rho) \leqslant C \|f|t\| \, |t - z|^{-1} \|_{L^{s'}(B_R)} + Cf(R) R^{1/s} \| \, |t - z|^{-1} \|_{L^{s'}(\partial B_R)},$$

where $|z| = \rho$, $1/s + 1/s' = 1$. Raise both sides to the power s', multiply by $|z|^{-s'}$, and integrate over $z \in B_R$, to derive

$$\int_{B_R} f(\rho)^{s'} \rho^{-s'} \leqslant CR^{2-s'} \int_{B_R} f(\rho)^{s'} \rho^{-s'} + Cf(R)^{s'} R^{2-s'}.$$

If $w(z_0) \neq 0$, $|z_0|$ small enough, then

$$\rho_0^{-n} < CR^{-n}, \qquad \rho_0 = |z_0|,$$

which is impossible if $n \to \infty$.]

6. Prove the following theorem: Let Ω be a strictly convex bounded domain in R^2 and let ϕ be a strictly concave function in $C^2(\overline{\Omega})$ [that is, $(\partial^2\phi/\partial x_i \, \partial x_j)$ is negative definite] with $\phi < 0$ on $\partial\Omega$, $\max_\Omega \phi = \phi(x^*) > 0$ $(x^* \in \Omega)$. Let u be the solution of the variational inequality

$$-\Delta u \geqslant 0, \qquad u \geqslant \phi, \qquad -\Delta u(u - \phi) = 0 \text{ in } \Omega,$$

$$u = 0 \qquad \text{on } \partial\Omega.$$

Then (a) the noncoincidence set N is connected; (b) the coincidence set Λ is connected.

[*Hint*: To prove (b), let $x_0 \in \partial\Omega$, π_{x_0} the plane tangent to $\partial\Omega$ at x_0 and to $z = \phi(x)$ $[x = (x_1, x_2)]$, say at $(\tilde{x}, \gamma(\tilde{x})) \in R^3$. By comparison $L(x) > u(x)$, where $x_3 = L(x)$ is the equation for π_{x_0}, and $|\nabla L(x_0)| > |\nabla u(x_0)|$; $\nabla L(x_0)$ and $\nabla u(x_0)$ are both normal to $\partial\Omega$ at x_0. The mappings

$$\Gamma_1 : x_0 \to \nabla L(x_0) \equiv \nabla\phi(\gamma(\tilde{x})),$$

$$\Gamma_0 : x_0 \to \nabla u(x_0)$$

are $1 - 1$ (since Ω is strictly convex) and $\Gamma_0 \subset$ interior of Γ_1. The mapping $x \to \nabla\phi(x)$ is $1 - 1$ (since ϕ is strictly concave; it maps γ onto Γ_1 and x^* into the origin 0. Hence it maps interior γ onto interior Γ_1; consequently, $\Gamma_0 = \nabla\phi(\tilde{\gamma})$ for some curve $\tilde{\gamma}$ in the interior of γ. Since $u < L$, the free boundary Γ lies outside γ and N is bounded by Γ and $\partial\Omega$. The mapping

$$\sigma = (\nabla\phi)^{-1} \nabla u \qquad (x \in N)$$

is open, $\sigma(\Gamma) = \Gamma$ and $\sigma(\partial\Omega) = \tilde{\gamma}$ lies inside γ and hence inside Γ. By (a), $\sigma(N)$ is connected. But then, $\Lambda \setminus \Gamma = \sigma(N) \cup$ (interior of $\tilde{\gamma}$) is also a connected open set.]

From Problem 6 we can easily deduce that N is homeomorphic to an annulus. An additional argument shows that Theorem 2.1 can be applied (observe that Γ is not assumed a priori to be a Jordan curve); for details, see Lewy and Stampacchia [138a].

3. GENERAL PROPERTIES OF THE FREE BOUNDARY

In this section we establish two basic facts for the obstacle problem:

(i)   The free boundary has measure zero;

(ii) If $y \in \Gamma$ (the free boundary), then

$$\liminf_{\substack{x \to y \\ x \in N}} D_{ii}(u(x) - \phi(x)) \geq 0,$$

where ϕ is the obstacle, N the noncoincidence set, and i is any direction.

We first work with the simple case of

$$(3.1) \qquad -\Delta u + f \geq 0, \qquad u \geq 0, \qquad (-\Delta u + f)u = 0 \text{ in } \Omega,$$

where Ω is a bounded domain in R^n, $f \in C^\alpha(\Omega)$ $(0 < \alpha < 1)$; we shall afterward deal with a general elliptic operator and a general obstacle.

As before, we set

$$N = \{x \in \Omega, u(x) > 0\},$$

$$\Lambda = \{x \in \Omega, u(x) = 0\},$$

$$\Gamma = \partial N \cap \Omega.$$

Recall that $u \in C^{1,1}(\Omega)$. For simplicity we may assume that $u \in C^{1,1}(\overline{\Omega})$.

Our first result is a simple lemma of nondegeneracy, asserting that u cannot be uniformly small in some neighborhood of a point of $\overline{N}$, provided that $f \geq \text{const.} > 0$.

Lemma 3.1.   *Suppose that $f \geq \lambda > 0$ in Ω and let x_0 be any point in $\overline{N}$. Then for any ball $B_r(x_0) \subset \Omega$,*

$$(3.2) \qquad \sup_{B_r(x_0)} [u(x) - u(x_0)] \geq \frac{\lambda r^2}{2n}.$$

Proof.   Suppose first that $x_0 \in N$. The function

$$w(x) = u(x) - u(x_0) - \frac{\lambda}{2n}|x - x_0|^2$$

satisfies $\Delta w \geq 0$ in N and $w(x_0) = 0$. Hence, by the maximum principle, sup w in $N \cap B_r(x_0)$ is nonnegative and it is attained on the boundary. But since

$w < 0$ on ∂N, there must then exist a point $x_1 \in \partial B_r(x_0) \cap \bar{N}$ such that $w(x_1) \geq 0$, and (3.2) follows.

If $x_0 \notin N$, we apply (3.2) to a sequence of points $x_m \in N$, $x_m \to x_0$.

The following Calculus type lemma will be useful.

Lemma 3.2. *Let u be any nonnegative function in $C^{1,1}(D)$, where D is an open set in R^n, and let $|D_{ij}u| \leq M$ for all i, j. If $u(y) < \delta^2$ for some $y \in D$ and if dist$(y, \partial D) > \max(\delta, h)$, where δ, h are some positive numbers, then*

$$|\nabla u(y)| \leq C_M \delta \qquad (C_M = M + 1)$$

and, for any direction, e_i,

(3.3)
$$\int_0^h \int_0^\tau D_{ii}u(y + te_i)\, dt\, d\tau \geq -C_M \delta(\delta + h).$$

Proof. In the inequality

(3.4) $$0 \leq u(y + se_i) = u(y) + u_i(y)s + \int_0^s \int_0^\tau D_{ii}u(y + te_i)\, dt\, d\tau$$

[where $s < $ dist$(y, \partial D)$], choose the e_i in the direction for which $u_i(y) = -|\nabla u(y)|$ and take $s = \delta$. This gives

$$0 \leq \delta^2 - |\nabla u(y)|\delta + \delta^2 M,$$

from which we deduce that $|\nabla u(y)| \leq C_M \delta$. Using this estimate we now apply (3.4) with $s = h$, and any direction e_i, and (3.3) follows.

Corollary 3.3. *If $x \in N$, dist$(x, \Gamma) < \varepsilon$, dist$(x, \partial\Omega) \geq \delta_0 > 0$, then*

(3.5) $$u(x) \leq C\varepsilon^2,$$

(3.6) $$|\nabla u(x)| \leq C\varepsilon,$$

where C depends only on the $C^{1,1}$ norm of u in Ω and on δ_0.

Indeed, (3.5) follows from the fact that $u = \nabla u = 0$ on Γ and the $C^{1,1}$ nature of u, and (3.6) is then a consequence of the preceding lemma.

Theorem 3.4. *Suppose that $f \geq \lambda > 0$ in Ω. If $x_0 \in \Gamma$, dist$(x_0, \delta\Omega) \geq \delta_0 > 0$, then*

(3.7) $$\frac{|B_\varepsilon(x_0) \cap N|}{|B_\varepsilon|} \geq \gamma > 0$$

for all $0 < \varepsilon \leq \varepsilon_0$, where γ and ε_0 depend only on λ, δ_0, and the $C^{1,1}$ norm of u.

Here we have used the notation

$$|A| = \text{Lebesgue measure of } A.$$

Proof. By Lemma 3.1 there exists a point $y \in \partial B_\varepsilon(x_0)$ such that

$$u(y) \geq \frac{\lambda \varepsilon^2}{2n}.$$

Using (3.6) we get, for small enough δ,

$$u(x) \geq \frac{\lambda \varepsilon^2}{2n} - \delta C \varepsilon^2 > 0$$

if $|x - y| < \delta \varepsilon$. Thus there exists a ball $B_{\delta \varepsilon}(y)$ contained in $B_{2\varepsilon}(x_0)$ in which $u > 0$. Since δ can be chosen independently of ε (for ε small enough), the assertion follows.

We recall that for any measurable set S,

$$\lim_{\varepsilon \to 0} \frac{|B_\varepsilon(x_0) \cap S|}{|B_\varepsilon|} = 1 \qquad \text{for a.a. } x_0 \in S;$$

that is, a.a. points of S have density 1. Since by Theorem 3.4, all points of Γ do not have density 1, we conclude:

Theorem 3.5. *If $f \geq \lambda > 0$, then the free boundary has Lebesgue measure zero.*

In the next theorem we study the behavior of the second pure derivatives $D_{ii}u$ near the free boundary; here we do not assume that $f > 0$, but for simplicity we take $f = $ const.

Theorem 3.6. *Let $f = $ const., $y \in \Gamma$, dist$(y, \partial \Omega) \geq \delta_0 > 0$. Then if $x \in N$,*

$$(3.8) \qquad D_{ii}u(x) \geq \frac{-C}{|\log |x - y||^\varepsilon}, \qquad \varepsilon = \frac{1}{2(n-1)},$$

where C is a positive number depending on n, δ_0 and the $C^{1,1}$ norm of u.

It follows that

$$(3.9) \qquad D_{ii}u(x) \geq \frac{-C}{|\log \text{dist}(x, \Gamma)|^\varepsilon}$$

and, in particular,

(3.10) $$\liminf_{\substack{x \to y \\ x \in N}} D_{ii}u(x) \geqslant 0 \qquad \text{if } y \in \Gamma.$$

Proof. Take for simplicity $y = 0$ and define

$$-M_k = \inf_{B_{2^{-k}}} D_{ii}u.$$

We shall estimate the M_k recursively for all k for which $M_k > 0$.

Let $x \in N$, $|x| \leqslant 2^{-(k+1)}$ and let $B_s(x)$ be the largest ball in N. Then $s \leqslant 2^{-(k+1)}$ and there exists a point $y_0 \in \partial B_s(x) \cap \Gamma$. Let y_1 be a point on the segment $\overline{xy_0}$ with distance δs to y_0; δ is small and still to be determined. If

$$|D_{ij}u| \leqslant M,$$

then

$$u(y_1) \leqslant \frac{M}{2}(\delta s)^2.$$

Suppose that $\langle e_i, x - y_0 \rangle \geqslant 0$ when e_i is the unit vector in the ith direction. Applying (3.3) with $y = y_1$, $h = \sqrt{\delta}\,s$, we get

$$\int_0^{\sqrt{\delta}s} \int_0^\tau D_{ii}u(y_1 + te_i)\, dt\, d\tau \geqslant -C\delta^{3/2}s^2.$$

If $\langle e_i, x - y_0 \rangle \leqslant 0$, then we replace e_i by $-e_i$ above and observe that $D_{-i,-i}u = D_{ii}u$.

We conclude that

(3.11) $$\sup_{0 < t < \sqrt{\delta}s} D_{ii}u(y_1 + t(\pm e_i)) \geqslant -C\delta^{1/2}$$

and the segment $y_1 + t(\pm e_i)$ stays a distance $\geqslant \frac{1}{2}\delta s$ from $\partial B_s(x)$ (for $0 < t < \sqrt{\delta}\,s$).

The function

$$w = D_{ii}u + M_k$$

is harmonic in $B_s(x)$ and is positive there [since $B_s(x) \subset B_{2^{-k}}$]; by (3.11)

$$w(\tilde{y}) \geqslant M_k - C\delta^{1/2}$$

for some point $\tilde{y} \in B_{s-\delta s/2}$. We now use Harnack's inequality (see Lemma 3.9):

(3.12) $$w(x) \geqslant w(\tilde{y})c\delta^{n-1} \qquad (c > 0).$$

It follows that

$$(3.13) \qquad D_{ii}u(x) \geqslant -M_k + c\left(M_k - C\delta^{1/2}\right)\delta^{n-1}.$$

Choosing δ such that $C\delta^{1/2} = \varepsilon_0 M_k$, where ε_0 is sufficiently small so that $\delta < \frac{1}{2}$ if $C\delta^{1/2} = \varepsilon_0 M$, we get

$$D_{ii}u(x) \geqslant -M_k + CM_k^{2n-1}.$$

Since x was an arbitrary point with $|x| < 2^{-(k+1)}$, we deduce that

$$(3.14) \qquad -M_{k+1} \geqslant -M_k + CM_k^{2n-1}.$$

One can now show by induction that

$$M_k \leqslant Ck^{-\varepsilon} \qquad \left[\varepsilon = \frac{1}{2(n-1)}\right]$$

with suitable C. Thus

$$-M_k \geqslant -C|\log|x||^{-\varepsilon} \qquad \text{if } 2^{-(k+1)} \leqslant |x| \leqslant 2^{-k},$$

and (3.8) follows.

We shall generalize the results above to non-constant f and to an elliptic operator

$$(3.15) \qquad Au = -\sum a_{ij}(x)\frac{\partial^2 u}{\partial x_i \partial x_j} + \sum b_i(x)\frac{\partial u}{\partial x_i} + c(x)u.$$

The variational inequality is

$$(3.16) \qquad (Au - f) \geqslant 0, \qquad u \geqslant \phi, \qquad (Au - f)(u - \phi) = 0 \text{ in } \Omega.$$

We assume that the solution is in $C^{1,1}(\overline{\Omega})$ and that $f \in C^\alpha$, $\phi \in C^{2+\alpha}$. Without loss of generality we may take $b_i = 0$, $c = 0$, for otherwise we can absorb the terms $\sum b_i u_{x_i} + cu$ into f. We may also assume that $\phi = 0$; for otherwise we consider $u - \phi$ instead of u. Thus without loss of generality we reduce (3.15), (3.16) to

$$(3.17) \qquad Au = -\sum_{i,j=1}^{n} a_{ij}(x)\frac{\partial^2 u}{\partial x_i \partial x_j},$$

$$(3.18) \qquad (Au - f) \geqslant 0, \qquad u \geqslant 0, \qquad (Au - f)u = 0 \text{ in } \Omega.$$

We shall always assume that

$$(3.19) \qquad a_{ij} \in C^{2+\alpha}(\Omega), \quad f \in C^\alpha(\Omega) \qquad \text{for some } 0 < \alpha < 1.$$

The proof of Lemma 3.1 remains valid if we take

$$w(x) = u(x) - u(x_0) - \lambda\gamma\,|\,x - x_0\,|^2$$

with γ positive and small enough (depending on the modulus of ellipticity of A). Thus

(3.20)
$$\sup_{B_r(x_0)} \left[u(x) - u(x_0)\right] \geqslant \lambda\gamma r^2 \quad \text{if } x_0 \in \overline{N}.$$

Theorems 3.4 and 3.5 remain unchanged. For the sake of convenient reference, we state:

Theorem 3.7. *Lemma* 3.1 *and Theorems* 3.4 *and* 3.5 *remain valid for any solution of* (3.18).

Some nontrivial changes occur in the proof of Theorem 3.6:

Theorem 3.8. *The assertion of Theorem* 3.6 *remains valid for any solution of* (3.18).

Proof. In order to proceed as in the proof of Theorem 3.6, we must be able to apply Harnack's inequality to some "small perturbation" of $D_{ii}u + M_k$. To accomplish this, we note first that

(3.21)
$$Au_{ii} = f_{ii} + 2\sum a_{jk,i}u_{ijk} + \sum a_{jk,ii}u_{jk},$$

where $a_{jk,i} = \partial a_{jk}/\partial x_i$, $u_{ijk} = \partial^3 u/\partial x_i\,\partial x_j\,\partial x_k$, and so on.
Let u^0 be the solution of

$$Au^0 = f \quad \text{in } B_s(x),$$
$$u^0 = 0 \quad \text{on } \partial B_s(x).$$

Then $u^0 \in C^{2+\alpha}$ and by writing the Schauder estimates for $\tilde{u}(y) = s^{-2}u^0(x + sy)$ in $\{|y| < 1\}$, we find that

$$|u^0|_{C^{2+\alpha}} \leqslant C|f|_{C^\alpha},$$

where the norms are taken in $B_s(x)$, and C is a positive constant independent of s.
Formally,

$$Au_{ii}^0 = f_{ii} + 2\sum a_{jk,i}u_{ijk}^0 + \sum a_{jk,ii}u_{jk}^0.$$

Let u^1 be the solution of

$$Au^1 = -2\sum a_{jk,i}u^0_{jk} \qquad \text{in } B_s(x),$$
$$u^1 = 0 \qquad \text{on } \partial B_s(x).$$

Then again

$$|u^1|_{C^{2+\alpha}} \leq C|f|_{C^\alpha},$$

and

$$Au^1_i = -2\sum a_{jk,i}u^0_{ijk} - 2\sum a_{jk,ii}u^0_{jk} + \sum a_{jk,i}u^1_{jk}.$$

Next let u^2 be the solution of

$$Au^2 = \sum a_{jk,ii}u^0_{jk} - \sum a_{jk,i}u^1_{jk} \qquad \text{in } B_s(x),$$
$$u^2 = 0 \qquad \text{on } \partial B_s(s).$$

Then

$$|u^2|_{C^{2+\alpha}} \leq C|f|_{C^\alpha}.$$

Finally, if

$$A\bar{v} = 0 \qquad \text{in } B_s(x),$$
$$\bar{v} = u^0_{ii} + u^1_i + u^2 \qquad \text{on } \partial B_s(x),$$

then the function $v = u^0_{ii} + u^1_i + u^2 - \bar{v}$ satisfies

(3.22)
$$Av = f_{ii} \qquad \text{in } B_s(x),$$
$$v = 0 \qquad \text{on } \partial B_s(x),$$

$$|v|_{C^\beta} \leq \text{const.} \qquad \text{for some } \beta > 0.$$

From (3.21) we see that

$$Au_{ii} = f_{ii} + g_0 + \sum_{j=1}^n D_j g_j,$$

where g_0, g_j are L^∞ functions. Following the preceding method, we can find functions v_i such that

$$Av_0 = g_0,$$
$$Av_j = D_j g_j \qquad (1 \leq j \leq n)$$

in $B_s(x)$, $v_i = 0$ on $\partial B_s(x)$, and

$$|v_i|_{C^\mu} \leqslant C \qquad \text{for any } 0 < \mu < 1.$$

Let

$$V = v + \sum_{j=0}^{n} v_j$$

and consider the function

$$(3.23) \qquad w = D_{ii}u - V + M_k + Cs^\beta \qquad \text{in } B_s(x).$$

It is a solution of $Aw = 0$. We recall the Harnack inequality [161e]:

Lemma 3.9. *Let u satisfy*

$$-\sum a_{ij}(x)\frac{\partial^2 u}{\partial x_i \partial x_j} + \sum b_i(x)\frac{\partial u}{\partial x_i} + c(x)u = 0$$

in B_ρ and assume that

$$\sum a_{ij}(x)\xi_i\xi_j \geqslant \lambda |\xi|^2, \qquad \sum |b_i| \leqslant M, \qquad 0 \leqslant c \leqslant M,$$

$$|a_{ij}(x) - a_{ij}(y)| \leqslant A |x - y|^\varepsilon \qquad (\lambda > 0, M > 0, A > 0, \varepsilon > 0).$$

If u is positive in B_ρ, then, for any $|x| < \rho/2, |y| < \rho$,

$$\frac{1}{\gamma} u(y)\frac{\rho}{\rho - |y|} \geqslant u(x) \geqslant \gamma u(y)\frac{(\rho - |y|)^{n-1}}{\rho^{n-1}},$$

where γ is a positive constant depending only on λ, M, A, ε.

Now, since

$$|V|_{C^\beta} \leqslant \text{const.}, \qquad V = 0 \text{ on } \partial B_s(x),$$

we can choose C large enough in (3.23) so that w is positive in $B_s(x)$. Applying the Harnack inequality as in the proof of Theorem 3.6 [see (3.12)], we obtain [see (3.13)]

$$D_{ii}u(x) \geqslant -M_k - Cs^\beta + \gamma \delta^{n-1}\left(M_k - C\delta^{1/2} - Cs^\beta\right).$$

Making the same choice of δ as before, we get [see (3.14)]

$$-M_{k+1} \geqslant -M_k + CM_k^{2n-1} - C2^{-\beta k},$$

and we can establish as before that $M_k \leqslant Ck^{-\varepsilon}$.

Definition 3.1. For any bounded set the *minimum diameter* of S, $MD(S)$, is the infimum of distances between pairs Π_1, Π_2 of parallel planes such that S is contained in the strip determined by Π_1, Π_2.

In the next two sections we shall study the behavior of the free boundary in a neighborhood of a point $x^0 \in \Gamma$. For simplicity we take $x^0 = 0$. The study will depend on the "thickness" of Λ at 0. This is defined in terms of

$$(3.24) \qquad\qquad \delta_r(\Lambda) = \frac{MD(\Lambda \cap B_r)}{r}.$$

One can easily verify that

$$(3.25) \qquad c_0 \frac{|\Lambda \cap B_r|}{|B_r|} \leqslant \delta_r(\Lambda) \leqslant 2 \qquad (c_0 \text{ positive constant}).$$

Indeed, this is obvious if $r = 1$ and for any r it then follows by scaling.

The following fundamental theorem of Caffarelli [56b] will be proved in the next two sections.

Theorem 3.10. *Suppose that $f \geqslant \lambda > 0$ in Ω. There exists a positive nondecreasing function $\sigma(r)$ $(0 < r < r_0)$ with $\sigma(0 +) = 0$ such that if for some $0 < r < r_0$,*

$$(3.26) \qquad\qquad \delta_r(\Lambda) > \sigma(r),$$

then for some $\tilde{r} > 0$, $\Gamma \cap B_{\tilde{r}}(0)$ is given by a C^1 surface

$$x_i = k(x_1, \ldots, x_{i-1}, x_{i+1}, \ldots, x_n) \qquad (k \in C^1)$$

and u is in $C^2((N \cup \Gamma) \cap B_{\tilde{r}}(0))$.

In view of (3.25) we also have:

Corollary 3.11. *If*

$$(3.27) \qquad\qquad \limsup_{r \to 0} \frac{|\Lambda \cap B_r|}{|B_r|} > 0,$$

then the assertions of Theorem 3.10 are valid.

The proof of Theorem 3.10 shows that $\sigma(r)$ can be chosen to be independent of $x^0 \in \Gamma$, provided that $\text{dist}(x^0, \partial\Omega) \geqslant \delta_0 > 0$.

In Theorem 3.10, u is the solution of a general elliptic obstacle problem [that is, (3.18)]. We shall first consider, however, the special case

$$(3.28) \qquad\qquad A = -\Delta, \qquad f = 1,$$

and only at the very end (in Section 5) point out the modifications needed for the general case.

4.   CONVEXITY PROPERTIES OF THE COINCIDENCE SET

The results of this section will be used in Section 5 in proving Theorem 3.10.

Definition 4.1.   A function u is said to belong to the class $P_r(M)$ if

$$(4.1) \qquad \begin{aligned} & u \in C^{1,1}(B_r) \text{ and } \sup_{B_r} |D_{ij}u| \leqslant M \\ & \text{for all derivatives } D_{ij} = D_i D_j, \end{aligned}$$

$$(4.2) \qquad u \geqslant 0 \quad \text{in } B_r,$$

$$(4.3) \qquad 0 \in \Gamma,$$

$$(4.4) \qquad \Delta u = 1 \quad \text{in } N$$

where

$$N = \{ x \in B_r, u(x) > 0 \}, \qquad \Gamma = \partial N \cap B_r;$$

we also set

$$\Lambda = \{ x \in B_r, u(x) = 0 \}$$

and write

$$N = N(u), \qquad \Lambda = \Lambda(u), \qquad \Gamma = \Gamma(u).$$

Notice that $u(0) = 0$, $\nabla u(0) = 0$.

For any fixed $s > 0$, the *scaled function*

$$u_s(x) = \frac{1}{s^2} u(sx)$$

is in $P_{r/s}(M)$ if $u \in P_r(M)$, and

$$\Lambda(u_s) = \Lambda_s(u), \qquad N(u_s) = N_s(u), \qquad \Gamma(u_s) = \Gamma_s(u),$$

where, by definition,

$$E_s = \{ x; sx \in E \} \qquad \text{for any set } E.$$

Suppose that

(4.5)
$$u^{(m)} \in P_{r_m}(M), \quad r_m \to r_0, \quad r_0 > 0, \quad u^{(m)} \to u_0$$
$$\text{uniformly in compact subsets of } B_{r_0}.$$

Then clearly

(4.6)
$$\varlimsup_{m \to \infty} \Lambda(u^{(m)}) \subset \Lambda(u_0),$$

where $\varlimsup \Lambda(u^{(m)})$ means the set of all limit points of sequences $\{x_m\}$, $x_m \in \Lambda(u^{(m)})$.

Lemma 4.1. *If (4.5) holds, then $u_0 \in P_{r_0}(M)$ and*

(4.7)
$$\varlimsup_{m \to \infty} N(u^{(m)}) \subset \overline{N(u_0)} .$$

Proof. Suppose that $y \in \varlimsup N(u^{(m)}) \setminus \overline{N(u_0)}$. Then $u_0 = 0$ in $B_\varepsilon(y)$ for some $\varepsilon > 0$ and at the same time there is a sequence $y_m \in N(u^{(m)})$ such that $y_m \to y$. By Lemma 3.1

$$\sup_{B_{\varepsilon/2}(y_m)} u^{(m)} \geqslant \frac{\varepsilon^2}{8n}$$

and since $u^{(m)} \to u_0$ uniformly in $B_\varepsilon(y)$, if ε is small enough, also

$$\sup_{B_\varepsilon(y)} u_0 \geqslant \frac{\varepsilon^2}{8n} > 0,$$

a contradiction.

It is clear that u_0 satisfies (4.1), (4.2), and (4.4). In view of (4.7), also (4.3) holds for u_0 and thus $u_0 \in P_{r_0}(M)$.

Definition 4.2. If

(4.8)
$$u^{(m)} \in P_r(M), \quad \varepsilon_m \downarrow 0, \quad u_{\varepsilon_m}^{(m)} \to u_0 \text{ uniformly}$$
$$\text{in compact subsets of } R^n$$

[recall that $u_{\varepsilon_m}^{(m)}$ is a scaled function of $u^{(m)}$], then we call u_0 a *blow-up limit* of $u^{(m)}$.

Lemma 4.2. *If u_0 is a blow-up limit of $u^{(m)}$, then u_0 is a convex function in R^n.*

Proof. By Theorem 3.6,

$$D_{ii}u_{\varepsilon_m}^{(m)}(x) = D_{ii}u^{(m)}(\varepsilon_m x) \geqslant -C |\log(\varepsilon_m|x|)|^{-\varepsilon}.$$

Hence, for fixed $R > 0$, the function

$$w_m(x) = u_{\varepsilon_m}^{(m)}(x) + C |\log(\varepsilon_m R)|^{-\varepsilon} |x|^2$$

is convex in B_R (i.e., $D_{ll}w \geqslant 0$ for any direction l) and $w_m \to u_0$ uniformly in B_R. Thus u_0 is convex.

Notation. If $u \in P_r(M)$ and u is convex in B_r, then we say that u belongs to $P_r^*(M)$.

Lemma 4.3. *Suppose that $u \in P_r^*(M)$, $\varepsilon_m \downarrow 0$, $u_{\varepsilon_m} \to u_0$ uniformly in compact subsets of B_r. Then in an appropriate system of coordinates u_0 has one of the following two forms:*

$$(4.9) \qquad u_0(x) = \sum_{i=1}^n a_i(x_i)^2, \qquad a_i \geqslant 0, \quad \sum_{i=1}^n a_i = \tfrac{1}{2},$$

$$(4.10) \qquad u_0(x) = \tfrac{1}{2}(x_n^+)^2.$$

Proof. Consider first the case

$$(4.11) \qquad \operatorname{int}\Lambda(u) = \varnothing.$$

Since u is convex, the set $\Lambda(u)$ is convex and therefore must lie in a hyperplane. But then $\Delta u = 1$ a.e. It follows that u_0 is a nonnegative solution of $\Delta u_0 = 1$ in R^n. Since

$$|D_i D_j u_0| \leqslant M, \qquad x \in R^n,$$

we can apply Liouville's theorem to $D_i D_j u_0$ and conclude that $D_i D_j u_0 = \text{const}$. Recalling also that $u_0(0) = 0$, $u_0 \geqslant 0$, (4.9) follows.
   Suppose next that

$$(4.12) \qquad \operatorname{int}\Lambda(u) \neq \varnothing.$$

Let $x \in \Lambda(u)$. Since $\Lambda(u)$ is convex, $\varepsilon_m tx \in \Lambda(u)$ for any $t > 0$ provided that $\varepsilon_m t < 1$. Thus $tx \in \Lambda(u_{\varepsilon_m})$ and

$$\Lambda(u_0) \supset \overline{\lim} \Lambda(u_{\varepsilon_m}) \supset C(\Lambda(u)),$$

where $C(K)$ denotes the cone generated by K (with vertex at the origin). If $x \notin C(\Lambda(u))$, then $x' \notin C(\Lambda(u))$ for all x' with $|x' - x| < \delta$ (δ small enough). But then $\varepsilon_m x' \in N(u)$ if $0 < \varepsilon_m < 1$ and thus $x' \in N(u_{\varepsilon_m})$. Using Lemma 4.1, we deduce that $x' \in N(u_0)$ and, consequently, $x \in \bar{N}(u_0)$. [Indeed, since $N(u_0)$ is convex and $\text{int}\,\Lambda(u_0) \supset \text{int}\,C(\Lambda(u)) \neq 0$, if $x \notin N(u_0)$, then there is an open set in $\{|x' - x| < \delta\}$ that is contained in $\text{int}\,\Lambda(u_0)$.]

We have thus proved that

$$\Lambda(u_0) = C(\Lambda(u)).$$

Take any direction $-e_i$ interior to $\Lambda(u_0)$. Then any line in the direction e_i intersects $N(u_0)$ in a half-line along which $D_{ii}u_0 \geqslant 0$ by convexity of u_0 (D_i = derivative in the direction e_i) and $D_i u_0 = 0$ at the initial point [which belongs to $\Gamma(u_0)$]. It follows that $D_i u_0 \geqslant 0$ in $N(u_0)$ and, by the maximum principle,

$$(4.13) \qquad D_i u_0 > 0 \qquad \text{in } N(u_0).$$

We next show that $\Lambda(u_0)$ is a half-space. Indeed, if not, then let the x_1-plane and x_2-plane be two planes of support. Writing

$$x = (\rho\cos\theta, \rho\sin\theta, x_3, \ldots, x_n),$$

we have

$$\Lambda(u_0) \subset \{x; \theta_0 < \theta < 2\pi - \theta_0\}$$

for some $\pi/2 < \theta_0 < \pi$. If $\pi/2 < \theta_1 < \theta_0$, then the function

$$h(x) = \rho^{\pi/(2\theta_1)}\cos\frac{\pi\theta}{2\theta_1}$$

is harmonic and vanishes on $\theta = \pm\theta_1$. It follows that for some small enough $c > 0$,

$$D_i u_0 \geqslant ch$$

if $\theta = \pm\theta_1$ or if $|\theta| \leqslant \theta_1, |x| = \delta > 0$ (c depends on δ). By the maximum principle, we then get

$$D_i u_0 \geqslant c\rho^{\pi/(2\theta_1)}\cos\frac{\pi\theta}{2\theta_1}$$

if $|\theta| \leqslant \theta_1, |x| < \delta$. Since $\pi/(2\theta_1) < 1$, we get a contradiction to the Lipschitz continuity of $D_i u_0$ at 0.

We have thus proved that $\Lambda(u_0)$ is a half-space, say $\{x_n > 0\}$. By the uniqueness to the Cauchy problem we deduce that u_0 must have the form (4.10).

Notation. We denote by $\alpha(x, y)$ the angle between the vectors x and y.

Lemma 4.4. *For any $\varepsilon > 0, \delta > 0$ there exists a $\lambda = \lambda(\varepsilon, \delta)$ such that if $u \in P_1^*(M)$ and $\delta_1(\Lambda(u)) > \varepsilon$ [$\delta_1(\Lambda)$ is defined in (3.24)], then in an appropriate system of coordinates*

(4.14)
$$\Lambda(u) \supset B_\lambda \cap \left\{ x; \alpha(x, -e_n) < \frac{\pi}{2} - \delta \right\},$$
$$N(u) \supset B_\lambda \cap \left\{ x; \alpha(x, e_n) < \frac{\pi}{2} - \delta \right\}.$$

Proof. If the assertion is not true, then there exists a sequence $\lambda_m \downarrow 0$ and $u^{(m)} \in P_1^*(M)$ for which (4.14) is contradicted in any system of coordinates. Thus for any coordinate system $e_1, \ldots, e_n$ at least one of the relations

(4.15)
$$\Lambda(u^{(m)}) \supset B_{\lambda_m} \cap \left\{ x; \alpha(x, -e_n) < \frac{\pi}{2} - \delta \right\},$$

(4.16)
$$N(u^{(m)}) \supset B_{\lambda_m} \cap \left\{ x; \alpha(x, e_n) < \frac{\pi}{2} - \delta \right\}$$

does not hold.

Notice that if (4.15) holds in a system of coordinates, then the same is true of (4.16). Indeed, otherwise there is a direction $e^{(m)}$ with $\alpha(e^{(m)}, -e_n) < \pi/2 - \delta$ such that $se^{(m)} \notin N(u^{(m)})$ for some $0 < s < \lambda_m$. Since [by (4.15)] a small neighborhood of $-se^{(m)}$ belongs to $\Lambda(u^{(m)})$, it follows from the convexity of $\Lambda(u^{(m)})$ that $0 \in \operatorname{int} \Lambda(u^{(m)})$, a contradiction.

From the remark above we conclude that already (4.15) is not true in any system of coordinates.

Take a subsequence $u^{(m)} \to u$ uniformly in any compact set. Then $u \in P_1^*(M)$.

For any sequence A_m of closed sets in B_1 there holds (see Problem 1)

(4.17)
$$\delta_1\left(\overline{\lim} A_m \right) \geqslant \overline{\lim} \, \delta_1(A_m).$$

Since $\Lambda(u) \supset \overline{\lim} \Lambda(u^{(m)})$, we then have

$$\delta_1(\Lambda(u)) \geqslant \varepsilon.$$

But then the convex set $\Lambda(u)$ has nonempty interior. It follows from the proof of Lemma 4.3 that in appropriate system of coordinates the cone $\Lambda(u)$ is the half-space $\{x_n \leqslant 0\}$.

On the other hand, since (4.15) is false, we can find in every convex cone of opening $< \pi/2 - \delta$ a unit vector $e^{(m)}$ contained in $N(u^{(m)})$ and thus $e = \lim e^{(m)}$ is in $\overline{N(u)}$ (by Lemma 4.1). This contradicts the previous conclusion that $\Lambda(u)$ is a half-space.

Lemma 4.5. *If* $u \in P_1^*(M)$ *and* $\delta_{1/4}(\Lambda(u)) \geqslant \varepsilon > 0$, *then there is an appropriate system of coordinates, a positive number* μ, *and a* C^1 *function* $g(x_1,\ldots,x_n)$ *such that*

(4.18) $$\Lambda(u) \cap B_\mu \subset \{x; \, x_n \leqslant g(x_1,\ldots,x_{n-1})\};$$

μ *and the modulus of continuity of* grad g *depend only on* M *and* ε.

Proof. Since

$$\Lambda(u) \cap B_{1/2}(y) \supset \Lambda(u) \cap B_{1/4} \quad \text{if } |y| < \tfrac{1}{4},$$

it follows that

$$\text{MD}\left(\Lambda(u) \cap B_{1/2}(y)\right) \geqslant \text{MD}\left(\Lambda(u) \cap B_{1/4}\right) = \tfrac{1}{4}\delta_{1/4}(\Lambda(u)) \geqslant \frac{\varepsilon}{4}.$$

Hence

(4.19)
$$\begin{array}{c} \text{Lemma 4.4 can be applied with respect} \\ \text{to any point } y \in \Gamma, |y| < \tfrac{1}{4}. \end{array}$$

By Lemma 4.4 with $y = 0 \in \Gamma$, for any $\delta = 1/m$ there is a system of coordinates (e_i^m) $(i = 1,\ldots,n)$ such that (4.14) holds with $e_n = e_n^m$. Since $\Lambda(u)$ is convex, any line $x^0 + te_n^m$ intersects $\Lambda(u)$ in a segment lying in $\{t < t^0\}$ and $N(u)$ in a segment lying in $\{t > t^0\}$; see Figure 2.1. If $x = \Sigma x_i e_i^m$, then we can represent $x^0 + t^0 e_n^m$ by

(4.20) $$x_n = g^m(x') \qquad [x' = (x_2,\ldots,x_n)]$$

and $\Lambda(u)$ by $x_n \leqslant g^m(x')$.

It is important to notice that if we use any coordinate system centered at 0 in which the direction e_n is near the direction e_n^m, then the remark above regarding $x^0 + te_n^m$ is valid also with respect to $x^0 + te_n$ provided that $|x^0|$ is small enough; see Figure 2.1.

Since (4.14) holds for (e_i^m) in B_{λ_m} (λ_m depends on $\delta = 1/m$ and ε; $\lambda_m \to 0$) it follows that for a suitable choice of the e_i^m, $1 \leqslant i \leqslant n - 1$, and for a subsequence

$$e_i^m \to e_i^0,$$

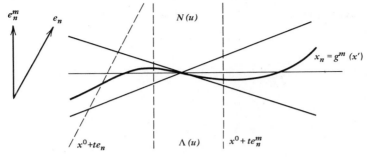

FIGURE 2.1

$$\left(e_i^0\right) = T^m\left(e_i^m\right), \qquad T^m \text{ an orthonormal matrix,}$$

and

(4.21) $$T^m \to I, \qquad I \text{ the identity matrix.}$$

From (4.14) we get

$$\frac{|g^m(x') - g^m(0)|}{|x'|} \le \frac{C}{m}, \qquad |x'| < \lambda_m,$$

and using (4.21) we easily deduce that

(4.22) $$\frac{|g^0(x') - g^0(0)|}{|x'|} \le \beta_m \qquad \text{if } |x'| < \lambda_m \text{ where } \beta_m \to 0;$$

here $x_n = g^0(x)$ is a representation of $\Gamma(u)$, which exists in the coordinate system (e_i^0) (with $x = \Sigma x_i e_i^0$) by the remark following (4.20) and by (4.21). From (4.22) we see that $g^0(x')$ is differentiable at 0 with zero gradient.

In view of (4.19) we can do the same about each point $y \in \Lambda(u)$ with $|y|$ small enough. Thus there is a system of coordinates $(e_i^{m, y})$ and a limiting one (as $m \to \infty$) $(e_i^{0, y})$, and $\Lambda(u)$ can be represented in a ρ_0-neighborhood of y (ρ_0 is independent of y) in the form

$$x_n \le g^{m, y}(x') \qquad \text{or} \qquad x_n \le g^{0, y}(x')$$

where $x = \Sigma x_i e_i^{m, y}$ or $x = \Sigma x_i e_i^{0, y}$, respectively. Furthermore,

(4.23) $$\frac{|g^{0, y}(x') - g^0(0)|}{|x'|} \le \beta_m \qquad \text{if } |x'| < \lambda_m, \beta_m \to 0.$$

The system of coordinates $e_i^{m,y}$ is related to $e_i^{0,y}$ by

$$e_i^{m,y} = T^{m,y}\left(e_i^{0,y}\right)$$

where $T^{m,y} \to I$ uniformly with respect to y. Notice also that $(e_i^{0,y})$ is related to (e_i^0) by

(4.24) $$\left(e_i^{0,y}\right) = T^{0,y}\left(e_i^0\right)$$

and

(4.25) $$T^{0,y} \to I \qquad \text{if } y \to 0$$

(with appropriate choice of $e_i^{0,y}$, $1 \le i \le n - 1$); this follows, in the same way as (4.21), from the assertion (4.14) for $y = 0$ and for y near 0.

We can rewrite the inequality (4.23) in terms of the systems of coordinates (e_i^0); taking into account the rotation (4.24) and (4.25), we obtain from (4.23)

$$g^0(y + h) - g^0(y) = h \cdot c(y) + h \cdot o(1),$$

where $o(1) \to 0$ as $|h| \to 0$, uniformly with respect to y, and $|c(y)| \le C$. This implies that $g^0 \in C^1$.

PROBLEM

1. Prove (4.17).

   [*Hint*: Set $A = \overline{\lim} A_m = \cap_{n=1}^{\infty} \cup_{m=n}^{\infty} A_m$, $D_n = \cup_{m=n}^{\infty} A_m$. Then $D_n \downarrow A$ and any ε-neighborhood of A contains the D_n, $n \ge n_0(\varepsilon)$.]

5. REGULARITY OF THE FREE BOUNDARY WHEN MD (Λ) IS POSITIVE

Lemma 4.4 has implications for implications for any u in $P_1(M)$:

Lemma 5.1. *Given $\varepsilon > 0$, $\delta > 0$, there exists a $\rho_0 = \rho_0(\varepsilon, \delta) > 0$ such that if $u \in P_1(M)$ and $\delta_\rho(\Lambda(u)) \ge \varepsilon$ for some $0 < \rho < \rho_0$, then in an appropriate system of coordinates*

(5.1)
$$\Lambda(u) \supset B_{\rho\lambda/2} \cap \{x; -x_n \ge 2\rho\lambda\delta\},$$
$$N(u) \supset B_{\rho\lambda/2} \cap \{x; x_n \ge 2\rho\lambda\delta\};$$

$\lambda = \lambda(\varepsilon, \delta)$ *is defined in Lemma 4.4.*

Note that the right-hand sides in (4.14) are cones intersected with a ball, whereas in (5.1) they are only strips intersected with a ball.

Proof. If the assertion is not true, then there are sequences $u^{(m)} \in P_1(M)$ and $\rho_m > 0$, $\rho_m \downarrow 0$ such that in some system of coordinates at least one of the relations

$$\Lambda(u^{(m)}) \supset B_{\rho_m\lambda/2} \cap \{x; -x_n > 2\rho_m\lambda\delta\},$$

$$N(u^{(m)}) \supset B_{\rho_m\lambda/2} \cap \{x; x_n > 2\rho_m\lambda\delta\}$$

is not satisfied.

A subsequence of $u_{\rho_m}^{(m)}$ is convergent to a function u in $P_1^*(M)$. Using the relations

$$\Lambda(u) \supset \overline{\lim} \Lambda\left(u_{\rho_m}^{(m)}\right), \qquad \overline{N(u)} \supset \overline{\lim} N\left(u_{\rho_m}^{(m)}\right)$$

we obtain a contradiction to Lemma 4.4.

We now choose $\varepsilon < \frac{1}{8}$, $\delta < \frac{1}{8}$. Then (5.1) implies that

$$\delta_{\rho\lambda/2}(\Lambda(u)) \geq 1 - 4\delta > \tfrac{1}{2},$$

and therefore the lemma can be applied again to $\rho\lambda/2$. Proceeding inductively, we obtain:

Lemma 5.2. *Let $\varepsilon < \frac{1}{8}$, $\delta < \frac{1}{8}$ and define $\lambda = \lambda(\varepsilon, \delta)$ as in Lemma 4.4 and $\rho_0 = \rho_0(\varepsilon, \delta)$ as in Lemma 5.1. If $u \in P_1(M)$ and $\rho < \rho_0$, then there exist systems of coordinates $(e_i^{(k)})$ $(1 \leq i \leq n)$ such that*

$$(5.2) \qquad \Lambda(u) \supset B_{\rho(\lambda/2)^k} \cap \left\{-x_n^{(k)} \geq \rho(\lambda/2)^k 4\delta\right\},$$

$$(5.3) \qquad N(u) \supset B_{\rho(\lambda/2)^k} \cap \left\{x_n^{(k)} \geq \rho(\lambda/2)^k 4\delta\right\}.$$

Here $(x_1^{(k)}, \ldots, x_n^{(k)})$ are the coordinates of a point with respect to $(e_i^{(k)})$. We would like to show that the systems $(e_i^{(k)})$ converge to a system $(e_i^{(0)})$ and then repeat the proof of Lemma 4.5. For this we need an estimate on $D_{ii}u$:

Lemma 5.3. *Under the hypotheses of Lemma 5.2 there exists a $\mu \in (0, 1)$ such that*

$$(5.4) \qquad D_{ii}u \geq -M\mu^{k-1} \qquad \text{in } N(u) \cap B_{\rho(\lambda/2)^k}.$$

Proof. Let h be the harmonic function in $B_1 \cap \{x_n < \frac{1}{2}\}$ such that

$$h = 0 \qquad \text{on } x_n = \tfrac{1}{2},$$
$$h = -1 \qquad \text{on } \partial B_1 \cap \{x_n < \tfrac{1}{2}\}.$$

Then

$$-\mu \equiv \inf_{B_{1/2}} h > -1$$

Suppose by induction that

(5.5) $$D_{ii}u \geqslant -M\mu^{k-1} \quad \text{in } N(u) \cap B_{\rho(\lambda/2)^k}.$$

Choosing $e_n^{(k+1)}$ in the vertical direction, we consider the function

$$w(x) = M\mu^{k-1}h\left(\frac{x}{\rho(\lambda/2)^k}\right) \quad \text{in } N(u) \cap B_{\rho(\lambda/2)^k}.$$

The functions w and $D_{ii}u$ are both harmonic in this set and $w \leqslant D_{ii}u$ on

$$N(u) \cap \partial B_{\rho(\lambda/2)^k}, \quad \text{by (5.5),}$$

and on

$$\partial N(u) \cap B_{\rho(\lambda/2)^k}, \quad \text{by (3.10).}$$

Applying the maximum principle, we obtain

$$D_{ii}u \geqslant -M\mu^k \quad \text{in } B_{\rho(\lambda/2)^{k+1}}.$$

Lemma 5.4. *If* $u \in C^{1,1}(B_R)$, $u \geqslant 0$, $u(0) = 0$, *and* $D_{ii}u \geqslant -\tau$ *on an interval* $\overline{0x_0}(|x_0| < R)$, $\tau \geqslant 0$, *then, for* $0 < t < 1$,

(5.6) $$u(tx_0) \leqslant |x_0|^2\tau + u(x_0).$$

Proof. Suppose that $u(t_0x_0) = \max_{0 \leqslant t \leqslant 1}u(tx)$. If $t_0 = 1$, then (5.6) is obvious. If, on the other hand, $t_0 < 1$, then $D_iu(t_0x_0) = 0$, where i is in the direction $x_0/|x_0|$. Writing

$$0 = u(0) = u(t_0x_0) + \int\int D_{ii}u \geqslant u(t_0x_0) - \tau|x_0|^2,$$

we get $u(tx_0) \leqslant u(t_0x_0) \leqslant \tau|x_0|^2 \leqslant \tau|x_0|^2 + u(x_0).$

Lemma 5.5. *Let the hypothesis of Lemma 5.2 hold and let*

(5.7)

$$\theta_k = (3Mn)^{1/2}\mu^{(k-2)/2}\rho(\lambda/2)^{k-1}, \quad x \in N(u) \cap \left(\overline{B_{\rho(\lambda/2)^k}} \setminus B_{\rho(\lambda/2)^{k+1}}\right).$$

Then there exists a point y such that $|x - y| \leqslant \theta_k$ and the points sy are contained in $N(u)$ for all s with

$$(5.8) \qquad s \geqslant 1, \quad |sy| \leqslant \rho(\lambda/2)^{k-1}.$$

Proof. By Lemma 3.1 there exists a point y such that $|x - y| \leqslant \theta_k$ and

$$(5.9) \qquad u(y) \geqslant \frac{\theta_k^2}{2n}.$$

Let s be as in (5.8). Taking $x_0 = sy$, $t = 1/s$ in the preceding lemma, and noting, by Lemma 5.3, that

$$D_{ii}u \geqslant -M\mu^{k-2} \qquad \text{on the interval } \overline{0x_0},$$

we get

$$u(y) \leqslant |sy|^2 M\mu^{k-2} + u(sy).$$

Substituting (5.9) and recalling (5.8) and the definition of θ_k in (5.7), we get $u(sy) > 0$.

Note that in the previous lemma

$$(5.10) \qquad \alpha(x, y) \leqslant C\frac{\mu^{(k-2)/2}}{\lambda^2}, \qquad C = C(n, M),$$

so that $\alpha(x, y) \to 0$ if $k \to \infty$.

Lemma 5.6. *Let the hypotheses of Lemma 5.2 hold. Then there exists a positive integer $k_0 = k_0(\varepsilon, \delta)$ and a constant $C = C(\varepsilon, \delta) > 0$ such that in an appropriate system of coordinates*

$$(5.11) \qquad \Lambda(u) \supset B_{\rho(\lambda/2)^{k_0}} \cap \left\{ x; \alpha(x, -e_n) < \frac{\pi}{2} - C\delta \right\},$$

$$(5.12) \qquad N(u) \supset B_{\rho(\lambda/2)^{k_0}} \cap \left\{ x; \alpha(x, e_n) < \frac{\pi}{2} - C\delta \right\}.$$

This is a major improvement over Lemma 5.2, where instead of the cones [in (5.11), (5.12)] we only had strips [in (5.1)].

Proof. Take e_n as $e_n^{(k)}$ with $k = k_0 - 1$ [see Lemma 5.2 for definition of $e_n^{(k)}$]. Suppose that (5.11) is not true. Then there is a point x_j, such that

$$x_j \in N(u) \cap (B_{\rho(\lambda/2)^j} \setminus B_{\rho(\lambda/2)^{j+1}}), \qquad j \geqslant k_0$$

and

(5.13) $$\alpha\left(x_j, -e_n^{(k_0)}\right) < \frac{\pi}{2} - C\delta.$$

By Lemma 5.5 there exists a point y_j and a segment sy_j in $N(u)$, where

$$s \geq 1, \qquad |sy_j| \leq \rho\left(\frac{\lambda}{2}\right)^{j-1}.$$

Take $x_{j-1} = s_j y_j$ with $|s_j y_j| = \rho(\lambda/2)^{j-1}$. Then

$$x_{j-1} \in N(u) \cap \left(\overline{B_{\rho(\lambda/2)^{j-1}}} \setminus B_{\rho(\lambda/2)^j} \right)$$

and

$$\alpha(x_j, x_{j-1}) \leq \frac{C}{\lambda^2} \mu^{(j-2)/2}, \qquad \text{by (5.10).}$$

Repeating this procedure step by step we find a sequence of points x_i, $k_0 \leq i < j$, such that

$$x_i \in N(u) \cap \partial B_{\rho(\lambda/2)^i}$$

and

$$\alpha(x_i, x_{i-1}) \leq \frac{C}{\lambda^2} \mu^{(i-2)/2}.$$

Thus $x = x_{k_0}$ satisfies

$$x \in N(u) \cap \partial B_{\rho(\lambda/2)^{k_0}}$$

and

$$\alpha(x, x_j) \leq \tilde{C}\mu^{(k_0-2)/2} \qquad \left[\tilde{C} = \tilde{C}(\lambda, \mu)\right].$$

If C (in (5.11) and (5.13)) is taken to be ≥ 5 and if k_0 is sufficiently large, then

$$\alpha(x, -e_n) < \frac{\pi}{2} - 4\delta$$

and we get a contradiction to Lemma 5.2.

To prove the assertion (5.12), we first note that the proof of (5.11) applies not only with respect to 0 but also with respect to any point $x \in \Gamma(u)$, $|x|$

small enough, say

$$x \in \Gamma, \qquad x \in B_{\rho(\lambda/2)^{k_0}};$$

k_0 and C may be taken independently of x.

If (5.12) is not true for any positive constant C and integer $k_0 > 0$, then for any large constant C^* there is a point x_0,

$$x_0 \in \Gamma(u) \cap B_{(1/2)\rho(\lambda/2)^{k_0}} \cap \left\{ x;\ \alpha(x, e_n) \leqslant \frac{\pi}{2} - C^*\delta \right\}.$$

By the previous remark we have that $\Lambda(u)$ contains [within $\frac{1}{2}\rho(\lambda/2)^{k_0}$-neighborhood of x_0] a cone with vertex x_0 and opening $\pi/2 - C\delta$. By (5.2) of Lemma 5.2 the axis of this cone must form an angle $\leqslant C_0\delta$ with $-e_n$ (C_0 is a universal constant). Hence if C^* is sufficiently large, then the cone contains 0 in its interior, a contradiction.

Proof of Theorem 3.10. We consider first the case $Au = -\Delta u$, $f \equiv 1$. Lemma 5.6 for $u \in P_1(M)$ asserts the same as Lemma 4.4 for $u \in P_1^*(M)$. Since the proof of Lemma 4.5 is based solely on the conclusions of Lemma 4.4, the same proof applies to $u \in P_1(M)$. It follows that Γ is C^1 in a neighborhood of 0.

It remains to prove that $u \in C^2((N \cup \Gamma) \cap B_\rho)$ for some small $\rho > 0$. Consider the functions $D_{ij}u$ in N. We have to show that if $x \to x_0 \in \Gamma$, $|x_0| < \rho$, then $D_{ij}u(x)$ converges to a limit, the limit being

$$\langle \nu, e_i \rangle \cdot \langle \nu, e_j \rangle$$

where $\langle a, b \rangle$ denote scalar product of a, b, and ν is the inward normal to Γ at x_0; note that ν is continuous along Γ, and set $\nu_0 = \nu(x_0)$.

It suffices to take $x_0 = 0$ and e_i tangent to Γ at 0 and prove that

(5.14)
$$\lim_{\substack{x \to 0 \\ x \in N}} D_{ij}u(x) = 0;$$

for then also, by the equation $-\Delta u + 1 = 0$,

$$\lim_{\substack{x \to 0 \\ x \in N}} D_{\nu\nu}u(x) = 1.$$

Suppose that the assertion (5.14) is false. Then there exists a sequence $x_m \to 0$, $x_m \in N$, such that

(5.15)
$$|D_{ij}u(x_m)| \geqslant \eta > 0.$$

Let y_m be the nearest point to x_m on Γ and introduce the functions

$$(5.16) \qquad u_m(x) = \frac{u(2\,|\,x_m - y_m\,|\,x + y_m)}{(2\,|\,x_m - y_m\,|)^2}.$$

Since $\overrightarrow{y_m x_m}$ is in the direction of the inward normal to $\Gamma(u)$ at y_m, $u(x) > 0$ in the region that is approximately the half space with axis $\overrightarrow{y_m x_m}$. It follows that $u_m(x) > 0$ in the region that is approximately the half-space $\{\langle x, x_m - y_m \rangle \geq 0\}$, or also approximately $\{\langle x, \nu_0 \rangle \geq 0\}$. It follows that for a subsequence

$$u_m \to w \text{ uniformly in compact sets;}$$

$w = 0$ if $\langle x, \nu_0 \rangle \leq 0$ and $\overline{N(w)}$ contains the half-space $\{\langle x, \nu_0 \rangle > 0\}$. Since w is also convex, it follows that $\Lambda(w)$ coincides with $\{\langle x, \nu_0 \rangle \leq 0\}$. By the uniqueness of the Cauchy problem, we then get

$$w(x) = \tfrac{1}{2}\langle x, \nu_0 \rangle^2 \qquad \text{if } x_n > 0.$$

Since

$$B_{|x_m - y_m|/2}(x_m) \subset N(u),$$

$\Delta u_m = 1$ in $B_{1/4}(\nu_0/2)$ for all m sufficiently large. It follows that

$$\lim D_{ij} u_m\left(\tfrac{1}{2}\nu_0\right) = D_{ij} w\left(\tfrac{1}{2}\nu_0\right) = 0$$

(since e_i is a tangential direction), contradicting (5.15).

The proof of Theorem 3.10 in the general case is similar; here we use Theorem 3.8 instead of Theorem 3.6. The only difference is that in the proof of Lemma 5.3 we have to make a modification similar to the one given in the proof of Theorem 3.8. Notice also that blow up of solutions of $Au = f$ are solutions of

$$\sum a_{ij}(0) D_{ij} u = f(0),$$

so that Lemma 4.3 and the proof that $u \in C^2$ are unchanged.

PROBLEM

1. Extend the proof of Lemma 5.3 to the general case of the variational inequality (3.18).

6. THE FREE BOUNDARY FOR THE FILTRATION PROBLEM

Let Ω be the cylinder in R^{n+1},

$$\Omega = B_{R_0} \times \{0 < y < H\},$$

where B_R denotes the ball $\{|x| < R\}$, $x = (x_1, \ldots, x_n)$; we shall write $X = (x, y)$.

Consider the variational inequality

$$(6.1) \qquad -\Delta u + f \geqslant 0, \quad u \geqslant 0, \quad (-\Delta u + f)u = 0 \qquad \text{in } \Omega$$

and set, as usual,

$$N = \{u > 0\}, \qquad \Lambda = \{u = 0\}, \qquad \Gamma = \partial N \cap \Omega.$$

We assume:

$$(6.2) \qquad f \in C^1(\overline{\Omega}); \qquad f_y \geqslant 0 \text{ and } f > 0 \text{ in } \overline{\Omega},$$

$$(6.3) \qquad u = u_y = 0 \qquad \text{on } B_{R_0} \times \{y = H\},$$

$$(6.4) \qquad u_y \leqslant 0 \qquad \text{in } \Omega.$$

It follows that for any $x \in B_{R_0}$,

$$(6.5) \qquad (x, y) \in N \qquad \text{if and only if } y < \psi(x),$$

where $\psi(x)$ is some function with values in $[0, H]$. Thus N is an x-subgraph and Γ is an x-graph.

Theorem 6.1. *If $X^0 = (x^0, y^0) \in \Gamma$, then there exists a neighborhood $V = B_\rho(x^0) \times \{|y - y^0| < \rho\}$ of X^0 such that $\Gamma \cap V$ is given by*

$$(6.6) \qquad y = \psi(x) \qquad (|x - x^0| < \rho)$$

and ψ is Lipschitz continuous.

Proof. Take for simplicity $x^0 = 0$. We recall that $u \in C^{1,1}(\Omega)$ and, without loss of generality we may assume that $u \in C^{1,1}(\overline{\Omega})$ (otherwise we replace Ω by $\Omega \cap \{y > \varepsilon\}$ for some small $\varepsilon > 0$ and use Theorem 4.3 of Chapter 1). The function u_y satisfies in N

$$-\Delta u_y = -f_y \leqslant 0, \qquad u_y \leqslant 0;$$

consequently, by the strong maximum principle,

(6.7) $u_y < 0$ in N.

For any $R > 0$, set

$$N_R = [B_R \times (0, H)] \cap N.$$

For any vector e in R^n, $|e| \leq 1$, $K > 0$, $A > 0$, consider the function

(6.8) $w = -Ku_y + e \cdot u_x - Au$

in N_{2R}, where R is such that $B_{2R} \subset B_{R_0}$. Since

$$\Delta w = -Kf_y + e \cdot f_x - Af,$$

it follows from (6.2) that if A is sufficiently large (independently of K, e), say $A \geq A_0$, then

(6.9) $\Delta w < -1$ in N_{2R};

from now on we take $A = A_0$.

Let δ be a small number to be determined later; it will depend only on R. Clearly,

$$w = 0 \qquad \text{on } \partial N_R \cap \Gamma$$

and, by (6.7),

$$w > 0 \text{ on } \partial N_{R'} \cap \{X; \text{dist}(X, \Gamma) > \delta\}$$
(6.10)
$$\text{for all } R \leq R' \leq 2R,$$

provided that $K \geq K(\delta)$. If we can also show that

(6.11) $w > 0$ on $\partial N_R \cap \{X; 0 < \text{dist}(X, \Gamma) \leq \delta\}$,

then we can apply the maximum principle [recall (6.9)] to deduce that

(6.12) $w > 0$ in N_R.

We shall now establish (6.11), provided that $\delta \leq \delta_0(R)$, by contradiction. Suppose there is a point $\overline{X} = (\overline{x}, \overline{y})$ such that

$$w(\overline{X}) \leq 0, \qquad \overline{x} \in \partial B_R, \qquad \text{dist}(\overline{X}, \Gamma) \leq \delta.$$

Consider the function

$$\tilde{w}(X) = w(X) + \frac{1}{2(n+1)} |X - \overline{X}|^2 \qquad \text{in } N_{2R}.$$

By (6.10)

$$\tilde{w}(X) > 0 \qquad \text{on } \partial N_{2R} \cap \{X; \text{dist}(X, \Gamma) > \delta\}.$$

Also $\tilde{w} \geq 0$ on Γ. Finally, if

$$X \in \partial N_{2R}, \qquad \text{dist}(X, \Gamma) \leq \delta,$$

then, since $u \in C^{1,1}$ and $u = 0$, $\nabla u = 0$ on Γ,

$$\tilde{w}(X) \geq -Ku_y - (A + 1)C_0\delta - \frac{1}{2(n + 1)}R^2 > 0 \qquad (C_0 > 0)$$

provided that $\delta \leq R^2/C$ for some constant C (independent of K, R, δ). Noting by (6.9) that $\Delta \tilde{w} < 0$ in N_{2R}, the maximum principle gives $\tilde{w} > 0$ in N_{2R}. This contradicts the assumption that

$$\tilde{w}(\overline{X}) = w(\overline{X}) \leq 0.$$

Having established (6.12) we shall now use only the weaker statement

$$-Ku_y + e \cdot u_x > 0 \qquad \text{in } N_R.$$

Since this is true for any vector e with $|e| \leq 1$ and a fixed $K > 0$, it follows that there is a cone M with vertex x^0, opening $\gamma > 0$, and with axis parallel to $-e_{n+1} = (0, 0, \ldots, 0, -1)$ such that $M \cap V \subset N$.

The same holds (with the same γ) with respect to any point $X^* \in \Gamma$, X^* in a small neighborhood of X^0. Since N is also an x-subgraph, it follows that for the cone M^+ with vertex X^0, opening γ, and axis parallel to e_{n+1}, $M^+ \cap V$ must belong to Λ. But then

$$\frac{|\psi(x) - \psi(x^0)|}{|x - x^0|} \leq C\gamma,$$

and the proof is complete.

Theorem 6.2. *Under the conditions of Theorem 6.1:*

(i) *ψ is in C^1 and u is in $C^2(N \cup \Gamma)$.*
(ii) *If further $f \in C^{m+\alpha}$ (m integer ≥ 1, $0 < \alpha < 1$), then $\psi \in C^{m+1+\alpha}$.*
(iii) *If f is also analytic, then $\psi(x)$ is analytic.*

Proof. The assertion (i) follows from Theorems 6.1 and 3.10. Notice that Theorem 3.10 asserts a C^1 representation

$$(6.13) \qquad x_i = h(x_1, \ldots, x_{i-1}, x_{i+1}, \ldots, x_n, x_{n+1}) \qquad (x_{n+1} = y).$$

Since, however, ψ is Lipschitz, $\partial h/\partial x_{n+1} \neq 0$ and the implicit function theorem can be used to invert (6.13) into $x_{n+1} = \psi(x)$ with ψ in C^1. The assertions (ii), (iii) follow from Theorem 1.1.

Consider now the filtration problem studied in Chapter 1, Section 5. As an immediate corollary of Theorem 6.2 we obtain:

Theorem 6.3. *Let $u = \phi(x)$ denote the free boundary for the filtration problem of Chapter 1, Section 5. Then $\phi(x)$ is analytic for $0 < x < a$, and $\phi'(x) < 0$ for $0 < x < a$.*

Indeed, since $w_y < 0$ in the noncoincidence set W, we find that $y = \phi(x)$ is analytic, and since $w_x < 0$ in W, also the inverse function $x = \phi^{-1}(y)$ is analytic, so that $\phi'(x) \neq 0$.

Theorem 6.2 can be applied also to three-dimensional filtration problems. Suppose that a dam Ω in R^3 (made up of homogeneous porous medium) is given by

$$\Omega = \{0 < x_3 < H, 0 < x_1 < a, g_1(x_1) < x_2 < g_2(x_1)\},$$

where x_3 is taken as the upward vertical axis. To the left of the face G_1: $x_2 = g_1(x_1)$, there is a reservoir of water at level H; to the right of the face G_2: $x_2 = g_2(x_1)$, there is a reservoir of water at a lower level h ($0 < h < H$); and the dam is impervious at the faces S_0: $x_1 = 0$ and S_a: $x_1 = a$ and at the bottom B: $x_3 = 0$. We denote by T the top face: $x_3 = H$.

Setting $u(x) = p(x) + x_3$ (p the pressure), one seeks functions $u(x), \phi(x_1, x_2)$ such that

$$\Delta u = 0 \qquad \text{in the wet part } \{p > 0\},$$

$$u = H \qquad \text{on } G_1,$$

$$u = h \qquad \text{on } G_2 \cap \{x_3 \leqslant h\},$$

$$u = x_3 \qquad \text{on } G_2 \cap \{x_3 > h\},$$

$$\frac{\partial u}{\partial x_3} = 0 \qquad \text{on } B,$$

and

$$u = x_3, \quad \frac{\partial u}{\partial \nu} = 0 \qquad \text{on the free boundary } \Gamma: x_3 = \phi(x_1, x_2).$$

Let $g(x_1, x_2)$ be the solution of

$$\Delta g = 0 \qquad \text{in } B,$$

$$g = \tfrac{1}{2}H^2 \qquad \text{on } \{x_2 = g_1(x_1)\},$$

(6.14)
$$g = \tfrac{1}{2}h^2 \qquad \text{on } \{x_2 = g_2(x_1)\},$$

$$\frac{\partial g}{\partial x_1} = 0 \qquad \text{on } \{x_1 = 0\} \cup \{x_1 = a\}.$$

Assuming that

(6.15) $$g_i'(0) = g_i'(a) = 0, \qquad g_1(x_1) < g_2(x_1),$$

one can establish the existence of a solution g in $C^1(\bar{B})$. We now define

$$w(x_1, x_2, x_3) = \int_{x_3}^{H} (u(x_1, x_2, t) - t)\, dt,$$

$$g(x) = \begin{cases} g(x_1, x_2) & \text{on } B \\ \tfrac{1}{2}(H - x_3)^2 & \text{on } G_1 \\ \tfrac{1}{2}(h - x_3)^2 & \text{on } G_2 \cap \{x_3 < h\} \\ 0 & \text{on } G_2 \cap \{x_3 > h\} \text{ and on } T. \end{cases}$$

Proceeding formally, we find that w is a solution of the variational inequality

$$w \in K,$$

(6.16) $$\int_{\Omega} \nabla w \cdot \nabla(v - w)\, dx \geq -\int_{\Omega} (v - w)\, dx \qquad \forall\, v \in K,$$

where

(6.17) $$K = \{v \in H^1(\Omega), v \geq 0, v = g \text{ on } \partial\Omega \setminus (S_0 \cup S_a)\}.$$

Theorem 6.4. *There exists a unique solution of the variational inequality* (6.16) *and* (i) $\partial w/\partial x_3 \leq 0$ *in* Ω, (ii) *the free boundary is given by* $x_3 = \phi(x_1, x_2)$, *where* ϕ *is analytic in* B.

Proof. The existence, uniqueness, and $W_{\text{loc}}^{2,p}$ regularity follow from the general theory of Chapter 1. To derive $W^{2,p}$ regularity near the corner points of the boundary, say near $x_3 = H$, we reflect across $x_3 = H$, by considering

$$\tilde{w}(x_1, x_2, x_3) = \begin{cases} w(x_1, x_2, x_3) & \text{if } x_3 < H, \\ w(x_1, x_2, 2H - x_3) & \text{if } x_3 > H; \end{cases}$$

for more details, see reference 167b.

The proof that $\partial w / \partial x_3 \leq 0$ follows by the maximum principle as in Chapter 1, Section 5.

Next, by comparison we find that

$$w \leq \tfrac{1}{2}(H - x_3)^2$$

and consequently $\partial w / \partial x_3 = 0$ on $x_3 = H$. We are now in a position to apply Theorem 6.2 in order to deduce the assertion (ii).

In the remaining part of this section we concentrate on the two-dimensional case and derive more information on the free boundary. One of the main results to be proved is the following:

Theorem 6.5. *Let* $y = \phi(x)$ *denote the free boundary for the filtration problem of Chapter* 1, *Section* 5. *Then*

(6.18) $\phi'(x)$ *is strictly monotone decreasing,*

(6.19) $\phi'(0) = 0, \qquad \phi'(a) = -\infty,$

(6.20) $\phi(a) > h.$

The last inequality asserts that the *seepage line* $\{x = a, \ h < y < \phi(a)\}$ is nonempty—a well-known fact in hydraulics engineering.

Lemma 6.6. *The solution* w *of the variational inequality* (*of Chapter* 1, *Section* 5) *satisfies*

(6.21) $0 \leq w_{xx} \leq 1, \qquad 0 \leq w_{yy} \leq 1.$

Proof. Let $H^* > H$, $\Omega_* = \{0 < x < a, 0 < y < H^*\}$ and define

$$g_* = \begin{cases} g & \text{on } \partial\Omega_* \cap \{y < H\} \\ 0 & \text{on } \partial\Omega_* \cap \{y > H\}, \end{cases}$$

$$K_* = \{v \in H^1(\Omega_*), v \geq 0, v = g_* \text{ on } \partial\Omega_*\}.$$

Let w_* be the solution of

$$w_* \in K_*,$$

$$\int_{\Omega_*} \nabla w_* \cdot \nabla (v - w_*) \geq - \iint_{\Omega_*} (v - w_*) \qquad \forall \, v \in K_*.$$

By uniqueness we find that $w = w_*$. Thus it suffices to establish (6.21) for w_*.
Consider the penalized problem.

(6.22) $$- \Delta w + \beta_\varepsilon(w) = -1 \qquad \text{in } \Omega_*,$$

(6.23) $$w = g_\delta \qquad \text{on } \partial \Omega_*$$

where g_δ differs from g_* only in a δ-neighborhood of the points $(0,0)$, $(0, H)$, $(a, 0)$, (a, h); g_δ is in C^4, and

$$\frac{\partial^2 g_\delta}{\partial x^2} = \frac{\partial^4 g_\delta}{\partial x^4} = 0 \qquad \text{at } (0,0) \text{ and at } (a, 0),$$

$$\left| \frac{\partial^2}{\partial y^2} g_\delta(0, y) \right| \leq C, \qquad \left| \frac{\partial^2}{\partial y^2} g_\delta(a, y) \right| \leq C,$$

$$\left| \frac{\partial^2}{\partial x^2} g_\delta(x, 0) \right| \leq C.$$

Denote the solution by $w_{\varepsilon\delta}$. Taking β_ε such that $\beta_\varepsilon'' \leq 0$, we find that $\zeta = \partial^2 w_{\varepsilon\delta}/\partial x^2$ satisfies

(6.24) $$- \Delta \zeta + \beta_\varepsilon'(w_{\varepsilon\delta})\zeta = -\beta_\varepsilon''(w_{\varepsilon\delta})\left(\frac{\partial w_{\varepsilon\delta}}{\partial x} \right)^2 \geq 0 \qquad \text{in } \Omega_*.$$

Next, on $\{ y = 0 \}$, $\zeta \geq -C$; on $\{ y = H^* \}$, $\zeta = 0$; on $\{ x = 0 \}$ and on $\{ x = a \}$,

$$\zeta = 1 - \frac{\partial^2}{\partial y^2} g_\delta - \beta_\varepsilon(g_\delta) \geq -C.$$

Hence, by the maximum principle, $\zeta \geq -C$ in Ω_*.
Similarly, $\partial^2 w_{\varepsilon\delta}/\partial y^2 \geq -C$ in Ω_*. Since

$$\Delta w_{\varepsilon\delta} = 1 + \beta_\varepsilon(w_{\varepsilon\delta}) \leq 1$$

(if we take $\beta_\varepsilon \leq 0$), we get

$$| D_x^2 w_{\varepsilon\delta} | \leq C, \qquad | D_y^2 w_{\varepsilon\delta} | \leq C.$$

We now let $\delta \to 0$ and obtain

(6.25) $|D_x^2 w_\varepsilon| \leqslant C, \qquad |D_y^2 w_\varepsilon| \leqslant C,$

where w_ε is the solution of (6.22) with

$$w_\varepsilon = g \qquad \text{on } \partial\Omega_*.$$

Letting $\varepsilon \to 0$, we conclude that w_{xx}, w_{yy} are bounded functions.
   If we can prove that

$$w_{xx} \geqslant 0,\ w_{yy} \geqslant 0 \qquad \text{in } W \equiv \{w > 0\},$$

then (6.21) follows by recalling that $\Delta w = 1$ in W. It will be enough to prove
that $w_{xx} \geqslant 0$ in W.
   Now by Theorem 3.6, $w_{xx} \geqslant 0$ on the free boundary. The harmonic function
w_{xx} is continuous and $\geqslant 0$ on the remaining part of ∂W, except possibly at the
points $(0,0)$, $(0, H)$, $(a, 0)$, (a, h), and $(a, \phi(a))$; designate these points by X_i
$(1 \leqslant i \leqslant 5)$. If we can show that

(6.26) $\displaystyle \liminf_{(x,\,y) \to X_i} w_{xx}(x, y) \geqslant 0,$

then the maximum principle gives $w_{xx} \geqslant 0$ in W.
   Since w_{xx} is bounded in W, (6.26) follows immediately from the following
Phragmén–Lindelöf type of lemma.

Lemma 6.7. *Let D be a bounded domain and let (x_0, y_0) be a point of ∂D.
Suppose that there is a line segment σ such that $\sigma \cap D = \varnothing$ and (x_0, y_0) belongs
to the boundary of σ. Let v be a bounded harmonic function in D and let*

$$\gamma_1 = \liminf_{(x,\,y)\in\partial D,\,(x,\,y)\to(x_0,\,y_0)} v(x, y),$$

$$\gamma_2 = \limsup_{(x,\,y)\in\partial D,\,(x,\,y)\to(x_0,\,y_0)} v(x, y).$$

Then

$$\gamma_1 \leqslant \liminf_{(x,\,y)\in D,\,(x,\,y)\to(x_0,\,y_0)} v(x, y) \leqslant \limsup_{(x,\,y)\in D,\,(x,\,y)\to(x_0,\,y_0)} v(x,y) \leqslant \gamma_2.$$

Proof. Suppose for simplicity that $(x_0, y_0) = (0,0)$ and that $\sigma = \{(\rho, \theta);$
$\theta = \pi, 0 < \rho < \rho_1\}$, where (ρ, θ) are polar coordinates. For any $\varepsilon > 0$, consider

the harmonic function

$$w = A\rho^{1/3}\cos\frac{\theta}{3} - \varepsilon \log \rho + (\gamma_2 + \varepsilon) - v$$

in $D_\eta = D \cap \{\eta < \rho < \rho_0\}$. If A is sufficiently large, then $w > 0$ on $D_\eta \cap \{\rho = \rho_0\}$. Since clearly $w > 0$ on the remaining boundary of D_η if η is positive and sufficiently small, we conclude that $w > 0$ in D_η and, hence, also in D_0. Letting $\varepsilon \to 0$ and then $\rho \to 0$, we get

$$\limsup_{(x, y)\in D,(x, y)\to(x_0, y_0)} v(x, y) \le \gamma_2.$$

In the same way one can prove the inequality regarding $\liminf v$.

Lemma 6.8. *For any point $(x_0, \phi(x_0))$ with $0 < x_0 < a$, the function w_{yy} cannot take an extremum at $(x_0, \phi(x_0))$ with respect to any $\overline{W}$-neighborhood of $(x_0, \phi(x_0))$.*

Here W is the noncoincidence set.

Proof. Differentiating the relations

$$w_x(x, \phi(x)) = 0, \qquad w_y(x, \phi(x)) = 0,$$

we get

$$(6.27) \qquad w_{xx} + w_{xy}\phi' = 0, \qquad w_{xy} + w_{yy}\phi' = 0.$$

Substituting $w_{xx} = 1 - w_{yy}$, we obtain

$$(6.28) \qquad w_{yy}(x, \phi(x)) = \frac{1}{1 + (\phi'(x))^2},$$

$$(6.29) \qquad w_{xy}(x, \phi(x)) = -\frac{\phi'(x)}{1 + (\phi'(x))^2}.$$

To prove the lemma by contradiction, assume that w_{yy} takes a local extremum at $(x_0, \phi(x_0))$. Then

$$(6.30) \qquad 0 = \frac{d}{dx}w_{yy}(x, \phi(x)) = -\frac{2\phi'(x)\phi''(x)}{\left(1 + (\phi'(x))^2\right)^2} \qquad \text{at } x = x_0.$$

By Theorem 6.3, $\phi'(x_0) \ne 0$ and, therefore,

$$\phi''(x_0) = 0.$$

From (6.29) we then obtain

(6.31) $$\frac{d}{dx} w_{xy}(x, \phi(x)) = 0 \qquad \text{at } x = x_0.$$

But on $y = \phi(x)$,

$$\frac{d}{dx} w_{xy} = w_{xyx} + w_{xyy}\phi' = -w_{yyy} + w_{yyx}\phi' = A\frac{\partial}{\partial\nu} w_{yy},$$

where $A \neq 0$. Thus $\partial w_{yy}/\partial\nu = 0$ at $(x_0, \phi(x_0))$, which contradicts the maximum principle.

We shall need the following general local result.

Lemma 6.9. *Suppose*

$$-\Delta w + 1 \geqslant 0, \qquad w \geqslant 0, \qquad (-\Delta w + 1)w = 0$$

in $\Omega = \{x \geqslant 0, y \geqslant 0, x + y \leqslant \varepsilon_0\}, |w_{xx}| \leqslant C, |w_{yy}| \leqslant C,$ *and*

$$w(x, 0) = 0, \qquad \frac{w(0, y)}{y^2} \to \frac{\gamma}{2} \quad \text{if } y \to 0, 0 \leqslant \gamma \leqslant 1.$$

Suppose also that the free boundary is a curve $y = \phi(x)$, *and* $\phi \in C^1[0, \varepsilon_1]$ *for some* $\varepsilon_1 > 0$. *Then*

(6.32) $$\gamma = \frac{1}{1 + (\phi'(0))^2}.$$

Proof. Set $\lambda = \phi'(0)$. Consider the family of functions

$$w_\varepsilon(x, y) = \frac{1}{\varepsilon^2} w(\varepsilon x, \varepsilon y) \qquad (0 < \varepsilon < 1).$$

They all solve the same variational inequality as w and, since

$$|D_x^2 w_\varepsilon| \leqslant C, \qquad |D_y^2 w_\varepsilon| \leqslant C,$$

there is a sequence $w_\varepsilon \to u$ uniformly. Clearly,

$$\Delta u = 1 \qquad \text{in } \{x > 0, y > 0, y > \lambda x\},$$

$$u(0, y) = \frac{\gamma}{2} y^2, \qquad u(x, \lambda x) = 0, \quad \frac{\partial}{\partial\nu} u(x, \lambda x) = 0.$$

The function

$$\bar{u} \equiv u - \frac{(\lambda x - y)^2}{2(\lambda^2 + 1)}$$

is harmonic for $\{x > 0, y > 0, y > \lambda x\}$ and vanishes on $y = \lambda x$ with its normal derivative. Hence it vanishes identically. But $\bar{u}(0, y) \equiv 0$ gives (6.32).

Lemma 6.10. *If* $\phi'(0) = \lim_{x \to 0} \phi'(x)$ *exists, then* $\phi'(0) = 0$; (ii) *if* $\phi'(a) \equiv \lim_{x \to a} \phi'(x)$ *exists, then* $\phi(a) > h$ *and* $\phi'(a) = -\infty$.

Proof. The assertion (i) follows from the preceding lemma. To prove (ii), suppose that

$$(6.33) \qquad \phi(a) = h.$$

Set $\gamma = -\phi'(a)$ and consider first the case $\gamma < \infty$. From (6.28) we get

$$w_{xx}(x, \phi(x)) \to \frac{\gamma^2}{1 + \gamma^2} \qquad \text{if } x \to a.$$

Also,

$$w_{xx}(a, y) = 0 \qquad \text{if } 0 < y < h.$$

Introducing the harmonic function,

$$z(x, y) = \frac{\gamma^2}{1 + \gamma^2} \frac{1}{\mu} \left(\arctan \frac{y - h}{a - x} + \frac{\pi}{2} \right), \qquad \mu = \frac{\pi}{2} + \arctan \gamma,$$

we see that

$$\begin{aligned} w_{xx} - z \to 0 \qquad &\text{if } (x, y) = (x, \phi(x)), x \uparrow a, \\ &\text{or if } (x, y) = (a, y), y \uparrow h. \end{aligned}$$

Applying Lemma 6.7, we find that $w_{xx} - z \to 0$ if $y = h, x \uparrow a$. It follows that

$$w_{xx}(x, h) \to \frac{\pi}{2\mu} \frac{\gamma^2}{1 + \gamma^2} \qquad \text{if } x \to a.$$

Applying now Lemma 6.9 in $\{x \leqslant a, y \leqslant h, (a - x) + (h - y) \leqslant \varepsilon_0\}$, we find that

$$\frac{\pi}{2\mu} \frac{\gamma^2}{3(1 + \gamma^2)} = \frac{1}{1 + (1/\gamma)^2},$$

which is impossible since $\gamma > 0$.

We conclude that $\phi'(a) = -\infty$ and the previous argument gives $w_{xx}(x, h) \to \frac{1}{2}$ if $x \to a$. Applying Lemma 6.9 again, we obtain

$$\frac{1}{2} = \frac{1}{1 + 0},$$

which is impossible. Thus the assumption (6.33) leads to a contradiction.

Having proved that $\phi(a) > h$, we now introduce the harmonic function

$$z(x, y) = A \arctan \frac{y - \phi(a)}{a - x} + B,$$

where

$$A \arctan \gamma + B = \frac{\gamma^2}{1 + \gamma^2}, \qquad -\frac{\pi}{2} A + B = 1.$$

Then

$$B = \left(\frac{\pi}{2} \frac{\gamma^2}{1 + \gamma^2} + \arctan \gamma \right) \Big/ \left(\frac{\pi}{2} + \arctan \gamma \right).$$

As in the preceding proof, we find that

$$w_{xx} - z \to 0 \qquad \begin{aligned} &\text{if } (x, y) = (x, \phi(x)), \, x \uparrow a, \\ &\text{or if } (x, y) = (a, y), \, y \uparrow \phi(a), \end{aligned}$$

so that, by Lemma 6.7,

$$w_{xx}(x, \phi(a)) - z(x, \phi(a)) \to 0 \qquad \text{if } x \to a.$$

Hence

$$w_{xx}(x, \phi(a)) \to B \qquad \text{if } x \to a.$$

If we now apply Lemma 6.9 in $\{x \leqslant a, y \leqslant \phi(a)\}$, we deduce that

$$B = \frac{1}{1 + (1/\gamma)^2}.$$

But this is impossible if $\gamma \neq \infty$. Thus either $\gamma = 0$ or $\gamma = \infty$.

Suppose next that $\gamma = 0$. To derive a contradiction, consider the function $V = w_x - \varepsilon w_y$ in a domain

$$D_\delta = \{a - \delta < x < a, \, h + \delta < y < \phi(x)\},$$

where δ is sufficiently small; recall that $\phi'(a - \delta) < 0$. Then

$$V_x(x, \phi(x)) = \frac{(\phi')^2 + \varepsilon\phi'}{1 + (\phi')^2} > 0$$

in some neighborhood of $(a - \delta, \phi(a - \delta))$ provided that ε is small enough. Since $V = 0$ on Γ, it follows that $V < 0$ if $x = a - \delta$, $\phi(a - \delta) - \eta < y < \phi(a - \delta)$ for some $\eta = \eta(\delta)$ small enough provided that $0 < \varepsilon \leq \varepsilon_0(\delta)$.

Recalling that $w_x < 0$ on the remaining part of $\partial D_\delta \setminus \Gamma$ and that $w_y = 0$ on $\{x = a, h + \delta < y < \phi(a)\}$, we conclude that

$$V < 0 \qquad \text{on } \partial D_\delta$$

if ε is small enough. By the maximum principle it follows that

(6.34) $$V < 0 \qquad \text{in } D_\delta.$$

Now, if $\gamma = 0$, then $\partial/\partial x - \varepsilon(\partial/\partial y)$ is an interior derivative to D_δ at $(x, \phi(x))$ with x near a, and (6.34) gives $w < 0$ in D_δ, at points near $(a, \phi(a))$, which is impossible. Thus $\gamma \neq 0$ and consequently $\gamma = -\infty$. This completes the proof of Lemma 6.10.

Lemma 6.11. *Let z be a bounded nonconstant harmonic function in a bounded domain D. Let $X_1, \ldots, X_k$ be boundary points of D such that for each X_j there is a segment σ_j, $\sigma_j \cap D = \varnothing$, $X_j \in \partial \sigma_j$. Assume that z is continuous on $\overline{D} \setminus \{X_1, \ldots, X_k\}$. If*

(6.35)
$$(x_0, y_0) \in D, \quad \alpha \equiv z(x_0, y_0) \neq z(x, y)$$
$$\textit{for all } (x, y) \in \partial D, \quad (x, y) \neq X_j \quad (1 \leq j \leq k),$$

then there exists a simple piecewise analytic curve S,

$$S \subset \{z = \alpha\}, \qquad (x_0, y_0) \in S,$$

such that S can be parametrized by $x = x(t), y = y(t)$ $(-\infty < t < \infty)$, and

(6.36)
$$(x(t), y(t)) \to X_j \qquad \textit{if } t \to \infty,$$
$$(x(t), y(t)) \to X_l \qquad \textit{if } t \to -\infty,$$

where $X_j \neq X_l$.

Proof. The level curves of $\{z = \alpha\}$ which initiate at (x_0, y_0) form locally m analytic curves, and the angle between two adjacent ones is π/m; if $m > 1$, we say that (x_0, y_0) is a branch point.

We construct S step by step by extending the endpoints. If an endpoint is not a branch point, then the extension is locally unique; if it is a branch point of order m, then we can pick up any one of the $2m - 1$ arcs.

As long as S remains in a compact subdomain D_1 of D, we can get in each step a new arc of length $\geq c$ (c depending on D_1). S cannot have self-intersection, for otherwise $z \equiv \alpha$ in a domain bounded by a part of S.

It follows that if $D_0 \subset \overline{D}_0 \subset D_1 \subset \overline{D}_1 \subset D$ (D_i open domains), then S must exit D_1 after a finite number of steps (or else there would be an infinite number of arcs of length $\geq c > 0$ which intersect any small neighborhood of some point of D_1). S may reenter D_0, but then it must again exit it; such excursions from ∂D_1 into D_0 are finite in number (or else again there would be an infinite number of arcs of length $\geq c$ which intersect any small neighborhood of some point of ∂D_0).

We parametrize S by $x = x(t)$, $y = y(t)$ in such a way that $-\infty < t < \infty$. Let V_δ be a δ-neighborhood of $X_1, \ldots, X_k$ and let D_0 be such that ∂D_0 lies in a δ-neighborhood of ∂D. If $|t| \geq T_\delta$, T_δ sufficiently large, then $\text{dist}((x(t), y(t)), \partial D) < \delta$ and, recalling (6.35), we conclude that $(x(t), y(t)) \in V_\delta$ if δ is small enough. Since δ is arbitrary, we obtain (6.36). Finally, Lemma 6.7 and the maximum principle implies that $X_j \neq X_l$.

Remark 6.1.   In applications we shall often not have the precise setting of Lemma 6.11 at hand, but rather some slightly different versions of it.

Since the free boundary is analytic, w has analytic extension across the free boundary. It follows that the function w_{yy} can be extended as a harmonic function into a domain containing $W \cup \Gamma$.

Lemma 6.12.   *The function $\phi'(x)$ cannot take a local minimum in the interval $0 < x < a$.*

Proof.   We argue by contradiction. Suppose that $\phi'(x)$ has a local minimum at some point x_0, $0 < x_0 < a$, and let $\alpha = w_{yy}(x_0, \phi(x_0))$. Then

$$(6.37) \qquad w_{yy}(x, \phi(x)) > \alpha \qquad \text{if } x \neq x_0, \ |x - x_0| \text{ small.}$$

It follows that there are at least two level curves of $\{w_{yy} = \alpha\}$ which initiate at $(x_0, \phi(x_0))$ and go into W. Thus there exists an open component G of the set $W \cap \{w_{yy} < \alpha\}$ such that $(x_0, \phi(x_0)) \in \partial G$.

Denote by S_1 and S_2 the two level curves of $\{w_{yy} = \alpha\}$ which start at $(x_0, \phi(x_0))$ and which bound ∂G, with S_1 being to the left of S_2 near $(x_0, \phi(x_0))$. These curves cannot intersect each other in $\overline{W}$ [except at $(x_0, \phi(x_0))$] since otherwise $w_{yy} \equiv \alpha$ in G (by Lemma 6.7 and the maximum principle).

Each S_i cannot intersect the free boundary, for otherwise Lemma 6.8 would give $w_{yy} \equiv \alpha$ in a region bounded by S_i and Γ.

Next, $w_{yy} = 0$ or 1 on $\partial W \setminus \Gamma$ [except possibly at $(0, H)$, (a, h), $(a, \phi(a))$ where w_{yy} may not be perhaps continuous]. Thus the proof of Lemma 6.11 can be applied in order to deduce that S_i is continuous up to its endpoint P_i, $P_1 \neq P_2$, and only the following cases can occur:

(i) $P_1 = (0, H)$.
(ii) $P_1 = (a, h)$, $P_2 = (a, \phi(a))$, and $\phi(a) > h$.

In case (i) we take any level curve T starting at $(x, \phi(x))$, $x < x_0$, $w_{yy}(x, \phi(x)) > \alpha$. Then

$$w_{yy} = \beta > \alpha \qquad \text{along } T.$$

The endpoint of T must also coincide with P_1 and then, after applying Lemma 6.7 near $(0, H)$ (noting that $w_{yy} = 1$ on $x = 0$), we obtain

$$1 \geqslant \alpha \geqslant \beta, \qquad \text{a contradiction.}$$

In case (ii), since $w_{yy}(a, y) = 0$ if $h < y < \phi(a)$, we obtain by the previous argument that $\beta \geqslant \alpha$ and similarly

$$w_{yy}(x, \phi(x)) \geqslant w_{yy}(x', \phi(x')) \qquad \text{if } x' < x.$$

It follows that $\phi'(x)$ is monotone and, in particular, $\phi'(a - 0)$ exists. But then Lemma 6.10(ii) gives

$$\alpha \leqslant w_{yy}(x, \phi(x)) \to 0, \qquad \text{if } x \to a,$$

a contradiction.

Proof of Theorem 6.5.   From Lemma 6.12 we have that $\phi'(x)$ is monotone. Hence Lemma 6.10 can be applied to deduce that $\phi'(0) = 0$, $\phi'(a) = -\infty$, $\phi(a) > h$. Finally, $\phi'(x)$ is clearly decreasing.

PROBLEMS

1. Prove that w_{xx} and w_{yy} are continuous in $\overline{W} \setminus \{(a, h)\}$; w as in Theorem 6.5.

2. Consider the filtration problem with permeable bottom; thus $u_y(x, 0) = 0$ is replaced by $u_y(x, 0) = l(x)$ for $0 < x < a$. In the variational inequality the only change that occurs is in the boundary condition on $y = 0$:

$$w(x, 0) = k(x) \qquad \text{where } k'' = l(x), \quad k(0) = \frac{H^2}{2}, \quad k(a) = \frac{h^2}{2}.$$

Suppose that $l(0) = l(a) = 0$, $k(x) > 0$. Prove:

(i) $w_y \leqslant 0$.

(ii)   If $l(x) > 0$, then $w_{xx} \geqslant 0$ and $\phi'(x) \leqslant 0$ for $0 \leqslant x \leqslant x_0$, $\phi'(x) \geqslant 0$ for $x_0 \leqslant x \leqslant a$.

(iii)   If $x_0 < a$, then $\phi'(a - 0) = 0$.

(iv)   If $l(x)$ changes sign m times, then $\phi'(x)$ changes sign at most $m + 1$ times.

3.   Let w be as in Theorem 6.5. Prove:

(i) $w_{xy} \to \infty$ if $(x, y) \to (a, h)$.

(ii) $w_{xy} > 0$ in W.

(iii) $\phi'(x) - (1/\pi)\log(a - x)$ is in $C^\infty(0, a]$.

[*Hint:* For (i) consider

$$z(x, y) = -\frac{a - x}{2\pi}\left[\log\left(y^2 + (a - x)^2\right) - \log\left((y - h)^2 + (a - x)^2\right)\right]$$

$$+ \frac{y - h}{2\pi}\left[\arctan\frac{y}{a - x} - \arctan\frac{y - h}{a - x}\right]$$

and show that $\zeta = w_y - z$ is harmonic and smooth in $\overline{W}$-neighborhood of (a, h). For (ii) consider the hodograph map $f = p + iq$, where $p = -w_{xy}$, $q = 1 - w_{yy}$, and show that f maps W 1–1 and conformally onto the domain bounded by the rays $q = 0$, $p < 0$; $q = -1$, $p < 0$, and the half circle $p^2 + q^2 + q = 0$, $p \leqslant 0$. For (iii) use the mapping $(\bar{x}, \bar{y}) = (\tilde{u}_x(x, y), \tilde{u}_y(x, y))$, where $\tilde{u} = w - x^2/2$. Show that it is 1–1 and that $v(\bar{x}, \bar{y}) = u(x, y)$ is harmonic, $v(\bar{x}, 0) = -\bar{x}^2/2$ if $\bar{x} > -a$, $= -a^2/2$ if $\bar{x} < -a$, and that $v(\bar{x}, \bar{y}) - z(\bar{x}, \bar{y})$ is C^∞ for $(\bar{x}, \bar{y})$ near $(-a, 0)$, $\bar{y} < 0$, where

$$z(x, y) = \frac{1}{2\pi}xy\left[\log(x^2 + y^2) - \log\left((x + a)^2 + y^2\right)\right]$$

$$+ (y^2 - x^2)\left[\arctan\frac{x}{y} - \arctan\frac{x + a}{y}\right]$$

$$- a^2\left[\arctan\frac{x + a}{y} + \frac{\pi}{2}\right].$$

Finally,

$$\phi'(x) = -\frac{w_{xy}}{w_{yy}}\bigg|_\Gamma = \frac{v_{\bar{y}}(u_x, 0)}{v_{\bar{x}}(u_x, 0)}\bigg|_\Gamma = -\frac{1}{\pi}\log\frac{x}{x - a} + \rho_0(x),$$

where $\rho_0 \in C^\infty[a - \varepsilon, a]$.]

4.   Consider the setting of Problem 5 of Chapter 1, Section 5, with

$$k(y) = \begin{cases} 1 & \text{if } y_0 < y < H \\ k & \text{if } 0 < y < y_0 \end{cases}$$

and $k > 1$, so that $k(y)$ is not increasing. Let

$$k_n(y) = k(y) \qquad \text{if } |y - y_0| > \frac{1}{n}, \quad 1 < k_n(y) < k.$$

Denote by w_n the solution of the variational inequality corresponding to k_n. Prove: $w_{n,x} \leqslant 0$ [so that the free boundary is given by $x = \psi_n(y)$] and $0 \leqslant w_{n,xx} \leqslant k$. Deduce that:

(i) $w_x \leqslant 0$, w_{xx} and w_{yy} are bounded if $y \neq y_0$.
(ii) $0 \leqslant w_x \leqslant k$, $w_{yy} \geqslant 0$.
(iii)   The free boundary is $x = \psi(y)$, and $\psi(y)$ is analytic if $y \neq y_0$.
(iv) $\psi(y_0 \pm 0)$ exist.

[*Hint*: see Theorem 7.2.]

[In reference 58b it is proved that (a) $\psi(y)$ has at most one local maximum and one local minimum [and there are examples where $\psi(y)$ is not monotone]; (b) ψ has at most one inflection point for $y > y_0$, and at most one inflection point for $y < y_0$; (c) denoting by α and β the angles from the negative x-axis to the tangents to Γ at $y = y_0$ going upward and downward, respectively, the following *refraction law* holds:

$$\text{either} \qquad \alpha = 0, \quad \beta = \pi, \quad \text{or } \alpha = \pi, \quad \beta = 0,$$

$$\text{or} \qquad \frac{\pi}{2} < \alpha < \pi, \quad \tan^2\alpha = k, \quad \beta = \alpha - \frac{\pi}{2},$$

$$\text{or} \qquad 0 < \alpha < \frac{\pi}{2}, \quad \tan^2\alpha = k, \quad \beta = \alpha + \frac{\pi}{2}.]$$

7.   REGULARITY OF THE FREE BOUNDARY FOR THE ELASTIC–PLASTIC TORSION PROBLEM

Lemma 7.1.   *Suppose that*

(7.1) $-\Delta u + f \geqslant 0, \qquad u \geqslant 0, \qquad (-\Delta u + f)u = 0 \text{ in } \Omega,$

(7.2) $f \geqslant \lambda > 0, \qquad |u|_{C^{1,1}(\Omega)} \leqslant M < \infty,$

where Ω is a domain in R^{n+1} intersecting the cylinder

$$K_R = B_R \times \{0 < y < b\}.$$

Let G be a domain in $K_R \cap N$ ($N = \{u > 0\}$) such that

$$\partial G \cap \left(\bar{B}_R \times \{0 < y \leqslant b\} \right) \text{ is contained in } \Lambda = \{u = 0\},$$

$$\partial G \cap \{y = b\} \text{ contains a point } (x^0, b),$$

$$\partial G \cap \{y = 0\} \text{ contains a point } (x^1, 0) \in \Gamma.$$

Then

(7.3)
$$b \leqslant 2 \left(\frac{(n+1)M}{\lambda} \right)^{1/2} R.$$

Proof.   We follow the proof of Lemma 3.1 with

$$w(x, y) = u(x, y) - \frac{\lambda}{2(n+1)} \left\{ |x - x_m|^2 + (y - y_m)^2 \right\},$$

where $x_m \to x^0$, $y_m \to b$ and conclude that

$$\sup_{\partial G \cap \{y=0\}} u \geqslant \frac{\lambda}{2(n+1)} b^2.$$

Since, however, $u(x^1, 0) \in \Gamma$, the left-hand side is $\leqslant \frac{1}{2}M(2R)^2$, and (7.3) follows.

This lemma is particularly useful in case $n = 1$. We shall present a version that will be readily applicable in the elastic–plastic problem.

We assume that

$$\Omega = \{a_1 < x < a_2, \ 0 < y < l(x)\}, \quad l(x) \text{ continuous},$$

(7.4) $N = \{a_1 < x < a_2, \ 0 < y < \psi(x)\},$

$$\varepsilon < \psi(x) \leqslant l(x) \qquad \text{for some } \varepsilon > 0.$$

Note that Γ contains points $(x, \psi(x))$ for x in some subset of $\{a_1 < x < a_2\}$ (not necessarily the entire interval).

Theorem 7.2.   *Assume that (7.1), (7.2), (7.4) hold and that $f \equiv 1$. Then $\psi(x)$ is continuous.*

Results of the type of Lemma 7.1 and Theorem 7.2 are called *nonoscillation lemmas*.

Proof.   Suppose that $a_1 < x_0 < a_2$, $\psi(x_0) < l(x_0)$; then $(x_0, \psi(x_0)) \in \Gamma$. There must exist sequences $x_m \downarrow x_0$, $\bar{x}_m \uparrow x_0$ such that $\psi(x_m) \to \psi(x_0)$, $\psi(\bar{x}_m)$

$\to \psi(x_0)$. Indeed, otherwise $u > 0$ on one side of an interval $\{x = x_0, \psi(x_0) < y < \psi(x_0) + \delta\}$ and we get a contradiction from the uniqueness to the Cauchy problem (here we use the fact that $f \equiv 1$).

We next claim that $\psi(x) < l(x)$ if $|x - x_0|$ is sufficiently small and $\psi(x) \to \psi(x_0)$ if $x \to x_0$.

Indeed, if this is not true, then there is a sequence $x_m^* \to x_0$ such that $\psi(x_m^*) \to \psi(x_0) + \delta_0$ for some $\delta_0 > 0$. We may assume that each x_m^* lies in some interval (x_j, x_k), x_m as in the first part of the proof, with $\psi(x_j) \to \psi(x_0)$, $\psi(x_k) \to \psi(x_0)$; here $j = j(m)$, $k = k(m)$. Let

$$x_m^{**} \in (x_j, x_k), \qquad \psi(x_m^{**}) = \max_{x_j \leqslant x \leqslant x_k} \psi(x) \equiv H_m.$$

Denote by G the interval $\{x = x_m^{**}, h_m < y < H_m\}$, where $h_m = \max\{\psi(x_j), \psi(x_k)\}$, and by K the rectangle $x_j < x < x_k, h_m < y < H_m$.

Applying Lemma 7.1 with $K_R = K$, we get

$$H_m - h_m \leqslant C|x_k - x_j| \to 0,$$

a contradiction to $\overline{\lim}(H_m - h_m) \geqslant \delta_0 > 0$.

We have established so far that the set $S = \{\psi(x) < l(x)\}$ is open and ψ is continuous in each of the intervals of this set.

Suppose next that $x_0 \in \bar{S} \setminus S$, $a_1 < x_0 < a_2$. If $x_m \to x_0$, then clearly $\liminf \psi(x_m) \geqslant \psi(x_0)$, and since $\psi(x_0) = l(x_0)$ it follows that $\lim \psi(x_m) = \psi(x_0)$. Finally, $\psi(x)$ is obviously continuous at any point $x_0 \notin \bar{S}$.

Consider now the obstacle problem

$$-\Delta u + f \geqslant 0, \qquad u \geqslant \phi, \qquad (-\Delta u + f)(u - \phi) = 0 \text{ in } \Omega,$$

(7.5)

$$|u|_{C^{1,1}(\Omega)} \leqslant M < \infty, \qquad \phi \in C^3$$

where Ω is a domain in R^n.

On the coincidence set Λ we have $f - \Delta\phi \geqslant 0$ a.e. On Γ everywhere we have $f - \Delta\phi \geqslant 0$; indeed, if $(f - \Delta\phi)(x_0) < 0$, $x_0 \in \Gamma$, then $v = u - \phi$ satisfies $-\Delta v > 0$, $v \geqslant 0$ in a neighborhood of x_0, and the strong maximum principle is contradicted since $v(x_0) = 0$.

If we wish to apply the nonoscillation lemma and results of Sections 1 to 5, then we must establish the inequality $f - \Delta\phi > 0$ in Γ. We shall give a useful sufficient condition for this to hold.

Lemma 7.3. *Let (7.5) hold and assume that*

(7.6) $f - \Delta\phi$ *and* $\nabla(f - \Delta\phi)$ *do not vanish simultaneously.*

Then $f - \Delta\phi > 0$ *on the free boundary of u.*

Proof. Let $\psi = f - \Delta\phi$. We suppose that $\psi(x_0) = 0$ for some point $x_0 \in \Gamma$ and derive a contradiction. Since $\nabla\psi(x_0) \neq 0$, for any small $\varepsilon > 0$ and $R > 0$,

$$\psi < 0 \qquad \text{in } K_\varepsilon \cap B_R(x_0)$$

where K_ε is the cone

$$K_\varepsilon = \{-\langle x - x_0, \nabla\psi(x_0)\rangle \geq \varepsilon \,|\, x - x_0|\}.$$

The function $v = u - \phi$ then satisfies

$$-\Delta v > 0 \qquad \text{in } K_\varepsilon \cap B_R(x_0),$$

and since $v \geq 0$, the strong maximum principle gives

$$v > 0 \qquad \text{in } K_\varepsilon \cap B_R(x_0).$$

By slightly increasing ε we may suppose that $v > 0$ on $\overline{K}_\varepsilon \cap \partial B_R(x_0)$.

Since the opening of K_ε can be made arbitrarily close to $\pi/2$, there exists (by Problem 6) a harmonic function of the form

$$h(x) = |x - x_0|^\lambda f_\lambda(\theta)$$

in K_ε, where $1 < \lambda < 2$ and θ is the angle between $\overrightarrow{x_0 x}$ and the axis of K_ε, and

$$h(x) > 0 \text{ in } K_\varepsilon, \qquad h(x) = 0 \text{ on } \partial K_\varepsilon.$$

By the maximum principle

$$v(x) \geq ch(x) \qquad \text{in } K_\varepsilon \cap B_R(x_0)$$

for some small constant $c > 0$. Hence

$$v(x) \geq cf_\lambda(0) \,|x - x_0|^\lambda$$

along the axis of K_ε. Since, however, $v(x_0) = 0$, $\nabla v(x_0) = 0$,

$$v(x) \leq M \,|x - x_0|^2,$$

which contradicts the previous inequality.

We are now ready to proceed with the study of the free boundary of the elastic–plastic torsion problem. For clarity we first consider the case where the domain Ω is simply connected; we restrict ourselves to the physical two-dimensional case.

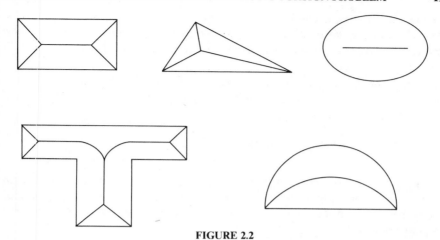

FIGURE 2.2

We assume that $\partial\Omega$ consists of a finite number of disjoint arcs $S_1, \ldots, S_m$ and that the endpoint V_i of S_i is the initial point of S_{i+1} $(S_{m+1} = S_1, V_m = V_0)$. We also assume that each S_i is in $C^{3+\alpha}$ up to the endpoints.

Definition 7.1. Denote by α_i the angle formed by S_i, S_{i+1} at V_i, which opens into Ω. If $\alpha_i > \pi$, then we say that V_i is a *reentrant corner*; if $\alpha_i < \pi$, then we say that V_i is a *nonreentrant corner*; for simplicity we shall always assume that $\alpha_i \neq \pi$.

Definition 7.2. The *ridge* of Ω is the set of all points $x_0 \in \Omega$ such that $d(x)$ is not in $C^{1,1}(V)$ for any neighborhood V of x_0; here

$$d(x) = \text{dist}\,(x, \partial\Omega).$$

In Figure 2.2 we indicate the ridge of several domains.
In order to characterize the ridge geometrically, we need a few lemmas.

Lemma 7.4. *If $d(x_0) = |x_0 - y| = |x_0 - z|$, where y and z are different points on $\partial\Omega$, then $d(x)$ is not differentiable at x_0.*

Proof. Along the segments $\overline{x_0 y}$ and $\overline{x_0 z}$ $d(x)$ is linear with derivative -1; the directions $\overrightarrow{x_0 y}, \overrightarrow{x_0 z}$ are distinct. If $d(x)$ is differentiable at x_0, then $|\nabla d(x_0)| = 1$ and there must be a unique direction l along which $\partial d(x_0)/\partial l = -1$, a contradiction.

Notation. We denote by R_0 the set of all points $x_0 \in \Omega$ such that $d(x_0) = |x_0 - y| = |x_0 - z|$ for at least two distinct points y, z on $\partial\Omega$.

Lemma 7.5. *Let $x_0 \in \Omega \setminus R_0$ and denote by $y_0 = y(x_0)$ the nearest point on $\partial\Omega$ to x_0. Denote by z_0 the center of the osculating circle at y_0 (suppose that y_0 is not a vertex.) Then d is in $C^{1,1}$ in some neighborhood of x_0 if and only if $x_0 \neq z_0$.*

Proof. The point z_0 cannot lie between x_0 and y_0, for otherwise y_0 is not the nearest point to x_0. If $x_0 \neq z_0$, then by reference 109, p. 382, d is in $C^{3+\alpha}$ in a neighborhood of x_0 and

$$(7.7) \qquad\qquad \Delta d(x) = -\frac{\kappa}{1 - \kappa d},$$

where $\kappa = \kappa(x)$ is the curvature of $\partial\Omega$ at the point $y = y(x)$, nearest to x on $\partial\Omega$. If z_0 is a finite point and x_0 lies in the interval $\overline{y_0 z_0}$, then $\kappa(x_0) > 0$ and $\Delta d(x)$ becomes unbounded as $x = x_0 \to z_0$; thus d is not in $C^{1,1}$ in any neighborhood of z_0.

Lemma 7.6. *If in the preceding lemma y_0 is a vertex (it is then necessarily a reentrant corner), then d is in $C^{1,1}$ in some neighborhood of x_0 if and only if $x_0 \neq z_1$, $x_0 \neq z_2$, where z_i are the centers of the osculating circles corresponding to the two arcs of $\partial\Omega$ which meet at y_0.*

In fact, if l_i is a segment connecting y_0 to z_i, then the regularity of d in one side of l_i is the same as before, whereas in the sector between l_1 and l_2 $d(x) = |x - y_0|$. Thus it remains to verify that the first derivatives of d from both sides of l_i agree. But since the derivative of d along l_i is 1, the derivative of d in a direction normal to l_i must tend to zero when the approach is from either side of l_i (since $|\nabla d| = 1$).

Notation. A point x_0 is said to belong to the set R_1 if there exists precisely one point $y_0 \in \partial\Omega$ such that $d(x_0) = |x_0 - y_0|$ and x_0 is the center of the osculating circle at y_0 (or the center of one of the two osculating circles, if y_0 is a vertex).

From the preceding three lemmas, we obtain:

Theorem 7.7. $R = R_0 \cup R_1$.

If there are no reentrant corners, then $R_1 \subset \overline{R_0}$, so that $R = \overline{R_0}$.

Theorem 7.8. $R \subset E$, *that is, the ridge is elastic.*

Proof. Let $x_0 \in R_0$. Then

$$d(x_0) = |x_0 - y_0| = |x_0 - y_1| \qquad \text{where } y_0 \in \partial\Omega, \quad y_1 \in \partial\Omega, \quad y_0 \neq y_1.$$

By Problem 2 of Chapter 1, Section 7, it follows that if $x_0 \in P$ (the plastic set),

then

$$u = d \qquad \text{on the line segments } l_i \colon \overline{x_0 y_i}\,.$$

It follows (see the proof of Lemma 7.4) that u is not differentiable at x_0, which is impossible.

Next suppose that $x_0 \in R_1$ and $x_0 \in P$, and denote by y_0 the nearest point to x_0 on $\partial\Omega$. Suppose first that y_0 is not a vertex. Then

$$u = d \qquad \text{on } l \colon \overline{x_0 y_0}\,;$$

$$\nabla(u - d) = 0 \qquad \text{on } l \text{ (since } u - d \leqslant 0 \text{ in } \Omega),$$

$$(7.8) \qquad \frac{\partial^2 d(x)}{\partial \eta^2} = -\frac{\kappa}{1 - \kappa d} \to -\infty \qquad \text{if } x = (0, \eta) \to x_0 = (0, 0)$$

where (ξ, y) are coordinates with η normal to l and ξ along l; (7.8) follows from the proof of (7.7). Since $u \leqslant d$, we then must also have

$$\frac{\partial^2 u}{\partial \eta^2} \to -\infty,$$

where the second derivative is taken in the sense of limit of second-order difference quotients of u. Thus $\partial u/\partial \eta$ is not Lipschitz in any neighborhood of x_0, which is impossible.

If y_0 is a vertex, then it is necessarily a reentrant corner and one easily finds that $\partial^2 d(x)/\partial \eta^2$, taken in the sense of limit of second-order difference quotients, tends to $-\infty$ as $x = (0, \eta) \to x_0$; thus the analysis above applies.

We now parametrize $\partial\Omega$ by $x_i = f_i(s)$ $(0 \leqslant s \leqslant L)$ and write $x = (x_1, x_2)$, $f(s) = (f_1(s), f_2(s))$. Denote the inward normal to $\partial\Omega$ at $f(s)$ by $\nu(s)$ $(s \neq s_j$, where s_j is the parameter corresponding to the vertex V_j). Let us assume first that there are no reentrant corners. Then, by Problem 2 of Chapter 1, Section 7, there exists a nonnegative function $\delta(s)$ such that

$$\Lambda = \{x;\, x = f(s) + t\nu(s),\quad 0 \leqslant t \leqslant \delta(s),\quad 0 \leqslant s \leqslant L\}.$$

By Theorem 7.8,

$$\text{dist}\,(f(s) + \delta(s)\nu(s), \partial\Omega) = \delta(s),$$

$$X_\varepsilon \equiv (f(s) + \delta(s)(\nu(s) + \varepsilon)) \in E \qquad \forall\, 0 < \varepsilon \leqslant \varepsilon_s,$$

where $X_\varepsilon \notin R$ if $0 < \varepsilon < \varepsilon_s$ and $X_{\varepsilon_s} \in R$; ε_s is uniformly positive in any closed interval of s that does not contain the parameters s_i of the vertices.

We are thus in a position where the proof of Theorem 7.2 can be applied provided we can prove that

(7.9) $$\mu + \Delta d < 0 \qquad \text{on } \Gamma.$$

(Recall, by Theorem 7.8, that $d \in C^{3+\alpha}$ in a neighborhood of Γ.)
Let $x_0 \in \Gamma$ and define

(7.10) $$\psi \equiv \mu + \Delta d = \mu - \frac{\kappa}{1 - \kappa d} \qquad [\text{by } (7.7)].$$

Denote by l the direction of $\overrightarrow{y_0 x_0}$, where y_0 is the nearest point to x_0 on $\partial \Omega$. We would like to apply Lemma 7.3 in order to deduce (7.9). Suppose then that $\psi(x_0) = 0$. Then by (7.10), $\kappa > 0$ at x_0 and, since $\partial \kappa / \partial l = 0$ at x_0,

$$-\frac{\partial \psi}{\partial l} = \frac{\kappa^2}{(1 - \kappa d)^2} \frac{\partial d}{\partial l} = \frac{\kappa^2}{(1 - \kappa d)^2} > 0$$

Thus Lemma 7.3 yields the assertion (7.9).

We are now in a position where Theorem 7.2 (or, rather, the proof of Theorem 7.2) can be applied. It yields the continuity of the function $\delta(s)$; notice that $\delta(s)$ may take the value zero.

Definition 7.3. If $\delta(s) > 0$ in an interval $a < s < b$ and $\delta(a) = \delta(b) = 0$, then we call the plastic set

$$P_{a,b} = \{ f(s) + t\nu(s); 0 \leqslant t \leqslant \delta(s), \quad a \leqslant s \leqslant b \}$$

a *plastic component* or a *plastic loop*.

Having proved that the free boundary of any plastic loop, that is,

$$\Gamma_{a,b} = \{ f(s) + \delta(s)\nu(s); a < s < b \},$$

is a continuous curve, Theorems 2.4 and 2.5 show that $\Gamma_{a,b}$ has a $C^{2+\alpha}$ nondegenerate parametrization. But $\Gamma_{a,b}$ may still have cusps! The next theorem excludes this possibility.

Theorem 7.9. *The coincidence set has positive density at each point x_0 of the free boundary such that $d \in C^{3+\alpha}$ in some neighborhood of x_0.*

Proof. Let $x_0 \in \Gamma$ and let y_0 be the nearest point to x_0 on $\partial \Omega$, $y_0 \neq$ vertex. (See Figure 2.3.) For simplicity we take $y_0 = (0, 0)$, $x_0 = (0, h)$, $h > 0$. Let

$$x_1 = \phi_1(t), \qquad x_2 = \phi_2(t)$$

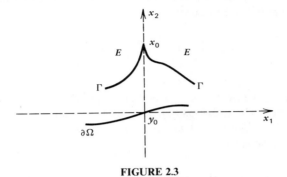

FIGURE 2.3

be a local $C^{2+\alpha}$ parametrization of Γ in a neighborhood of x_0, with $\phi_1(0) = 0$, $\phi_2(0) = h$. We only have to consider the case when $\phi_1'(0) = \phi_2'(0) = 0$, $\phi_1''(0) = 0$, $\phi_2''(0) \neq 0$; by Theorems 2.5 and 3.4, this is the only possible case in which the assertion of the theorem may not hold. Γ has a cusp as in Figure 2.3.

For $t \neq 0$, t near 0, $\phi_2'(t) \neq 0$ and thus the free boundary is in C^2 and u is in C^2 up to the boundary.

The function $w = u_{x_2} - 1$ is negative in E and $w(x_0) = 0$. Since the inside ball property holds at x_0, the strong maximum principle gives $w_{x_2} < 0$ at x_0, that is,

$$(7.11) \qquad\qquad u_{x_2 x_2} < 0 \qquad \text{at } x_0;$$

the derivative is taken in the sense of lim sup of difference quotients of u_{x_2}.

Differentiating the relation

$$(u - d)_{x_2}(\phi_1(t), \phi_2(t)) = 0$$

with respect to t, we get

$$(u - d)_{x_1 x_2}\phi_1'(t) + (u - d)_{x_2 x_2}\phi_2'(t) = 0.$$

Dividing by $\phi_2'(t)$ and letting $t \to 0$, we obtain (since u and d are in $C^{1,1}$)

$$(u - d)_{x_2 x_2}(\phi_1(t), \phi_2(t)) \to 0.$$

Since d is linear along $\overline{y_0 x_0}$,

$$d_{x_2 x_2} \to 0 \qquad \text{if } (x_1, x_2) \to x_0;$$

consequently,

$$u_{x_2 x_2} \to 0 \qquad \text{as } (x_1, x_2) \to x_0 \text{ along } \Gamma.$$

Applying Lemma 6.7, we obtain

$$u_{x_2 x_2}(x) \to 0 \qquad \text{if } x \in E, \quad x \to x_0$$

which contradicts (7.11).

Consider finally the portion of a free boundary whose nearest point on $\partial\Omega$ is a reentrant corner. The previous considerations extend to this case with small changes (again a variant of Theorem 7.2 is needed here). Notice that the obstacle in this case is analytic.

Definition 7.4. If Γ_0 is a connected component of Γ and if for some $x_0 \in \Gamma_0$ the nearest point to $\partial\Omega$ is a reentrant corner V_i, then call the corresponding plastic set (bounded by Γ_0 and $\partial\Omega$) a *reentrant plastic component* or a *reentrant plastic loop*.

Applying Theorem 7.9 and Corollary 3.11, we can now state the following fundamental result.

Theorem 7.10. *The free boundary is locally $C^{m+\alpha}$ ($m \geqslant 3$, $0 < \alpha < 1$) or analytic if the part of $\partial\Omega$ which parametrizes it is in $C^{m+\alpha}$ or is analytic, respectively. Any open portion of Γ whose nearest points to $\partial\Omega$ coincide with a reentrant corner is analytic.*

PROBLEMS

1. In the following problems Ω is a multiply connected two-dimensional domain and u is the solution of the variational inequality (1.6.4). With ϕ, ψ defined by (1.6.6), (1.6.7), we define: The *upper ridge* R^+ is the set of points x_0 in Ω such that ψ is not $C^{1,1}$ in any neighborhood of x_0. Similarly, one defines the *lower ridge* R^- with respect to ϕ. The set $R = R^+ \cup R^-$ is called the *ridge*. (This definition depends on the solution u, that is, on the constants c_j.) Prove: $R \subset E$.

2. Extend Theorem 7.10 to the case where $P^+ \cap P^- = \varnothing$ ($P^\pm$ are defined in Problem 1 of Chapter 1, Section 7).

3. Let $v \in H_0^1(\Omega^*)$, $\phi_\varepsilon \leqslant v \leqslant \psi_\varepsilon$, where ϕ_ε, ψ_ε are defined as ϕ, ψ but with c_j replaced by $c_j + \varepsilon$. Show that

$$\int_{\Omega^*} |\nabla v|^2 - 2\mu \int_{\Omega^*} v \geqslant \int_{\Omega^*} |\nabla u|^2 - 2\mu \int_{\Omega^*} u.$$

4. Suppose that Ω has only one "hole" Ω_1 and $u = C_\mu$ in Ω_1. Set $C_\infty = \text{dist}(\partial\Omega_1, \partial\Omega^*)$. Show that if $C_\mu < C_\infty$ and $\mu > |\Omega_1|/|\partial\Omega_1|$, then

$$P^- \setminus P^+ = \Omega \cap \{u = \phi, u < \psi\}$$

is nonempty.

[*Hint*: Suppose that $u > \phi$ a.e., so that $\Delta u \geq -\mu$. For small $\varepsilon > 0$, if $\zeta = \max\{0, \phi + \varepsilon - u\}$, then $u + \zeta \in H_0^1(\Omega^*)$, $\phi_\varepsilon \leq u + \zeta \leq \psi_\varepsilon$. Use Problem 3.]

5. Assume that $\partial\Omega_1$ has a vertex W such that $|W - Z_1| = |W - Z_2| = C_\infty$, $Z_j \in \partial\Omega^*$; the rays $\gamma_j = \overrightarrow{WZ_j}$ form angle θ, $0 < \theta < \pi$, and the sector obtained by rotating $\overrightarrow{WZ_1}$ by an angle θ to position $\overrightarrow{WZ_2}$ is contained in $\bar{\Omega}$. Denote by Σ the domain bounded by $\overrightarrow{WZ_1}$, $\overrightarrow{WZ_2}$ and $\partial\Omega^*$. Prove:

If $C_\mu = C_\infty$, then, for any neighborhood G of W,

$G \cap \Sigma$ intersects $P^- \setminus P^+$ in a set of positive measure.

[*Hint*: If not, $\Delta u \geq -\mu$ in $G \cap \Sigma$. Let $\Delta v = -\mu$ in $G \cap \Sigma$, $v = u$ on $\partial(G \cap \Sigma)$; l the linear function with $l = u$ on $\overrightarrow{WZ_1} \cup \overrightarrow{WZ_2}$ ($|\nabla l| > 1$); $\tilde{v} = v - l + \mu r^2/4$, where $r = r(x) = |x - W|$. Then $\Delta\tilde{v} = 0$; deduce that $\tilde{v} \in C^{1+\epsilon}$ in $G \cap \bar{\Sigma}$ so that $\partial v/\partial\gamma < -1$ at W, γ the bisector of Σ.]

6. Consider a cone $K_\psi = \{x; x_n > 0, \cos^{-1}(x_n/|x|) < \psi\}$. Prove that if $\psi < \pi/2$ and $\pi/2 - \psi$ is small, then there exists a $1 < \lambda < 2$ and a function $f_\lambda(\theta)$ [where $\theta = \cos^{-1}(x_n/|x|)$] such that the function $v = |x|^\lambda f_\lambda(\theta)$ is harmonic and positive in K_ψ and $v = 0$ on ∂K_ψ.

[*Hint*: If $f_\lambda(\theta) = g(z)$, $z = \cos\theta$,

$$(z^2 - 1)g'' + (n - 1)zg' - \lambda(n - 2 + \lambda)g = 0,$$

and then $g(z) = C_\lambda^{(n-2)/2}(z)$, where the C_λ^k are the Gegebauer functions [84, p. 178]. If $\lambda \downarrow 1$, $C_\lambda^k(z) \to \text{const.} z$ and $f_\lambda(\theta) \to \text{const.} \cos\theta$. The first zero of $f_\lambda(\theta)$ (for $\lambda > 1$) is $\theta = \psi_\lambda < \pi/2$ [otherwise, $|x|^\lambda f_\lambda(\theta) > cx_n$ ($c > 0$) in $\{x_n > 0, |x| < 1\}$ by the maximum principle] and, by continuity, $\psi_\lambda \to \pi/2$ if $\lambda \downarrow 1$.]

8. THE SHAPE OF THE FREE BOUNDARY FOR THE ELASTIC–PLASTIC TORSION PROBLEM

In this section Ω is a simply connected two-dimensional domain; $\partial\Omega$ is made up of $C^{3+\alpha}$ arcs $S_1, \ldots, S_m$ and the vertices are denoted by V_i ($V_i = \bar{S}_i \cap \bar{S}_{i+1}$). The angle of $\partial\Omega$ at V_i is denoted by α_i and it is assumed that $\alpha_i \neq \pi$ (see Section 7).

We begin by studying the free boundary in a neighborhood of a reentrant corner V_j. Set $\alpha = \alpha_j/2$ (so that $\alpha > \pi/2$), $\beta = \alpha - \pi/2$, and introduce polar coordinates (r, θ) about V_j so that the tangents to S_j, S_{j+1} at V_j are the rays $\theta = -\alpha$ and $\theta = \alpha$, respectively.

Theorem 8.1. *There exists a continuous positive-valued function $\rho(\theta)$, defined for $-\beta < \theta < \beta$, such that*

$$(8.1) \qquad \{(r, \theta); \quad 0 < r < \rho(\theta), -\beta < \theta < \beta\} \subset P.$$

Thus a neighborhood of a reentrant corner always contains a plastic sector of opening $\alpha_j - \pi$.

Proof. We begin with a lemma that is a weaker version of the theorem in case Ω is a sector D.

Lemma 8.2. *Let $D = \{(r, \theta); 0 < r < r_0, |\theta| < \gamma\}$, $\gamma > \pi/2$ and denote by E, P the elastic and plastic sets of D. Then there is a segment $\sigma = \{(r, \theta); 0 < r < r_1, \theta = 0\}$ which is contained in P.*

Proof. Consider the function

$$u_\theta = xu_y - yu_x$$

in the region $E^+ = E \cap \{\theta > 0\}$. This function is harmonic [since $\Delta u_\theta = (\Delta u)_\theta = (-\mu)_\theta = 0$] and vanishes on the part of ∂E^+ where $r = r_0$ (since $u = 0$ if $r = r_0$) and on $\theta = 0$ (by symmetry). On the segment $\theta = \gamma$, $u_\theta \leqslant 0$ (since $u = 0$ on $\theta = \gamma$, $u \geqslant 0$ elsewhere). Finally, on the free-boundary part of ∂E^+, $u_\theta = d_\theta \leqslant 0$ since d is decreasing in θ along any circular arc in E^+. By the maximum principle we then conclude that $u_\theta \leqslant 0$ in E^+.

In the smaller set

$$E_0^+ = E \cap \left\{0 < \theta < \gamma - \frac{\pi}{2}, r < \frac{r_0}{2}\right\},$$

P is a subgraph $\{r \leqslant f(\theta)\}$; $f(\theta)$ may a priori be equal to zero for some values of θ. Since $u_\theta \leqslant 0$ in E^+ and since $d(r, \theta) = r$ in E_0^+, we have

$$(u - d)_\theta \leqslant 0 \qquad \text{in } E_0^+.$$

It follows that $f(\theta)$ is monotone decreasing in θ, $0 < \theta < \gamma - \pi/2$.

Now, if the assertion of the lemma is not true, then $f(0) = 0$, and from the monotonicity of f it follows that $f(\theta) = 0$ if $0 \leqslant \theta \leqslant \gamma - \pi/2$. By symmetry we also have $f(\theta) = 0$ if $-\gamma + (\pi/2) \leqslant \theta \leqslant 0$. Let

$$G = \left\{(r, \theta); \, |\theta| < \gamma - \frac{\pi}{2}, 0 < r < \frac{r_0}{2}\right\},$$

$$B = D \cap \left\{0 < r < \frac{r_0}{2}\right\}.$$

Then G is contained in E and therefore $\Delta u = -\mu$ in G. On the other hand, if $x \in B \setminus G$, then either $x \in E$, in which case $\Delta u = -\mu < 0$ or $x \in P$, in which case a.e. $\Delta u \leqslant \Delta d = 0$ (since d is linear in $B \setminus G$). Thus

$$\Delta u \leqslant 0 \qquad \text{a.e. in } B.$$

Consider the harmonic function

$$v = cr^{\pi/2\gamma}\cos\frac{\pi\theta}{2\gamma} \qquad (c > 0)$$

in B. Suppose that

(8.2) $$\frac{\partial u}{\partial \nu} \neq 0 \qquad \text{at } \left(\frac{r_0}{2}, \pm\gamma\right),$$

where ν is the normal. Then, if c is sufficiently small, $v \leqslant u$ on ∂B. Since also $\Delta v \geqslant \Delta u$ a.e. in B, the maximum principle gives $u \geqslant v$ in B. In particular,

$$u(r, 0) \geqslant v(r, 0) = cr^{\pi/2\gamma}.$$

Since, however, $u(r, 0) \leqslant d(r, 0) = r$ and $(\pi/2\gamma) < 1$, we get a contradiction if $r \to 0$. This completes the proof of the lemma in case (8.2) is satisfied.

The assertion (8.2) is true (by the strong maximum principle) if there exists a D-neighborhood of $(r_0/2, \pm\gamma)$ which is elastic. On the other hand, if no such neighborhood exists, then there is a plastic segment initiating at a point $(r_0^*/2, \pm\gamma)$ and orthogonal to $\{\theta = \pm\gamma\}$, where r_0^* is arbitrarily close to r_0. We then have

$$\frac{\partial u}{\partial \nu}\left(\frac{r_0^*}{2}, \pm\gamma\right) = 1.$$

Thus by replacing, if necessary, B by a domain

$$B^* = \left\{0 < r < \frac{r_0^*}{2}, -\gamma < \theta < \gamma\right\},$$

we can continue as before to deduce that $u \geqslant v$ in B^* and, consequently, derive a contradiction.

Proof of Theorem 8.1.   Take any direction $\theta = \theta_0$ in $(-\beta, \beta)$ and construct a domain D as in the preceding lemma so that it is contained in Ω, its axis of symmetry is $\{\theta = \theta_0\}$, and

$$\gamma = \frac{\pi}{2} + \frac{1}{2}(\beta - |\theta_0|).$$

Denote by u_D the solution u corresponding to the domain D. By comparison, in D we have that $u_D \leqslant u$. Also, $u \leqslant d = r$ along the ray $\theta = \theta_0$, if r is small enough. By Lemma 8.2,

$$u_D(r, \theta_0) = r \qquad \text{if } r \text{ is small enough, say } r < \bar{r}.$$

It follows that $u(r, \theta_0) = r$ if $r < \bar{r}$, that is, $(r, \theta_0) \in P$. Since $\bar{r}$ can be taken independently of θ_0 provided that θ_0 is restricted to a closed subinterval of $(-\beta, \beta)$, the proof is complete.

We next study the free boundary near a nonreentrant corner V_k; here $\alpha_k < \pi$.

Theorem 8.3.   *If V_k is a nonreentrant corner, then there exists a neighborhood $B_R(V_k)$ of V_k such that $\Omega \cap B_R(V_k)$ is elastic.*

Proof.   Consider first the case where S_k, S_{k+1} are line segments and introduce polar coordinates with center at V_k so that some $\overline{\Omega}$-neighborhood of V_k is given by

$$G = \{(r, \theta); 0 < r < r_0, -\alpha < \theta < \alpha\}, \qquad \alpha_k = 2\alpha.$$

Recall that $\Delta u = -\mu$ in E and $\Delta u = \Delta d$ a.e. in P. Since $P \cap R = \varnothing$ and Δd is a bounded function in $G \setminus R$ (in fact, with bounded derivative), we conclude that

(8.3) Δu is bounded in G.

Suppose now that $\alpha > \pi/4$, and consider the function

$$v = r^{\pi/2\alpha}\cos\frac{\pi\theta}{2\alpha} + \frac{1}{2}(y^2 - x^2\tan^2\alpha) \qquad \text{in } G.$$

It satisfies

$$\Delta v = 1 - \tan^2\alpha < 0 \qquad \text{in } G.$$

Also, $v = 0$ on $\theta = \pm\alpha$ and $v > 0$ if $r = r_0$ provided that r_0 is sufficiently small, and the normal derivative of v is different from zero at $(r_0, \pm\alpha)$.

Recalling (8.3) we conclude, by the maximum principle, that

$$u \leqslant Cv \qquad \text{in } G$$

for some positive constant C. Hence

$$u(r, \theta) \leqslant Cdr^\nu < \tfrac{1}{2}d \qquad \left(\nu = \frac{\pi}{2\alpha} - 1 > 0\right)$$

if r is small enough, say $r < \delta$. Thus $\overline{G} \cap \{r < \delta\} \subset E$.

Consider next the case where $\alpha \leqslant \pi/4$. It will be enough to prove that

$$(8.4) \qquad \{(r, \theta) \in G; \, -\alpha \leqslant \theta \leqslant 0, r < \delta\} \subset E,$$

provided that δ is sufficiently small; the proof for $0 \leqslant \theta \leqslant \alpha$ is similar.

Let Ω_1 be a domain containing Ω such that $\partial\Omega_1$ has vertex V_k and $\Omega_1 \cap \{r < r_0\}$ is a sector of opening $\gamma > \pi/2$, with $\gamma < \pi$, and such that the two arcs of $\partial\Omega_1$ which meet at V_k are line segments and one of them lies on $\theta = -\alpha$. If we denote by w the elastic–plastic solution corresponding to Ω_1, then, by comparison, $u \leqslant w$ in Ω.

By the proof above,

$$w \leqslant \tfrac{1}{2}d_1 \qquad \text{in some } \Omega_1\text{-neighborhood of } V_k,$$

where d_1 is the distance function for Ω_1. Since $d_1 = d$ if $-\alpha \leqslant \theta \leqslant 0$, r small enough, we obtain that $u < d$ in this set, and (8.4) follows.

We have now completed the proof of the theorem in case S_k, S_{k+1} are line segments. In the general case we transform an Ω-neighborhood of V_k onto a sector G as above by conformal mapping f. Denote by $z = g(\tilde{z})$ the inverse of $\tilde{z} = f(z)$. Then we can write

$$g(\tilde{z}) = \left(h(\tilde{z}^\gamma)\right)^{1/\gamma}, \qquad \gamma = \frac{\pi}{\alpha} > 1.$$

where $\eta = h(\zeta)$ maps a domain $\tilde{\Omega}_1$ bounded in part by two smooth curves T_1, T_2 which form angle π at their common endpoint, the origin, into $\operatorname{Im}\eta > 0$; T_1 and T_2 are mapped into $\{\operatorname{Im}\eta = 0, \eta > 0\}$ and $\{\operatorname{Im}\eta = 0, \eta < 0\}$, respectively.

The inverse function k of h takes $C^{1+\varepsilon}$ values on $\{\operatorname{Im}\eta = 0, |\eta| < \varepsilon_0\}$ for some $\varepsilon_0 > 0$ and thus by general regularity results for elliptic equations (see reference 180), it is in $C^{1+\varepsilon}$ up to the boundary. We shall show that for some $\eta_0 > 0$

$$(8.5) \qquad |\nabla k| \geqslant c > 0 \qquad \text{if } \operatorname{Im}\eta \geqslant 0, \, |\eta| \leqslant \eta_0.$$

Using (ρ, θ) as polar coordinates of ζ, we note that, for any small $\lambda > 0$, the function

$$v = \rho \sin \theta + \rho^{2-\lambda} \cos (2 - \lambda) \theta$$

is positive on the boundary of $\tilde{\Omega}_1 \cap \{ |\zeta| = \rho_0 \}$ if ρ_0 is small enough (in fact, $\rho \sin \theta \geqslant -\text{const.} \, \rho^2$ in $\tilde{\Omega}_1$, near $\zeta = 0$). Hence $Cv \geqslant |\operatorname{Im} h|$ on this boundary if C is large enough and, by the maximum principle,

$$|\operatorname{Im} h| \leqslant Cv \leqslant \text{const.} \, \rho \qquad \text{in } \tilde{\Omega}_1.$$

By interior estimates for harmonic functions

$$|\nabla(\operatorname{Im} h)| \leqslant C$$

along the bisector of T_1, T_2. It follows that $|\nabla k| \geqslant c > 0$ along the bisector $\operatorname{Im} \eta = 0$. Since $k \in C^{1+\varepsilon}$, we also get (8.5).

From (8.5) it follows that

$$(8.6) \qquad |\nabla h| \leqslant C \qquad \text{in some } \tilde{\Omega}_1\text{-neighborhood of } 0,$$

and, in fact $h \in C^{1+\varepsilon}$ up to the boundary.

Setting $\tilde{u}(\tilde{z}) = u(z)$, where $z = g(\tilde{z})$, we have

$$\Delta \tilde{u} = |g'|^2 \, \Delta u,$$

and the right-hand side is a bounded function [by (8.6)]. We can then apply to $\tilde{u}$ the proof of the preceding special case and conclude that

$$\tilde{u} \leqslant C \tilde{d} \tilde{r}^{\nu} \qquad \text{near the origin,}$$

where $\nu = (\pi/2\alpha) - 1$, $\tilde{d}$ is the distance function for the transformed region, and $\tilde{r} = |\tilde{z}|$. Since $\tilde{d} \leqslant Cd$, $\tilde{r} \leqslant Cr$, we get $u \leqslant \frac{1}{2} d$ in some Ω-neighborhood of V_k, and the proof is complete.

It is not known whether the number of plastic components in Ω is finite. However, in case $\partial\Omega$ contains a linear segment λ, one can prove that the number of plastic components connected to λ is finite:

Theorem 8.4. *Let λ be a closed interval contained in one of the open arcs S_l. Then the number of plastic components connected to λ is finite.*

Proof. Take for simplicity $\lambda = \{ a \leqslant x_1 \leqslant b, \, x_2 = 0 \}$ and suppose that there are infinite number of plastic components P_i, given by

$$P_i = \{ a_i \leqslant x_1 \leqslant b_i, \quad 0 \leqslant x_2 \leqslant \phi(x_1) \}.$$

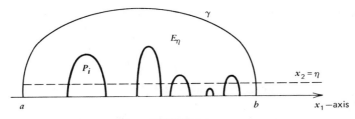

FIGURE 2.4

Set

$$H_i = \max_{a_i \leqslant x \leqslant b_i} \phi(x).$$

If $i \to \infty$, then $b_i - a_i \to 0$ and, by the nonoscillation lemma, also $H_i \to 0$. Hence any line $x_2 = \eta$ ($\eta > 0$ and small) intersects only a finite number $\lambda(\eta)$ of the plastic components, and $\lambda(\eta) \to \infty$ if $\eta \to 0$.

Consider a subdomain E_η of $E \cap \{x_2 > \eta\}$ as in Figure 2.4. It is bounded by a piecewise analytic curve γ in E, parts of the ∂P_j, and some intervals lying on $x_2 = \eta$. We choose γ such that $d(x) = x_2$ in E_η for all η small; γ is independent of η. [If the P_i accumulate at $x_1 = a$, then we take γ to start on $x_1 = a - \varepsilon$, for some small $\varepsilon > 0$, if $(a - \varepsilon, 0)$ has elastic Ω-neighborhood; otherwise, γ will start from a free boundary point on $\{x_1 = a - \varepsilon\}$; a similar modification of γ is made about $x_1 = b$.]

On every interval

$$\{x_2 = \eta, \quad d_i \leqslant x_1 \leqslant e_i\}$$

on ∂E_η with endpoints on Γ the function $(u - d)_{x_1} = u_{x_1}$ is analytic and changes sign (since $u - d = 0$ at the endpoints and $u - d < 0$ inside). We can choose points $d_i < \bar{d}_i < \bar{e}_i < e_i$ such that

$$(8.7) \quad u_{x_1}(x_1, \eta) < 0 \text{ if } d_i < x_1 \leqslant \bar{d}_i, \quad u_{x_1}(x_1, \eta) > 0 \text{ if } \bar{e}_i \leqslant x_1 < e_i.$$

We now consider the level curves Δ_i and Γ_i of $\{u_{x_1} = \text{const}\}$, which initiate at $(\bar{d}_i, \eta)$ and $(\bar{e}_i, \eta)$, respectively. By Lemma 6.11 these curves can be continued until they exit $E_0 \equiv \cup_{\eta > 0} E_\eta$. Notice that $u_{x_1} = (u - d)_{x_1} = 0$ on ∂P_j, and thus Δ_i, Γ_i cannot exit E_0 on the free boundary. They also cannot come arbitrarily close to $\{x_2 = 0\}$ since $u_{x_1} = 0$ on $\{x_2 = 0\}$. Finally, because of (8.7) they must exit E_0 on γ at points M_i (in which the sign of u_{x_1} alternates) and M_i precedes M_j along γ if the same holds for the initial points of the curves on $x_2 = \eta$.

It follows that u_{x_1} alternates signs on γ at least $2\lambda(\eta) - 1$ times, a number that grows to infinity as $\eta \to 0$. But since u_{x_1} is analytic in a neighborhood of $\bar{\gamma}$ and γ is piecewise analytic, this is impossible.

We next proceed to give an estimate on the number of plastic components connected to λ. Take

$$\lambda = \overline{S}_1 = \{(x_1, 0); 0 \leqslant x_1 \leqslant b\}, \qquad V_0 = (b, 0), \quad V_1 = (0, 0)$$

and set $\mu = \overline{S}_2 \cup \cdots \cup \overline{S}_m$. We assume that some Ω-neighborhood of S_1 lies in $\{x_2 > 0\}$.

Let $x = f(x)$ $(0 \leqslant s \leqslant L)$ be a parametrization of $\partial \Omega$ with $f(0) = V_0$, $f(s_1) = V_1$. We consider first the case where

(8.8)   there are no reentrant corners.

The distance function $d(x)$ is differentiable along $x = f(s)$, $s \neq s_j$ [$f(s_j) = V_j$]. Set

$$d_1(s) = \frac{\partial d}{\partial x_1}(f(s)) \qquad \text{if } s \neq s_j,$$

$$d_1(s_j) = d_1(s_j + 0) \qquad [d_1(0) = d_1(L - 0)],$$

and assume that

(8.9)   the set $\{s; d_1(s) = 0, 0 \leqslant s \leqslant L\}$ consists of a finite number of points A_i and intervals J_k.

It follows that $d_1(s)$ changes sign (from positive to negative, or vice versa) a finite number of times. Set

(8.10) $k = $ number of times $d_1(s)$ changes sign from positive to negative as s increases from $s = s_1$ to $s = L$.

Denote by $P_1, \ldots, P_\tau$ the plastic components connected to λ; by Theorems 8.3 and 8.4 and by assumption (8.8), $\tau < \infty$. We can write

$$P_j = \{(x_1, x_2); 0 \leqslant x_2 \leqslant \phi(x_1), a_j \leqslant x_1 \leqslant b_j\},$$

where $a_1 < b_1 \leqslant a_1 < b_2 \leqslant \cdots < b_{\tau-1} \leqslant a_\tau < b_\tau$; notice that

$$a_1 > 0, \qquad b_\tau < b.$$

By Section 7, $\phi(x_1)$ is positive and analytic in each interval $a_j < x_1 < b_j$; it is continuous and equal to zero at $x_1 = a_j$, $x_1 = b_j$. We define $\phi(x_1) = 0$ if $x_1 \notin \cup_{j=1}^\tau (a_j, b_j)$.

In each interval $\{a_j \leqslant x_1 \leqslant b_j\}$, ϕ has a certain number N_j of points where it achieves a local maximum (these are strict local maxima since ϕ is analytic).

Theorem 8.5. *Each N_j is finite and*

(8.11) $$\sum_{j=1}^\tau N_j \leqslant k.$$

From this result we immediately obtain:

Corollary 8.6. (i) *The total number of plastic components connected to λ is finite and $\leqslant k$; (ii) if the number of plastic loops connected to λ is precisely k, then in* each interval $\{a_j < x < b_j\}$ there is a point c_j such that $\phi(x_1)$ is strictly increasing if $a_j < x_1 < c_j$ and strictly decreasing if $c_j < x_1 < b_j$.

Corollary 8.7. *If Ω is a convex polygon, then for any side S_j there is at most one plastic loop connected to S_j; if it is given by $\{f(s) + \nu(s)t, 0 \leqslant t \leqslant h(s), A_j \leqslant s \leqslant B_j\}$, then $h(s)$ is strictly increasing in some interval $A_j \leqslant s \leqslant C_j$ and strictly decreasing in the interval $C_j \leqslant s \leqslant B_j$.*

To facilitate the proof of Theorem 8.5 we begin with some auxiliary results. For the proof of the first lemma assume for definiteness that $d_1(s) > 0$ if $s > s_1$, $s - s_1$ small and $d_1(s) < 0$ if $s < L$, $L - s$ small; the proof in the other cases is similar.

Setting $l = 2k$, there is a partition

$$s_1 = s_{1,1} \leqslant s_{1,2} \leqslant s_{1,3} \leqslant \cdots \leqslant s_{1,m_1}$$
$$= s_{2,1} \leqslant s_{2,2} \leqslant \cdots \leqslant s_{2,m_2} = s_{3,1} \leqslant s_{3,2} \leqslant \cdots \leqslant s_{l,m_l}$$
$$= s_{l+1,1} \leqslant s_{l+1,2} \leqslant \cdots \leqslant s_{l+1,m_{l+1}} = L$$

such that

$$d_1(s) > 0 \text{ if } s_{1,1} < s < s_{1,2} \quad (s_{1,1} < s_{1,2}),$$

$$d_1(s) = 0 \text{ if } s_{1,2} \leqslant s \leqslant s_{1,3} \quad (s_{1,2} \leqslant s_{1,3}),$$

$$d_1(s) > 0 \text{ if } s_{1,3} < s < s_{1,4} \quad (s_{1,3} < s_{1,4}),$$

$$d_1(s) = 0 \text{ if } s_{1,4} \leqslant s \leqslant s_{1,5} \quad (s_{1,4} \leqslant s_{1,5}),$$

$$d_1(s) > 0 \text{ if } s_{1,5} < s < s_{1,6} \quad (s_{1,5} < s_{1,6}),$$

$$\vdots$$

(8.12)
$$d_1(s) < 0 \text{ if } s_{2,1} < s < s_{2,2} \quad (s_{2,1} < s_{2,2}),$$

$$d_1(s) = 0 \text{ if } s_{2,2} \leqslant s \leqslant s_{2,3} \quad (s_{2,2} \leqslant s_{2,3}),$$

$$d_1(s) < 0 \text{ if } s_{2,3} < s < s_{2,4} \quad (s_{2,3} < s_{2,4}),$$

$$\vdots$$

$$d_1(s) < 0 \text{ if } s_{l+1,1} < s < s_{l+1,2} \quad (s_{l+1,1} < s_{l+1,2}),$$

$$d_1(s) = 0 \text{ if } s_{l+1,2} \leqslant s \leqslant s_{l+1,3} \quad (s_{l+1,2} \leqslant s_{l+1,3}),$$

Thus $d_1(s)$ changes sign at the points $s_{1,m_1}, s_{2,m_2}, \ldots, s_{l,m_l}$. The vertices V_i $(2 \leqslant i \leqslant m - 1)$ need not necessarily coincide with points $f(s_{j,m_j})$; that is, s_i may be a point in some interval $(s_{r,p}, s_{r,p+1})$. Some of the intervals $[s_{i,j}, s_{i,j+1}]$ where $d_1 \equiv 0$ may consist of a single point.

We recall that the free boundary can be written in the form

$$x = f(s) + \nu(s)h(s),$$

where $\nu(s)$ is the normal to $\partial\Omega$ at $f(s)$ and $h(s) > 0$; for the points where $h(s) = 0$, there is no plastic interval initiating at $f(s)$ and perpendicular to $\partial\Omega$.

Consider the function

$$(8.13) \qquad \tilde{d}_1(s) \equiv \frac{\partial d}{\partial x_1}(f(s) + \nu(s)h(s)) \qquad (0 \leqslant s < L)$$

at a point $s = s^0$ such that $s^0 \neq s_i$ for $1 \leqslant i \leqslant m$, and

$$(8.14) \qquad\qquad\qquad d_1(s^0) > 0.$$

By continuity, $d_1(s) > 0$ in some interval $|s - s^0| < \delta_0$. Hence some Ω-neighborhood of $f(s^0)$ is given by

$$g(x_2) < x_1 < g(x_2) + \delta_1, \qquad |x_2 - x_2^0| < \delta_1,$$

with $x_1 = g(x_2)$ being a part of $\partial\Omega$, and $f(s^0) = (g(x_2^0), x_2^0) \equiv x^0$.

Set $y^0 = f(s^0) + \nu(s^0)h(s^0) \equiv (y_1^0, y_2^0)$. Suppose that $h(s^0) > 0$. Then x^0 is the nearest point on $\partial\Omega$ to y^0. It is clear that, for any η positive and sufficiently small, the ray initiating at $y^\eta = (y_1^0 - \eta, y_2^0)$ and in the direction $\overline{y^0 x^0}$ intersects $\partial\Omega$ at $(x_1^\eta, x_2^\eta) \equiv x^\eta$ and $|x^\eta - y^\eta| < |x^0 - y^0|$; in fact, since $g'(x_2)$ is a bounded function,

$$|x^\eta - y^\eta| \leqslant |x^0 - y^0| - c\eta \qquad (c > 0).$$

This implies that $d(y^\eta) < d(y^0) - c\eta$, so that

$$(8.15) \qquad\qquad\qquad \tilde{d}_1(s^0) = d_{x_1}(y^0) \geqslant c.$$

Since $\nabla(u - d) = 0$ on the free boundary, we conclude that

$$(8.16) \quad u_{x_1}(f(s^0) + \nu(s^0)h(s^0)) = \tilde{d}_1(s^0) \geqslant c > 0 \qquad \text{if } h(s^0) > 0.$$

By continuity, the relation (8.16) is satisfied also at any point s^0 such that

$$(8.17) \qquad \begin{array}{l} h(s^0) = 0, \text{ and there exists a sequence} \\ s^k \text{ such that } s^k \to s^0, h(s^k) > 0. \end{array}$$

Suppose next that $h(s^0) = 0$ but (8.17) is not satisfied. Then $h(s) = 0$ in a neighborhood of s^0. Consequently, there exists an Ω-neighborhood W of x^0 which is elastic. Since $-\Delta u = \mu > 0$ in W, $u > 0$ in W, $u = 0$ on $\partial\Omega$, the strong maximum principle gives $(\partial u/\partial \nu(s^0))(x^0) > 0$. Recalling (8.14), we deduce that $u_{x_1}(f(s^0)) > 0$. We have thus proved, in general, that

(8.18) if $d_1(s^0) > 0$, then $\tilde{d}_1(s^0) > 0$, $u_{x_1}(f(s^0) + \nu(s^0)h(s^0)) > 0$.

Similarly, one can show that if $d_1(s^0) < 0$, then

$$\tilde{d}_1(s^0) < 0, \qquad u_{x_1}(f(s^0) + \nu(s^0)h(s^0)) < 0.$$

Finally, if $d_1(s) = 0$ in an open interval $\alpha < s < \beta$, then $\{f(s), \alpha < s < \beta\}$ is a line segment parallel to the x_1-axis; it follows that

$$\tilde{d}_1(s) = 0, \qquad u_{x_1}(f(s) + \nu(s)h(s)) = 0 \qquad \text{if } \alpha < s < \beta.$$

Set

(8.19) $$u_1(s) = \frac{\partial u}{\partial x_1}(f(s) + \nu(s)h(s)), \qquad 0 < s < L,$$

if $s \neq s_i$, and $u_1(s_i) = u_1(s_i + 0)$. By Theorem 8.3, $h(s) = 0$ in a neighborhood of $s = s_i$.

We can now summarize:

Lemma 8.8.   *The assertion (8.12) holds with $d_1(s)$ replaced by either $\tilde{d}_1(s)$ or $u_1(s)$; further,*

(8.20) $$\tilde{d}_1(s) = u_1(s) = 0 \qquad \text{if } 0 < s < s_1.$$

Definition 8.1.   The points $f(s) + \nu(s)h(s)$ for which $u_1(s) = 0$ will be called *flat points*. The set of flat points consist of a finite number of points and a finite number of arcs (called *flat intervals*).

Notice that the flat intervals correspond to $[0, s_1]$ and to those intervals $[s_{1,2}, s_{1,3}]$, $[s_{1,4}, s_{1,5}], \ldots$ in (8.12) where $d_1(s) \equiv 0$ and which do not reduce to single points.

The function u_{x_1} is harmonic in E and it has harmonic continuation into a neighborhood of the interior of each flat interval.

The proof of Theorem 8.5 will employ level-curves analysis. We shall need a variant of Lemma 6.11.

Lemma 8.9.   *Through each point $x^0 \in E$ where $u_{x_1}(x^0) = 0$ there passes a piecewise analytic curve $\gamma : \{x = \gamma(t), -\infty < t < \infty\}$, contained in E and along* *which* $u_{x_1} = 0$. γ *has no self-intersections and*

$$\gamma(+\infty) = \lim_{t \to \infty} \gamma(t), \qquad \gamma(-\infty) = \lim_{t \to -\infty} \gamma(t)$$

exist and belong to ∂E.

The proof is similar to the proof of Lemma 6.11. It uses the conclusion of Lemma 8.8, which implies that $\gamma(t)$ must stay away from all the points of ∂E except for the flat points and the flat intervals. Notice that γ cannot oscillate near a flat interval I (as $t \to \infty$ or $t \to -\infty$) since u_{x_1} has harmonic continuation into a neighborhood of the interior of I.

Notation.   Denote by λ_E the part of λ which is not in the plastic components connected to λ, and denote by $\partial_0 P^\lambda$ the union of the free boundaries of the plastic components connected to λ. Set $\lambda_0 = \bar{\lambda}_E \cup \partial_0 P^\lambda$. Similarly, we define for $\mu = \cup_{j=2}^m \bar{S}_j$ the sets μ_E, $\partial_0 P^\mu$, and $\mu_0 = \bar{\mu}_E \cup \partial_0 P^\mu$.

Consider now a plastic loop P_j $(1 \leqslant j \leqslant \tau)$; it is given by

$$0 \leqslant x_2 \leqslant \phi(x_1), \qquad a_j \leqslant x_1 \leqslant b_j.$$

Suppose that β is a point of local maximum of ϕ. Then

(8.21) $\phi(x_1) < \phi(\beta)$ if $a_j < \beta - \delta_0 < x_1 < \beta + \delta_0 < b_j$, $x_1 \neq \beta$

for some $\delta_0 > 0$.

Lemma 8.10.   *There exists a level curve γ of $\{u_{x_1} = 0\}$, given by $\gamma : \{x = \gamma(t), 0 \leqslant t < \infty\}$ such that $\gamma(0) = (\beta, \phi(\beta))$, $\gamma(t) \in E$ if $0 < t < \infty$, and $\gamma(\infty) = \lim_{t \to \infty} \gamma(t)$ belongs to $\mu_0 \setminus \lambda_0$.*

Proof.   For any small $\delta > 0$, $\delta < \delta_0$,

$$u(\beta - \delta, \phi(\beta)) < d(\beta - \delta, \phi(\beta)) = \phi(\beta),$$

$$u(\beta + \delta, \phi(\beta)) < d(\beta - \delta, \phi(\beta)) = \phi(\beta).$$

Hence $u_x(\beta - \delta_1, \phi(\beta)) > 0$, $u_x(\beta + \delta_2, \phi(\beta)) < 0$ for some $0 < \delta_1, \delta_2 < \delta$. By continuity,

$$u_x(\beta - \delta_1, \phi(\beta) + \varepsilon) > 0, \qquad u_x(\beta + \delta_2, \phi(\beta) + \varepsilon) < 0$$

if $\varepsilon > 0$ is sufficiently small. Consequently, there is a number $\tilde{\delta}_\varepsilon$, $-\delta_1 < \tilde{\delta}_\varepsilon < \delta_2$, such that

(8.22) $$u_x(\beta + \tilde{\delta}_\varepsilon, \phi(\beta) + \varepsilon) = 0.$$

It follows that there is a level curve of $\{u_{x_1} = 0\}$, from those initiating at

$(\beta, \phi(\beta))$, which is contained in E (except for its initial point). We now continue to extend this curve as in the proof of Lemmas 6.11 and 8.9. It remains to show that $\gamma(\infty) \in \mu_0 \setminus \lambda_0$.

If this is not the case, then $\gamma(\infty) \in \lambda_0$ and there exists a subdomain E_0 of E bounded by γ and a part of λ_0 such that $u_{x_1} = 0$ on its boundary. It follows by Lemma 6.7 and the maximum principle that $u_{x_1} \equiv 0$, which is impossible.

Proof of Theorem 8.5.   Consider first the case where $k = 1$ and take for definiteness $d_1(s)$ positive for $s > s_1$, $s - s_1$ small and the negative for $s < L$, $L - s$ small.

Denote by $\mu_0(s', s'')$ the portion of μ_0 corresponding to $s' \leqslant s \leqslant s''$, and denote by $\mu_0(s)$ the point of μ_0 corresponding to s.

Suppose there is a curve γ as in Lemma 8.9 such that $\gamma(-\infty)$ and $\gamma(+\infty)$ both lie on μ_0. Thus

$$\gamma(-\infty) = \mu_0(\sigma_1), \quad \gamma(\infty) = \mu_0(\sigma_2), \qquad s_1 < \sigma_1 < \sigma_2 < L.$$

Denote by μ_1 the curve obtained from μ_0 by replacing $\mu_0(\sigma_1, \sigma_2)$ by γ, and parametrize γ by the parameter s, $\sigma_1 < s < \sigma_2$.

Denote by E_1 the subdomain of E bounded by λ_0, μ_1 and suppose that there exists another curve of the type γ, say γ_1, in E_1 such that $\gamma_1(-\infty)$ and $\gamma_1(+\infty)$ lie in μ_1. We then modify μ_1 into a curve μ_2, replacing the part of μ_1 from $\gamma_1(-\infty)$ to $\gamma_1(+\infty)$ by γ_1. Again we parametrize μ_2 by s so that s varies in $s_1 \leqslant s < L$. We can repeat this process step by step. Note that in the first step the points $\gamma(-\infty)$, $\gamma(+\infty)$ cannot coincide and cannot lie in the same flat interval (otherwise, $u_{x_1} \equiv 0$). Similarly, in the second step the points $\gamma_1(-\infty)$, $\gamma_1(+\infty)$ cannot lie on an arc of μ_1 along which $u_{x_1} = 0$, and so on. Recalling Lemma 8.8, we conclude that the process of constructing the curves $\mu_1, \mu_2, \ldots$ must end at some j. Thus there are no level curves γ_j (as in Lemma 8.9) which lie in the subdomain E_j bounded by λ_0, μ_j for which $\gamma_j(-\infty) \in \mu_j, \gamma_j(+\infty) \in \mu_j$.

We shall assume for simplicity that $\mu_j = \mu_0$; but the proof given below extends almost word by word (with a slightly more complicated notation) to the general case where $\mu_j \neq \mu_0$. Thus we assume that

(8.23)   there is no level curve γ as in Lemma 8.9
            such that $\gamma(t) \in E$ and $\gamma(-\infty), \gamma(+\infty)$ lie in μ_0.

Consider a curve γ as in Lemma 8.10 and let $\gamma(+\infty) = \mu_0(s^0), s_1 < s^0 < L$. Denote by λ_{01} the part of λ_0 from $(0,0)$ to $(\beta, \phi(\beta))$ and by λ_{02} the arc $\lambda_0 \setminus \lambda_{01}$. Denote by D_1 the domain bounded by $\mu_0(s_1, s^0), \lambda_{01}$, and γ, and by D_2 the domain bounded by $\mu_0(s^0, L), \lambda_{02}$, and γ.

Consider first the case where

(8.24) $$u_1(s) \geqslant 0 \qquad \text{if } s_1 < s < s^0,$$
           $$u_1(s) \leqslant 0 \qquad \text{if } s^0 < s < L.$$

By the maximum principle

$$(8.25) \qquad u_{x_1} > 0 \qquad \text{in } D_1,$$

$$(8.26) \qquad u_{x_1} < 0 \qquad \text{in } D_2.$$

From (8.25) it follows that $(u - d)_{x_1} > 0$ in the subset of D_1 consisting of all points x for which $d(x) = x_2$. This inequality easily implies that

$$(8.27) \qquad \begin{array}{c} \text{the function } x_2 = \phi(x_1) \text{ is} \\ \text{monotone increasing for } 0 < x_1 < \beta. \end{array}$$

Similarly, we deduce from (8.26) that

$$(8.28) \qquad \begin{array}{c} \text{the function } x_2 = \phi(x_1) \text{ is} \\ \text{monotone decreasing for } \beta < x_1 < b. \end{array}$$

It follows that there can be at most one loop and that $\phi(x_1)$ has a unique strict maximum. Thus the proof of the theorem is complete in this case.

Suppose next that (8.24) is not satisfied. Assume for definiteness that for some s^*, $s^0 < s^* < L$,

$$(8.29) \quad \begin{array}{ll} u_1(s) \geqslant 0 & \text{if } s_1 < s < s^*, \qquad u_1(s) \not\equiv 0 \text{ if } s^0 < s < s^*, \\ u_1(s) < 0 & \text{if } s^* < s < s^* + \delta \qquad \text{for some small } \delta > 0, \\ u_1(s) \leqslant 0 & \text{if } s^* < s < L. \end{array}$$

Then the conclusion (8.27) follows as before.

Since $u_1(s)$ changes sign along $\mu_0(s^0, L)$, there is a level curve γ_1 in D_2 of $\{u_{x_1} = 0\}$. By (8.23), $\gamma_1(-\infty)$ and $\gamma_1(+\infty)$ cannot both lie in $\mu_0(s^0, L)$. They also cannot both lie in λ_{02}. Hence we may take $\gamma_1(-\infty) \in \lambda_{02}$ and $\gamma_1(+\infty) \in \mu_0(s^0, L)$. Let σ_1 be defined by $\mu_0(\sigma_1) = \gamma_1(+\infty)$. If $u_1(s)$ changes sign along $\mu_0(s^0, \sigma^1)$ [or along $\mu_0(\sigma^1, L)$], then we can construct a level curve γ_2 of $\{u_{x_1} = 0\}$ lying in the region bounded by γ, γ_1 and $\mu_0(s^0, \sigma^1)$ [or $\mu_0(\sigma^1, L)$] and a part of λ_{02}. For this curve,

$$\gamma_1(-\infty) \in \lambda_{02}, \qquad \gamma_2(+\infty) = \mu_0(\sigma^2) \in \mu_0(s^0, \sigma^1)$$
$$\left[\text{or } \mu_0(\sigma^2) \in \mu_0(\sigma^1, L)\right].$$

Again if $u_1(s)$ changes sign along one of the arcs $\mu_0(s^0, \sigma^2)$, $\mu_0(\sigma^2, L)$ [or $\mu_0(\sigma^1, \sigma^2)$, $\mu_0(\sigma^2, L)$], then we can construct a level curve γ_3 of $\{u_{x_1} = 0\}$ which starts on λ_{02} and ends on the corresponding arc, say at σ^3, and so on. Since no two terminal points of curves γ_j can lie in the same flat interval, the process of constructing the curves γ_j must terminate. It follows that there exists a level curve, say γ_j, with $\gamma_j(-\infty) \in \lambda_{02}$ and $\gamma_j(+\infty) = \mu_0(\sigma_j)$ such that $s^0 < \sigma_j \leqslant s^*$ and $u_1(s) \equiv 0$ on $\mu_0(\sigma_j, s^*) = 0$. Consequently, $u_1(s) \leqslant 0$ on $\mu_0(\sigma_j, L)$ and $u_1(s) \geqslant 0$ on $\mu_0(s_1, \sigma_j)$.

The argument following (8.24) now shows that $\phi(x_1)$ is monotone increasing for $0 < x_1 < \delta$ and monotone decreasing for $\delta < x_1 < b$, where $\gamma_j(-\infty) = (\delta, \phi(\delta))$. This completes the proof, in case $k = 1$, and incidentally shows that (since $\delta = \beta$) γ_j must coincide with γ, that is, $s_0 = \sigma_j$.

Consider now the case of general k. Let $\beta_j (1 \leqslant j \leqslant N)$ be points of strict local maximum for $\phi(x_1)$. Construct level curves γ^j of $\{u_{x_1} = 0\}$ as in Lemma 8.10, which initiate at $(\beta_j, \phi(\beta_j))$; each γ^j must terminate on μ_0, at some point $\mu_0(\sigma_j)$ and $\bar{\gamma}^j \cap \bar{\gamma}^l = \varnothing$ if $j \neq l$ (otherwise, $u_{x_1} \equiv 0$). Set $\gamma^0 = \mu_0(s_1, \sigma_1)$, $\gamma^{N+1} = \mu_0(\sigma_N, L)$ and denote by D_j the domain bounded by γ_j, γ_{j+1}, $\mu_0(\sigma_j, \sigma_{j+1})$ and λ_{0j}, where λ_{0j} is the part of λ_0 from β_j to β_{j+1}.

Along each arc $\mu_0(\sigma_j, \sigma_{j+1})$ $(1 \leqslant j \leqslant N - 1)$ the sign of $u_1(s)$ cannot be fixed [otherwise, ϕ would be monotone in (β_j, β_{j+1})]. Also, it cannot happen that the sign changes precisely once from positive to negative, for this would imply that $\phi(x_1)$ is increasing in some interval $(\beta_j, \beta_j + \eta)$ and decreasing in the interval $(\beta_j + \eta, \beta_{j+1})$ (by the proof in the special case $k = 1$ when applied to D_j; notice that $u_{x_1} = 0$ on γ^j, γ^{j+1}).

It follows that $u_1(s)$ changes sign from negative to positive in each interval (σ_j, σ_{j+1}). Finally, $u_1(s)$ cannot be $\leqslant 0$ in the entire interval $(0, \sigma_1)$ [for otherwise, $\phi(x_1)$ would be decreasing for $0 < x_1 < \beta_1$]; similarly, $u_1(s)$ cannot be $\geqslant 0$ in the entire interval (σ_{N+1}, L). Thus the number of changes of sign, from positive to negative, of $u_1(s)$ along μ_0 is at least k. This completes the proof.

Theorem 8.11. *Theorem 8.5 remains valid even if the assumption (8.8) is dropped.*

Here, in counting the N_j, we do not include the possibly two reentrant loops which are connected to the endpoints of λ.

For proof, see Problem 1.

Consider the case where Ω is a rectangle

$$(8.30) \qquad \Omega = \{-a < x_1 < a, -b < x_2 < b\}.$$

By Corollary 8.7 there are at most four plastic loops:

$$P_1 = \{-\alpha \leqslant x_1 \leqslant \alpha, -b \leqslant x_2 \leqslant -b + \phi(x_1)\},$$

$$P_2 = \{-\beta \leqslant x_2 \leqslant \beta, -a \leqslant x_1 \leqslant -a + \psi(x_2)\},$$

$$(8.31) \qquad P_3 = \text{the reflection of } P_1 \text{ with respect to the } x_1\text{-axis},$$

$$P_4 = \text{the reflection of } P_2 \text{ with respect to the } x_2\text{-axis}.$$

Denote by l the bisector to $\partial\Omega$ at $(-a, -b)$, that is,

$$l: x_2 = x_1 + a - b.$$

Denote by $\rho(x)$ the reflection of a point $x = (x_1, x_2)$ with respect to l where $x_2 \leqslant x_1 + a - b$, and by ρ^* the inverse reflection to ρ.

Theorem 8.12. *If $b < a$, then $\rho^*(P_2) \subset P_1$.*

Proof.   Let

$$D = \{(x_1, x_2); x_2 < x_1 + a - b, -a < x_1 < -a + 2b\},$$

$$D_1 = D \cap E.$$

Consider the function

$$w(x) = u(\rho x) - u(x) \qquad \text{in } D_1.$$

Since

$$-\Delta u(x) \leqslant \mu \text{ in } \rho(D_1), \qquad -\Delta u(x) = \mu \text{ in } D_1,$$

we have

$$-\Delta w \leqslant 0 \qquad \text{in } D_1.$$

The boundary of D_1 consists of five arcs γ_i: γ_1 lying on the line l, γ_2 lying on the line $x_2 = -b$, γ_3 lying on ∂P_1, γ_4 lying on ∂P_4, and γ_5 lying on the line $x_1 = -a + 2b$. On γ_1 we have

$$w(x) = u(\rho x) - u(x) = u(x) - u(x) = 0.$$

On γ_2,

$$w(x) = u(\rho x) - u(x) = u(-a, \tilde{x}_2) - u(\tilde{x}_1, -b) = 0 - 0 = 0$$

(where $\tilde{x}_1 + a = \tilde{x}_2 + b$). On $\gamma_3 \cup \gamma_4$,

$$w(x) = u(\rho x) - d(x) \leqslant 0,$$

since $u(x) = d(x)$, $d(\rho x) \leqslant d(x)$. Finally, on γ_5,

$$w(x) = u(\rho x) - u(x) = 0 - u(x) \leqslant 0.$$

Thus $w(x) \leqslant 0$ on the boundary of D_1. Applying the maximum principle, it follows that

$$w(x) < 0 \qquad \text{in } D_1.$$

Notice that for every point $x^0 \in D_1$ with $\rho x^0 \in \partial P_2$,

$$d(\rho x^0) = \text{dist}(\rho x^0, \gamma) \qquad \gamma \text{ the line } x_2 = -b;$$

hence $d(\rho x^0) = d(x^0)$. Since $u(\rho x^0) = d(\rho x^0)$, it follows that

$$u(\rho x^0) = d(x^0).$$

Consequently,

$$w(x^0) = u(\rho x^0) - u(x^0) = d(x^0) - u(x^0).$$

Since $w(x^0) < 0$, we conclude that $u(x^0) > d(x^0)$, which is impossible. This contradiction implies that there are no points x^0 in D_1 for which $\rho x^0 \in \partial P_2$; that is, $\rho^*(\partial P_2)$ is contained in P_1. This completes the proof.

Consider next the case where

$$(8.32) \qquad Q \text{ is a triangle with sides } \gamma_1, \gamma_2, \gamma_3$$

and denote the length of γ_i by $|\gamma_i|$. By Corollary 8.7, the plastic set consists of three loops P_i; P_i is based on γ_i (and may be the empty set).

Denote by l the bisector to γ_1, γ_2 and denote by ρx the reflection of x across l, when x varies in the half plane that contains γ_1. Denote by ρ^* the inverse reflection to ρ.

Theorem 8.13. *If $|\gamma_2| < |\gamma_1|$, then $\rho^*(P_2) \subset P_1$.*

Proof. Denote by D^* the triangle with sides γ_2, $\tilde{\gamma}_3$, $\tilde{l}$ where $\tilde{\gamma}_3 \subset \gamma_3$ and $\tilde{l} \subset l$ and set

$$D = \rho^*(D^*), \qquad D_1 = D \cap E.$$

Since $|\gamma_2| < |\gamma_1|$, $\rho^*(\tilde{\gamma}_3)$ [and hence also $\rho^*(D^*)$] is contained in Ω. The angles of the triangles D^*, D at the point $l \cap \gamma_3$ are equal and hence they are smaller than $\pi/2$. It follows that

$$(8.33) \qquad \text{if } x \in \partial P_3 \cap D, \qquad \text{then } d(\rho x) - d(x) \leq 0.$$

Consider the function

$$w(x) = u(\rho x) - u(x)$$

in D_1. By (8.33), if $x \in \partial P_3 \cap D_1$, then

$$w(x) = u(\rho(x)) - d(x) \leqslant d(\rho x) - d(x) \leqslant 0.$$

On the remainder part of ∂D_1 we also have $w(x) \leqslant 0$ (as in the proof of Theorem 8.12). Since $-\Delta w \leqslant 0$ in D_1, the maximum principle implies that $w(x) < 0$ in D_1. From this inequality we can now deduce, as in the proof of Theorem 8.12, that $\rho^*(P_2) \subset \rho(P_1)$.

The method of proving Theorems 8.12 and 8.13 is based on the principle of reflection. This principle can be applied also in other situations. We shall consider a case where the boundary of Ω is not necessarily polygonal.

Definition 8.2. Let $x^0 \in \partial\Omega$, $x^0 \neq$ vertex, and denote by $N_0(x^0)$ the line normal to $\partial\Omega$ at x^0. Denote by σ the reflection with respect to $N_0(x^0)$. Suppose that

(8.34) $$\sigma(\Omega \cap T_-) \subset \Omega \cap T_+$$

where T_-, T_+ are the two half-planes into which $N_0(x^0)$ divides the plane. Then we say that x^0 has the *reflection property*.

Theorem 8.14. *Let $x^0 = f(s^0) \neq$ vertex, and suppose that*

$$y^0 = f(s^0) + \nu(s^0)\rho_0(s^0)$$

is a point of the free boundary, that is, $f(s^0) + \nu(s^0)\rho \in P$ if $0 < \rho \leqslant \rho_0(s^0)$, $f(s^0) + \nu(s^0)\rho \in E$ if $\rho = \rho_0(s^0) + \varepsilon$ for any $\varepsilon > 0$ and small. If x^0 has the reflection property with T_+ containing the points $f(s)$, $s > s^0$, $s - s^0$ small, then

(8.35) $$\frac{d}{ds}\rho_0(s) \geqslant 0 \qquad at \ s = s^0.$$

Proof. Consider the function

$$w(x) = u(\sigma^{-1}x) - u(x) \qquad in \ E_+ \equiv E \cap \sigma(\Omega \cap T_-).$$

It satisfies

$$-\Delta w \leqslant \mu - \Delta u = 0 \qquad in \ E_+.$$

On ∂E_+ we have,

$$w(x) = 0 \qquad if \ x \in N(x^0),$$

$$w(x) = 0 - u(x) \leqslant 0 \qquad if \ x \in \sigma(\partial\Omega \cap T_-),$$

$$w(x) = u(\sigma^{-1}x) - d(x) \leqslant d(\sigma^{-1}x) - d(x) \leqslant 0$$

$$if \ x \in \partial E_+ \cap \sigma(\Omega \cap T_-),$$

[where (8.34) was used in deriving the last inequality]. Hence, by the maximum principle, $w \leq 0$ in E_+, that is,

$$(8.36) \qquad u(\sigma^{-1}x) \leq u(x) \qquad \text{if } x \in E_+ .$$

Denote by d_-, d_+ the distance functions from the curves $x = f(s)$, $s \leq s^0$ and $x = f(s)$, $x \geq s^0$, and let

$$\gamma = \{ f(s^0) + \rho\nu(s^0), 0 < \rho < \rho_0(s^0) \}.$$

Then

$$d_1 = d_- , \text{grad } d_- = \text{grad } d_+ = \nu(s^0) \qquad \text{on } \gamma.$$

Introducing orthogonal coordinates (ρ, s) [where $x = f(s) + \nu(s)\rho$], we then have

$$\frac{\partial}{\partial s} d_\pm = 0, \qquad \frac{\partial}{\partial \rho} d_\pm = 1 \qquad \text{on } \gamma.$$

Also [see (7.7), (7.8)],

$$\frac{\partial^2}{\partial \rho^2} d_\pm = \frac{\partial^2}{\partial s \, \partial \rho} d_\pm = 0, \qquad \frac{\partial^2}{\partial s^2} d_+ = \frac{\partial^2}{\partial s^2} d_- \qquad \text{on } \gamma.$$

It follows that

$$(8.37) \qquad |d_+(x_\varepsilon) - d_-(x_{-\varepsilon})| = O(\varepsilon^3)$$

if x_ε lies on the normal to $N(x^0)$ at $f(s^0) + \nu(s^0)\rho_0(s^0)$ and

$$x_\varepsilon \in T_+ , \qquad \text{dist}(x_\varepsilon, \gamma) = \varepsilon, \qquad x_{-\varepsilon} = \sigma^{-1}x_\varepsilon.$$

Suppose now that (8.35) is not true; that is, suppose that

$$(8.38) \qquad \frac{d}{ds} \rho_0(s^0) < 0;$$

we shall derive a contradiction. From (8.38) it follows that $x_\varepsilon \in E$, $x_{-\varepsilon} \in P$ for small $\varepsilon > 0$; consequently,

$$u(x_{-\varepsilon}) = d_-(x_{-\varepsilon}).$$

On the other hand, if σ and τ are the tangent and normal directions to the free boundary at a point y, then

$$(u - d)_{\sigma\sigma} = (u - d)_{\sigma\tau} = 0, \qquad (u - d)_{\tau\tau} \neq 0 \text{ at } y.$$

It follows that

$$u(x_\varepsilon) < d_+(x_\varepsilon) - C\varepsilon^2 \qquad (C > 0).$$

Using (8.37) we deduce that

$$u(x_\varepsilon) < d_-(x_{-\varepsilon}) + O(\varepsilon^3) - C\varepsilon^2 = u(x_{-\varepsilon}) + O(\varepsilon^3) - C\varepsilon^2,$$

which contradicts (8.36) if ε is small enough.

Denote by R_1 the first quadrant of the plane. We shall assume that Ω satisfies the following conditions:

(8.39) $\partial\Omega$ is in $C^{3+\alpha}$; Ω is symmetric with respect to the x-axis and the y-axis;

the set $\partial\Omega \cap R_1$ is given by $y = f(x)$, $0 \leq x \leq a$, where

$$(8.40) \qquad\qquad\qquad\qquad f'(x) \leq 0,$$

$$(8.41) \qquad\qquad\qquad\qquad f''(x) \leq 0,$$

$$(8.42) \qquad\qquad -\frac{f''(x)}{\left[1 + (f'(x))^2\right]^{3/2}} \text{ is nondecreasing.}$$

The condition (8.41) means that Ω is convex. The condition (8.42) means that the curvature at the points $(x, f(x))$ of $\partial\Omega \cap R_1$ is nondecreasing as x increases.

Lemma 8.15. *If Ω satisfies the conditions (8.39)–(8.42), then the reflection property is satisfied at every point of $\partial\Omega$.*

For a proof of this geometric lemma, we refer the reader to reference 59.

Corollary 8.16. *Let Ω be a domain satisfying (8.39)–(8.42). Then the free boundary consists of two curves:*

$$\Gamma_+ : (x, f(x)) + \rho_0(x)\nu_+(x), \qquad -\alpha < x < \alpha,$$

$$\Gamma_- : (x, -f(x)) + \rho_0(x)\nu_-(x), \qquad -\alpha < x < \alpha,$$

where $0 \leqslant \alpha \leqslant a$, $\rho_0(-x) = \rho_0(x)$, $\nu_{\pm}(x)$ is the inward normal to $\partial\Omega$ at $(x, \pm f(x))$; the function $\rho_0(x)$ is monotone decreasing in x for $0 < x < \alpha$.

Note that if $\alpha = 0$, then there are no free-boundary points, whereas if $\rho_0(a - 0) > 0$, then the two curves form one connected curve.

Example. If Ω is the interior of an ellipse,

$$\frac{x^2}{a^2} + \frac{y^2}{b^2} = 1, \qquad a > b,$$

then the conditions (8.39)–(8.42) are satisfied.

PROBLEMS

1. Prove Theorem 8.11.

    [*Hint*: Use Theorem 8.1 in order to extend Lemma 8.8.]

2. Extend Theorem 8.5 to the case where λ is a circular arc $\{r = r_0, \theta_1 < \theta < \theta_2\}$ and u_{x_1} is replaced by u_θ; note that $\Delta u_\theta = 0$ in E.

3. If $r = h(\theta)$ is the free boundary near a reentrant corner, estimate the number of local maxima of h in terms of $\partial d/\partial\theta$ along $\partial\Omega$; see Problem 2.

4. If Ω is T-shaped with the x_2-axis as the axis of symmetry, then there are no plastic components based on the open sides of $\partial\Omega$ which are parallel to the x_1-axis and one of whose endpoints is a reentrant corner.

5. Consider a pentagon all of whose vertices have the same size angle. Generalize Theorem 8.12 by proving that if P_i is a plastic loop based on a side l_i and if $|l_i| \leqslant |l_j|$, then $\rho_{ij}(P_i) \subset P_j$, where ρ_{ij} is a suitable reflection.

6. Prove the same for a hexagon with equal angles.

7. Prove that E is contained in an ε_μ-neighborhood of the ridge R, where $\varepsilon_\mu \to 0$ if $\mu \to \infty$ (actually $\varepsilon_\mu = C/\mu$, $C > 0$; see reference 58f). (This result indicates that Theorem 8.5 gives a sharp estimate for μ large.)

8. If Ω is a rectangle, then for all μ sufficiently large E contains c/μ neighborhood of the ridge ($c > 0$).

9. With the notation (8.30), (8.31), prove that (a) $\phi(x_1)$ has precisely one point of inflection in the interval $0 < x_1 < \alpha$; (b) $\phi'(\alpha - 0) = 0$.

    [*Hint*: Use the method of Section 7.]

9. THE FREE BOUNDARY FOR THE STEFAN PROBLEM

In this section we use the notation

$$Hu = \Delta u - u_t$$

for the heat operator. A function u satisfying $Hu = 0$ is called *caloric*; if it satisfies $Hu \geq 0$ (≤ 0) in the sense of distributions then it is called *subcaloric* (*supercaloric*).

In Chapter 1, Section 9, we studied the Stefan problem. We recall that the conditions on the domain are stated in the second paragraph of that section, and the conditions on the data are given there in (9.9). In this section we shall require precisely the same conditions.

In Chapter 1, Section 9, it was proved that

$$(9.1) \qquad\qquad \overline{N(0)} \subset N(t) \qquad \text{if } t > 0,$$

where

$$N(t) = \{x \in \Omega; u(x, t) > 0\},$$

and that

$$(9.2) \qquad\qquad 0 \leq \frac{\partial u}{\partial t} \leq C,$$

$$(9.3) \qquad\qquad D_x^2 u \in L^\infty_{\text{loc}}.$$

We recall that the proof of (9.2) utilizes the penalized problem (9.12) of Chapter 1. Differentiating the parabolic equation with respect to t we obtained the equation

$$-H\left(\frac{\partial u_\varepsilon}{\partial t}\right) + \beta'_\varepsilon(u_\varepsilon)\frac{\partial u_\varepsilon}{\partial t} = 0,$$

from which we deduced, by the maximum principle, that $\partial u_\varepsilon / \partial t \geq 0$; (9.2) then follows by taking $\varepsilon \to 0$. Notice also that

$$-H\left(\frac{\partial u_\varepsilon}{\partial t}\right) \leq 0,$$

and consequently

$$(9.4) \qquad\qquad H(u_t) \geq 0;$$

that is, the temperature u_t is subcaloric.

In this section we shall prove that u_t is continuous and then extend the results of Sections 3 to 5 to the present parabolic case.

We introduce the *parabolic distance*

$$d((x, t), (x_0, t_0)) = \{|x - x_0|^2 + |t - t_0|\}^{1/2}$$

and the moduli of continuity

(9.5) $$\omega_\varepsilon(r) = C|\log r|^{-\varepsilon} \qquad (C > 0, \varepsilon > 0),$$

(9.6) $$\sigma_\gamma(r) = C2^{-|\log r|^\gamma} \qquad (C > 0, \gamma > 0).$$

Theorem 9.1

(i) *The temperature u_t is continuous in $\overline{\Omega} \times [0, T]$;*

(ii) *for any compact set K in $\Omega \times (0, T)$, u_t is uniformly continuous with modulus of continuity $\omega_\varepsilon(r)$, for any $0 < \varepsilon < 2/(n - 2)$ if $n \geqslant 3$, and $\sigma_\gamma(r)$, for any $0 < \gamma < \frac{1}{2}$ if $n = 2$.*

In the modulus of continuity, r is understood to be the parabolic distance. To prove the theorem, we need several lemmas.

Lemma 9.2. *Let w be a bounded measurable subcaloric function in a cylinder $D \times (0, T)$. Then there exists a function $\tilde{w}$ such that (i) $\tilde{w} = w$ a.e. in $D \times (0, T)$, (ii) $\tilde{w}$ is upper semicontinuous in $D \times (0, T)$, (iii) for any ball K, $K \subset D$, if z satisfies*

$$\Delta z - z_t = 0 \qquad \text{in } K \times (t_0, t_1) \ (0 < t_0 < t_1 < T),$$
$$z = \tilde{w} \qquad \text{on the parabolic boundary of } K \times (t_0, t_1),$$

then $z \geqslant \tilde{w}$ in $K \times (t_0, t_1)$.

This is a fairly standard result: For proof we refer to reference 58c. From now on we understand by u_t the version for which the properties (i) to (iii) are satisfied.

We extend h by 0 into $\Omega \setminus \overline{G}$.

Lemma 9.3. *u_t is continuous at $t = 0$.*

Proof. We claim that

(9.7) $$\begin{array}{c} N(t) \text{ is contained in a } \delta(t)\text{-neighborhood} \\ \text{of } G, \text{ where } \delta(t) \to 0 \text{ if } t \to 0. \end{array}$$

Indeed, otherwise there are points (x^m, t^m) such that $u(x^m, t^m) > 0$, $t_m \to 0$,

$\text{dist}(x^m, G) \geqslant 2c > 0$. Consider the function

$$v^m(x, t) = u(x, t) - \frac{1}{8n}\left(|x - x^m|^2 + (t^m - t)\right)$$

in the set R of points (x, t) such that $u(x, t) > 0$ and

$$|x - x^m| < c, \qquad 0 < t < t^m.$$

Clearly, $Hv^m \geqslant 0$ in R, and $v^m(x^m, t^m) > 0$. By the maximum principle, v must take a positive maximum on the parabolic boundary of R, say at (x_1^m, t_1^m). Thus

$$(9.8) \qquad u(x_1^m, t_1^m) > \frac{1}{8n}\left(|x^m - x_1^m|^2 + (t^m - t_1^m)\right).$$

Since $t_1^m = 0$ is impossible and since (x_1^m, t_1^m) cannot lie on the free boundary, we must have $|x^m - x_1^m| = c$. Hence (9.8) gives

$$u(x_1^m, t_1^m) > \frac{c^2}{8n}, \qquad t_1^m \to 0, \qquad \text{dist}(x_1^m, G) > c,$$

which is again impossible since $h(x) = 0$ if $x \notin G$.

Since $u_t \geqslant 0$,

$$(9.9) \qquad u(x, t) > 0 \qquad \text{if } x \in G.$$

From (9.7), (9.9) and the boundedness of u_t we deduce that

$$(9.10) \qquad \int_\Omega |u_t(x, t) - h(x)|\, dx \to 0 \qquad \text{if } t \to 0;$$

here we used the fact that Γ_1 is in C^1.

We now proceed to prove the continuity of u_t at a point $(y, 0)$; it suffices to take y in Γ_1. Let K be a small ball with center y, and let w^ε ($\varepsilon \geqslant 0$) be the solution of

$$\Delta w^\varepsilon - w_t^\varepsilon = 0 \qquad \text{in } K \times (\varepsilon, 1),$$

$$w^\varepsilon = u_t \qquad \text{on the parabolic boundary of } K \times (\varepsilon, 1).$$

By (9.10),

$$(9.11) \qquad w^\varepsilon(x, t) \to w^0(x, t) \qquad \text{if } \varepsilon \to 0.$$

From Lemma 9.2 we have, for any $\varepsilon > 0$,

$$u_t(x, t) \leqslant w^\varepsilon(x, t).$$

Hence

$$(9.12) \qquad u_t(x, t) \leqslant w^0(x, t).$$

Since $w^0(x, 0) = h(x)$ is continuous at y, $w^0(x, t) \to h(y) = 0$ if $x \to y$, $t \to 0$. Recalling that $u_t \geqslant 0$ and using (9.12), we find that $u_t(x, t) \to 0$ if $x \to y$, $t \to 0$.

We shall consider now the case $n \geqslant 3$.

Lemma 9.4. *Let $0 < \theta < \frac{1}{2}$ and let $f_\theta(x, t)$ be a continuous and nonnegative function defined for $|x| \leqslant 1$, $0 \leqslant t \leqslant \lambda$, and satisfying:*

$$f_\theta \geqslant 0 \qquad if \, |x| = 1, 0 < t < \lambda,$$

$$f_\theta \geqslant 0 \qquad if \, |x| \leqslant 1, t = 0,$$

$$f_\theta \geqslant 1 \qquad if \, |x| \leqslant \theta, 0 < t < \frac{\lambda_0}{2},$$

$Hf_\theta \leqslant 0$ in the distribution sense, in $\{|x| < 1, 0 < t < \lambda\}$, where $0 < \lambda_0 < \lambda$. Then, for any $0 < \beta < 1$, if λ_0 is sufficiently large (independently of θ, λ),

$$(9.13) \qquad f_\theta(x, t) \geqslant C\theta^{n-2} \qquad for \, |x| < \beta, \, \lambda_0 < t < \lambda,$$

where C is a positive constant depending only on β, λ_0, λ.

Proof. Let g be the solution of

$$\Delta g - g_t = 0 \qquad if \, |x| < 1, \quad 0 < t < \infty,$$

$$g = 0 \qquad if \, |x| = 1, \quad 0 \leqslant t \leqslant \infty,$$

$$g(x, 0) = |x|^{2-n} - 1 \qquad if \, |x| \leqslant 1.$$

Then $g \geqslant 0$ and

$$(9.14) \qquad g\left(x, \frac{\lambda_0}{2}\right) < \frac{1}{2}\left(|x|^{2-n} - 1\right) \qquad if \, |x| \leqslant \beta_0, \quad \beta_0 < 1,$$

provided that λ_0 is sufficiently large.

Applying the maximum principle to the function

$$z(x, t) \equiv f_\theta(x, t) - \left(\frac{|x|^{2-n} - 1}{\theta^{2-n} - 1} - \frac{g(x, t)}{\theta^{2-n} - 1} \right)$$

in the domain $\theta < |x| < 1$, $0 < t < \lambda_0/2$, we find that $z \geqslant 0$ in this domain.

Using (9.14), we conclude that

$$f_\theta\left(x, \frac{\lambda_0}{2}\right) \geqslant \frac{1}{2} \frac{|x|^{2-n} - 1}{\theta^{2-n} - 1} \geqslant C\theta^{n-2} \qquad (C > 0)$$

if $|x| < \beta_0$, where $\beta < \beta_0 < 1$. We now represent $f_\theta(x, t)$ in $|x| \leqslant \beta_0$, $(\lambda_0/2)$ $\leqslant t \leqslant \lambda$, by Green's function. Using the last inequality, (9.13) follows (with a different C).

Proof of Theorem 9.1 for $n \geqslant 3$. Let (x^0, t^0) to be a point in $\Omega \times (0, T)$ such that $u(x^0, t^0) = 0$. We shall prove the continuity of u_t at this point. For simplicity we make a translation of coordinates so that $x^0 = 0$, $t^0 = 0$; thus u is a solution of the variational inequality in a neighborhood V of the origin.

For any positive integer k, $k > k_0$, let

$$\Gamma_k = \left\{(x, t); |x| \leqslant 4^{-k}, -2\lambda_0 4^{-2k} \leqslant t \leqslant 0\right\},$$

where λ_0 is a sufficiently large constant to be determined later, and k_0 is sufficiently large so that $\Gamma_{k_0} \subset V$. In the sequel we denote by C, C_1 generic constants independent of k.

Set

$$m_k = \sup_{\Gamma_k} u_t,$$

$$M_k = \max\left\{m_k, Ck^{-\varepsilon}\right\},$$

where ε is a positive constant to be determined later. We shall prove:

Lemma 9.5. *For any $\delta > \varepsilon/2$, there holds*

$$(9.15) \qquad m_{k+1} \leqslant M_k\left(1 - \frac{C}{k^{\delta(n-2)}}\right) \qquad \text{for all } k \geqslant k_0.$$

Proof. Let

$$w^*(x, t) = \frac{1}{4^{-2k}} \int_{-4^{-2k}}^0 u_t(x, t + h)\, dh.$$

Since w^* is a convolution of the subcaloric function u_t with a positive kernel, it is also subcaloric.

Since $u(0, t) = 0$, $D_x u(0, t) = 0$ if $t \leqslant 0$, and since $D_x^2 u$ is a bounded function,

$$\frac{1}{4^{-2k}} \int_{-4^{-2k}}^0 u_t(x, t + h)\, dh \leqslant \frac{1}{4^{-2k}} u(x, t) \leqslant \frac{1}{4^{-2k}} C|x|^2.$$

Hence

$$(9.16) \qquad w^*(x, t) \leqslant \frac{C}{k^{2\delta}} \qquad \text{if } |x| = \frac{1}{4^k k^\delta}, \quad \delta > 0.$$

We introduce the coordinates

$$(9.17) \qquad x' = 4^k x, \qquad t' = 4^{2k} t$$

and the function

$$w(x', t') = w^*(x, t).$$

Γ_k is mapped onto $|x'| \leqslant 1$, $-2\lambda_0 \leqslant t \leqslant 0$, and

$$\Delta w - w_t \geqslant 0 \qquad \text{if } |x'| < 1, \quad -2\lambda_0 + 1 < t' < 0,$$

$$w \leqslant m_k \qquad \text{if } |x'| \leqslant 1, \quad -2\lambda_0 + 1 < t' \leqslant 0,$$

$$w(x', t') \leqslant \frac{C}{k^{2\delta}} \qquad \text{if } |x'| = k^{-\delta}.$$

Let

$$(9.18)$$

$$v = \frac{M_k}{M_k - Ck^{-2\delta}} (w - Ck^{-2\delta}) \qquad \text{in } |x'| \leqslant 1, \quad -2\lambda_0 + 1 \leqslant t' \leqslant 0.$$

Then we can apply Lemma 9.4 to the function $1 - v/M_k$ with $\theta = k^{-\delta}$, and thus obtain

$$1 - \frac{v(x', t')}{M_k} \geqslant C_1 k^{-\delta(n-2)} \qquad (C_1 > 0),$$

if $|x'| \leqslant \beta$, $-\lambda_0 \leqslant t' \leqslant 0$, provided that λ_0 is sufficiently large. Thus

$$v(x', t') \leqslant M_k(1 - C_1 k^{-\delta(n-2)}).$$

Hence, from (9.18),

$$w \leqslant (M_k - Ck^{-2\delta})(1 - C_1 k^{-\delta(n-2)}) + Ck^{-2\delta}$$

$$= M_k(1 - C_1 k^{-\delta(n-2)}) + CC_1 k^{-2\delta} k^{-\delta(n-2)}.$$

Choosing $\varepsilon < 2\delta$, we conclude that

$$(9.19) \quad w(x', t') \leq M_k(1 - Ck^{-\delta(n-2)}) \qquad \text{if } |x'| \leq \beta, \quad -\lambda_0 \leq t' \leq 0.$$

Let

$$z(x', t') = u_t(x, t)$$

and consider the function

$$h(x', t') = M_k - z(x', t') \qquad \text{for } |x'| \leq \beta, \quad -\lambda_0 + 1 \leq t' \leq 0.$$

In view of (9.19),

$$\int_{t'-1}^{t'} h(x', s)\, ds \geq M_k - \int_{t'-1}^{t'} z(x', s)\, ds \equiv M_k - w(x', t') \geq CM_k k^{-\delta(n-2)}.$$

Also $h \geq 0$, $\Delta h - h_t \leq 0$. Let $\hat{h}$ be the caloric function defined in $|x'| \leq \beta$, $-\lambda_0 + 1 < t' \leq 0$, having the same boundary values as h on the parabolic boundary. By Lemma 9.2, $h \geq \hat{h}$. Representing $\hat{h}$ in terms of Green's function in $|x'| \leq \beta$, $-\lambda_0 + 1 \leq t' \leq 0$, we then find that

$$h(x', t') \geq C_1 CM_k k^{-\delta(n-2)} \qquad (C_1 > 0)$$

in a smaller region $|x'| \leq \beta_0$, $-(\lambda_0/2) \leq t' \leq 0$, where $\beta_0 < \beta$. Therefore, in this region

$$z(x', t') \leq M_k - C_1 CM_k k^{-\delta(n-2)}.$$

Choosing $\beta_0 = \frac{1}{4}$, (9.15) follows.

We now choose δ so that $\delta(n-2) < 1$ [and then $\varepsilon < 2/(n-2)$] and proceed to prove by induction that

$$m_k \leq C^* k^{-\varepsilon} \qquad \text{if } k \geq k_1.$$

This inequality certainly holds for any given $k_1 \geq k_0$ if C^* is large enough (depending on k_1). Assuming this inequality for k, we get, by definition of M_k,

$$M_k = C^* k^{-\varepsilon}, \qquad \text{provided that we take } C^* > C.$$

Next, by Lemma 9.5,

$$m_{k+1} \leq C^* k^{-\varepsilon}\left[1 - \frac{C}{k^{\delta(n-2)}}\right] \leq C^*(k+1)^{-\varepsilon},$$

since

$$1 - \frac{C}{k^{\delta(n-2)}} \leq \left(\frac{k+1}{k}\right)^{-\varepsilon} \qquad \text{if } k \geq k_1;$$

k_1 is determined so that this inequality holds if $k \geq k_1$; k_1 is independent of C^*.

Having completed the inductive proof, we can now write that

$$u_t(x, t) \leq C^* k^{-\varepsilon} \qquad \text{if } |x| \leq 4^{-k}, \quad -2\lambda_0 4^{-2k} \leq t \leq 0.$$

This means that

$$(9.20) \qquad u_t(x, t) \leq C \left| \log \left(|x|^2 + |t| \right) \right|^{-\varepsilon}$$

provided that (x, t) is in a small neighborhood of $(0, 0)$ and $t \leq 0$. This inequality shows that u_t is continuous from below at the point $(x^0, t^0) = (0, 0)$ of the free boundary, and $u_t(0, 0 - 0) = 0$, that is, if $x \to x^0$ $t \uparrow t^0$, then $u_t(x, t) \to u_t(x^0, t^0 - 0) = 0$.

In order to prove continuity from above at any point (x^0, t^0) of the free boundary, take for simplicity $(x^0, t^0) = (0, 0)$. Let ζ be the solution of

$$\Delta \zeta - \zeta_t = 0 \qquad \text{if } |x| < \mu, \quad 0 < t < \mu,$$

$$\zeta(x, t) = u_t(x, t - 0) \qquad \text{on } |x| = \mu,$$

$$0 < t < \mu \text{ and on } |x| < \mu, \quad t = 0.$$

Since $u_t(x, 0 - 0)$ satisfies (9.20) with $t = 0$, the function ζ is continuous at $(0, 0)$ and, in fact,

$$\zeta(x, t) \leq C \left| \log \left(|x|^2 + t \right) \right|^{-\varepsilon}$$

(as seen by representing ζ by a fundamental solution). Since u_t is subcaloric, we have (by Lemma 9.2) that

$$u_t \leq \zeta.$$

It follows that (9.20) holds for $t \geq 0$. This completes the proof of part (i) of Theorem 9.1.

To prove the assertion (ii), let $P = (x, t)$, $\overline{P} = (\overline{x}, \overline{t})$ be two points in $K \times (\delta, T]$. Denote by $d(P, \overline{P})$ the parabolic distance from P to $\overline{P}$. Denote by d_p the parabolic distance from P to the parabolic free boundary, that is, to the part of the free boundary whose points (y, s) satisfy $s \leq t$. Define $d_{\overline{P}}$ in the same way, and let

$$d_{P\overline{P}} = \min \left(d_P, d_{\overline{P}} \right).$$

Since u_t is a bounded solution of the heat equation outside the free boundary, the interior Schauder estimates give

$$(9.21) \qquad |u_t(P) - u_t(\bar{P})| \leq C(d(P, \bar{P}))^{1/2}$$

if $d(P, \bar{P}) \leq (d_{P\bar{P}})^2$. Suppose next that

$$(9.22) \qquad (d_{P\bar{P}})^2 < d(P, \bar{P}).$$

In view of (9.20),

$$(9.23) \qquad u_t(\bar{P}) \leq C|\log d_{\bar{P}}|^{-\varepsilon},$$

$$(9.24) \qquad u_t(P) \leq C|\log d_P|^{-\varepsilon}.$$

Suppose for definiteness that $d_{P\bar{P}} = d_{\bar{P}}$. Then

$$d_{\bar{P}} < d(P, \bar{P})^{1/2}$$

by (9.22), and

$$d_P \leq d(P, \bar{P}) + d_{\bar{P}} \leq C(d(P, \bar{P}))^{1/2}.$$

Using this in (9.23), (9.24), we find that

$$|u_t(P) - u_t(\bar{P})| \leq C|\log d(P, \bar{P})|^{-\varepsilon}.$$

Recalling (9.21), the proof of the theorem is complete. [Notice that ε can be taken to be any positive number smaller than $2/(n - 2)$.]

Proof of Theorem 9.1 in case $n = 2$. We modify Lemma 9.4 taking

$$g(x, 0) = \log \frac{1}{|x|}.$$

Instead of (9.13) we now get

$$f_\theta(x, t) \geq C \log \frac{1}{\theta}.$$

We change (9.16) into

$$w^*(x, t) \leq \frac{C}{2^{2k^\varepsilon}} \qquad \text{if } |x| < \frac{1}{4^k 2^{k^\varepsilon}}.$$

This enables us to prove the inequality

$$m_{k+1} \leqslant M_k \left(1 - \frac{C}{k^\varepsilon} \right)$$

provided that

$$M_k = \max \left\{ m_k, 2^{-2k^{\varepsilon'}} \right\}, \qquad \varepsilon' < \varepsilon,$$

where m_k is defined as before.

We can next show, by induction, that for any $0 < \delta < \varepsilon'$,

$$m_k \leqslant C2^{-k^\delta}$$

provided that $\varepsilon < 1 - \delta$. Thus we obtain

$$u_t(x, t) \leqslant Ce^{-k^\gamma} \qquad \text{if } |x| \leqslant 4^{-k}, \quad -2\lambda_0 4^{-2k} \leqslant t \leqslant 0,$$

where γ can be taken to by any positive number $< \frac{1}{2}$.

The rest of the proof is the same as in the case $n \geqslant 3$.

We proceed to study the free boundary. The following notation will be used:

$$N = \text{the noncoincidence set,}$$

$$N_t = N(t) = \{x : (x, t) \in N\},$$

$$\Lambda = \text{the coincidence set,}$$

$$\Lambda_t = \{x; (x, t) \in \Lambda\},$$

$$\Gamma = \text{the free boundary,}$$

$$\Gamma_t = \{x; (x, t) \in \Gamma\}.$$

The function u satisfies $Hu = k$ in a neighborhood of any free boundary point (x_0, t_0) $(t_0 > 0)$; for simplicity we take $k = 1$.

Since $\Delta u = 1 + u_t \geqslant 0$, Lemma 3.1 gives

$$(9.25) \qquad \text{if } (x, t) \in N \cup \Gamma \qquad \text{then} \qquad \sup_{x' \in N_t \cap \partial B_r(x)} u(x', t) \geqslant \frac{r^2}{2n}.$$

Theorem 9.6. *The Lebesgue measure of Γ_t is equal to zero.*

The proof is similar to the proof in the elliptic case (Theorem 3.5); it utilizes (9.25).

We shall need the following Harnack inequality.

Lemma 9.7. *Let u be a positive caloric function in a cylinder $Z_\gamma = \{|x| < 1, 0 < t < \gamma\}$ and let $(y, t) \in Z_\gamma$, $|y| = 1 - \delta$, $t = \delta^2$, where $0 < \delta < \min(\frac{1}{2}, \gamma/2)$. Then*

$$u(0, \gamma) \geq c\delta^{n+2}u(y, t) \qquad (c > 0),$$

where c is a constant depending only on n, γ.

For proof, see Problem 1.

We return to the variational inequality for the Stefan problem.

Lemma 9.8. *For any pure second spatial derivative u_{ii},*

$$(9.26) \qquad u_{ii}(x, t) \geq -C|\log d(x, t)|^{-\varepsilon} \qquad \left[\varepsilon = \frac{1}{2(n-1)}\right],$$

where $(x, t) \in N$, $d(x, t) = $ parabolic distance to the parabolic part of the free boundary.

Proof. The proof is similar to the proof of Theorem 3.6. We denote by $-M_k$ the infimum of u_{ii} in a cylinder

$$Z_k = \{|x| < 2^{-k}, 0 < t < C2^{-2k}\}$$

[we take $(0, 0) \in \Gamma$] and replace the ball $B_s(x)$ in the proof of Theorem 3.6 by a cylinder about (x, t):

$$\tilde{Z}_s(x, t) = B_s(x) \times (t - Cs^2, t);$$

the cylinder is contained in N and its parabolic boundary contains a free boundary point (y_0, t_0). We now define y_1 as before; that is, y_1 is obtained by going a distance δs from y_0 into the center x of $B_s(x)$ [notice that $u(y_1, t_0) \leq C(\delta s)^2$]. Next we choose $t_1 \geq t_0$ such that $t_1 \geq (\delta s)^2$ and $t_1 \leq t_0 + (\delta s)^2$ [then $u(y_1, t_1) \leq C(\delta s)^2$].

Proceeding as in the proof of Theorem 3.6, we derive

$$u_{ii}(\tilde{y}, t_1) \geq -C\delta^{1/2}$$

for some $\tilde{y}$ inside $B_{s(1-\delta)}(x)$. We now apply Harnack's inequality (Lemma 9.7) and continue as in the proof of Theorem 3.6.

Let (x_0, t_0) be a free-boundary point and define

$$\delta_r(\Lambda) = \frac{\mathrm{MD}(\Lambda_{t_0} \cap B_r(x_0))}{r}.$$

We then have the analog of Theorem 3.10:

Theorem 9.9.   *Let (x_0, t_0) be a free-boundary point; $t_0 > 0$. Then there exists a positive nondecreasing function $\sigma(r)$ $(0 < r < r_0)$ with $\sigma(0 +) = 0$ such that if, for some $0 < r < r_0$,*

$$\text{(9.27)} \qquad\qquad\qquad \delta_r(\Lambda) > \sigma(r),$$

then there exists a neighborhood V of (x_0, t_0) such that $V \cap \Gamma$ can be represented in the form

$$x_i = k(x_1, \ldots, x_{i-1}, x_{i+1}, \ldots, x_n, t)$$

with $k \in C^1$, and all the second derivatives of u $(D_x^2 u, D_x D_t u, D_t^2 u)$ are continuous in $(N \cup \Gamma) \cap V$.

From the results of Section 1 it then follows that $k \in C^\infty$ and u is in $C^\infty((N \cup \Gamma) \cap V)$.

Corollary 9.10.   *If*

$$\text{(9.28)} \qquad\qquad \limsup_{r \to 0} \frac{|\Lambda_{t_0} \cap B_r(x_0)|}{B_r} > 0,$$

then the assertions of Theorem 9.9 are valid.

The rest of this section is devoted to the proof of Theorem 9.9. We begin with a simple observation regarding the rate of growth of N_t:

$$\text{(9.29)} \qquad \begin{array}{c} \text{if } B_\rho(x_1) \subset \Lambda_{t_1}, \text{ then } B_{\rho - Ch} \subset \Lambda_{t_1 + h^2} \text{ provided that} \\ C \text{ is a sufficiently large positive constant.} \end{array}$$

Indeed, if $y_0 \in B_{\rho - Ch}(x_1) \cap \overline{N}_{t_1 + h^2}$, then, by (9.25),

$$\sup_{B_\rho(x_1)} u(x, t_1 + h^2) \geqslant cC^2 h^2 \qquad (c > 0).$$

Using the boundedness of u_t, we get

$$\sup_{B_\rho(x_1)} u(x, t_1) > 0$$

if C is large enough, a contradiction.

We take for simplicity (x_0, t_0) as the origin.

For each t we may consider u as a solution of the elliptic variational inequality

$$-\Delta u \geqslant f \qquad (f = 1 + u_t),$$

$$u \geqslant 0,$$

$$u(-\Delta u - f) = 0.$$

Since u_t is continuous, we can apply the results of Sections 4 and 5. [Note that the Hölder continuity of f is required in the proof of (3.8); in Sections 4 and 5 only the continuity of f is required.] We conclude that there exists a system of coordinates $(x_1, \ldots, x_n)$ and a neighborhood V of $(x_0, t_0) = (0, 0)$, such that $\Gamma_{t_0} \cap V$ can be represented in the form

$$(9.30) \qquad x_n = g(x_1, \ldots, x_{n-1}, t_0)$$

with $D_{x_i} g$ continuous, and $D_x^2 u(x, t_0)$ continuous in $(N_{t_0} \cup \Gamma_{t_0}) \cap V$.

From (9.29) we then see that the condition (9.27) remains valid also at any point $(x_1, t_1) \in \Gamma$ near (x_0, t_0), with $\sigma(r)$ replaced by $c\sigma(r)$ with sufficiently small constant $c > 0$. Thus for any t_1 near t_0 there is a system of coordinates and a representation (9.30) (with t_0 replaced by t_1) of $\Gamma_{t_1} \cap V$. The property (9.29) shows that one can actually use the same system of coordinates $(x_1, \ldots, x_n)$ for all t_1 near t_0; furthermore, g is in $C^{1/2}$ in t. We summarize:

Lemma 9.11. *There exists a system of coordinates $(x_1, \ldots, x_n)$ and a neighborhood V of $(x_0, t_0) = (0, 0)$ such that $\Gamma \cap V$ is given by*

$$(9.31) \qquad x_n = g(x_1, \ldots, x_{n-1}, t),$$

where $D_{x_i} g$ is continuous in $(x_1, \ldots, x_{n-1})$, uniformly with respect to t, g is $C^{1/2}$ in t, and $N \cap V$ is given by

$$x_n < g(x_1, \ldots, x_{n-1}, t);$$

finally, $D_x^2 u$ is continuous in x, uniformly with respect to t, in $(N \cup \Gamma) \cap V$.

We shall henceforth assume (as we may) that

$$g_{x_i}(x_0, t_0) = 0 \qquad \text{for } 1 \leqslant i \leqslant n - 1.$$

We shall write

$$x' = (x_1, \ldots, x_{n-1}), \qquad y = x_n.$$

Lemma 9.12. *Let v be a continuous subcaloric function in a cylinder $Z = \{|x| < 1, 0 < t < \gamma\}$ and suppose that $v \leqslant M$ in Z, and $v \leqslant \mu$ in a subcylinder*

$Z_0 = \{x \in G_0, 0 < t < \gamma\}$ with meas$(G_0) \geq \theta > 0$, and $\mu = M/2$, $M > 0$. Then there exists a number λ, $0 < \lambda < 1$, such that

$$v(0, \gamma) \leq \lambda M;$$

λ depends only on γ, θ.

Indeed, if $\bar{v}$ is caloric in Z and $\bar{v} = v$ on the parabolic boundary, then we represent $\bar{v}$ by Green's function in a cylinder $\{|x| < \rho, 0 < t < \gamma\}$ and choose ρ such that

$$\rho_0 < \rho < 1, \qquad \text{meas}\left(\partial B_\rho \cap G_0\right) \geq c_0,$$

where c_0 and ρ_0 are positive numbers depending only on θ.

Using Lemmas 9.11 and 9.12, we can now show that

$$(9.32) \qquad u_t \leq C d^\delta(x, \Gamma_t) \qquad \text{for some } \delta > 0,$$

where $d(x, A) = \text{dist}(x, A)$.

Indeed, the proof is similar to the proof of Lemma 5.3 with $D_{ii}u$ replaced by u_t and $h(x)$ replaced by a caloric function $h(x, t)$; Lemma 9.12 is applied in order to get an upper bound for $-h$, smaller than 1.

Since $g \in C^{1/2}$ in t, (9.32) gives

$$(9.33) \qquad u_t(x, t) \leq C d^\delta((x, t), \Gamma),$$

where $d((x, t), \Gamma)$ is the parabolic distance from (x, t) to the parabolic part of the free boundary.

From (9.33) and Schauder's interior estimates we get [see the argument following (9.21)],

$$(9.34) \qquad \begin{array}{c} u_t \text{ is Hölder continuous in } (x, t) \text{ in} \\ (N \cup \Gamma) \cap V, \text{ with some exponent } \delta > 0. \end{array}$$

Since $u_{yy} \to 1$ as $(x, t) \to (x_0, t_0)$, we have that $u_{yy} \geq \frac{1}{2}$ in $(N \cup \Gamma) \cap V$, if V is small enough; consequently,

$$(9.35) \quad -u_y(x', g(x', t) - \eta, t) \geq \frac{1}{2}\eta \qquad \text{if } \eta \text{ is small enough, } \eta > 0.$$

The next lemma will enable us to estimate $D_x D_t u$ in terms of $D_x^2 u$ and $D_t^2 u$ in terms of $D_x D_t u$.

Lemma 9.13. *There exist positive constants C_1, C_2 such that*

$$(9.36) \qquad u_t(x', y, t) \leq -C_1 u_y(x', y, t) + C_2 d^2((x', y), \Gamma_t)$$

in $N \cap V$.

Proof. Introduce the cylinders

$$(9.37) \qquad Z_{\rho, \varepsilon}(x_1, t_1) = \{(x, s); |x - x_1| < \rho, t_1 - \varepsilon < s < t_1\}$$

for small ρ, ε. For any $x_1 = (x_1', y_1) \in \Gamma_{t_1}$ we compare in $N \cap Z_{\rho, \varepsilon}(x_1, t_1)$ the caloric functions u_t and $w = -Au_y - u + [|x - x_1|^2 + (t_1 - t)]/(2n + 1)$. It is easily seen that $w > 0$ on the parabolic boundary [except for (x_1, t_1)] provided that A is large enough, and that $Cw \geqslant u_t$ on the parabolic boundary provided that C is sufficiently large; here (9.35) has been used. Applying the maximum principle, we obtain

$$Cw \geqslant u_t \qquad \text{in } N \cap Z_{\rho, \varepsilon}(x_1, t_1).$$

Since (x_1, t_1) can be any point of the free boundary, (9.36) follows.

Lemma 9.14. *The derivatives $D_x D_t u$, $D_t^2 u$ are bounded in $N \cap V$.*

Proof. Consider mollifiers in space

$$(J_\varepsilon v)(x, t) = \frac{1}{\varepsilon^n} \int \rho\left(\frac{x - y}{\varepsilon}\right) v(y, t) \, dy$$

where $\rho \in C^\infty$, $\rho \geqslant 0$, $\rho(x) = 0$ if $|x| > 1$ and $\int \rho(x) \, dx = 1$. We shall estimate $D_x D_t u$ by the Bernstein method. For this we consider the function

$$(9.38) \qquad z(x, t) = \zeta^2(x, t) |\nabla_x(J_\varepsilon u_t)|^2 + A(J_\varepsilon u_t)^2 \qquad (A > 0),$$

where $\zeta \in C_0^\infty$ in a neighborhood V of (x_0, t_0), $0 \leqslant \zeta \leqslant 1$, and $\zeta = 1$ in a smaller neighborhood V_0 of (x_0, t_0). Denote by D_ε the set of all points $(x, t) \in N \cap V$ such that $\text{dist}(x, \Lambda_t) > \varepsilon$. Setting $\theta = J_\varepsilon u_t$, we compute, in D_ε,

$$\Delta z - z_t = 2\zeta^2 \sum_{i, j} \theta_{x_i x_j}^2 + 8 \sum \zeta \zeta_{x_j} \theta_{x_i} \theta_{x_i x_j} + 2\left\{A + \sum \left(\zeta \zeta_{x_j}\right)_{x_j} - \zeta \zeta_t\right\} \sum_i \theta_{x_i}^2$$

$$\geqslant 2\zeta^2(1 - \varepsilon) \sum \theta_{x_i x_j}^2 + 2\left\{A - \frac{2}{\varepsilon} |\nabla \zeta|^2 + \sum \left(\zeta \zeta_{x_j}\right)_{x_j} - \zeta \zeta_t\right\} > 0$$

for $0 < \varepsilon < 1$, provided that A is large enough. Thus z cannot take a maximum in the interior.

From Lemma 9.13 it follows that if $(x, t) \in D_\varepsilon$, $\text{dist}(x, \Gamma_t) < 2\varepsilon$, then $0 \leqslant u_t \leqslant C$. Hence $0 < \theta \leqslant C\varepsilon$ and also, as easily computed

$$|\theta_{x_j}| \leqslant C$$

if $(x, t) \in \partial D_\varepsilon$, $\text{dist}(x, \Gamma_t) = \varepsilon$. It follows that $z \leqslant C$ on ∂D_ε; consequently,

$z \leqslant C$ in D_ε and, in particular,

$$| \nabla_x J_\varepsilon(u_t) | \leqslant C.$$

Taking $\varepsilon \to 0$, we obtain

$$| D_x D_t u | \leqslant C$$

in some N-neighborhood of (x_0, t_0).

Next we introduce a mollifier J_ε in time, denote it by $\tilde{J}_\varepsilon$, and the function

$$\tilde{z}(x, t) = \zeta^2(x, t) | D_t(\tilde{J}_\varepsilon u_t) |^2 + A | \nabla_x(\tilde{J}_\varepsilon u_t) |^2 \qquad (A > 0)$$

in a set $\tilde{D}_\varepsilon$ whose points (x, t) are characterized by: $(x, t - \varepsilon)$ belongs to some N-neighborhood of (x_0, t_0); here the function ζ is similar to the function ζ used in (9.38), but it has a smaller support.

Setting $\tilde{\theta} = \tilde{J}_\varepsilon u_t$, we compute

$$\Delta \tilde{z} - \tilde{z}_t = 2\zeta^2 \sum_j \tilde{\theta}_{tx_j}^2 + 8 \sum_j \zeta \zeta_{x_j} \tilde{\theta}_t \tilde{\theta}_{tx_j} - 2\zeta \zeta_t \tilde{\theta}_t^2$$

$$+ 2 \sum_j \left(\zeta \zeta_{x_j} \right)_{x_j} \tilde{\theta}_t^2 + 2A \sum_{i,j} \tilde{\theta}_{x_i x_j}.$$

Using the relation $\tilde{\theta}_t = \Delta \tilde{\theta}$ and the boundedness of $\tilde{\theta}_{tx_i}$, we find that $\Delta \tilde{z} - \tilde{z}_t > 0$ if A is sufficiently large.

Next, if $(x, t - \gamma \varepsilon) \in \partial D_\varepsilon \cap \Gamma$ for some $0 < \gamma \leqslant 2$, then by Lemma 9.13, the boundedness of u_{yt}, and the fact that $g \in C_t^{1/2}$, we find that

$$0 \leqslant u_t \leqslant C\varepsilon.$$

It follows that $0 \leqslant \tilde{\theta}_t \leqslant C$ on the part of $\partial \tilde{D}_\varepsilon$ that lies near Λ. Hence $\tilde{z} \leqslant C$ on $\partial \tilde{D}_\varepsilon$ and, by the maximum principle, $\tilde{z} \leqslant C$ in $\tilde{D}_\varepsilon$. We conclude that

$$\left| D_t(\tilde{J}_\varepsilon u_t) \right| \leqslant C$$

in $\tilde{D}_\varepsilon \cap V_1$, where V_1 is some neighborhood of (x_0, t_0); taking $\varepsilon \to 0$, we obtain the boundedness of u_{tt}.

Corollary 9.15. *If $B_\rho(x) \subset \Lambda_t$ $[(x, t)$ near (x_0, t_0), ρ small$]$, then, for a suitable positive constant C, $B_{\rho - Ch}(x) \subset \Lambda_{t+h}$ for all small $h > 0$; consequently, g is Lipschitz continuous in t.*

The proof is the same as that of (9.29), but exploits the boundedness of u_{tt}.

Now that Lemma 9.14 has been proved, one can complete the proof of Theorem 9.9 by some modifications in the hodograph–Legendre approach;

this is outlined in Problems 7 to 10. Here we proceed with a method that is in the spirit of the analysis of Sections 4 and 5.

Lemma 9.16. *For any pure second derivative u_{jj} (in space or time),*

$$(9.39) \qquad u_{jj}(x, t) \geqslant -Cd^{\varepsilon}(x, t) \qquad (\text{for some } \varepsilon > 0, C > 0)$$

in $N \cap V$, where $d(x, t)$ is the parabolic distance to the parabolic part of the free boundary.

Proof. Since Γ is Lipschitz (in V) and u_{jj} is a bounded caloric function, the nontangential limits of u_{ii} on Γ exist a.e. and they form a function f_{ii} in $L^{\infty}(\mu)$; μ is the caloric measure (see reference 121). Let $-M = \text{ess sup } f_{ii}$ on Γ_0, where Γ_0 is a fixed open Γ-neighborhood of (x_0, t_0), and let K be a compact subset of Γ_0. Let D be an N-neighborhood of (x_0, t_0) with $\partial D \cap \Gamma = \Gamma_0$.

For any small $h > 0$, let

$$D_h = \{(x, t); (x, t) \in D, (x, t) + \lambda e_j \in D \quad \forall 0 < \lambda < h\},$$

where e_j is the unit vector in the direction j. Consider the functions

$$F_h^1 = \frac{2}{h^2} \int_0^h \int_0^s u_{jj}((x, t) + re_j) \, dr \, ds,$$

$$F_h^2 = \frac{2}{h^2} \int_0^h \int_s^h u_{jj}((x, t) + re_j) \, dr \, ds.$$

Then $HF_h^i = 0$ in D_h and $F_h^i \to v_{jj}$ in D if $h \to 0$. We claim that

$$(9.40) \qquad v_{jj}(x, t) \geqslant -M - C\rho_K^{\varepsilon}(x, t)$$

where ρ_K is the parabolic distance to the parabolic part of the boundary of K. Indeed, the proof is similar to the proof of Lemma 5.3; it uses Lemma 9.7 and the Lipschitz nature of g. Notice that the maximum principle (which is needed in the proof) is still valid when the boundary values are taken in the sense of nontangential limits; in fact, this follows from the unique representation of bounded caloric functions by means of Green's function [121].

From (9.40) we deduce that

$$(9.41) \qquad F_h^i \geqslant -M - C(\rho_K^{\varepsilon} + h^{\varepsilon})$$

with another C.

Let

$$\partial_1 D_h = \{(x, t) \in \partial D_h; (x, t) + \lambda e_j \in K \text{ for some } 0 < \lambda < h\}.$$

If $\lambda \leqslant h/2$, then

$$\int_\lambda^t \int_\lambda^s u_{jj}\big((x,t) + re_j\big)\, dr\, ds = u\big((x,t) + he_j\big) \geqslant 0$$

and consequently,

$$F_h^1 \geqslant -\tfrac{3}{4}M.$$

If $\lambda < h/2$, then $F_h^2 \geqslant -3M/4$. Thus, in either case,

$$F_h^1 + F_h^2 \geqslant -\tfrac{7}{4}M \qquad \text{on } \partial_1 D_h$$

and, by the proof of (9.40) applied to $F_h^1 + F_h^2$,

$$F_h^1 + F_h^2 \geqslant -\tfrac{7}{4}M + C\rho_K^\varepsilon \qquad \text{in } D_h.$$

Taking $h \to 0$, we obtain

$$v_{jj}(x,t) \geqslant -\tfrac{7}{8}M + C\rho_K^\varepsilon(x,t)$$

and, as (x,t) tends to K,

$$\operatorname*{ess\ sup}_K f_{jj} \geqslant -\tfrac{7}{8}M.$$

Taking $K \uparrow \Gamma_0$, we get $-M \geqslant -7M/8$; hence $M \leqslant 0$.

We can now apply the proof of (9.40) once again to deduce (9.39).

Lemma 9.17. *Let η be any unit vector in the space of variables $(x',t) = (x_1,\ldots,x_{n-1},t)$ and set $X_\lambda = \lambda\eta$, $\lambda > 0$. Then*

$$(9.42) \qquad g_\nu(0) \equiv \lim_{\lambda \to 0} \frac{g(X_\lambda) - g(0)}{\lambda} \quad \text{exists}$$

and, for some $\varepsilon > 0$, $C > 0$ (independent of η),

$$(9.43) \qquad \frac{g(X_\lambda) - g(0)}{\lambda} - g_\eta(0) \geqslant -C\lambda^\varepsilon.$$

Proof. Set

$$\bar{g}_\eta(0) = \limsup_{\lambda \to 0} \frac{g(X_\lambda) - g(0)}{\lambda}.$$

If $0 < \lambda_1 < \lambda_2/2$, then

(9.44)
$$\frac{g(X_{\lambda_1}) - g(0)}{\lambda_1} < \frac{g(X_{\lambda_2}) - g(0)}{\lambda_2} + C\lambda_2^\varepsilon.$$

Indeed, consider the ball $[X_{\lambda_2} = (x'_{\lambda_2}, t_{\lambda_2})]$

$$B = B_{C''\lambda_2^{1+\varepsilon}}\left(x'_{\lambda_2}, g(X_{\lambda_2}) + C'\lambda_2^{1+\varepsilon}, t_{\lambda_2}\right).$$

Since g is Lipschitz, if C'/C'' is large enough, then $B \subset \Lambda$. Consequently, if (9.44) is not valid for some large enough C, then we can apply the caloric analog of Lemma 5.5 to deduce the following: There is an interval analogous to $\{sy\}$, contained in N, which either intersects B or passes above it; in either case we get a contradiction. Here we point out that Lemma 5.5 is based on Lemma 5.3 (whose caloric analog is Lemma 9.16) and it extends to the parabolic case with minor changes.

Having proved (9.44), we take a sequence $\lambda_1 = \lambda_{1j} \to 0$ such that

$$\frac{g(X_{\lambda_1}) - g(0)}{\lambda_1} \to \bar{g}_\eta(0)$$

and obtain from (9.44) that

$$\liminf_{\lambda_2 \to 0} \frac{g(X_{\lambda_2}) - g(0)}{\lambda_2} \geq \bar{g}_\eta(0).$$

It follows that the limit (9.42) exists. Finally, (9.43) follows from (9.44) by taking $\lambda_1 \to 0$.

Denote by $C(x_0, t_0)$ the cone generated by all the tangents $g_\eta(0)$, which opens into N. Similarly, one defines the cone $C(x, t)$ for any $(x, t) \in \Gamma \cap V$.

Lemma 9.18. $C(x, t)$ *is a convex cone; since $g \in C_x^1$, the cone consists of two hyperplanes.*

Proof. It suffices to prove that $C(x_0, t_0)$ is convex. Introduce directions η_1, η_2 in the (x', t) space and the tangent rays γ_j in the direction $g_{\eta_j}(0)$. Let $(x'_j, g(x'_j, t_j), t_j)$ be a point of Γ, with distance λ from (x_0, t_0), and lying within distance $o(\lambda)$ of γ_j (as $\lambda \downarrow 0$). We apply Lemma 9.17 with (x_0, t_0) replaced by $(x'_1, g(x'_1, t_1), t_1)$ and η the direction from (x_1, t_1) to (x_2, t_2). Denoting by $(\tilde{x}_1, \tilde{t}_1)$ a point in the interval from (x_1, t_1) to (x_2, t_2) which divides it by ratio $\theta : (1 - \theta)$, (9.43) yields

$$u\left(\tilde{x}_1, g(\tilde{x}_1, \tilde{t}_1) - C\lambda^{1+\varepsilon}, \tilde{t}_1\right) > 0.$$

Taking $\lambda \to 0$, we find that

$$\theta g_{\eta_1}(0) + (1 - \theta)g_{\eta_2}(0) \in C(x_0, t_0),$$

and the proof is complete.

Proof of Theorem 9.9.   We begin by proving that the tangent cone $C(x_0, t_0)$ is a hyperplane:

$$(9.45) \quad C(x_0, t_0) = \{y = g(x_0', t_0) + A_0(x' - x_0') + A_1(t - t_0)\}.$$

We already know that $C(x_0, t_0)$ consists of two hyperplanes

$$y = a(t - t_0) + g(x_0', t_0), \qquad t \geqslant t_0,$$

$$y = b(t - t_0) + g(x_0', t_0), \qquad t < t_0$$

with $b \leqslant a$. We have to show that $a = b$.

Recall that

$$(9.46)$$
$$|\nabla g(x_1', t) - \nabla g(x_2', t)| \leqslant \sigma(|x_1' - x_2'|),$$
$$|u_{ij}(x_1, t) - u_{ij}(x_2, t)| \leqslant \sigma(|x_1 - x_2|),$$

where u_{ij} are any second-order spatial derivatives and $\sigma(\lambda) \downarrow 0$ if $\lambda \downarrow 0$.

Take three points on a vertical segment,

$$A = (x_0', g(x_0', t_0) - \alpha, t_0),$$

$$B = (x_0', g(x_0', t_0) - \alpha, t_0 + \beta),$$

$$C = (x_0', g(x_0', t_0) - \alpha, t_0 - \beta) \qquad (\alpha > 0, \beta > 0).$$

Since g is Lipschitz, if $\beta = o(\alpha)$, then all the points lie in N. Since $u_{yy}(x_0, t_0) = 1$, we have, using (9.46),

$$(9.47) \qquad\qquad |u(A) - \tfrac{1}{2}\alpha^2| \leqslant \sigma(\alpha)\alpha^2.$$

By Lemma 9.17 and the Lipschitz character of g,

$$g(x_0', t_0 + \beta) = g(x_0', t_0) + a\beta + \varepsilon_1(\beta),$$

$$(9.48) \qquad g(x_0', t_0 - \beta) = g(x_0', t_0) - b\beta + \varepsilon_2(\beta),$$

$$-C\beta^{1+\varepsilon} \leqslant \varepsilon_i(\beta) \leqslant C\beta.$$

Notice next that

$$(9.49) \qquad u_{yy}\big(x_0', g(x_0', t_0 + \beta), t_0 + \beta\big) = (\cos\theta)^2$$

where θ is the angle between the y-axis and the spatial normal to g at $(x_0', g(x_0', t_0 + \beta), t_0 + \beta)$. To estimate θ we use the fact that g is increasing in t:

$$0 \leqslant g(x', t_0 + \beta) - g(x', t_0)$$

$$= g(x_0', t_0 + \beta) - g(x_0', t_0) + g_{x'}(x_0', t_0 + \beta) \cdot (x' - x_0')$$

$$+ \sigma(|x' - x_0'|)|x' - x_0'| \, .$$

For $x' - x_0'$ in the direction of $\pm\nabla_{x'} g(x_0', t_0 + \beta)$, we have

$$g_{x'}(x_0', t_0 + \beta) \cdot (x' - x_0') = \pm\tan\theta \, |x' - x_0'| \, .$$

Taking $|x' - x_0'| = \beta^{1/2}$, we obtain, after using (9.46),

$$|\sin\theta| \leqslant \sigma(\beta^{1/2}) + \frac{a\beta + \varepsilon_1(\beta)}{\beta^{1/2}} \leqslant \sigma(\beta^{1/2}) + C\beta^{1/2}.$$

We now choose

$$(9.50) \qquad \beta = \sigma^{1/2}\big(\sqrt{\alpha}\,\big)\alpha.$$

Then we obtain

$$|\sin\theta| \leqslant C\sigma^{1/4}\big(\sqrt{\alpha}\,\big)\sqrt{\alpha} + \sigma\big(\sqrt{\beta}\,\big).$$

It follows from (9.49) that

$$u_{yy}\big(x_0', g(x_0', t_0 + \beta), t_0 + \beta\big) \geqslant 1 - C\sigma^{1/2}\big(\sqrt{\alpha}\,\big)\alpha + \sigma^2\big(\sqrt{\beta}\,\big).$$

Recalling (9.46), (9.48), we find that

$$u(B) - \tfrac{1}{2}\big(a\beta + \varepsilon_1(\beta) + \alpha\big)^2 \geqslant -C\alpha^2\Big[\sigma(M\alpha) + \sigma^2\big(\sqrt{\alpha}\,\big) + \sigma^{1/2}\big(\sqrt{\alpha}\,\big)\alpha\Big]$$

for some constant $M > 0$. Similarly,

$$u(C) - \tfrac{1}{2}\big(-b\beta + \varepsilon_2(\beta) + \alpha\big)^2 \geqslant -C\alpha^2\Big[\sigma(M\alpha) + \sigma^2\big(\sqrt{\alpha}\,\big) + \sigma^{1/2}\big(\sqrt{\alpha}\,\big)\alpha\Big].$$

Thus

$$(9.51) \quad \begin{aligned} I \equiv u(B) + u(C) - 2u(A) &\geq (a - b)\alpha^2\left(\sigma^{1/2}\left(\sqrt{\alpha}\right)\right) \\ &+ (\varepsilon_1(\beta) + \varepsilon_2(\beta))\alpha - C\alpha^2\sigma\left(\sqrt{\alpha}\right) - C\alpha^3\sigma^{1/2}\left(\sqrt{\alpha}\right). \end{aligned}$$

On the other hand,

$$I = (u(B) - u(A)) + (u(C) - u(A)) \leq \left(\sup_{J_1} u_t - \inf_{J_2} u_t\right)\sigma^{1/2}\left(\sqrt{\alpha}\right)\alpha$$

where $J_1 = \overline{AB}$, $J_2 = \overline{AC}$. Since u_{tt} is bounded, we obtain

$$I \leq C\sigma\left(\sqrt{\alpha}\right)\alpha^2,$$

which contradicts (9.51), unless $a = b$. We have thus completed the proof of (9.45).

The convergence in (9.42) is uniform with respect to both the choice of the free-boundary point (x_0, t_0) and the direction η. In view of (9.45) we conclude that

$$g(x', t') - g(x_0', t_0) = (x' - x_0')g_{x'}(x_0', t_0) + (t' - t_0)g_t(x_0', t_0)$$
$$+ \gamma(|x' - x_0'| + |t' - t_0|),$$

where $\gamma(\lambda) \to 0$ if $\lambda \to 0$. It follows that $g \in C^1$ in (x', t).

We next prove that

$$(9.52) \quad u_{ij} \to 0 \quad \text{if } (x, t) \to (x_0, t_0)$$

for any second-order derivative (in space and time) such that the direction i is tangential to the free boundary at (x_0, t_0).

We form the difference quotients

$$F_h = \frac{1}{h}\left(u_j((x, t) + he_i) - u_j(x, t)\right)$$

in the open set

$$D_h = \{(x, t) \in N \cap V, (x, t) + he_i \in N \cap V\}.$$

Since $g \in C^1$,

$$\frac{\partial}{\partial e_i} g \to 0 \quad \text{if } (x', t) \to (x_0', t_0).$$

Noting that $u_j(x', g(x', t), t) = 0$ and that $|D^2 u| \leqslant C$, we get

$$F_h = \varepsilon(\delta)$$

in a δ-neighborhood of (x_0, t_0) intersected with D_h, where $\varepsilon(\delta) \to 0$ if $\delta \to 0$. We also have

$$|F_h| \leqslant C \qquad \text{in } D_h.$$

We now represent F_h by means of Green's function (by reference 121, for instance) and conclude that

$$|F_h| \leqslant 2\varepsilon(\delta)$$

in a δ'-neighborhood of (x_0, t_0) intersected with D_h, where δ' is small enough. Taking $h \to 0$, we get

$$|u_{ij}| \leqslant 2\varepsilon(\delta)$$

in this neighborhood, and (9.52) follows.

It remains to prove that $\lim u_{\nu\nu}$ exists, where ν is the normal to the free boundary at (x_0, t_0). In view of (9.52), it suffices to prove this when ν is just one particular nontangential direction. But for ν in the direction of the y-axis the limit of $u_{\nu\nu}$ exists.

PROBLEMS

1. Prove Lemma 9.7.

    [*Hint*: Representing u by Green's function $G_{x,t}(y, s)$ of the cylinder $\{|y| < 1\}$ with pole at (x, t), we have to (i) estimate $\dfrac{\partial}{\partial\nu} G_{0,0}$ (ν inward normal) from below and $(\partial/\partial\nu) G_{y, \gamma-t}$ from above on the lateral boundary $\partial_l Z_\gamma$ of Z_γ, and (ii) estimate $G_{0,0}$ from below and $G_{y, \gamma-t}$ from above on the top $\partial_T Z_\gamma$ of Z_γ. With $K(|x|, t) = K(x, t)$ the fundamental solution of the heat equation,

    $$K(|x|, t) - K(1, t_0) \leqslant G_{0,0}(|x|, t) \qquad (|x| < 1, t < t_0)$$

    since $K_t(1, t) > 0$ if $\gamma < 2/n$. Hence

    $$\frac{\partial}{\partial\nu} G_{0,0}(1, t_0) \geqslant \frac{\partial}{\partial\nu} K(1, t_0) = \frac{C}{t_0^{(n+2)/2}} e^{-1/4t_0} \qquad (C > 0),$$

    $$G_{0,0}(|x|, \gamma) \geqslant C\big(e^{-|x|^2/4\gamma} - e^{-1/4\gamma}\big) \qquad (C > 0).$$

To estimate $\partial G_{y,t}(e_1, s)/\partial \nu$ $(e_1 = (1, 0, \ldots, 0))$ for $s \geqslant t$, let Σ be the half-space tangent to $\partial_t Z_\gamma$ along e_1 and $G^*_{y,t}(x, s)$ the Green function for Σ. Since $G^*_{y,t} \geqslant G_{y,t}$,

$$\frac{\partial}{\partial \nu} G_{y,t}(e_1, s) \leqslant \frac{C y_1}{(s-t)^{(n+2)/2}} \exp\left[-\frac{|e_1 - y|^2}{4(s-t)} \right]$$

$$\leqslant \frac{C}{\delta^{n+1}} \qquad \text{since } |e_1 - y| > \delta, \, s - t < \delta^2.$$

On $\partial_T Z_\gamma$, if $y_n = \text{dist}(y, \partial\Sigma)$, $x_n = \text{dist}(x, \partial\Sigma)$,

$$G_{y,t}(x, \gamma) \leqslant \frac{C}{\delta^n}\left[\exp\left(\frac{x_n y_n}{2\delta^2}\right) - \exp\left(-\frac{x_n y_n}{2\delta^2}\right) \right]$$

$$\leqslant \frac{C}{\delta^n}\frac{x_n y_n}{\delta^2} \leqslant \frac{C}{\delta^{n+2}}(1 - |x|);$$

finally, if $\gamma \geqslant 2/n$, repeat the above several times.]

2. Suppose that Γ_0 is the boundary of a ball B_{r_0} and write Δw in polar coordinates,

$$\Delta w = r^{1-n}(r^{n-1} w_r)_r + r^{-2} \Lambda w.$$

Assume that

$$g(x, t) - g(x, 0) - r_0^{-2}\Lambda\left(\int_0^t g(x, s)\, ds \right) > 0 \qquad \text{on } \Gamma_0 \times [0, T],$$

$$(r^2 h(x))_r < 0 \qquad \text{in } \overline{G}.$$

[The last condition implies (since $h > 0$ in G) that $\overline{G}_0 \cup G$ is star-shaped with respect to any point near the origin.] Let $v = r^{n-1} u_r$. Prove that:

(a) $\qquad -\Delta v + \dfrac{2(n-2)}{r} v_r + v_t = r^{n-3}(r^2 f)_r - 2r^{n-2} u_t$

in the distribution sense in N.

(b) $\qquad\qquad\qquad\qquad v < 0 \qquad \text{in } N$.

[Hint: Take $w = e^{-\alpha t} v$ and a test function $\zeta = e^{-\alpha t} \max(w - M, 0)$, where $M = \max w$.];

(c) $N(t)$ is star-shaped with respect to any point near the origin; therefore, Γ_t is Lipschitz;

(d) N is given locally by $x_i = g(x_1, \ldots, x_{i-1}, x_{i+1}, \ldots, x_n, t)$ with $g \in C^\infty$.

3. The noncoincidence set for the Stefan problem is given by $t < \phi(x)$, where ϕ is Lipschitz continuous.

    [*Hint*: Use the method of proof of Theorem 6.1 to deduce that $Au_t + \Sigma c_i u_{x_i} - u$ is positive if $\Sigma c_i^2 \leqslant 1$ and A is sufficiently large.]

4. Consider the one-dimensional Stefan problem:

$$u_t - u_{xx} = 0 \qquad \text{if } 0 < x < s(t), t > 0,$$

$$u_x(0, t) = f(t) \qquad \text{if } t > 0 [f(t) \leqslant 0],$$

$$u(x, 0) = h(x) \qquad \text{if } 0 < x < b [b > 0, \quad h(x) > 0],$$

$$u(s(t), t) = 0 \qquad \text{if } t > 0,$$

$$u_x(s(t), t) = -\frac{ds}{dt} \qquad \text{if } t > 0,$$

where $f(0) = h'(0)$, $h(b) = 0$, and f, h are continuous. Perform a transformation $\tilde{u}(x, t) = \int_{s(t)}^{x} u(x, \tau) \, d\tau$ to reduce it to a variational inequality with Neumann condition at $x = 0$. Prove that there exists a unique solution with $s'(t) > 0$, $s \in C^\infty$.

5. Suppose that in Problem 4

$$h'(x) < 0, \qquad h''(x) > 0, \qquad \frac{h''(x)}{h'(x)} \text{ is monotone increasing,}$$

and $f \equiv 0$. Prove that $s'(t)$ is strictly monotone increasing.

    [*Hint*: (a) Show that $u_x < 0$, $u_t > 0$; (b) $z = u_t/u_x$ satisfies a parabolic equation and does not take a local extremum on the free boundary; (c) consider the level curves $\{z = \alpha\}$; they are regular for a.a. α (by Sard's lemma).]

6. Extend the result of Problem 5 to the case where $f \not\equiv 0$ but $f'(t) < 0$.
    [*Hint*: Show that z cannot take minimum on $x = 0$.]

7. Use Lemma 9.14 to deduce that $\Gamma_t \in C^{1+\lambda}$ for any $0 < \lambda < 1$.
    [*Hint*: Let $\xi_\alpha = x_\alpha - x_{\alpha 0}$, $\alpha < n$, $\xi_n = -u_x(x, t)$, $x_0 = (x_{01}, \ldots, x_{0, n}) \in \Gamma_t$, $v(\xi) = x_n \xi_n + u(x, t)$. Then

$$F(D^2 v) \equiv -\frac{1}{v_{\xi_n \xi_n}} - \frac{1}{v_{\xi_n \xi_n}} \sum_{\alpha < n} v_{\xi_\alpha \xi_n}^2 + \sum_{\alpha < n} v_{\xi_\alpha \xi_\alpha} = f(\xi)$$

in $U = \{\xi, \xi_n < 0, |\xi| < \varepsilon\}$, $v = 0$ on $\xi_n = 0$, where $f = 1 + u_t \in C^{0,1}$. The function $w = v_{\xi_\beta}$ $(\beta < n)$ satisfies an elliptic equation in U, $w = 0$ on

$\xi_n = 0$; apply L^p elliptic estimates to deduce that $v \in C^{2+\lambda}$. Finally, Γ_t is given by $x_n = v_{\xi_n}$.]

8. Let G be a domain in R^n and S a $C^{1+\lambda}$ open hypersurface of ∂G. If

$$\Delta w = f \quad \text{in } G, w \in C^2(G) \cap C^0(\overline{G}),$$

$$w = g \quad \text{on } S,$$

where $f \in L^\infty(G)$, $g \in C^{1+\lambda}(S)$, then $w \in C^{1+\lambda}(G \cup S)$.
[*Hint*: Suppose that $f = 0$. Set $y_\alpha = x_\alpha \ (\alpha < n)$, $y_n = x_n - \phi(x')$, where S is given by $x_n = \phi(x')$. Then $z(y) = w - g$ satisfies

$$Lz = -\sum \frac{\partial}{\partial y_\alpha} g_\alpha \quad (y_n < 0), z = 0 \text{ on } y_n = 0.$$

The proof of Lemma 4.2 of Chapter 1 gives $z \in C^\mu$ for some $0 < \mu < 1$ and Theorem 9.2 of reference 3a gives $z \in C^{1+\lambda}$.]

9. By Problems 7 and 8, $u_t \in C^{1+\lambda}((N_t \cup \Gamma_t) \cap U)$, U a neighborhood of x_0, with Hölder coefficients for u_{tx_i}, $u_{x_i x_j}$ which are independent of t. Use this, the continuity of $g(x', t)$, and the monotonicity of g in t to prove that u_{tx_i}, $u_{x_i x_j}$ are continuous in t.

10. Prove that under the setting of Problem 9, Γ is in C^1 in a neighborhood of (x_0, t_0).
[*Hint*: Take finite differences in the relation $u_{x_n}(x', g(x', t), t) = 0$.]

11. Show that, in general, the temperature in the Stefan problem is not Hölder continuous if $n \geq 3$.
[*Hint*: Consider a radially symmetric solution $u(x, t) = \tilde{u}(r, t) \ (r = |x|)$ such that $(0, 1)$ is a free-boundary point and

$$N = \{(x, t); t > \sigma(r)\}, \qquad \Lambda = \{(x, t); t < \sigma(r)\}, \qquad \sigma(0) = 1.$$

By nondegeneracy, $\tilde{u}(r, 1) \geq cr^2 \ (c > 0)$. If u_t is Hölder continuous, then, for some $\varepsilon > 0$,

$$\tilde{u}_t(r, t) \leq Cr^\varepsilon \quad \text{if } 1 - \delta < t < 1 \text{ if } \delta = 0(r).$$

Hence $\tilde{u}(r, t) > 0$ if $\delta = c_0 r^{2-\varepsilon} \ (c_0 > 0)$ and thus $\Lambda \subset \{(x, t); 0 < 1 - t < c_0 r^{2-\varepsilon}\}$. Since u_t is a barrier for N at $(0, 1)$, we get a contradiction to the well-known fact that a barrier for the heat equation in N does not exist at $(0, 1)$ when Λ lies in the parabolic-type region above; see Dvoretzky and Erdös [82] and Spitzer [165] for a probabilistic proof (another proof should follow from a recent Wiener criterion of Evans and Gariepy [86]).

10. STABILITY OF FREE BOUNDARIES

In this section we consider both elliptic and parabolic variational inequalities and establish an estimate on the measure of the set $\{u > \varepsilon\}$ (when the obstacle is $\phi \equiv 0$). As an application we shall prove (a) that the free boundary has finite Hausdorff measure H_{n-1}, and (b) if the free boundary is smooth for a solution u of a variational inequality, then it is also smooth for any other solution u with $|\tilde{u} - u|_{L^\infty}$ small enough.

For simplicity we take the elliptic operator to be the Laplacian. Thus we consider

$$(10.1) \qquad \left.\begin{array}{r} -\Delta u + f \geqslant 0 \\ u \geqslant 0 \\ (-\Delta u + f)u = 0 \end{array}\right\} \qquad \text{a.e. in } \Omega$$

where Ω is a bounded domain in R^n. We assume that

$$(10.2) \qquad f \in C^{0,1}(\overline{\Omega}), \qquad u \in C^{1,1}(\overline{\Omega}),$$

$$(10.3) \qquad f \geqslant \lambda > 0 \qquad \text{in } \Omega.$$

Set

$$M = |u|_{C^{1,1}(\overline{\Omega})}, \qquad F = |f|_{C^{0,1}(\overline{\Omega})}$$

and as usual denote by N, Λ, Γ the noncoincidence set, the coincidence set, and the free boundary, respectively.

For any set $E \subset R^m$ and $\varepsilon > 0$, we write

$$E_{(\varepsilon)} = \{x \in R^m; \operatorname{dist}(x, E) < \varepsilon\},$$

$$E_{(-\varepsilon)} = \{x \in E; \operatorname{dist}(x, R^m \setminus E) > \varepsilon\}.$$

Let

$$(10.4) \qquad O_\varepsilon = \{x \in N; |\nabla u(x)| < \varepsilon\}.$$

We denote by $\mu(K)$ the Lebesgue measure of a set K in R^n and by $A(\partial K)$ the surface area of the boundary ∂K of K.

The following lemma is fundamental for all the results of this section.

Lemma 10.1. *For any compact domain $K \subset \Omega$ with C^1 boundary S,*

$$(10.5) \qquad \mu(O_\varepsilon \cap K) \leqslant C\varepsilon(\mu(K) + A(S))$$

for all $\varepsilon > 0$; C depends only on M, F, λ, and K.

Proof. Let $O_\varepsilon^i = \{x \in N; |D_i u| < \varepsilon\}$ and set

$$h^i = \begin{cases} -\varepsilon & \text{if } D_i u < -\varepsilon \\ D_i u & \text{if } -\varepsilon < D_i u < \varepsilon \\ \varepsilon & \text{if } D_i u > \varepsilon. \end{cases}$$

Then

$$\int_K \nabla h^i \cdot \nabla D_i u \, dx + \int_K h^i \Delta D_i u \, dx = \int_S h^i D_\nu(D_i u) \, dS$$

where D_ν is the derivative along the outward normal. Hence

$$\int_{O_\varepsilon^i \cap K} |\nabla D_i u|^2 \, dx \leqslant \int_K |h^i D_i f| \, dx + \int_S |h^i D_\nu D_i u| \, dS$$

$$\leqslant C\varepsilon(\mu(K) + A(S)).$$

Summing over i and using the fact that $O_\varepsilon^i \supset O_\varepsilon$ for each i, we get

$$\int_{O_\varepsilon \cap K} |\nabla^2 u|^2 \, dx \leqslant C\varepsilon(\mu(K) + A(S)).$$

Since finally

$$|\nabla^2 u|^2 \geqslant \frac{1}{n^2} |\Delta u|^2 \geqslant \frac{\lambda^2}{n^2} \qquad \text{in } N,$$

(10.5) follows.

Corollary 10.2. *Let K be any domain in $\Omega_{(-\varepsilon)}$ with C^1 boundary S. Then*

$$(10.6) \qquad \mu(\{0 < u < \varepsilon^2\} \cap K) \leqslant C\varepsilon(\mu(K) + A(S)).$$

Indeed, by Lemma 3.2,

$$\{0 < u < \varepsilon^2\} \cap K \subset O_{C\varepsilon} \cap K$$

for some constant C, and Lemma 10.1 yields (10.6).

Corollary 10.3. *If $x \in \Gamma \cap \Omega_{(-\delta)} (\delta > 0)$, then, for all $\varepsilon > 0$,*

$$(10.7) \qquad \frac{\mu(B_\varepsilon(x) \cap O_{C\varepsilon})}{\mu(B_\varepsilon)} \geqslant \gamma$$

where C and γ are positive constants depending only on M, F, λ, δ.

Proof. By Lemma 3.1, there exists a point $y \in \overline{B_\varepsilon(x)}$ such that

$$c_1 \varepsilon^2 < u(y) < c_2 \varepsilon^2 \qquad (c_i > 0).$$

By Lemma 3.2, $|\nabla u(y)| \leqslant c_3 \varepsilon$ and therefore there exists a ball $B_{c\varepsilon}(y)$ $(c > 0)$ in which $u > 0$. Since also $u < c_4 \varepsilon^2$ in this ball, we have $|\nabla u| \leqslant C\varepsilon$ (by Lemma 3.2) and thus the ball is contained in $O_{C\varepsilon}$, and (10.7) follows.

Theorem 10.4. *If K is any domain with C^1 boundary S such that $K \cup S \subset \Omega$, then*

$$(10.8) \qquad H_{n-1}(\Gamma \cap K) \leqslant C(\mu(K) + A(S)).$$

Here H_{n-1} is the $(n - 1)$-dimensional Hausdorff measure. We recall that, by definition, for any set $E \cap R^n$,

$$(10.9) \qquad H_{n-k}(E) = \lim_{\varepsilon \to 0\{B_i\}} \inf \varepsilon^{-k} \sum \mu(B_i),$$

where $\{B_i\}$ is any set of the open balls B_i with centers in E and radius ε which cover the set E. One may restrict the coverings $\{B_i\}$ to be such that each point of E is contained only in a finite number N of the balls; N is independent of the covering and of ε.

Proof. Take a covering of $\Gamma \cap K_{(-\varepsilon)}$ by balls B_i with centers in $\Gamma \cap K_{(-\varepsilon)}$ and radius ε, having at most N overlappings at each point. By Corollary 10.3,

$$\sum \mu(B_i) \leqslant \frac{C}{\gamma} \sum \mu(B_i \cap O_{C\varepsilon}) \leqslant \frac{NC}{\gamma} \mu(O_{C\varepsilon} \cap K).$$

Estimating the right-hand side by Lemma 10.1, we get

$$H_{n-1}(\Gamma \cap K_0) \leqslant C$$

for any $K_0 \subset K_{(-\delta)}$ for some $\delta > 0$; C is independent of δ. Taking $K_0 \uparrow K$, (10.8) follows.

In the next theorem we deal with two solutions u_1, u_2 of variational inequalities

$$(10.10) \qquad \left. \begin{aligned} -\Delta u_i + f_i &\geqslant 0 \\ u_i &\geqslant 0 \\ (-\Delta u_i + f_i)u_i &= 0 \end{aligned} \right\} \qquad \text{a.e. in } \Omega,$$

where

$$f_i \in C^{0,1}(\overline{\Omega}), \qquad u_i \in C^{1,1}(\overline{\Omega}),$$

(10.11)

$$f_i \geq \lambda > 0 \qquad \text{in } \Omega.$$

We define

$$N(u_i) = \text{the noncoincidence set for } u_i,$$

$$\Lambda(u_i) = \text{the coincidence set for } u_i,$$

$$\Gamma(u_i) = \text{the free boundary for } u_i.$$

Theorem 10.5. *Let* (10.10), (10.11) *hold. If*

$$|u_1 - u_2|_{L^\infty(\Omega)} < \varepsilon^2,$$

then (i)

(10.12) $\qquad \mu\big((\Lambda(u_1) \Delta \Lambda(u_2)) \cap \Omega_{(-\varepsilon)}\big) \leq C\varepsilon \qquad (C > 0)$

where C depends only on the $C^{0,1}$ norms of the f_i, the $C^{1,1}$ norm of u_i and λ, and (ii)

(10.13) $\qquad (\Lambda(u_2))_{(-C_0\varepsilon)} \subset \Lambda(u_1) \subset \{u_2 < \varepsilon^2\} \qquad (C_0 > 0)$

where C_0 depends only on the $C^{1,1}$ norm of the u_i and on λ.

Proof. Since $\Lambda(u_1) \subset \{u_2 < \varepsilon^2\}$, Corollary 10.2 shows that

$$\mu\big((\Lambda(u_1) \setminus \Lambda(u_2)) \cap \Omega_{(-\varepsilon)}\big) < C\varepsilon$$

and (10.12) follows. To prove the first part of (10.13), notice that if $x \in \overline{N(u_1)}$, then, by Lemma 3.1,

$$\sup_{B_{C_0\varepsilon}(x)} u_1 \geq \frac{\lambda}{2n} C_0^2 \varepsilon^2 > \varepsilon^2$$

if C_0 is large enough, and then

$$\sup_{B_{C_0\varepsilon}(x)} u_2 > 0,$$

which implies that $x \notin (\Lambda(u_2))_{(-C_0\varepsilon)}$.

Theorem 10.6. *Let* (10.10), (10.11) *hold and let* K', K *be domains with* C^1 *boundary such that* $K' \subset K$, $\bar{K} \subset \Omega$, $c_0 = \text{dist}(K', \partial K) > 0$. *If* $\Gamma(u_1) \cap K$ *is a* C^1 *surface, then* $\Gamma(u_2) \cap K'$ *is a* C^1 *surface provided that*

$$(10.14) \qquad\qquad |u_1 - u_2|_{L^\infty(\Omega)} < \varepsilon_0;$$

ε_0 *is sufficiently small, depending on the* $C^{0,1}$ *norms of the* f_i, *the* $C^{1,1}$ *norms of the* u_i, λ, c_0, *and a* C^1 *bound on* $\Gamma(u_1) \cap K$.

Proof. Let $x_0 \in \Gamma(u_2) \cap K$. By (10.12), if $\varepsilon^2 = \varepsilon_0$ is small enough, then the condition

$$\frac{MD(\Lambda(u_2) \cap B_r(x_0))}{r} > \sigma(r)$$

of Theorem 3.10 is satisfied. Hence $\Gamma(u_2)$ is C^1 in a neighborhood of x_0.

By Theorem 1.1, if $f_i \in C^{m+\alpha}$ $(0 < \alpha < 1)$ for some integer $m \geq 1$, then Theorem 10.6 yields $C^{m+1+\alpha}$ smoothness of the free boundaries.

We shall next extend the preceding results to the Stefan problem. We recall that

$$(10.15) \qquad \left.\begin{array}{rl} -\Delta u + u_t &\geq -1 \\ u &\geq 0 \\ (-\Delta u + u_t + 1)u &= 0 \end{array}\right\} \quad \text{a.e. in } \Omega^*,$$

where Ω^* is some neighborhood of the free boundary, and

$$(10.16) \qquad\qquad u_t \geq 0,$$

$$(10.17) \qquad\qquad u_t, D_x^2 u \text{ are in } L^\infty_{\text{loc}}.$$

It follows that for each $t > 0$,

$$(10.18) \qquad \Delta u = f, \qquad f = 1 + u_t \geq 1 \text{ in } \Omega^* \cap N.$$

Defining

$$O_{\varepsilon, t} = \{x; |\nabla_x u(x, t)| < \varepsilon\},$$

$$N_t = \{x; (x, t) \in N\},$$

and similarly Λ_t, Γ_t, the proof of the Lemma 10.1 extends to yield

$$(10.19) \qquad \mu(O_{\varepsilon, t} \cap K) \leq C\varepsilon(\mu(K) + A(S)),$$

where K is any domain in R^n contained in a small neighborhood of a point x; $x \in \Gamma_t$.

Similarly,

$$\mu(\{x; 0 < u(x, t) < \varepsilon^2\} \cap K) \leqslant C\varepsilon(\mu(K) + A(S)),$$

$$\frac{\mu(B_\varepsilon(x) \cap O_{C\varepsilon, t})}{\mu(B_\varepsilon(x))} \geqslant \gamma > 0 \qquad (x \in \Gamma_t),$$

$$H_{n-1}(\Gamma_t \cap K) \leqslant C(\mu(K) + A(K)).$$

Next

$$\mu((\Lambda_t(u_1) \Delta \Lambda_t(u_2)) \cap \Omega_t^{**}) \leqslant C\varepsilon$$

if u_i are two solutions of the Stefan problem with

$$|u_1 - u_2|_{L^\infty(\Omega^*)} < \varepsilon^2$$

and Ω^{**} is a domain, $\overline{\Omega^{**}} \subset \Omega^*$, $\Omega_t^{**} = \{x; (x, t) \in \Omega^{**}\}$. Also,

$$(\Lambda_t(u_2))_{(-C\varepsilon)} \subset \Lambda_t(u_1), \qquad \text{for some } C > 0.$$

Finally:

Theorem 10.7. *Let u_1, u_2 be solutions of the Stefan problem and suppose that the free boundary of u_1 is given in some cylinder D (in $\{t > 0\}$) by*

$$x_n = g_1(x_1, \ldots, x_{n-1}, t), \qquad g_1 \in C^1.$$

Let δ be any small positive number. If

$$|u_1 - u_2|_{L^\infty(D_{(\delta)})} < \varepsilon_0,$$

where ε_0 is sufficiently small (depending on δ), then the free boundary of u_2 in $D_{(-\delta)}$ is given by

$$x_n = g_2(x_1, \ldots, x_{n-1}, t), \qquad g_2 \in C^\infty.$$

Here $D_{(\delta)} = \{(x, t); \text{dist}((x, t), D) < \delta\}$, and $D_{(-\delta)}$ is similarly defined.

Indeed, the proof is similar to the proof of Theorem 10.6 and makes use of Theorem 9.9.

PROBLEMS

1. If

$$-\Delta u_i + f_i \geqslant 0, \qquad u_i \geqslant 0, \qquad (-\Delta u_i + f_i)u_i = 0$$

a.e. in Ω, $u_i = 0$ on $\partial\Omega$, where $\partial\Omega \in C^{2+\alpha}$, $f_i \in L^\infty(\Omega)$, then for any $\varepsilon > 0$, $|u_1 - u_2|_{L^\infty(\Omega)} < \varepsilon$ if $|f_1 - f_2|_{L^\infty(\Omega)}$ is sufficiently small.

2. If the free boundary of u_1 in Theorem 10.6 is locally (say in K) given by $x_n = f_1(x_1, \dots, x_{n-1})$, $f \in C^1$, then the free boundary of u_2 is given [in any subset K' of K, $c_0 = \mathrm{dist}(K', \partial K) > 0$] by $x_n = f_2(x_1, \dots, x_{n-1})$, where $f_2 \in C^1$ and

$$|f_1 - f_2| \leqslant C\varepsilon_0^{1/2};$$

C depends on the same constants on which ε_0 depends.

3. Theorem 10.6 and the preceding results are false if the nondegeneracy condition $f \geqslant \lambda$ is not satisfied. Check this in case $n = 1$ with

$$-u'' \geqslant 0, \qquad u \geqslant \phi, \qquad u''(u - \phi) = 0 \text{ in } -1 < x < 1,$$

with $u(-1) = u(1) = 0$ and $\phi(x) = \varepsilon x^2$, $-1 < \varepsilon < 1$.

4. Extend Theorem 10.6 and all the preceding results to the elliptic operator

$$Au = -\sum a_{ij}(x)\frac{\partial^2 u}{\partial x_i \partial x_j} + \sum b_i(x)\frac{\partial u}{\partial x_i}.$$

[*Hint*: For Lemma 10.1, integrate

$$\int_{O_\varepsilon} h_\varepsilon^i D_i f \qquad \text{by parts.}]$$

11. FREE BOUNDARIES WITH SINGULARITIES

Let Ω be a convex domain in R^n with C^2 boundary, symmetric with respect to $x_n = 0$. We identify R^{n-1} with the hyperplane $\{(x', 0); x' \in R^{n-1}\}$, where $x' = (x_1, \dots, x_{n-1})$. Let E be an open set in $R^{n-1} \cap \Omega$ and let F be a closed set in $R^{n-1} \cap \Omega$ such that

$$E \subset F.$$

We write these sets as disjoint unions of their open and closed components, respectively,

(11.1) $$E = \cup E_i, \qquad F = \cup F_i$$

and assume that

$$E_i \subset F_i \qquad \forall i.$$

Let ϕ be a C^2 function in $\overline{\Omega}$, $\phi < 0$ on $\partial\Omega$ and denote by u the solution of the obstacle problem

(11.2) $$\int_\Omega \nabla u \cdot \nabla(v - u)\, dx \geq 0 \qquad \forall v \in K; u \in K,$$

where $K = \{v \in H_0^1(\Omega), v \geq \phi \text{ a.e.}\}$. We denote by Λ the coincidence set and by $\mathring{\Lambda}$ the interior of Λ. In the next theorem we establish for some ϕ a disjoint decomposition

(11.3)
$$\Lambda = \cup \Lambda_j, \qquad \Lambda_j \text{ closed components,}$$

$$\mathring{\Lambda} = \cup \Lambda_{oj}, \qquad \Lambda_{oj} \text{ open components,}$$

such that

(11.4) $$E_j = \Lambda_{oj} \cap R^{n-1}, \qquad F_j = \Lambda_j \cap R^{n-1}.$$

Theorem 11.1. *Let E, F be as above. There exists a C^∞ strictly superharmonic function ϕ (that is, $\Delta\phi < 0$) in $\overline{\Omega}$, with $\phi < 0$ on $\partial\Omega$, such that for the solution u there holds the decomposition (11.3) with the relations (11.4).*

It follows that whenever $(x_0', 0)$ is a point of accumulation of components of F_i, the free boundary has infinite number of components in any neighborhood of $(x_0', 0)$.

It is interesting to compare this situation with the positive result for $n = 2$ obtained by reference 138a (see also Section 2, Problem 6) in case ϕ is a strictly concave function: N is homeomorphic to an annulus and, if ϕ is analytic, Γ is a Jordan curve having analytic parametrization.

Proof. For any open set O in R^m there exists a nonnegative C^∞ function $\alpha(x)$ such that $O = \{x \in R^m; \alpha(x) > 0\}$. Indeed, if $B_j = B_{r_j}(x_j)$ form a covering of O by balls contained in O, with $r_j < \frac{1}{2}$, and $\zeta \in C^\infty$, $\zeta(x) > 0$ if $|x| < 1$, $\zeta(x) = 0$ if $|x| \geq 1$, then we can take

$$\alpha(x) = \sum \alpha_j \zeta\left(\frac{x - x_j}{r_j}\right) \qquad \left(\alpha_j = (r_j)^j\right).$$

Take $O = E \subset R^{n-1}$ and construct $\alpha(x')$ as above. Let

$$\Gamma_+ = \{(x', \alpha(x')); x' \in \Omega \cap R^{n-1}\},$$

$$\Gamma_- = \{(x', -\alpha(x')); x' \in \Omega \cap R^{n-1}\},$$

and split Ω into three sets

$$\Omega_+ : x_n > \alpha(x'),$$

$$\Omega_- : x_n < -\alpha(x'),$$

$$\Lambda : -\alpha(x') \leqslant x_n \leqslant \alpha(x').$$

Notice that $E \subset \text{int}(\Lambda) \cap R^{n-1}$.

Consider the Cauchy problem

(11.5) $$\Delta v = 1 + h \qquad \text{in } \Omega_+,$$

(11.6) $$v = 0, \qquad \frac{\partial v}{\partial x_n} = 0 \text{ on } \Gamma_+.$$

Lemma 11.2. *There exist functions* v, h *in* $C^\infty(\overline{\Omega}_+)$ *for which* (11.5), (11.6) *hold and* h *vanishes on* Γ_+ *with all its derivatives; further,*

(11.7) $$v > 0, \qquad \Delta v > 0 \text{ in } \Omega_+.$$

Proof. Use the relation

$$\Delta v = 1 \qquad \text{in } \Omega_+$$

and (11.6) in order to compute formally the derivatives

$$v_k(x') = \frac{\partial^k}{\partial x_n^k} v \qquad \text{on } \Gamma_+$$

(assuming that $v \in C^\infty$), and then define

(11.8) $$v(x', x_n) = \sum_{k=2}^\infty v_k(x')\zeta(c_k(x_n - \alpha(x')))\frac{(x_n - \alpha(x'))^k}{k!};$$

here ζ is a C_0^∞ cutoff function in one variable, $\zeta \equiv 1$ in a neighborhood of the origin. If we choose, for instance,

$$c_k = 2 + \sum_{|\beta| \leqslant k} \left|D^\beta v_k\right|_{L^\infty},$$

then the series in (11.8) is uniformly convergent in Ω_+ together with each of its derivatives, and, by the definition of the v_k,

(11.9)
$$\Delta v(x) - 1 \text{ and all its derivatives}$$
$$\text{converge to zero as dist}(x, \Gamma_+) \to 0.$$

We take $c_2 = 0$ and consider the first term in (11.8)

$$T = \tfrac{1}{2} v_2(x')(x_n - \alpha(x'))^2.$$

One easily computes that

$$v_2 = \left(1 + |\nabla\alpha|^2\right)^{-1}$$

and thus T is positive in Ω_+. By increasing the c_k if necessary we can make the remainder of the series arbitrarily small, say

$$|v - T| < \tfrac{1}{2}T.$$

Consequently, $v > 0$ in Ω_+. Writing

$$T = \tfrac{1}{2}x_n^2 - \frac{x_n^2 |\nabla\alpha|^2 + 2x_n\alpha - \alpha^2}{2(1 + |\nabla\alpha|^2)},$$

we see that $\Delta T - 1$ can be made arbitrarily small if $\alpha(x')$ is replaced at the beginning by $\varepsilon\alpha(x')$ with ε small enough. Thus $\Delta T > \tfrac{1}{2}$ in Ω_+. By increasing the c_k, if necessary, we get $\Delta v > 0$ in Ω_+. Defining $h = 1 - \Delta v$ and recalling (11.9), the proof of the lemma is complete.

We now extend the definition of v into Ω by

$$v(x) = \begin{cases} v(x', -x_n) & \text{if } x \in \Omega_- \\ 0 & \text{if } x \in \Lambda \end{cases}$$

and let

$$f = \Delta v \quad \text{in } \Omega_+ \cup \Omega_-$$
$$= 1 \quad \text{in } \Lambda.$$

Then $f \in C^\infty(\overline{\Omega})$, $f > 0$ in $\overline{\Omega}$.

Let ψ be the solution of

$$\Delta\psi = -f \quad \text{in } \Omega,$$
$$\psi = -v \quad \text{on } \partial\Omega.$$

Then $u = v + \psi$ satisfies

$$\Delta u = 0 \qquad \text{in } \Omega \setminus \Lambda,$$
$$u = 0 \qquad \text{on } \partial\Omega,$$
$$u = \psi, \quad \nabla u = \nabla \psi \qquad \text{on } \partial\Lambda,$$
$$u > \psi \qquad \text{in } \Omega \setminus \Lambda.$$

Thus u is a solution of the variational inequality with obstacle ψ, and the first relation in (11.4) holds.

To satisfy the second relation in (11.4) we change ψ into $\phi(x) = \psi(x) - \beta(x')$, where $\beta \in C^\infty$, $\beta \geq 0$, and

$$F = \{x'; \beta(x') = 0\}$$

[β is constructed in the same way as α, with E replaced by $(R^{n-1} \cap \Omega) \setminus F$]. The sets $\{u = \psi\}$, $\{u = \phi\}$ can differ only on the set F of measure zero. Hence u is the solution of the obstacle problem corresponding to the obstacle ϕ. Since $\psi = -v$ on $\partial\Omega$, $\psi < 0$ on $\partial\Omega$ except possibly on $\partial\Omega \cap R^{n-1}$ and (since $F \subset \Omega \cap R^{n-1}$) $\beta(x') < 0$ on this set; thus $\phi < 0$ on $\partial\Omega$.

Next, $\Delta\psi < -\varepsilon < 0$ for some ε, and by replacing $\beta(x')$ by $\delta\beta(x')$ with δ small enough, we can achieve $\Delta\phi < 0$. Finally, since $\phi < \psi$ on $(\Omega \cap R^{n-1}) \setminus F$, the second condition in (11.4) is also satisfied; this completes the proof.

We shall next give examples of analytic superharmonic obstacles for which the free boundary has a cusp.

Take any analytic curve Γ with a cusp at one point A, having the figure "∞" as in Figure 2.5 (such a curve is given in Problem 2), and denote by Λ the closed set enclosed by Γ, and by V some neighborhood of Γ.

Consider the Cauchy problem:

$$\Delta\tilde{u} = 1 \qquad \text{in } N,$$
$$\tilde{u} = 0, \quad \frac{\partial\tilde{u}}{\partial\nu} = 0 \qquad \text{along } \Gamma.$$

By the Cauchy–Kowalewsky theorem this problem has a unique analytic

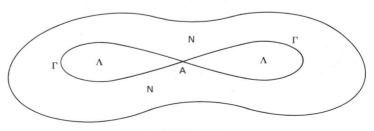

FIGURE 2.5

solution $\tilde{u}$ in $N \cup \Gamma$ provided that the neighborhood V is small enough. Since

$$\frac{\partial^2 \tilde{u}}{\partial \nu^2} = 1 > 0 \qquad \text{on } \Gamma,$$

$\tilde{u} > 0$ in N if V is further decreased. Extending $\tilde{u}$ into Λ by zero and setting

$$u = \tilde{u} - \frac{|x|^2}{2}, \qquad \phi = -\frac{|x|^2}{2}$$

we conclude:

Theorem 11.3. *The function u is a solution in Ω of the obstacle problem*

$$-\Delta u \geq 0, \qquad u \geq \phi, \qquad \Delta u(u - \phi) = 0 \text{ in } \Omega$$

with strictly concave analytic obstacle ϕ, and the free boundary Γ has a cusp at A.

One can also construct examples in which the domain N is not restricted to belong to a sufficiently small neighborhood of Γ; this is done in the following problems.

PROBLEMS

1. Let Ω be a domain in R^2, Λ a closed subset bounded by a piecewise C^1 curve Γ, $N = \Omega \setminus \Lambda$. Let $f(z)$ be holomorphic function in N, continuous in $N \cup \Gamma$, with $f = \phi_x - i\phi_y$ on Γ, where $\phi \in C^1(\Gamma)$. Then the function $u(z) = \operatorname{Re} \int f(z)\, dz$ is a single-valued harmonic function in N, and $u = \phi$, $\nabla u = \nabla \phi$ on Γ.

2. Let

$$\sigma(z) = \tfrac{1}{2}(z + z^{-1}) + \frac{\varepsilon}{2}(z - z^{-1})P(z), \qquad P(z) = z^2 + 2 + z^{-2}.$$

Then $\sigma(e^{i\theta}) = \cos\theta + \varepsilon i\, |e^{2i\theta} + 1|^2$ and traces a curve Γ of shape "∞" as $0 \leq \theta < 2\pi$, with a cusp corresponding to $\theta = \pi/2$. Show that for small $\varepsilon > 0$ the annulus $A = \{1 < |z| < 2\}$ is mapped by σ in a 1-1 way onto a set N; the inner boundary of N is Γ and the outer boundary is $\sigma\{|z| = 2\}$. Let $\Lambda =$ the set enclosed by Γ, $\Omega = N \cup \Lambda$. This notation will be used in the next two problems.

3. Let $\phi = -\tfrac{1}{2}(x^2 + y^2)$. Then $f = \phi_x - i\phi_y$ on Γ if $f(z) = -\bar{z}$ on Γ, or $f(\sigma(z)) = -\overline{\sigma(z)}$ on $|z| = 1$, or yet if

$$f(z) = -\phi\left(\frac{1}{\phi^{-1}(z)}\right)$$

$[\overline{\sigma(z)} = \sigma(\bar{z}), \bar{z} = 1/z$ on $|z| = 1]$. The right-hand side is holomorphic in N. If u, defined by Problem 1, satisfies

(*) $\hspace{3cm} u > \phi \hspace{1cm}$ in $N \cup \partial\Omega$,

then check that u solves the obstacle problem in Ω with obstacle ϕ, and Γ is the free boundary.

4. Prove (*).
   [*Hint*: Since

$$\int_{\sigma(\gamma)} f(z)\, dz = \int_{\gamma} f(\sigma(z))\sigma'(z)\, dz,$$

(*) reduces to

$$W \equiv u - \phi = \operatorname{Re}\int_{z_0}^{z}\left[\overline{\sigma(z)} - \sigma(z^{-1})\right]\sigma'(z)\, dz > 0$$

if $1 < |z| \leqslant 2$, with any $|z_0| = 1$. Write $\sigma = \sigma_+ + \varepsilon\sigma_-$; then $\sigma_\pm(1/z) = \pm\sigma(z)$. If

$$W = a + \varepsilon b + \varepsilon^2 c,$$

then

$$a = (\operatorname{Im}\sigma_+)^2, \hspace{1cm} a(re^{i\theta}) = \tfrac{1}{4}(r - r^{-1})^2 \sin^2\theta > 0$$

if $|z| = 2$, $z \neq \pm 2$. At $z = \pm 2$, $b > 0$, so that $W > 0$ on $|z| = 2$. W cannot take minimum in $\{1 < |z| < 2\}$ since

$$W_x - iW_y = \left(\overline{\sigma(z)} - \sigma\!\left(\frac{1}{z}\right)\right)\sigma'(z) \neq 0.]$$

5. Take $P(z) = (z - 2 + z^{-1})^2$ in Problems 2 to 4 in order to produce an example (with the same ϕ) of a free boundary Γ given by

$$\sigma(e^{i\theta}) = \cos\theta + \varepsilon i\, |e^{2i\theta} - 1|^4 \sin\theta \hspace{1cm} (0 \leqslant \theta < 2\pi);$$

it has a cusp at $\sigma(1)$; see Figure 2.6.

FIGURE 2.6

12. BIBLIOGRAPHICAL REMARKS

The hodograph–Legendre transformation and the treatment of Section 1 are due to Isakov [189a, b] and to Kinderlehrer and Nirenberg [123a]. Schaeffer [158b] has given the method in Problem 3 of Section 1. The hodograph–Legendre transformations have also been applied to elliptic systems of equations and to minimal surfaces by Kinderlehrer, Nirenberg, and Spruck [124a, b].

Theorem 2.1 is due to Lewy and Stampacchia [138a]. Theorem 2.4 was established by Caffarelli and Riviere [63a]. All the other results of Section 2 are due to Kinderlehrer [122b].

The results of Sections 3 to 5 are due to Caffarelli [56b, e]; see also [56a]. The original proof of Theorem 3.10 appears in reference 56b; the proof presented in Sections 4 and 5 is taken from reference 56e.

Theorem 6.1 is due to Alt [5a]. The three-dimensional dam problem was first introduced by Stampacchia [167b] and Gilardi [106a]. Theorem 6.5 was established by Friedman and Jensen [95b]; the results of Problem 3 of Section 6 are also taken from reference 95b. The results of Problem 2 are due to Friedman [94i] for $m = 1$ and to Jensen [116a] for $m \geqslant 1$. Another proof of Theorem 6.5 was given by Cryer [75]; the convexity of the free boundary for another filtration problem was established by Boieri and Gastaldi [42]. The filtration problem with several layers was studied by Baiocchi, Comincioli, Magenes, and Pozzi [21], Baiocchi [19a], Baiocchi and Friedman [22], and Benci [28], and by Caffarelli and Friedman [58a, b]. In reference 58a and b, the shape of the free boundary is studied and the asymptotic behavior is derived when one of the layers becomes very thin; the limit problem is a free-boundary problem with mixed Dirichlet and Neumann conditions.

The dam problem with time-dependent boundary conditions (also for slightly compressible fluids) was treated by Friedman and Jensen [95a, c], Friedman and Torelli [99], and Torelli [174a–c].

The Baiocchi transformation (Section 5 of Chapter 1) can be extended to establish existence for some nonrectangular dams in two dimensions; see Baiocchi [19b]. The filtration problem for general dams with any number of reservoirs and in any number of dimensions will be studied in Chapter 5; the methods will be entirely different from the methods developed for the rectangular dam. We shall prove existence, uniqueness, and analyticity of the free boundary. For the corresponding time-dependent problem Gilardi [106b] established the existence of a "weak" solution.

The regularity of the free boundary for the elastic–plastic torsion problem was proved by Caffarelli and Riviere [63c]. Lemma 7.3 is taken from reference 63a. Theorem 7.7 is due to Caffarelli and Friedman [58f]; related results were proved by Ting [172a, b]. Problems 1 to 5 of Section 7 are taken from reference 58f; related work is due to Chipot [71b].

Theorems 8.1, 8.3, and 8.4 are due to Caffarelli and Friedman [58f]. Theorems 8.5 and 8.11 were established by Friedman and Pozzi [98], and the remaining part of Section 8 is taken from Caffarelli, Friedman, and Pozzi [59].

Earlier work for regular polygonal domains was done by Ting [172a, b].

Elastic–plastic problems with several materials were considered by Caffarelli and Friedman [58n]; see also Brezis, Caffarelli, and Friedman [46]. Problems of unloading have been studied by Caffarelli and Friedman [58q] and by Ting [172c, d].

Theorem 9.1 is due to Cafarelli and Friedman [58c]. Theorems 9.6, 9.8, and 9.9 are taken from Caffarelli [56b]. In Lemma 9.14 we have replaced the original proof by a proof given by Kinderlehrer and Nirenberg [123b]; the completion of the proof of Theorem 9.9 as outlined in Problems 7 to 10 of Section 9 is given in reference 123b.

For one-dimensional parabolic free-boundary problems, Friedman [94h] proved that the free boundary is analytic if the boundary conditions are analytic. Another proof was given by Kinderlehrer and Nirenberg [123c].

In Section 9, Problem 2 is based on Friedman and Kinderlehrer [97], Problems 4 and 5 are based on Friedman and Jensen [95b], Problem 6 is taken from Friedman [94j], and Problem 11 is due to Caffarelli (oral communication). Some results on the free boundary in two space dimensions are given in Caffarelli [56c].

Theorem 10.6 and all the preceding developments of Section 10 are due to Caffarelli [56f]. Schaeffer [158a] had previously proved the weaker result: If $\Gamma(u_1)$ is C^∞ and $u_1 - u_2$ is small in C^∞ sense, then $\Gamma(u_2)$ is C^∞. His method uses the Nash–Moser implicit function theorem, and was extended to parabolic variational inequalities by Hanzawa [111].

The examples in Section 11 (including the problems) are due to Şchaeffer [158c]. Positive results for $n = 2$ were proved by Lewy and Stampacchia [138a] for $-\Delta$ and by Kinderlehrer [122a] for minimal surfaces. Caffarelli and Riviere [63b] studied the nature of the singularities of free boundaries for $n = 2$. A two-dimensional free boundary problem was studied by Kinderlehrer and Stampacchia [126a].

Variational inequalities have been used in problems of flow past a profile: see Brezis and Stampacchia [53b, c], Brezis and Duvaut [190], Hummel [115], Shimborsky [162a, b], and Tomarelli [173]; see also Elliott and Janovsky [83]. They also appear in problems of lubrication; see Capriz and Cimatti [65] and Cimatti [72].

In stochastic control, problems of optimal stopping time can be reduced to variational inequalities; see Bensoussan and Lions [34], Friedman [94g]. Finally, problems with partial observations, such as some problems in sequential analysis, can also be reduced to variational inequalities and give rise to questions regarding qualitative behavior of the free boundary; see Caffarelli and Friedman [58l, o] and Friedman [94j, k].

3

JETS AND CAVITIES

If we minimize the functional

$$J_1(v) = \int_\Omega |\nabla v|^2 + \int_\Omega fv^+$$

over the class of functions $v \in K$, where

$$K = \{v; \, v - u^0 \in H_0^1(\Omega), \, v \geq 0\},$$

where $u^0 \in H^1(\Omega)$, $u^0 \geq 0$, then the minimizer u is a solution of the variational inequality

$$-\Delta u \geq f, \qquad u \geq 0, \qquad (\Delta u - f)u = 0 \text{ a.e. in } \Omega.$$

In this chapter we minimize functionals

$$J(v) = \int_\Omega |\nabla v|^2 + \int_\Omega fI_{\{v>0\}} \qquad (f > 0)$$

over the same set K, where I_A = characteristic function of a set A. Notice that $J(v)$ is no longer continuous, although it is lower semicontinuous.

Such functionals (with f = const.) give rise to free-boundary problems

$$\Delta u = 0 \qquad \text{in } \{u > 0\},$$

$$u = 0, \frac{\partial u}{\partial \nu} = \text{const. on the free boundary } \Omega \cap \partial\{u > 0\},$$

which arise in flow problems of jets and cavities. We shall develop a general theory for the minimizers u and then proceed to solve problems of jets and cavities.

1. EXAMPLES OF JETS AND CAVITIES

In this chapter we consider only fluids that are incompressible, irrotational, and inviscid. That means that the velocity vector is $\nabla \phi$, where ϕ, the *velocity potential*, satisfies $\Delta \phi = 0$ in the fluid. If the flow is two-dimensional, then the harmonic conjugate ψ is called a *stream function*, and if the flow is three-dimensional axially symmetric (with the x-axis as the axis of symmetry), then the function ψ determined by

$$(1.1) \qquad \phi_x = \frac{1}{y} \psi_y, \qquad \phi_y = -\frac{1}{y} \psi_x$$

is called the *stream function*. In either case (ϕ_x, ϕ_y) gives the velocity in the fluid and the stream lines (i.e., the lines along which the tangent is in the direction of the velocity) are given by $\psi = $ const.

We shall consider two situations: (a) there is a "cavity" in the fluid; (b) a part Γ of the fluid's boundary is surrounded by air.

In case (a), if the fluid is fast moving, the cavity consists of a mixture of vapor and gas and the pressure p in the cavity is constant (we are dealing here with stationary problems only). By Bernoulli's law,

$$(1.2) \qquad p + \tfrac{1}{2} |\nabla \phi|^2 + gy = \text{const.}$$

throughout the fluid, provided gravity is in the negative y-axis. Since the pressure is continuous across the fluid's boundary, $p = $ const. on the boundary of the cavity and (1.2) gives

$$(1.3) \qquad \tfrac{1}{2} |\nabla \phi|^2 + gy = \text{const. on } \Gamma,$$

where Γ is the boundary of the cavity. Further, since Γ is a stream line, we also have

$$(1.4) \qquad \psi = \text{const. on } \Gamma.$$

In case (b), the pressure p outside the fluid is again constant, so that Bernoulli's law again gives (1.3); also (1.4) is valid.

Problem (a) is called a *cavity problem* and problem (b) is called a *jet problem*; there are of course also flow problems which involve both jets and cavities [as in Figure 3.1(c)]. In Figure 3.1, bold curves indicate fixed boundaries.

Unless otherwise explicitly stated, we do not include gravity in our discussion; that is, we take $g = 0$. (Problems with gravity are studied in Sections 18

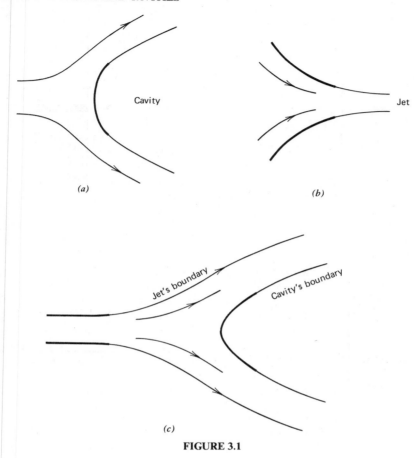

Cavity

Jet

(a)

(b)

Jet's boundary

Cavity's boundary

(c)

FIGURE 3.1

and 19.) Thus

$$\frac{\partial \psi}{\partial \nu} = \text{const. on } \Gamma \quad \text{if } n = 2,$$

$$\frac{1}{y} \frac{\partial \psi}{\partial \nu} = \text{const. on } \Gamma \text{ if } n = 3 \text{ and the flow is symmetric with respect to the } x\text{-axis.}$$

We shall not treat here three-dimensional flows that are not axially symmetric.

We shall now compute explicitly some examples, using conformal mappings; the method (called the *hodograph method*) is restricted to two-dimensional flows.

We begin with the example of Kirchoff (1869) of a symmetric flow of unit velocity past a vertical flat plate with free stream line detachment at the ends (see Figure 3.2a). We introduce the velocity magnitude $q = |\nabla\phi|$, the complex velocity vector $u + iv$ ($u = \phi_x, v = \phi_y$), the complex potential $f = \phi + i\psi$,

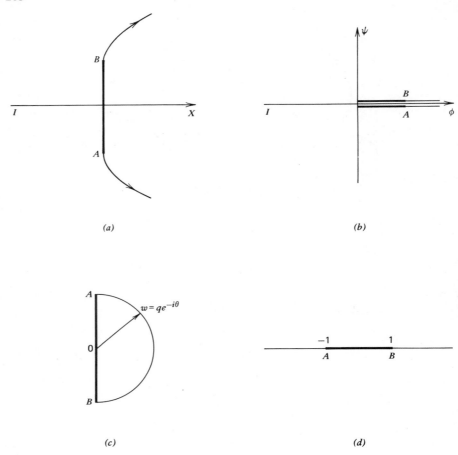

FIGURE 3.2

and the *hodograph variable*

$$w = \frac{df}{dz} = u - iv = qe^{-i\theta}.$$

The images of the physical plane in the f-plane and in the w-plane are described in Figure 3.2b and c.

The idea now is to parameterize both f and w by the same parameter t; t varies in the complex plane with the real interval $(-1, 1)$ deleted (Figure 3.2d):

$$f = kt^2, \qquad \frac{i}{2}\left(w - \frac{1}{w}\right) = t^{-1},$$

where $t = \pm 1$ correspond to the ends of the plate. Using the relation $w = df/dz$ it should then be possible to find a parameterization $z = z(t)$ for the free stream line. Indeed,

$$\frac{df}{dz} = w = i\left(-\frac{1}{t} + \left(\frac{1}{t^2} - 1\right)^{1/2}\right),$$

where the branch of the square root is chosen so that $w \to 0$ if $t \to 0$. Then

$$z = \int \frac{dz}{df} df = 2ik \int_0^t \left(1 + \sqrt{1 - t^2}\right) dt$$

and $k = l/(4 + \pi)$ is the length of the plate. The upper free-stream line is then given by

(1.5)
$$x = \frac{1}{4 + \pi}\left[t\sqrt{t^2 - 1} - \log\left(1 + \sqrt{t^2 - 1}\right)\right],$$

$$y = \frac{l}{2} + \frac{2l}{4 + \pi}(t - 1) \qquad \text{for } t \geq 1.$$

We shall next use the hodograph method to study the jet problem with symmetric nozzle formed by two line segments which form an opening of size $\alpha\pi$ (see Figure 3.3a). The pictures in the f and w planes are outlined in Figure 3.3b and c and in the t-plane (Figure 3.3d) we exclude the intervals $(-\infty, -1)$, $(1, \infty)$. Then

$$f = \frac{h}{\pi}\left(i\frac{\pi}{2} - \log t\right),$$

$$w = f'(z) = \left(\sqrt{1 - t^2} + it\right)^\alpha$$

and the free stream lines are given by

(1.6)
$$z(t) = -\frac{h}{\pi}\int_1^t \frac{\left(\sqrt{1 - \tau^2} - i\tau\right)^\alpha}{\tau} d\tau, \qquad 0 < t < 1.$$

Denote by y_0 the y-coordinate of B and by y_∞ the limiting height of the upper free boundary as $x \to \infty$. Then the number

$$C_c = \frac{y_\infty}{y_0}$$

is called the *contraction coefficient* of the nozzle. One can easily compute from

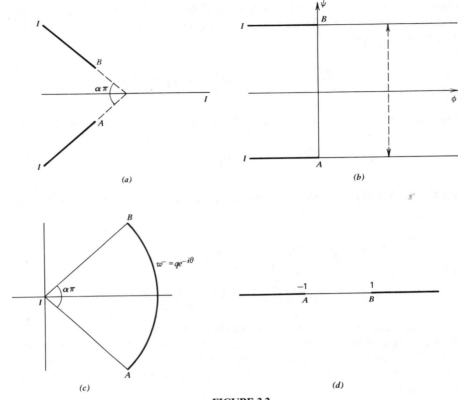

FIGURE 3.3

(1.6) that

(1.7)
$$C_c = \tfrac{1}{2} \qquad \text{if } \alpha = 2,$$
$$C_c = \frac{\pi}{\pi + 2} \qquad \text{if } \alpha = 1.$$

The nozzle with $\alpha = 2$ is called *Borda's mouthpiece*; (1.7) was first established by Borda (1766) using physical considerations.

Later we shall establish existence theorems for jets and cavities and prove that ψ_x, ψ_y have constant signs (for $y > 0$) for some geometric shapes of the nozzles and the obstacles (for cavities); we shall also establish asymptotic behavior of ψ at infinity. Using these facts one can rigorously justify the hodograph method for some specific geometries; in particular, (1.5) and (1.6) can be justified for our solutions.

PROBLEMS

1. The *drag D* of the plate (in Figure 3.2) is calculated by integrating the pressure difference on both sides:

   (1.8) $$D = \int_{-l/2}^{l/2} (p - p_c) \, dy, \qquad p_c = \text{cavity pressure.}$$

   (a) Show that $D = \pi l/(4 + \pi)$ (the drag coefficient is $C_D = D/l = \pi/(4 + \pi)$).
   (b) Show that $y \sim c\sqrt{x}$ along the upper free boundary as $x \to \infty$ (c constant).

2. (a) Verify (1.5), (1.6).
   (b) Prove that for general $\alpha \in (0,2)$

   $$\frac{1}{C_c} = 2 - \frac{1}{\pi} \sin \frac{\alpha \pi}{2} \left[\Phi\left(\frac{1}{2} + \frac{\alpha}{4}\right) - \Phi\left(\frac{\alpha}{4}\right) - \frac{2}{\alpha} \right],$$

   where $\Phi(x)$ is the logarithmic derivative of the gamma function.

3. Prove in both the plane and axially symmetric flow that

   (1.9) $$\frac{\partial q}{\partial \nu} + \kappa q = 0 \qquad \text{in the fluid,}$$

   where $q = |\nabla \phi|$, ν is the normal to the stream line, and κ is the curvature of the stream line, taken positive when the stream line is convex to the region on the side of the positive normal ν.

2. THE VARIATIONAL PROBLEM

Let Ω be a domain in R^n, not necessarily bounded, and assume that $\partial\Omega$ is locally piecewise C^1 graph. Let S be a nonempty open subset of $\partial\Omega$ with piecewise C^1 boundary ∂S. Let u^0 be a function satisfying:

$$u^0 \in L^1_{\text{loc}}(\Omega), \qquad \nabla u^0 \in L^2(\Omega), \qquad u^0 \geqslant 0.$$

We introduce a convex set

$$K = \left\{ v \in L^1_{\text{loc}}(\Omega), \nabla v \in L^2(\Omega), v = u^0 \text{ on } S \right\}$$

and a functional

(2.1)
$$J(v) = \int_\Omega \left[|\nabla v|^2 + I_{\{v>0\}} Q^2 \right] dx,$$

where Q is a given nonnegative measurable function and I_A denotes the characteristic function of a set A.

Theorem 2.1. *If $J(u^0) < \infty$, then there exists a $u \in K$ such that*

$$J(u) = \min_{v \in K} J(v).$$

Proof. Since J is nonnegative, there exists a minimizing sequence u_m, that is,

$$u_m \in K, \qquad J(u_m) \to \alpha \equiv \inf_{v \in K} J(v) \qquad (0 \leq \alpha < \infty).$$

Since $\{\nabla u_m\}$ is bounded in $L^2(\Omega)$ and $u_m - u^0 = 0$ on S, it follows that $u_m - u^0$ is bounded in $L^2(\Omega \cap B_R)$ for any ball B_R with large R. Therefore, there exists a u in K such that, for a subsequence,

$$\nabla u_m \to \nabla u \qquad \text{weakly in } L^2(\Omega),$$
$$u_m \to u \qquad \text{a.e. in } \Omega.$$

Moreover, there is a function $\gamma \in L^\infty(\Omega)$ such that

$$I_{\{u_m>0\}} \to \gamma \qquad \text{weakly star in } L^\infty(\Omega).$$

Now, for any large R,

$$\int_{\Omega \cap B_R} \left[|\nabla u|^2 + \gamma \min(Q, R)^2 \right]$$

$$\leq \liminf_{m \to \infty} \int_{\Omega \cap B_R} |\nabla u_m|^2 + \lim_{m \to \infty} \int_{\Omega \cap B_R} I_{\{u_m>0\}} \min(Q, R)^2$$

$$\leq \lim J(u_m).$$

Letting $R \to \infty$ and noting that $\gamma = 1$ a.e. in $\{u > 0\}$, we conclude that

$$J(u) \leq \int_\Omega \left(|\nabla u|^2 + \gamma Q^2 \right) \leq \lim J(u_m).$$

From now on we assume that

(2.2) $$0 < Q_{min} \leqslant Q \leqslant Q_{max} < \infty.$$

Definition 2.1. A function $u \in K$ is called a *local minimum* (or a *local minimizer*) for J if, for some small $\varepsilon > 0$, $J(u) \leqslant J(v)$ for any $v \in K$ such that

$$\|\nabla(v - u)\|_{L^2(\Omega)} + \|I_{\{v>0\}} - I_{\{u>0\}}\|_{L^1(\Omega)} < \varepsilon.$$

Lemma 2.2. *If u is a local minimum, then u is subharmonic and hence, by Lemma 1.10.1, there is a version for which*

(2.3) $$u(x) = \lim_{r \downarrow 0} \fint_{B_r(x)} u \qquad \forall x \in \Omega,$$

and u is upper semicontinuous.

In the sequel we always work with this version.

Proof. For a nonnegative $\zeta \in C_0^\infty(\Omega)$,

$$J(u - \varepsilon\zeta) - J(u) \geqslant 0 \qquad \forall \varepsilon > 0,$$

from which it follows that $\int \nabla\zeta \cdot \nabla u \leqslant 0$, that is, u is subharmonic.

Lemma 2.3. *If u is a local minimum, then*

$$0 \leqslant u \leqslant \sup_\Omega u^0.$$

Proof. Use $J(u_\varepsilon) \geqslant J(u)$ with $u_\varepsilon = u - \varepsilon \cdot \min(u, 0)$ and $u_\varepsilon = u + \varepsilon \cdot \min(\sup_\Omega u^0 - u, 0)$.

In Section 3 it will be proved that a local minimum is continuous (in fact, Lipschitz continuous). It follows that the set $\{u > 0\}$ is open. We shall use this fact in establishing the next two results.

Lemma 2.4. *If u is a local minimum, then u is harmonic in the open set $\{u > 0\}$.*

Proof. Use $J(u + \varepsilon\zeta) \geqslant J(u)$ with $\zeta \in C_0^\infty(\{u > 0\})$, $|\varepsilon|$ sufficiently small.

Theorem 2.5. *Let u be a local minimum and $Q^2 \in H^{1,1}_{\text{loc}}(\Omega)$. Then*

(2.4)
$$\lim_{\varepsilon \downarrow 0} \int_{\partial\{u>\varepsilon\}} (|\nabla u|^2 - Q^2)\eta \cdot \nu = 0$$

for any n-vector η with components in $C^1_0(\Omega)$, where ν is the normal.

Proof. For any real ε, $|\varepsilon|$ small, let $\tau_\varepsilon(x) = x + \varepsilon\eta(x)$ and define u_ε by $u_\varepsilon(\tau_\varepsilon(x)) = u(x)$. Then $u_\varepsilon \in K$ and

$$0 \leqslant J(u_\varepsilon) - J(u)$$

$$= \int_{\{u>0\}} \left\{ \left[\left| \nabla u (D\tau_\varepsilon)^{-1} \right|^2 + Q^2 \circ \tau_\varepsilon \right] \det D\tau_\varepsilon - \left[|\nabla u|^2 + Q^2 \right] \right\}$$

$$= \varepsilon \int_{\{u>0\}} (|\nabla u|^2 + Q^2) \nabla \cdot \eta$$

$$+ \varepsilon \int_{\{u>0\}} \left[-2\nabla u D\eta \, \nabla u + \nabla Q^2 \cdot \eta \right] + o(\varepsilon).$$

It follows that the linear term in ε vanishes and, since u is harmonic in $\{u > 0\}$,

$$0 = \int_{\{u>0\}} \nabla \cdot \left[(|\nabla u|^2 + Q^2)\eta - 2(\eta \cdot \nabla u) \nabla u \right]$$

$$= \lim_{\varepsilon \downarrow 0} \int_{\partial\{u>\varepsilon\}} \left[(|\nabla u|^2 + Q^2)\eta - 2(\eta \cdot \nabla u) \nabla u \right] \cdot \nu$$

$$= \lim_{\varepsilon \downarrow 0} \int_{\partial\{u>\varepsilon\}} (Q^2 - |\nabla u|^2)\eta \cdot \nu.$$

Theorem 2.5 shows that if the free boundary $\partial\{u > 0\}$ in C^1 and if u is in C^1 in $\{u > 0\}$ uniformly up to the free boundary, then

(2.5)
$$|\nabla u| = Q \qquad \text{on the free boundary.}$$

Remark. Since $J(u^+) \leqslant J(u) \forall u \in K$, $\min\limits_{v \in K} J(v) = \min\limits_{\substack{v \in K \\ v \geqslant 0}} J(v)$.

PROBLEMS

1. Use spherical coordinates

$$x = (r \cos \phi \sin \theta, \, r \sin \phi \sin \theta, \, r \cos \theta)$$

to define a function

$$u(x) = r \max \left\{ \frac{f(\theta)}{f'(\theta_0)}, 0 \right\}$$

where

$$f(\theta) = 2 + \cos\theta \log\frac{1 - \cos\theta}{1 + \cos\theta}$$

is a solution of

$$((\sin\theta)f')' + 2(\sin\theta)f = 0, \qquad f'\left(\frac{\pi}{2}\right) = 0$$

and θ_0 is the unique zero of f in $(0, \pi/2)$.

(a) Show that u is harmonic in $\{u > 0\}$, $|\nabla u| = 1$ on $\partial\{u > 0\}\setminus\{0\}$, $\partial\{u > 0\}$ has a singularity at the origin, and $|\nabla u| < 1 - c\varepsilon$ $(c > 0)$ if $\theta_0 + \varepsilon < \theta < \pi - \theta_0 - \varepsilon$.

(b) Show that u is not a minimum with respect to J, when $\Omega = B_1(0)$, $S = \partial\Omega$, $u^0 = u$, $Q = 1$.

[*Hint*: $\theta_0 \sim 33°$, $\cos^2\theta_0 > \frac{1}{2}$, and for $\theta_0 < \theta < \pi/2$,

$$(\sin\theta)f'(\theta_0) > (\sin\theta_0)f'(\theta)\left(1 + (\cos^2\theta_0)f(\theta)\right).$$

Use this to show that if $\zeta_\varepsilon(x) = 4\varepsilon(1/r - 1)$, $u_\varepsilon = \max(u - \zeta_\varepsilon, 0)$, then

$$\varepsilon^{-3/2}\left(J(u_\varepsilon) - J(u)\right) \to \frac{32}{3}\pi\sqrt{f'(\theta_0)}\left(\sqrt{2}\,\sin\theta_0\right)$$

$$- \int_{\theta_0}^{\pi/2} (f(\theta))^{-3/2}\left[(\sin\theta)f'(\theta_0) - (\sin\theta_0)f'(\theta)\right] d\theta < 0.]$$

2. Consider the functional

$$J_\alpha(v) = \int_\Omega \left[|\nabla v|^2 + Q^2(v^+)^\alpha\right] dx \qquad (0 < \alpha < 1)$$

and define $K_\alpha = \{v \in K, Q^2(v^+)^\alpha \in L^2(\Omega)\}$. Extend Theorem 2.1 and Lemma 2.2 to J_α, K_α.

3. REGULARITY AND NONDEGENERACY

From Theorem 2.4 it follows that if the free boundary is smooth then, for any point x_0 of the free boundary,

(3.1) $u(x + x_0) = Q(x_0)\max\{-x \cdot \nu(x_0), 0\} + o(|x|)$

as $|x| \to 0$, where ν is the outward normal (with respect to $\{u > 0\}$). In this section we derive a weak formulation of (3.1), in terms of averages of u. An estimate from above is given in Lemma 3.1 and an estimate from below is given in Lemma 3.3. The first estimate is sufficient for deducing that u is Lipschitz continuous (Theorem 3.2).

We shall often use scaling: If u is a local minimum in $B_\rho(x_0) \subset \Omega$, then

$$(3.2) \qquad u_\rho(x) = \frac{1}{\rho} u(x_0 + \rho x)$$

is a local minimum for $Q_\rho(x) = Q(x_0 + \rho x)$ in $B_1 = B_1(0)$.

Lemma 3.1. *There exists a positive constant* $C^* = C_n Q_{\max}$ *(C_n depending only on n) such that any (local) minimum u has the following property for every (small) ball $B_r(x_0) \subset \Omega$:*

$$(3.3) \qquad \frac{1}{r} \fint_{\partial B_r(x_0)} u \geq C^* \qquad implies\ that\ u > 0\ in\ B_r(x_0).$$

Proof. The idea of the proof is that if the average of u on $\partial B_r(x_0)$ is large, then replacing u in $B_r(x_0)$ by the harmonic function v with boundary values u will decrease the functional, unless of course $u = v$.

Let v be harmonic in $B_r(x_0)$, $v = u$ on $\partial B_r(x_0)$, and extend v by u into $\Omega \setminus B_r(x_0)$. Since $u \geq 0$, v is positive in $B_r(x_0)$. Clearly $v \in K$ and, therefore,

$$J(u) \leq J(v),$$

by the definition of minimum; for local minimum we must take r small. It follows that

$$\int_{B_r(x_0)} \left(|\nabla u|^2 + I_{\{u>0\}} Q^2 \right) \leq \int_{B_r(x_0)} \left(|\nabla v|^2 + Q^2 \right), \qquad \bullet$$

or

$$(3.4) \qquad \int_{B_r(x_0)} |\nabla(u - v)|^2 \leq Q_{\max}^2 \int_{B_r(x_0)} I_{\{u=0\}}.$$

We wish to estimate the measure of $B_r(x_0) \cap \{u = 0\}$ from above by the left-hand side of (3.4). By scaling we may assume that $B_r(x_0) = B_1$.

For any $\xi \in \partial B_1$ define

$$r_\xi = \inf\{r; \tfrac{1}{4} \leq r \leq 1 \quad \text{and} \quad u(r\xi) = 0\}$$

if the set of r's is nonempty and $r_\xi = 1$ if this set is empty. For a.a. ξ, $u(r\xi)$ is in H^1 in r, $r \in [\frac{1}{4}, 1]$ (and thus in $C^{1/2}$ in r) so that, if $r_\xi < 1$,

$$v(r_\xi \xi) = v(r_\xi \xi) - u(r_\xi \xi) = \int_{r_\xi}^1 \frac{d}{dr}(u - v)(r\xi)dr$$

$$\leq \sqrt{1 - r_\xi} \left[\int_{r_\xi}^1 |\nabla(v - u)(r\xi)|^2 dr \right]^{1/2}.$$

On the other hand, by Poisson's formula

$$v(r_\xi \xi) \geq c_n(1 - r_\xi) \fint_{\partial B_1} u, \qquad c_n > 0.$$

It follows that

$$(1 - r_\xi) \left(\fint_{\partial B_1} u \right)^2 \leq C_n \int_{r_\xi}^1 |\nabla(v - u)(r\xi)|^2 dr.$$

This inequality is obviously valid also if $r_\xi = 1$. Integrating over ξ and recalling the definition of r_ξ we obtain

$$\int_{B_1 \setminus B_{1/4}} I_{\{u=0\}} \left(\fint_{\partial B_1} u \right)^2 \leq C_n \int_{B_1} |\nabla(v - u)|^2 \leq \tilde{C}_n Q_{\max}^2 \int_{B_1} I_{\{u=0\}}, \qquad \text{by (3.4).}$$

Replacing $B_{1/4}$ in the above analysis by $B_{1/4}(\bar{x})$ where $\bar{x} \in \partial B_{1/2}$, we obtain a similar estimate with $B_{1/4}$ replaced by $B_{1/4}(\bar{x})$. Adding the two inequalities and using (3.3), we find that

$$(C^*)^2 \int_{B_1} I_{\{u=0\}} \leq C_n Q_{\max}^2 \int_{B_1} I_{\{u=0\}}.$$

Hence, if C^* is sufficiently large then $I_{\{u=0\}} = 0$ a.e. in B_1, that is, $u > 0$ a.e. in B_1. From (3.4) we then further deduce that $u = v$ a.e. and, by (2.3), $u = v > 0$ everywhere in B_1.

Remark 3.1. If in Lemma 3.1 $B_r(x_0)$ is not contained in Ω but $B_r(x_0) \cap \partial\Omega$ is in $C^{2+\alpha}$ and $u = 0$ on $B_r(x_0) \cap \partial\Omega$, then the assertion (3.3) is still valid (u is extended by zero into $R^n \setminus \Omega$); more precisely, for any $0 < \kappa < 1$, $\varepsilon > 0$, there is a positive constant $C_{\kappa, \varepsilon}$ such that if $C_{\kappa, \varepsilon}^* = C_{\kappa, \varepsilon} Q_{\max}$, then

$$\frac{1}{r^{n+1}} \int_{\partial B_r(x_0) \setminus B_{\varepsilon r}(\partial\Omega)} u \geq C_{\kappa, \varepsilon}^* \text{ implies that } u > 0 \text{ in } B_{\kappa r}(x_0) \cap \Omega,$$

where $B_{\varepsilon r}(\partial\Omega)$ is the (εr)-neighborhood of $\partial\Omega$.

Theorem 3.2. $u \in C^{0,1}(\Omega)$.

Proof. If $x \in \Omega$, $u(x) > 0$, then by Lemma 2.2,

$$\fint_{\partial B_r(x)} u \geq u(x) > 0,$$

and from Lemma 3.1 we conclude that $u > 0$ in some neighborhood of x. Hence the set $\{u > 0\}$ is open.

Let D, D' be bounded open sets with $\bar{D} \subset D'$, $\bar{D}' \subset \Omega$. Let $\rho_0 = \text{dist}(D, \partial D')$. For any $x \in D$ with $u(x) > 0$, let $B_r(x)$ be the largest ball in $D' \cap \{u > 0\}$. If $r < \rho_0$, then $\partial B_r(x)$ contains a free-boundary point. Hence by Lemma 3.1, if r is small enough, say $r \leq r_0$, then

$$\frac{1}{r + \varepsilon} \fint_{B_{r+\varepsilon}(x)} u \leq C^*$$

for all small $\varepsilon > 0$ and, by subharmonicity of u, also

$$(3.5) \qquad \frac{1}{r} \fint_{\partial B_r(x)} u \leq C^*.$$

If $r > \min(\rho_0, r_0)$, then, since u is upper semicontinuous and therefore bounded in $\bar{D}'$, we have

$$(3.6) \qquad \frac{1}{r} \fint_{\partial B_r(x)} u \leq \bar{C}$$

with another constant $\bar{C}$. Finally, if $C = \max(C^*, \bar{C})$,

$$|Du(x)| \leq C \frac{1}{r} \fint_{\partial B_r(x)} u \leq C \qquad \text{in } D \cap \{u > 0\}.$$

Since $Du(x) = 0$ a.e. on $\{u = 0\}$, it follows that $u \in C^{0,1}(\Omega)$.

We next establish a "nondegeneracy" estimate from below on averages of u.

Lemma 3.3. *For any $0 < \kappa < 1$, there exists a positive constant $C_* = C_{n,\kappa} Q_{\min}$ ($C_{n,\kappa}$ depending only on n, κ) such that for any (local) minimum and for every (small) ball $B_r(x_0) \subset \Omega$:*

$$(3.7) \qquad \text{If } \frac{1}{r} \fint_{\partial B_r(x_0)} u \leq C_*, \qquad \text{then } u = 0 \text{ in } B_{\kappa r}(x_0).$$

Proof.   The idea of the proof is that if the average of u on $\partial B_r(x_0)$ is small, then replacing u by a function w which vanishes in $B_{\kappa r}(x_0)$ will decrease J.

By scaling we may take $B_r(x_0) = B_1$. We choose a function

$$v(x) = \frac{\varepsilon\sqrt{\kappa}}{-\Phi_\kappa(\sqrt{\kappa})}\max\left\{-\Phi_\kappa(|x|), 0\right\},$$

where

(3.8) $$\Phi_R(r) = \begin{cases} \dfrac{R}{n-2}\left[\left(\dfrac{r}{R}\right)^{2-n} - 1\right] & \text{if } n \geqslant 3 \\[2ex] R\log\dfrac{R}{r} & \text{if } n = 2 \\[2ex] R - r & \text{if } n = 1 \end{cases}$$

is the fundamental solution and

(3.9) $$\varepsilon = \frac{1}{\sqrt{\kappa}}\sup_{B_{\sqrt{\kappa}}} u \leqslant C_{n,\kappa}\fint_{\partial B_1} u$$

since u is subharmonic.

Notice that $v = 0$ in B_κ. Since $v \geqslant u$ on $\partial B_{\sqrt{\kappa}}$, $\min(u, v)$ is admissible function if extended by u outside $B_{\sqrt{\kappa}}$. Thus $J(u) \leqslant J(\min(u, v))$, which implies that

$$\int_{B_\kappa}\left(|\nabla u|^2 + I_{\{u>0\}}Q^2\right) \leqslant \int_{B_{\sqrt{\kappa}}\setminus B_\kappa}\left(|\nabla\min(u, v)|^2 - |\nabla u|^2\right)$$

$$= \int_{B_{\sqrt{\kappa}}\setminus B_\kappa}\nabla(\min(u, v) - u)\cdot\nabla(\min(u, v) + u)$$

(3.10) $$= -\int_{B_{\sqrt{\kappa}}\setminus B_\kappa}\nabla\max(u - v, 0)\cdot\nabla(v + u)$$

$$\leqslant -2\int_{B_{\sqrt{\kappa}}\setminus B_\kappa}\nabla\max(u - v, 0)\cdot\nabla v = 2\int_{\partial B_\kappa} u\nabla v\cdot\nu$$

$$\leqslant C_{n,\kappa}\varepsilon\int_{\partial B_\kappa} u.$$

We wish to estimate $\int_{\partial B_\kappa} u$ by the left-hand side of (3.10). To do this we write, using the definition of ε in (3.9),

$$\int_{\partial B_\kappa} u \leqslant C_{n,\kappa}\left(\int_{B_\kappa} u + \int_{B_\kappa} |\nabla u|\right)$$

$$\leqslant C_{n,\kappa}\left[\frac{\varepsilon}{Q_{\min}^2}\int_{B_\kappa} I_{\{u>0\}}Q^2 + \frac{1}{Q_{\min}}\int_{B_\kappa}\left(|\nabla u|^2 + I_{\{u>0\}}Q^2\right)\right]$$

$$\leqslant C_{n,\kappa}\frac{1}{Q_{\min}}\left(\frac{\varepsilon}{Q_{\min}}+1\right)\int_{B_\kappa}\left(|\nabla u|^2 + I_{\{u>0\}}Q^2\right).$$

Substituting this estimate into the right-hand side of (3.10), we find that $u = 0$ in B_κ if $\varepsilon/Q_{\min}$ is small enough, that is, if

$$\fint_{\partial B_1} u \text{ is small enough} \qquad [\text{by (3.9)}].$$

Remark 3.2. Lemma 3.3 remains valid if $B_r(x_0)$ is not contained in Ω and $u = 0$ on $B_r(x_0) \cap \partial\Omega$ (u is extended by zero into $R^n \setminus \Omega$).

Corollary 3.4. *For any bounded domain D with $\overline{D} \subset \Omega$ and for some small enough $d_0 > 0$ there exist positive constants c, C such that if a ball $B_d(x)$ with $x \in D$, $d < d_0$ is contained in $\{u > 0\} \cap \Omega$ and $\partial B_d(x)$ contains a free-boundary point, then*

$$cd(x) \leqslant u(x) \leqslant Cd(x).$$

The next result may be compared with Theorem 3.4 of Chapter 2.

Theorem 3.5. *There exist positive constants $0 < c < c_0 < 1$ such that for any local minimum u and for every small ball $B_r(x_0) \subset \Omega$ with $x_0 \in \partial\{u > 0\}$,*

$$(3.11) \qquad c \leqslant \frac{|B_r(x_0) \cap \{u > 0\}|}{|B_r|} \leqslant c_0.$$

It follows that $\partial\{u > 0\}$ has Lebesgue measure zero.

Proof. Take $x_0 = 0$. By Lemma 3.3 there is a point $y \in \partial B_{r/2}$ such that $u(y) \geqslant cr$. Since u is subharmonic, we conclude that

$$\frac{1}{\kappa r}\fint_{\partial B_{\kappa r}(y)} u \geqslant \frac{1}{\kappa r}u(y) \geqslant \frac{c}{\kappa} \qquad (0 < \kappa < 1).$$

Choosing κ small enough, we can apply Lemma 3.1 to deduce that $u > 0$ in $B_{\kappa r}(y)$, which yields the lower bound in (3.11).

To establish the upper bound, choose v as in the proof of Lemma 3.1. Then

$$\int_{B_r} I_{\{u=0\}} \geqslant \int_{B_r} |\nabla(u-v)|^2 \geqslant \frac{c}{r^2} \int_{B_r} |u-v|^2$$

$$\geqslant \frac{c}{r^2} \int_{B_{\kappa r}} |u-v|^2 \qquad (\kappa \text{ small}).$$

If $y \in B_{\kappa r}$, then, by Theorem 3.2 and Lemma 3.3,

$$(v-u)(y) = v(y) - u(y) \geqslant (1 - C\kappa) \fint_{\partial B_r} u - C\kappa r$$

$$\geqslant (1 - C\kappa) C_* r - C\kappa r \geqslant cr \qquad (c > 0)$$

for small enough κ. Thus

$$\fint_{B_r} I_{\{u=0\}} \geqslant c_0 > 0$$

and the upper estimate in (3.11) follows.

Definition 3.1. Let u be a local minimum, and let $B_{\rho_k}(x_k)$ be a sequence of balls in Ω with $\rho_k \to 0$, $x_k \to x_0 \in \Omega$, and $u(x_k) = 0$. The sequence of functions

$$u_k(x) = \frac{1}{\rho_k} u(x_k + \rho_k x)$$

is called a *blow-up sequence*.

Since $|\nabla u_k(x)| \leqslant C$ in every compact set of R^n, if k is large enough, and since $u_k(0) = 0$, it follows that for a subsequence

$$u_k \to u_0 \qquad \text{in } C_{\text{loc}}^\alpha(R^n) \qquad \forall \, 0 < \alpha < 1,$$

$$\nabla u_k \to \nabla u_0 \qquad \text{weakly star in } L_{\text{loc}}^\infty.$$

The function u_o is called a *blow-up limit*.

Recall that the *Hausdorff distance* $d(A, D)$ between two sets A, D is the infimum of the numbers ε such that

$$\bigcup_{x \in A} B_\varepsilon(x) \supset D, \qquad \bigcup_{x \in D} B_\varepsilon(x) \supset A.$$

Lemma 3.6. *The following properties hold*:

(a) $\partial\{u_k > 0\} \to \partial\{u_0 > 0\}$ *in the Hausdorff distance*;

(b) $I_{\{u_k > 0\}} \to I_{\{u_0 > 0\}}$ *in L_{loc}^1*;

(c) If $x_k \in \partial\{u > 0\}$ then $0 \in \partial\{u_0 > 0\}$;

(d) $\nabla u_k \to \nabla u_0$ a.e.;

(e) If Q is continuous, then u_0 is an absolute minimum for $Q(x_0)$ in every bounded domain.

Proof. To prove (a), notice that if $B_r(y) \cap \partial\{u_0 > 0\} = \varnothing$ and $u_0 = 0$ in $B_r(y)$, then the u_k are uniformly small in $B_r(y)$ for k large and, by nondegeneracy (Lemma 3.3) $u_k = 0$ in $B_{r/2}$; if $u_0 > 0$ in $B_r(y)$ then $u_k > 0$ in $B_{r/2}(y)$. Hence $B_{r/2}(y) \cap \partial\{u_k > 0\} = \varnothing$ if k is large enough. On the other hand, if $B_r(y) \cap \partial\{u_k > 0\} = \varnothing$, then the u_k are harmonic in $B_r(y)$ and the same holds for u_0.

Lemmas 3.1 and 3.3 remain valid for u_0, by passage to the limit with u_k; hence the proof of Theorem 3.5 is also valid for u_0, so that

(3.12) $\partial\{u_0 > 0\}$ has Lebesgue measure zero.

Hence an r-neighborhood V_r of this set intersected with any ball B_R has measure $\mu_{r, R}$, where $\mu_{r, R} \downarrow 0$ if $r \downarrow 0$. By the first argument of the preceding section we have that

$$\int_{B_R} \left| I_{\{u_k > 0\}} - I_{\{u_0 > 0\}} \right| \leqslant \int_{V_r \cap B_R} = \mu_{r, R}$$

if k is large enough, and (b) follows.

To prove (c), note, by Lemmas 3.1 and 3.3, that if x_k is a free-boundary point for u_k, then for all small r, say $0 < r < r_0$,

$$c \leqslant \frac{1}{r} \fint_{\partial B_r(x_k)} u_k \leqslant C \qquad \text{for some } c > 0, C > 0;$$

r_0 is independent of k. Taking $k \to \infty$, we obtain

$$c \leqslant \frac{1}{r} \fint_{\partial B_r(x_0)} u_0 \leqslant C$$

which implies that $0 \in \partial\{u_0 > 0\}$.

By (a), for any compact subset E of $\{u_0 > 0\} \cup \text{int }\{u_0 = 0\}$, u_k is harmonic in E if k is large enough, so that $\nabla u_k \to \nabla u_0$ uniformly in any such set as $k \to \infty$. Recalling (3.12), (d) follows.

To prove (e), choose a bounded domain D with $\bar{D} \subset \Omega$. For any function v with $v - u_0 \in H_0^{1,2}(D)$ and $\eta \in C_0^1(D)$, $0 \leqslant \eta \leqslant 1$, set

$$v_k = v + (1 - \eta)(u_k - u_0), \qquad Q_k(x) = Q(x_k + \rho_k x).$$

Since u is a local minimum, if k is large enough, then

$$\int_D \left(|\nabla u_k|^2 + I_{\{u_k>0\}}Q_k^2 \right) \leqslant \int_D \left(|\nabla v_k|^2 + I_{\{v_k>0\}}Q_k^2 \right),$$

Taking $k \to \infty$ and using (d), we obtain

$$\int_D \left[|\nabla u_0|^2 + I_{\{u_0>0\}}Q^2(x_0) \right] \leqslant \int_D \left[|\nabla v|^2 + I_{\{\eta=1\}}I_{\{v>0\}}Q^2(x_0) \right]$$

$$+ \int_D I_{\{0<\eta<1\}}Q^2(x_0).$$

Choosing η such that $\eta = 1$ in D_j, $D_j \uparrow D$, we obtain

$$\int_D \left[|\nabla u_0|^2 + I_{\{u_0>0\}}Q^2(x_0) \right] \leqslant \int_D \left[|\nabla v|^2 + I_{\{v>0\}}Q^2(x_0) \right]$$

for any v with $v - u_0 \in H_0^{1,2}(D)$; this completes the proof.

PROBLEMS

1. Prove Remark 3.1.

2. Extend Theorem 3.2 by proving that $u \in C^{0,1}$ up to any open subset S_1 of $\partial\Omega$, provided that $\bar{S}_1$ is in $C^{2+\alpha}$ and $u^0 \in C^{2+\alpha}(S_1)$.
   [*Hint*: Cf. proof of Theorem 8.5.]

3. Prove Remark 3.2.

4. Consider

$$J_\lambda(v) = \int_{B_1} \left(|\nabla v|^2 + \lambda I_{\{v>0\}} \right) \qquad \text{with } u^0 \equiv 1, \quad \lambda > 0.$$

   By symmetrization (see Section 7) any absolute minimum u must be a function $u = u(r)$, $r = |x|$, and by Section 4, $u(r)$ has at most a finite number of free-boundary points. Prove: (a) the minimizer u_λ is unique; (b) if $0 < \lambda < \lambda_0$, then $u_\lambda \equiv 1$; (c) if $\lambda > \lambda_0$, then $u_\lambda(r) = 0$ if and only if $0 < r < R_\lambda$, and $R_\lambda > 0$, $R_\lambda \uparrow$ if $\lambda \uparrow$.

5. Suppose that $\partial\Omega \setminus S$ is bounded. If u^0 has bounded support, then every local minimum u has bounded support.
   [*Hint*: Let $\partial\Omega \setminus B_R \subset S$, $u^0 = 0$ in $\partial\Omega \setminus B_R$, $D = B_{3R} \setminus B_R$, (trace u) $\neq 0$ on $\partial D \cap \Omega$. Extend u by 0 into $R^n \setminus (B_R \cup \Omega)$ (then it is subharmonic) and let

$$\Delta v = 0 \text{ in } D, \qquad v = u \text{ on } \partial D.$$

The function

$$u_\varepsilon = \begin{cases} u + \varepsilon \min (v - u, 0) & \text{in } \Omega \cap D \\ u & \text{in } \Omega \setminus D(\varepsilon > 0) \end{cases}$$

is in K and $\{u_\varepsilon > 0\} = \{u > 0\}$. Show that

$$\int_{\Omega \cap D} |\nabla \min (v - u, 0)|^2 = -\frac{d}{d\varepsilon} J(u_\varepsilon)\Big|_{\varepsilon = 0} \leq 0,$$

so that $u \leq v$ in D; hence

$$C \equiv \sup_{\partial B_{2R}} u < \infty$$

and, by the maximum principle, $u \leq C$ in $R^n \setminus B_{2R}$. Now proceed as in Lemma 3.3 (see Remark 3.2).]

4. REGULARITY OF THE FREE BOUNDARY

We shall define a concept of flatness at the origin 0 ($0 \in \partial\{u > 0\}$) in the direction e_n of the positive x_n-axis; the definition will depend on parameters σ_+, σ_-, τ.

Definition 4.1. Let $0 \leq \sigma_+, \sigma_- \leq 1$, $\tau > 0$. We say that u belongs to the *flatness class* $F(\sigma_+, \sigma_- ; \tau)$ in B_ρ if

$$0 \in \partial\{u > 0\},$$

$$u(x) = 0 \qquad \text{for } x_n \geq \sigma_+ \rho,$$

$$u(x) \geq -Q(0)(x_n + \sigma_- \rho) \qquad \text{if } x_n \leq -\sigma_- \rho,$$

$$|\nabla u| \leq Q(0)(1 + \tau) \qquad \text{in } B_\rho,$$

$$\operatorname*{osc}_{B_\rho} Q \leq Q(0)\tau.$$

If the origin is replaced by x_0 and the direction e_n by a unit vector ν, then we say that u belongs to the *flatness class* $F(\sigma_+, \sigma_- ; \tau)$ in $B_\rho(x_0)$ *in the direction ν.*

Theorem 4.1. *Suppose that Q is Hölder continuous and u is a local minimizer. Then for any bounded domain D with $\overline{D} \subset \Omega$, there exist positive constants α, β, σ_0, τ_0 such that the following holds: If*

$$(4.1) \qquad u \in F(\sigma, 1; \infty) \text{ in } B_\rho(x_0) \subset D \qquad \text{in direction } \nu$$

with $\sigma \leqslant \sigma_0$ and $\rho \leqslant \tau_0 \sigma^{2/\beta}$, then

$$B_{\rho/4}(x_0) \cap \partial\{u > 0\} \text{ is a } C^{1+\alpha} \text{ surface};$$

it is given by $x_n = \phi(x_1, \ldots, x_{n-1})$, $\phi \in C^{1+\alpha}$ if the direction of the x_n-axis is taken in the direction ν.

The condition (4.1) is called the *flatness* condition.

Theorem 4.2. *If $n = 2$ and Q is Hölder continuous, then for any local minimum u the free boundary $\partial\{u > 0\}$ is locally a $C^{1+\alpha}$ curve.*

These two fundamental theorems are due to Alt and Caffarelli [6a]; the proofs are lengthy and will not be given here.

Combining Theorems 4.1 and 4.2 with Theorem 1.4 of Chapter 2 we can obtain further regularity of the free boundary. In particular:

Corollary 4.3. *Under the assumptions of Theorem 4.1 or 4.2, if Q is analytic, then the free boundary is analytic.*

5. THE BOUNDED GRADIENT LEMMA AND THE NONOSCILLATION LEMMA

In the proof of Theorem 3.2 the bound on $|Du|$ may depend on u; in fact, from (3.5), (3.6) we have that

$$(5.1) \qquad\qquad |Du(x)| \leqslant C\left(1 + \sup_{B_r(x)} u\right).$$

In some applications it is important to obtain a bound on Du which does not depend on $\sup u$. This is done in the following *bounded gradient* lemma.

Lemma 5.1. *Let D be a bounded domain, $\overline{D} \subset \Omega$. There exists a constant $C > 0$ depending only on D, Ω, Q_{max} such that for any minimum u, if D contains a free-boundary point x_0 then*

$$(5.2) \qquad\qquad |\nabla u(x)| \leqslant C \qquad \text{if } x \in D \cap \{u > 0\}.$$

Proof. Let $\rho_0 = \text{dist}(D, \partial\Omega)$. For every $x \in D \cap \{u > 0\}$ there are points $x_1, \ldots, x_k = x$ in D (k depending on ρ_0) such that

$$x_j \in B_{r_0}(x_{j-1}) \qquad \text{for } j = 1, \ldots, k, \ r_0 = \frac{\rho_0}{4}.$$

Choose $k_0 \in \{1, 2, \ldots, k\}$ the largest such that $B_{2r_0}(x_{k_0})$ contains a point $y_0 \in \partial\{u > 0\}$; such a k_0 exists since $B_{2r_0}(x_1)$ contains x_0. Then u is harmonic in $B_{2r_0}(x_{j-1})$ for $j \geqslant k_0 + 2$ and, by Harnack's inequality,

$$u(x_j) \leqslant C \int_{\partial B_{2r_0}(x_{j-1})} u = Cu(x_{j-1}).$$

Since u is subharmonic, we also have

$$u(x_{k_0+1}) \leqslant C \int_{\partial B_{4r_0}(y_0)} u \leqslant 4CC^* r_0$$

by Lemma 3.1. Therefore, $u(x) \leqslant C$ for any $x \in D$, that is,

$$\sup_D u \leqslant C,$$

where C depends only on Ω, D, $Q_{\max}$. Using this estimate in (5.1), the proof of the lemma is complete.

Remark 5.1. If u is only a local minimum, then the proof above is valid provided that r_0 is chosen sufficiently small (depending on the ε in Definition 2.1); C then depends also on ε.

Remark 5.2. If we make use of Remarks 3.1 and 3.2, we find that Lemma 5.1 extends to the case where $\bar{D}$ intersects $\partial\Omega$, provided that $\partial D \cap \partial\Omega$ is contained in a $C^{2+\alpha}$ open subset T of $\partial\Omega$ and $u = 0$ on T.

The next result, essentially unrelated to the previous lemma, is concerned with the behavior of the free boundary near the fixed boundary, and is stated in case Ω is a two-dimensional domain. We denote the points of Ω by $X = (x, y)$.

Let G be a domain in Ω bounded by four curves:

$$y = y_1, \qquad y = y_1 + h \quad (h > 0),$$

$$\gamma_1 : X = X^1(t) \qquad \text{and} \qquad \gamma_2 : X = X^2(t),$$

where $0 \leqslant t \leqslant T$. We assume: If $X^j(t) = (x^j(t), y^j(t))$, then

$$y_1 < y^j(t) < y_1 + h \qquad \text{if } 0 < t < T,$$

$$y^j(0) = y_1, \qquad y^j(T) = y_1 + h,$$

$$x_0 \leqslant x^j(t) \leqslant x_0 + \delta.$$

Further, γ_2 lies to the right of γ_1; more precisely, $x^1(0) < x^2(0)$ and γ_1, γ_2 do not intersect. We finally assume that

(5.3)
$$\gamma_1, \gamma_2 \text{ belong to } \partial\{u > 0\},$$
$$u > 0 \text{ in some } G\text{-neighborhood of } \gamma_1 \cup \gamma_2,$$

$$S_1 \text{ is a } C^{2+\alpha} \text{ open subset of } \partial\Omega,$$

(5.4)
$$u = 0 \text{ on } S_1,$$

$$C_* = \operatorname{dist}(G, \partial\Omega \setminus S_1) > 0.$$

The *nonoscillation* lemma asserts:

Lemma 5.2.   Under the foregoing assumptions,

(5.5)
$$h \leqslant C\delta,$$

where C is a constant depending only on C_*, $\partial\Omega$, $Q_{\max}$, $Q_{\min}$, and $\sup u^0$.

   Proof.   By Section 3, Problem 2 we obtain (5.1) in G, with $B_r(x)$ replaced by Ω; recalling Lemma 2.3, we conclude that

(5.6)
$$|Du(X)| \leqslant C \qquad \text{in } G,$$

where C depends on the same constants as in the assertion of the lemma. Set

$$G_0 = \{X \in G; u(X) > 0\},$$

$$a = y_1, \qquad b = y_1 + h$$

and apply Green's formula to $y - a$ and u in G_0:

$$\int_{\partial G_0} u \frac{\partial y}{\partial \nu} \, ds - \int_{\partial G_0} \frac{\partial u}{\partial \nu} (y - a) \, ds = 0.$$

Since $u = 0$ on $\partial G_0 \cap \{a < y < b\}$ and $y - a = 0$ on $\{y = a\}$, and since $-\partial u/\partial \nu \geqslant 0$ on $\partial G_0 \cap \{a < y < b\}$, we get

(5.7) $$\int_{\gamma_1 \cup \gamma_2} Q(y - a) \, ds \leqslant \int_{\partial G_0 \cap \{y=b\}} u \, dx + h \int_{\partial G_0 \cap \{y=b\}} \frac{\partial u}{\partial \nu} \, dx.$$

Using (5.6), we find that the second term on the right-hand side is bounded by

$Ch\delta$, whereas the first term is bounded by $C\delta^2$ (since $u \leqslant C\delta$). Noting that

$$\int_{\gamma_1 \cup \gamma_2} Q(y - a)\, ds \geqslant Q_{\min} h^2,$$

we obtain from (5.7)

$$Q_{\min} h^2 \leqslant Ch\delta + C\delta^2,$$

and the assertion (5.5) follows.

Remark 5.3. The previous proof is valid also if γ_1 lies on $\partial\Omega$ and

$$(5.8) \qquad\qquad\qquad -\frac{\partial u}{\partial \nu} \geqslant Q \qquad \text{on } \gamma_1.$$

Remark 5.4. It is well known [108] that if

$$\Delta u = 0 \qquad \text{in } \Omega,$$

$$u \in C^0(\bar{\Omega}),$$

$$u = \phi \qquad \text{on } \partial\Omega,$$

and if $S_0 = B_R(x_0) \cap \partial\Omega$ is a $C^{1+\alpha}$ surface and $\phi \in C^{1+\alpha}(S_0)$, then $u \in C^{1+\alpha}(S_0)$. The same is valid also for general elliptic operators with smooth coefficients. Using this fact, we can replace the $C^{2+\alpha}$ condition in Section 3 Problem 2, and Remarks 5.2 and (5.4) by a $C^{1+\alpha}$ condition.

6. CONVERGENCE OF FREE BOUNDARIES

In this section we establish for $n = 2$ a general theorem on the convergence of free boundaries corresponding to a sequence of local minimizers u_m. The theorem can be used also when the free boundaries of u_m converge to the fixed boundary.

Let U be an open set in R^2 such that $U \cap \Omega \neq \varnothing$; U is not necessarily contained in Ω. We extend u_m by 0 into $U \setminus \Omega$ and assume that $u_m \in C^0(U)$, Q_m are constants λ_m; thus $u_m \geqslant 0$ and

$$\Delta u_m = 0 \qquad \text{in } U \cap \{u_m > 0\},$$

$$(6.1) \qquad U \cap \partial\{u_m > 0\} \text{ is in } C^{1+\alpha} \qquad (0 < \alpha < 1),$$

$$-\frac{\partial u_m}{\partial \nu} = \lambda_m \text{ on } U \cap \partial\{u_m > 0\} \qquad (\lambda_m > 0),$$

where ν is the outward normal. We assume that as $m \to \infty$,

$$\lambda_m \to \lambda \qquad (\lambda > 0),$$

$$u_m \to u \qquad \text{uniformly in } U, u \in H^1(U),$$

(6.2)
$$\nabla u_m \to \nabla u \qquad \text{weakly in } L^2(U),$$

$$U \cap \{u_m > 0\} \to U \cap \{u > 0\} \qquad \text{in measure},$$

$$U \cap \partial\{u > 0\} \qquad \text{is in } C^{1+\alpha}.$$

Theorem 6.1. *Let* u_m *be local minimizers satisfying* (6.1), (6.2). *Then*

(6.3)
$$\Delta u = 0 \qquad \text{in } U \cap \{u > 0\},$$

(6.4)
$$-\frac{\partial u}{\partial \nu} = \lambda \qquad \text{on } U \cap \partial\{u > 0\}.$$

Proof. For any nonnegative $\zeta \in C_0^\infty(U)$,

(6.5)
$$\lambda_m \int_{\partial\{u_m > 0\}} \zeta = -\int_{\{u_m > 0\}} \nabla\zeta \cdot \nabla u_m \to -\int_{\{u > 0\}} \nabla\zeta \cdot \nabla u,$$

using the assumption that $U \cap \{u_m > 0\} \to U \cap \{u > 0\}$. Using this last property, we can also deduce that

(6.6)
$$\int_{\partial\{u > 0\}} \zeta \leqslant \liminf_{m \to \infty} \int_{\partial\{u_m > 0\}} \zeta.$$

Indeed, for any vector $f = (f_1, f_2)$ with $f_i \in C_0^1(U)$, $f_1^2 + f_2^2 \leqslant 1$, we have

$$\int_{\partial\{u > 0\}} \zeta f \cdot \nu = \int_{\{u > 0\}} \nabla \cdot \zeta f \leftarrow \int_{\{u_m > 0\}} \nabla \cdot \zeta f = \int_{\partial\{u_m > 0\}} \zeta f \cdot \nu.$$

Taking $f = f^j$ such that $f = \nu$ on $U_j \cap \partial\{u > 0\}$, where $U_j \uparrow U$, we find that

$$\int_{U_j \cup \partial\{u > 0\}} \zeta \leqslant \liminf_{m \to \infty} \int_{\partial\{u_m > 0\}} \zeta$$

and (6.6) follows.

From (6.5), (6.6) we get

$$\lambda \int_{\partial\{u > 0\}} \zeta \leqslant -\int_{\{u > 0\}} \nabla\zeta \cdot \nabla u.$$

Since $\partial\{u > 0\}$ is in $C^{1+\alpha}$, u is also in $C^{1+\alpha}$ up to this boundary (see Remark 5.4) and we obtain

$$\int_{\partial\{u>0\}} \left(\lambda + \frac{\partial u}{\partial \nu}\right)\zeta \leq 0,$$

and since ζ is arbitrary,

$$\lambda \leq -\frac{\partial u}{\partial \nu}.$$

To prove the reverse inequality, take any point $X_0 = (x_0, y_0)$ in $U \cap \partial\{u > 0\}$. Then there is a sequence of points $X_m \in U$ such that $X_m \to X_0$, $u_m(X_m) = 0$. Indeed, otherwise we shall have $\Delta u_m = 0$ in some neighborhood V of X_0 and therefore also $\Delta u = 0$ in V. By the maximum principle it then follows that $u \equiv 0$ in V, contradicting the assumption that $X_0 \in \partial\{u > 0\}$.

Choose r small and domains $D_m \subset B_r(X_0) \subset U$ with

$$B_r(X_0) \cap \partial D_m \to B_r(X_0) \cap \partial\{u > 0\} \text{ in } C^{1+\beta} \qquad (0 < \beta < \alpha),$$

(6.7) $$u_m > 0 \qquad \text{in } D_m,$$

$$X_m \notin D_m.$$

For definiteness we may assume that

$$B_r(X_0) \cap \partial\{u > 0\} \text{ is given by } y = f(x), \quad f'(x_0) = 0,$$

$$D_m = B_r(X_0) \cap \{y < f_m(x)\},$$

$$f_m(x) = f(x) - \varepsilon_m, \qquad \varepsilon_m \downarrow 0 \text{ if } m \to \infty.$$

Let

(6.8) $$\eta(x) = \begin{cases} \exp\left(-\dfrac{x^2}{1-x^2}\right) & \text{if } |x| < 1, \\ 0 & \text{if } |x| > 1 \end{cases}$$

and define

$$f_{s,m}(x) = f_m(x) + s\eta\left(\frac{x - x_0}{r/2}\right), \qquad s > 0.$$

Define a domain $D_{s,m}$ by

$$D_{s,m} = B_r(X_0) \cap \{y < f_{s,m}(x)\}.$$

Then $D_{0, m} = D_m$ and, in view of (6.7), there is a value $s = s_m$ such that

$$u_m > 0 \quad \text{in } D_{s_m, m},$$

$$B_r(X_0) \cap \partial D_{s_m, m} \text{ contains a free-boundary point}$$

$$\text{of } u_m, \text{ say } \tilde{X}_m = (\tilde{x}_m, \tilde{y}_m), \text{ where } \tilde{y}_m = f_{s_m, m}(\tilde{x}_m),$$

and

(6.9) $$s_m \to 0 \quad \text{if } m \to 0.$$

Set $\tilde{D}_m = D_{s_m, m}$.

Let v_m be the solution of

$$\Delta v_m = 0 \quad \text{in } \tilde{D}_m,$$

$$v_m = 0 \quad \text{on } \partial \tilde{D}_m \cap B_r(X_0),$$

$$v_m = u_m \quad \text{on } \partial \tilde{D}_m \cap \partial B_r(X_0).$$

Since $\Delta u_m = 0$ in $\tilde{D}_m$, the maximum principle gives $u_m \geqslant v_m$ in $\tilde{D}_m$ and, therefore,

(6.10) $$\lambda_m = -\frac{\partial u_m(\tilde{X}_m)}{\partial \nu} \geqslant -\frac{\partial v_m(\tilde{X}_m)}{\partial \nu}.$$

In view of (6.9),

$$f_{s_m, m}(x) \to f(x) \quad \text{in } C^{1+\beta}.$$

Hence, by elliptic estimates, v_m in $\tilde{D}_m \cap B_{r/2}(X_0)$ converges to u in $\{u > 0\} \cap B_{r/2}(X_0)$ in the $C^{1+\beta}$ sense. We may assume that $\tilde{X}_m \to \tilde{X}$ as $m \to \infty$ and, by (6.9),

$$\tilde{X} \in B_{r/2}(X_0) \cap \partial\{u > 0\}.$$

We then have

$$\frac{\partial v_m(\tilde{X}_m)}{\partial \nu} \to \frac{\partial u(\tilde{X})}{\partial \nu}.$$

Recalling (6.10), the inequality

(6.11) $-\dfrac{\partial u(\tilde{X})}{\partial \nu} \leqslant \lambda$

follows. Taking $r \to 0$, we obtain (6.11) at $\tilde{X} = X_0$, and the proof is complete.

PROBLEMS

1. Extend all the results of Sections 2, 3, 5, and 6 to the functional

$$J(v) = \int_\Omega \left(\left| \frac{\nabla v}{y} \right|^2 + Q^2 I_{\{v>0\}} \right) y \, dx \, dy,$$

where Ω is a two-dimensional domain; here

$$\frac{\partial^2 u}{\partial x^2} + \frac{\partial^2 u}{\partial y^2} - \frac{1}{y}\frac{\partial u}{\partial y} = 0 \quad \text{in } \{u > 0\},$$

$$-\frac{1}{y}\frac{\partial u}{\partial \nu} = Q \quad \text{on the free boundary,}$$

where ν is the outward normal. [Theorem 4.2 also extends to this case and, in fact, all the results of Sections 2 to 6 extend to functionals

$$\int_\Omega \left[a_{ij}(x) v_{x_i} v_{x_j} + Q^2 I_{\{v>0\}} \right] dx$$

with $\Omega \subset R^n$ provided that (a_{ij}) is positive definite and smooth.]
[*Hint:* See Lemmas 8.3 and 8.6.]

2. Do the same for

$$J_0(v) = \int_\Omega | \nabla v + eQ I_{\{v>0\}} |^2 \, dx,$$

where $\Omega \subset R^n$ and e is a unit vector in R^n, and for

$$J_1(v) = \int_\Omega \left| \frac{\nabla v}{y} + eQ I_{\{v>0\}} \right|^2 y \, dx \, dy,$$

where $\Omega \subset R^2$.

7. SYMMETRIC REARRANGEMENTS

Later in this chapter and in the following chapter we shall use some well-known facts about symmetrization of functions.

Let E be any open set in the two-dimensional strip $\{(x, y); -a < x < a\}$ and let $E_\rho = E \cap \{x = \rho\}$. Denoting by $|E_\rho|$ the one-dimensional Lebesgue measure of E_ρ, let

$$(7.1) \qquad E_\rho^* = \left\{(\rho, y); -\tfrac{1}{2}|E_\rho| < y < \tfrac{1}{2}|E_\rho|\right\}.$$

The set

$$(7.2) \qquad E^* = \bigcup_{-a < \rho < a} E_\rho^*$$

is called the *Steiner symmetrization* of E about the line $y = 0$. If E is closed, we define

$$E_\rho^* = \left\{(\rho, y); -\tfrac{1}{2}|E_\rho| \leqslant y \leqslant \tfrac{1}{2}|E_\rho|\right\}$$

and then E^*, defined by (7.2), is again called the Steiner symmetrization of E.

Clearly, meas E = meas E^*. One can show that if E is open (closed), then E^* is open (closed).

Let $u(x, y)$ be a continuous nonnegative function defined in a rectangle

$$Q_{a,b} = \{(x, y); -a < x < a, -b < y < b\}$$

such that $u(x, y) = u(x, -y)$. For any fixed x, consider the function $v(y) = u(x, y)$ and consider the sets

$$v^{-1}[c, \infty), \qquad v^{-1}(c, \infty) \text{ for any } c \geqslant 0,$$

where $v^{-1}(A) = \{y \in [-b, b]; v(y) \in A\}$ for any set A. Denote by $(v^{-1}[c, \infty))^*$ and $(v^{-1}(c, \infty))^*$ the segments

$$\left\{y; |y| \leqslant \tfrac{1}{2}|v^{-1}[c, \infty)|\right\} \text{ and } \left\{y; |y| < \tfrac{1}{2}|v^{-1}(c, \infty)|\right\}$$

respectively. Define

$$u^*(x, y) = c \qquad \text{if and only if } y \in \left(v^{-1}[c, \infty)\right)^* \setminus \left(v^{-1}(c, \infty)\right)^*.$$

It follows that $u^*(x, y) = u^*(x, -y)$, $u^*(x, y)$ is monotone decreasing in y for

$y > 0$, and

$$\text{meas} \{y; u^*(x, y) \in (c, d)\} = \text{meas} \{y; u(x, y) \in (c, d)\}$$

for any $0 < c < d$. One can also show that

$$(7.3) \qquad \int g(u^*(x, y), x) \, dx \, dy = \int g(u(x, y), x) \, dx \, dy$$

for any continuous function g and that

$$(7.4) \quad \int u^*(x, y) w^*(x, y) h(x) \, dx \, dy \geqslant \int u(x, y) w(x, y) h(x) \, dx \, dy$$

if w satisfies the same properties as u and h is a nonnegative continuous function. If $h > 0$, $w > 0$, then strict inequality holds in (7.4) unless $u^* = u$ a.e.

Using (7.3), one can now define u^* also when u is any L^1 function, by approximation; it is an even function in y, decreasing in y for $y > 0$, and it satisfies (7.3), (7.4).

Definition 7.1. The function $u^*(x, y)$ is called (*symmetric*) *decreasing rearrangement* of $u(x, y)$ in the variable y or a *Steiner symmetrization* of $u(x, y)$ with respect to the line $y = 0$.

Theorem 7.1. *If* $u \in H^1(Q_{a, b})$, $u \geqslant 0$, $u(x, y) = u(x, -y)$, $u(x, b) = 0$ *then*

$$(7.5) \qquad \int_{Q_{a, b}} |\nabla u^*|^2 h(x) \, dx \, dy \leqslant \int_{Q_{a, b}} |\nabla u|^2 h(x) \, dx \, dy$$

for any continuous nonnegative function $h(x)$.

This result is basically contained in reference 154; see also references 92 and 103.

One can define similarly the concept of *increasing rearrangement* u^*, assuming that $u \geqslant 0$, $u(x, 0) = 0$ and $u(x, y) = u(x, -y)$. This function is, for any x, monotone increasing in y, for $y > 0$, and it has the same distribution function in y as $u(x, y)$, that is,

$$\text{meas} \{y; u(x, y) \in A\} = \text{meas} \{y; u^*(x, y) \in A\}.$$

One can also define increasing rearrangement in y with respect to the measure $y \, dy$ instead of dy. The following result is proved in reference 102:

Theorem 7.1'. *If u^* is an increasing rearrangement of u in y with respect to the measure $y\,dy$, then*

(7.6)
$$\int_{Q_{a,b}} |\nabla u^*|^2 y\,dx\,dy \leqslant \int_{Q_{a,b}} |\nabla u|^2 y\,dx\,dy.$$

Let (r, θ, z) be cylindrical coordinates for $x \in R^3$ and suppose that

$$u(x) = \rho(r, z), \qquad \rho \geqslant 0, \qquad \rho(r, z) = \rho(r, -z).$$

Let u^* be a decreasing rearrangement of u in the variable z. Then we have:

Theorem 7.2. *There holds*

(7.7)
$$\int_{R^3}\int_{R^3} \frac{u^*(x)u^*(y)}{|x-y|}\,dx\,dy \geqslant \int_{R^3}\int_{R^3} \frac{u(x)u(y)}{|x-y|}\,dx\,dy$$

and equality holds if and only if $u^* = u$ a.e.

The inequality (7.7) is a consequence of the following result:

Theorem 7.3. *If $f(x)$, $g(x)$, $h(x)$ are even nonnegative functions and if f^*, g^*, h^* are their decreasing symmetric rearrangements (about $x = 0$), then*

(7.8) $\displaystyle\int_{R^1}\int_{R^1} f^*(x)g^*(y)h^*(x-y)\,dx\,dy \geqslant \int_{R^1}\int_{R^1} f(x)g(x)h(x-y)\,dx\,dy.$

The proof is given in reference 112; the assertion about equality in (7.8) follows from the proof.

We next introduce a spherically symmetric decreasing rearrangement. Let $u(x)$ be defined in a bounded open set $\Omega \subset R^n$ and let

$$\mu(t) = \text{meas}\,\{x \in \Omega;\, |u(x)| > t\}.$$

Define

$$u^*(r) = \inf\,\{t \geqslant 0;\, \mu(t) < r\}$$

and

$$u^*(x) = u^*(\gamma_n r^n), \qquad r = |x|, \qquad \gamma_n = \frac{\pi^{n/2}}{\Gamma(1 + n/2)}.$$

Then $u^*(r)$ is a decreasing function and

$$(7.9) \qquad \int_{R^n} g(u^*(x)) \, dx = \int_\Omega g(u(x)) \, dx$$

for any continuous positive function g; also,

$$\int_{R^n} u^* v^* \, dx \geqslant \int_G uv \, dx,$$

u and v are in $L^2(G)$. Next:

Theorem 7.4.   *If* $u \in W^{1,p}(G)$, $1 \leqslant p < \infty$, *then*

$$(7.10) \qquad \int_{R^n} |\nabla u^*|^p \, dx \leqslant \int_G |\nabla u|^p \, dx.$$

This result for $p = 2$ is given in reference 154 and for general p in reference 169. We finally mention a generalization of Theorem 7.3, given in reference 163.

Theorem 7.5.   *Let* f, g, h *be nonnegative functions in* R^n *and let* f^*, g^*, h^* *be their spherically symmetric decreasing rearrangements, respectively. Then*

$$\int_{R^n} \int_{R^n} f^*(x) g^*(x) h^*(x - y) \, dx \, dy \geqslant \int_{R^n} \int_{R^n} f(x) g(x) h(x - y) \, dx \, dy.$$

PROBLEMS

1.   Prove (7.3).

2.   Prove (7.4).

3.   Show that Theorem 7.2 follows from Theorem 7.3.

4.   Prove Theorem 7.4 for $p = 2$, $n = 1$.

5.   Prove that if $u(x, y)$ is continuous, then $u^*(x, y)$ is continuous.

8. AXIALLY SYMMETRIC JET FLOWS

Let N be a continuous curve $X = X(t) = (x(t), y(t))$ $(0 \leqslant t < \infty)$ such that

$$
\begin{aligned}
&X(0) = A, \quad \text{where } A = (0,1); \\
&X = X(t) \quad \text{is a } y\text{-graph}; \\
&y(t) \geqslant 1; \\
&|X(t)| \to \infty \quad \text{if } t \to \infty; \\
&X(t) \quad \text{is piecewise } C^{1+\alpha} \quad (0 < \alpha < 1); \\
&\nabla X(t \pm 0) \neq 0 \quad \text{for all } 0 \leqslant t < \infty;
\end{aligned}
$$
(8.1)

The second condition in (8.1) means that any line $y = y_0$ which does intersect N intersects it in either one point or one segment. We have assumed for simplicity that $X(0) = (0, 1)$, but all the subsequent results hold for the general case of $X(0) = (x(0), y(0))$, $y(0) > 0$.

Set

$$
A' = (-a, 1), \qquad A'' = (-a, 0)
$$

for some $a > 1$, and let

l_1 : the ray $y = 1$, $x > 0$,

l_0 : the curve consisting of $\overline{AA'}$, $\overline{A'A''}$ and

the interval $y = 0$, $-\infty < x < -a$.

Let Ω be the domain in R^2 bounded by N, the x-axis and l_1 and let D be the subdomain of Ω bounded by N and l_0; see Figure 3.4.

Definition 8.1. N is called a *nozzle*.

Definition 8.2. Set $R_+ = R^2 \cap \{y > 0\}$. The *axially symmetric jet problem* is the problem of finding a function $u(x, y)$ in $C^0(\overline{R}_+)$, a C^1 *curve* $\Gamma : X = X_0(t) = (x_0(t), y_0(t))$ $(0 \leqslant t < \infty)$ in R_+ and positive parameters λ, Q such that

(8.2) $$X_0(0) = A,$$

(8.3) $$\Gamma_0 = \Gamma \setminus \{A\} \quad \text{does not intersect } N,$$

(8.4) $$x_0(t) \to \infty, \quad y_0(t) \to h \quad (h > 0),$$

$$x_0'(t) \to 1, \quad y_0'(t) \to 0 \quad \text{if } t \to \infty,$$

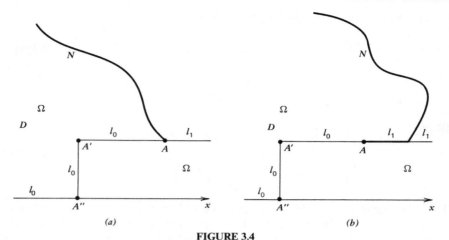

FIGURE 3.4

(8.5)   the curve $N \cup \Gamma$ is C^1 in a neighborhood of A,

(8.6) $$0 \leqslant u \leqslant Q \quad \text{in } \overline{R_+},$$

(8.7) $$u(x,0) = 0 \quad \text{if } -\infty < x < \infty,$$

(8.8) $$u = Q \quad \text{on } N,$$

(8.9) $$Lu = 0 \quad \text{in } R_+ \cap \{u < Q\},$$

where

$$Lu = \frac{\partial^2 u}{\partial x^2} + \frac{\partial^2 u}{\partial y^2} - \frac{1}{y}\frac{\partial u}{\partial y},$$

(8.10) $$\left[\partial\{u < Q\} \setminus N\right] \cap R_+ \quad \text{coincides with } \Gamma_0,$$

so that $u = Q$ on Γ,

(8.11)

$u \in C^1$ in $\{u < Q\}$ neighborhood of any point of Γ_0 and $\dfrac{1}{y}\dfrac{\partial u}{\partial \nu} = \lambda \quad$ on Γ_0,

where ν is the outward normal to Γ_0, and

(8.12) u is in $C^1\left(B_R(A) \cap \{\overline{u < Q}\}\right) \quad$ for some $R > 0$.

Definition 8.3. u is called the *stream function*, Γ (or Γ_0) is called the *free* *boundary*, λ is called the *velocity* on the free boundary, and $2\pi Q$ is called the *flux*.

The flux of a flow across a planar section T in R^3 is defined by

$$F = \int_T \nabla\phi \cdot \nu \, dS \qquad (\nu = \text{normal to } T),$$

where ϕ is the velocity potential. In the axially symmetric case (see Section 1), if $T = \{u < Q\} \cap \{x = x_0\}$,

$$F = \int_T \phi_x 2\pi y \, dy = \int 2\pi u_y \, dy = 2\pi Q.$$

Notice next that since $u = Q$ on $N \cup \Gamma$, $N \cup \Gamma$ is a stream line.

From some of the conditions in (8.2)–(8.12), one can deduce (see Problem 1) that

(8.13)

$$u_x \to 0, \ \frac{1}{y} u_y \to \lambda \qquad \text{if } x \to \infty,$$

$$\text{uniformly in } \{u < Q\} \cap \left\{ y > \frac{h}{2} \right\}.$$

It follows that

(8.14)
$$h = \left(\frac{2Q}{\lambda} \right)^{1/2}.$$

If (u, Γ, λ, Q) is a solution of the jet problem, then also $(\beta u, \Gamma, \beta\lambda, \beta Q)$ is a solution of the jet problem, for any $\beta > 0$. (Notice that h does not change by this scaling.) Therefore, from now on we take Q fixed and seek a solution (u, Γ, λ) corresponding to this Q.

Under the assumption of (8.1) we anticipate the free boundary Γ to lie in $\{0 < y < 1\}$, that is, $u = Q$ in $R_+ \setminus \Omega$. This is why in the future we shall replace R_+ by Ω.

Definition 8.4. The condition (8.2) is called the *continuous-fit* condition, and the conditions (8.5), (8.12) are called the *smooth-fit* conditions.

Theorem 8.1. *If the nozzle N satisfies (8.1), then there exists a solution of the axially symmetric jet problem.*

The proof of Theorem 8.1 depends on the variational approach of the preceding sections. We need, however, to truncate Ω by domains Ω_μ bounded on the left and above.

Consider first the case where either $y(t) \to \infty$ if $t \to \infty$, or $y(t) \to y_\infty < \infty$ and $x(t) \to -\infty$ if $t \to \infty$.

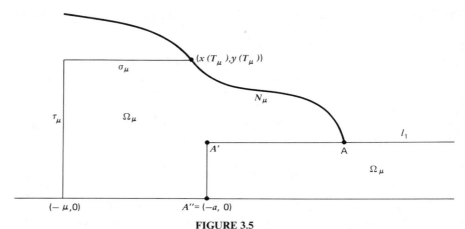

FIGURE 3.5

Let N_μ (for any large μ, $\mu \geqslant a$) be the part of N corresponding to $X(t)$, $0 \leqslant t \leqslant T_\mu$, and introduce the horizontal segment

$$\sigma_\mu = \left\{ \left(x, y(T_\mu) \right); -\mu < x < x(T_\mu) \right\}$$

and the vertical segment

$$\tau_\mu = \left\{ (-\mu, y); 0 < y < y(T_\mu) \right\};$$

T_μ is chosen so that $x(t) > -\mu$ if $0 \leqslant t \leqslant T_\mu$ and $T_\mu \to \infty$ if $\mu \to \infty$. Let Ω_μ be the domain bounded by N_μ, σ_μ, τ_μ, l_1 and the line $\{ y = 0, -\mu < x < \infty \}$; see Figure 3.5.

Suppose next that $y(t) \to y_\infty < \infty$, $x(t) \to +\infty$ if $t \to \infty$. We then define T_μ as before and let

$$\sigma_\mu^1 = \left\{ \left(x(T_\mu), y \right); y(T_\mu) < y < y(T_\mu) + \mu \right\},$$

$$\sigma_\mu = \left\{ \left(x, y(T_\mu) + \mu \right); -\mu < x < x(T_\mu) \right\},$$

$$\tau_\mu = \left\{ (-\mu, y); 0 < y < y(T_\mu) + \mu \right\}.$$

We now define Ω_μ to be the domain bounded by N_μ, σ_μ, τ_μ, l_1 when N_μ is the union of $\{ X(t), 0 \leqslant t \leqslant T_\mu \}$ with σ_μ^1. We define a class of admissible functions

$$K_\mu = \left\{ v; v \in H^{1,2}(\Omega_\mu \cap B_R) \qquad \forall R > 0, 0 \leqslant v \leqslant Q, \right.$$

$$v = Q \qquad \text{on } N_\mu \cup l_1,$$

(8.15) $\qquad v = 0 \qquad \text{on } \tau_\mu \text{ and on } \{ -\mu < x < \infty, y = 0 \},$

$$\left. v = Q \frac{x + \mu}{x(T_\mu) + \mu} \qquad \text{on } \sigma_\mu \right\}$$

and a functional

$$(8.16) \qquad J_{\lambda,\mu}(v) = \int_{\Omega_\mu} \left| \frac{1}{y} \nabla v - \lambda I_{\{v<Q\}\setminus D} e \right|^2 y \, dx \, dy,$$

where e is the unit vector $(0, 1)$.

Problem $J_{\lambda,\mu}$. Find a solution $u = u_{\lambda,\mu}$ in K_μ of

$$(8.17) \qquad J_{\lambda,\mu}(u) = \min_{v \in K_\mu} J_{\lambda,\mu}(v).$$

The term $I_{\{v<Q\}\setminus D}$ in $J_{\lambda,\mu}$ needs some explanation. Since the free boundary is $\partial\{u < Q\}$, $\{v < Q\}$ must replace $\{v > 0\}$ in the functional in (2.1). Next, our purpose is to find a free boundary that does not go "infinitely deep" leftward into Ω, that is, which does not enter some domain D with a sufficiently large a; for this reason we have included the complement of D in I. Of course, we shall have to prove later that the free boundary does not touch ∂D.

The functional

$$(8.18) \qquad \tilde{J}_{\lambda,\mu}(v) = \int_{\Omega_\mu} \left(\left| \frac{1}{y} \nabla v \right|^2 + \lambda^2 I_{\{v<Q\}\setminus D} \right) y \, dx \, dy$$

is infinite for every admissible function. For this reason we work with $J_{\lambda,\mu}(v)$; this functional (as we shall see) remains finite for flows that are nearly horizontal as $x \to \infty$.

Since we anticipate $h \leq 1$, we shall henceforth always assume that

$$(8.19) \qquad \lambda \geq 2Q.$$

The proof of Theorem 8.1 begins in this section and ends in Section 12.

The outline of the proof is as follows. We solve the problem $J_{\lambda,\mu}$ for any λ, μ, and then show that for some $\lambda = \lambda(\mu)$ the solution solves the jet problem in the truncated domain Ω_μ. Letting $\mu \to \infty$, we shall obtain the desired solution.

We cannot deal directly with the functional

$$(8.20) \qquad J_\lambda(v) = \int_\Omega \left| \frac{1}{y} \nabla v - \lambda I_{\{v<Q\}\setminus D} e \right|^2 y \, dx \, dy$$

since in general $J_\lambda(v) = \infty$ for any admissible function.

In the remaining part of this section we establish for $J_{\lambda,\mu}$ results analogous to results established in Sections 2 and 3.

Lemma 8.2. *Problem $J_{\lambda,\mu}$ has a solution.*

Proof. Since $\lambda \geqslant 2Q$, the function

(8.21)
$$u_0 = \min\left\{\frac{\lambda}{2}y^2, Q\right\},$$

for $x > 1$, can be extended into a function in K_μ, and

$$J_{\lambda,\mu}(u_0) < \infty.$$

Now take a minimizing sequence u_k with

$$u_k \to u \quad \text{weakly in } H^{1,2}(\Omega_\mu \cap B_R) \quad \forall R > 0,$$
$$u_k \to u \quad \text{a.e. in } \Omega_\mu,$$

$$I_{\{u_k < Q\} \setminus D} \to \gamma \quad \text{weakly star in } L^\infty(\Omega_\mu \cap B_R) \quad \forall R > 0.$$

Then

$$\sqrt{y}\left(\frac{\nabla u_k}{y} - \lambda I_{\{u_k < Q\} \setminus D}e\right) \to f \quad \text{weakly in } L^2(\Omega_\mu \cap B_R) \quad \forall R > 0,$$

where

$$f = \sqrt{y}\left(\frac{\nabla u}{y} - \lambda \gamma e\right).$$

Since $\gamma = 1$ a.e. in $\{u < Q\} \setminus D$, $\gamma = 0$ in D,

$$J_{\lambda,\mu}(u) \leqslant \int_{\Omega_\mu} |f|^2 \leqslant \liminf_{k \to \infty} J_{\lambda,\mu}(u_k).$$

Lemma 8.3. *There exists a positive constant C independent of μ, λ such that for any ball $B_r(X^0) \subset \Omega_\mu$ with $X^0 = (x^0, y^0)$, $r < y^0/2$,*

$$\frac{1}{r}\fint_{\partial B_r(X^0)} (Q - u) \geqslant C\lambda y^0 \quad \text{implies that } u < Q \text{ in } B_r(X^0);$$

here $u = u_{\lambda,\mu}$ is any solution of problem $J_{\lambda,\mu}$.

Proof. Let

$$Lv = 0 \quad \text{in } B_r(X^0),$$
$$v = u \quad \text{on } \partial B_r(X^0),$$

and define $v = u$ in $\Omega_\mu \setminus B_r(X^0)$. Then $J_{\lambda,\mu}(u) \leqslant J_{\lambda,\mu}(v)$, so that

$$0 \geqslant \int_{B_r(X^0)} \frac{1}{y} \nabla(u-v) \cdot \nabla(u+v)$$

$$- 2\lambda \int_{B_r(X^0)} \frac{\partial}{\partial y}(u-v) - \lambda^2 \int_{B_r(X^0)} y I_{\{u=Q\} \setminus D}.$$

The first integral is equal to

$$\int_{B_r(X^0)} \frac{1}{y} |\nabla(u-v)|^2$$

(see the proof of Lemma 3.1). The second integral is equal to zero. Hence

$$\int_{B_r(X^0)} |\nabla(u-v)|^2 \leqslant 3\lambda^2 (y^0)^2 \int_{B_r(X^0)} I_{\{u=Q\}}.$$

Introducing

$$\tilde{u}(X) = \frac{Q - u(X^0 + rX)}{r}, \qquad \tilde{v}(X) = \frac{Q - v(X^0 + rX)}{r},$$

we have

$$\tilde{L}\tilde{v} \equiv \Delta\tilde{v} - \frac{1}{y + y^0/r} \frac{\partial}{\partial y} \tilde{v} = 0 \qquad \text{in } B_1(0),$$

$$\tilde{L}\tilde{u} \geqslant 0 \qquad \text{in } B_1(0)$$

and

(8.22) $$\int_{B_1(0)} |\nabla(\tilde{u} - \tilde{v})|^2 \leqslant 3\lambda^2 (y^0)^2 \int_{B_1(0)} I_{\{\tilde{u}=0\}}.$$

We can now continue as in the proof of Lemma 3.1, noting that $\tilde{L}$ is uniformly elliptic and smooth in $B_1(0)$, independently of r.

Lemma 8.4. *There holds, for $u = u_{\lambda,\mu}$,*

(8.23) $$u(x, y) \leqslant Q \frac{y^2}{h_\lambda^2}, \qquad \text{where } h_\lambda = \left(\frac{2Q}{\lambda}\right)^{1/2}.$$

Proof. Define u_0 in Ω_μ by (8.21). Then $\min(u, u_0)$ is admissible. Hence

$$0 \geqslant J_{\lambda, \mu}(u) - J_{\lambda, \mu}(\min(u, u_0))$$

$$= \int_{\Omega_\mu} \frac{1}{y} \nabla(u - \min(u, u_0)) \cdot \nabla(u + \min(u, u_0))$$

$$- 2\lambda \int_{\Omega_\mu \setminus D} \frac{\partial}{\partial y}(u - \min(u, u_0))$$

$$+ \lambda^2 \int_{\Omega_\mu \setminus D} y\left(I_{\{u<Q\}} - I_{\min\{(u, u_0)<Q\}}\right)$$

$$= I_1 + I_2 + I_3.$$

We can write

$$I_1 = \int_{\Omega_\mu} \frac{1}{y} \nabla(\max(u - u_0, 0)) \cdot \nabla(u + u_0)$$

and, since we may integrate only over $\{u > u_0\}$,

$$I_1 = \int_{\Omega_\mu} \frac{1}{y} |\nabla(\max(u - u_0, 0))|^2 + 2\int_{\Omega_\mu} \frac{1}{y} \nabla(\max(u - u_0, 0)) \cdot \nabla u_0.$$

The last integral is equal to

$$\int_{\Omega_\mu \cap \{u_0<Q\}} \lambda \frac{\partial}{\partial y} \max(u - u_0, 0) = \int_{\bar{\Omega}_\mu \cap \{y>0\} \cap \partial\{u_0<Q\}} \lambda \max(u - u_0, 0) = 0$$

since $u_0 = Q \geqslant u$ in the last integrand. Thus

$$I_1 = \int_{\Omega_\mu} \frac{1}{y} |\nabla(\max(u - u_0, 0))|^2.$$

Next,

$$I_2 = -2\lambda \int_{(\Omega_\mu \setminus D) \cap \{u_0<Q\}} \frac{\partial}{\partial y} \max(u - u_0, 0)$$

$$= -2\lambda \int_{(\bar{\Omega}_\mu \setminus D) \cap \{y>0\} \cap \partial\{u_0<Q\}} \max(u - u_0, 0) = 0.$$

Finally,

$$I_3 = \lambda^2 \int_{\Omega_\mu \setminus D} y \left(I_{\{u<Q\}} - I_{\{u<Q\} \cup \{u_0<Q\}} \right)$$

$$= -\lambda^2 \int_{\Omega_\mu \setminus D} y I_{\{u=Q\} \cap \{u_0<Q\}}.$$

Combining these estimates, we obtain

$$0 \geqslant \int_{\Omega_\mu} \frac{1}{y} |\nabla \max(u-u_0,0)|^2 - \lambda^2 \int_{\Omega_\mu \setminus D} y I_{\{u=Q\} \cap \{u_0<Q\}}$$

$$\geqslant \int_{(\Omega_\mu \setminus D) \cap \{u=Q\} \cap \{u_0<Q\}} y \left(\left| \frac{\nabla u_0}{y} \right|^2 - \lambda^2 \right) + \int_{\Omega_\mu \cap \{u_0<u<Q\}} \frac{1}{y} |\nabla(u-u_0)|^2$$

$$= \int_{\Omega_\mu \cap \{u_0<u<Q\}} \frac{1}{y} |\nabla(u-u_0)|^2.$$

It follows that $u \leqslant u_0$ in $\Omega_\mu \cap \{u_0 < Q\} \cap \{u < Q\}$.

Now the proof of Theorem 3.2 applies with obvious changes in compact subsets of Ω_μ. Thus u is continuous in Ω_μ and clearly also on τ_μ. Thus $u(x, y) < Q$ if $y > 0$, y small, and x is near $-\mu$. By continuity we now deduce that $u \leqslant u_0$ in Q_μ, wherever $u_0 < Q$.

Theorem 8.5. $u = u_{\lambda, \mu}$ *is Lipschitz continuous in every compact subset of* $\overline{\Omega}_\mu$ *that does not contain A or points where $\partial\Omega_\mu$ is not $C^{1+\alpha}$.*

Proof. The Lipschitz continuity in compact subsets of Ω follows as in Theorem 3.2. To prove Lipschitz continuity near l_1, note that if the distance r from X to $\partial\{u < Q\}$ is less than the distance s from X to l_1, then we can argue as in Theorem 3.2. If, however, $s < r$, set $X^0 = (x, 1)$ if $X = (x, y)$ and $s^0 = \min(\frac{1}{2}, x)$, $B_0 = B_{s^0}(X^0)$. Let

$$Lv = 0 \quad \text{in } B_0 \cap \{y < 1\},$$
$$v = Q \quad \text{on } \partial B_0 \cap \{y < 1\},$$
$$v = 0 \quad \text{on } B_0 \cap \{y = 1\}.$$

Since $Q - u$ is L-subsolution and $Q - u \leqslant 0$ on $\partial B_0 \cap \{y < 1\}$, we have $Q - u \leqslant v$. To the function $\tilde{v}(X') = v(X^0 + s^0 X')$ we can apply elliptic estimates and conclude that

$$v(X^0 + s^0 X') \leqslant C |y'| \quad \text{if } X' = (x', y') \in B_{1/2}(0) \cap \{y' < 0\}.$$

Hence, if $s = 1 - y < s^0/4$,

$$(8.24) \qquad |\nabla(Q - u)(X)| \leqslant \frac{C}{s} \fint_{\partial B_s(X)} (Q - u) \leqslant \frac{C}{s} \fint_{\partial B_s(X)} v \leqslant \frac{C}{s^0};$$

the first inequality in (8.24) follows by representing $w(X') \equiv (Q - u)(X + sX')$ in $|X'| < 1$ in terms of Green's function for the corresponding uniformly elliptic operator $\tilde{L}$, and then applying ∇.

It remains to prove Lipschitz continuity near $y = 0$. If $X^0 = (x^0, y^0)$ with $0 < y^0 < \frac{1}{2}h_\lambda$ and $r_0 = \frac{1}{2}y^0$, then by Lemma 8.4,

$$\left. \begin{array}{r} u \leqslant \dfrac{\lambda}{2} y^2 \\[2mm] Lu = 0 \end{array} \right\} \quad \text{in } B_{r_0}(X^0).$$

Hence the normalized $\tilde{u}(X) = u(X^0 + r_0 X)/r_0^2$ satisfies

$$\left. \begin{array}{r} \tilde{u} \leqslant C \\[2mm] \Delta\tilde{u} - \dfrac{1}{2 + y}\tilde{u}_y = 0 \end{array} \right\} \quad \text{in } B_1(0),$$

which implies that

$$|\nabla u(X^0)| = r_0 |\nabla\tilde{u}(0)| \leqslant Cy^0.$$

Lemma 8.6. *There is a constant $c > 0$ independent of λ, μ such that for any ball $B_r(X^0)$ with center $X^0 = (x^0, y^0) \in \Omega \setminus D$ and radius $r \leqslant y^0/2$, the following holds:*

$$(8.25) \qquad \frac{1}{r} \fint_{\partial B_r(X^0)} (Q - u) \leqslant c\lambda y^0$$

implies that $u = Q$ in $B_{r/8}(X^0) \cap (\Omega \setminus (D \cup B_r(A')))$;

here $u = u_{\lambda, \mu}$ and $Q - u$ is extended by zero into $B_r(X^0) \setminus \Omega$.

Proof. The set

$$B_{r/8}(X^0) \cap (\Omega \setminus (D \cup B_r(A')))$$

can be covered by balls of the form

$$B_{r_1}(X^1) \subset B_{r/4}(X^0) \cap (\Omega \setminus \bar{D}) \qquad \text{with } r_1 = \frac{r}{16};$$

thus it suffices to prove that $u = Q$ in each such ball.

Introducing

$$\tilde{u}(X) = \frac{Q - u(X^1 + r_1 X)}{r_1},$$

we have

$$\tilde{L}\tilde{u} = 0 \quad \text{in } B_1,$$

$$\tilde{L}\tilde{u} \geqslant 0 \quad \text{in } \left\{ y > -\frac{y^1}{r_1} \right\} \quad \text{where } X^1 = (x^1, y^1);$$

our assumption is that

(8.26) $$\fint_{\partial B_0} \tilde{u} \leqslant \delta\lambda y^0, \qquad B_0 = B_{16}\left(\frac{X^0 - X^1}{r_1} \right),$$

where δ is to be chosen small enough. This implies, since $B_2(0) \subset B_8((X^0 - X^1)/r_1)$, that

(8.27) $$\sup_{B_2} \tilde{u} \leqslant C\delta\lambda y^0.$$

Let v be the solution of

$$Lv = 0 \quad \text{in } B_{2r_1}(X^1) \setminus B_{r_1}(X^1),$$
$$v = Q \quad \text{on } \partial B_{r_1}(X^1),$$
$$v = u \quad \text{on } \partial B_{2r_1}(X^1),$$

and define $v = Q$ in $B_{r_1}(X^1)$ and $v = u$ outside $\partial B_{2r_1}(X^1)$. Then $\max(v, u)$ is an admissible function. Proceeding analogously to Lemma 3.3, we have

$$0 \geqslant J_{\lambda, \mu}(u) - J_{\lambda, \mu}(\max(v, u))$$

$$= \int_\Omega \frac{1}{y} \nabla(\min(u - v, 0)) \cdot \nabla(u + v) - 2\lambda \int_{\Omega \setminus D} \frac{\partial}{\partial y} \min(u - v, 0)$$

$$+ \lambda^2 \int_{\Omega \setminus D} y I_{\{u < Q\} \cap \{v = Q\}} \equiv I_1 + I_2 + I_3.$$

By integration of parts, $I_2 \geqslant 0$, since $\min(\ldots) \leqslant 0$ on the top and $= 0$ at the bottom. Next,

$$I_1 = \int_\Omega \frac{1}{y} |\nabla \min(u-v,0)|^2 + 2\int_\Omega \frac{1}{y} \nabla(\min(u-v,0)) \cdot \nabla v$$

$$\geqslant \int_{B_{r_i}(X^1)} \frac{1}{y} |\nabla u|^2 - 2\int_{\partial B_{r_i}(X^1)} (Q-u)\left|\frac{\nabla v}{y} \cdot \nu\right|,$$

$$I_3 \geqslant c\lambda^2 y^0 \int_{B_{r_i}(X^1)} I_{\{u<Q\}}.$$

Combining these estimates, we obtain

$$\int_{B_{r_i}(X^1)} \left(\left|\frac{\nabla u}{y}\right|^2 + c\lambda^2 I_{\{u=Q\}}\right) \leqslant \frac{C}{y^0}\int_{\partial B_{r_i}(X^1)} (Q-u)\left|\frac{\nabla v}{y} \cdot \nu\right|$$

or, in terms of the normalized $\tilde{u}$ and the corresponding $\tilde{v}$,

$$(8.28) \qquad \int_{B_1} \left(|\nabla\tilde{u}|^2 + c\lambda^2 y_0^2 I_{\{\tilde{u}>0\}}\right) \leqslant C\int_{\partial B_1} \tilde{u}|\nabla\tilde{v}\cdot\nu|.$$

By (8.27), $\tilde{v}\leqslant\tilde{u}\leqslant C\delta\lambda y^0$ on ∂B_2 and by interior elliptic estimates in $B_{3/2}$,

$$|\nabla\tilde{v}\cdot\nu|\leqslant C\delta\lambda y^0,$$

and the right-hand side in (8.28) is then estimated by (see the proof of Lemma 3.3)

$$C\delta\lambda y^0\left\{\left(C\delta\lambda y^0 + \frac{1}{\varepsilon}\right)\int_{B_1} I_{\{u>0\}} + \frac{1}{2}\varepsilon\int_{B_1}|\nabla\tilde{u}|^2\right\}.$$

Choosing $\varepsilon = (C\delta\lambda y^0)^{-1}$, we then obtain from (8.28)

$$(1 - \delta^2 C)(\lambda y^0)^2\int_{B_1} I_{\{\tilde{u}>0\}} \leqslant 0,$$

so that $\tilde{u}=0$ in B_1. This completes the proof.

Corollary 8.7. *For any small $\delta>0$ there is a constant $C_\delta>0$ independent of μ, λ such that*

$$u = Q \qquad in \ (\Omega\setminus D)\cap\{\{y\geqslant C_\delta h_\lambda\}\setminus B_\delta(A')\},$$

where $h_\lambda = (2Q/\lambda)^{1/2}$.

Proof. Since

$$\frac{1}{r} \int_{\partial B_r(X^0)} (Q - u) \le \frac{y^0}{r} \frac{Q}{y^0}, \qquad X^0 \in \Omega \setminus D,$$

choosing $r = y^0/2$, $c(y^0)^2 = Q/\lambda$, Lemma 8.6 gives $u = Q$ in $B_r(X^0)$.

Lemma 8.8. $u < Q$ *in some neighborhood of A'.*

For proof, see Problem 2.

PROBLEMS

1. Prove (8.13) if $X_0(t)$ is in $C^{2+\alpha}$.
    [*Hint*: Consider $u_n(x, y) = u(x + n, y)$ and apply Theorem 1.4 of Chapter 2.]

2. Prove Lemma 8.8.
    [*Hint*: Let $Lv = 0$ in $B_\delta(A') \cap D$, $v = Q$ on $B_\delta(A') \cap \partial D$, $v = u$ on $\partial B_\delta(A') \cap D$. Show that $v \le Q - Cr^{2/3}$ by considering

$$w(z') = (Q - v)(z), z = x + iy, z' = z^{2/3}.]$$

9. THE FREE BOUNDARY IS A CURVE $x = k(y)$

Since the results of Section 4 apply also to general elliptic equations with smooth coefficients, the free boundary

$$\Gamma_{\lambda, \mu} = \partial\{u_{\lambda, \mu} < Q\} \cap (\Omega \setminus D)$$

is locally analytic. In this section we show that $\Gamma_{\lambda, \mu}$ is given by a continuous curve $x = k_{\lambda, \mu}(y)$ or, briefly, $x = k(y)$.

Lemma 9.1. *Let $\lambda/2Q \le \Lambda$ and let σ be a segment joining the points $(x_0 \pm b, y_0) \in \Omega \setminus \overline{D}$, $b > 0$. If a solution $u = u_{\lambda, \mu}$ of problem $J_{\lambda, \mu}$ satisfies*

$$u = Q \qquad \text{in a neighborhood of } \sigma,$$

$$(x_0, y_1) \text{ is a free-boundary point with } y_1 > y_0,$$

then

$$b \le C,$$

where C is a constant depending only on Λ.

Proof. By Lemma 8.4 we have $y_0 \geq c_0 > 0$, where c_0 depends only on Λ. For any $s_0 \geq 0$, define

$$\Gamma_{s_0} = \left\{ (x, y); |x - x_0| \leq b, \, y - y_0 = s_0 \eta\left(\frac{x - x_0}{\sqrt{b}} \right) \right\},$$

where $\eta(x)$ is defined in (6.8), and let s be the smallest value s_0 for which Γ_{s_0} touches the free boundary at some point, say $(\tilde{x}, \tilde{y})$. Notice that

$$|\tilde{x} - x_0| \leq \sqrt{b}, \qquad s \leq C_0,$$

where C_0 is a universal constant.

For fixed large m, consider the domain E bounded by Γ_s, $y = y_0 + m$ and $x = x_0 \pm b$, and let w be the solution of

$$Lw = 0 \qquad \text{in } E,$$

$$w = Q \qquad \text{on } \Gamma_s,$$

$$w = 0 \qquad \text{on } \partial E \setminus \Gamma_s.$$

Since u is a supersolution in $\{y > 0\}$ (when extended by Q), $u \geq w$ and

$$(9.1) \qquad \lambda = \frac{1}{y} \frac{\partial u}{\partial \nu} \leq \frac{1}{y} \frac{\partial w}{\partial \nu} \qquad \text{at } (\tilde{x}, \tilde{y}).$$

Suppose that the assertion of the lemma is not true. Then the setting above holds for a sequence of b's with $b \to \infty$. Choose a subsequence such that the corresponding values of

$$\frac{\lambda}{Q}, \quad x_0, y_0, s, \tilde{y}, \frac{\tilde{x} - x_0}{\sqrt{b}}$$

converge, and let

$$x_\infty = \lim \frac{\tilde{x} - x_0}{\sqrt{b}}; \qquad \text{then } x_\infty \in [-1, 1].$$

After a translation in the x-direction such that $(\tilde{x}, \tilde{y})$ lies on the y-axis, we see that E converges to a strip

$$E_\infty = \{ y_0 + s\eta(x_\infty) < y < y_0 + m \}$$

and $(\tilde{x}, \tilde{y})$ converges to $(0, y_0 + s\eta(x_\infty))$, where x_0, y_0, s designate the values of x_0, y_0, s in the construction above for each b. Furthermore, the functions w converge to a solution W of

$$
\begin{aligned}
LW &= 0 && \text{in } E_\infty, \\
W &= Q && \text{on } y = y_0 + s\eta(x_\infty) \equiv y_1, \\
W &= 0 && \text{on } y = y_0 + m.
\end{aligned}
$$

By uniqueness (see Corollary 9.10),

$$
W = Q \frac{(y_0 + m)^2 - y^2}{(y_0 + m)^2 - y_1^2}
$$

and then (9.1) gives

$$
\lambda \leqslant \lim_{b \to \infty} \frac{1}{y} \frac{\partial W}{\partial \nu} \bigg|_{(\bar{x}, \bar{y})} = \frac{1}{y} \frac{\partial W}{\partial \nu} \bigg|_{(0, \, y_1)}
$$

$$
= \frac{2Q}{(y_0 + m)^2 - y_1^2} < 2Q
$$

if m is large enough, contradicting (8.19).

Lemma 9.2. *There exists a minimizer $u = u_{\lambda, \mu}$ such that $u_x \geqslant 0$.*

Proof. The proof is by symmetrization. Since, however, Ω_μ is unbounded, we introduce another truncation. For any $R > 0$, let

$$
(9.2) \qquad J_{\lambda, \mu, R}(v) = \int_{\Omega_\mu \cap \{x < R\}} \left| \frac{1}{y} \nabla v - \lambda I_{\{v < Q\} \setminus De} \right|^2 y \, dx \, dy.
$$

By the proof of Lemma 8.1 there exists a solution $u = u_{\lambda, \mu, R}$ of the problem

$$
u \in K_\mu,
$$

$$
(9.3)
$$

$$
J_{\lambda, \mu, R}(u) = \min_{v \in K_\mu} J_{\lambda, \mu, R}(v).
$$

Denote by u^* the increasing rearrangement of $u(x, y)$ in the direction x. Since $u(-\mu, y) = 0$, if we define $u(x, y) = u(-2\mu - x, y)$, then Theorem 7.1 can be applied. Hence

$$
\int_{\Omega_\mu \cap \{x < R\}} \frac{1}{y} |\nabla u^*|^2 \leqslant \int_{\Omega_\mu \cap \{x < R\}} \frac{1}{y} |\nabla u|^2.
$$

The term

$$J_1 = -\lambda \int_{(\Omega_\mu \setminus D) \cap \{u<Q\} \cap \{x<R\}} \frac{\partial u}{\partial y}$$

$$= -\lambda \int_{(\Omega_\mu \setminus D) \cap \{x<R\}} \frac{\partial u}{\partial y} = -\lambda \int_{-a}^0 u(x,1) - \lambda QR$$

also decreases by the rearrangement, since the rearrangement increases $u(x,1)$ for $-a < x < 0$. It follows that there is a minimum $u_{\lambda,\mu,R}$ rearranged monotonically in x, and since (Problem 1)

(9.4) $u_{\lambda,\mu,R} \to u_{\lambda,\mu}$ a.e.

for a sequence $R \to \infty$, where $u_{\lambda,\mu}$ is a minimum of $J_{\lambda,\mu}$, the assertion follows.

From now on we take only solutions $u_{\lambda,\mu}$ that satisfy

(9.5) $\dfrac{\partial}{\partial x} u_{\lambda,\mu}(x,y) \geq 0.$

[In Section 10 it will be shown that Problem $J_{\lambda,\mu}$ has a unique solution $u_{\lambda,\mu}$ satisfying (9.5).] From (9.5) we deduce that

(9.6)

$$\{u_{\lambda,\mu}(x,y) < Q\} \cap \{0 < y < 1\} = \{(x,y); -\mu < x < k_{\lambda,\mu}(y), 0 < y < 1\},$$

for some function $k_{\lambda,\mu}(y)$ which we briefly designate by $k(y)$.
    Lemma 8.4 shows that

$$k(y) = \infty \qquad \text{if } 0 < y < h_\lambda.$$

Theorem 9.3. *If $\lambda > 2Q$, then $k(y) = k_{\lambda,\mu}(y)$ is a bounded continuous function for $h_\lambda < y < 1$; further,*

(9.7) $x_{\lambda,\mu} \equiv \lim\limits_{y \uparrow 1} k_{\lambda,\mu}(y)$ *exists and* $-a < x_{\lambda,\mu} < \infty$,

(9.8) $\lim\limits_{y \downarrow h_\lambda} k_{\lambda,\mu}(y) = \infty,$

the inverse $y = m_{\lambda,\mu}(x)$ of $x = k_{\lambda,\mu}(y)$ exists and is differentiable for all x sufficiently large, and

(9.9) $\lim\limits_{x \to \infty} m'_{\lambda,\mu}(x) = 0.$

The proof is given in a sequence of lemmas.

Lemma 9.4. $k(y)$ is continuous in $(0, 1)$ with values in $[-a, \infty]$.

Proof. $k(y)$ is clearly lower semicontinuous. Let $0 < y_0 < 1$ with $x_0 = k(y_0) < \infty$ and suppose that there is a sequence $y_n \to y_0$ such that $k(y_n) \geq x_0 + \varepsilon$ for some $\varepsilon > 0$. Then the segment $\sigma = \{(x, y_0); 0 < x - x_0 < \varepsilon\}$ belongs to the free boundary. Therefore, $Lu = 0$ in an upper or lower neighborhood V of σ and

$$u = Q, \qquad \frac{1}{y_0} \frac{\partial u}{\partial y} = \lambda \text{ on } \sigma.$$

By uniqueness for the Cauchy problem it follows that

$$u = \frac{\lambda}{2} y^2 + \left(Q - \frac{\lambda}{2} y_0^2 \right) \qquad \text{in } V$$

and by unique continuation we get $u = Q$ on the segment $\{(x, y_0), -\mu < x < x_0\}$, which is impossible.

Lemma 9.5. If (x_0, y_0) is a free-boundary point in $\Omega \setminus \bar{D}$, then

(9.10) $\qquad\qquad u = Q \qquad in \ (x_0 + C, \infty) \times (y_0, 1),$

where C is as in Lemma 9.1.

Proof. By Lemma 9.4, for any $\varepsilon > 0$,

$$u(x, y) = Q \qquad \text{if } x > x_0 + \varepsilon, \quad |y - y_0| < \delta$$

provided that $\delta = \delta(\varepsilon)$ is sufficiently small. It follows that $u = Q$ in a neighborhood of the segment

$$\{y = y_0, |x - (x_0 + b)| < b\}$$

for any $b > 0$. Now apply Lemma 9.1 to deduce (9.10).

Lemma 9.6. There exists a number $h \in [h_\lambda, 1]$ such that

$$k(y) < \infty \qquad if \ h < y < 1,$$
$$k(y) = \infty \qquad if \ 0 < y < h.$$

Proof. Let

$$h = \inf \{y; k(y) < \infty\}.$$

By Lemma 9.4, $k(h) = \infty$. Take a sequence $y_n \downarrow h$ such that $-a < k(y_n) < \infty$. Applying Lemma 9.5 with $y_0 = y_n$, we find that $k(y) < \infty$ if $y_n < y < 1$, and the assertion follows.

Lemma 9.7. $k(1) \equiv \lim_{y \uparrow 1} k(y)$ *exists and* $k(1) > -a$.

Proof. If the first assertion is not true, then we get a situation contradicting the nonoscillation lemma (Lemma 5.2; see Problem 1, Section 6). The second assertion follows from Lemma 8.8.

Lemma 9.8. (*i*) *If* $k(1) > 0$, *then*

$$(9.11) \qquad \lambda \leqslant \frac{1}{y} \frac{\partial u}{\partial \nu} \qquad on \ y = 1, 0 < x < k(1)$$

where ν *is the outward normal.*
  (*ii*) *If* $k(1) < 0$, *then*

$$(9.12) \qquad \lambda \geqslant \frac{1}{y} \frac{\partial u}{\partial \nu} \qquad on \ y = 1, k(1) < x < 0,$$

where ν *is the outward normal to D.*
  (*iii*) *If* $k(y_0) = -a$ *for some* $y_0 \in (0, 1)$, *then*

$$(9.13) \qquad \lambda \geqslant \frac{1}{y} \frac{\partial u}{\partial \nu} \qquad at \ (-a, y_0),$$

where ν *is the outward normal with respect to D.*

Note that

$$\frac{\partial}{\partial \nu} = \frac{\partial}{\partial y}, \qquad -\frac{\partial}{\partial y}, \qquad \frac{\partial}{\partial x}$$

respectively in (9.11), (9.12), (9.13) and that the last two derivatives are taken as lim inf of quotient differences.
   For proof of the lemma, see Problem 2; only part (i) will be needed later.
   We shall need the following Phragmén–Lindelöf type of result.

Lemma 9.9. *Suppose that*

$$\begin{aligned} L\psi &\geqslant 0 & if \ x > 0, 0 < y < 1, \\ \psi &\leqslant M & if \ x > 0, 0 < y < 1 \ (M > 0), \\ \psi(x, 0) &\leqslant 0, & \psi(x, 1) \leqslant 0 \ if \ x > 0. \end{aligned}$$

Then $\lim \sup_{x \to \infty} \psi \leqslant 0$.

Proof. For any $\varepsilon > 0$, consider the function

$$w = \psi - \frac{My^2}{r^3} - \varepsilon x \qquad (r^2 = x^2 + y^2)$$

in $\{0 < x < N_\varepsilon, 0 < y < 1\}$. It satisfies $Lw \geqslant 0$, and $w \leqslant 0$ on the boundary if $N_\varepsilon > M/\varepsilon$. By the maximum principle we get $w \leqslant 0$, that is,

$$\psi(x, y) \leqslant \frac{My^2}{r^3} + \varepsilon x \qquad \text{if } 0 < x < N_\varepsilon, 0 < y < 1.$$

Taking $\varepsilon \to 0$, we get $\psi(x, y) \leqslant My^2/r^3$, and the assertion follows.

Corollary 9.10. *Suppose that*

$$
\begin{aligned}
L\psi &\geqslant 0 & &\text{if } -\infty < x < \infty, \quad 0 < y < 1, \\
\psi &\leqslant M & &\text{if } -\infty < x < \infty, \quad 0 < y < 1, \\
\psi(x, 0) &\leqslant 0, & &\psi(x, 1) \leqslant 0 \text{ if } -\infty < x < \infty.
\end{aligned}
$$

Then $\psi(x, y) \leqslant 0$ for $-\infty < x < \infty, 0 < y < 1$.

Indeed, by the preceding lemma, for any $\delta > 0$,

$$\psi(\pm N_\delta, y) < \delta \qquad \text{for } 0 < y < 1$$

provided that N_δ is large enough; hence, by the maximum principle,

$$\psi(x, y) < \delta \qquad \text{if } -N_\delta < x < N_\delta, 0 < y < 1.$$

Lemma 9.11. *If $\lambda > 2Q$, then $k(0) < \infty$ (and consequently $h < 1$).*

Proof. Suppose that $k(0) = \infty$. Consider the functions $u_n(x, y) = u(x + n, y)$ $(n = 1, 2, \ldots)$. They satisfy

$$\left. \begin{aligned} Lu_n &= 0 \\ |\nabla u_n| &\leqslant C \end{aligned} \right\} \qquad \text{in } \{x > 1 - n, 0 < y < 1\},$$

$$u_n(x, 0) = 0 \qquad \text{if } x > -n,$$

$$u_n = Q, \qquad \frac{1}{y}\frac{\partial u_n}{\partial y} \geqslant \lambda \text{ if } y = 1, x > -n$$

where (9.11) was used. Therefore, for a subsequence

$$u_n \to v$$

uniformly in compact subsets of $\{0 < y \leq 1\}$, and

$$Lv = 0 \quad \text{in } \{0 < y < 1\},$$

$$v(x, 0) = 0 \quad \text{if } -\infty < x < \infty,$$

$$v = Q, \quad \frac{1}{y}\frac{\partial v}{\partial y} \geq \lambda \text{ if } -\infty < x < \infty, \quad y = 1.$$

By Corollary 9.10 we have that $v \equiv Qy^2$, and therefore

$$\frac{1}{y}\frac{\partial v}{\partial y} = 2Q.$$

Thus $2Q \geq \lambda$, a contradiction.

Completion of the Proof of Theorem 9.3. We already know that $0 < h < 1$ and that

$$k(y) \to \infty \quad \text{if } y \downarrow h.$$

But then the free boundary satisfies the flatness condition of Section 4 for all values $x = k(y)$ with $x \geq x_0$, x_0 large. It follows that, for $x \geq x_0$, the free boundary can be written in the form

$$y = m(x)$$

with $-\varepsilon_0 \leq m'(x) \leq 0$, where ε_0 is arbitrarily small, positive number, provided that x_0 is large enough depending on ε_0. Thus (9.9) holds.

Using Theorem 1.4 of Chapter 2 we also deduce

$$|m''(x)| \leq C \quad \text{if } x \geq x_*,$$

(9.14) $$|Du| + |D^2u| \leq C \quad \text{in } \{u > 0\} \cap \{x > x_*\}.$$

Recall also that

(9.15) $$\left.\begin{array}{r} u = Q \\[4pt] \dfrac{1}{y}\dfrac{\partial u}{\partial \nu} = \lambda \end{array}\right\} \quad \text{on } y = m(x).$$

For any sequence $x_n \to \infty$, consider the functions

$$u_n(x, y) = u(x + n, y).$$

Then for a subsequence

$$u_n \to v$$

uniformly in compact subsets of $\{0 < y < 1\}$ and by (9.9), (9.14), (9.15), and elliptic estimates,

$$Lv = 0 \qquad \text{in } \{0 < y < h\},$$

$$v(x,0) = 0 \qquad \text{if } -\infty < x < \infty,$$

$$v = Q, \qquad \frac{1}{y}\frac{\partial v}{\partial y} = \lambda \text{ if } -\infty < x < \infty, \quad y = h.$$

It follows (using Corollary 9.10) that

$$v = Q\frac{y^2}{h^2}$$

and therefore $2Q/h^2 = \lambda$. Thus $h = h_\lambda$ and the proof of Theorem 9.3 is complete.

Lemma 9.12. *If* $\lambda \downarrow 2Q$, *then* $x_{\lambda,\mu} \to \infty$.

Proof. Suppose that $x_{\lambda,\mu} \to \bar{x} < \infty$ for $\lambda = \lambda_n \downarrow 2Q$. Thus for a subsequence

$$u_{\lambda_n,\mu} \to v$$

uniformly in compact subsets of Ω, and v is a minimizer corresponding to $\lambda = 2Q$. By Lemma 8.4, $v < Q$ in Ω and therefore Theorem 6.1 (adapted to L) gives

$$v = Q, \qquad \frac{1}{y}\frac{\partial v}{\partial \nu} = 2Q \text{ on } y = 1, x > \bar{x}.$$

By uniqueness for the Cauchy problem, we then get

$$v(x, y) = Qy^2 \qquad \text{throughout } \Omega,$$

which is impossible.

PROBLEMS

1. Prove (9.4).

   [*Hint*: See Lemma 3.6.]

2. Prove Lemma 9.8.

   [*Hint*: Use the method of proof of Theorem 2.5.]

10. MONOTONICITY AND UNIQUENESS

From now on we choose $a = \mu$ in Figure 3.5.

Theorem 10.1. *Problem $J_{\lambda,\mu}$ has a unique solution $u_{\lambda,\mu}$ satisfying* (9.5).

Proof. Suppose that $\tilde{u}_{\lambda,\mu}$ is another solution and set

$$u_1 = u_{\lambda,\mu}, \qquad u_2 = \tilde{u}_{\lambda,\mu}.$$

Introduce the functions

$$v_1 = \min(u_1, u_2) = u_1 \wedge u_2,$$

$$v_2 = \max(u_1, u_2) = u_1 \vee u_2.$$

We claim that

(10.1) $$J_{\lambda,\mu}(u_1) + J_{\lambda,\mu}(u_2) = J_{\lambda,\mu}(v_1) + J_{\lambda,\mu}(v_2).$$

Indeed, if we develop both sides, we see that it suffices to prove that, for any $0 < R < \infty$,

(10.2)

$$\int_{\Omega_\mu \cap \{x<R\}} \frac{1}{y}\left(|\nabla u_1|^2 + |\nabla u_2|^2\right) = \int_{\Omega_\mu \cap \{x<R\}} \frac{1}{y}\left(|\nabla v_1|^2 + |\nabla v_2|^2\right),$$

(10.3) $$\lambda \int_{(\Omega_\mu \setminus D) \cap \{x<R\}} \left(\frac{\partial u_1}{\partial y} + \frac{\partial u_2}{\partial y}\right) = \lambda \int_{(\Omega_\mu \setminus D) \cap \{x<R\}} \left(\frac{\partial v_1}{\partial y} + \frac{\partial v_2}{\partial y}\right),$$

(10.4) $$\lambda^2 \int_{(\Omega_\mu \setminus D) \cap \{x<R\}} y\left(I_{\{u_1<Q\}} + I_{\{u_2<Q\}}\right)$$

$$= \lambda^2 \int_{(\Omega_\mu \setminus D) \cap \{x<R\}} y\left(I_{\{v_1<Q\}} + I_{\{v_2<Q\}}\right).$$

Now (10.2) and (10.4) are obvious, whereas (10.3) reduces to the obvious relation

$$\int_{-a}^{R} (u_1(x,1) + u_2(x,1))\, dx = \int_{-a}^{R} (v_1(x,1) + v_2(x,1))\, dx.$$

Since $v_i \in K_\mu$, we have

$$J_{\lambda,\mu}(u_i) \leq J_{\lambda,\mu}(v_i),$$

so that, by (10.1),

$$(10.5) \qquad J_{\lambda,\mu}(u_i) = J_{\lambda,\mu}(v_i).$$

We claim that

$$(10.6) \qquad \begin{aligned} &\text{if } 0 < u_1(X^0) = u_2(X^0) < Q, \text{ then either} \\ &u_1 \geqslant u_2 \text{ or } u_2 \geqslant u_1 \text{ in a neighborhood of } X^0. \end{aligned}$$

Indeed, let V be a disc with center X^0 such that $0 < u_i < Q$ in V for $i = 1, 2$ and let w be defined by

$$\begin{aligned} Lw &= 0 &&\text{in } V, \\ w &= v_1 &&\text{outside } V. \end{aligned}$$

If (10.6) does not hold in V, then v_1 is not a solution of $Lv_1 = 0$ in V and, consequently,

$$\int_V \frac{|\nabla w|^2}{y} < \int_V \frac{|\nabla v_1|^2}{y}.$$

It follows that $J_{\lambda,\mu}(w) < J_{\lambda,\mu}(v_1) = J_{\lambda,\mu}(u_1)$, a contradiction since $w \in K_\mu$. From (10.6) it follows that if $u_1 \not\equiv u_2$, then we cannot have

$$0 < u_1(X^0) = u_2(X^0) < Q$$

at any point $X^0 \in \Omega_\mu$; otherwise, the maximum principle would give $u_1 \equiv u_2$ in a neighborhood of X^0 and, by unique continuation, also in Ω_μ; here we use the fact [which follows from (9.5)] that the sets $\{0 < u_i < Q\}$ are connected.
We conclude that

$$(10.7) \qquad \text{either } u_1 \geqslant u_2 \text{ in } \Omega_\mu \qquad \text{or} \qquad u_2 \geqslant u_1 \text{ in } \Omega_\mu.$$

The argument above does not give uniqueness, but it does apply to general nozzles [not just to those satisfying (8.1)]. We shall now use a slightly modified argument to prove uniqueness when the nozzle is a y-graph.
If u_1, u_2 are two solutions, extend u_i by Q to the right of N and by 0 to the left of $x = -\mu$. For any small $\varepsilon > 0$, define

$$u_1^\varepsilon(x, y) = u_1(x - \varepsilon, y),$$

$$v_1 = u_1^\varepsilon \wedge u_2, \qquad v_2 = u_1^\varepsilon \vee u_2$$

and denote by $J_{\lambda,\mu}^\varepsilon$ the functional $J_{\lambda,\mu}$ corresponding to Ω_μ^ε, D^ε, the translations of Ω_μ, D, respectively, by $x \to x + \varepsilon$. We claim that

(10.8) $$J^\varepsilon_{\lambda, \mu}(u^\varepsilon_1) + J_{\lambda, \mu}(u_2) = J^\varepsilon_{\lambda, \mu}(v_1) + J_{\lambda, \mu}(v_2).$$

The proof in fact is similar to the proof of (10.1) and depends on the choice $a = \mu$. Notice next that u^ε_1 is a minimizer for $J^\varepsilon_{\lambda, \mu}$, that v_1 is in the corresponding class of admissible functions, and that $v_2 \in K_\mu$. Hence

$$J^\varepsilon_{\lambda, \mu}(u^\varepsilon_1) = J^\varepsilon_{\lambda, \mu}(v_1), \qquad J_{\lambda, \mu}(u_2) = J_{\lambda, \mu}(v_2).$$

Proceeding as before we conclude that either $u^\varepsilon_1 \geqslant u_2$ everywhere or $u^\varepsilon_1 \leqslant u_2$ everywhere. Since, however, $u^\varepsilon_1 < u_2$ at some points of N_μ, it follows that

(10.9) $$u_1(x - \varepsilon, y) \leqslant u_2(x, y).$$

Taking $\varepsilon \to 0$, we obtain $u_1 \leqslant u_2$. Similarly, one shows that $u_2 \geqslant u_1$.

Theorem 10.2. *If* $\lambda_1 < \lambda_2$, *then*

(10.10) $$\frac{1}{\lambda_1}(Q - u_{\lambda_1, \mu}) \geqslant \frac{1}{\lambda_2}(Q - u_{\lambda_2, \mu}).$$

Proof. Set

$$u_i = \frac{1}{\lambda_i}(Q - u_{\lambda_i, \mu}),$$

$$v_1 = u_1 \vee u_2, \qquad v_2 = u_1 \wedge u_2,$$

$$J_{\lambda_i, \mu}(u_{\lambda_i, \mu}) = \lambda_i^2 J(u_i).$$

Then, as in the proof of (10.1),

(10.11) $$J(u_1) + J(u_2) = J(v_1) + J(v_2).$$

Since $0 \leqslant \lambda_i u_i \leqslant Q$, we have $0 \leqslant v_i \leqslant Q/\lambda_i$, so that $Q - \lambda_i v_i$ is in K_μ. Therefore,

$$J_{\lambda_i, \mu}(u_{\lambda_i, \mu}) \leqslant J_{\lambda_i, \mu}(Q - \lambda_i v_i) = \lambda_i^2 J(v_i),$$

and from (10.11) we deduce that

$$J(u_i) = J(v_i).$$

We can now argue as before and show that either $u_1 \geq u_2$ or $u_2 \geq u_1$ throughout Ω_μ. Since (by Theorem 9.3) $u_1 > u_2$ at some points (x, y) with $x \to \infty$, (10.10) follows.

Corollary 10.3. *If* $2Q < \lambda_1 < \lambda_2$, *then*

$$(10.12) \qquad k_{\lambda_1, \mu}(y) > k_{\lambda_2, \mu}(y) \qquad if \, h_{\lambda_1} < y < 1, \quad k_{\lambda_2}(y) > -a.$$

Proof. Theorem 10.2 gives

$$k_{\lambda_1, \mu}(y) \geq k_{\lambda_2, \mu}(y).$$

To prove strict inequality, suppose that

$$x_0 \equiv k_{\lambda_1, \mu}(y_0) = k_{\lambda_2, \mu}(y_0) > -a.$$

Then, by the strong maximum principle (using Problem 1),

$$\left| \frac{\partial u_{\lambda_2, \mu}}{\partial \nu} \right| > \left| \frac{\partial u_{\lambda_1, \mu}}{\partial \nu} \right| \qquad at \, (x_0, y_0).$$

Since (x_0, y_0) lies on both free boundaries, we have

$$\frac{1}{\lambda_i} \frac{1}{y} \frac{\partial}{\partial \nu} u_{\lambda_i, \mu} = 1 \qquad at \, (x_0, y_0) \, (i = 1, 2),$$

a contradiction.

Lemma 10.4. *If* $\lambda_n \to \lambda$, $\lambda > 2Q$, *then*

$$(10.13) \qquad u_{\lambda_n, \mu} \to u_{\lambda, \mu} \qquad weakly \ in \ H_{\text{loc}}^{1,2} \ and \ a.e.,$$

and

$$(10.14) \qquad k_{\lambda_n, \mu}(y) \to k_{\lambda, \mu}(y) \qquad for \ each \ y \in (h_\lambda, 1]$$

Proof. By the proof of Lemma 3.6, for any subsequence of $u_{\lambda_n, \mu}$ such that

$$u_{\lambda_n, \mu} \to w \qquad weakly \ in \ H_{\text{loc}}^{1,2} \ and \ a.e.,$$

we have: w is a minimizer for $J_{\lambda, \mu}$. By uniqueness, we then obtain the assertion (10.13). Lemma 3.6(c) also gives (10.14) for $y < 1$.

To prove (10.14), for $y = 1$ suppose it is not true. Then for a subsequence

$$k_{\lambda_n, \mu}(1) \to k_{\lambda, \mu}(1) + \beta, \qquad \beta \neq 0.$$

If $\beta < 0$, then Theorem 6.1 gives

$$\frac{1}{y} \frac{\partial u_{\lambda, \mu}(x, 1 - 0)}{\partial y} = \lambda \qquad \text{if} \qquad k_{\lambda, \mu}(1) + \beta < x < k_{\lambda, \mu}(1),$$

so that, by uniqueness for the Cauchy problem,

$$u_{\lambda, \mu} \equiv Qy^2 \qquad \text{with } Q = 2\lambda,$$

a contradiction. If $\beta > 0$ and $k_{\lambda, \mu}(1) < 0$ then the proof of Theorem 6.1 applies again and yields

$$\frac{1}{y} \frac{\partial u_{\lambda, \mu}(x, 1 + 0)}{\partial y} = 0 \qquad \text{if } k_{\lambda, \mu}(1) < x < 0,$$

and we again get a contradiction. If $\beta > 0$ and $k_{\lambda, \mu}(1) > 0$, then we obtain a contradiction to the nonoscillation lemma (Lemma 5.2 and Remark 5.3).

By Lemma 9.11,

$$x_{\lambda, \mu} \equiv k_{\lambda, \mu}(1) \in (0, \infty)$$

if $\lambda > 2Q$, $\lambda - 2Q$ is small enough, and by Corollary 8.7,

$$x_{\lambda, \mu} < 0 \qquad \text{if } \lambda \geq \tilde{\lambda},$$

where $\tilde{\lambda}$ is independent of a, μ.

Definition 10.1.

$$\Sigma_\mu^+ = \{\lambda; \lambda > 2Q, \ x_{\lambda, \mu} \geq 0\},$$

$$\Sigma_\mu^- = \{\lambda; \lambda > 2Q, \ x_{\lambda, \mu} < 0\},$$

$$\lambda_\mu = \sup \{\lambda; \lambda \in \Sigma_\mu^+\}.$$

As noted above,

(10.15) $$2Q < \lambda_\mu < \tilde{\lambda}; \qquad \tilde{\lambda} \text{ independent of } \mu.$$

From Corollary 10.3 we have

$$\Sigma_\mu^+ = \{\lambda; 2Q < \lambda < \lambda_\mu\},$$

(10.16)

$$\Sigma_\mu^- = \{\lambda; \lambda_\mu < \lambda < \infty\}.$$

Choose

$$\Lambda = \frac{\tilde\lambda}{2Q}$$

in Lemma 9.1 and denote the corresponding C by C^*. If $\lambda \in \Sigma_\mu^+$, then $\lambda < \tilde\lambda$, $x_{\lambda,\mu} > 0$ and Lemma 9.1 then shows that the free boundary must lie in $\{x > -C^*\}$. Choosing

(10.17) $$a_0 = 1 + C^*, \quad \mu > a_0,$$

we then have:

Lemma 10.5. *If* $\lambda \in \Sigma_\mu^+$, *then the free boundary lies in* $\{x > 1 - a_0\}$.

Thus the free boundary does not intersect ∂D.

We conclude this section with a gradient estimate near the initial point of the free boundary.

Lemma 10.6. *Set* $X_{\lambda,\mu} = (x_{\lambda,\mu}, 1)$. *Then for some* $R > 0$,

(10.18) $$|\nabla u_{\lambda,\mu}(X)| \le C \qquad in \ \Omega \cap B_R(X_{\lambda,\mu}),$$

where C *is a constant independent of* μ.

Proof. Set $r = |X - X_{\lambda,\mu}|$. It suffices to establish the estimate

(10.19) $$|\nabla u(X)| \le C \qquad in \ \Omega \cap \{\beta < r < 2\beta\}$$

for any sufficiently small β (with C independent of β). Set

$$u^\beta(X) = \frac{1}{\beta} u(X_{\lambda,\mu} + \beta X) \qquad in \ \tilde B = \{X; \tfrac{1}{2} < |X| < \tfrac{5}{2}, X_{\lambda,\mu} + \beta X \in \Omega\}$$

and let

$$\tilde G = \{X; 1 < |X| < 2, X_{\lambda,\mu} + \beta X \in \Omega\}.$$

Then

$$u^\beta = Q_\beta, \qquad \frac{1}{y}\frac{\partial u^\beta}{\partial \nu} = \lambda \text{ on the free boundary}$$

and

$$0 \leqslant u^\beta \leqslant Q_\beta, \qquad \text{where } Q_\beta = \frac{Q}{\beta}.$$

We can now apply the proof of the bounded gradient lemma (Lemma 5.1 and Remark 5.2) to $Q_\beta - u^\beta$ with D, Ω replaced by $\tilde{B}$ and $\tilde{G}$, respectively. We deduce that

$$|\nabla u^\beta| \leqslant C \qquad \text{in } \tilde{B},$$

and (10.19) follows.

PROBLEM

1. Let Ω_0 be a subdomain of Ω bounded by $\{y = 0\}$, a curve N_0 in $\{y > 0\}$, and a line segment $x = x_0$ with possibly $x_0 = \pm\infty$, and let

    $$Lu \geqslant 0, \qquad u \leqslant M \text{ in } \Omega_0,$$

    $$u \leqslant 0 \qquad \text{on } \{y = 0\} \text{ and on } N_0;$$

    then $u \leqslant 0$ in Ω_0.

11. THE SMOOTH-FIT THEOREMS

Theorem 11.1. *If $x_{\lambda,\mu} = 0$, then $N \cup \Gamma_{\lambda,\mu}$ is continuously differentiable in a neighborhood of A.*

Proof. The proof consists of several steps. In the first step we study the blow-up limit of

$$(11.1) \qquad u^\gamma(X) = \frac{1}{\gamma}(Q - u(A + \gamma X)), \qquad u = u_{\lambda,\mu}$$

for any sequence $\gamma = \gamma_n \to 0$. By Lemma 10.6,

$$|\nabla u^\gamma(X)| \leqslant C$$

in every compact subset G of R^2 which is contained in $\Omega^\gamma \equiv \{X; u^\gamma(X) > 0\}$, provided that γ is sufficiently small (C is independent of G).

By the proof of Lemma 3.6, for a subsequence

$$u^{\gamma_n} \to u^0 \qquad \text{uniformly in compact sets,}$$

where $u^0 \not\equiv 0$. Further, u^0 is an absolute minimizer for the functional

$$(11.2) \qquad\qquad J(v) = \int \left(|\nabla v|^2 + \lambda^2 I_{\{v > 0\}} \right)$$

is any bounded domain. Set

$$\Omega^0 = \{X; u^0(X) > 0\}.$$

Notice that $u_x^0 \leq 0$.

If

$$(11.3) \qquad \begin{array}{l} \text{the tangent line to } N \text{ at } A \text{ is} \\ \text{given by } \alpha x + \beta y - \beta = 0, \quad \alpha^2 + \beta^2 > 0, \end{array}$$

then, for $\alpha \neq 0$, the boundary of Ω^0 is a y-graph consisting of the ray

$$(11.4) \qquad\qquad N^0 : \alpha x + \beta y = 0, \quad y \geq 0$$

and of the free boundary

$$\Gamma^0 = \partial\{u^0 > 0\} \setminus N^0.$$

Lemma 11.2. *If $\alpha \neq 0$ then Γ^0 is the ray*

$$\Gamma^0 : \alpha x + \beta y = 0, \quad y < 0$$

and $\nabla u^0 \equiv const.$

Proof. By Theorem 4.2, Γ^0 is analytic. We also have

$$\begin{array}{ll} \Delta u^0 = 0 & \text{in } \Omega^0, \\ u^0 = 0 & \text{on } N^0 \cup \Gamma^0, \\ -\dfrac{\partial u^0}{\partial \nu} = \lambda & \text{on } \Gamma^0. \end{array}$$

We shall consider first the case where

$$(11.5) \qquad\qquad \Gamma^0 \text{ is a connected curve initiating at } 0.$$

We claim that

$$(11.6) \qquad\qquad |\nabla u^0| < \lambda \qquad \text{in } \Omega^0.$$

To prove it, consider the mapping

$$h: X \to X' = \nabla u^0(X), \qquad X \in \overline{\Omega^0}.$$

Since u^0 is analytic, h is an open mapping and consequently $h(\Omega^0)$ is a domain whose boundary $\partial h(\Omega^0)$ consists of all points

$$\lim h(X^n), \qquad \text{where } X^n \in \Omega^0, \text{dist}(X^n, \partial\Omega^0) \to 0.$$

Observe that

$$(11.7) \qquad \begin{array}{l} \nabla u^0(X) \in N' \text{ if } X \in N^0, \text{ where } N' \text{ is} \\ \text{the straight line } N': -\alpha x' + \beta y' = 0, \end{array}$$

and

$$(11.8) \qquad\qquad |\nabla u^0(X)| = \lambda \qquad \text{if } X \in \Gamma^0.$$

We shall use these facts to show that

$$(11.9) \qquad\qquad \partial h(\Omega^0) \text{ is contained in the disc } |X'| \le \lambda.$$

Indeed, if (11.9) is not true then, since $|\nabla u^0| \le C$, there is a closed disc $B': |X' - X^*| \le \delta_0$ such that

$$B' \cap h(\Omega^0) = \varnothing,$$
$$\begin{array}{ll} B' \cap \overline{h(\Omega^0)} & \text{is nonempty and does not} \\ & \text{intersect } N' \text{ and } \{|X'| \le \lambda\}. \end{array}$$

Introduce the notation

$$z = x + iy \text{ if } X = (x, y), \qquad z^* = x^* + iy^* \text{ if } X^* = (x^*, y^*)$$

and consider the analytic function

$$\theta(z) = \frac{\delta_0}{u_x^0 + iu_y^0 - z^*} \qquad \text{in } \Omega^0.$$

Then there is a sequence of points z^n such that

$$(11.10) \qquad\qquad z^n \in \Omega^0, \qquad |\theta(z^n)| \to 1.$$

On the other hand, $|\theta(z)|$ is a bounded subharmonic function and $|\theta(z)| \leqslant \theta_0 < 1$ on $(N^0 \cup \Gamma^0) \setminus 0$. Since $N^0 \cup \Gamma^0$ is a y-graph, there is a line segment initiating at 0 which lies outside Ω^0. Hence Lemma 6.7 of Chapter 2 can be applied to deduce that

$$\limsup |\theta(z)| \leqslant \theta_0 \qquad \text{as } z \to 0 + 0i.$$

Thus $\limsup |\theta(z)| < \theta_0$ as $\text{dist}(z, \partial\Omega^0) \to 0$, whereas $|\theta(z)| \leqslant \text{const.}$ in Ω^0. By the Phragmén–Lindelöf theorem it follows that $|\theta(z)| \leqslant \theta_0$ in Ω^0 and (11.10) is contradicted.

Having proved (11.9), we have from (11.8) that the subharmonic function $|\nabla u^0(X)|$ in Ω^0 achieves its maximum in $\overline{\Omega}^0$ on Γ^0. Recalling the relation

$$\frac{\partial q}{\partial \nu} + \kappa q = 0 \qquad \text{along a stream line}$$

where $q = |\nabla u^0|$, κ curvature (see Problem 3 of Section 1), we deduce that

(11.11) Γ^0 is convex to the fluid.

We next introduce the conformal mapping σ from the unit disc E onto Ω^0 mapping the complex numbers i, $-i$ into 0 and ∞, respectively. Since $\Gamma^0 \cup N^0$ is a Jordan curve, σ is continuous and 1-1 from $\overline{E}$ onto $\overline{\Omega}^0$ (by a standard theorem in conformal mappings; see reference 27, Chap. 4, Sec. 8). We can choose σ so that it maps the part $\partial_1 E$ of ∂E going clockwise from $-i$ to i onto Γ^0, and $\partial_2 E \equiv \partial E \setminus \partial_1 E$ into N^0.

Suppose now that $\nabla u^0 \not\equiv \text{const.}$ Then the mapping $k = u_z^0 \circ \sigma$ is a conformal mapping from E into the set

B : the intersection of $\{|X'| < \lambda\}$ with one of the two
half-planes determined by the line $-\beta x' + \alpha y' = 0$.

Since

$$k_j^{\pm} = \lim k(z) \qquad \text{as } z \to \pm i, \quad z \in \partial_j E$$

exists, the set of limit points of $k(z)$ as $z \to \pm i$, $z \in E$ consists of the interval $\gamma^{\pm}$ joining $k_1^{\pm}$ to $k_2^{\pm}$. [This follows by applying Lemma 6.7 of Chapter 2 to functions $\tilde{k}_j \pm c \arctan(x/(\pm i - y))$, $c > 0$ where $k = \tilde{k}_1 + i\tilde{k}_2$.] Hence, if $\gamma^{\pm}$ does not lie on ∂B, then $\overline{k(E)}$ is a proper subset of $\overline{B}$. But then the proof of (11.9) can be applied [with a suitable disc B' in $B \setminus k(E)$] in order to derive a contradiction.

We have thus proved that $\partial k(E) \subset \partial B$ and therefore k maps E onto B. Since ∂E and ∂B are Jordan curves it follows, by conformal mappings (reference 27, Chap. 4, Sec. 8) that k is continuous in $\overline{E}$. But then

$$(11.12) \qquad u_z^0(X) \to \delta\lambda \left(\frac{\alpha}{\sqrt{\alpha^2 + \beta^2}} + \frac{i\beta}{\sqrt{\alpha^2 + \beta^2}} \right) \qquad \text{if } X \to A$$

$$(11.13) \quad u_z^0(X) \to -\delta\lambda \left(\frac{\alpha}{\sqrt{\alpha^2 + \beta^2}} + \frac{i\beta}{\sqrt{\alpha^2 + \beta^2}} \right) \qquad \text{if } |X| \to \infty,$$

where $\delta = 1$ or $\delta = -1$.

Since $u_x^0 \leqslant 0$, we must have $\alpha = 0$, contradicting the assumption $\alpha \neq 0$.

We have thus established that $\nabla u^0 \equiv$ const. The lemma immediately follows, provided (11.5) holds.

It remains to prove (11.15). We first show that

$$(11.14) \qquad\qquad \overline{\lim_{y \uparrow 1}} \ \frac{|k(y)|}{1 - y} \qquad \text{is finite.}$$

Indeed, suppose first that

$$(11.15) \qquad\qquad \frac{k(y_n)}{1 - y_n} \to +\infty \qquad \text{for a sequence } y_n \uparrow 1.$$

Set $X_n = (k(y_n), y_n)$, $r_n = |X_n|$, and take a blow-up sequence u_{r_n} of $Q - u$ with respect to $B_{r_n}(A)$; for a subsequence $u_{r_n} \to U$. By Theorem 6.1

$$U_y(x, 0 - 0) = -\lambda, \qquad \text{if } 0 < x < 1$$

and, therefore, $U(x, y) = -\lambda y$ if $y \leqslant 0$. Also $U(x, y) \equiv 0$ if $y > 0$ (since U is harmonic to the left of the nozzle and vanishes on the nozzle and on $\{y = 0\}$). Replacing U in $B_r(-1, 0)$ (r small) by a harmonic function $\tilde{U}$ [$\tilde{U} = U$ on $\partial B_r(-1, 0)$], we decrease the functional $J(U)$, thereby contradicting the fact that U is a local minimum. If

$$\frac{k(y_n)}{1 - y_n} \to -\infty \qquad \text{for a sequence } y_n \uparrow 1,$$

then we use Theorem 6.1 to derive, for a blow-up limit U,

$$U = \lambda y \qquad \text{if } y > 0,$$

which is impossible, since $U = 0$ on N^0.

From (11.14) and nondegeneracy it follows that, for any small $r > 0$, $\partial B_r(A)$ intersects the free boundary of u at a point $(k(y_r), y_r)$ such that

$$(11.16) \qquad (k(y_r), y_r) \subset \{c|x| < 1 - y\} \qquad \text{for some } c > 0.$$

Since $u_x^0 \leqslant 0$, $u^0(x, y) > 0$ if and only if $x < k_0(y)$ for some function $k_0(y)$. From (11.16) we see that, for a sequence $z_n \uparrow 0$,

$$(11.17) \qquad\qquad k_0(z_n) \text{ is finite} \quad \text{and} \quad k_0(z_n) \to 0.$$

Suppose $\{\gamma < y < \beta\}$ is a maximal interval where $k_0(y)$ is finite valued. We claim that

$$(11.18) \qquad\qquad\qquad \beta = 0.$$

Indeed, if $\beta < 0$ then either $k_0(\beta - 0) = -\infty$ or $k_0(\beta - 0) = +\infty$. Consider the first case and set

$$\Gamma': x = k_0(y), \qquad \gamma < y < \beta.$$

Below Γ' there is a connected component D of $\{u^0 > 0\}$. The function $|\nabla u^0|$ is subharmonic in D, $= \lambda$ on ∂D, and it is bounded in D. By the Phragmén–Lindelöf theorem it follows that $|\nabla u^0| \leqslant \lambda$ in D. But then [see (11.11)] Γ' is convex to the fluid, a contradiction to $k_0(\beta - 0) = -\infty$.

If $k_0(\beta - 0) = \infty$ we get a contradiction to the nonoscillation lemma. Thus (11.18) is proved.

From (11.16) and (11.18) it follows that the free boundary Γ^0 consists of Γ' [with $\beta = 0$, $k_0(0) = 0$] and perhaps one horizontal line $\tau_\delta = \{y = \delta\}$ with $u^0 = 0$ in the $\{y > \delta\}$ neighborhood of τ_δ. If in fact such τ_δ is a part of the free boundary, then $k_0(\gamma + 0) = -\infty$. [If $k_0(\gamma + 0) = +\infty$ then there must be free-boundary points between Γ' and τ_δ.] However, the proof of (11.11) shows that Γ' is convex to the fluid, contradicting $k_0(\gamma + 0) = -\infty$. Consequently, Γ^0 coincides with Γ' and (11.5) is established.

Lemma 11.2 extends to the case $\alpha = 0$; see Problem 3.

Completion of the Proof of Theorem 11.1. Consider first the case when the slope of N at A is finite, that is,

$$(11.19) \qquad\qquad \tau \equiv -\frac{\beta}{\alpha} \quad \text{is finite.}$$

We shall prove that

$$(11.20) \qquad\qquad \frac{k(y) - k(1)}{y - 1} \to \tau, \qquad \text{if } y \uparrow 1.$$

If (11.20) is not true, then there is a sequence $y_n \uparrow 1$ such that

$$(11.21) \qquad\qquad \lim_{n \to \infty} \left| \frac{k(y_n) - k(1)}{y_n - 1} - \tau \right| > 0.$$

Taking $\gamma = \gamma_n = 1 - y_n$, we may assume that

$$u^\gamma(X) = u^{\gamma_n}(X) \to u^0(X),$$

with u^γ as in (11.1). The free boundary of u^γ is given by

$$x = k^\gamma(y) = \frac{1}{\gamma}(k(1 + \gamma y) - k(1)),$$

so that

$$k^{\gamma_n}(-1) = \frac{k(y_n) - k(1)}{1 - y_n}.$$

Thus (11.21) implies that

$$\lim_{n \to \infty} |k^{\gamma_n}(-1) + \tau| > 0.$$

By Lemma 11.2, $k^0(-1) = -\tau$, where k^0 is the free boundary of u^0. Consequently,

$$\lim_{n \to \infty} |k^{\gamma_n}(-1) - k^0(-1)| > 0,$$

which is impossible by the proof of Lemma 3.6(a).

Finally, if $\tau = \pm\infty$, then by Problem 3

$$\frac{k(y) - 1}{y - 1} \to \pm\infty \qquad \text{if } y \uparrow 1.$$

In order to complete the proof of Theorem 11.1, consider first the case

(11.22) $\tau = 0.$

From (11.20) we obtain

$$\frac{1}{\gamma y}|k(1 + \gamma y) - k(1)| = \varepsilon(\gamma y), \qquad \varepsilon(t) \downarrow 0 \text{ if } \varepsilon \downarrow 0.$$

It follows that if $-2 < y_1, y_2 < 0$, then

$$|k^\gamma(y_1) - k^\gamma(y_2)| \le \frac{1}{\gamma}|k(1 + \gamma y_1) - k(1)| + \frac{1}{\gamma}|k(1 + \gamma y_2)$$

$$-k(1)| \le 4\varepsilon(2\gamma).$$

Consequently, for any $\delta > 0$ there is a $\gamma_0(\delta) > 0$ such that

$$|k^\gamma(y_1) - k^\gamma(y_2)| < \delta \qquad \text{if } -2 < y_1, y_2 < 0, \gamma < \gamma_0(\delta).$$

Thus the flatness condition of Section 4 is satisfied. It follows that

$$\left| \frac{d}{dy} k^\gamma(y) \right| < \eta \qquad \text{at } y = -1,$$

where η can be chosen arbitrarily small provided that $\delta \leqslant \delta_0(\eta)$ and $\gamma \leqslant \gamma_0(\delta)$. We conclude that

$$|k'(1 - \gamma)| < \eta \qquad \text{if } \gamma \leqslant \gamma_0(\delta_0(\eta)),$$

that is,

$$k'(y) \to 0 \qquad \text{if } y \uparrow 1.$$

In case $\tau \neq 0$, we first perform an orthogonal transformation $(x, y) \to (x', y')$ in a neighborhood of A so that in the new coordinates the slope of N at A is equal to zero. Applying the previous proof and then returning to the original coordinates, we obtain the relation

$$k'(y) \to \tau \qquad \text{if } y \uparrow 1.$$

This completes the proof of Theorem 11.1.

Theorem 11.3. *If $x_{\lambda, \mu} = 0$, then $u_{\lambda, \mu}$ is continuously differentiable in the closure of $\{u_{\lambda, \mu} < Q\} \cap B_\delta(A)$, for some $\delta > 0$.*

Proof. Set

$$d(X) = \text{dist}(X, N).$$

Let N_1 be a smooth curve initiating at A and lying in $\{u < Q\}$, which forms an angle θ with the tangent to N at A, $0 < \theta < \pi/2$. We claim: If $v = Q - u$, then

$$(11.23) \qquad v - \lambda d(X) = o(r) \qquad (r = |X - A|, \quad X \in N_1).$$

Indeed, otherwise there is a sequence

$$X_n \in N_1, \qquad r_n = |X_n - A| \to 0$$

such that

$$(11.24) \qquad |v(X_n) - \lambda d(X_n)| > c r_n, \qquad c > 0.$$

We shall derive a contradiction using a blow-up sequence

$$u_n(X) = \frac{1}{r_n}(Q - u(A + r_n X)).$$

Notice that

$$(11.25) \quad \frac{1}{r_n}v(X_n) = u_n(Y_n), \quad \text{where } Y_n = \frac{X_n - A}{r_n}, \quad |Y_n| = 1.$$

As in the proof of Theorem 11.1,

$$u_n(X) \to u^0(X) \quad \text{uniformly on compact sets of } R^2,$$

and

$$u^0(X) = \lambda \tilde{d}(X),$$

where $\tilde{d}(X) = \text{dist}(N^0 \cup \Gamma^0)$. Thus if $Y_n \to Y$, then

$$u_n(Y_n) \to \lambda \tilde{d}(Y), \quad |Y| = 1.$$

From (11.25) we then get

$$v(X_n) = \lambda r_n \tilde{d}(Y)(1 + o(1)).$$

Also, since $X_n \in N_1$,

$$\tilde{d}(Y) - \frac{d(X_n)}{r_n} \to 0$$

and consequently,

$$v(X_n) = \lambda d(X_n)(1 + o(1)),$$

contradicting (11.24).

Having proved (11.23), now extend N into a C^2 curve $N \cup \hat{N} \equiv \tilde{N}$ and denote by D_R (R positive and small) the domain bounded by $\tilde{N}$ and $\partial B_R(A)$ which contains a part of N_1 initiating at A. Let w be a solution of

$$Lw = 0 \quad \text{in } D_R,$$

$$w = 0 \quad \text{on } \tilde{N},$$

$$\frac{\partial w}{\partial \nu} = \lambda \quad \text{at } A \; (\nu \text{ inward normal}).$$

Then (see Remark 5.4) $w \in C^{1+\alpha}$ and consequently

$$(11.26) \qquad w(X) = \lambda \hat{d}(X) + o(r) \qquad (r = |X - A|),$$

where $\hat{d}(X) = \operatorname{dist}(X, \tilde{N})$.

Denote by G_R the subdomain of D_R bounded by N, N_1 and $\partial B_R(A)$ and let W be defined by

$$LW = 0 \qquad \text{in } G_\rho,$$

$$W = 0 \qquad \text{on } (N \cup N_1) \cap \partial G_\rho,$$

$$W = \overline{C} \qquad \text{on } \partial B_\rho(A) \cap \partial G_\rho$$

for some $\rho < R$. By Problem 1,

$$(11.27) \qquad \qquad \nabla W(X) \to 0 \qquad \text{if } X \to A.$$

Consider the function

$$Z = v - (1 + \varepsilon)w - W \qquad \text{(for any } \varepsilon > 0)$$

in G_ρ. From (11.23), (11.26) we see that $Z < 0$ on $\partial G_\rho \cap N_1$ if ρ is small enough. Clearly also, $Z = 0$ on $\partial G_\rho \cap N$ and $Z < 0$ on $\partial G_\rho \cap \partial B_\rho(A)$ if $\overline{C}$ is large enough. Hence, by the maximum principle, $Z < 0$ in G_ρ.

In view of (11.26), (11.27),

$$\varepsilon w > W \qquad \text{in } G_\rho \cap B_\delta(A)$$

if δ is small enough, say if $\delta \leqslant \delta_0(\varepsilon)$. Therefore,

$$v - (1 + 2\varepsilon)w < 0 \qquad \text{in } G_\delta.$$

Since $v - (1 + 2\varepsilon)w = 0$ on $N \cap \partial G_\delta$, we deduce that

$$\frac{\partial v}{\partial \nu} \leqslant (1 + 2\varepsilon)\frac{\partial w}{\partial \nu} \qquad \text{on } N \cap B_\delta(A).$$

Similarly,

$$\frac{\partial v}{\partial \nu} \geqslant (1 - 2\varepsilon)\frac{\partial w}{\partial \nu}$$

and, recalling that $w \in C^{1+\alpha}$, we conclude that

$$\lim_{\substack{x \in N \\ X \to A}} \frac{\partial u}{\partial \nu} \quad \text{exists and is equal to } -\lambda.$$

The same is of course true for $X \in \Gamma_{\lambda, \mu}$, $X \to A$. Since $N \cup \Gamma_{\lambda, \mu}$ is a C^1 curve, it follows that the derivatives u_x, u_y along $N \cup \Gamma_{\lambda, \mu}$ are continuous functions. Since these derivatives are also bounded in $B_\delta(A) \cap \{u < Q\}$, for some $\delta > 0$, we can apply the Phragmén–Lindelöf theorem (see Problem 2) to u_x and to u_y/y (both satisfy homogeneous elliptic equations with no lowest-order terms) and deduce that u_x, u_y are continuous at A.

Corollary 11.4. *If $x_{\lambda, \mu} \neq 0$, then the curve $\overline{A X}_{\lambda, \mu} \cup \Gamma_{\lambda, \mu}$ is continuously differentiable in a neighborhood of $X_{\lambda, \mu}$ and $\nabla u_{\lambda, \mu}$ is uniformly continuously differentiable in $\{u_{\lambda, \mu} < Q\} \cap B_\delta(X_{\lambda, \mu})$ for some $\delta > 0$.*

The proof is the same as the proof of Theorems 11.1 and 11.3; see Problem 3.

Corollary 11.5. *Under the conditions of Corollary 10.3,*

$$k_{\lambda_1, \mu}(1) > k_{\lambda_2, \mu}(1).$$

Proof. If $x_0 \equiv k_{\lambda_1, \mu}(1) = k_{\lambda_2, \mu}(1)$, then for some small $\delta > 0$,

$$(11.28) \quad u_1 > u_2(1 + \varepsilon) \quad \text{on } \{u_2 > 0\} \cap B_\delta(X_0) \quad (u_i = u_{\lambda_i, \mu})$$

where $X_0 = (x_0, 1)$ and ε is some small positive number. In fact, (11.28) follows by the maximum principle, noting that $u_1 > u_2$ on the closure of $\{u_2 > 0\} \cap \partial B_\delta(X_0)$, for some $\delta > 0$.

From (11.28) we obtain

$$(11.29) \qquad \left| \frac{\partial u_1}{\partial \nu} \right| > \left| \frac{\partial u_2}{\partial \nu} \right| (1 + \varepsilon) \qquad \text{at } X_0,$$

where each derivative exists, by Corollary 11.4, and is equal to 1; a contradiction.

PROBLEMS

1. Prove (11.27).

[*Hint*: Make a conformal mapping and use Remark 5.4.]

2. Extend Lemma 6.7 of Chapter 2 to elliptic operators

$$L = \sum a_{ij} u_{x_i x_j} + \sum b_i u_{x_i}.$$

3. Prove Theorem 11.1 in case $\alpha = 0$.

[*Hint*: If $N^0 = \{x \geqslant 0, y = 0\}$, we have to show

$$\frac{k(y)}{1 - y} \to -\infty, \qquad \text{if } y \uparrow 1.$$

If (11.15) holds then by Theorem 6.1 we obtain $U = -\lambda y$ ($y < 0$) for a blow-up limit, and $J(U)$ can be decreased by replacing U in $B_r(-x_0, 0)$ by a harmonic function $\tilde{U}$ [$\tilde{U} = U$ on $\partial B_r(-x_0, 0)$]. If

(*) $$\frac{k(y_n)}{1 - y_n} \to \gamma, \qquad -\infty < \gamma < \infty,$$

take the blow-up limit U of

$$u_{r_n}(X) = \frac{1}{r_n}(Q - u(A + r_n X)), \qquad r_n = |X_n - A|$$

where $X_n = (k(y_n), y_n)$. Since (11.15) was excluded, by nondegeneracy, ∂B_r intersects the free boundary of U for any $r < 1$. Now we can proceed as in the proof of (11.15), and then show that $N^0 \cup \Gamma^0$ is linear, a contradiction. If $N^0 = \{x \leqslant 0, y = 0\}$ and if

$$\frac{k(y_n)}{1 - y_n} \to -\infty,$$

then apply the proof of the nonoscillation lemma in the region G bounded by N, $x = k(y)$, $\{x = 0\}$ and $x = \{k(y_n)\}$, noting that $\partial u/\partial \nu = \lambda y$ on the free boundary and $\partial u/\partial \nu \geqslant 0$ on $\partial G \cap N$. Next, (*) again leads to a contradiction.]

12. EXISTENCE AND UNIQUENESS FOR AXIALLY SYMMETRIC JETS

Let λ_μ be as in Definition 10.1 and set

$$u_\mu = u_{\lambda_\mu, \mu}, \qquad \Gamma_\mu = \text{free boundary of } u_\mu.$$

Lemma 12.1. *The free boundary Γ_μ initiates at A.*

Proof. Take $\lambda_n \uparrow \lambda_\mu$. Then

$$b = \lim_{n \to \infty} k_{\lambda_n, \mu}(1)$$

is $\geqslant 0$ by continuity (Lemma 10.4). If $b > 0$, then $k_{\lambda,\mu}(1) > 0$ for $\lambda > \lambda_\mu$, $\lambda - \lambda_\mu$ small enough (again by Lemma 10.4), and we get a contradiction to the definition of λ_μ.

Recalling Lemma 10.5 and the smooth fit theorems (Theorems 11.1 and 11.3) we see that $(u_\mu, \Gamma_\mu, \lambda_\mu)$ satisfies all the properties of a solution of the jet problem with Ω replaced by Ω_μ.

We now take a sequence $\mu = \mu_n \to \infty$ such that

$$\lambda_{\mu_n} \to \lambda,$$

$$u_{\mu_n} \to u \qquad \text{weakly in } H^{1,2}_{\text{loc}} \text{ and a.e. in } \Omega.$$

Since $0 \leqslant u_\mu \leqslant Q$, elliptic estimates hold for u_μ in bounded subsets of $\Omega \setminus D$, uniformly with respect to μ. Since $u_x \geqslant 0$, the free boundary of u is a y-graph. Theorem 9.3 remains valid with the same proof for u; thus the free boundary is given by

$$\Gamma : x = k(y), \qquad h_\lambda < y \leqslant 1.$$

Lemma 10.5 implies that Γ does not intersect ∂D (except initially at A).

We next claim that

$$(12.1) \qquad\qquad\qquad\qquad k(1) = 0.$$

Indeed, the proof is similar to the proof of (10.14) for $y = 1$.

Having proved the continuous fit condition (12.1) for u, Γ, we can now apply Theorems 11.1 and 11.3 to deduce the smooth fit; these theorems are local and apply to u as well as to $u_{\lambda,\mu}$.

We have thus established that (u, Γ, λ) forms a solution of the jet problem; this completes the proof of Theorem 8.1.

Corollary 12.2.   *For the solution (u, Γ, λ) of the jet problem that was constructed above, the free boundary has the form $x = k(y)$, $h_\lambda < y < 1$.*

For uniqueness we require, in addition to (8.1), that

$$(12.2) \qquad \begin{array}{l} \text{there is a point } O' = (-b, 0) \ (b \geqslant 0) \text{ such} \\ \text{that } N \text{ is star-shaped with respect to } O'. \end{array}$$

Theorem 12.3.   *If N satisfies (8.1), (12.2), then the solution of the jet problem is unique.*

Proof.   Suppose that (u, Γ, λ) and $(\tilde{u}, \tilde{\Gamma}, \tilde{\lambda})$ are two different solutions. By Problem 21.7, $N \cup \Gamma$ and $N \cup \tilde{\Gamma}$ are star-shaped with respect to O'. However, the uniqueness proof need not rely upon this fact. Indeed, because of (12.2) and the smooth-fit, there is a point $O'' = (-b'', 0)$ (with $b'' \geq b$) such that (i) N is star-shaped with respect to O'', and (ii) any segment from O'' to N does not intersect Γ and $\tilde{\Gamma}$. For simplicity we take $A = (b'', 1)$, $O'' = (0, 0)$. Properties (i) and (ii) will be sufficient for carrying out the uniqueness proof.

Without loss of generality we may take $\lambda \geq \tilde{\lambda}$. Consider first the case

$$(12.3) \qquad\qquad\qquad \lambda > \tilde{\lambda}.$$

By (8.4), (8.14) it follows that Γ lies below $\tilde{\Gamma}$ for x large enough. Setting

$$\tilde{u}_\rho(X) = \tilde{u}(\rho X) \qquad (\rho > 0),$$

it follows that there exist numbers $\rho \leq 1$ such that

$$(12.4) \qquad\qquad\qquad \{u < Q\} \subset \{\tilde{u}_\rho < Q\}.$$

Let ρ be the largest number with this property. If $\rho < 1$, then there is a point X^0 which belongs to the boundary of each of the sets in (12.4) (since $\tilde{u}_\rho < u$ in $\{u < Q\}$, by the maximum principle). Also, X^0 must belong to the free boundary of both functions; here and in the preceding sentence we make use of the properties (i), (ii).

It follows, by the maximum principle, that

$$(12.5) \qquad\qquad \left| \frac{\partial \tilde{u}_\rho}{\partial \nu} \right| > \left| \frac{\partial u}{\partial \nu} \right| \qquad \text{at } X^0$$

that is, $\rho\tilde{\lambda} > \lambda$, which contradicts (12.3).

If $\rho = 1$, then we again argue as before with $X^0 = A$ and obtain $\tilde{\lambda} > \lambda$, a contradiction.

Consider next the case

$$(12.6) \qquad\qquad\qquad \tilde{\lambda} = \lambda.$$

If Γ and $\tilde{\Gamma}$ are given for large x by $y = m(x)$ and $y = \tilde{m}(x)$, respectively, then

$$\lim_{x \to \infty} m(x) = \lim_{x \to \infty} (\tilde{m})(x).$$

It follows that there exists a $\rho \leq 1$ such that (12.4) holds. If, for the largest such ρ, $\rho < 1$, then there exists a point X^0 as before and we obtain $\rho\lambda > \lambda$, a contradiction. If, however, $\rho = 1$, then we take $X^0 = A$ and obtain, as in the proof of (11.24),

$$\left|\frac{\partial \tilde{u}}{\partial \nu}\right| > (1 + \varepsilon)\left|\frac{\partial u}{\partial \nu}\right| \qquad \text{at } A, \text{ for some } \varepsilon > 0,$$

that is, $\lambda > (1 + \varepsilon)\lambda$, which is impossible.

PROBLEMS

1. Assume that

$$(12.7) \qquad \begin{array}{l} N \text{ satisfies } (8.1), \ x(t) \rightarrow -\infty \text{ if} \\ t \rightarrow \infty, \text{ and } N \text{ is uniformly } C^1 \text{ near } \infty. \end{array}$$

If $\mu_n \rightarrow \infty$, $u_{\lambda, \mu_n} \rightarrow u_\lambda$ weakly in $H_{\text{loc}}^{1,2}$ and a.e. in Ω, then we say that u_λ is a *solution of problem* J_λ. Prove: If u_1, u_2 are two solutions of problem J_λ, then for $x < -x_0$ (for some $x_0 > 0$),

$$(12.8) \qquad |\nabla u_i| \leq C \min(y, 1),$$

$$(12.9) \qquad |u_1 - u_2| \leq \frac{Cy^2}{|X|^3},$$

$$(12.10) \qquad |\nabla(u_1 - u_2)| \leq \begin{cases} \dfrac{Cy}{|X|^3} & \text{if } y \leq \varepsilon_0 |X| \\[2mm] \dfrac{C}{|X|} & \text{if } y > \varepsilon_0 |X|, \end{cases}$$

where ε_0 is a small positive number.

[*Hint*: To prove (12.8) near $y = 0$, use Lemma 8.4 and consider $(1/r^0)u(X^0 + r_0 X)$. To prove (12.9), apply the maximum principle to

$$\frac{Cy^2}{|X|^3} - \varepsilon x \pm (u_1 - u_2).]$$

2. Suppose that u_1, u_2 are two solutions of problem J_λ. Let $\phi(x) \in C^\infty$, $\phi(x) = 0$ if $x > -\mu + 2$, $\phi(x) = 1$ if $x < -\mu + 1$, $\phi \geq 0$, $|\phi'| \leq 2$. Define

$$\tilde{u}_2 = u_2 + (u_1 - u_2)\phi,$$

$$v_1 = u_1 \vee \tilde{u}_2, \qquad v_2 = u_1 \wedge \tilde{u}_2$$

$$\tilde{v}_2 = v_2 + (u_2 - v_2)\phi.$$

Prove: If

$$\tilde{J}_{\lambda,\kappa}(v) = \int_{\Omega\cap\{x>-\kappa\}} \left|\frac{1}{y}\nabla v - \lambda I_{\{v<Q\}\setminus D}e\right|^2 y\, dx\, dy,$$

then for any $\kappa > \mu$,

$$\tilde{J}_{\lambda,\kappa}(u_1) + \tilde{J}_{\lambda,\kappa}(u_2) \geqslant \tilde{J}_{\lambda,\kappa}(v_1) + J_{\lambda,\kappa}(\tilde{v}_2) - \eta(\kappa),$$

where $\eta(\kappa) \to 0$ if $\kappa \to \infty$.

3. Use Problem 2 to prove the uniqueness of a solution u_λ of problem J_λ.

4. Use Problem 3 to establish the continuity of u_λ in λ, that is, if $\lambda_n \to \lambda$, then $u_{\lambda_n} \to u$ weakly in $H^{1,2}_{\mathrm{loc}}$ (see Theorem 10.4) and give a proof of Theorem 8.1 [assuming that (12.7) holds] by showing that, for some λ, u_λ satisfies the continuous- and smooth-fit conditions.

13. CONVEXITY OF THE FREE BOUNDARY

We shall study the shape of the free boundary for the axially symmetric jet problem. We assume that

(13.1)
$$N \text{ is concave to the fluid, that is,}$$
$$\Omega \cap \{y > 1\} \text{ is a convex domain.}$$

Theorem 13.1. *If N satisfies (8.1), (13.1), then the free boundary is convex to the fluid.*

This means that $k''(y) \geqslant 0$, and consequently there is a $y_0 \in (h, 1)$ such that

(13.2)
$$k(y) \text{ is monotone decreasing for } h < y < y_0,$$
$$k(y) \text{ is monotone increasing for } y_0 < y < 1.$$

In particular, if N is given by $y = g(x)$, $g'(x) \leqslant 0$, $g''(x) \leqslant 0$, then Γ is given by $y = f(x)$, $f'(x) \leqslant 0$, $f''(x) \geqslant 0$.

Proof. Suppose first that

(13.3)
$$N \text{ lies in a half-plane } \{x < B\},$$
$$X(t) \text{ is uniformly } C^{1+\alpha} \text{ for } t \geqslant t_0,$$

and consider the speed function

(13.4)
$$q = \frac{|\nabla u|}{y}.$$

Then

$$\nabla\left(y\nabla\left(yq^2\right)\right) \geqslant 0 \qquad \text{in } \{u < Q\},$$

that is, q^2 is a subsolution; thus it cannot take a local maximum in the interior of the fluid. We also have

$$q = \lambda \qquad \text{on the free boundary,}$$

$$q(X) \to \lambda \qquad \text{if } X \to A \text{ (by the smooth fit).}$$

Further, q cannot take maximum at any point X^0 of N $(X^0 \neq A)$ since otherwise $\partial q/\partial \nu < 0$ at X^0 (ν inward normal), so that, by (1.9),

$$\kappa q = -\frac{\partial q}{\partial \nu} > 0,$$

which contradicts (13.1).

Suppose we can show that

(13.5)
$$\begin{aligned} &\text{if } q(x_n, y_n) \to \lambda_0, \lambda_0 = \sup q, \quad (x_n, y_n) \to (x_0, y_0) \\ &\text{and either } y_0 = 0 \text{ or } x_0 = \pm\infty \text{ or } y_0 = \pm\infty, \text{ then } \lambda_0 \leqslant \lambda. \end{aligned}$$

Then it would follow that

(13.6)
$$\max_{\bar{\Omega}} q = \lambda$$

and then

$$\kappa q = -\frac{\partial q}{\partial \nu} > 0 \qquad \text{on } \Gamma \quad (\nu \text{ inward normal}),$$

which is the conclusion of the theorem.

To prove (13.5), consider first the case $y_0 = 0$ and let

$$u_n(x, y) = \frac{u(x_n + y_n x, y_n y)}{y_n^2}.$$

Then $Lu_n = 0$ and, by Lemma 8.4,

$$u_n(x, y) \leqslant \frac{Qy^2}{h^2} = \frac{\lambda}{2}y^2, \qquad h = \left(\frac{2Q}{\lambda}\right)^{1/2}.$$

Further, by the assumptions of (13.5),

$$|\nabla u_n(x, y)| \leq \lambda_0 y,$$

$$|\nabla u_n(0, 1)| \to \lambda_0.$$

For a subsequence,

$$u_n \to w \qquad \text{uniformly in } B_2(0) \cap \{y > 0\},$$

and then

$$Lw = 0,$$

(13.7)
$$w(x, y) \leq \frac{\lambda}{2} y^2,$$

$$|\nabla w(x, y)| \leq \lambda_0 y,$$

$$|\nabla w(0, 1)| = \lambda_0.$$

Setting

$$e_0 = \frac{\nabla w(0, 1)}{\lambda_0},$$

$$W = e_0 \cdot \nabla w - \lambda_0 y,$$

we compute

$$LW = \frac{1}{y}\left(\lambda_0 - \frac{w_y}{y} e_0 \cdot e\right) \geq \frac{1}{y}\left(\lambda_0 - \frac{|\nabla w|}{y}\right) \geq 0$$

in $B_2(0) \cap \{y > 0\}$, where $e = (0, 1)$. Since also W takes its maximum (zero) at the point $(0, 1)$, the maximum principle gives $W \equiv 0$; that is,

$$\nabla w \cdot e_0 = \lambda_0 y.$$

Recalling $|\nabla w| \leq \lambda_0 y$, we see that $\nabla w \equiv \lambda_0 y e_0$ and consequently

$$w(x, y) = \frac{\lambda_0}{2} y^2.$$

Thus, by the second relation in (13.7), $\lambda_0 \leq \lambda$.

Consider next the case $x_0 = \infty$, $y_0 > 0$. Since we assume that (13.3) holds, we must have $0 < y_0 < 1$. Recalling that

$$u(x, y) - \min\left\{\frac{\lambda}{2}y^2, Q\right\} \to 0 \qquad \text{if } x \to \infty, \quad 0 < y < 1,$$

and that Γ is given, for large x, by

$$y = f(x), \qquad f'(x) \to 0, \qquad |f''(x)| \le C \qquad (x \to \infty),$$

we deduce that, for any $\varepsilon > 0$,

$$\nabla u(x, y) - \lambda ye \to 0 \qquad \text{uniformly if } \varepsilon < y < f(x), \quad x \to \infty.$$

Hence

$$\frac{1}{y_n}|\nabla u(x_n, y_n)| \to \lambda,$$

so that $\lambda_0 = \lambda$.

Consider next the case in (13.5) where $x_0 = -\infty$, $0 < y_0 < \infty$, and introduce

$$u_n(x, y) = u(x + x_n, y).$$

Then for a subsequence

$$u_n(x, y) \to w(x, y) \qquad \text{in } C^1(B),$$

for some neighborhood B of $(0, y_0)$, and

$$Lw = 0 \qquad \text{in } \{0 < y < H\},$$

$$w(x, 0) = 0 \qquad \text{if } -\infty < x < \infty,$$

$$w(x, y) \le Q \qquad \text{in } \{0 < y < H\},$$

where $H \ge h$. Here H is the asymptotic height of N. If H is finite, then actually $w(x, H) = Q$ and, by Corollary 9.10, $w(x, y) \equiv Qy^2/H^2$. It follows that

$$\frac{1}{y}|\nabla w(x, y)| = \frac{2Q}{H^2} \le \frac{2Q}{h^2} = \lambda,$$

so that $\lambda_0 \leqslant \lambda$. If $H = \infty$, then, by Corollary 9.10,

$$w(x, y) \leqslant \varepsilon y^2 \quad \text{in } \left\{ 0 < y < \left(\frac{Q}{\varepsilon} \right)^{1/2} \right\}$$

for any $\varepsilon > 0$. Thus $w \equiv 0$ and $\lambda_0 = 0$.

It remains to consider the case in (13.5) where $y_0 = +\infty$. Since $X(t)$ is uniformly in $C^{1+\alpha}$ for $t \geqslant t_0$, by elliptic estimates we have (see Remark 5.4)

$$|\nabla u| \leqslant C \quad \text{in } \Omega \cap \{ y \geqslant C_0 \}$$

for some $C_0 > 1$. Hence

$$\frac{1}{y_n} |\nabla u(x_n, y_n)| \to 0,$$

so that $\lambda_0 = 0$.

We have thus completed the proof of (13.5) and thereby also the proof of the theorem in case (13.3) holds. The proof for a general nozzle N follows by truncating N by nozzles N_ε satisfying (13.3). If we replace $x(t)$ by $\min(x(t), 1/\varepsilon)$, then the first condition in (13.3) holds. To satisfy the second condition we replace the part of the nozzle with $t \geqslant T_\varepsilon$ by a horizontal ray, where $T_\varepsilon \to \infty$ if $\varepsilon \to 0$.

PROBLEMS

1. If $A = (0, b)$, the *contraction coefficient* of N is defined by

$$C_c = \frac{\pi h^2}{\pi b^2},$$

where h is the asymptotic height of the jet. Thus $C_c = 2/\lambda$ if $b = 1, Q = 1$. Prove that $C_c = \frac{1}{2}$ for the Borda mouthpiece.
    [*Hint*: Integrate

$$\text{div} \left(\frac{u_y^2 - u_x^2}{y}, -\frac{2u_x u_y}{y} \right) = 0$$

over $\Omega \cap B_R$; by scaling, $|\nabla u(X)| \leqslant C/|X|$ over ∂B_R. Take $R \to \infty$.]

2. Let N, N^0 be two nozzles satisfying (8.1), (12.2) such that N^0 lies to the left of N. Choose $K_\mu, J_{\lambda, \mu}$ for N and a suitable truncation N_μ^0 for N such that with the corresponding $K_\mu^0, J_{\lambda, \mu}^0$ there holds: If u_1, u_2 are minimizers for

$J_{\lambda,\mu}$, $J^0_{\lambda,\mu}$ respectively, then $u_1 \wedge u_2 \in K_\mu$, $u_1 \vee u_2 \in K^0_\mu$. Show that

$$J_{\lambda,\mu}(u_1) + J^0_{\lambda,\mu}(u_2) = J_{\lambda,\mu}(u_1 \wedge u_2) + J^0_{\lambda,\mu}(u_1 \vee u_2)$$

and deduce that $u_1 \leqslant u_2$ everywhere.

3. If N, N^0 are as in Problem 2, show that for the corresponding solutions (u, Γ, λ), $(u^0, \Gamma^0, \lambda^0)$, there holds $\lambda^0 \leqslant \lambda$, with strict inequality if $N \neq N^0$.
    [*Hint*: If $u_{\lambda_\mu,\mu}$ has continuous fit at A, then apply Problem 2 to $u_{\lambda_\mu,\mu}$, $u^0_{\lambda_\mu,\mu}$ to deduce that if $u^0_{\lambda,\mu}$ has continuous fit, then $\lambda \leqslant \lambda_\mu$.]

4. If N is given by $y = g(x)$, $g'(x) \leqslant 0$, $g''(x) \leqslant 0$, then the solution u satisfies $u_y \leqslant 0$.
    [*Hint*: Apply the maximum principle to u_y/y.]

14. THE PLANE SYMMETRIC JET FLOW

The plane symmetric jet problem is the two-dimensional analog of the axially symmetric jet problem. In the conditions (8.2)–(8.12) we make the following changes:

$$Lu = \Delta u = \frac{\partial^2 u}{\partial x^2} + \frac{\partial^2 u}{\partial y^2},$$

and (8.11) is replaced by

$$\frac{\partial u}{\partial v} = \lambda \qquad \text{on } \Gamma_0.$$

Theorem 14.1. *If N satisfies (8.1), then there exists a solution of the plane symmetric jet problem.*

Theorem 14.2. *The solution is unique if N satisfies also the star-shape condition (12.2).*

The proofs are similar to the proof of Theorems 8.1 and 12.2. Notice that in the present case

$$u(x, y) \sim \lambda y \qquad \text{as } x \to \infty \qquad \text{in the set } \{u < Q\},$$

and the asymptotic height of the jet is

$$h = \frac{Q}{\lambda}.$$

PROBLEMS

1. Prove that the number of inflection points of the free boundary does not exceed the number of inflection points of N plus a possible inflection at the detachment point. Here an inflection point means a point of maximum or minimum of the angle θ between the tangent to the curve and the x-axis. [*Hint*: $\theta = \arctan(u_x/u_y)$, $\log u_z = \log|\nabla u| + i\theta$, so that θ cannot take local extremum at any point of the free boundary. If $\theta(A_i)$ is a local maximum for θ restricted to Γ, study the region $\{\theta < \theta(A_i)\}$ connected to A_i, and do the same for local minima.]

2. If N is given by $y = g(x)$ with $g(x)$ monotone decreasing, then Γ is given by $y = f(x)$ with $f(x)$ monotone decreasing.
   [*Hint*: Use symmetrization in the y-direction, employing

$$u(x, y) < Q\frac{x + \mu}{x(\tau_\mu) + \mu}$$

below σ_μ (see Figure 3.5).]

15. ASYMMETRIC JET FLOWS

In this section we consider plane flows which are not necessarily symmetric. The nozzle is made up of two curves N_1, N_2 satisfying:

$$N_i: \quad X = X_i(t) = (x_i(t), y_i(t)), \quad 0 \leqslant t < \infty,$$
$$\text{where } x_i(t), y_i(t) \text{ are piecewise } C^{1+\alpha} \quad (0 < \alpha < 1)$$
$$x_i(t) \leqslant x_i(0), \quad \nabla X_i(t \pm 0) \neq 0;$$

(15.1) every line $x = \text{const.} = c$ either does not intersect N_i, or intersects it at one point, or intersects it at one segment, and $y_2 < y_1$ if $(c, y_i) \in N_i$;

$x_1(0) = x_2(0) = 0$ and the open interval $\overline{X_1(0)X_2(0)}$ does not contain any points of $N_1 \cup N_2$; $x_i(t) \to -\infty$ if $t \to \infty$.

Thus N_i is an x-graph and N_1 lies above N_2. We take for simplicity

$$A = X_1(0) = (0, 1), \qquad B = X_2(0) = (0, -1).$$

Definition 15.1. The *asymetric jet problem* seeks a function u, nonintersecting C^1 curves Γ_1, Γ_2, a positive number λ, and a direction $e = (e_1, e_2)$ with $e_1 > 0$ such that

(15.2)
$$\Gamma_i \text{ initiates at } X_i(0), \quad \Gamma_i \cup N_i \text{ is } C^1$$
$$\text{in a neighborhood of } X_i(0);$$
$$\Gamma_i \text{ is given by } y_i = f_i(x), \quad 0 < x < \infty,$$

(15.3)
$$\Delta u = 0 \text{ in the flow domain bounded by}$$
$$N_1 \cup \Gamma_1 \text{ and } N_2 \cup \Gamma_2;$$

(15.4)
$$u \text{ is uniformly continuously differentiable}$$
$$\text{in a neighborhood of } \Gamma_i \cup \{X_i(0)\} \text{ intersected}$$
$$\text{with the flow domain;}$$

(15.5)
$$\left| \frac{\partial u}{\partial \nu} \right| = \lambda \quad \text{on } \Gamma_i;$$

(15.6)
$$u = 1 \text{ on } N_1 \cup \Gamma_1, \quad u = -1 \text{ on } N_2 \cup \Gamma_2;$$

finally,

$$f_1(x) - f_2(x) \to d\sqrt{1 + \alpha^2} \quad (d \text{ const.})$$

(15.7)
$$f_i'(x) \to 0 \quad \text{if } x \to \infty, \text{ where } \tan a = \frac{e_2}{e_1}.$$

From these conditions it follows that

(15.8) $\lambda d = 2$,   where d is the asymptotic width of the jet.

Note that in general the stream function can be taken to satisfy $u = \pm Q$ on Γ_i; for simplicity, however, we normalize it by taking $Q = 1$.

Theorem 15.1. *For any nozzle N satisfying (15.1), there exists a solution (u, Γ, λ) of the asymmetric jet problem.*

This result is valid also if the last condition in (15.1) is dropped out; see Section 17, Problem 1.

The proof of Theorem 15.1 begins in this section and ends in Section 17.

As a corollary of the proof we shall obtain the following result in case of a symmetric nozzle.

Theorem 15.2. *If N_1 satisfies (15.1) and lies above the x-axis, then there exists a solution of the plane symmetric jet problem with nozzle $N = N_1$.*

[Here again the last condition of (15.1) may be dropped out.]
For proof, see Section 17, Problem 2.

In order to introduce a variational formulation, we denote by D the domain bounded by N_1, N_2 and the segment $\overline{AB}$ and by Ω the domain

$$D \cup \overline{AB} \cup R_+^2 ,$$

where $R_+^2 = \{(x, y); x > 0\}$; $\overline{AB}$ is called the *mouth* of the jet.
Define functions ϕ, Φ by

$$\Delta\phi = 0 \quad \text{in } D,$$

$$\phi(x, y) = 1 \quad \text{if } x < 0, (x, y) \text{ lies above } N_1,$$

(15.9)
$$= -1 \quad \text{if } x < 0, (x, y) \text{ lies below } N_2,$$

$$= -1 \quad \text{if } (x, y) \in \overline{AB} \cup R_+^2 ;$$

$$\Delta\Phi = 0 \quad \text{in } D,$$

$$\Phi(x, y) = 1 \quad \text{if } x < 0, (x, y) \text{ lies above } N_1,$$

(15.10)
$$= -1 \quad \text{if } x < 0, (x, y) \text{ lies below } N_2,$$

$$= 1 \quad \text{if } (x, y) \in \overline{AB} \cup R_+^2 .$$

We require that both functions be bounded and this uniquely determines the functions (by the Phragmén–Lindelöf theorem).
Notice that

$$-1 < \phi < \Phi < 1 \quad \text{in } D,$$
$$\phi \text{ is subharmonic in } \Omega,$$
$$\Phi \text{ is superharmonic in } \Omega.$$

By the maximum principle $\phi_y \geqslant 0$, $\Phi_y \geqslant 0$ in D and thus for any $\mu > 0$ there is a C^1 function $h_\mu(y)$ satisfying

$$h_\mu(y) \text{ is increasing in } y,$$

$$\phi(-\mu, y) \leqslant h_\mu(y) \leqslant \Phi(-\mu, y).$$

Introduce classes of admissible functions

$$K = \{v \in H^{1,2}_{\text{loc}}(R^2), \phi \leqslant v \leqslant \Phi\},$$

$$K_\mu = \{v \in K, v(-\mu, y) = h_\mu(y)\},$$

and let

$$\Omega_\mu = \Omega \cap \{x > -\mu\}.$$

For any unit vector $e = (e_1, e_2)$, denote by $e^\perp$ the vector obtained from e by rotation counterclockwise through an angle $\pi/2$.

For any $\lambda > 0$, $\mu > 0$ and unit vector $e = (e_1, e_2)$ with $e_1 \geqslant 0$, we introduce the functional

$$(15.11) \qquad J_{\lambda, e, \mu}(v) = \int_{\Omega_\mu} |\nabla v - \lambda e^\perp I_{\{|v|<1\}}|^2 \, dx \, dy$$

and consider:

Problem $J_{\lambda, e, \mu}$.   Find $u = u_{\lambda, e, \mu}$ in K_μ such that

$$(15.12) \qquad J_{\lambda, e, \mu}(u) = \min_{v \in K_\mu} J_{\lambda, e, \mu}(v).$$

Remark 15.1. If we take in the definition of K $\phi \equiv -1$, $\Phi \equiv 1$ and replace $I_{\{|v|<1\}}$ in (15.11) by $I_{\{|v|<1\} \cap \{x \geqslant 0\}}$, then run into difficulty in the proof of Lemma 17.1.

If S is an infinite strip in the direction e with width $2/\lambda$ and if

$$\hat{u}(X) = -1 + \text{dist}(X, \partial_1 S) \qquad \text{for } X \in S,$$

where $\partial_1 S$ is the lower of the two lines bounding S, then we can define a function u_0 in K_μ which coincides with $\hat{u}$ in $S \cap \{x > 1\}$ such that

$$J_{\lambda, e, \mu}(u_0) < \infty.$$

Consequently, there exists a solution to problem $J_{\lambda, \mu, e}$.

By increasing rearrangements in the y-direction we can construct a solution u for which $u_y \geqslant 0$ (see Lemma 9.2); we shall henceforth deal only with such solutions. (It will be shown in Section 17 that problem $J_{\lambda, \mu, e}$ has a unique such solution.)

We also note (see Problem 1) that

$$(15.13) \qquad u_{\lambda, \mu, \nu} \text{ is harmonic in } D.$$

We define

$$(15.14) \qquad u(X) = \lim_{r \to 0} \fint_{B_r(X)} u$$

if the limit exists and $u = 0$ on the set T (of measure zero) where the limit does not exist; it will be shown below that T is the empty set.

Lemma 15.3. *There exists a positive constant c (independent of λ, μ) such that if $B_r(X^0) \subset \Omega_\mu$, then*

$$(15.15) \qquad \frac{1}{r} \fint_{\partial B_r(X^0)} (1 + u) \geq \lambda c \qquad \textit{implies that } 1 + u > 0 \textit{ in } B_r(X^0).$$

Proof. Consider the class $\tilde{K}$ of functions $\tilde{h}$ in $H^{1,2}(B_r(X^0))$ satisfying: $\tilde{h} = u$ on $\partial B_r(X^0)$, $\tilde{h} \geq u$ a.e., and let h be the solution of the variational inequality:

$$\min_{\tilde{h} \in \tilde{K}} \int_{B_r(X^0)} |\nabla \tilde{h}|^2 = \int_{B_r(X^0)} |\nabla h|^2, \qquad h \in \tilde{K}.$$

Then h is superharmonic (take $\tilde{h} = h + \varepsilon \zeta$ where $\zeta \in C_0^\infty(B_r(X^0))$, $\zeta \geq 0$) and $h \leq \Phi$ [otherwise, the superharmonic function $\min(h, \Phi)$ would give a smaller value for the functional]. Clearly also, $h \geq \phi$ and $h = u$ a.e. where $u = 1$. Extending h by u outside $B_r(X^0)$, we obtain an admissible function. It follows that

$$\int_{B_r(X^0)} |\nabla u - \lambda e^\perp I_{\{|u| < 1\}}|^2 \leq \int_{B_r(X^0)} |\nabla h - \lambda e^\perp I_{\{|h| < 1\}}|^2,$$

and thus

$$(15.16) \qquad \int_{B_r(X^0)} |\nabla(h - u)|^2 \leq \int_{B_r(X^0)} |\nabla(1 + u)|^2 - \int_{B_r(X^0)} |\nabla(1 + h)|^2$$

$$\leq \lambda^2 \int_{B_r(X^0)} I_{\{1 + u = 0\}}$$

since $h > -1$ in $B_r(X^0)$. We can now proceed to estimate the left-hand side of (15.16) as in Lemma 3.1, using the fact that $h + 1$ is superharmonic, and thus deduce that

$$\int_{B_r(X^0)} I_{\{1 + u = 0\}} = 0.$$

From (15.16) we then deduce that $u = h$ a.e. in $B_r(X^0)$. Since h is super-harmonic, it follows that

$$\fint_{B_\rho(X)} h \uparrow h(X) \qquad \forall\, x \in B_r(X^0);$$

hence [recalling definition (15.14)] the same is true for u and $u = h$ everywhere in $B_r(X^0)$. In particular, $u > -1$ in $B_r(X^0)$, and the lemma follows.

We claim:

(15.17) $\{u = 1\}$ and $\{u = -1\}$ are closed subsets in Ω_μ.

Indeed, if

$$X^n \to X^0, \qquad u(X^n) = -1$$

then setting $r_n = 2\,|X^n - X^0|$ and applying Lemma 15.3, we get

$$\fint_{\partial B_r(X^0)} (1 + u) < cr \qquad \text{for } r_n < r < r_0.$$

Hence

$$\fint_{B_r(X^0)} (1 + u) \to 0 \qquad \text{if } r \to 0,$$

so that $1 + u(X^0) = 0$. Thus $\{u = -1\}$ is closed in Ω_μ, and the proof for $\{u = 1\}$ is similar.

Corollary 15.4. $u \in C^{0,1}(\Omega_\mu)$.

The proof is similar to the proof of Theorem 3.2. Since $\partial u_{\lambda,\,e,\,\mu}/\partial y \geqslant 0$,

$$-1 < u_{\lambda,\,e,\,\mu}(x, y) < 1 \qquad \text{if and only if } f_2(x) < y < f_1(x).$$

Definition 15.2.   The set

$$\Gamma_1 = \{X \in R_+^2 \,;\, X \in \partial\{|u| = 1\}, u(X) = 1\}$$

is called the *upper free boundary*, and the set

$$\Gamma_2 = \{X \in R_+^2 \,;\, X \in \partial\{|u| = 1\}, u(X) = -1\}$$

is called the *lower free boundary*.

As in Section 9 we have

$$f_i(x) \text{ is continuous wherever real, and}$$

$$\Gamma_i = \{(x, f_i(x)); f_i(x) \text{ is real valued}\};$$

notice also that Γ_1 lies above Γ_2.

The nondegeneracy lemma (Lemma 3.3) extends to $1 + u$, in any set where $u < 1$. It implies:

Lemma 15.5. *If $X^0 \in \Gamma^1$ and* dist$(X^0, \overline{AB}) > R$ *for some $R > C/\lambda$ (C is a positive constant independent of λ, μ), then $B_R(X^0)$ intersects $\Gamma_2 \cup \{(0, y);$ $-\infty < y < -1\}$.*

Indeed, if $u < 1$ in $B_R(X^0)$, then by nondegeneracy,

$$\sup_{\partial B_R(X^0)} (1 + u) \geqslant c\lambda R \qquad (c > 0).$$

Since the left-hand side is $\leqslant 2$, we get $R \leqslant 2/(c\lambda)$.

Suppose that the conditions $u = \pm 1$ on Γ_i are replaced by $u = \pm M$ and ϕ, Φ are replaced accordingly by $M\phi, M\Phi$. Let $u^M_{\lambda, e, \mu}$ be the corresponding minimizers.

Lemma 15.6. *Let X^0 be a free-boundary point in $\Omega \setminus \overline{D}$ and let*

$$G = B_r(X^0) \subset B_R(X^0) \subset \Omega_\mu \setminus \overline{D} \qquad (0 < r < R).$$

Then

$$|\nabla u^M_{\lambda, e, \mu}| \leqslant C \qquad \text{in } G,$$

where C is a positive constant independent of M.

This is the bounded gradient lemma; for proof, see Problem 2.

Lemma 15.7. *If the free boundaries Γ_i are connected unbounded curves with initial point $X_i = (0, y_i)$, and if $\delta = |X_1 - X_2| > 0$, then*

$$|\nabla u(X)| \leqslant C\lambda \qquad \text{if } \left|X - \frac{X_1 + X_2}{2}\right| > 2\delta, \quad x > 0$$

where C is a universal constant.

For proof, see Problem 3. The proof also shows that if $X^1 = X^2$, then

$$|\nabla u| \leqslant C\lambda$$

in the region bounded by Γ_1, Γ_2 and $\partial B_R(X_1)$. But this is impossible since $u = 1$ on Γ_1 and $u = -1$ on Γ_2. Thus we must have $X^1 \neq X^2$. Further, if we connect

$$Y^1 = X^1 + \frac{X^1 - X^2}{2} \quad \text{to} \quad Y^2 = X^2 - \frac{X^1 - X^2}{2}$$

by half a circle γ lying in $x > 0$, then

$$2 = u(Y^1) - u(Y^2) \leqslant \int_\gamma |\nabla u| \leqslant C\lambda |Y^1 - Y^2|.$$

Thus:

Lemma 15.8. *Under the assumptions of Lemma* 15.6,

(15.18) $$|X^1 - X^2| \geqslant \frac{c}{\lambda},$$

where c is a positive universal constant.

PROBLEMS

1. Prove (15.13).
   [*Hint*: For any ball B in D,

   $$\min_v \int_B |\nabla v|^2 = \int_B |\nabla u|^2 \quad \text{if } v = u \text{ on } \partial B, \phi \leqslant v \leqslant \Phi,$$

   and by $W^{2,p}$ regularity,

   $$\Delta u = I_{\{u=\phi\}}\Delta\phi + I_{\{u=\Phi\}}\Delta\Phi = 0 \quad \text{in } B.]$$

2. Prove Lemma 15.6.
   [*Hint*: See the proof of Lemma 5.1 and take $B_r(X_{k_0+1})$ with the smallest r $(r \geqslant r_0)$ such that $\partial B_R(x_{k_0+1})$ contains a free boundary point y_0, say with $u = -1$. Then

   $$\fint_{\partial B_r(x_{k_0+1})} (1 + u) \leqslant cr.]$$

3. Prove Lemma 15.7 by the method of Lemma 10.6, using a variant of Lemma 15.3.

16. THE FREE BOUNDARY FOR THE ASYMMETRIC CASE

Definition 16.1. By a *constant flow in direction e with speed* λ, we mean the function w given by

$$w(te + se^{\perp}) = \lambda s \qquad \text{if } 0 < |s| < \frac{1}{\lambda}, t \text{ real,}$$

$$= -1 \qquad \text{if } s < -\frac{1}{\lambda}, t \text{ real,}$$

$$= 1 \qquad \text{if } s > \frac{1}{\lambda}, t \text{ real.}$$

Lemma 16.1. *Let $X_n = (x_n, y_n)$ belong to Γ_2 and $x_n \to \infty$. Then for a subsequence*

$$u_n(X) \equiv u(X_n + X) \to w(X - e)$$

uniformly in compact subsets of R^2, where w is the constant flow in the direction e with speed λ; moreover, the free boundaries of u_n converge to the free boundary of the constant flow in C^1. The same conclusion holds if $X_n \in \Gamma_1$.

Proof. Since

$$\int_{\Omega_\mu} |\nabla u(X) - \lambda e^{\perp} I_{|u(X)|<1}|^2 \leq C_0 < \infty,$$

we have, for any $C > 0$,

$$I_n \equiv \int_{\{|X|<C\}} |\nabla u_n(X) - \lambda e^{\perp} I_{\{|u_n|<1\}}|^2 \to 0 \qquad \text{if } n \to \infty.$$

Without loss of generality we may assume that $u_n \to u_0$ weakly in $H^{1,2}_{\text{loc}}$. It follows that

$$\int_{\{|X|<C\}} |\nabla u_0(X) - \lambda e^{\perp} I_{\{|u_0|<1\}}|^2 \leq \lim I_n = 0.$$

Thus

(16.1) $$\nabla u_0(X) = \lambda e^{\perp} I_{\{|u_0|<1\}} \qquad \text{a.e.}$$

Since $|\nabla u(X_n + X)| \le c_1$ if $|X| < C$, for a suitable c_1 independent of C (provided that n is sufficiently large), we conclude that for a subsequence $u_n \to u_0$ uniformly in compact subsets, and that

$$u(X + X_n) < 1 \qquad \text{if } |X| < \frac{1}{c_1}.$$

Hence by the nondegeneracy lemma [recall that $u(X_n) = -1$],

$$\frac{1}{r} \oint_{\partial B_r} (1 + u_n) \ge c\lambda \qquad \text{if } r < \frac{1}{c_1}.$$

We conclude that $u_0 \not\equiv -1$ in B_r, with $u_0(0) = -1$.

Introduce orthogonal coordinates (x', y') such that e is in the direction of the positive x'-axis and $e^\perp$ is in the direction of the positive y'-axis. Set $X' = (x', y')$, $w'(X') = u_0(X)$. From (16.1) we have

(16.2)
$$\frac{\partial w'}{\partial x'} = 0, \qquad \frac{\partial w'}{\partial y'} = \lambda I_{\{|w'| < 1\}} \qquad \text{a.e.}$$

Thus $w'(X')$ is a function $w_0(y')$ of y' only and it is monotone nondecreasing in y'. Since $w_0 \not\equiv -1$ in any neighborhood of 0, $w_0(0) = -1$ and $-1 \le w_0 \le 1$, it follows that

$$w'(X') = \begin{cases} -1 + \lambda y' & \text{if } 0 < y' < \dfrac{2}{\lambda}, \\ -1 & \text{if } y' < 0, \\ 1 & \text{if } y' > \dfrac{2}{\lambda}, \end{cases}$$

and the first statement of the lemma follows.

Denote by $\{n'\}$ the subsequence for which $u_{n'} \to u_0$.

To prove C^1 convergence of the free boundaries, consider first the case of e nonvertical. By nondegeneracy it follows that the upper free boundaries of u_n in B_R and the upper free boundary of u_0 lie each within a δ-neighborhood of one another provided that n is large enough; here R is arbitrarily large and δ is arbitrarily small.

This yields the flatness condition of Section 4 and thus implies that the upper free boundary of u_n in B_R converges to the upper free boundary of u_0 in the C^1 norm. Thus

$$f_1'(x + x_{n'}) \to \frac{e_2}{e_1} \qquad \text{if } |x| < R, n' \to \infty.$$

A similar analysis applies to f_2 and to the case where e is vertical.

Lemma 16.2. *Let $X_n = (x_n, y_n)$ belong to Γ_2 and $x_n \to \xi$, ξ a finite nonnegative number, $|y_n| \to +\infty$; then for a subsequence*

$$u_n(X) \equiv u(X_n + X) \to w(x - e)$$

uniformly in compact subsets of $\{x > -\xi\}$, where w is the constant flow with speed λ and direction $e = (0, -1)$. The same assertion holds for $X_n \in \Gamma_1$.

Proof. In the present case $u_0(X)$ is defined in the strip $R_\xi = \{(x, y); x \geq -\xi\}$. Also, $u_0 \not\equiv -1$ in any R_ξ-neighborhood of $(0,0)$, as before. Further,

$$(16.3) \qquad\qquad u_0(-\xi, y) \equiv 1 \qquad \text{if } y_n \to +\infty$$

and

$$(16.4) \qquad\qquad u_0(-\xi, y) \equiv -1 \qquad \text{if } y_n \to -\infty.$$

Since $u_0(0,0) = -1$, (16.3) cannot occur. Next, e must be the vector $(0, -1)$ for otherwise we shall have, by (16.4) and $\partial w/\partial x' = 0$ (cf. (16.2)) that $u_0 \equiv -1$ in R_ξ. We can now complete the proof of the lemma as before.

Lemma 16.3. *There cannot exist a line*

$$l: y = \alpha x + \beta, \qquad -\infty < \alpha < 0,$$

which is tangent to the lower free boundary and stays above it, and which intersects $\{x = 0\}$ at $\beta > 1$ (that is, above A).

Proof. Suppose that such a line l exists. Denote by w the constant flow with speed λ and direction $e_l = (1, \alpha)/\sqrt{1 + \alpha^2}$, and define

$$w^\sigma(x, y) = w(x, y - \sigma), \qquad \sigma \geq \sigma_0$$

where the lower free boundary of w^{σ_0} is l. In view of Lemma 15.5, Γ_1 must lie entirely within some distance R of l, and therefore there exists a smallest value of σ, $0 \leq \sigma - \sigma^0 \leq R\sqrt{1 + \alpha^2}$, such that

$$w^\sigma(X) \leq u(X) \qquad \text{everywhere in } \overline{R_+^2},$$

and we may assume that equality holds at some point $\overline{X}$ in the closure of $\{|u| < 1\}$. (If not, we argue with a line $y = \alpha_\varepsilon x + \beta - \varepsilon$ having the same properties as l.) Suppose first that $u(\overline{X}) > -1$. By the strong maximum principle $\overline{X}$ cannot belong to the set $\{|u| < 1\}$. Thus it must lie on the boundary of the strip S^σ (supporting w^σ) as well as on $\overline{\Gamma}_1$. By Theorem 11.1 (which applies here as well), if $\overline{\Gamma}_1 \cap \{x = 0, y > 1\}$ is nonempty, then Γ_1 is C^1

at $x = 0$ and its tangent is in the direction of the positive y-axis. It follows that $\overline{X}$ cannot in fact lie in $\{x = 0\}$ and thus $\overline{X} \in \Gamma_1$.

The strong maximum principle now gives

$$\lambda = \frac{\partial w^\sigma}{\partial \nu} > \frac{\partial u}{\partial \nu} = \lambda \qquad \text{at } \overline{X},$$

which is impossible. If $u(\overline{X}) = -1$, then the last relation holds at the point $\overline{X}$ of $l \cap \Gamma_2$ and this is again impossible.

Lemma 16.4. *There is a constant $R > 0$ such that each ball $B_R(X^0) \subset \Omega \setminus \overline{D}$ with $|u(X^0)| < 1$ contains a free-boundary point.*

This implies that $\Gamma_1 \cup \Gamma_2$ is nonempty.

Proof. If $\Gamma_1 \cup \Gamma_2$ does not intersect $B_R(X^0)$, then by the nondegeneracy lemma,

$$\sup_{\partial B_R(X^0)} (1 + u) \geqslant c\lambda R$$

for any R sufficiently large. This is impossible since the left-hand side is $\leqslant 2$.

Lemma 16.5. *If $e \neq (0, 1)$, then Γ_1 is nonempty.*

Proof. Suppose that Γ_1 is empty. Then $u < 1$ in $\Omega_\mu \setminus D$ and by Lemma 16.4 there is a sequence $X_n = (x_n, y_n)$ in Γ_2 such that $x_n \leqslant C$, $y_n \to +\infty$. But then, by Lemma 16.2, $e = (0, 1)$, contradicting our assumption on e.

Lemma 16.6. *If $e = (0, -1)$, then Γ_2 is empty.*

Proof. If Γ_2 is nonempty, then, by Lemmas 16.1 and 16.2, for any sequence $X_n = (x_n, y_n) \in \Gamma_2$ with $|X_n| \to \infty$ we must have

$$y_n \to -\infty, \qquad \frac{x_n}{|y_n|} \to 0.$$

Indeed, otherwise there exists a sequence of points Y_n on Γ_n with slope $\geqslant \alpha$ (α some negative finite number) such that $|Y_n| \to \infty$. But this contradicts Lemmas 16.1 and 16.2. It follows that Γ_2 lies below any line $y = \alpha x + \beta$ with arbitrary α, $-\infty < \alpha < 0$, and a suitable β. It is easy to see that a line l as in Lemma 16.3 can then be constructed, contradicting Lemma 16.3.

Lemma 16.7. *If $e \neq \pm(0, 1)$, then Γ_1 is connected and $f_1(x)$ is finite valued for all $0 < x < \infty$.*

Proof. By Lemma 16.5, Γ_1 is nonempty. Consider a maximal interval $\{a < x < b\}$ where $f_1(x)$ is finite valued. If $b < \infty$, then f_1 cannot have a finite limit for $x \uparrow b$, since Γ_1 is a smooth curve. Hence

$$|f_1(x)| \to \infty \qquad \text{if } x \uparrow b,$$

thus implying (by Lemma 16.2) that e is vertical.

We have thus proved that $b = \infty$. The proof that $a = 0$ is similar.

Lemma 16.8. *If $e = (0, -1)$, then Γ_1 is connected, $f_1(x)$ is finite on some interval $0 < x < b$ ($b < \infty$) and $f_1(x) \to -\infty$ if $x \uparrow b$.*

Proof. By Lemma 16.5, Γ_1 is nonempty. For any maximal interval $\{c < x < d\}$ where $f_1(x)$ is finite, we must have, by Lemma 16.2,

$$f_1(x) \to -\infty \qquad \text{if } x \uparrow d, \quad \text{provided that } d < \infty,$$

$$f_1(x) \to -\infty \qquad \text{if } x \downarrow c, \quad \text{provided that } c > 0.$$

Also, since Γ_2 is empty (by Lemma 16.6), Lemma 15.5 implies that $d < \infty$. If $c > 0$, then we obtain a contradiction to the nonoscillation lemma in a region

$$\{c < x < d, -\infty < y < -y_0\},$$

where y_0 is sufficiently large. We conclude that $c = 0$, and this completes the proof.

We summarize:

Theorem 16.9.

(i) *If e is nonvertical, then Γ_1, Γ_2 are nonempty and are given by*

$$y = f_i(x), \qquad 0 < x < \infty,$$

where $f_i(x)$ is continuous,

$$f_i'(x) \to \frac{e_2}{e_1} \qquad \text{as } x \to \infty,$$

$$f_i(0) \equiv \lim_{x \downarrow 0} f_i(x) \text{ exists,}$$

and $f_1(0) - f_2(0) \geqslant c/\lambda > 0$ ($c > 0$).

(ii) *If $e = (0, 1)$, then Γ_1 is empty and Γ_2 is nonempty and is given by*

$$y = f_2(x), \qquad 0 < x < b_2 \quad (b_2 < \infty),$$

where $f_2(x)$ is continuous,

$$f_2(x) \to +\infty, \; f_2'(x) \to +\infty \qquad as \; x \uparrow b_2$$

and

$$f_2(0) \equiv \lim_{x \downarrow 0} f_2(x) \; exists;$$

(iii)   *If $e = (0, -1)$, then Γ_2 is empty and Γ_1 is nonempty and is given by*

$$y = f_1(x), \qquad 0 < x < b_1 \qquad (b_1 < \infty),$$

where $f_1(x)$ is continuous,

$$f_1(x) \to -\infty, \; f_1'(x) \to -\infty \qquad as \; x \uparrow b_1,$$

and

$$f_1(0) \equiv \lim_{x \downarrow 0} f_1(x) \; exists.$$

Using Theorem 1.4 of Chapter 2, we further deduce that, in case (i),

$$\left| \frac{d^j f_i(x)}{dx^j} \right| \leq C \qquad (1 \leq x < \infty)$$

for any $j \geq 2$.

The smooth-fit theorems of Section 11 are valid also in the present case with essentially the same proof.

17.   MONOTONICITY, CONTINUITY, AND EXISTENCE FOR THE ASYMMETRIC JET PROBLEM

Lemma 17.1.   *Let $e = (e_1, e_2)$, $\tilde{e} = (\tilde{e}_1, \tilde{e}_2)$ be unit vectors with $e_1 \geq 0$, $\tilde{e}_1 \geq 0$ such that $\tilde{e}$ is obtained from e by rotation counterclockwise by a nonnegative angle $< \pi$. Then*

(17.1) $$u_{\lambda, e, \mu} \geq u_{\lambda, \tilde{e}, \mu}.$$

Here $u_1 = u_{\lambda, e, \mu}$ and $u_2 = u_{\lambda, \tilde{e}, \mu}$ are any minimizers of Problems $J_{\lambda, e, \mu}$ and $J_{\lambda, \tilde{e}, \mu}$, respectively

Proof. Set

$$v_1 = u_1 \vee u_2, \qquad v_2 = u_1 \wedge u_2.$$

We claim that

$$(17.2) \qquad \tilde{J}_{\lambda, e, \mu}(u_1) + J_{\lambda, \tilde{e}, \mu}(u_2) = J_{\lambda, e, \mu}(v_1) + J_{\lambda, \tilde{e}, \mu}(v_2).$$

Indeed, if $\tilde{e} = e$, then the proof is as in Section 10. Suppose then that $\tilde{e} \neq e$. Denote by l_1, l_2 the straight lines in the directions of $e^\perp$ and $\tilde{e}^\perp$, respectively, passing through a point $X^0 = (x^0, y^0)$; X^0 is taken on a ray from the origin in the direction of $e + \tilde{e}$. Denote by Π_i the half-plane bounded by l_i which contains the origin, and set

$$S = R_+^2 \cap \Pi_1 \cap \Pi_2, \qquad S^0 = R_+^2 \setminus S.$$

By Theorem 16.9 we can choose $|X^0|$ sufficiently large so that the sets

$$\{|u_1| < 1\} \cap S^0, \qquad \{|u_2| < 1\} \cap S^0$$

are disjoint and

$$\{|u_i| < 1\} \cap S \text{ is bounded and has positive distance}$$
$$\text{to } l_j, \text{ where } (i, j) = (1, 2) \text{ or } (i, j) = (2, 1).$$

Let

$$S_\mu = (D \cup S) \cap \{x > -\mu\}.$$

Then we can write

$$J_{\lambda, e, \mu}(w) = \int_{S_\mu} + \int_{S^0} \equiv J_1(w) + J_2(w).$$

Noting that $v_1 = u_1$, $v_2 = u_2$ in S^0, it remains to prove (17.2) for J_1. Note that for $w = u_i, v_i$

$$J_1(w) = \int_{S_\mu} \left(|\nabla w|^2 + \lambda^2 I_{\{|w| < 1\}} \right) - 2 \int_{S_\mu} \frac{\partial}{\partial l} w$$

$$\equiv \tilde{J}(w) + J_0(w),$$

where l is in the direction l_1 of $e^\perp$ if $w = u_1$ or $w = v_1$, whereas l is in the direction l_2 of $\tilde{e}^\perp$ if $w = u_2$ or $w = v_2$. Since (17.2) holds for the functional $\tilde{J}$, it remains to prove that

(17.3)
$$\int_{S_\mu} \left(\frac{\partial v_1}{\partial l_1} + \frac{\partial v_2}{\partial l_2} \right) = \int_{S_\mu} \left(\frac{\partial u_1}{\partial l_1} + \frac{\partial v_2}{\partial l_2} \right).$$

Writing

$$\int_{S_\mu} \frac{\partial}{\partial l_i} v_i = \int [v_i], \qquad \int_{S_\mu} \frac{\partial}{\partial l_i} u_i = \int [u_i],$$

where the right-hand side in each equality is a one-dimensional integral, and noting that $[v_i] = [u_i]$ since $v_i = u_i$ at those points of S_μ where $[\cdots]$ is evaluated, (17.3) follows.

Having proved (17.2), we can now complete the proof of Lemma 17.1 by the argument used in the proof of Theorem 10.1.

Lemma 17.2. *There is a unique solution $u_{\lambda, e, \mu}$ to problem $J_{\lambda, e, \mu}$ satisfying $\partial u_{\lambda, e, \mu}/\partial y \geqslant 0$.*

Proof. If u_1, u_2 are two solutions, then set $u_1^\varepsilon(x, y) = u_1(x, y - \varepsilon)$ and

$$v_1 = u_1^\varepsilon \wedge u_2, \qquad v_2 = u_1^\varepsilon \vee u_2.$$

We can establish the analog of (10.8) for $J_{\lambda, e, \mu}$, by essentially the same proof, and then proceed as in the proof of Theorem 10.1.

Denote the free boundaries for $u_{\lambda, e, \mu}$ by $\Gamma^i_{\lambda, e, \mu}$; thus

$$\Gamma^i_{\lambda, \mu} : y = f^i_{\lambda, e, \mu}(x)$$

and $f^1_{\lambda, e, \mu}(x) > f^2_{\lambda, e, \mu}(x)$.

Lemma 17.3. *If $e = (1, 0)$, then there exists a positive constant c such that for all λ sufficiently small (say $\lambda < \lambda_0$)*

(17.4)
$$f^1_{\lambda, e, \mu}(0) \geqslant \frac{c}{\lambda}, \qquad f^2_{\lambda, e, \mu}(0) \leqslant -\frac{c}{\lambda};$$

c and λ_0 are independent of μ.

Proof. From Theorem 16.9, we have

$$f^1_{\lambda, e, \mu}(0) - f^2_{\lambda, e, \mu}(0) > \frac{c}{\lambda}.$$

therefore, if the assertion is not true, say for f_2, then for any small $\varepsilon > 0$ the lower free boundary starts on $x = 0$, $y \geqslant -\varepsilon/\lambda$ and the upper free boundary starts on $x = 0$, $y > 2\varepsilon/\lambda$.

Let

$$G' = \left\{ (x, y); \frac{\varepsilon^2}{\lambda^2} < x^2 + y^2 < \frac{4\varepsilon^2}{\lambda^2}, x > 0 \right\},$$

$$G = G' \backslash \{ u = -1 \}.$$

Then a part $\partial_0 G$ of ∂G consists of subarcs of the lower free boundary, and the remaining part $\partial_1 G$ of ∂G consists of circular parts of $\partial G'$ and of the segment

$$x = 0, \qquad \frac{\varepsilon}{\lambda} < y < \frac{2\varepsilon}{\lambda}.$$

By the bounded gradient lemma (applied to u/λ), we then have

$$|Du| \leqslant C\lambda \qquad \text{in } G,$$

where C is independent of λ.

Take a point $(0, \bar{y}) \in \partial G$, i.e.,

$$\frac{\varepsilon}{\lambda} < \bar{y} < \frac{2\varepsilon}{\lambda},$$

and connect it to a point X^0 on $\partial_0 G$ by a circular arc γ lying in G. Then

$$2 = u(0, \bar{y}) - u(X^0) \leqslant \int_\gamma |Du| \leqslant C\lambda \frac{\varepsilon}{\lambda} = C\varepsilon,$$

which is impossible if ε is small enough.

Lemma 17.4. *If e is nonvertical, then either*

(17.5) $$-1 < f^2_{\lambda, e, \mu}(0) < f^1_{\lambda, e, \mu}(0) < 1,$$

or

(17.6) $$\min \left\{ 1 - f^2_{\lambda, e, \mu}(0), f^1_{\lambda, e, \mu}(0) + 1 \right\} \leqslant \frac{C}{\lambda}$$

for all λ sufficiently large; C is a constant independent of e, μ.

Proof. Set $u = u_{\lambda, e, \mu}$, $\Gamma^i = \Gamma^i_{\lambda, e, \mu}$. Since

$$\frac{1}{r} \int_{\partial B_r(X^0)} (1 + u) \leqslant \frac{2}{r},$$

the nondegeneracy lemma implies that if $X^0 \in \Gamma^2$, then $B_{c/\lambda}(X^0)$ must intersect $\{u = 1\}$ (otherwise, $1 + u = 0$ in a neighborhood of X^0); here c is positive and is independent of μ, e.

Similarly, if $X^0 \in \Gamma^1$, then $B_{c/\lambda}(X^0)$ intersects $\{u = -1\}$. Hence, if the assertion of the lemma is not true, then

$$B_{c/\lambda}(A_2) \text{ intersects } \Gamma^1, \qquad \text{say at } X^1,$$

$$B_{c/\lambda}(A_1) \text{ intersects } \Gamma^2, \qquad \text{say at } X^2,$$

where A_i is the initial point of Γ^i. But this gives a situation contradictory to the bounded gradient lemma.

Proof of Theorem 15.1. We now have all the ingredients necessary for completing the proof of Theorem 15.1. We say that λ belongs to the set Σ_μ if there is a nonvertical e such that

$$f^1_{\lambda, e, \mu}(0) > 1, \qquad f^2_{\lambda, e, \mu}(0) < -1.$$

Let

$$\lambda_\mu = \sup \{\lambda; \lambda \in \Sigma_\mu\}.$$

By Lemma 17.3, Σ_μ is nonempty and contains all λ sufficiently small. By Lemma 17.4, $\lambda_\mu \leq C$ for some positive constant C independent of μ.

Using Lemma 17.2, we can establish as in Lemma 10.4 that $u_{\lambda, \mu, e}$ depends continuously on λ, μ, e. More precisely,

(17.7)
$$\begin{array}{c} \text{if } \lambda_n \to \lambda, \mu_n \to \mu, e_n \to e, \text{ then} \\ u_{\lambda_n, e_n, \mu_n} \to u_{\lambda, e, \mu} \text{ weakly in } H^{1,2}_{\text{loc}} \text{ and a.e.,} \end{array}$$

(17.8)
$$f^i_{\lambda_n, e_n, \mu_n}(x) \to f^i_{\lambda, e, \mu}(x) \qquad \text{for each } x \geq 0.$$

If e is vertical, say $e = (0, 1)$, then (17.8) for $i = 1$ is taken in the sense that

$$f^1_{\lambda_n, e_n, \mu}(x) \to +\infty.$$

Now let $\lambda_n \in \Sigma_\mu$, $\lambda_n \uparrow \lambda_\mu$ and choose e_n such that $f^2_{\lambda_n, e_n, \mu}(0) < -1$ and $f^1_{\lambda_n, e_n, \mu}(0) > 1$, $e_n \to e_\mu$. By continuity

(17.9)
$$f^1_{\lambda_\mu, e_\mu, \mu}(0) \geq 1, \qquad f^2_{\lambda_\mu, e_\mu, \mu}(0) \leq -1.$$

We claim that

(17.10)
$$e_\mu \text{ is nonvertical,}$$

and thus $f^i_{\lambda_\mu, e_\mu, \mu}(0)$ are finite. Indeed, suppose for instance that $e_\mu = (0, -1)$. Then for any $\tilde{e} = (\tilde{e}_1, \tilde{e}_2)$ with $\tilde{e}_1 > 0$, we have

(17.11) $$f^1_{\lambda_\mu, \tilde{e}, \mu}(x) > f^1_{\lambda_\mu, e_\mu, \mu}(x) \qquad \text{if } x \geqslant 0.$$

Indeed, the inequality " $\geqslant$ " follows from the monotonicity of Lemma 17.1, whereas the strict inequality is deduced as in the proofs of Corollaries 10.3 and 11.5.

We now choose $|\tilde{e} - e_\mu|$ small enough so that, by continuity,

$$f^2_{\lambda_\mu, \tilde{e}, \mu}(0) < -2.$$

Since also $f^1_{\lambda_\mu, \tilde{e}, \mu}(0) > 1$ [by (17.9), (17.11)], we can choose $\lambda > \lambda_\mu$ with $\lambda - \lambda_\mu$ small so that

$$f^2_{\lambda, \tilde{e}, \mu} < -1, \qquad f^1_{\lambda, \tilde{e}, \mu}(0) > 1$$

(by continuity again). Thus $\lambda \in \Sigma_\mu$, a contradiction since $\lambda > \lambda_\mu$.

Having proved (17.10) we can now easily show that equalities must hold in (17.9). Indeed, suppose for instance that

$$f^1_{\lambda_\mu, e_\mu, \mu}(0) > 1.$$

Let $\tilde{e} = (\tilde{e}_1, \tilde{e}_2)$ (with $\tilde{e}_1 > 0$) be a vector obtained by rotating e clockwise. If $|\tilde{e} - e|$ is small, then, by continuity,

$$f^1_{\lambda_\mu, \tilde{e}, \mu}(0) > 1.$$

By strict monotonicity [see (17.11)] we also have

$$f^2_{\lambda_\mu, \tilde{e}, \mu}(0) < -1.$$

But then, as before, $\lambda \in \Sigma_\mu$ if $\lambda > \lambda_\mu$ and $\lambda - \lambda_\mu$ is small enough; a contradiction.

We have thus completed the proof of the continuous fit for $u_{\lambda_\mu, e_\mu, \mu}$. We then also have (as in Section 11) a smooth fit at the detachment points A and B.

To complete the proof, we take a sequence $\mu = \mu_n \to \infty$ and set

$$\lambda_n = \lambda_{\mu_n}, \qquad e_n = e_{\mu_n}.$$

Then, for a subsequence, $\lambda_n \to \lambda$, $e_n \to e$ and

$$u_{\lambda_n, \mu_n, e} \to u \qquad \text{weakly in } H^{1,2}_{\text{loc}} \text{ and a.e.}$$

We can now prove that there is a continuous fit for u [by the same argument as used to prove (17.8)]. It also follows that e is nonvertical. The proof that u is a solution of the jet problem can now be completed by already familiar arguments (see Section 12).

PROBLEMS

1. Extend Theorem 15.1 to the case where the last condition in (15.1) is dropped out.

   [*Hint*: Approximate the nozzle by nozzles satisfying (15.1).]

2. Prove Theorem 15.2.

   [*Hint*: Let N_2 denote the reflection of N_1 about the x-axis. By uniqueness $u_{\lambda, e, \mu}(x, y) = -u_{\lambda, e, \mu}(x, -y)$ provided that $e = (1, 0)$ and $h_\mu(-y) = -h_\mu(y)$. Choose λ with continuous fit at A.]

3. Extend the results of Problems 1 to 4 of Chapter 12, to the asymmetric case assuming that $N_1 \cup N_2$ lies in a sector of opening $\theta_0 < \pi$.

   [*Hint*: Consider $\pm(u_1 - u_2) + (C \cos \alpha\theta)/r^\alpha$ to deduce

   $$|u_1 - u_2| \leqslant \frac{C}{|X|^\alpha} \qquad \left(\alpha = \frac{\pi}{\theta_0} > 1\right),$$

   $$|\nabla(u_1 - u_2)| \leqslant \frac{C}{|X|^\alpha}.]$$

4. Denote by N_1^* the reflection of N_1 with respect to the x-axis, and assume that

   $$N_1^* \text{ lies above } N_2.$$

   Prove that for any solution (u, Γ, λ, e) of the jet problem obtained by the procedure of this section,

   $$e_2 \geqslant 0 \qquad \text{if } e = (e_1, e_2).$$

   [*Hint*: Let $u_1 = u_{\lambda, e_0, \mu}$, $e_0 = (1, 0)$. Then $u^*(x, y) = -u_1(x, -y)$ solves Problem $J_{\lambda, e_0, \mu}$ for the reflected nozzle N^*; denote this problem by $J_{\lambda, e_0, \mu}^*$. Then $u^* \wedge u_1$ is in K_μ and $u^* \vee u_1$ in the class K_μ corresponding to N^*. Show (see proof of Lemma 17.1) that

   $$u^* \wedge u_1 = u^*;$$

   next compare with Problem 2 of Section 13.]

5. In Problems 5 to 10 we outline a solution of the *impinging jet problem* by the methods of Sections 15 to 17. The problem is similar to the asymmet-

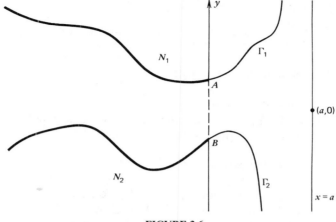

FIGURE 3.6

ric jet problem except for the presence of a wall $x = a$ $(a > 0)$, which blocks the path of the jet. Thus we anticipate that the free boundary will consist of two curves (see Figure 3.6).

(17.12)
$$\Gamma_i: y = f_i(x) \ (0 < x < b_i) \text{ where } b_i < a \text{ and }$$
$$f_1(x) \to \infty \text{ if } x \to b_1, f_2(x) \to -\infty \text{ if } x \to b_2.$$

The unknown parameters are λ and the value θ of u on the wall $x = a$ $(-1 < \theta < 1)$. We shall use the same notation as in Section 15 except for replacing Ω by $\Omega \cap \{x < a\}$ (we denote this new domain again by Ω) and taking a fixed unit vector e, namely

$$e = \left(\frac{x - a}{r}, \frac{y}{r} \right), \qquad \text{where } r^2 = (x - a)^2 + y^2.$$

The variational functional is

$$J_{\lambda, \mu}(v) = \int_{\Omega_\mu} | \nabla v + \lambda e^\perp I_{\{|v|<1\} \setminus D} |^2 \, dx \, dy,$$

where $D = \Omega \cap \{x < 0\}$, and the admissible class is

$$K_{\theta, \mu} = \{ v \in H_{loc}^{1,2}, v = 1 \text{ on } N_1, v = -1 \text{ on } N_2,$$

$$v = \theta \text{ on } \{x = a\}, v(-\mu, y) = h_\mu(y) \},$$

where $h_\mu(y)$ is monotone increasing, $h_\mu = 1$ if $(-\mu, y) \in N_1$ and $h_\mu = -1$ if $(-\mu, y) \in N_2$, and $-1 \leq \theta \leq 1$. Prove that there exists a unique solution $u = u_{\lambda, \theta, \mu}$ of problem $J_{\lambda, \theta, \mu}$:

$$u \in K_{\theta, \mu}, \qquad J_{\lambda, \mu}(u) = \min_{v \in K_{\theta, \mu}} J_{\lambda, \mu}(v).$$

[*Hint*: Compare $u_1(x, y)$ with $u_2(x, y - \varepsilon)$.]

6. If Γ_1, Γ_2 are nonempty, then (17.12) holds.
   [*Hint*: If both Γ_1, Γ_2 go to $-\infty$, then u solves the jet problem without a wall and Γ_2 must be empty (using Lemma 16.3).]

7. $u_{\lambda, \theta}$ is monotone increasing in θ.

8. Set $\lambda_{\theta, \mu} = \sup_{\lambda \in \Sigma} \lambda$; λ belongs to Σ if for some $\theta \in (-1, 1)$, Γ_1 starts above A and Γ_2 starts below B. Let $\lambda_i \in \Sigma$, $\lambda_i \uparrow \lambda_{\theta, \mu}$, $\theta_i \to \theta$ (θ_i corresponds to λ_i in the definition of $\lambda_i \in \Sigma$). Prove that $\theta \neq \pm 1$.
   [*Hint*: If $\theta = \pm 1$, take a constant flow in vertical direction and speed $\lambda_{\theta, \mu}$ and argue as in Lemma 16.3.]

9. Prove that the impinging jet problem has a solution with free boundaries satisfying (17.12).

10. If the nozzle is symmetric with respect to the x-axis and if N_1 is star shaped with respect to $(0, 0)$, then there exists a unique symmetric solution u, Γ, θ to the impinging jet problem [that is, $\theta = 0$ and $u(x, -y) = -u(x, y)$].

18. JETS WITH GRAVITY

The theory developed in Sections 8 to 17 can be extended to jets in a gravity field. As is customary, we take the gravity force to be in the negative y-direction. We shall consider here only the axially symmetric case, with the axis of symmetry in the direction of the gravity force. The nozzle N satisfies:

(18.1)

N is a piecewise $C^{1+\alpha}$ curve given by $X = X(t) = (x(t), y(t))$,

$0 \leq t < \infty \qquad (0 < \alpha < 1)$;

$X(0) = A = (a, 0) \quad (a > 0)$;

$x(t) \geq a$ and $X(t)$ is an x-graph;

If $x(t) \equiv a$ in some interval $0 < t < t_0$, then

$y(t) > 0$ for $0 < t < t_0$;

$\nabla X(t \pm 0) \neq 0$ and $X(t)$ is in $C^3[0, t]$ for some $t_1 > 0$;

$y(t) \to \infty$ if $t \to \infty$;

and finally,

(18.2) N is star-shaped with respect to
               some point $O^* = (0, y^*)$; $y^* > 0$.

Set

$$Lu = \frac{\partial^2 u}{\partial x^2} + \frac{\partial^2 u}{\partial y^2} - \frac{1}{x}\frac{\partial u}{\partial x}.$$

Definition 18.1. The *jet problem* consists of finding a stream function $u(x, y)$, a curve $\Gamma: y = \phi(x)$ (the free boundary) and constants $\lambda > 0$, $Q > 0$ such that

$y = \phi(x)$ is defined and continuous for $0 < x \leqslant a$,

$\phi(a) = 0$,

$N \cup \Gamma$ is a C^1 curve in a neighborhood of A,

Γ is a C^1 curve,

(18.3) $Lu = 0$ in the flow region J bounded by $N \cup \Gamma$ and the y-axis,

$u(0, y) = 0$ for $-\infty < y < \infty$,

$u = Q$ on $N \cup \Gamma$,

$\dfrac{1}{x}\dfrac{\partial u}{\partial \nu} = \sqrt{\lambda - y}$ on Γ,

u is in C^1 in $J \cup (N \cup \Gamma)$, except at the points

where N is not in $C^{1+\alpha}$;

finally,

for some large $y_0 > 0$, $\Gamma \cap \{y < -y_0\}$ is given by

(18.4) $x = f(y)$ and

$$f(y) = \frac{\sqrt{2Q}}{|y|^{1/4}}(1 + o(1))   \text{ as } y \to -\infty.$$

Denote by α_0 the angle formed by the segment γ_0 connecting A to the origin and the tangent ray to N at A.

Theorem 18.1

(i) *Let N satisfy (18.1), (18.2). Then there exists a continuous and strictly monotone increasing function $k(\lambda)$ $(0 \leqslant \lambda < \infty)$ with $k(0) \geqslant 0$, such that a solution to the jet problem with parameters Q, λ exists if and only if $Q = k(\lambda)$.*

(ii) *The solution $u_{\lambda, Q}$ $(Q = k(\lambda))$ is unique.*

(iii) *If $\pi/2 \leqslant \alpha_0 < 2\pi/3$ then $k(0) > 0$, and if $2\pi/3 \leqslant \alpha_0 \leqslant 3\pi/2$ then $k(0) = 0$.*

The function $Q = k(\lambda)$ is called the *solution curve*.

The proof is given in this and in the following section.

One can further show (see reference 7c) that if N is also uniformly C^1 at ∞, then

$$(18.5) \qquad \lim_{\lambda \to \infty} \frac{k(\lambda)}{\sqrt{\lambda}} \equiv c_0 \qquad \text{exists,}$$

and $2c_0$ is the contraction coefficient of the nozzle N.

In order to introduce a variational approach, we need some notation.

Let μ be any positive number, $\mu > \lambda$. Denote by l_μ the segment on $y = \mu$ from $x = 0$ until the first intersection with N; the point of intersection is denoted by (a_μ, b_μ). We further introduce (see Figure 3.7)

$$E^+ = \{x > 0\};$$

N_μ, the union of the part of N from A to (a_μ, μ) and the

segment $\{x = a_\mu, \mu < y < \mu + 1\}$;

γ_0, the interval $\{y = 0, 0 < x < a\}$;

D_μ^*, the region bounded by γ_0, N_μ, the segment $\{y = \mu + 1, 0 < x < a_\mu\}$

$(18.6) \qquad$ and the y-axis from $y = 0$ to $y = \mu + 1$;

M, the ray $\{x = a, -\infty < y \leqslant 0\}$;

E_λ, the half strip $\{0 < x < a, -\infty < y < \lambda\}$;

$$D_\mu = D_\mu^* \setminus \overline{E}_\lambda;$$

$$\Omega_\mu = \text{int } \overline{D_\mu \cup E_\lambda}.$$

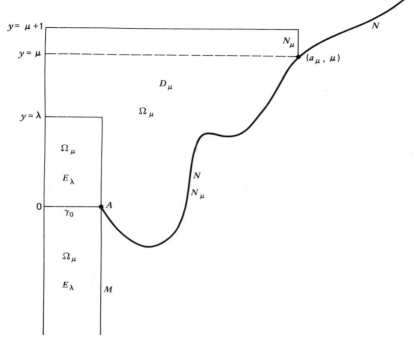

FIGURE 3.7

We define a function Q_μ on N_μ by

$$Q_\mu(x, y) = Q \qquad \text{if } 0 \leqslant y \leqslant \mu,$$
$$= Q(\mu + 1 - y) \qquad \text{if } \mu < y \leqslant \mu + 1.$$

We next introduce a class of admissible functions, depending on Q and the truncating parameter μ:

$$K_{Q,\mu} = \{v; v \in H^{1,2}(E^+ \cap B_R) \text{ for all } R > 0, 0 \leqslant v \leqslant Q \text{ a.e.},$$

and

(18.7)
$$v(0, y) = 0 \text{ if } -\infty < y < \infty,$$
$$v(x, y) = 0 \text{ if } y \geqslant \mu + 1, \quad x \geqslant 0,$$
$$v(x, y) = Q(\mu + 1 - y) \text{ if } \mu \leqslant y \leqslant \mu + 1, \quad x \geqslant a_\mu,$$
$$v(x, y) = Q \text{ if } -\infty < y \leqslant \mu, (x, y) \notin \Omega_\mu\}.$$

We finally introduce the functional J depending on λ, Q, and μ:

$$(18.8) \qquad J_{\lambda, Q, \mu}(v) = \int_{\Omega_\mu} \left| \frac{1}{x} \nabla v - \sqrt{\lambda - y}\, I_{\{v < Q\} \cap E_\lambda} e \right|^2 x\, dx\, dy,$$

where e is the vector $e = (1, 0)$, and $\lambda \geq 0$, $Q > 0$.

Problem $J_{\lambda, Q, \mu}$. Find $u = u_{\lambda, Q, \mu}$ in $K_{Q, \mu}$ such that

$$(18.9) \qquad\qquad J_{\lambda, Q, \mu}(u) = \min_{v \in K_{Q, \mu}} J_{\lambda, Q, \mu}(v).$$

Lemma 18.2. *There exists a function v in $K_{Q, \mu}$ such that $J_{\lambda, Q, \mu}(v) < \infty$.*

Proof. Take $y_0 = \lambda - (\sqrt{2Q}/a)^4$ and define

$$v(x, y) = \begin{cases} \dfrac{x^2}{2}\sqrt{\lambda - y} & \text{if } 0 < x \leqslant \dfrac{\sqrt{2Q}}{(\lambda - y)^{1/4}}, \quad y \leqslant y_0 \\[3mm] Q & \text{if } x > \dfrac{\sqrt{2Q}}{(\lambda - y)^{1/4}}, \quad y \leqslant y_0. \end{cases}$$

Then

$$\int_{\{y < y_0\}} \left| \frac{1}{x} \nabla v - \sqrt{\lambda - y}\, I_{\{v < Q\} \cap E_\lambda} e \right|^2 x\, dx\, dy$$

$$= \int_{-\infty}^{y_0} \int_0^{\sqrt{2Q}/(\lambda - y)^{1/4}} \left(\frac{x}{4(\lambda - y)^{1/2}} \right)^2 x\, dx\, dy = \int_{-\infty}^{y_0} \frac{Q^2}{16(\lambda - y)^2}\, dy < \infty.$$

We can now clearly extend the definition of v into $\{y > y_0\}$ so that it belongs to $K_{Q, \mu}$ and $J_{\lambda, Q, \mu}(v) < \infty$.

Using Lemma 18.2, we can now proceed to prove that problem $J_{\lambda, Q, \mu}$ has a solution $u = u_{\lambda, Q, \mu}$. Notice that $Lu = 0$ in D_μ.

Lemma 18.3. *There exists a constant C depending on Q but not on λ, μ, such that*

$$(18.10) \qquad u_{\lambda, Q, \mu}(x, y) \leqslant \begin{cases} Cx^2 \sqrt{\lambda + 2 - y} & \text{in } E_\lambda, \\[2mm] Cx^2 & \text{in } \Omega_\mu \cap \{y > \lambda\}. \end{cases}$$

Proof. Let $u = u_{\lambda, Q, \mu}$ and choose large constants $c_Q > 0$, $c_\lambda > \lambda$ (to be defined below). Let

(18.11)
$$G = \{0 < x < a, \, y < c_\lambda\},$$
$$v(x, y) = c_Q x^2 \left(\sqrt{c_\lambda - y} + \frac{1}{r^3} \right) \qquad [(x, y) \in G],$$

where

$$r^2 = x^2 + (c_\lambda - y)^2.$$

Then

$$v \geqslant c_Q a^2 \qquad\qquad \text{if } x = a, \, y < c_\lambda - 1,$$

$$v \geqslant c_Q \frac{a^2}{(1 + a^2)^{3/2}} \qquad \text{if } x = a, \, c_\lambda - 1 \leqslant y \leqslant c_\lambda,$$

$$v \geqslant \frac{c_Q}{a} \qquad\qquad \text{if } 0 \leqslant x \leqslant a, \, y = c_\lambda.$$

Therefore, if c_Q is large enough, then

(18.12)
$$v > Q \qquad \text{on } \partial G \setminus \{x = 0\}.$$

Since $L(x^2 \sqrt{c_\lambda - y}) \leqslant 0$, whereas $L(x^2/r^3) = 0$, we have

(18.13)
$$Lv \leqslant 0 \qquad \text{in } G.$$

Moreover, if

(18.14)
$$c_\lambda \geqslant \lambda + 2 \qquad \text{and} \qquad c_Q \geqslant 1,$$

then we have for $y \leqslant \lambda$ the estimate

(18.15)
$$\frac{1}{x} \left| \frac{\partial v}{\partial x} \right| = c_Q \left| 2 \left(\sqrt{c_\lambda - y} + \frac{1}{r^3} \right) - \frac{3x^2}{r^5} \right|$$
$$\geqslant 2 c_Q \sqrt{c_\lambda - y} \left(1 - \frac{3/2}{(c_\lambda - y)^{7/2}} \right) \geqslant \sqrt{c_\lambda - y} \geqslant \sqrt{\lambda - y}.$$

We now choose temporarily

(18.16)
$$c_\lambda = \mu + 2$$

(later we shall choose a better value for c_λ, which will not depend on μ). Since

$|\nabla v|^2/x$ is smooth below $\{y = c_\lambda\}$ [here we make use of (18.16)] and since (18.12) holds, we can compare u with

$$\tilde{u} = \begin{cases} \min(u, v) & \text{in } \Omega_\mu \cap G, \\ u & \text{in } \Omega_\mu \setminus G. \end{cases}$$

We have

(18.17) $J_{\lambda, Q, \mu}(u) \leq J_{\lambda, Q, \mu}(\tilde{u}).$

Using (18.13), (18.15), we obtain as in the proof of Lemma 8.4 (recall that $u < Q$ in $\{y > \lambda\}$) that

(18.18) $u(x, y) \leq v(x, y) \leq 2c_Q\sqrt{\mu + 2 - y}\, x^2$ for $(x, y) \in \Omega_\mu.$

We shall use (18.18) in order to compare u with $\tilde{u}$, with a choice of c_λ independent of μ, namely with c_λ as in (18.14).

For $X \in G \cap \{u > v\}$, we have

(18.19) $c_Q\dfrac{x^2}{r^3} \leq v(X) \leq u(X) \leq 2c_Q\sqrt{\mu + 2 - y}\, x^2;$

that is, $\tilde{u} = u$ in a neighborhood of the singular points of $\nabla v/x$. Therefore, we can again use $\tilde{u}$ as a comparison function and repeat the argument based on (18.17). We thus obtain (18.18) with $\mu + 2$ replaced by c_λ; that is,

(18.20) $u(x, y) \leq Cx^2\sqrt{c_\lambda - y}$ for $y \leq \lambda,$

where C is a constant depending on Q.

Finally, we use the function $\tilde{v}(x, y) = Cx^2$ (we assume that $Ca^2 > Q$) as a comparison function for u in

$$\tilde{G} = \{\lambda < y < \mu + 1, 0 < x < a\}.$$

Since by (18.20), $u < \tilde{v}$ on $\partial\tilde{G}$, we obtain $u(x, y) \leq Cx^2$ for $y > \lambda$.

Lemma 18.4.   *There exist positive constants C, C_0 depending on Q but not on λ, μ, such that if $X^0 = (x^0, y^0)$ and*

$$0 < x^0 < \frac{Q^{1/2}}{C_0}(\lambda + 2 - y^0)^{-1/4} \qquad \text{for } -\infty < y^0 < \lambda,$$

$$x^0 < \frac{Q^{1/2}}{C_0} \qquad \text{for } \lambda < y^0 < \mu + 1,$$

then

$$(18.21) \quad |\nabla u_{\lambda,Q,\mu}(X^0)| \leqslant \begin{cases} Cx^0\sqrt{\lambda + 2 - y^0} & \text{if } -\infty < y^0 < \lambda, \\ Cx^0 & \text{if } \lambda < y^0 < \mu + 1. \end{cases}$$

Proof. Take $r_0 = x^0/2$. By Lemma 18.3,

$$u_{\lambda,Q,\mu} < Q \quad \text{in } B_{r_0}(X^0)$$

and, consequently,

$$Lu_{\lambda,Q,\mu} = 0 \quad \text{in } B_{r_0}(X^0).$$

Hence the normalized function

$$\tilde{u}(X) = \frac{1}{r_0} u_{\lambda,Q,\mu}(X^0 + r_0 X)$$

satisfies

$$\tilde{u} \leqslant Cx^0\sqrt{\lambda + 2 - y^0} \quad (\text{if } y^0 < \lambda),$$

$$\Delta\tilde{u} - \frac{1}{2 + x}\tilde{u}_x = 0 \quad \text{in } B_1(0),$$

which implies

$$|\nabla u_{\lambda,Q,\mu}(X^0)| = |\nabla\tilde{u}(0)| \leqslant Cx^0\sqrt{\lambda + 2 - y^0}.$$

The proof of (18.21) for $y^0 > \lambda$ is similar.

From the general results of Section 3 we have that the solution $u = u_{\lambda,Q,\mu}$ is Lipschitz continuous in $\overline{\Omega}_\mu \setminus (\{x = 0\} \cup \{A\})$; the Lipschitz continuity in a neighborhood of $\{x = 0\}$ was proved in Lemma 18.4. By Section 4, the free boundary

$$\Gamma \equiv \partial\{u < Q\} \cap \{u = Q\}$$

is analytic in E_λ.

Lemma 18.5. *There exists a unique solution u of problem $J_{\lambda,Q,\mu}$, and $u_y \leqslant 0$.*

Proof. Using decreasing rearrangements in the y-direction, we get a solution u with $u_y \leqslant 0$. Suppose that u_1 and u_2 are two such solutions of problem $J_{\lambda,Q,\mu}$ and set

$$u_2^\varepsilon(x, y) = u_2(x, y - \varepsilon) \quad \text{for } 0 < \varepsilon < 1.$$

Denote by Ω_μ^ε the domain obtained by translating Ω_μ upward by ε, and denote the corresponding $K_{Q,\mu}$, $J_{\lambda,Q,\mu}$ by $K_{Q,\mu}^\varepsilon$ and $J_{\lambda,Q,\mu}^\varepsilon$. Thus

$$J_{\lambda,Q,\mu}^\varepsilon(v) = \int_{\Omega_\mu^\varepsilon} \left| \frac{\nabla v}{x} - \sqrt{\lambda - y + \varepsilon}\, I_{\{v<Q\}} \cap E_{\lambda+\varepsilon} e \right|^2 x\, dx\, dy.$$

Since N is an x-graph, the function $v_1 = u_1 \wedge u_2^\varepsilon$ belongs to $K_{Q,\mu}$ and the function $v_2 = u_1 \vee u_2^\varepsilon$ belongs to $K_{Q,\mu}^\varepsilon$. One can check that

(18.22) $J_{\lambda,Q,\mu}(v_1) + J_{\lambda,Q,\mu}^\varepsilon(v_2) \leqslant J_{\lambda,Q,\mu}(u_1) + J_{\lambda,Q,\mu}^\varepsilon(u_2^\varepsilon).$

This implies

$$u_2^\varepsilon \geqslant u_1$$

in the connected component D of $\{u_1 < Q\}$ containing the top of Ω_μ. In particular,

$$u_2^\varepsilon \geqslant u_1 \qquad \text{in } D.$$

Since $\partial u_1/\partial y \leqslant 0$, we conclude that D has the form $\{y > \psi(x)\}$ for some ψ. As a part of the free boundary, the graph of ψ is smooth. Therefore the function

$$\tilde{u} = \begin{cases} u_1 & \text{in } D \\ Q & \text{in } \Omega_\mu \setminus D \end{cases}$$

is admissible; hence $J(u_1) \geqslant J(\tilde{u})$ which implies $\partial u_1/\partial y = 0$ in $\Omega_\mu \setminus D$. Consequently the open set $\{u_1 < Q\}$ is connected and thus

$$u_2^\varepsilon \geqslant u_1 \qquad \text{in } \{u_1 < Q\}.$$

Taking $\varepsilon \to 0$ we obtain $u_2 \geqslant u_1$ in $\{u_1 < Q\}$. Similarly, $u_1 \geqslant u_2$ in $\{u_2 < Q\}$. It follows that $u_1 \equiv u_2$. This completes the proof of the lemma.

Since the free boundary is analytic and $u_y \leqslant 0$, we can write the free boundary in the form $y = \phi(x)$, where $\phi(x)$ is continuous and finite valued on some open subset of $\{0 < x < a\}$.

Lemma 18.6. *The free boundary Γ in $\{0 < x < a\}$ is given by a continuous function $y = \phi(x)$, $0 < x < a$.*

Proof. We first show that the domain of definition of $\phi(x)$ is precisely one interval. Suppose that this is not true. Then there is a value $\bar{x} \in (0, a)$ such that Γ does not intersect $\{x = \bar{x}\}$, but Γ contains points in $\{x < \bar{x}\}$ and in

$\{x > \bar{x}\}$. It follows that there are curves

$$\Gamma_i: y = \phi(x), \ x_i < x < \bar{x}_i \qquad (i = 1, 2; \ x_1 < \bar{x}_1 < \bar{x} < x_2 < \bar{x}_2)$$

such that

$$\phi(x) \to -\infty \qquad \text{if } x \to \bar{x}_1 \text{ and if } x \to x_2.$$

Take r_0 small and $X^0 = (\bar{x}_1, y^0)$, where $y^0 < 0$, $|y^0|$ sufficiently large. Then $B_{r_0/2}(X^0)$ contains free boundary points; hence by the nondegeneracy lemma (Lemma 3.3 with Q replaced by $\sqrt{\lambda - y}$)

$$Q \geqslant \fint_{B_{r_0}(X^0)} (Q - u) \geqslant c x_1 r_0 \min_{B_{r_0/2}(X^0)} \sqrt{\lambda - y} \to \infty$$

as $y_0 \to -\infty$, a contradiction.

We have thus proved that the domain of definition of $\phi(x)$ is one interval. The right endpoint must be $x = a$, for otherwise we can derive a contradiction as before. Similarly, the left endpoint must be $x = 0$.

Lemma 18.7. $\lim_{x \to a} \phi(x)$ *exists and is finite* [*we shall denote it by* $\phi(a)$].

Proof. If $\lim_{x \to a} \phi(x) = -\infty$, then we get a contradiction by the argument of the previous lemma. If $\phi(x)$ oscillates as $x \to a$, then we can derive a contradiction to the nonoscillation lemma.

Notice that $-\infty < \phi(a) \leqslant \lambda$.

Lemma 18.8. *The free boundary* Γ *does not intersect the line* $y = \lambda$, *that is*,

$$\phi(x) < \lambda \qquad \text{if } 0 < x < a.$$

Proof. Suppose that $X^0 = (x^0, \lambda)$ is a free-boundary point where $0 < x^0 < a$. By Lemma 3.1 (see also Lemma 8.3),

$$\fint_{\partial B_r(X^0)} (Q - u) \leqslant C r x^0 \max_{B_r(X^0)} \sqrt{\lambda - y}$$

for small r, where we define

$$\sqrt{t} = 0 \qquad \text{if } t < 0.$$

Hence

$$(18.23) \qquad \fint_{\partial B_r(X^0)} (Q - u) \leqslant C r^{3/2}.$$

Consider the function w satisfying

$$Lw = 0 \qquad \text{in } D_\mu,$$
$$w = 0 \qquad \text{on } y = \lambda, \quad 0 < x < a,$$
$$w = Q - u \qquad \text{on the remaining boundary of } \partial D_\mu.$$

Then $0 < w < Q - u$ in D_μ, by the maximum principle.

We now introduce Green's function G_r for L in the half disc

$$B = B_r(X^0) \cap \{y > \lambda\}.$$

Since $Lw = 0$ in B, we can write, for X in $B_{r/4}(X^0) \cap \{y > \lambda\}$,

$$w(X) = -\int_{\partial B} \frac{\partial G_r}{\partial \nu} w \leq -\int_{\partial B \cap \{y > \lambda\}} \frac{\partial G_r}{\partial \nu}(Q - u) \leq C\!\!\!\fint_{\partial B_r(X^0)}(Q - u) \leq Cr^{3/2}$$

by (18.23). If we take $X = (x^0, \lambda + r/8)$, we obtain for $r \to 0$

(18.24)
$$\frac{\partial w}{\partial y}(X^0) = 0,$$

a contradiction to the maximum principle.

The proof above extends to the case $x^0 = a$, provided that N does not contain the point (a, λ).

Lemma 18.9. *There exists a y_0 sufficiently large such that $\Gamma \cap \{y < -y_0\}$ is given by $x = f(y)$, where*

(18.25)
$$f(y) = \frac{\sqrt{2Q}}{|y|^{1/4}}(1 + o(1)), \qquad f'(y) = o(1)$$

for $y \to -\infty$.

Proof. Take any sequence $(x_n, y_n) \in \Gamma$ with $y_n \to -\infty$ and take any $R > 0$. Then

$$\int_{\{|y - y_n| < R\}} \left| \frac{\nabla u}{x} - \sqrt{\lambda - y}\, I_{\{u < Q\}} e \right|^2 dx\, dy \to 0 \qquad \text{if } n \to \infty.$$

Consider the transformation

$$\tilde{u}_n(\tilde{x}, \tilde{y}) = u\!\left(\frac{\tilde{x}}{r_n}, y_n + \frac{\tilde{y}}{r_n}\right), \qquad r_n = (\lambda - y_n)^{1/4}.$$

The integral above becomes

$$\int_{\{|\tilde{y}|<R(\lambda-y_n)^{1/4}\}}$$

$$\left[\left((\lambda-y_n)^{1/2}\frac{1}{\tilde{x}}\frac{\partial\tilde{u}_n}{\partial\tilde{x}} - (\lambda-y_n)^{1/2}\left(1 - \frac{\tilde{y}}{(\lambda-y_n)^{5/4}}\right)^{1/2} I_{\{\tilde{u}_n<Q\}}\right)^2\right.$$

$$\left. + (\lambda-y_n)\left(\frac{1}{\tilde{x}}\frac{\partial\tilde{u}_n}{\partial\tilde{y}}\right)^2\right]\tilde{x}(\lambda-y_n)^{-3/4}\,d\tilde{x}\,d\tilde{y}$$

(18.26)

$$= (\lambda-y_n)^{1/4}\int_{\{|y|<R(\lambda-y_n)^{1/4}\}}\left[\left(\frac{1}{\tilde{x}}\frac{\partial\tilde{u}_n}{\partial\tilde{x}} - \left(1 - \frac{\tilde{y}}{(\lambda-y_n)^{5/4}}\right)^{1/2} I_{\{\tilde{u}_n<Q\}}\right)^2\right.$$

$$\left. + \left(\frac{1}{\tilde{x}}\frac{\partial\tilde{u}_n}{\partial\tilde{y}}\right)^2\right]\tilde{x}\,d\tilde{x}\,d\tilde{y}.$$

Therefore, the last integral tends to zero as $n\to\infty$; hence we have, for a subsequence,

$$\tilde{u}_n\to\tilde{u}\qquad\text{weakly in }H^{1,2}_{\text{loc}},$$

$$\tilde{u}_n\to\tilde{u}\qquad\text{a.e.,}$$

$$I_{\{\tilde{u}_n<Q\}}\to\gamma\qquad\text{weakly star in }L^{\infty}_{\text{loc}}\text{, and }0\leqslant\gamma\leqslant 1;$$

further,

$$\frac{1}{\tilde{x}}\frac{\partial\tilde{u}_n}{\partial\tilde{x}} - I_{\{\tilde{u}_n<Q\}}\to 0\qquad\text{in }L^2_{\text{loc}},$$

$$\frac{\partial\tilde{u}_n}{\partial\tilde{y}}\to 0\qquad\text{in }L^2_{\text{loc}};$$

that is,

(18.27) $$\frac{\partial\tilde{u}}{\partial\tilde{x}} = \gamma\tilde{x}\qquad\text{and}\qquad\frac{\partial\tilde{u}}{\partial\tilde{y}} = 0.$$

Since $\gamma = 1$ a.e. on $\{\tilde{u}<Q\}$, we conclude that

(18.28) $$\gamma = I_{\{\tilde{u}<Q\}}.$$

But since $\tilde{u} = 0$ on the $\tilde{y}$-axis (recall that $\tilde{u}_n = 0$ on the $\tilde{y}$-axis and $\tilde{u}_n\to\tilde{u}$

weakly in $H^{1,2}_{\text{loc}}$), it follows from (18.27) and (18.28) that

$$\tilde{u}(\tilde{x}, \tilde{y}) = \begin{cases} \dfrac{\tilde{x}^2}{2} & \text{for } 0 \le \tilde{x} \le \sqrt{2Q} \\ Q & \text{for } \tilde{x} \ge \sqrt{2Q}. \end{cases}$$

Let $|\tilde{y}| \le 1$ and $\tilde{x} > \sqrt{2Q}$. Then for a.a. small $r > 0$,

$$\lim_{n \to \infty} \frac{1}{r} \fint_{\partial B_r(\tilde{X})} (Q - \tilde{u}_n) = 0 \qquad (\tilde{X} = (\tilde{x}, \tilde{y})).$$

Therefore, by nondegeneracy $\tilde{X}$ cannot be a free-boundary point for $\tilde{u}_n$ for large n.

Similarly, for $\tilde{x} < \sqrt{2Q}$,

$$\lim_{n \to \infty} \frac{1}{r} \fint_{\partial B_r(\tilde{X})} (Q - \tilde{u}_n) \to \infty \qquad \text{for } r \to 0,$$

which implies that $\tilde{X}$ is not in $\partial\{\tilde{u}_n < Q\}$ for large n. This shows that $\partial\{\tilde{u}_n < Q\}$ converges to $\{\tilde{x} = \sqrt{2Q}\}$ locally in the Hausdorff metric. In particular,

(18.29) $$(\lambda - y_n)^{1/4} x_n = \tilde{x}_n \to \sqrt{2Q}.$$

Note that $\tilde{u}_n$ is a minimizer of the functional

$$\int \left| \frac{\nabla v}{x} - \left(1 - \frac{y}{(\lambda - y_n)^{1/4}}\right)^{1/2} I_{\{v < Q\}} e \right|^2 x \, dx \, dy$$

in $B_R(0)$ for large n, and the free boundary is given by

$$x = f_n(y) = r_n f\left(y_n + \frac{y}{r_n}\right)$$

$$= \frac{\sqrt{2Q}}{|(y_n/r_n^4) + (y/r_n^5)|^{1/4}} (1 + o(1)) \to \sqrt{2Q} \quad \text{as } n \to \infty$$

[where f is not necessarily single valued]. Thus for large n the free boundary for u_n satisfies the flatness condition of Section 4, which implies that f_n is single valued and

$$|f_n'(y)| < \varepsilon_n \qquad \text{for } |y| \le R$$

for each R, where $\varepsilon_n \to 0$; this and (18.29) complete the proof of (18.25).

Lemma 18.10

(i)   *If* $\lambda_1 < \lambda_2$, *then* $u_{\lambda_1, Q, \mu} \leqslant u_{\lambda_2, Q, \mu}$.

(ii) *If* $Q_1 > Q_2$, *then*

$$\frac{1}{Q_1} u_{\lambda, Q_1, \mu} \leqslant \frac{1}{Q_2} u_{\lambda, Q_2, \mu}.$$

For proof, see Problem 2.

From the monotonicity for u we obtain strict monotonicity for the corresponding free boundaries. We also obtain continuity of $u_{\lambda, Q, \mu}$ with respect to Q, λ, μ.

PROBLEMS

1.   Prove (18.22). (Note that the inequality is a consequence of the monotonicity of $\sqrt{\lambda - y}$ in y.)

2.   Prove Lemma 18.10.
[*Hint*: See the proof of (18.22).]

3.   In case of symmetric plane flow under gravity.

$$u = Q, \qquad \frac{\partial u}{\partial \nu} = \sqrt{\lambda - y} \qquad \text{on the free boundary,}$$

$$\Delta u = 0 \qquad \text{in the fluid region.}$$

The variational functional is

$$J_{\lambda, Q, \mu}(v) = \int_{\Omega_\mu} | \nabla v - \sqrt{\lambda - y} \, I_{\{u < Q\} \cap E_\lambda} e |^2 \, dx \, dy.$$

(a)   Extend Lemma 18.2, taking $v = x\sqrt{\lambda - y}$ if $y < y_0$.
(b)   Extend Lemma 18.9; here

$$f(y) = \frac{2Q}{|y|^{1/2}} (1 + o(1)) \qquad \text{as } y \to -\infty.$$

19. THE CONTINUOUS FIT FOR THE GRAVITY CASE

Let γ_1, γ_2 be two C^2 arcs initiating at A and forming angle α. Denote by G the domain bounded by γ_1, γ_2 and $B_R(A)$, for some small $R > 0$.

We introduce polar coordinates (r, θ) with center at A, such that $\theta = 0$ corresponds to the tangent to γ_1 at A.

Lemma 19.1. *Let w be a bounded positive solution of*

$$Lw = 0 \qquad in \ G,$$

$$w = 0 \qquad on \ G \cap (\gamma_1 \cup \gamma_2).$$

If $0 < \alpha \le \pi$, then for $x \in G$

(19.1) $w(X) \le Cr^{(\pi/\alpha)-1}\mathrm{dist}(X, \gamma_1 \cup \gamma_2), \ r = |X - A|,$

for some $C > 0$, and

(19.2) $|\nabla w(X)| \le Cr^{(\pi/\alpha)-1} \qquad for \ X \in \gamma_1 \cup \gamma_2.$

   Proof.   By reference 128, p. 236, w has the form

$$w(X) = a_0 r^{\pi/\alpha} \sin \frac{\pi}{\alpha} \theta + w_0(X)$$

for some constant a_0, where w_0 is of class $\mathring{W}_0^{k+2}$ (see reference 128, p. 231) provided that $k + 1 \ne \pi/\alpha$, k a positive integer and $\pi/\alpha < k + 1 \le 2\pi/\alpha$. Thus we have

$$\int_G \frac{|w_0|^2}{r^{2(k+2)}} \, dx \, dy < \infty.$$

(Notice that we can choose $k + 1 = 3$ if $\alpha = \pi/2$ and $k + 1 = 2$ if $\pi/2 < \alpha \le \pi$.) Now let $X^0 \in \gamma_1 \cup \gamma_2$, $r = |X^0 - A| > 0$, $X \in B_{(\varepsilon/2)r}(X^0)$, and $d = \mathrm{dist}(X, \gamma_1 \cup \gamma_2)$. Then, if ε is small, we have for $\varepsilon \le \varepsilon' \le 2\varepsilon$, the estimate

$$w(X) \le \frac{Cd}{r} \oint_{\partial B_{\varepsilon'r}(X^0)} w \le Cdr^{(\pi/\alpha)-1} + \frac{Cd}{r} \oint_{\partial B_{\varepsilon'r}(X^0)} w_0.$$

Integrating over ε', we obtain

$$w(X) - Cdr^{(\pi/\alpha)-1} \le \frac{Cd}{r^3} \int_{B_{2\varepsilon r}(X^0) \setminus B_{\varepsilon r}(X^0)} |w_0|$$

$$\le \frac{Cd}{r^2} \left\{ \int_{B_{2\varepsilon r}(X^0) \setminus B_{\varepsilon r}(X^0)} |w_0|^2 \right\}^{1/2}$$

$$\le \frac{Cd}{r^2} \left\{ \int_{B_{2r}(A)} \left(\frac{r}{|\tilde{X} - A|} \right)^{2(k+2)} |w_0(\tilde{X})|^2 \, d\tilde{X} \right\}^{1/2}$$

$$\le \frac{Cd}{r} r^{k+1}.$$

Dividing by d and letting $d \to 0$, we get (19.2). The estimate of $w(X)$ for all X follows by Harnack's inequality (see Problem 1).

Lemma 19.2. *If $\alpha_0 \geqslant 2\pi/3$, then for any $Q > 0$ there exists a small enough $\lambda > 0$ such that*

$$\phi_{\lambda, Q, \mu}(a) < 0.$$

Proof. If the assertion is not true, then, by continuity, it follows that the free boundary $y = \phi(x)$ of $u_0 \equiv u_{0, Q, \mu}$ starts at A [by Lemma 18.8, $\phi(x) < 0$ if $0 < x < a$]. We proceed to derive a contradiction.

Let l be a line segment initiating at A and contained in Ω_μ such that there are no free boundary points converging to A and lying above l. Denote the endpoint of l by $A_l = (a_l, b_l)$; let $B_l = (0, b_l)$. Denote by D^* the domain bounded by N_μ, l, $\overline{A_l B_l}$ and the lines $y = \mu + 1$ and $x = 0$, and by α the angle between N_μ and l at A.

Let w be the solution of

$$
\begin{aligned}
Lw &= 0 && \text{in } D^*, \\
w &= Q - u_0 && \text{on } \partial D^* \setminus \gamma^* \ \left(\gamma^* = l \cup \overline{A_l B_l} \right), \\
w &= 0 && \text{on } \gamma^*.
\end{aligned}
$$

Since $\phi(x) < 0$ if $0 < x < a$, we can choose l short enough so as to ensure that there are no free-boundary points above $\overline{A_l B_l}$; thus

$$u_0 < Q, \quad Lu_0 = 0 \qquad \text{in } D^*.$$

By the maximum principle we then deduce that

$$(19.3) \qquad\qquad\qquad w < Q - u_0 \qquad \text{in } D^*.$$

By Lemma 3.1 and Remark 3.1, we have:

$$(19.4) \qquad \begin{aligned} &\text{if there exists a free boundary point in } B_{\varepsilon r_0/2}(X_0), \\ &\text{for some } X_0 \in l \text{ near } A \ (r_0 = |X_0 - A|, 0 < \varepsilon < 1), \end{aligned}$$

then

$$(19.5) \qquad \fint_{\partial B_{\varepsilon r_0}(X_0) \cap D^* \setminus B_{\varepsilon r_0/2}(l)} (Q - u_0) \leqslant C\varepsilon r_0 \max_{B_{\varepsilon r_0}(X_0)} \sqrt{0 - y}$$

$$\leqslant \begin{cases} C\varepsilon r_0 \sqrt{\varepsilon r_0} & \text{if } \alpha = \alpha_0 \\ C\varepsilon r_0 \sqrt{r_0} & \text{if } \alpha > \alpha_0; \end{cases}$$

here C is a universal constant and $B_r(l) = r$-neighborhood of l.

Consider first the case where $\alpha \leqslant \pi$ and N is a y-graph near A. Let w_0 be the harmonic function in $D_\delta^* = D^* \cap \{y < \delta\}$ (for some small $\delta > 0$) with

$$w_0 = Q \qquad \text{on the } y\text{-axis},$$

$$w_0 = 0 \qquad \text{elsewhere on } \partial D_\delta^*.$$

Then $w_{ox} \leqslant 0$, so that

$$Lw_0 = -\frac{1}{x} w_{ox} \geqslant 0.$$

Thus, by the maximum principle,

$$w \geqslant w_0.$$

By comparison,

$$w_0 \geqslant c_0 r^{\pi/\alpha} \sin \frac{\pi\theta}{\alpha} \qquad (c_0 > 0),$$

which gives a lower bound for w in D_δ^*.

If $\alpha > \pi$ or if N is not a y-graph, let γ^{**} be an arc in D^* initiating at A and forming angle $\tilde{\alpha} = \min(\alpha, \pi)$ with γ^* at A. Let D^{**} be a domain bounded by γ^*, γ^{**} and $\partial B_R(A)$, for some small enough R, so that $D^{**} \subset D^*$. We can now repeat the construction of w_0 with respect to D^{**}, and again obtain a lower bound for w.

From these lower bounds and from (19.3) we conclude that if r_0 is small,

$$\fint_{\partial B_{\varepsilon r_0}(X_0) \cap D^* \setminus B_{\varepsilon r_0/2}(l)} (Q - u_0) \geqslant c_1 \varepsilon r_0^{\pi/\tilde{\alpha}} \qquad [c_1 > 0]$$

and therefore, by (19.5),

$$C r_0^{3/2} \sqrt{\varepsilon} \geqslant r_0^{\pi/\tilde{\alpha}} \qquad \text{if } \alpha = \alpha_0,$$

$$C r_0^{3/2} \geqslant r_0^{\pi/\tilde{\alpha}} \qquad \text{if } \alpha > \alpha_0$$

for some constant C. Since $\alpha_0 \geqslant 2\pi/3$, we obtain a contradiction in both cases provided that ε is small enough if $\alpha = \alpha_0$ and provided that r_0 is small enough (and ε fixed, say $\varepsilon = \frac{1}{4}$) in case $\alpha > \alpha_0$. [Notice that N is a y-graph near A if $\alpha_0 < \pi$.]

Thus (19.4) cannot be true. But if we choose l with slope

$$k \equiv \lim_{x \to a} \frac{-\phi(x)}{a - x} - \delta \qquad (\delta > 0),$$

then the fact that (19.4) is not true for some $\varepsilon > 0$ and all r small implies that

$$\lim_{x \to a} \frac{-\phi(x)}{a - x} < k - \varepsilon,$$

that is $\varepsilon < \delta$, which is impossible if δ is small enough (since ε is independent of δ). This contradiction completes the proof of the lemma.

Lemma 19.3. *If $\pi/2 \leqslant \alpha_0 < 2\pi/3$, then there exists an $\varepsilon > 0$ (depending on N and η below, but not on μ) such that if $Q < \varepsilon$, then the free boundary of $u_{0, Q, \mu}$ starts at A; furthermore, in a small neighborhood of A, the free boundary lies above a curve*

$$(19.6) \qquad y = -\eta(a - x)^\delta \qquad \left[\delta = 2\left(\frac{\pi}{\alpha_0} - 1 \right) \right],$$

where η is any positive constant, that is,

$$(19.7) \qquad \phi_{0, Q, \mu}(x) \geqslant -\eta(a - x)^\delta \qquad if \ a - \varepsilon' < x < a$$

for some small $\varepsilon' > 0$.

Note that $1 < \delta \leqslant 2$ [$\delta = 2$ if $\alpha_0 = \pi/2$ and $\delta \downarrow 1$ if $\alpha_0 \uparrow (2\pi/3)$].

Proof. Construct a smooth curve $\gamma : x = h(y)$ $(-\infty < y < 0)$ which coincides with (19.6) in some neighborhood of A and which decreases to $a/2$ as $y \downarrow -\infty$. If suffices to show:

$$(19.8) \qquad \begin{array}{l} \text{if } Q \text{ is small enough then for any } 0 < \lambda \leqslant 1 \\ \text{the free boundary of } u_{\lambda, Q, \mu} \text{ lies above } \gamma. \end{array}$$

To prove (19.8), introduce the domain

$$\tilde{\Omega}_\mu = \Omega_\mu \setminus \{ h(y) \leqslant x < a \}.$$

Let v be the solution of

$$\begin{array}{ll} Lv = 0 & \text{in } \tilde{\Omega}_\mu, \\ v = 1 - u_{1, 1, \mu} & \text{on } \partial\tilde{\Omega}_\mu \setminus \gamma, \\ v = 0 & \text{on } \gamma. \end{array}$$

From Lemma 19.4 (below) it follows that (19.8) holds for $\lambda = 1$, if $Q < \varepsilon$ and ε is small enough.

If follows by continuity and monotonicy of $u_{\lambda, Q, \mu}$ in λ that if (19.8) is not true for some $0 < \lambda < 1$, then there exists a smallest value λ with $0 < \lambda < 1$ for which (19.8) holds. The free boundary of $u_{\lambda, Q, \mu}$ then lies above γ and it is tangent to it at some finite point $\tilde{X} = (\tilde{x}, \tilde{y})$ (the fact that $\tilde{X}$ is a finite point is a consequence of Lemma 18.9 and the definition of γ, and $\tilde{X} = A$ would be a contradiction to the fact that continuous fit implies smooth fit).

By the maximum principle,

$$Q - u_{\lambda, Q, \mu} < Qv$$

in the set $\{u_{\lambda, Q, \mu} < Q\}$. By nondegeneracy we have

$$(19.9) \qquad ca\sqrt{\lambda - \tilde{y} - r} \leqslant \frac{1}{r} \fint_{\partial B_r(\tilde{X})} (Q - u_{\lambda, Q, \mu}) \leqslant \frac{1}{r} \fint_{\partial B_r(\tilde{X})} Qv,$$

where $c > 0$ is a universal constant and $r < \min \{r_0, |A - \tilde{X}|\}$ (r_0 is small but fixed). Since the right-hand side is $\leqslant CQ/r$, we see that

$$(19.10) \qquad \begin{array}{l} \text{if } Q \text{ is small enough then } \tilde{X} \text{ lies} \\ \text{in a small neighborhood of } A. \end{array}$$

We now apply Lemma 19.1 to v to deduce that provided $r \to 0$, the right-hand side of (19.9) is bounded by

$$CQr^{(\pi/\alpha_0)-1} \leqslant C\varepsilon \left(\frac{|\tilde{y}|}{\eta} \right)^{((\pi/\alpha_0)-1)/\delta};$$

we have used here the fact that $\tilde{X}$ lies on the curve (19.6). Using this inequality in (19.9) with $r \to 0$, we obtain

$$1 \leqslant C\varepsilon \left(\frac{1}{\eta} \right)^{1/2}$$

provided that (we recall) Q is small enough, say $Q \leqslant \varepsilon$. If we now choose ε small enough, then we arrive at a contradiction.

Lemma 19.4. *There exists a universal constant $c > 0$ such that if*

$$(19.11) \qquad Q \leqslant \begin{cases} ca^2\sqrt{\lambda} & \text{for } \lambda > a \\ ca\lambda^{3/2} & \text{for } 0 < \lambda \leqslant a, \end{cases}$$

then the free boundary of $u_{\lambda, Q, \mu}$ in $\{a/2 < x < a\}$ lies above the line $y = \lambda/2$.

Proof. Suppose that $\tilde{X} = (\tilde{x}, \tilde{y})$ is a free-boundary point with $a/2 < \tilde{x} < a$. By nondegeneracy,

$$(19.12) \qquad Q \geqslant \fint_{\partial B_r(\tilde{X})} (Q - u_{\lambda, Q, \mu}) \geqslant c_o \tilde{x} r \min_{B_r(\tilde{X})} \sqrt{\lambda - y}$$

if $r < \tilde{x}/2$, where c_0 is a universal positive constant. If $\tilde{y} < \lambda/2$, then we obtain a contradiction to (19.11) by choosing $r = a/4$ if $\lambda > a$ and $r = \lambda/4$ if $\lambda < a$.

Lemma 19.5. *There exist positive universal constants c, C such that*

$$(19.13) \qquad \lambda - \tilde{y} \geqslant \left(\frac{cQ}{\tilde{x}^2} \right)^2 - \tilde{x}$$

for every free boundary point $(\tilde{x}, \tilde{y})$; in particular, if

$$(19.14) \qquad Q > Ca^2 \sqrt{\lambda + a},$$

then the free boundary of $u_{\lambda, Q, \mu}$ starts below the line $y = 0$, that is,

$$(19.15) \qquad \phi_{\lambda, Q, \mu}(a) < 0.$$

Proof. Suppose $\tilde{X} = (\tilde{x}, \tilde{y})$ belongs to the free boundary Γ of $u_{\lambda, Q, \mu}$. Denote by $D(y_0)$ the half disk

$$\left\{ x > 0, x^2 + (y - y_0)^2 < \tilde{x}^2 \right\}.$$

If $y_0 > \lambda + \tilde{x}$ then $\overline{D(y_0)} \cap \Gamma = \varnothing$. Since $\overline{D(\tilde{y})} \cap \Gamma$ is nonempty there exists a number $y^* \in [\tilde{y}, \lambda + \tilde{x})$ such that

$$D(y^*) \cap \Gamma = \varnothing, \qquad \partial D(y^*) \cap \Gamma \neq \varnothing.$$

Let $\hat{X} = (\hat{x}, \hat{y})$ be a point in $\partial D(y^*) \cap \Gamma$.

We introduce the solution v of

$$Lv = 0 \qquad \text{in } D(y^*),$$

$$v(0, y) = 0 \qquad \text{on } \partial D(y^*) \cap \{x = 0\},$$

$$v = Q \qquad \text{on } \partial D(y^*) \cap \{x > 0\}.$$

By the maximum principle $v > u_{\lambda, Q, \mu}$ in $D(y^*)$ and

$$(19.16) \qquad \sqrt{\lambda - \hat{y}} = \frac{1}{\hat{x}} \left| \frac{\partial u_{\lambda, Q, v}}{\partial \nu} \right| \geqslant \frac{1}{\hat{x}} \left| \frac{\partial v}{\partial \nu} \right| \qquad \text{at } \hat{X}.$$

To estimate the right-hand side let w be the solution of

$$\Delta w = 0 \quad \text{in } D(y^*),$$

$$w = v \quad \text{on } \partial D(y^*).$$

It is easily seen that $w_x \geq 0$ so that

$$Lw = -\frac{1}{x} w_x \leq 0.$$

Thus, by the maximum principle $w \geq v$ in $D(y^*)$ and

$$(19.17) \qquad\qquad\qquad \left| \frac{\partial w}{\partial \nu} \right| \leq \left| \frac{\partial v}{\partial \nu} \right| \quad \text{at } \hat{X}.$$

Introduce polar corrdinates (ρ, θ) about $(0, y^* + \tilde{x})$ with $\theta = 0$ in the direction of the negative y axis. Then the harmonic function $w - 2Q\theta/\pi$ is in $C^{1,1}$ on $\partial D(y^*) \cap \{y > y^*\}$, and therefore it is C^1 in $D(y^*) \cap \{y > y^*\}$. It follows that if $\tilde{x} = 1$ then

$$\left| \frac{\partial w}{\partial \nu} - \frac{2Q}{\pi} \frac{\partial \theta}{\partial \nu} \right| \leq C,$$

and, consequently,

$$\left| \frac{\partial w}{\partial \nu} \right| \geq cQ \quad \text{on } \partial D(y^*) \cap \{x > 0\} \quad (c > 0)$$

at points near $(0, y^* + \tilde{x})$. The same inequality holds near $(0, y^* - \tilde{x})$. Finally, this inequality also holds (by using barriers) on the remaining part of $\partial D(y^*) \cap \{x > 0\}$. By scaling we find that, for any $\tilde{x}$,

$$\left| \frac{\partial w}{\partial \nu} \right| \geq c \frac{Q}{\tilde{x}} \quad \text{on } \partial D(y^*) \cap \{x > 0\}.$$

Using this in (19.17), we obtain from (19.16)

$$\sqrt{\lambda - \hat{y}} \geq \frac{c}{\tilde{x}} \frac{Q}{r} \geq \frac{cQ}{\tilde{x}^2}.$$

Since $\hat{y} \geq \tilde{y} - \tilde{x}$ we find that (19.13) holds.

Lemma 19.6. *For any $\lambda > 0$, there exists a $Q > 0$ such that for $u_{\lambda, Q, \mu}$ there is a smooth fit at A.*

Proof. For Q sufficiently large (19.14) holds and for Q sufficiently small

$$\phi_{\lambda, Q, \mu}(a) > 0$$

by Lemma 19.4. By continuity there exists an intermediate value of Q for which $u_{\lambda, Q, \mu}$ has continuous fit at A; the smooth fit follows from Section 11.

Lemma 19.7. *Let $2\pi/3 \leqslant \alpha_0 \leqslant 3\pi/2$. Then for any $Q > 0$ there exists a $\lambda > 0$ such that for $u_{\lambda, Q, \mu}$ there is a smooth fit at A.*

Proof. By Lemma 19.2, if λ is small enough, then $\phi_{\lambda, Q, \mu}(a) < 0$, and by Lemma 19.4, if λ is large enough, then $\phi_{\lambda, Q, \mu}(a) < 0$. Now proceed as in the preceding proof.

Lemma 19.8. *Let $\pi/2 \leqslant \alpha_0 < 2\pi/3$. Then there exists a $\tilde{Q} > 0$ such that if $Q < \tilde{Q}$, then there is no $\lambda \geqslant 0$ for which $u_{\lambda, Q, \mu}$ has smooth fit at A.*

This follows immediately from Lemmas 18.10. and 19.3.

Lemma 19.9. *Let $u_{\lambda, Q, \mu}$ and $u_{\lambda', Q', \mu}$ be two solutions with smooth fit at A. Then:*

(i)   *If $\lambda > \lambda' > 0$, then $Q > Q'$.*
(ii) *If $\lambda = \lambda' > 0$, then $Q = Q'$.*
(iii) *If $Q = Q'$, then $\lambda = \lambda'$.*

This follows easily from Lemma 18.10.

Definition 19.1. For each $\lambda > 0$, denote by $k_\mu(\lambda)$ the value of Q for which $u_{\lambda, Q, \mu}$ has a smooth fit at A. The function

$$Q = k_\mu(\lambda)$$

is called the *solution curve* for Ω_μ. From the previous lemmas it follows:

Theorem 19.10. *The function $k_\mu(\lambda)$ is continuous and strictly monotone increasing in λ, $\lambda > 0$, and*

$$k_\mu(0) \equiv k_\mu(0+) = 0 \qquad if \ \frac{2\pi}{3} \leqslant \alpha_0 \leqslant \frac{3\pi}{2},$$

$$k_\mu(0) > 0 \qquad if \ \frac{\pi}{2} \leqslant \alpha_0 < \frac{2\pi}{3} \ ;$$

further,

(19.18) $c\sqrt{\lambda} \leqslant k_\mu(\lambda) \leqslant C\sqrt{\lambda}$ *if* $\lambda > 1$,

where c, C are positive constants depending only on a.

It is easily seen that for any sequence $\mu_j \to \infty$ and a sequence $Q_j, c_1 \leqslant Q_j \leqslant c_2$, where c_i are positive constants, there exist subsequences such that $Q_j \to Q$ and

(19.19)
$$u_{\lambda, Q_j, \mu_j} \to u_{\lambda, Q} \quad \text{weakly in } H^{1,2}(E^+ \cap B_R) \text{ for any } R > 0,$$
$$u_{\lambda, Q_j, \mu_j} \to u_{\lambda, Q} \quad \text{a.e.}$$

Let

Ω = the domain bounded by $N \cup M$ and the *y*-axis,

$\tilde{K}_Q = \{v;\ v \in H^1(E^+ \cap B_R) \text{ for all } R > 0; 0 \leqslant v \leqslant Q \text{ a.e.}$

$v = Q$ on $N \cup M$ and, as before, on $E^+ \setminus \Omega$; $v = 0$ on the *y*-axis$\}$,

$$\tilde{J}_{\lambda, Q, \mu}(v) = \int_{\Omega \cap \{y < \mu\}} \left| \frac{1}{x} \nabla v - \sqrt{\lambda - y}\, I_{\{v < Q\} \cap E_\lambda} e \right|^2 x\, dx\, dy.$$

Then $u_{\lambda, Q}$ is a minimizer in the following sense:

(19.20) $\tilde{J}_{\lambda, Q, \mu}(u_{\lambda, Q}) \leqslant \tilde{J}_{\lambda, Q, \mu}(v)$

for any $\mu > \lambda$ and for any $v \in \tilde{K}_Q$ such that $v = u_{\lambda, Q}$ on $\{y = \mu\}$.

All the properties established above for $u_{\lambda, Q, \mu}$ follow also for $u_{\lambda, Q}$ (either with the same proof or by passage to limit with μ); this in particular is true of Lemma 19.2 and 19.3.

If we take $Q_j = k_{\mu_j}(\lambda)$, then $u_{\lambda, Q}$ has a smooth fit at A. Similarly, we can take $\lambda_j = k_{\mu_j}^{-1}(Q)$ (Q large enough) and obtain (for $\mu_j \to \infty$) a solution $u_{\lambda, Q}$ with smooth fit at A.

The function $u_{\lambda, Q}$ (with smooth fit at A) thus constructed is then a solution of the jet problem. Notice that if $\lambda_1 > \lambda_2$, then solutions $u_{\lambda_1, Q_1}, u_{\lambda_2, Q_2}$ with smooth fit at A can be constructed using the same sequence μ_j; from this we infer that $Q_1 \geqslant Q_2$.

Lemma 19.11. *If* $u_{\lambda_1, Q}$ *and* $u_{\lambda_2, Q}$ *are two solutions of the jet problem, then* $\lambda_1 = \lambda_2$ *and* $u_{\lambda_1, Q} = u_{\lambda_2, Q}$.

The proof uses a comparison argument as in the case without gravity.

Proof. Let $u_i = u_{\lambda_i, Q}$. For any small $\delta > 0$ there is a $y_\delta < 0$ such that the free boundary Γ_i of u_i is given by

$$(19.21) \quad x = f_i(y) \equiv \frac{\sqrt{2Q}}{|y|^{1/4}} (1 + \delta_i(y)), \qquad |\delta_i(y)| \leqslant \delta \text{ for } y \leqslant y_\delta.$$

We assume for definiteness that $\lambda_2 \leqslant \lambda_1$. Take any $0 < k < 1$ and $\sigma \geqslant 0$ and [recalling (18.2)] consider

$$u_1^\sigma(x, y) = u_1\left(\frac{x}{k}, y^* + \frac{y - y^* - \sigma}{k} \right).$$

Its free boundary Γ_1^σ is given by

$$x = kf_1\left(y^* + \frac{y - y^* - \sigma}{k} \right) \qquad \text{for } y < k(y_\delta - y^*) + y^* + \sigma$$

and

(19.22)

$$kf_1\left(y^* + \frac{y - y^* - \sigma}{k} \right) = \frac{k^{5/4}\sqrt{2Q}\,(1 + \delta_1(y))}{(-y + \sigma + y^*(1 - k))^{1/4}} < k^{1/4} f_2(y) < f_2(y)$$

if

$$k = \frac{1 - \delta}{1 + \delta} < 1 \qquad (k \uparrow 1 \text{ if } \delta \downarrow 0).$$

Thus the free boundary Γ_1^σ of u_1^σ lies above the free boundary Γ_2 of u_2 in the region $\{y < y_\delta\}$ (y_δ is independent of σ).

From (19.22) we also infer that

$$kf_1\left(y^* + \frac{y - y^* - \sigma}{k} \right) \to 0 \qquad \text{if } \sigma \to \infty$$

uniformly for $y \leqslant y_0$, for any y_0. It follows that

(19.23) $\qquad$ $\Gamma_1^\sigma \cup N^\sigma$ lies above $\Gamma_2 \cup N$ everywhere if σ is large enough, where N^σ is the transformed nozzle corresponding to u_1^σ.

Now let σ_k be the smallest nonnegative value of σ for which (19.23) holds. By the maximum principle we have

(19.24) $\qquad\qquad\qquad\qquad$ $u_1^{\sigma_k} \geqslant u_2.$

[Indeed, for any large $M > 0$ and small $\varepsilon > 0$, the function

$$U = u_1^{\sigma_k} - u_2 + \frac{Cx^2}{\left(x^2 + (y + M)^2\right)^{3/2}} + \varepsilon(y + M) \qquad [C = Q(1 + a)]$$

satisfies $U \geqslant 0$ on $y = -M$ and on the free boundary of $u_1^{\sigma_k}$ in $\{y > -M\}$, and $U \to \infty$ if $y \to \infty$; since $LU \leqslant 0$ in $\{0 < u_1^{\sigma_k} < Q\}$, the maximum principle yields $U(x, y) \geqslant 0$ if $y \geqslant -M$. Now take $\varepsilon \to 0$ and then $M \to \infty$.] Moreover, if $\sigma_k > 0$, $\Gamma^{\sigma_k} \cup N^{\sigma_k}$ touches Γ_2 at some finite point $\overline{X} = (\overline{x}, \overline{y})$; otherwise [since (19.22) holds for all $\sigma \geqslant 0$], we can further decrease σ_k keeping the condition (19.23).

If we assume that $\sigma_k \geqslant \text{const.} > 0$ for all k near 1, the point $\overline{X}$ cannot belong to N^{σ_k} for k near 1, for Γ_2 detaches smoothly from A. Hence $\overline{X} \in \Gamma_1^\sigma \cap \Gamma_2$ and (19.24) implies that

$$\frac{1}{k}\sqrt{\lambda_1 - y^* - \frac{\overline{y} - y^* - \sigma_k}{k}} = \frac{1}{\overline{x}}\left|\frac{\partial u_1^{\sigma_k}(\overline{X})}{\partial \nu}\right| \leqslant \frac{1}{\overline{x}}\left|\frac{\partial u_2(\overline{X})}{\partial \nu}\right| = \sqrt{\lambda_2 - \overline{y}}\,.$$

This gives $\lambda_1 < \lambda_2$ if y^* was chosen large enough such that Γ_1 lies in $\{y < y^*\}$, contradicting the assumption $\lambda_2 \leqslant \lambda_1$. [Observe that N is star-shaped also with respect to any point $(0, y)$ with $y > y^*$.]

Therefore, $\sigma_k \to 0$ for a sequence $k \uparrow 1$, and we then obtain from (19.24) that

$$u_1(x, y) \geqslant u_2(x, y).$$

Now, if $u_1 \not\equiv u_2$, we easily deduce that for some small $R > 0$ and $\varepsilon > 0$,

$$(1 + \varepsilon)(Q - u_1) \leqslant Q - u_2$$

on the boundary of $G \equiv B_R(A) \cap \{u_1 < Q\}$; hence also in G. Using the smooth fit at A for u_i, we get

$$(1 + \varepsilon)\sqrt{\lambda_1} = (1 + \varepsilon)\left|\frac{\partial}{\partial \nu}(Q - u_1)(A)\right| \leqslant \left|\frac{\partial}{\partial \nu}(Q - u_2)(A)\right| = \sqrt{\lambda_2}\,,$$

which is impossible since $\lambda_1 \geqslant \lambda_2$. This completes the proof.

Similarly, we can prove:

Lemma 19.12. *If u_{λ, Q_1} and u_{λ, Q_2} are solutions of the jet problem, then $Q_1 = Q_2$ and $u_{\lambda, Q_1} = u_{\lambda, Q_2}$.*

From Lemmas 19.11 and 19.12 it follows that the function $k(\lambda)$ is continuous and strictly monotone increasing. This completes the proof of Theorem 18.1.

PROBLEMS

1. Show that (19.2) implies (19.1).

    [*Hint*: Suppose that $\gamma_1 = $ positive x-axis, $\gamma_2 = $ positive y-axis. If $X^0 = (x^0, y^0), 0 < y^0 < \frac{1}{3}x^0, r = 2y^0, B = B_r(X^0) \cap \{y > 0\}$, and if $w(X^0) > Mr$, then (by Harnack's inequality) $w \geqslant cMr$ in $B_{r/4}(X^0)$ and, by scaling, $w_y(x^0, 0) \geqslant c_1 M$, implying by (19.2) a bound on M. Use Harnack's inequality when X^0 is not near the axes.]

2. Prove Lemma 19.9.

3. Prove (19.19).

4. Prove that $c\sqrt{\lambda} \leqslant k(\lambda) \leqslant C\sqrt{\lambda}$ as $\lambda \to \infty$ $(c > 0, C > 0)$.

5. Extend Theorem 18.1 to the symmetric plane jet with gravity; see Section 18, Problem 3.

6. Consider the symmetric plane jet with gravity and assume that $x(t)$, $y(t)$ in (18.1) are both increasing functions of t. Show that the free boundary is given by a continuous and strictly monotone increasing function $x = g(y)$, $-\infty < y < 0$.

    [*Hint*: See Section 14, Problem 2.]

20. AXIALLY SYMMETRIC FINITE CAVITIES

In this section we consider the Riabouchinsky model of an axially symmetric finite cavity; this will be used in the next section to study axially symmetric infinite cavities.

In both models there is an obstacle N' lying in an ideal axially symmetric fluid; we denote by N the curve in $\{x < 0, y < 0\}$ which generates N' by rotation about the axis of symmetry (the x-axis). We assume:

N is a piecewise $C^{1+\alpha}$ curve given by

$$X = X(t) = (x(t), y(t)), 0 \leqslant t \leqslant \bar{t}, \text{ with } \nabla X(t \pm 0) \neq 0;$$

(20.1) $x(t)$ and $y(t)$ are monotone nondecreasing and $y(t) > 0$ if $t > 0$;

$$X(0) = (-b, 0), \quad X(\bar{t}) = (-a, y(\bar{t})), \quad 0 < a < b.$$

For simplicity we take $y(\bar{t}) = 1$ and set

$$A \equiv X(\bar{t}) = (-a, 1).$$

In the Riabouchinsky model we introduce another obstacle T, which is the reflection of N about the y-axis.

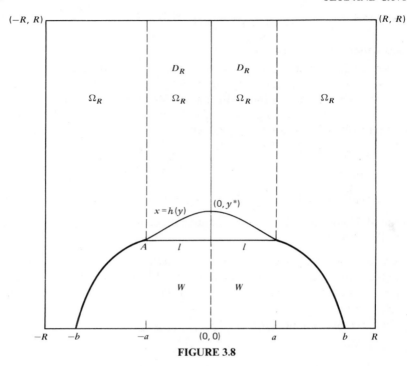

FIGURE 3.8

N is called the *nose* or *projectile* and T the *tail*.

Connect $(-a, 1)$ to $(a, 1)$ by a line segment l and introduce the sets (see Figure 3.8)

$$E_R = \{(x, y); -R < x < R, 0 < y < R\},$$

$$E = R^2 \cap \{y > 0\},$$

$$D_R = \{(x, y); -a < x < a, 1 < y < R\},$$

$W =$ the domain bounded by N, T, l, and the segment $\{(x, 0); -b < x < b\}$,

$$\Omega_R = E_R \setminus \overline{W},$$

$$\Omega = E \setminus \overline{W}.$$

Definition 20.1. The (*axially symmetric*) *finite cavity problem* consists of finding a function $u(x, y)$ (the stream function), a curve Γ (the free boundary) and a positive number λ such that the following conditions are satisfied:

Γ is C^1, Γ is symmetric with

respect to the y-axis;

$\Gamma_- \equiv \Gamma \cap \{x \le 0\}$ is given by $x = h(y)$, where h is

strictly monotone decreasing and continuous for

(20.2) $\quad 1 \le y \le y^*, h(1) = -a, h(y^*) = 0, -a < h(y) < 0$

for $1 < y < y^*$;

$\Gamma \cup N$ is C^1 in a neighborhood of A,

and

u is symmetric with respect to
the y-axis and $u \in C(\bar{E})$;

(20.3) $\quad$ $\begin{aligned} u(x,0) &= 0 & &\text{if } -\infty < x \le -b, \\ u(x, y) &= 0 & &\text{if } (x, y) \in W \text{ or if } |x| \le h(y), 1 \le y \le y^*, \\ u(x, y) &> 0 & &\text{elsewhere in } E, \\ Lu &= 0 & &\text{in } \{u > 0\}, \end{aligned}$

where

$$Lu = u_{xx} + u_{yy} - \frac{1}{y} u_y;$$

further,

(20.4) $\qquad u = 0, \qquad \frac{1}{y}\left|\frac{\partial u}{\partial \nu}\right| = \lambda \qquad \text{on } \Gamma;$

(20.5) $\qquad \begin{aligned} &\nabla u \text{ is uniformly continuously differentiable} \\ &\text{in } \Omega \cap \{u > 0\}\text{-neighborhood of } A; \end{aligned}$

(20.6) $\qquad u - \frac{y^2}{2} \to 0 \qquad \text{as } (x, y) \to \infty.$

The set

$$K = W \cup \{(x, y); |x| \le h(y), 1 \le y \le y^*\}$$

is called the *cavity*.

Theorem 20.1. *There exists a unique solution of the finite cavity problem.*

The proof is given in several steps. We first introduce a variational approach. For any $R > 0$, let

(20.7)

$$K_R = \left\{ v \in H_{\text{loc}}^{1,2}(E); \, v = \Phi \text{ on } \partial\Omega_R \setminus \partial W, \, v = 0 \text{ in } W, 0 \leqslant v \leqslant \Phi \text{ in } \Omega_R \right\},$$

where

$$\Phi = \frac{y^2}{2}.$$

Let

(20.8) $$J_{\lambda, R}(v) = \int_{\Omega_R} \left[\left| \frac{\nabla v}{y} \right|^2 + \lambda^2 I_{\{v > 0\} \cap D_R} \right] y \, dx \, dy$$

and consider:

Problem $J_{\lambda, R}$. Find $u = u_{\lambda, R}$ in K_R such that

(20.9) $$J_{\lambda, R}(u) = \min_{v \in K_R} J_{\lambda, R}(v).$$

This problem clearly has a solution u. Set $u = yU$ and write

$$J_{\lambda, R}(u) = \tilde{J}_{\lambda, R}(U).$$

Then U is a minimizer of $\tilde{J}(V)$ in the class $\tilde{K}_R$ of functions $V = v/y$, $v \in K_R$. Since Φ/y is monotone increasing in y, if U^* is the symmetric increasing rearrangement of U in the y-direction, then $U^* \in \tilde{K}_R$. Using Theorem 7.1′ [and, in particular, (7.6)], we easily see that

$$\tilde{J}_{\lambda, R}(U^*) \leqslant \tilde{J}_{\lambda, R}(U),$$

with strict inequality if $U^* \neq U$ on a set of positive measure. Hence U is monotone increasing in y, and the same is then true of u.

We can also rearrange $\Phi - u$ symmetrically and decreasing in $|x|$; here even the second term in (20.8) may decrease by the rearrangement. It follows that u may be assumed to be decreasing in x for $x < 0$. Thus

$$u_y \geqslant 0 \text{ in } \Omega_R, \qquad u_x \leqslant 0 \text{ in } \Omega_R \cap \{x < 0\}.$$

Since the free boundary is locally analytic, it follows that is has the form

$$x = \pm h(y) \qquad (1 \leqslant y \leqslant y^*), \quad h \leqslant 0,$$

where h is monotone increasing and continuous; h and y^* depend of course on λ, R; say $h = h_{\lambda, R}$, $y^* = y^*_{\lambda, R}$.

The free boundary, if nonempty, may either start on the horizontal line $\{y = 1\}$ or on the vertical line $\{x = -a\}$; in both cases it satisfies the smooth fit property (Section 11), since the proof of this property is local.

Now take a sequence $R \to \infty$ such that

$$u_{\lambda, R} \to u_\lambda \qquad \text{weakly in } H^{1,2}_{\text{loc}}(E) \text{ and a.e.}$$

Lemma 20.2. *If $\lambda \leq \Lambda < \infty$, then there exists a positive constant C^* depending only on Λ such that*

(20.10) $$y^*_{\lambda, R} \leq C^*$$

for all R sufficiently large.

Proof. Set

$$\tilde{\Omega}_R = \{u_{\lambda, R} > 0\} \cap \{x < 0\} \cap \{y < y^*_{\lambda, R}\}.$$

By the bounded gradient lemma,

$$|\nabla u(x, y)| \leq C \qquad \text{if } y = y^*_{\lambda, R}, \quad -\frac{R}{2} \leq x \leq 0,$$

where C is independent of λ, R. Hence, for any $\delta > 0$,

$$u(x, y) \leq \delta y \qquad \text{if } y = y^*_{\lambda, R}, \quad -\frac{\delta}{C} y \leq x \leq 0.$$

On the remaining part of $\partial\tilde{\Omega}_R$,

$$u \leq -\varepsilon x y^2$$

for any $\varepsilon > 0$, provided that $y^*_{\lambda, R}$ is sufficiently large. Hence, by the maximum principle,

$$u \leq \delta y - \varepsilon x y^2 \qquad \text{in } \tilde{\Omega}_R,$$

where δ and ε can be taken arbitrarily small if $y^*_{\lambda, R}$ is sufficiently large.

Consequently, if the assertion of the lemma is not true, then $u = u_{\lambda, R} \to 0$ for a sequence $R \to 0$, and this contradicts the nondegeneracy lemma (Lemmas 3.3 and 8.6).

From Lemma 20.2 it follows that

$$(20.11) \qquad -\frac{Cy^2}{r^3} \leqslant u_{\lambda, R} - \frac{y^2}{2} \leqslant 0 \qquad \text{on } \partial\{u_{\lambda, R} > 0\}$$

where $r^2 = x^2 + y^2$ and C is a positive constant independent of λ, R (provided that $\lambda \leqslant \Lambda$). By the maximum principle we obtain the same inequality in $\{u_{\lambda, R} > 0\}$; therefore, also

$$(20.12) \qquad -\frac{Cy^2}{r^3} \leqslant u_\lambda - \frac{y^2}{2} \leqslant 0 \qquad \text{in } \{u_\lambda > 0\}.$$

Definition 20.2. The function u_λ is called a *solution of problem* J_λ.

Note that u_λ satisfies all the properties enlisted in (20.2)–(20.6) except for the continuous and smooth-fit conditions.

Definition 20.3. Any function u_λ satisfying all the properties (20.2)–(20.6) except for the continuous and smooth fit conditions will be called a λ-*solution*.
For any λ-solution u_λ we consider the flow region $Q_\lambda = \{u_\lambda > 0\}$. The domain

$$\alpha Q_\lambda \equiv \{\alpha X; \ X \in Q_\lambda\} \qquad (0 < \alpha < 1)$$

where $X = (x, y)$ is called a *magnification* of Q_λ. It is again a flow region of a λ-solution with N replaced by αN; this λ-solution is

$$u_\lambda^\alpha(x, y) \equiv \alpha^2 u_\lambda\left(\frac{x}{\alpha}, \frac{y}{\alpha}\right).$$

Lemma 20.3. *Suppose that* u_{λ_1}, u_{λ_2} *are* λ_1- *and* λ_2-*solutions,* $u_{\lambda_1} \not\equiv u_{\lambda_2}$, *and* Q_{λ_1} *can be magnified by a factor* $\alpha < 1$ *in such a way that*

$$\alpha Q_{\lambda_1} \supset Q_{\lambda_2}, \qquad \partial\{\alpha Q_{\lambda_1}\} \cap \partial Q_{\lambda_2} \cap \{y > 0\} \neq \varnothing.$$

Then $\lambda_1 > \lambda_2$.

Proof. Set

$$\tilde{u}_1(x, y) = \alpha^2 u_{\lambda_1}\left(\frac{x}{\alpha}, \frac{y}{\alpha}\right), \qquad u_2(x, y) = u_{\lambda_2}(x, y).$$

Then

$$U(x, y) \equiv \tilde{u}_1(x, y) - u_2(x, y) \to 0 \qquad \text{if } r \to \infty.$$

Since $U \geqslant 0$ on the boundary of Q_{λ_2} the maximum principle gives $U > 0$ in Q_{λ_2} and

$$\frac{1}{y}\left|\frac{\partial \tilde{u}_1}{\partial \nu}\right| > \frac{1}{y}\left|\frac{\partial u_2}{\partial \nu}\right| \qquad \text{at } X^0,$$

where X^0 is a point in $\{y > 0\}$ where ∂Q_{λ_2} and $\partial(\alpha Q_{\lambda_1})$ intersect. Since such a point must belong to both free boundaries, we deduce that $\lambda_1 > \lambda_2$.

Corollary 20.4. *If $\lambda_1 \geqslant \lambda_2$, then $\{u_{\lambda_1} > 0\} \subset \{u_{\lambda_2} > 0\}$.*

Indeed, otherwise we can magnify $\{u_{\lambda_2} > 0\}$ so that it contains $\{u_{\lambda_1} > 0\}$ and thus deduce that $\lambda_2 > \lambda_1$.

Corollary 20.5. *For any $\lambda > 0$, there is precisely one λ-solution (which is then the solution of problem J_λ).*

Indeed, if u_1, u_2 are two solutions, then $\{u_1 > 0\} = \{u_2 > 0\}$, by Corollary 20.4, and, by the maximum principle, $u_1 - u_2 \equiv 0$.

Corollary 20.5 implies that u_λ varies continuously in the parameter λ (see Lemma 10.4).

Lemma 20.6. *If $\lambda \leqslant 1$, then the free boundary of u_λ is empty.*

Proof. Let u_0 be the flow past W satisfying (20.6) (it is simply a solution of problem J_λ with $\lambda = 0$) and set

$$\lambda_0 = \frac{1}{y}\left|\frac{\partial u_0}{\partial \nu}\right| \qquad \text{at } (0, 1).$$

If u_λ has free boundary points, then by magnifying $\{u_\lambda > 0\}$ until it contains W we obtain, using the argument of Lemma 20.3,

$$\lambda > \lambda_0.$$

Since by the maximum principle

$$u_0 \geqslant \frac{1}{2}(y^2 - 1) \qquad \text{in } \{y > 1\},$$

we have $\lambda_0 \geqslant 1$; thus $\lambda > 1$, contradicting our assumption on λ.

Lemma 20.7. *If λ is sufficiently large, then the free boundary for u_λ starts on the vertical line $x = -a$.*

Proof. Take a domain $\tilde{W}$ bounded by $y = 0$, $x = a'$, $x = a' + 2\varepsilon$ and half a circle

$$\left\{ (x - (a' + \varepsilon))^2 + (y - \delta)^2 < \varepsilon^2, \ y > \delta \right\}$$

with ε small and δ near 1. Denote by $\tilde{u}$ the flow past $\tilde{W}$.

If the free boundary initiates on $\{y = 1\}$, then in view of the smooth-fit property we can choose the parameters ε, δ in such a way that $\tilde{W}$ lies inside $\{u = 0\}$ but $\partial\tilde{W} \cap \{y > \delta\}$ touches the free boundary, say at X^0. But then

$$(20.13) \qquad \lambda = \frac{1}{y}\left|\frac{\partial u}{\partial \nu}\right| \leq \frac{1}{y}\left|\frac{\partial \tilde{u}}{\partial \nu}\right| \equiv \tilde{\lambda} \qquad \text{at } X^0.$$

The choice of ε, δ depends only on a; therefore, using the inner ball property to construct a barrier for $\tilde{u}$ at X^0, we get

$$(20.14) \qquad\qquad\qquad \tilde{\lambda} \leq C,$$

where C depends only on a. Thus, if $\lambda > C$, then the free boundary must start on the line $x = -a$.

Combining Lemmas 20.6 and 20.7 with the continuous dependence of u_λ (and its free boundary) upon λ we now deduce by familiar arguments the existence of a parameter λ with smooth fit at A.

This parameter is unique; in fact, the proof is similar to the proof of Corollary 11.5. We have thus completed the proof of Theorem 20.1.

One can obtain a standard series development for u near infinity. Thus, if we introduce the velocity potential ϕ by

$$(20.15) \qquad\qquad \begin{aligned} \phi_x &= \frac{1}{y} u_y, \\[2mm] \phi_y &= -\frac{1}{y} u_x \end{aligned}$$

in $\{y > 0\}$, then the function

$$\tilde{\phi}(x, y, z) = \phi\left(x, \sqrt{y^2 + z^2}\right)$$

is harmonic in the region Ω^* obtained by rotating $\{u > 0\}$ about the x-axis. Developing $\tilde{\phi}$ near infinity in terms of zonal harmonics ([120, p. 254]), we find that

$$(20.16) \qquad \left. \begin{array}{l} \nabla\phi(x, y) \to (1,0) \\[2mm] \dfrac{1}{y} \nabla u(x, y) \to (0, 1) \end{array} \right\} \qquad \text{as } x^2 + y^2 \to \infty.$$

PROBLEMS

1. Derive the expansion (for any $u = u_\lambda$)

$$u(x, y) = -\frac{y^2}{2} - \frac{cy^2}{r^3} + O\left(\frac{1}{r^2}\right).$$

2. Extend Theorem 20.1 to the case where $N = \{(-a, y); 0 \leqslant y \leqslant 1\}$.

3. Extend Theorem 20.1 to plane symmetric finite cavities.

4. Extend Theorem 20.1 to the case where $x(t)$ is not a monotone nondecreasing function (that is, N is just a y-graph) provided that N is star-shaped with respect to the origin.

5. Let G be a bounded star-shaped domain (with respect to the origin) in R^2 and let $\Omega = R^2 \setminus \overline{G}$. Consider the functional

$$J(v) = \int_\Omega \left(|\nabla v|^2 + \lambda^2 I_{\{v>0\}} \right) dx$$

over the class

$$K = \left\{ v \in H^{1,2}_{\text{loc}}(R^2), v = 1 \text{ on } \partial G, v \geqslant 0 \text{ in } \Omega \right\}.$$

Prove:
(a) There exists a unique minimum $u = u_\lambda$.
(b) If $\lambda_1 > \lambda_2$, then $\{u_{\lambda_1} > 0\} \subset \{u_{\lambda_2} > 0\}$.
(c) $\partial\{u_\lambda > 0\}$ is star-shaped with respect to the origin.
(d) If G is convex, then the free boundary is convex.

[*Hint*: By Problem 5 of Section 3, any (local) minimum has a bounded support. For (a)–(c) use the method of Lemma 20.3, and for (d) notice that q takes minimum on the free boundary (see Theorem 13.1).]

6. Consider the axially symmetric cavity problem in a pipe $|z| < Z$ with obstacle $y = g(x)$, $0 \leqslant x \leqslant a$, where $g(x)$ is monotone increasing. Construct the variational functional and establish the existence of a solution with free boundary $y = f(x)$, $f(x)$ monotone increasing, $f'(x) \to 0$ if $x \to \infty$.

2.1   AXIALLY SYMMETRIC INFINITE CAVITIES

Let N be a nose satisfying:

N is a piecewise $C^{1+\alpha}$ curve given by

$$X = X(t) = (x(t), y(t)), 0 \leq t \leq \bar{t}, \text{ with } \nabla X(t \pm 0) \neq 0;$$

(21.1) $x(t)$ and $y(t)$ are monotone nondecreasing and $y(t) > 0$

if $t > 0$; $X(0) = (0,0)$ and $X(\bar{t}) = (a, y(\bar{t})), a > 0.$

For simplicity we take $y(\bar{t}) = 1$ and set

$$A \equiv X(\bar{t}) = (a,1).$$

Definition 21.1.   Let N (the nose) be as in (21.1). The (*axially symmetric*) *infinite cavity problem* consists of finding a function $u(x, y)$ and a curve Γ such that:

Γ is C^1 and is given by a strictly monotone

(21.2)   increasing and continuous function $y = f(x), a \leq x < \infty;$

$f(a) = 1$ and $\Gamma \cup N$ is C^1 in a neighborhood of A,

and

$u \in C(\bar{E}),$

$u(x,0) = 0$ if $-\infty < x \leq 0,$

$u = 0$ on $N \cup \Gamma$ and in the region K bounded by $N \cup \Gamma$

and the segment $\{(x,0); a \leq x < \infty\};$

(21.3) $u > 0$ elsewhere in $E;$

$Lu = 0$ in $\{u > 0\};$

u is uniformly continuously differentiable

in some $\{u > 0\}$-neighborhood of $A;$

further,

(21.4) $\dfrac{1}{y}\left|\dfrac{\partial u}{\partial \nu}\right| = 1 \qquad \text{on } \Gamma,$

and u, Γ possess the asymptotic behavior:

$$(21.5) \qquad\qquad f'(x) \to 0 \qquad \text{if } x \to \infty,$$

$$(21.6) \quad \frac{1}{y}\nabla u(x, y) \to (0, 1) \qquad \text{uniformly in } \{u > 0\} \text{ if } x^2 + y^2 \to \infty.$$

K is called the *cavity*.

Theorem 21.1. *There exists a solution of the infinite cavity problem.*

To prove the theorem we first consider the finite cavity problem with tail $T = T_\mu$ which is symmetric to N with respect to the line $x = \mu$, $\mu > a$. By Section 20 there exists a unique solution u_μ, Γ_μ, λ_μ to this problem and

$$\Gamma_\mu : y = f_\mu(x) \qquad (a < x < 2\mu - a),$$

where

$$f_\mu(x) = f_\mu(2\mu - x),$$

$$f_\mu(x) \text{ is strictly monotone increasing for } a < x \leqslant \mu.$$

From Lemmas 20.6 and 20.7 [see (20.13), (20.14)], we have

$$(21.7) \qquad\qquad 1 \leqslant \lambda_\mu \leqslant C \qquad (C \text{ independent of } \mu).$$

Lemma 21.2. *Set $I = \{(x, 0); \; -\infty < x \leqslant 0\}$. If a straight line l does not intersect $I \cup N$, then it can intersect $\Gamma_\mu^- \equiv \Gamma_\mu \cap \{x < \mu\}$ in at most one point.*

Proof. Since Γ_μ^- is given by a monotone-increasing curve $y = f_\mu(x)$, it suffices to take l with positive slope. If the assertion is not true, then by parallel translation of l we arrive at the situation (see Figure 3.9) where the segment $\overline{BM}$ of l lies in the cavity and is tangent to Γ_μ^- at M, and on the arc $\widehat{MH}$ of Γ_μ^- there is no other point with the same properties. It follows that

$$(21.8) \qquad \begin{array}{l} \text{if } \widehat{MH} \text{ enters the half-plane to the right side} \\ \text{of } l \text{ at a point } F, \text{ then the slope of } \Gamma_\mu^- \text{ at} \\ \text{each point of } \widehat{FH} \text{ is less than the slope of } l \end{array}$$

(otherwise, the first point M' of $\widehat{FH}$ where the slope of Γ_μ^- equals that of l has the same properties as M, a contradiction).

If there is a point F as in (21.8), then we set $\delta = \widehat{MF}$. If, on the other hand, $\widehat{MH}$ lies entirely to the left of l, then we set $\delta = \widehat{MH}$.

Denote by E the point in δ which is farthest from the line l, and denote by γ the ray going from E to $x = \mu$ parallel to l. In view of (2.18), γ lies above Γ_μ^-.

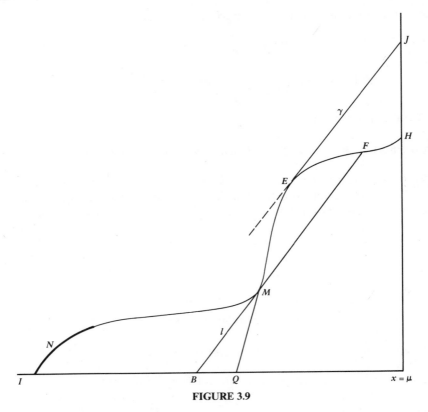

FIGURE 3.9

Denote by Q the intersection of the line EM with the x-axis. Clearly, Q lies to the right of B.

Denote by R the region bounded by $y = 0$ from $x = -\infty$ to B, the segment $\overline{BM}$ the subarc $\widehat{ME}$ of Γ_μ^- , the arc γ, and the segment on the line $x = \mu$ from J to $+\infty$.

Denote by $\tilde{u}$ the solution of

$$L\tilde{u} = 0 \qquad \text{in } R,$$

$$\tilde{u} = 0 \qquad \text{on } \partial R \cap \{x < \mu\},$$

(21.9) $$\tilde{u}_x(x, y) = 0 \qquad \text{if } y_J < y < \infty \qquad [J = (x_J, y_J)],$$

$$\tilde{u} - \frac{y^2}{2} \to 0 \qquad \text{as } r^2 = x^2 + y^2 \to \infty.$$

The solution can be obtained by reflecting the curve $\partial R \cap \{x < \mu\}$ about $x = \mu$ and applying the method of Section 20 with $\lambda = 0$; see also Problem 1.

We shall prove below that

(21.10)
$$\frac{|\nabla u(M)|}{|\nabla u(E)|} \leq \frac{|\nabla \tilde{u}(M)|}{|\nabla \tilde{u}(E)|}.$$

Assuming this inequality for the moment, consider the similarity transformation

$$(x, y) \to (\alpha(x - x_0), \alpha y) \qquad [Q = (x_0, 0)],$$

which maps E into M. Thus if

$$M = (x_M, y_M), \qquad E = (x_E, y_E),$$

then

$$\alpha y_E = y_M.$$

It is easily seen that the inverse mapping takes R onto a subset R' of R. The function

$$\hat{u}(x, y) = \frac{1}{\alpha^2} \tilde{u}(\alpha(x - x_0), \alpha y)$$

satisfies $L\hat{u} = 0$ in R', $\hat{u} \leq \tilde{u}$ on $\partial R'$ and $\hat{u} = \tilde{u} = 0$ at E ($E \in \partial R'$). By the maximum principle we get $\hat{u} < u$ in R' and

$$|\nabla \hat{u}(E)| < |\nabla \tilde{u}(E)|;$$

hence

$$\frac{1}{\alpha} |\nabla \tilde{u}(M)| < |\nabla \tilde{u}(E)|.$$

Substituting this into (21.10), we obtain

$$\frac{|\nabla u(M)|}{|\nabla u(E)|} < \alpha.$$

Since, however, M and E lie on the free boundary,

$$\frac{|\nabla u(M)|}{|\nabla u(E)|} = \frac{y_M}{y_E} = \alpha, \qquad \text{a contradiction.}$$

It remains to prove (21.10). We may assume that u, $\tilde{u}$ are smooth up to the boundary, for the general case then follows by approximating the flow region by regions with smooth boundaries.

For any point $Z = (\zeta, \eta)$ in $\widehat{ME}$, we introduce the function

$$\Omega_Z(X) = V(Z)\tilde{u}(X) - \tilde{V}(Z)u(X) \qquad (X = (x, y)),$$

where

$$V(Z) = \frac{1}{\zeta}|\nabla u(Z)|, \qquad \tilde{V}(Z) = \frac{1}{\zeta}|\nabla \tilde{u}(Z)|,$$

and the level curve

$$\Delta_Z : \Omega_Z(X) = 0$$

initiating at Z and going into the domain R_* where both u and $\tilde{u}$ are positive. We note that the velocity potentials ϕ, $\tilde{\phi}$ corresponding to u, $\tilde{u}$ possess expansions in terms of zonal harmonics of the form (see reference 120, p. 254)

$$\sum c_k P_k(\cos \theta) r^k;$$

the same is true of u, $\tilde{u}$ (see also reference 148).

Therefore, at each point X^0 where u, $\tilde{u}$ are analytic, there is a finite number m of simple branches of Δ_Z and if $m > 1$, then two adjacent branches form nonzero angles at X^0. This is valid also about $y = 0$, $-\infty < x < 0$, since $|\nabla u| \leqslant Cy$, $|\nabla \tilde{u}| \leqslant Cy$ (by the proof of Theorem 8.5), which implies that $\phi\left(x, \sqrt{y^2 + z^2}\right)$ $\tilde{\phi}\left(x, \sqrt{y^2 + z^2}\right)$ are harmonic about the x-axis, for $-\infty < x < 0$.

Since $\partial \Omega_Z(X)/\partial \nu = 0$ at $X = Z$, it follows that Δ_Z does in fact initiate from Z into R_*. It also follows that if Δ_Z exits the region $R_\mu = R_* \cap \{x < \mu\}$ at a point of I, then it is nontangential at the exit point (since $\Omega_Z = 0$ along I). If Δ_Z exists R_μ at $x = \mu$, then, because of symmetry with respect to the line $x = \mu$, it must intersect $x = \mu$ nontangentially. Notice that Δ_Z cannot exit R_μ on $\widehat{ME}$ (otherwise, $\Omega_Z \equiv 0$) and on the arcs γ (where $\Omega_Z < 0$), N (where $\Omega_Z > 0$) and the subarc Γ' of Γ_μ from A to M (where again $\Omega_Z > 0$).

From the considerations above it follows that the subsets $\mathcal{A}$, $\mathcal{B}$ of $\widehat{MF}$ defined below are open:

$$Z \in \mathcal{A} \qquad \text{if } V(Z) \neq \tilde{V}(Z) \text{ and } \Delta_Z \text{ does not exit } R_\mu \text{ on } x = \mu;$$

$$Z \in \mathcal{B} \qquad \text{if } V(Z) \neq \tilde{V}(Z) \text{ and } \Delta_Z \text{ exits } R_\mu \text{ on } x = \mu.$$

If Δ_Z remains in R_μ, then, if we parametrize it by the length s, the points $X(s)$ of Δ_Z must satisfy: $|X(s)| \to \infty$ if $s \to \infty$. It follows that

$$(21.11) \qquad \frac{1}{y}\frac{\partial}{\partial x}\Omega_Z \to 0, \qquad \frac{1}{y}\frac{\partial}{\partial y}\Omega_Z \to V(Z) - \tilde{V}(Z)$$

at $x = X(s)$, $s \to \infty$. If also $V(Z) \neq \tilde{V}(Z)$, then Δ_Z becomes asymptotically horizontal, going to $-\infty$.

Consider now the case

$$(21.12) \qquad\qquad \mathcal{C} = \widehat{ME}.$$

For any $Z \in \widehat{ME}$, if Δ_Z exits R_μ on I (say at X^0), then $\Omega_Z \geq 0$ on the boundary of the region G_Z bounded by Δ_Z and the part of ∂R_μ from X^0 to Z. By the maximum principle it follows that $\Omega_Z > 0$ in G_Z and

$$(21.13) \qquad\qquad \frac{\partial}{\partial \nu}\Omega_Z > 0 \text{ at } M, \qquad \nu \text{ inward normal.}$$

Suppose next that Δ_Z does not exit R_μ. Then, by the remark containing (21.11), Δ_Z is asymptotically horizontal. Hence, for any $\varepsilon > 0$,

$$\Omega_Z(X) + \varepsilon(\mu - x)y^2 \geq 0$$

on ∂R_μ from $(-\infty, 0)$ to Z, on Δ_Z, and for $X \in G_Z$ as $|X| \to \infty$. By the maximum principle we conclude that $\Omega_Z + \varepsilon(\mu - x)y^2 > 0$ in G_Z. Letting $\varepsilon \to 0$, we again arrive at (21.13).

Taking $Z \to E$ in (21.13), we obtain

$$V(E)|\nabla \tilde{u}(M)| \geq \tilde{V}(E)|\nabla u(M)|$$

and (21.10) follows.

In case

$$(21.14) \qquad\qquad \mathcal{B} = \widehat{ME},$$

we can argue in a similar way. Here $\Omega_Z < 0$ in the region $\tilde{G}_Z$ bounded by Δ_Z and the part of ∂R_μ from Z to the exit point X^1 of Δ_Z on $\{x = \mu\}$. In fact, since $\Omega_Z(x, y) = \Omega_Z(2\mu - x, y)$, we can apply the maximum principle in the closure of $\tilde{G}_Z \cup \{(2\mu - x, y); (x, y) \in \tilde{G}_Z\}$. We conclude that

$$\frac{\partial}{\partial \nu}\Omega_Z < 0 \qquad \text{at } E,$$

and taking $Z \to M$ (21.10) follows.

Consider finally the case

$$(21.15) \qquad\qquad \mathcal{C} \neq \varnothing, \qquad \mathcal{B} \neq \varnothing.$$

Since $\mathcal{C}, \mathcal{B}$ are open and disjoint, there exists a point $Z \in \widehat{ME}$ such that

$Z \notin \mathcal{C} \cap \mathcal{B}$. Therefore, Δ_Z does not exit R_μ and, since

$$| \nabla u(Z) | = | \nabla \tilde{u}(Z) | ,$$

we have

$$| \Omega_Z(X) | \leqslant C \qquad \text{if } | X | \to \infty.$$

It follows that for any $\varepsilon > 0$,

$$\Omega_Z(X) + \varepsilon y^2 + \varepsilon(\mu - x)$$

is positive in G_Z and

$$\Omega_Z(X) - \varepsilon y^2 - \varepsilon(\mu - x)$$

is negative in $\tilde{G}_Z$. Letting $\varepsilon \to 0$, we can then derive as before

$$\frac{\partial}{\partial \nu} \Omega_Z(M) \geqslant 0, \qquad \frac{\partial}{\partial \nu} \Omega_Z(E) \leqslant 0,$$

that is,

$$| \nabla \tilde{u}(M) | \geqslant | \nabla u(M) | ,$$

$$| \nabla \tilde{u}(E) | \leqslant | \nabla u(E) | ,$$

and (21.10) follows.

Having completed the proof of Lemma 21.2 we now choose any $0 \leqslant a_0 \leqslant a$ such that the straight-line segment connecting $(a_0, 0)$ to A does not intersect N. From Lemma 21.2 we then deduce that Γ_μ^- is star-shaped with respect to $(a_0, 0)$. Consequently, the function

(21.16) $\dfrac{f_\mu(x)}{x - a_0}$ is monotone decreasing, $a < x < \mu$.

Hence its derivative exists a.e. and is nonpositive. This implies that

$$f_\mu'(x) \leqslant \frac{f_\mu(x)}{x - a_0} \qquad \text{a.e.}$$

Since $y = f_\mu(x)$ has analytic representation, we easily deduce:

Lemma 21.3. *There holds*:

(21.17) $$0 \leqslant f_\mu'(x) \leqslant \frac{f_\mu(x)}{x - a_0} \qquad \text{if } a < x < \infty.$$

We now choose a sequence $\mu = \mu_n \to \infty$ such that [recall (21.7)]

$$u_{\mu_n} \to u, \qquad f_{\mu_n} \to f, \qquad \lambda_{\mu_n} \to \lambda \quad (\lambda \geqslant 1)$$

in the appropriate sense. Then u, f, λ satisfy (21.2), (21.3). From (21.16) and Lemma 21.3, we obtain:

Lemma 21.4. *The function $f(x)$ satisfies the following properties*:

(21.18) $$\frac{f(x)}{x - a_0} \qquad \text{is monotone decreasing}, \, a < x < \infty,$$

(21.19) $$f'(x) \leqslant \frac{f(x)}{x - a_0} \qquad \text{for } a < x < \infty.$$

If follows that

(21.20) $$\beta \equiv \lim_{x \to \infty} \frac{f(x)}{x} \qquad \text{exists.}$$

Lemma 21.5. $\beta = 0$.

Proof. Set

$$u_R(X) = \frac{1}{R^2} u(RX).$$

Since $u(X) \leqslant y^2/2$, also $u_R(X) \leqslant y^2/2$. Hence there exists a sequence $R_n \to \infty$ such that

$$u_{R_n}(X) \to u_0(X)$$

uniformly in compact subsets of $\{y \geqslant 0\}$. The free boundary Γ_R of u_R is given by

$$y = f_R(x) = \frac{1}{R} f(Rx).$$

By nondegeneracy, if $X^0 \in \Gamma_R$ and $B(X^0, h)$ does not intersect the fixed boundary of u_R, then

$$\sup_{B(X^0, h) \cap \{u_R > 0\}} u_R(X) \geq ch \qquad (c > 0);$$

it follows that $u_0 \not\equiv 0$. From (21.20) we deduce that the free boundary of u_0 is

$$y = f_0(x) = \beta x.$$

We have

$$Lu_0 = 0 \qquad \text{in the sector } \Sigma : \theta_0 < \theta < \pi$$

where $0 \leq \theta_0 < \pi$, $\tan \theta_0 = \beta$, and

$$u_0 = 0 \qquad \text{on } \partial\Sigma,$$

$$u_0 \leq \frac{y^2}{2} \qquad \text{in } \Sigma.$$

Consider the function

$$(21.21) \qquad v(r, \theta) = -r^{\alpha+1} \sin^2 \theta P^1_{\alpha-1}(\cos \theta), \qquad \alpha > 1$$

where $P^\lambda_\nu(z)$ are the Legendre functions as defined in reference 84, p. 122 [$P^0_n(z)$ is the Legendre polynomial of order n]. One can check (see also reference 148, where the P^ν_μ coincide with the $P^{\nu-1/2}$ of reference 84) that

$$(21.22) \qquad\qquad\qquad\qquad Lv = 0.$$

Lemma 21.6. *There holds*:

$$(21.23) \qquad P^1_{\alpha-1}(\cos \theta) < 0 \qquad if \ 0 < \theta < \frac{\pi}{2\alpha - 1}.$$

Proof. From reference 84, p. 159, formula (27),

$$(21.24) \qquad P^\mu_\nu(\cos \theta) > 0 \qquad if \ \mu < \frac{1}{2}, \ 0 < \left(\nu + \frac{1}{2}\right)\theta < \frac{\pi}{2}.$$

Recall that $P^\mu_\nu(z)$ is a single-valued analytic function defined in the complex plane cut along the negative real axis. By reference 84, p. 160, formula (1),

$$(21.25) \qquad P^1_\nu(z) = (\nu + 1)\nu P^{-1}_\nu(z), \qquad z \text{ complex}$$

and by reference 84, p. 143, formula (1)

$$(21.26) \qquad P_\nu^\mu(x) = \lim\left[e^{i\mu\pi/2}P_\nu^\mu(x+i0) - e^{-i\mu\pi/2}P_\nu^\mu(x-i0)\right]$$

if x is real, $-1 < x < 1$. Therefore,

$$P_\nu^1(x) = \left[e^{i\pi/2}P_\nu^1(x+i0) + e^{-i\pi/2}P_\nu^{-1}(x-i0)\right]$$
$$= (\nu+1)\nu i\left[P_\nu^{-1}(x+i0) - P_\nu^{-1}(x-i0)\right],$$

and the right-hand side is equal to

$$(\nu+1)\nu\left(-P_\nu^{-1}(x)\right)$$

by (21.26) with $\mu = -1$. Hence

$$P_\nu^1(x) = -(\nu+1)P_\nu^{-1}(x).$$

Since the right-hand side is positive (by (21.24)) if $0 < \theta < \pi/(2(\nu + \frac{1}{2}))$, the assertion (21.23) follows by taking $\nu = \alpha - 1$.

We can now complete the proof of Lemma 21.5. Indeed, if $\beta > 0$, then we choose $\alpha > 1$ such that

$$\frac{\pi}{2\alpha - 1} > \pi - \beta$$

and set $u(r, \theta) = v(r, \pi - \theta)$, v as in (21.21). In view of (21.22),

$$Lu = 0 \qquad \text{in the sector } \Sigma,$$

and by Lemma 21.6,

$$u \geqslant 0 \qquad \text{on } \partial\Sigma.$$

Since $\alpha > 1$, we also have

$$\frac{u}{y^2} \to \infty \qquad \text{if } X \in \Sigma, \ |X| \to \infty.$$

We can now apply the maximum principle to deduce that for any $\varepsilon > 0$,

$$u_0(X) < \varepsilon u(r, \theta) \qquad \text{if } X \in \Sigma, \ |X| < \rho$$

provided that ρ is sufficiently large. It follows that $u_0 \equiv 0$, a contradiction.

From Lemmas 21.4 and 21.5 we get

Corollary 21.7. *There holds*:

$$(21.27) \qquad\qquad f'(x) \to 0 \qquad if\ x \to \infty.$$

We next prove:

Lemma 21.8. $\lambda = 1$.

Proof. From Corollary 21.7 it follows that there exists a sequence $\alpha_n \uparrow \infty$ such that

$$(21.28) \qquad\qquad f'(x) < \frac{1}{n} \qquad if\ x > \frac{\alpha_n}{2}.$$

Define functions $u_n(X)$ by

$$u_n(x, y) = \frac{1}{f^2(\alpha_n)} u\big(xf(\alpha_n) + \alpha_n, \, yf(\alpha_n)\big)$$

$$= \frac{1}{f^2(\alpha_n)} u\big((X + X_n)f(\alpha_n)\big)$$

where

$$X_n = \left(\frac{\alpha_n}{f(\alpha_n)}, 0 \right).$$

The free boundary for $u_n(X)$ is given by

$$y = f_n(x) = \frac{1}{f(\alpha_n)} f\big(xf(\alpha_n) + \alpha_n\big);$$

in particular,

$$(21.29) \qquad\qquad f_n(0) = 1.$$

For any $C > 0$,

$$(21.30) \qquad f_n'(x) = f'\big(xf(\alpha_n) + \alpha_n\big) < \frac{1}{n} \qquad if\ |x| < C$$

provided that n is sufficiently large; indeed, if $|x| < C$, then

$$xf(\alpha_n) + \alpha_n > \frac{\alpha_n}{2}$$

[since $(f(t)/t) \to 0$ if $t \to \infty$] and thus (21.30) is a consequence of (21.28).

Since $u(X) \leqslant y^2/2$ we also have

$$u_n(X) \leqslant \tfrac{1}{2}y^2.$$

Hence for a subsequence

$$f_n(x) \to f_*(x)$$

pointwise, and

$$u_n(X) \to u_*(X)$$

uniformly in compact subsets of $\{(x, y) \in R^2; y > f_*(x)\}$, and

$$Lu_* = 0 \qquad \text{if } y > f_*(x),$$

$$u_* = 0, \quad \frac{1}{y}\frac{\partial u_*}{\partial \nu} = \lambda \qquad \text{on } y = f_*(x).$$

From (21.29), (21.30) we also have $f_*(x) = 1$. Hence by uniqueness, for the Cauchy problem

$$u_*(X) = \frac{\lambda y^2}{2} - \frac{\lambda}{2}.$$

Since further $u_*(X) \leqslant y^2/2$, it follows that $\lambda \leqslant 1$. Recalling that $\lambda \geqslant 1$, the proof is complete.

Let B be a ball in R^2 containing N and denote by G_1 the intersection of B with $\{(x, y); x < \mu, y > 0\}$. Denote by G_2 the reflection of G_1 with respect to $x = \mu$. Let $G = G_1 \cup G_2$ and denote by D_μ^* the domain in R^3 obtained by rotating $D_\mu \setminus G$ about the x-axis; here $D_\mu = \{u_\mu > 0\}$.

Let ϕ_μ be the velocity potential corresponding to u_μ. Then the function

$$\tilde{\phi}_\mu(x, y, z) = \phi_\mu\left(x, \sqrt{x^2 + y^2}\right)$$

is harmonic in D_μ^* and its gradient is bounded by 1 at infinity. From the bounded gradient lemma we have

(21.31) $$\left|\frac{\partial \tilde{\phi}_\mu}{\partial x}\right| \leqslant C, \qquad \left|\frac{\partial \tilde{\phi}_\mu}{\partial y}\right| \leqslant C$$

on ∂D_μ^*, where C is a constant independent of μ. By the maximum principle we then obtain the same estimate in D_μ^*.

Now, for any $\varepsilon > 0$ there is $x_0 > a$ such that

$$\frac{f(x_0)}{x_0 - a_0} < \varepsilon.$$

Hence

$$\frac{f_\mu(x_0)}{x_0 - a_0} < \varepsilon \qquad \text{if } \mu \geqslant \mu_0.$$

Since $f_\mu(x)/(x - a_0)$ is monotone decreasing for $a < x < \mu$, we then also have

(21.32) $\qquad\qquad \dfrac{f_\mu(x)}{x - a_0} < \varepsilon \qquad \text{if } x_0 < x < \mu, \quad \mu \geqslant \mu_0,$

and, by Lemma 21.3,

(21.33) $\qquad\qquad f_\mu'(x) < \varepsilon \qquad \text{if } x_0 < x < \mu, \quad \mu \geqslant \mu_0.$

Recalling that

$$\frac{1}{y}\frac{\partial u_\mu}{\partial \nu} = \lambda_\mu, \qquad \lambda_{\mu_n} \to 1$$

and using (21.31), we deduce that

(21.34) $\qquad \left| \dfrac{1}{y} \dfrac{\partial u_\mu\big(x, f_\mu(x)\big)}{\partial y} - 1 \right| < \eta(\varepsilon), \qquad \left| \dfrac{1}{y} \dfrac{\partial u_\mu\big(x, f_\mu(x)\big)}{\partial x} \right| < \eta(\varepsilon)$

$$\text{if } x_0 < x < \mu_n, \mu = \mu_n \geqslant \tilde\mu(\varepsilon),$$

where $\eta(\varepsilon) \to 0$ if $\varepsilon \to 0$.

Let R_0 be a positive constant independent of μ such that the arcs

$$y = f_\mu(x), \qquad a < x < x_0$$

and N are all contained in the ball B_{R_0}. Denote by $B_{R_0}(O_1)$ the reflection of B_{R_0} with respect to the line $x = \mu$, that is, $O_1 = (2\mu, 0)$.

Denote by $\tilde D_\mu$ the set in R^3 obtained by rotating $D_\mu \setminus (\overline{B_{R_0}} \cup \overline{B_{R_0}(O_1)})$ about the x-axis. Consider in $\tilde D_\mu$ the harmonic functions

$$u_1 \equiv \frac{\partial \tilde\phi_\mu}{\partial x} - 1, \qquad u_2 \equiv \frac{\partial \tilde\phi_\mu}{\partial y}.$$

In view of (21.31),

(21.35) $\qquad\qquad\qquad\qquad |u_i| \leqslant C$

on the part of $\partial \tilde D_\mu$ obtained by rotating ∂B_{R_0} and $\partial B_{R_0}(O_1)$ about the x-axis.

On the remaining portion of $\partial \tilde{D}_\mu$ we have, by (21.34),

$$|u_i| \le \eta(\varepsilon).$$

Notice also that

$$u_i \to 0 \qquad \text{if } x^2 + y^2 \to \infty.$$

Thus we can compare u_i with the harmonic function in R^3,

$$v(\xi) = \eta(\varepsilon) + \frac{C_0}{|\xi|} + \frac{C_0}{|\xi - \xi_0|} \qquad [\xi = (x, y, z)],$$

where $\xi_0 = (2\mu, 0, 0)$ and C_0 is a sufficiently large constant depending only on C in (21.35) and on R_0, and deduce that

$$|u_i| \le v;$$

thus, if $\mu = \mu_n \ge \tilde{\mu}(\varepsilon)$,

$$\left| \frac{1}{y} \frac{\partial u_\mu}{\partial y} - 1 \right| + \left| \frac{1}{y} \frac{\partial u_\mu}{\partial x} \right| \le \eta(\varepsilon) + \frac{C_0}{(x^2 + y^2)^{1/2}} + \frac{C_0}{((x - 2\mu)^2 + y^2)^{1/2}}$$

in $\tilde{D}_\mu \cap \{z = 0\}$. Taking $\mu = \mu_n \to \infty$, we obtain

$$\left| \frac{1}{y} u_y - 1 \right| + \left| \frac{1}{y} u_x \right| \le \eta(\varepsilon) + \frac{C_0}{(x^2 + y^2)^{1/2}}.$$

Taking $x^2 + y^2 \to \infty$ and noting that $\eta(\varepsilon)$ can be made arbitrarily small, the assertion (21.6) follows. This completes the proof of Theorem 21.1.

PROBLEMS

1. Solve (21.9) by introducing the velocity potential ϕ and the harmonic function $\tilde{\phi}(x, y, z) = \phi(x, \sqrt{y^2 + z^2})$; $\tilde{\phi}$ is then a solution of the exterior Neumann problem.

2. Assume that $\lim_{\theta \to 0} [r(\theta)/\tilde{r}(\theta)]$ exists for any two solutions u, $\tilde{u}$ of the infinite cavity problem, where $r(\theta)$ ($\tilde{r}(\theta)$) is the distance from $(a, 0)$ to the free boundary of u ($\tilde{u}$) along the ray $y = (x - a)\tan\theta$, $x > 0$. Prove uniqueness for the infinite cavity problem; here the Phragmén–Lindelöf theorem needed for a difference ψ of solutions follows by the proof of Corollary 9.10 applied to $\psi + \eta y^2$, η small.

3. Extend Theorem 21.1 to the case where $x(t)$ is not monotone nondecreasing (that is, N is just a y-graph) provided that N is star-shaped with respect to some point $(b, 0)$, $b > 0$.

FIGURE 3.10

4. Let R_1, R_2 be two flow regions in $\{y > 0\}$ of axially symmetric stream functions u_1 and u_2, respectively, and denote by S_i the stream line $u_i = 0$; $S_i = \partial R_i$. Assume that $u_i \leqslant y^2/2$ and

$$ u_i = \frac{y^2}{2}(1 + o(1)) \qquad \text{as } r^2 = x^2 + y^2 \to \infty, \, (x, y) \in R_i. $$

Suppose that S_1, S_2 have an arc $\overparen{MN}$ in common and S_2 lies below S_1 from $-\infty$ to M, whereas S_2 lies above S_1 from N to ∞; see Figure 3.10. Suppose finally that u_i is C^1 in $\overline{R}_i$-neighborhoods of M and N. Prove:

$$ \frac{|\nabla u_1(M)|}{|\nabla u_1(N)|} \leqslant \frac{|\nabla u_2(M)|}{|\nabla u_2(N)|}; $$

this is called the *under–over theorem*.

5. Let u be a solution of the axially symmetric infinite cavity problem for a general smooth nose; the free boundary Γ is a C^1 curve connecting A to infinity (not necessarily a y- or x-graph) which is asymptotically horizontal, and $(1/y)\,|\partial u/\partial \nu| = 1$ on Γ. Let I be the line segment $\{(x, 0); -\infty < x < 0\}$, where $(0, 0)$ is the initial point of N. Prove *the single intersection property*: Any infinite straight line that does not intersect $I \cup N$ can intersect Γ in at most one point.

6. Theorem 21.1 can be extended to the plane symmetric infinite cavity problem. Denote the solution by ψ and the velocity potential by ϕ and set $f = \phi + i\psi$, $\omega = i\log(df/dz) = \theta + i\log q$ (see Section 1). Then the flow region R_z is mapped by f into a region R_f that contains a neighborhood of infinity slit along the positive real axis. R_f is mapped by $f = 1/t^2$ into a region R_t which near $t = 0$ consist of half a disc $\{|t| < \varepsilon, \text{Im } t > 0\}$.
   (a) Show that $\omega(t)$ has analytic extension into a neighborhood of $t = 0$.
   (b) Verify the formula

$$ z = \int e^{i\omega(t)} f'(t) \, dt $$

and use it to deduce that

$$z(t) = \frac{1}{t^2} + \frac{2i\omega'(0)}{t} + \left[\omega'(0)^2 - i\omega''(0)\right] \log t + \sum_{n=0}^{\infty} a_n t^n$$

near $t = 0$ [notice that $\omega(0) = 0$].

(c)   Show that if $\omega'(0) \neq 0$, then the free boundary has the form

$$y \sim 2\,|\,\omega'(0)\,|\,\sqrt{x} \qquad \text{for } x \to \infty.$$

(d)   If $\omega'(0) = 0$ and $\omega''(0) \neq 0$, then the free boundary has the form

$$y \sim |\,\omega''(0)\,|\,\log x \qquad \text{for } x \to \infty.$$

7.   The single intersection property (Lemma 21.5 and Problem 5) extend to axially symmetric jets: If a line l does not intersect the nozzle N then it can intersect the free boundary in at most one point. Prove it!

8.   If a nozzle satisfying (8.1) lies in $\{x \leqslant 0\}$ then the free boundary is given by a monotone decreasing function $y = h(x)$.

22. BIBLIOGRAPHICAL REMARKS

Surveys of the classical theory of jets and cavities can be found in references 41, 107b, 110, and 182. The classical approach to these problems is based on a reduction of the problem to a nonlinear functional or integral equation by means of a hodograph-related transformation. We briefly mention the works of Finn [90], Leray [135], and Weinstein [179] on jets and the works of Lavrentieff [134], Leray [135], and Serrin [161b] on cavities: these works establish existence theorems. Uniqueness results were previously proved by Lavrentieff [134] and more simply by Gilbarg [107a] and Serrin [161b].

Variational approach to cavities was developed by Garabedian, Lewy, and Schiffer [102] and by Garabedian and Spencer [103]; these authors use symmetrization, but their admissible classes are rather restricted and the analysis (which employs complex methods) is complicated; reference 102 is the only paper in the classical literature to establish existence for axially symmetric flows.

Various properties of solutions (such as comparison and monotonicity) are surveyed in references 41 and 107b. We mention specifically the under–over theorem and the single intersection property due to Serrin [161b, c, d], both of which are used in Section 21.

The material of this chapter (with the exception of Sections 1 and 7) is based on recent papers by Alt and Caffarelli [6a], Alt, Caffarelli, and Friedman [7a, b, c], and Caffarelli and Friedman [58p]. Sections 2 to 4 are taken from reference 6a; Sections 6 and 8 to 19 are based on reference 7 a, b, c; the results on the impinging jet problem (at the end of Section 17) are due to Caffarelli and Friedman (unpublished).

The paper [102] asserts a weaker version of Theorem 20.1, namely, the smoothness of the continuous fit is not established. The presentation of Section 20 given here is new.

Asymmetric jets with gravity have been studied in reference 7c. For some special geometries there are the results on existence and uniqueness that use hodograph-related transformations to reduce the problem to a nonlinear integral equation; see Budden and Norbury [54], Carter [68a, b], Keady [118a], Keady and Norbury [119a], and Larock and Street [133].

The results in Problem 5 of Section 20 were originally proved by Tepper [171a, b]. Related variational problems arising in optimization of heat flow have been studied by Acker [1a–c].

Phillips [187a, b] studied minimization problems for $\int(|\nabla u|^2 + (u^+)^\gamma)$ and obtained results analogous to those of Sections 2, 3. Alt, Caffarelli, and Friedman [7d, e] have studied jet problems with two fluids using variational techniques; here the condition $|\nabla u^+|^2 - |\nabla u^-|^2 = \lambda$ holds on the interface between two fluids.

4

VARIATIONAL
PROBLEMS WITH
POTENTIALS

In this chapter we study variational problems where the functional has the form

$$\int_D \int_D \frac{\rho(x)\rho(y)}{|x-y|}\, dx\, dy + \int_D A(\rho(x))\, dx + \int_D \Phi(\rho)(x)\, dx \qquad (D \subset R^3)$$

and $A(t)$ is a given function, $\Phi(\rho)$ is a given functional. The admissible functions ρ satisfy $\rho \geq 0$, $\int \rho(x)\, dx = M$ (M is given), and possibly some other constraints. The variational equations for these problems will have the form $\Delta u = g(x, u)$ and $\{u = 0\}$ is the free boundary $\partial\{\rho > 0\}$. Such variational problems arise in physical models, classical as well as recent ones. We shall deal here with four topics:

Sections 1 to 5: Self-gravitating axially symmetric rotating fluids.
Sections 6 to 10: Vortex rings.
Sections 11 to 13: The problem of confined plasma.
Sections 14 to 17: The Thomas–Fermi atomic model.

In studying these problems, we shall develop some general methods. Thus in Section 5 we derive results on the expansion of solutions of differential inequalities $|\Delta u| \leq C|u|^\beta$ ($\beta \geq 1$) near $x = x^0$, provided that $u(x^0) = 0$; in Section 9 we establish estimates on capacities. These general results are used in the analysis of several of the models.

As in Chapter 3 we shall be concerned with questions of existence, uniqueness, and regularity of solutions of the variational problem, with the smoothness and shape of the free boundary, and with asymptotic estimates.

1. SELF-GRAVITATING AXISYMMETRIC ROTATING FLUIDS

Consider inviscid and irrotational fluid with density ρ rotating about the z axis. It is assumed to be axisymmetric; that is, $\rho(x) = \rho(r, z)$, where $x = (r, \theta, z)$. The rotation law is

$$\mathbf{v} = s(r)\mathbf{i}_\theta$$

where $s(r)$ is a given function and

$$\mathbf{i}_\theta = (-\sin \theta, \cos \theta)$$

is tangent to the path of the particles. We introduce the unit vector

$$\mathbf{i}_r = -(\cos \theta, \sin \theta)$$

pointing into the origin. One can check that the law of conservation of mass

$$(1.1) \qquad\qquad\qquad \nabla(\rho\mathbf{v}) = 0$$

is satisfied.

Next, the Euler equations are

$$(1.2) \qquad\qquad\qquad (\mathbf{v} \cdot \nabla)\mathbf{v} = -\frac{1}{\rho}\nabla p + \nabla V$$

where p is the pressure in the fluid and V is the gravitational potential,

$$V(x) = \int_{R^3} \frac{\rho(y)\, dy}{|x - y|}.$$

One easily verifies that

$$(\mathbf{v} \cdot \nabla)\mathbf{v} = -\frac{s^2(r)}{r}\mathbf{i}_r = -\nabla f(r)$$

where

$$f(r) = -\int_r^\infty \frac{s^2(t)}{t}\, dt.$$

Thus Euler's equation becomes

$$(1.3) \qquad\qquad\qquad \frac{1}{\rho}\nabla p = \nabla(V + f).$$

If we restrict the fluid to be compressible and assume the polytropic law,

$$p = K\rho^\gamma, \qquad \gamma = 1 + \frac{1}{\beta}, \qquad 0 < \beta < 3, \ K \text{ const.},$$

then (1.3) becomes

(1.4) $\qquad\qquad c\rho^{1/\beta} = V + f + \lambda \text{ in } G \qquad (\lambda \text{ const.}),$

where $c = K(\beta + 1)$ and

$$G = \{x : \rho(x) > 0\}.$$

Similarly, in the incompressible case $\rho = 1$,

$$\nabla p = \nabla(V + f),$$

so that Euler's equations become

(1.5) $\qquad\qquad p = V + f + \lambda \text{ in } G \qquad (\lambda \text{ const.}).$

We have assumed that $\rho = \rho(r, z)$. We shall further assume that $\rho(r, z) = \rho(r, -z)$ and

$$\int_{R^3} \rho \, dx = M \qquad (M > 0 \text{ is given}).$$

Define

$$m(r) = \frac{1}{M} \int_{\{r(y)<r\}} \rho(y) \, dy,$$

where $r(y) = r$ if $y = (r, \theta, z)$. We now introduce a function $j(m)$ $(0 \leqslant m \leqslant 1)$, the angular momentum per unit mass, such that

(1.6) $j(0) = 0, \qquad j(m)$ monotone nondecreasing and $j^2(m) \in C^1[0, 1]$.

Formally, for $0 \leqslant m \leqslant 1$,

(1.7) $\qquad\qquad j(m(r)) = rs(r),$

so that

(1.8) $\qquad\qquad f(r) = -\int_r^\infty \frac{j^2(m(r))}{r^3} \, dr.$

(1.7) is clearly valid if we accept the intuitive assumption that the rotating fluid, during the evolution which led to equilibrium, was moving in such a way that the total angular momentum did not change for any fraction of the mass lying within a distance r from the axis of rotation.

Define a function u by

$$(1.9) \qquad\qquad u = V + f + \lambda.$$

Then we seek an equilibrium figure G for which $u > 0$ in G and

$$(1.10) \qquad \left.\begin{array}{l} u = c\rho^{1/\beta} \text{ in } G = \{\rho > 0\} \\[4pt] u \leqslant 0 \text{ in } R^3 \setminus G \end{array}\right\} \qquad \begin{array}{l}\text{equilibrium}\\ \text{conditions}\end{array}$$

in the compressible case, and

$$(1.11) \qquad \left.\begin{array}{l} u \geqslant 0 \text{ in } G \quad (\rho = I_G) \\[4pt] u \leqslant 0 \text{ in } R^3 \setminus G \end{array}\right\} \qquad \begin{array}{l}\text{equilibrium}\\ \text{conditions}\end{array}$$

in the incompressible case.

We shall derive these equations from a variational principle.

Definition 1.1. A function $\rho(x)$ is said to belong to class $\mathcal{C}$ if $\rho(x) = \rho(r, z)$, $\rho(r, z) = \rho(r, -z)$, $\rho \geqslant 0$, and

$$\int_{R^3} \rho(x) \, dx = M;$$

here M is a fixed positive number.

Definition 1.2. A function ρ is said to belong to class $\mathcal{C}_0$ if $\rho \in \mathcal{C}$ and $0 \leqslant \rho \leqslant 1$.

Set

$$(1.12) \quad E(\rho) = -\frac{1}{2} \int_{R^3} \int_{R^3} \frac{\rho(x)\rho(y)}{|x - y|} \, dx \, dy + \frac{1}{2} \int_{R^3} \frac{j^2(m(r(x)))}{r^2(x)} \rho(x) \, dx$$

$$+ K\beta \int_{R^3} \rho^\gamma(x) \, dx;$$

the terms on the right represent, respectively, the gravitational potential energy, the rotational kinetic energy, and the internal energy. Consider the variational problem:

Problem (E).   Find ρ such that

$$(1.13) \qquad E(\rho) = \min_{\tilde{\rho} \in \mathcal{Q}} E(\tilde{\rho}), \qquad \rho \in \mathcal{Q}.$$

We introduce the formal derivative of $E(\rho)$ (see Problem 1):

$$(1.14) \qquad E'(\rho) = -\int_{R^3} \frac{\rho(y)\, dy}{|x - y|} + \int_{r(x)}^{\infty} \frac{j^2(m(r))}{r^3}\, dr + c\rho^{1/\beta}(x).$$

Lemma 1.1. *If ρ is a solution of Problem (E) which belongs to L^∞, then a.e.*

$$(1.15) \qquad \begin{aligned} E'(\rho) &= \lambda && in\ G = \{\rho > 0\}, \\ E'(\rho) &\geqslant \lambda && in\ R^3 \setminus G, \end{aligned}$$

where λ is a constant. If further $V(x)$ is continuous, then ρ is continuous and (1.15) *holds everywhere.*

Proof.   Since $\{\rho > 0\}$ has positive measure we can choose $\eta_0 \geqslant 0$, supp $\eta_0 \subset \{\rho > \delta_0\}$ with some $\delta_0 > 0$ such that $\eta_0(x) = \eta_0(r, z) = \eta_0(r, -z)$ and $\int \eta_0(x)\, dx = 1$. For any $\delta > 0$ and for any $\eta(x) = \eta(r, z) = \eta(r, -z)$ such that $\eta \geqslant 0$ in the set $\{\rho < \delta\}$, the function

$$\rho_\varepsilon = \rho + \varepsilon\left(\eta - \left(\int \eta\right)\eta_0\right)$$

belongs to $\mathcal{Q}$ if $\varepsilon > 0$, ε sufficiently small (depending on δ). It follows that

$$(1.16) \qquad \frac{d}{d\varepsilon} E(\rho_\varepsilon)\bigg|_{\varepsilon = 0+} \geqslant 0,$$

which gives (see Problem 1)

$$(1.17) \qquad \int\left[E'(\rho)\eta - \left(\int E'(\rho)\eta_0\right)\eta\right] \geqslant 0.$$

Since δ and η are arbitrary, (1.15) follows with

$$\lambda = \int E'(\rho)\eta_0.$$

From (1.14), (1.9) we see that (1.15) reduces to

$$\begin{aligned} u &= c\rho^{1/\beta} && in\ G \\ u &\leqslant 0 && in\ R^3 \setminus G \end{aligned}$$

a.e., so that the equilibrium conditions (1.10) hold a.e.

If for a solution ρ of (1.13) the gravitational potential V is continuous, then u^+ is continuous. We can then take a version

$$\rho = \left(\left(\frac{u}{c} \right)^+ \right)^\beta$$

of ρ that is continuous in R^3, and the equilibrium conditions hold everywhere.

Consider next the incompressible case and set

$$(1.18) \quad E_0(\rho) = -\frac{1}{2} \int_{R^3} \int_{R^3} \frac{\rho(x)\rho(y)}{|x-y|} \, dx \, dy + \frac{1}{2} \int_{R^3} \frac{j^2(m(r(x)))}{r^2(x)} \rho(x) \, dx.$$

Problem (E_0).   Find ρ such that

$$(1.19) \qquad\qquad E_0(\rho) = \min_{\tilde{\rho} \in \mathcal{C}_0} E_0(\tilde{\rho}), \qquad \rho \in \mathcal{C}_0.$$

Introducing the derivative

$$(1.20) \qquad\qquad E_0'(\rho) = -\int_{R^3} \frac{\rho(y)\,dy}{|x-y|} + \int_{r(x)}^\infty \frac{j^2(m(r))}{r^3} \, dr,$$

we have:

Lemma 1.2.   *If ρ is a solution of Problem (E_0) and if ρ has compact support, then $\rho = I_G$ and*

$$(1.21) \qquad\qquad \begin{array}{ll} E_0'(\rho) \leqslant \lambda & \text{in } G, \\ E_0'(\rho) \geqslant \lambda & \text{in } R^3 \setminus G, \end{array}$$

where λ is a constant.

Proof.   Assume first that the set $\{0 < \rho < 1\}$ has positive measure; we shall derive a contradiction. Choosing η_0 such that for some small $\delta_0 > 0$,

$$\sup \eta_0 \subset \{\delta_0 < \rho < 1 - \delta_0\}, \qquad \int \eta_0 \, dx = 1,$$

we can proceed as before to establish that a.e.

$$(1.22) \qquad\qquad \begin{array}{ll} E_0'(\rho) \geqslant \lambda & \text{in } \{0 \leqslant \rho < 1\}, \\ E_0'(\rho) \leqslant \lambda & \text{in } \{0 < \rho \leqslant 1\}. \end{array}$$

By Theorem 7.2 of Chapter 3, the solution ρ must be already rearranged increasingly in z, for $z > 0$. But then one can compute that V is strictly decreasing in z, for $z > 0$ (see Problem 2). From (1.20) we then infer that the function $E_0'(\rho)$ is strictly monotone increasing in z, for $z > 0$. Thus the set

$$(1.23) \qquad \{(r, z); E_0'(\rho) = \lambda\}$$

has measure zero. Since, by (1.22), the set in (1.23) differs from the set $\{0 < \rho < 1\}$ by a null set, it follows that meas $\{0 < \rho < 1\} = 0$, contradicting our assumption.

We have thus proved that $\rho = I_G$ for some set G.

Notice (see Problem 2) that $E_0'(\rho)$ is a function of (r, z) only, say $h(r, z)$, and h is continuous and strictly decreasing in z for $z \geq 0$. It follows that

$$G = \{(r, z); |z| < \phi(r), 0 \leq r < \infty, \text{where } h(r, \phi(r)) = \lambda,$$

$$(1.24) \qquad \text{and } \phi(r) \text{ is positive and continuous}$$

$$\text{on the open set } \{r; h(r, 0) < \lambda\}\}.$$

Choose (r_0, z_0) with $r_0 > 0$, $z_0 = \phi(r_0) > 0$, and let $z_\delta = z_0 - \delta$, for any small $\delta > 0$. For any $\gamma > 0$, denote by B_γ the disc with center (r_0, z_δ) and radius γ; we choose γ sufficiently small so that $B_{2\gamma} \subset G$. Let $\eta_0(r, z) = \eta_0(r, -z)$ be defined by

$$\eta_0 = CI_{B_\gamma} \qquad \text{if } z > 0, \quad \int \eta_0 \, dx = 1.$$

Choosing $\eta(r, z) = \eta(r, -z) \geq 0$ with support in $R^2 \setminus G$, the function

$$\rho + \varepsilon \left(\eta - \left(\int \eta \right) \eta_0 \right)$$

is then in $\mathcal{C}_0$ for any small $\varepsilon > 0$, and we find as before that

$$(1.25) \qquad E_0'(\rho) \geq \lambda_0 \qquad \text{a.e. in } R^2 \setminus G,$$

where

$$\lambda_0 = \int E_0'(\rho) \eta_0.$$

Notice that

$$\lambda_0 - h(r_0, z_0) = 2C \int_{B_\gamma} h(r, z) \, dx - h(r_0, z_0)$$

$$= h(\tilde{r}, \tilde{z}) - h(r_0, z_0) \to 0 \qquad \text{if } \delta, \gamma \to 0,$$

where $(\tilde{r}, \tilde{z}) \in B_\gamma$.

Similarly, denoting by B^γ the disc with radius γ and center $(r_0, z_0 + \delta)$, we choose γ small and $\eta_1(r, z) = \eta_1(r, -z)$ such that

$$\eta_1 = CI_{B^\gamma} \qquad \text{if } z > 0, \quad \int \eta_1 \, dx = 1.$$

Then for any $\eta(r, z) = \eta(r, -z) \leq 0$ with support in G, the function

$$\rho + \varepsilon\left(\eta - \left(\int \eta\right)\eta_1\right)$$

is in $\mathcal{C}_0$ for any small $\varepsilon > 0$. We then find that

$$E_0'(\rho) \leq \lambda_1 \qquad \text{a.e. in } G$$

where

$$\lambda_1 = \int E_0'(\rho)\eta_1 \to h(r_0, z_0) \qquad \text{if } \delta, \gamma \to 0.$$

Thus (1.21) holds with $\lambda = h(r_0, z_0)$.

In the next section we establish estimates on the gravitational potential V of a function ρ in either $\mathcal{C}$ or $\mathcal{C}_0$. These estimates will be used in the existence proof for the variational problems (E) and (E_0) (in Section 3).

PROBLEMS

1. Let ρ, σ be bounded with compact support and $\rho, \rho_\varepsilon = \rho + \varepsilon\sigma$ belong to $\mathcal{C}$ for all small $\varepsilon > 0$. Prove that

$$\frac{d}{d\varepsilon} E(\rho_\varepsilon)\bigg|_{\varepsilon=0} = \int E'(\rho)\sigma.$$

[*Hint*: Set $m_\varepsilon(r) = m_{\rho_\varepsilon}$. Then $m_\varepsilon(r) - m_0(r) = \varepsilon m_\sigma(r)$, $m_\varepsilon(r) \leq Cr^2$. If $M = 1$, $\hat{\rho}(r) = 2\pi r \int_{-\infty}^{\infty} \rho(r, z) \, dz$, then $\int_0^s \hat{\rho}(r) \, dr = m_\rho(r)$, so that $m_\rho'(r) = \hat{\rho}(r)$. To compute the derivative at $\varepsilon = 0$ for the term involving j, set $L = j^2$. Then

$$\varepsilon^{-1}\left[\int \rho_\varepsilon L(m_\varepsilon)r^{-2} - \int \rho L(m_0)r^{-2}\right] = \int \sigma L(m_\varepsilon)r^{-2}$$

$$+ \varepsilon^{-1}\int \rho(L(m_\varepsilon) - L(m_0))r^{-2}$$

$$= J_1 + J_2;$$

since $m_\varepsilon \to m_0$ uniformly, $J_1 \to \int \hat{\sigma} L(m_0) r^{-2}$. Next

$$J_2 = \varepsilon^{-1} \int_0^\infty \hat{\rho}(L(m_\varepsilon) - L(m_0)) r^{-2} \, dr$$

$$= \varepsilon^{-1} \int_0^\infty \int_0^\varepsilon (\hat{\rho} + t\hat{\sigma}) L'(m_\varepsilon(r)) m_\sigma(r) \, dt r^{-2} \, dr$$

$$- \varepsilon^{-1} \int_0^\infty \int_0^\varepsilon t\hat{\sigma} L'(m_\varepsilon(r)) m_\sigma(r) \, dt r^{-2} \, dr$$

$$\equiv I_1 - I_2.$$

Since $(\hat{\rho} + t\hat{\sigma}) L'(m_t) = \dfrac{d}{dr} L(m_t)$,

$$I_1 = -\varepsilon^{-1} \int \int L(m_t) \left[\hat{\sigma} r^{-2} - 2 m_\sigma(r) r^{-3} \right]$$

$$\to -\int_0^\infty \hat{\sigma} L(m_0) r^{-2} \, dr + 2 \int_0^\infty m_\sigma(r) L(m_0) r^{-2}$$

$$= -K_1 + K_2$$

and $K_1 = \lim J_1$,

$$K_2 = 2 \int \sigma(x) \int_{r(x)}^\infty L(m_0) r^{-3} \, dr \, dx.$$

Finally,

$$I_2 = \varepsilon^{-1} \int_0^\infty \int_0^\varepsilon r^{-2} \hat{\sigma} t \frac{d}{dt} L(m_t) \, dt \, dr$$

$$= \int_0^\infty r^{-2} \hat{\sigma} \left[L(m_\varepsilon) - \varepsilon^{-1} \int_0^\infty L(m_t) \, dt \right] \to 0.$$

2. If $\rho(x) = \rho(r, z) = \rho(r, -z) \geqslant 0$, ρ has compact support, then:
   (i) The gravitational potential

$$V(x) = \int_{R^3} \frac{\rho(y)}{|x - y|} \, dy$$

is independent of θ.
   (ii) If ρ is bounded and decreasing in z, $z \geqslant 0$, then V is strictly decreasing in z, $z > 0$.

[*Hint*: For (i) consider $V(\tau_\phi x)$, where $\tau_\phi(r, z, \theta) = (r, z, \theta + \phi)$.]

2. ESTIMATES OF GRAVITATIONAL POTENTIALS

For any $\rho \in \mathcal{C}$ or $\rho \in \mathcal{C}_0$, set

$$\rho_0 = \sup_{x \in R^3} \rho(x);$$

clearly, $\rho_0 \leqslant 1$ if $\rho \in \mathcal{C}_0$.

Lemma 2.1.   *If $\rho \in \mathcal{C}$ or $\rho \in \mathcal{C}_0$ and if $\rho(r, z)$ is nonincreasing in z for $z > 0$, then*

$$(2.1) \qquad \sup_{x \in R^3} V(x) \leqslant c^* M^{2/3} \rho_0^{1/3} \qquad \left[c^* = 2\pi(3/4\pi)^{2/3} \right],$$

$$(2.2) \qquad V(x) \leqslant \frac{CM}{|x|} \log\left(1 + \frac{C|x|}{a_0} \right) \qquad (x \in R^3),$$

where a_0 is defined by

$$(2.3) \qquad \frac{4\pi}{3} \rho_0 a_0^3 = M$$

and C is a sufficiently large positive universal constant.

Proof.   To prove (2.1), consider the problem of maximizing

$$J(\rho) = \int_{R^3} \frac{\rho(y)}{|x - y|} dy \qquad (x \text{ fixed})$$

subject to the constraints

$$0 \leqslant \rho(y) \leqslant \rho_0, \qquad \int_{R^3} \rho(y)\, dy = M.$$

Clearly, the solution is given by $\rho = \rho_0 I_B$ for a ball B with center x and radius a_0. Calculating $J(\rho)$ for this density, we obtain

$$J(\rho) = 4\pi\rho_0 \int_0^{a_0} \frac{\xi^2\, d\xi}{\xi} = 2\pi\rho_0 a_0^2 = c^* \rho_0^{1/3} M^{2/3}.$$

We now proceed to prove that

$$(2.4) \qquad V(x) \leqslant \frac{CM}{R} \log\left(1 + \frac{CR}{a_0} \right) \qquad \text{if } x = (R, 0, z).$$

Suppose that $R \leqslant A a_0$, where A is a positive constant to be determined later.

Since

$$V(x) \leq \rho_0 \int_{\{|x-y|<a_0\}} \frac{dy}{|x-y|} = C\rho_0 a_0^2,$$

we have

$$V(x) \leq \frac{CM}{a_0},$$

which implies (2.4). Thus it remains to consider the case $R > Aa_0$.

Let $S = \{y = (r, \theta, z); \ |r - R| < R/2, \ |z - Z| < R\}$ and $S' = R^3 \setminus S$. Since for $y \in S'$ we have $|x - y| > R/2$, it follows immediately that

$$\int_{S'} \frac{\rho(y)}{|x-y|} dy \leq \frac{CM}{R}.$$

Thus it remains to estimate

$$I = \int_S \frac{\rho(y)}{|x-y|} dy$$

(2.5)
$$= \iint_{\substack{\{|r-R|<R/2\} \\ \{|z-Z|<R\}}} \rho(r, z) \, dr \, dz$$

$$\times \int_{-\pi}^{\pi} \frac{r \, d\theta}{\left[(z - Z)^2 + r^2 + R^2 - 2rR \cos \theta\right]^{1/2}}.$$

Writing $\lambda = z - Z$, $a = \lambda^2 + r^2 + R^2$, $b = 2rR$, we have

$$J(\lambda, r, R) \equiv \int_{-\pi}^{\pi} \frac{r \, d\theta}{[\lambda^2 + r^2 + R^2 - 2rR \cos \theta]^{1/2}}$$

(2.6)
$$= 2r \int_0^{\pi} \frac{d\theta}{(a - b \cos \theta)^{1/2}}$$

$$= 2r \cdot \frac{2K(k)}{(a + b)^{1/2}},$$

where $k^2 = 2b/(a + b)$ and $K(k)$ is the complete elliptic integral of the first kind; see formula 291.00 of reference 55. Since $k^2 = 4rR/(\lambda^2 + (r + R)^2)$, $k \to 1$ as $\lambda^2 + (r - R)^2 \to 0$. The function $K(k)$, which is singular at $k = 1$, possesses the asymptotic property (see formula 112.01 of reference 55)

$$K(k) \sim \log \frac{4}{k'} \qquad (k'^2 = 1 - k^2) \text{ as } k \to 1.$$

Using $k'^2 = (\lambda^2 + (r - R)^2)/(\lambda^2 + (r + R)^2)$, we now conclude from (2.6)
that

$$J(\lambda, r, R) \leqslant \frac{Cr}{\left[\lambda^2 + (r + R)^2\right]^{1/2}} \log C \frac{\lambda^2 + (r + R)^2}{\lambda^2 + (r - R)^2}$$

for some sufficiently large absolute positive constant C. Finally, since $|\lambda| < R$
and $|r - R| < R/2$, we have

(2.7) $$J(\lambda, r, R) \leqslant C \log \frac{CR}{\left[\lambda^2 + (r - R)^2\right]^{1/2}}.$$

Now applying estimate (2.7) to equation (2.5), we find

(2.8) $$I \leqslant F(\rho) \equiv C \iint\limits_{\substack{\{|r - R| < R/2\} \\ \{|z - Z| < R\}}} \rho(r, z) \log \frac{CR}{\left[(z - Z)^2 + (r - R)^2\right]^{1/2}} \, dr \, dz.$$

Also, there holds

$$M = 2\pi \iint\limits_{R^3} \rho(r, z) r \, dr \, dz \geqslant CR \iint\limits_{\substack{\{|r - R| < R/2\} \\ \{|z - Z| < R\}}} \rho(r, z) \, dr \, dz.$$

We consider, therefore, the problem of maximizing $F(\rho)$ subject to the constraints

$$0 \leqslant \rho(r, z) \leqslant \rho_0,$$

$$\iint\limits_{\substack{\{|r - R| < R/2\} \\ \{|z - Z| < R\}}} \rho(r, z) \, dr \, dz \leqslant CM/R.$$

Clearly, the maximum occurs for $\rho = \rho_0 I_D$, where $D = \{(r, z); (z - Z)^2 + (r - R)^2 < s^2\}$ and $\pi \rho_0 s^2 = CM/R$. Computing F for this density, we obtain

$$F = C \rho_0 \int_0^s \log \frac{CR}{\varepsilon} \xi \, d\xi$$

$$\leqslant C \rho_0 s^2 \left(1 + \log \frac{CR}{s}\right)$$

$$\leqslant \frac{CM}{R} \left(1 + \log \frac{C \rho_0 R^3}{M}\right)$$

$$\leqslant \frac{CM}{R} \log \frac{CR}{a_0};$$

we assume that the constant A is now fixed sufficiently large so that $R > Aa_0$ implies that $s < R/2$. The estimate for F above yields, by (2.8), the required estimate for I. This completes the proof of (2.4).

We next establish the estimate

$$(2.9) \qquad V(x) \leqslant \frac{CM}{Z} \log\left(1 + \frac{CZ}{a_0}\right) \qquad \text{if } x = (R, 0, Z), \quad Z > 0.$$

As in the proof of (2.4), it suffices to consider $Z > Aa_0$. The proof given below does not depend on the axisymmetry of ρ; therefore, we may assume that $R = 0$.

We claim that $V(x)$ is majorized by the potential due to a certain rearrangement of ρ; namely

$$(2.10) \qquad V(x) \leqslant \int_{R^3} \frac{\tilde{\rho}(y)}{|x - y|}\, dy,$$

where $\tilde{\rho} = \rho_0 I_{\tilde{G}}$, $\tilde{G} = \{y = (r, \theta, z); 0 \leqslant r < \sigma(z)\}$, and $\sigma(z)$ is given by

$$(2.11) \qquad \pi \rho_0 \sigma^2(z) = \int_{R^2} \rho(y_1, y_2, z)\, dy_1\, dy_2.$$

To prove the claim we express $V(x)$ in the form

$$(2.12) \qquad V(x) = \int_{-\infty}^{\infty} dz \int_{R^2} \frac{\rho(y_1, y_2, z)}{\left[(z - Z)^2 + y_1^2 + y_2^2\right]^{1/2}}\, dy_1\, dy_2.$$

Now, for fixed λ and σ, consider the problem of maximizing

$$F(\rho) = \int_{R^3} \frac{\rho(y_1, y_2)}{\left[\lambda^2 + y_1^2 + y_2^2\right]^{1/2}}\, dy_1\, dy_2$$

subject to the constraints

$$0 \leqslant \rho(y_1, y_2) \leqslant \rho_0,$$

$$\int_{R^2} \rho(y_1, y_2)\, dy_1\, dy_2 = \pi \rho_0 \sigma^2.$$

Clearly the maximum occurs for $\rho = \rho_0 I_D$, where D is a disc of radius σ. This reasoning, applied to the inner integral in (2.12) for each z, establishes (2.10).

By virtue of the obvious inequality [writing $y = (r, \theta, z)$]

$$\int_{\{|z - Z| \geqslant Z/2\}} \frac{\tilde{\rho}(y)}{|x - y|}\, dy \leqslant \frac{CM}{Z},$$

it remains only to estimate

$$I = \int_{\{|z-Z|<Z/2\}} \frac{\tilde{\rho}(y)}{|x-y|}\,dy.$$

Since $\rho(y_1, y_2, z)$ is a nonincreasing function of z for $z > 0$ and $\rho(y_1, y_2, -z) = \rho(y_1, y_2, z)$, it is evident from (2.11) that $\sigma(z)$ is a nonincreasing function of z for $z > 0$ and $\sigma(-z) = \sigma(z)$. Writing $\sigma_1 = \sigma(Z/2)$, we then have

$$\left\{ y = (r, \theta, z); 0 \leqslant r < \sigma_1, |z| < \frac{Z}{2} \right\} \subseteq \tilde{G}.$$

Hence $M/\rho_0 = \text{meas}\,\tilde{G} \geqslant c\sigma_1^2 Z$ for some absolute positive constant c. Thus for $\sigma_0 > 0$ defined by

$$\sigma_0^2 = \frac{CM}{\rho_0 Z} \qquad (C > 0),$$

it follows that $\sigma_1 \leqslant \sigma_0$. Furthermore,

$$\tilde{G} \cap \{z > Z/2\} \subseteq \{0 \leqslant r < \sigma_1\} \subseteq \{0 \leqslant r < \sigma_0\}.$$

Thus we can estimate I as follows:

$$I \leqslant \rho_0 \int_{\substack{\{0 \leqslant r < \sigma_0\} \\ \{|z-Z|<Z/2\}}} \frac{dy}{|x-y|}$$

$$= C\rho_0 \int_{-Z/2}^{Z/2} \int_0^{\sigma_0} \frac{r\,dr\,d\lambda}{(\lambda^2 + r^2)^{1/2}}$$

$$= C\rho_0 \sigma_0^2 \int_{-Z/2}^{Z/2} \frac{d\lambda}{\left(\lambda^2 + \sigma_0^2\right)^{1/2} + \lambda}$$

$$\leqslant C\rho_0 \sigma_0^2 \log \frac{Z}{\sigma_0}$$

$$= \frac{CM}{Z} \log \frac{CZ}{a_0};$$

here we use the fact that $Z > 2\sigma_0$, which is implied by the assumption $Z > Aa_0$ provided that the constant A is fixed sufficiently large. This completes the proof of (2.9). The assertion (2.2) is obviously a consequence of (2.4), (2.9).

Remark 2.1. The assumption that ρ is nonincreasing in z, $z > 0$, was not used in the proofs of (2.1) and (2.4).

Lemma 2.2. *If ρ is a solution of Problem (E) and if ρ has compact support, then*

$$(2.13) \qquad \sup_{x \in R^3} \rho(x) \leqslant \rho_*$$

where ρ_* is defined by

$$(2.14) \qquad \rho_*^{(1/\beta)-(1/3)} = CM^{2/3}$$

and C is a positive constant depending only on K, β.

Proof. From Lemma 1.1 we obtain

$$(2.15) \qquad C\rho^{1/\beta} = V + f + \lambda \qquad \text{in } G.$$

Since ρ has compact support and $V \to 0$, $f \to 0$ as $x \to \infty$, (1.10) and (1.11) yield

$$\lambda = \lim_{x \to \infty} u(x) \leqslant 0.$$

Noting also that $f \leqslant 0$, we obtain from (2.15)

$$C\rho^{1/\beta} \leqslant V.$$

Estimating the right-hand side by (2.1), the assertion follows.

PROBLEMS

1. Let T be the (solid) torus

$$T = \left\{ y = (r, \theta, z); (z - Z)^2 + (r - R)^2 < s^2 \right\}$$

with $0 < s < R/2$. Prove that

$$\int_T \frac{dy}{|x - y|} \geqslant Cs^2 \log \frac{R}{s} \qquad [x = (R, 0, Z)]$$

where C is a positive constant independent of R, Z, s.
[*Hint:* Use the proof of (2.4).]

2. Another common model for the rotating fluid is obtained when one prescribes the angular velocity $\Omega(r)$ instead of the angular momentum $j(m)$. In this case the equilibrium equation is

$$(2.16) \qquad \rho^{-1} \nabla p = \nabla V + r \Omega^2(r) \mathbf{i}_r$$

and in the functionals E, E_0 the kinetic energy term is replaced by

$$-\int \rho(x)J(r(x))\, dx,$$

where

$$(2.17) \qquad\qquad J(r) = \int_0^r s\Omega^2(s)\, ds.$$

In the incompressible case the equilibrium condition is

$$(2.18) \qquad \begin{array}{ll} V + J + \lambda \geqslant 0 & \text{in the fluid,} \\ V + J + \lambda \leqslant 0 & \text{outside the fluid.} \end{array}$$

If $\Omega(s) = \text{const.} = \Omega$, then $J(r) = \Omega^2 r^2/2$. Maclaurin (1742) discovered a family of solutions corresponding to incompressible oblate spheroids

$$\frac{r^2}{a_1^2} + \frac{z^2}{a_2^2} < 1 \qquad (a_1 > a_2).$$

For such a body,

$$V(x) = \pi\left(I - A_1 r^2 - A_2 z^2\right), \qquad I = 2a_1^2 A_1^2 + a_2^2 A_2^2$$

and

$$A_1 = \frac{\sqrt{1 - e^2}}{e^3}\sin^{-1} e - \frac{1 - e^2}{e^2}, \qquad A_2 = \frac{2}{e^2} - \frac{2\sqrt{1 - e^2}}{e^3}\sin^{-1} e,$$

where $e = (1 - a_2^2/a_1^2)^{1/2}$ is the eccentricity. Show that (2.18) is satisfied if

$$\frac{\Omega^2}{\pi} = 2\left(A_1 - \frac{a_2^2}{a_1^2}A_2\right) = \frac{\sqrt{1 - e^2}}{e^3}2(3 - 2e^2)\sin^{-1} e - \frac{6}{e^2}(1 - e^2),$$

which has the graph shown in Figure 4.1. Other families of solutions can be found in reference 69.

3. The rotating spheroids of Problem 2 are also solutions of the model (1.3) with $\rho = 1$, where f is given by (1.8); find the functions $j(m)$.

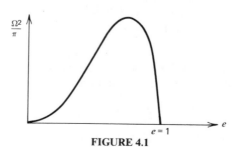

FIGURE 4.1

3. EXISTENCE OF SOLUTIONS

Theorem 3.1. *Problem (E) has a solution ρ which is continuous and has compact support, and $\rho(r, z)$ is nonincreasing in z for $z \geqslant 0$.*

In order to prove the theorem we shall first consider a family of truncated problems.

For any $N > 1$, denote by $\mathcal{Q}_N$ the class of functions $\tilde{\rho} \in \mathcal{Q}$ such that

$$\operatorname{supp} \tilde{\rho} \subset \left\{ x \in R^3; |x| \leqslant Na_* \right\},$$

$$0 \leqslant \tilde{\rho} \leqslant N\rho_* \qquad \text{a.e.,}$$

where ρ_* is defined by (2.14) and a_* is defined by (2.3) with $\rho_0 = \rho_*$, $a_0 = a_*$; that is

$$(3.1) \qquad\qquad \tfrac{4}{3}\rho_* a_*^3 = M.$$

Lemma 3.2. *There exists $\rho_N \in \mathcal{Q}_N$ satisfying*

$$(3.2) \qquad\qquad E(\rho_N) = \min_{\tilde{\rho} \in \mathcal{Q}_N} E(\tilde{\rho});$$

ρ_N is continuous in $\{|x| < Na_*\}$, $\rho_N(r, z)$ is nonincreasing in z for $z \geqslant 0$, and

$$(3.3) \qquad \begin{aligned} E'(\rho_N) &\leqslant \lambda_N && \text{on } \{\rho_N > 0\}, \\ E'(\rho_N) &\geqslant \lambda_N && \text{on } \{\rho_N < N\rho_*\} \end{aligned}$$

where λ_N is some constant.

Proof. Since $\mathcal{Q}_N$ is a bounded, closed, and convex subset of $L^p\{x; |x| < Na_*\}$, $1 \leqslant p < \infty$, we conclude that $\mathcal{Q}_N$ is compact in the weak topology of L^p. We claim: In the weak topology of L^p,

$$(3.4) \qquad \tilde{\rho} \to E(\tilde{\rho}) \qquad \text{is lower semicontinuous on } \mathcal{Q}_N.$$

Indeed, if $\rho_m \to \rho$ weakly in L^p, then

$$m_{\rho_m}(r) \to m_\rho(r) \qquad \text{pointwise}$$

[where $m_\rho(r)$ is the function $m(r)$ for the density ρ]. Since further $m_{\rho_m}(r)$, $m_\rho(r)$ are monotone, continuous, and uniformly bounded, the convergence is uniform. Consequently,

$$j^2\big(m_{\rho_m}(r(x))\big) \to j^2\big(m_\rho(r(x))\big)$$

uniformly in x. Noticing also that

$$r^{-2}(x)j^2\big(m_{\tilde{\rho}}(r(x))\big) \leqslant C \qquad \forall \tilde{\rho} \in \mathcal{Q}_N,$$

where C is independent of $\tilde{\rho}$, it follows that

$$\int_{R^3} r^{-2}(x)j^2\big(m_{\rho_m}(r(x))\big)\rho_m(x)\,dx \to \int_{R^3} r^{-2}(x)j^2\big(m_\rho(r(x))\big)\rho(x)\,dx.$$

Next, the functions

$$\int \frac{\rho_m(y)}{|x-y|}\,dy$$

are uniformly bounded and equicontinuous and thus converge uniformly to

$$\int \frac{\rho(y)}{|x-y|}\,dy.$$

Hence

$$\iint \frac{\rho_m(x)\rho_m(y)}{|x-y|}\,dx\,dy \to \iint \frac{\rho(x)\rho(y)}{|x-y|}\,dx\,dy.$$

Since finally

$$\liminf_{m\to\infty} \int \rho_m^\gamma(x)\,dx \geqslant \int \rho^\gamma(x)\,dx,$$

the proof of (3.4) is complete. It follows that there exists a solution $\rho_N \in \mathcal{Q}_N$ of (3.2).

Since ρ_N is bounded and has compact support, we can now proceed to prove (3.3) a.e. as in the proof of Lemma 1.1. We shall now prove that ρ_N is continuous.

Set

(3.5) $$u_N = V_{\rho_N} + f + \lambda_N$$

when V_ρ is the gravitational potential of a density function ρ. Then

(3.6) $$E'(\rho_N) - \lambda_N = c\rho_N^{1/\beta} - u_N.$$

We claim that

$$\rho_N < N\rho_* \text{ a.e. on the set } \big\{u_N < c(N\rho_*)^{1/\beta}\big\}.$$

Indeed, if $\rho_N = N\rho_*$ then a.e. $E'(\rho_N) \leqslant \lambda_N$, by (3.3), and thus, by (3.6),

$$c(N\rho_*)^{1/\beta} = c\rho_N^{1/\beta} \leqslant u_N < c(N\rho_*)^{1/\beta},$$

a contradiction. Similarly, $\rho_N = N\rho_*$ a.e. on the set $\{u_N > c(N\rho_*)^{1/\beta}\}$, and

$$\rho_N > 0 \qquad \text{a.e. on the set } \{u_N > 0\},$$
$$\rho_N = 0 \qquad \text{a.e. on the set } \{u_N < 0\}.$$

It follows that $0 < \rho_N < N\rho_*$ a.e. on the set

$$\left\{0 < u_N < c(N\rho_*)^{1/\beta}\right\}$$

and thus, by (3.3), (3.6),

$$c\rho_N^{1/\beta} - u_N = 0 \qquad \text{a.e. on this set.}$$

Consequently, a.e.

$$(3.7) \qquad \rho_N = \begin{cases} N\rho_* & \text{if } u_N > c\left(N\rho_*\right)^{1/\beta} \\ \left(\dfrac{1}{c}u_N\right)^{\beta} & \text{if } 0 < u_N < c\left(N\rho_*\right)^{1/\beta} \\ 0 & \text{if } u_N < 0. \end{cases}$$

By Theorem 7.2 of Chapter 3, ρ_N is monotone nonincreasing in z for $z \geqslant 0$ and then u_N is strictly decreasing in z for $z > 0$ (Section 1, Problem 2). It follows that the level sets

$$\{(r, z); u_N(r, z) = \alpha\}$$

have measure zero. Hence the right-hand side of (3.6) gives a continuous version of ρ_N.

Lemma 3.3. *There exists a sufficiently large $\overline{N}$ such that*

$$(3.8) \qquad\qquad\qquad \lambda_N < 0 \qquad \text{if } N \geqslant \overline{N}.$$

Proof. From (3.3) we have

$$(3.9) \qquad \lambda_N \leqslant E'(\rho_N) = -V_{\rho_N} - f + c\rho_N^{1/\beta} \qquad \text{on } \{\rho_N < N\rho_*\}.$$

Take x_0 such that $|x_0| < \frac{1}{2}Na_*$, $r(x_0) \geqslant \frac{1}{4}Na_*$ and

$$\rho_N(x_0) = \inf_{\substack{|x| < \frac{1}{2}Na_* \\ r(x) > \frac{1}{4}Na_*}} \rho_N(x).$$

Then $\rho_N(x_0)(Na_*)^3 \leqslant CM$ and hence $\rho_N(x_0) \leqslant CN^{-3}\rho_*$. Thus, since $\beta < 3$,

$$(3.10) \qquad c\rho_N^{1/\beta}(x_0) \leqslant \varepsilon(N)N^{-1}\rho_*^{1/\beta}, \qquad \varepsilon(N) \to 0 \text{ if } N \to \infty.$$

We also have

$$(3.11) \qquad\qquad\qquad V_{\rho_N}(x_0) \geqslant \frac{M}{2Na_*},$$

$$(3.12) \qquad\qquad -f(x_0) = \int_{r(x_0)}^{\infty} \frac{j^2(m(r))}{r^3}\, dr \leqslant \frac{C}{N^2 a_*^2}.$$

Substituting (3.10)–(3.12) into (3.9), we find that (3.8) holds if N is sufficiently large.

Lemma 3.4. *If $N \geqslant \bar{N}$, then*

$$(3.13) \qquad\qquad\qquad \sup_{x \in R^3} \rho_N(x) \leqslant \rho_*.$$

Indeed, since $\lambda_N < 0$ and ρ_N has compact support, the assertion (3.13) follows by the same proof as in Lemma 2.2.

Lemma 3.5. *There is a positive constant $\bar{C}$ independent of N such that*

$$(3.14) \qquad\qquad E(\rho_N) \leqslant -\bar{C}\rho_*^{1/3}M^{5/3} \qquad \forall\, N \geqslant \bar{N},$$

provided that $\bar{N}$ is sufficiently large.

Proof. Let $\tilde{\rho} = \theta\rho_* I_B$, where B is a ball about the origin, $0 < \theta < 1$, and $\int_{R^3} \tilde{\rho}(x)\, dx = M$. Then the radius R of B is given by

$$\tfrac{4}{3}\pi R^3 \theta\rho_* = M$$

and

$$\int_B \frac{j^2(m(r(x)))}{r^2(x)}\tilde{\rho}(x)\, dx \leqslant 4\pi R\theta\rho_* \int_0^R \frac{j^2(m(r))}{r^2} r\, dr,$$

$$m(r) \leqslant 2M^{-1}\pi r^2 R\theta\rho_* = \frac{3}{2}\frac{r^2}{R^2}.$$

Recalling that $j^2(m)$ is monotone nondecreasing and making a change of variables $m = 3r^2/2R^2$ in the last integral, we obtain

(3.15) $\quad \int_B \dfrac{j^2(m(r(x)))}{r^2(x)} \tilde{\rho}(x)\, dx \leqslant 4\pi R\theta\rho_* j_0^2 \leqslant C(\theta\rho_*)^{2/3} M^{1/3} j_0^2,$

where $j_0^2 = \int_0^1 (j^2(m)/m)\, dm$.

Next, since

$$\int_B \dfrac{\tilde{\rho}(y)}{|x-y|}\, dy \geqslant \dfrac{M}{2R},$$

we have

(3.16) $\quad\quad \displaystyle\int_B\int_B \dfrac{\tilde{\rho}(x)\tilde{\rho}(x)}{|x-y|}\, dx\, dy \geqslant \dfrac{M^2}{2R} \geqslant C(\theta\rho_*)^{1/3} M^{5/3}$

for some $C > 0$. Finally,

(3.17) $\quad\quad K\beta \displaystyle\int_B \tilde{\rho}^\gamma(x)\, dx \leqslant C(\theta\rho_*)^{1/\beta} M = C\theta^{1/\beta}\rho_*^{1/3} M^{5/3},$

where we have used $\gamma = 1 + 1/\beta$ and (2.14).

Combining the estimates (3.15)–(3.17), we get

$$E(\tilde{\rho}) \leqslant -C(\theta\rho_*)^{1/3} M^{5/3}\left[1 - \dfrac{C(\theta\rho_*)^{1/3} j_0^2}{M^{4/3}} - C\theta^{1/\beta - 1/3}\right]$$

with positive constants C. Choosing θ sufficiently small (depending only on M and the function j) the expression in braces becomes larger than $\frac{1}{2}$. Since finally $\tilde{\rho} \in \mathcal{Q}_N$ if N is large enough, $E(\rho_N) \leqslant E(\tilde{\rho})$ and (3.14) follows.

Lemma 3.6. *Set*

$$\mu_A = M^{-1}\int_{\{|x|<Aa_*\}} \rho_N(x)\, dx.$$

Then there exist positive constants A (sufficiently large) and δ (sufficiently small), both independent of N, such that

(3.18) $\quad\quad\quad\quad\quad \mu_A \geqslant \delta \quad\quad if \quad\quad N \geqslant \bar{N},$

where $\bar{N}$ is sufficiently large.

Proof. From Lemma 3.5 it follows that

$$\overline{C}\rho_*^{1/3}M^{5/3} \leqslant \iint \frac{\rho_N(x)\rho_N(y)}{|x-y|}$$

$$\leqslant \int_{\{|x|,|y|<Aa_*\}} \frac{\rho_N(x)\rho_N(x)}{|x-y|} + 2\int_{\{|x|>Aa_*\}} V(x)\rho_N(x)$$

$$\leqslant C\rho_*^{1/3}(\mu_A M)^{5/3} + C\rho_*^{1/3}M^{5/3}\frac{1}{A}\log A;$$

here we used Lemma 3.4 and the estimate (2.2). Taking A sufficiently large, the assertion (3.18) follows.

Lemma 3.7. *There exists a positive constant B (sufficiently large) such that*

$$(3.19) \qquad\qquad \operatorname{supp}\rho_N \subset \{x; |x|< Ba_*\} \qquad \forall N \geqslant \overline{N},$$

provided that $\overline{N}$ is sufficiently large.

Proof. Suppose that $x_2 \in \operatorname{supp}\rho_N$ and $|x_2|> Ba_*$; we shall derive a contradiction. Take x_1 such that $|x_1|\leqslant Aa_*, r(x_1) > \frac{1}{2}Aa_*$, and

$$\rho_1 = \rho(x_1) = \inf_{\substack{|x|<Aa_* \\ r(x)>(1/2)Aa_*}} \rho_N(x).$$

Then $\rho_1(Aa_*)^3 \leqslant CM$, hence $\rho_1 \leqslant CA^{-3}\rho_* < N\rho_*$ if $N > \overline{N}$. From the variational conditions (3.3), we have

$$c\rho_1^{1/\beta} \geqslant u(x_1) \geqslant u(x_1) - u(x_2) \qquad (u = u_N)$$

so that (with $V = V_{\rho_N}$)

$$(3.20) \qquad\qquad V(x_1) \leqslant V(x_2) + c\rho_1^{1/\beta} + \int_{r_1}^{r_2} \frac{j^2(m(r))}{r^3}\,dr,$$

where $r_i = r(x_i)$. Also

$$V(x_1) \geqslant \frac{\delta M}{2Aa_*} \qquad\qquad \text{(by Lemma 3.6),}$$

$$V(x_2) \leqslant \frac{CM}{a_*B}\log\frac{1}{B} \qquad \text{[by (2.2)],}$$

and

$$\int_{r_1}^{r_2} \frac{j^2(m(r))}{r^3} dr \leqslant \frac{j_1^2}{2r_1^2} \leqslant \frac{Cj_1^2}{A^2 a_*^2}$$

where

$$j_1 = \max_{0 \leqslant m \leqslant 1} j(m).$$

Thus (3.20) implies that

$$\frac{\delta M}{2Aa_*} \leqslant \frac{CM}{a_* B} \log \frac{1}{B} + \frac{Cj_1^2}{A^2 a_*^2} + C(\rho_* A^{-3})^{1/\beta}.$$

If A and B are sufficiently large (depending on M and j only), then we obtain, using (2.14) and (3.1),

$$\frac{\delta M}{4Aa_*} \leqslant \frac{C\rho_*^{1/3}}{A} A^{(1/3)-(1/\beta)},$$

which is impossible if A is sufficiently large.

Proof of Theorem 3.1. Choose $\rho = \rho_{\bar{N}}$, $\bar{N}$ as in the previous lemmas. It suffices to show that this ρ is a minimizer of $E(\tilde{\rho})$, for all $\tilde{\rho} \in \mathcal{Q}$. Let $\tilde{\rho} \in \mathcal{Q}$ and define

$$\sigma_N(x) = \begin{cases} \tilde{\rho}(x) & \text{if } |x| < Na_* \text{ and } \tilde{\rho}(x) < \frac{1}{2}N\rho_* \\ 0 & \text{otherwise.} \end{cases}$$

Set

$$\tau_N(x) = \theta_N \rho_* I_B(x),$$

where B is a ball of radius a_* about the origin and θ_N is defined by

$$\int_{R^3} \tau_N(x) \, dx = M - \int_{R^3} \sigma_N(x) \, dx.$$

Further, let $\tilde{\rho}_N = \sigma_N + \tau_N$. Since $\theta_N \to 0$ if $N \to \infty$, $\tilde{\rho}_N \in \mathcal{Q}_N$ if N is sufficiently large, say $N \geqslant N(\tilde{\rho})$. Also, it is straightforward to verify that

(3.21) $$E(\tilde{\rho}_N) \to E(\tilde{\rho}) \quad \text{if } N \to \infty.$$

By Lemmas 3.4 and 3.7 we have $\rho_N \in \mathcal{Q}_N^-$ for all $N \geq \overline{N}$, and therefore

$$E(\rho_N^-) \leq E(\rho_N).$$

Since $\tilde{\rho}_N \in \mathcal{Q}_N$, we also have

$$E(\rho_N) \leq E(\tilde{\rho}_N).$$

Thus

$$E(\rho_N^-) \leq \lim_{N \to \infty} E(\tilde{\rho}_N) = E(\tilde{\rho}).$$

Theorem 3.8. *Problem* (E_0) *has a solution* ρ *such that* $\rho = I_G$, *G has bounded support and is given by*

$$G = \{(r, z); -\phi(r) < z < \phi(r), 0 \leq r < \infty\},$$

where $\phi(r)$ *is positive and continuous on an open set.*

The proof is similar to the proof of Theorem 3.1. In fact, it is somewhat simpler since the estimate $\rho_* = 1$ is trivial; for the last assertion, see (1.24).

Consider now compressible fluid with general barotropic pressure–density relationship

$$p = P(\rho).$$

Set

$$\Psi(t) = t \int_0^t \frac{P(s)}{s^2} ds,$$

so that

$$\Psi''(t) = \frac{P'(t)}{t}.$$

The energy functional is

$$E_\Psi(\rho) = -\frac{1}{2} \int_{R^3} \int_{R^3} \frac{\rho(x)\rho(y)}{|x - y|} dx\, dy$$

(3.22)

$$+ \frac{1}{2} \int_{R^3} \frac{j^2(m(r(x)))}{r^2(x)} \rho(x)\, dx + \int_{R^3} \Psi(\rho(x))\, dx$$

and the equilibrium conditions are

(3.23)
$$u = \psi(\rho) \qquad \text{in } G = \{\rho > 0\},$$
$$u \leq 0 \qquad \text{in } R^3 \setminus G,$$

where $\psi(t) = \Psi'(t)$. Assume that $P(t)$ is positive, strictly increasing, and smooth for $t > 0$, and that

(3.24)
$$\liminf_{t \to \infty} \frac{P(t)}{t^{1/3}} = K \qquad (0 < K \leqslant \infty),$$

$$\lim_{t \to 0} \frac{P(t)}{t^{1/3}} = 0.$$

Then $\psi(t)$ is a positive, strictly increasing smooth function for $t > 0$, $\psi(0) = 0$, and

(3.25)
$$\liminf_{t \to \infty} \frac{\psi(t)}{t^{1/3}} = 4K \qquad (0 < K \leqslant \infty),$$

$$\lim_{t \to 0} \frac{\psi(t)}{t^{1/3}} = 0.$$

Set

(3.26)
$$M_0 = \left(\frac{4K}{c^*} \right)^{3/2},$$

where c^* is the universal constant defined in (2.1).

Consider the problem: Find ρ such that

(3.27)
$$E_\Psi(\rho) = \min_{\tilde{\rho} \in \mathcal{Q}} E_\Psi(\tilde{\rho}), \qquad \rho \in \mathcal{Q}.$$

Theorem 3.9. *If $M < M_0$, then there exists a solution ρ of (3.27); ρ is a continuous function with compact support and $\rho(r, z)$ is nonincreasing in z, $z \geqslant 0$.*

The proof is similar to the proof of Theorem 3.1 once we have established an a priori bound of a solution ρ, analogously to Lemma 2.2. From (3.25), (3.26) it follows that there exists a positive number ρ_* such that

(3.28)
$$\psi(\rho_*) = c^* M^{2/3} \rho_*^{1/3}.$$

We can now proceed as in Lemma 2.2 to establish that $\psi(\rho(x)) \leqslant \psi(\rho_*)$.

From now on when we speak of the compressible case we mean a solution of problem (E), although most of the results extend to solutions of problem (3.27), as in Theorem 3.9.

Definition 3.1.　Set

$$q(m) = \frac{j^2(m)}{ma_*M} \qquad (0 < m \leq 1),$$

(3.29)

$$Q = \int_0^1 q(m)\, dm.$$

The fluid is said to be *slowly rotating* if

(3.30)　　　$Q < C_0, q(1) < C_0$　　　for some positive constant C_0

in the compressible case, and

(3.31)　　　$Q < 1, q(1) < C_0$　　　for some positive constant C_0

in the incompressible case.

The physical interpretation of Q and $q(m)$ is given in Problem 4. Introducing

(3.32)　　　$j_0^2 = \int_0^1 \frac{j^2(m)}{m}\, dm, \qquad j_1 = \max_{0 \leq m \leq 1} j(m),$

(3.30) implies that

(3.33)　　　$j_0^2 \leq C_0 a_* M, \qquad j_1^2 \leq C_0 a_* M.$

Theorem 3.10.　*If the fluid is slowly rotating, then there exists a positive constant B such that for all $0 < M < \infty$,*

(3.34)　　　　　　　　$\operatorname{supp} \rho \subset \{x; |x| \leq Ba_*\}.$

This result is valid both for the compressible and the incompressible case. Since a_* is the radius of a ball with density ρ_* and mass M, we also have

(3.35)　　　　　　　　$\operatorname{supp} \rho \supset \{x; |x| \leq a_*\}.$

The proof of Theorem 3.10 is obtained simply by reviewing the proof of Theorem 3.1 and observing that the various constants C are independent of M, j provided that (3.33) holds.

Remark 3.1.　The notation in the proof of Theorem 3.1 can be slightly simplified if we do not keep track of the explicit dependence of the estimates on M and j. However, we did keep the dependence fairly explicit, having in mind applications to asymptotic estimates, such as in Theorem 3.10. In the next section we consider further applications to rapidly rotating fluids.

PROBLEMS

1. Prove (3.21).

2. Prove Theorem 3.9.

3. For degenerate gas ("white dwarfs")

$$P(t) = AF(x) \text{ where } x = (t/B)^{1/3} \text{ and}$$

$$F(x) = x(2x^2 - 3)(x^2 + 1)^{1/2} + 3\sinh^{-1}x$$

and A, B are positive constants. Verify (3.24) with $K < \infty$.

4. Imagine a (hypothetical) ball of radius a_*, uniform density ρ_*, and total mass M rotating with a distribution of angular momentum per unit mass given by $j(m)$. Let T and W be the total kinetic and gravitational potential energies, respectively; and, for $0 < \mu \leq 1$, let T_μ and W_μ be the kinetic and gravitational potential energies of a unit mass situated on the lateral surface of the cylinder of radius r such that $m(r) = \mu$. Show that

$$c_1 Q \leq \frac{T}{|W|} \leq c_2 Q, \qquad c_1 q(\mu) \leq \frac{T_\mu}{|W_\mu|} \leq c_2 q(\mu),$$

where c_1, c_2 are absolute positive constants.

5. Extend Theorems 3.1 and 3.8 to the model in Section 2, Problem 2, assuming $J(r)$ to be nondecreasing, $J(0) = 0$, $J(\infty) < \infty$ and

$$\lim_{r \to \infty} r(J(\infty) - J(r)) = 0.$$

4. RAPIDLY ROTATING FLUIDS

We now drop the assumptions (3.30), (3.31) and obtain a lower bound on the support of the solution, simultaneously for the compressible [problem (E)] and incompressible [problem (E_0)] case.

For any $0 < \mu \leq 1$, set

$$(4.1) \qquad R_\mu = \inf\left\{ R; \int_{\{r(x) < R\}} \rho(x)\, dx = \mu M \right\}.$$

Theorem 4.1. *There exists a positive constant C independent of M, j such that for any $0 < \mu \leq 1$,*

$$(4.2) \qquad C\frac{R_\mu}{a_*}\log\left(1 + C\frac{R_\mu}{a_*}\right) \geq \mu q(\mu).$$

Proof. Set $x_\mu = (R_\mu, 0, 0)$. Then

$$0 \leq u(x_\mu) = V(x_\mu) + f(x_\mu) + \lambda.$$

Since $\lambda \leq 0$, we have

$$(4.3) \qquad\qquad -f(x_\mu) \leq V(x_\mu).$$

By monotonicity of $j(m)$,

$$-f(x_\mu) = \int_{R_\mu}^\infty \frac{j^2(m(r))}{r^3}\,dr \geq \frac{j^2(\mu)}{2R_\mu^2}$$

and, by Lemma 2.1,

$$V(x_\mu) \leq C\frac{M}{R_\mu}\log\left(1 + C\frac{R_\mu}{a_*}\right).$$

Substituting these inequalities into (4.3), the assertion (4.2) follows.

Denote by a_1 the radius of the smallest ball about the origin which contains the fluid domain G. Thus, for a slowly rotating fluid,

$$C_1 \leq \frac{a_1}{a_*} \leq C_2,$$

where C_i are positive constants independent of M, j.

Corollary 4.2. *If $q(1) \geq c_0 Q$ for some positive constant c_0, then for any fixed $\varepsilon_0 > 0$ there holds*

$$(4.4) \qquad \frac{a_1}{a_*} \geq C_1\frac{Q}{\log(1 + Q)} \qquad \text{provided that } Q \geq \varepsilon_0,$$

where C_1 is a positive constant depending only on c_0, ε_0.

Definition 4.1. A fluid is said to be *rapidly rotating* if

$$(4.5) \qquad Q \geqslant \varepsilon_0 \qquad \text{for some positive constant } \varepsilon_0,$$

$$(4.6) \quad c_0 Q \leqslant q(m) \leqslant c_0^{-1} Q \qquad \text{for } 0 < m \leqslant 1 \text{ and a positive constant } c_0.$$

The lower estimate (4.4) is complemented by the following upper estimate:

Theorem 4.3. *If (4.5), (4.6) hold, then*

$$(4.7) \qquad \frac{a_1}{a_*} \leqslant C_2 Q \log (1 + Q),$$

where C_2 is a positive constant depending only on c_0, ε_0.

We shall first prove the theorem in the incompressible case. In this case, (4.5) and the first inequality in (4.6) can be rewritten in the form

$$(4.8) \qquad \frac{j_0^2}{M^{4/3}} \geqslant \varepsilon_0,$$

$$(4.9) \qquad \inf_{0 < m < 1} \frac{j^2(m)}{m} \geqslant c_0 j_0^2,$$

where j_0, j_1 are defined in (3.32).

Lemma 4.4. *In the incompressible case, if (4.8), (4.9) hold, then there exists a positive constant B depending only on ε_0, c_0 such that*

$$(4.10) \qquad \frac{a_1}{M^{1/3}} \leqslant B \log \left(1 + C \frac{a_1}{M^{1/3}} \right) \cdot \max \{ \hat{Q}, q(1) \},$$

where $\hat{Q}$ is determined by

$$(4.11) \qquad C\hat{Q} \log (1 + c\hat{Q}) = A_0 Q$$

and C, A_0 are some positive constant (A_0 is sufficiently large) independent of M, j, ε_0, c_0.

Note, by (4.6), that

$$\max \{ \hat{Q}, q(1) \} \leqslant CQ.$$

Substituting this into (4.10), the assertion (4.7) for the incompressible case follows (recall that $a_* = cM^{1/3}$, $c > 0$). Thus it remains to establish Lemma 4.4.

The proof is based on several lemmas. We let R^* denote the solution of the equation

$$(4.12) \qquad \frac{CR^*}{M^{1/3}} \log\left(1 + \frac{CR^*}{M^{1/3}}\right) = A_0 \frac{j_0^2}{M^{4/3}}.$$

We shall establish the required estimate (4.10) in the form

$$(4.13) \qquad a_1 \leq B \log\left(1 + C\frac{a_1}{M^{1/3}}\right) \cdot \max\{R^*, M^{1/3}q(1)\}.$$

Lemma 4.5. *Under assumptions (4.8), (4.9),*

$$(4.14) \quad \int_G\int_G \frac{dx\,dy}{|x-y|} \geq \int_G \frac{j^2(m(r))}{r^2}\,dx + \frac{CM^2}{R^*}\log\left(1 + \frac{CR^*}{M^{1/3}}\right).$$

Proof. Take $\tilde{\rho} = I_{\tilde{G}}$, where $\tilde{G}$ is a torus

$$z^2 + (r - R)^2 < s^2, \qquad 2\pi^2 s^2 R = M, \qquad s < \frac{R}{2}.$$

By Section 2, Problem 1,

$$\int_{\tilde{G}} \frac{dy}{|x-y|} \geq Cs^2 \log\frac{R}{s} \geq \frac{CM}{R}\log\left(1 + \frac{CR}{M^{1/3}}\right)$$

provided that

$$x = (r, \theta, z), \qquad z^2 + (r - R)^2 < \frac{s}{2}.$$

Hence,

$$\int_{\tilde{G}}\int_{\tilde{G}} \frac{dx\,dy}{|x-y|} \geq \frac{CM^2}{R}\log\left(1 + \frac{CR}{M^{1/3}}\right).$$

Also,

$$\int_{\tilde{G}} \frac{j^2(m(r))}{r^2}\,dx \leq \frac{CM}{R^2}\int_0^1 j^2(m)\,dm \leq \frac{CM}{R^2} j_0^2.$$

It follows that

$$E_0(\tilde{\rho}) \leq -\frac{CM^2}{R}\log\left(1 + \frac{CR}{M^{1/3}}\right) + \frac{CM}{R^2} j_0^2.$$

Choosing $R = R^*$, with A_0 fixed sufficiently large in (4.12), we find that

$$E_0(\tilde{\rho}) \leq -\frac{CM^2}{R^*}\log\left(1 + \frac{CR^*}{M^{1/3}}\right).$$

Using the inequality $E_0(\rho) \leqslant E_0(\tilde{\rho})$, (4.14) follows.

Lemma 4.6. *Assume that* (4.8), (4.9) *hold and set*

$$\mu_A = M^{-1} \operatorname{meas}(G \cap \{x; |x| < AR^*\}).$$

Then there exist positive constants A, δ $(0 < \delta \leqslant 1)$ *independent of* M, j *such that*

(4.15) $$\mu_A \geqslant \delta.$$

Proof.   The proof's outline is the same as for Lemma 3.6. However, in order to carry out the details we must use the kinetic energy in order to control a portion of the gravitational energy corresponding to a mass in a neighborhood of the origin. More specifically, we shall need to establish the inequality

(4.16) $$\iint_{\substack{\{r(x), r(y) < \eta R^*\} \\ \{x, y \in G\}}} \frac{dx\,dy}{|x - y|} \leqslant \int_{\substack{\{r(x) < \eta R^*\} \\ \{x \in G\}}} \frac{j^2(m(r))}{r^2}\,dx$$

for some $\eta > 0$ independent of M, j.
Let

$$0 = r_0 < r_1 < r_2 < \cdots < r_n = \eta R^*$$

be a partition of $[0, R^*]$ such that

(4.17) $$g(r_1) \equiv \operatorname{meas}\{G \cap (r < r_1)\} > 0.$$

Define

$$\Omega_i = \{(x, y); r_{i-1} < r(x) \vee r(y) < r_i\},$$

where $a \vee b = \max\{a, b\}$, and

$$m_i = m(r_i), \qquad \Delta_i m = m_i - m_{i-1}.$$

We can write the left-hand side of (4.16) in the form

$$\sum_{i=1}^n \int_{\Omega_i} \frac{dx\,dy}{|x - y|} = J_1 + \sum_{i=2}^n J_i$$

where

$$J_1 = \iint_{\substack{\{r(x), r(y) < r_1\} \\ \{x, y \in G\}}} \frac{dx\,dy}{|x - y|},$$

$$J_i < 2 \int_{\substack{\{r_{i-1} < r(x) < r_i\} \\ \{x \in G\}}} dx \int_{\substack{\{0 < r(y) < r_i\} \\ \{y \in G\}}} \frac{dy}{|x - y|}, \qquad i \geqslant 2.$$

By (2.4), if $i \geqslant 2$,

$$J_i \leqslant C \int_{\substack{\{r_{i-1}<r(x)<r_i\} \\ \{x \in G\}}} \frac{m_i M}{r(x)} \log \left(1 + \frac{Cr(x)}{M^{1/3}} \right) dx$$

$$\leqslant C \frac{m_i M}{r_{i-1}} (\Delta_i m) M \log \left(1 + \frac{Cr_i}{M^{1/3}} \right).$$

On the other hand, the right-hand side of (4.16) can be written in the form

$$\sum_{i=1}^{n} K_i,$$

where

$$K_i = \int_{\substack{\{r_{i-1}<r(x)<r_i\} \\ \{x \in G\}}} \frac{j^2(m(r))}{r^2} dx \geqslant \frac{j^2(m_{i-1})}{r_0^2} (\Delta_i m) M.$$

Choosing the partition sufficiently refined so that

$$\frac{r_i}{r_{i-1}} \leqslant 2, \qquad \frac{m_i}{m_{i-1}} < 2$$

[note that $m_1 > 0$ by (4.17)], we conclude that $K_i \geqslant J_i$ for $2 \leqslant i \leqslant n$ provided that

$$\frac{r_i}{M^{1/3}} \log \left(1 + \frac{Cr_i}{M^{1/3}} \right) \leqslant c_1 \frac{j^2(\frac{1}{2}m_i)}{m_i M^{4/3}}$$

for some sufficiently small positive constant c_1. In view of (4.9) it suffices to show that

$$\frac{r_i}{M^{1/3}} \log \left(1 + \frac{Cr_i}{M^{1/3}} \right) \leqslant c_2 \frac{j_0^2}{M^{4/3}}$$

with another sufficiently small positive constant c_2. Since $r_i \leqslant \eta R^*$, the last inequality follows from the definition of R^* provided that η is chosen sufficiently small (independent of M, j). We conclude that the left-hand side of (4.16) is smaller than the right-hand side plus the term J_1. Now taking $r_1 \downarrow \tilde{r}$ where

$$\tilde{r} = \inf \{r; g(r) > 0\}$$

(so that $J_1 \downarrow 0$), the proof of (4.16) is complete.

By (2.2) we have

$$\int_{\{|x| \geqslant AR^*\}} V(x)\, dx \leqslant \frac{CM^2}{AR^*} \log\left(1 + \frac{CAR^*}{M^{1/3}}\right).$$

Using both this inequaltiy with A fixed sufficiently large and (4.16) in the inequality (4.14), we deduce that

$$\frac{CM^2}{R^*} \log\left(1 + \frac{CR^*}{M^{1/3}}\right) \leqslant \int_{\substack{\{|x| \vee |y| < AR^*\} \\ \{r(x) \vee r(y) > \eta R^*\} \\ \{x, y \in G\}}} \frac{dx\, dy}{|x - y|}.$$

By (2.4), the right-hand side is smaller than

$$C\frac{(\mu_A M)^2}{\eta R^*} \log\left(1 + \frac{CAR^*}{M^{1/3}}\right).$$

Noting that

$$\log\left(1 + \frac{CAR^*}{M^{1/3}}\right) \leqslant B_1 \log\left(1 + \frac{CR^*}{M^{1/3}}\right),$$

where B_1 is a constant depending only on A, we conclude that $\mu_A > \delta$, where δ is a positive constant independent of $M, j(m)$.

Lemma 4.7. *Under assumptions* (4.8), (4.9),

(4.18) $$\lambda \geqslant \frac{M}{B_0}\left[\max\left\{R^*, M^{1/3}q(1)\right\}\right]^{-1} \qquad [\lambda = u(\infty)],$$

where B_0 is a sufficiently large positive constant depending only on ε_0, c_0.

Proof. Let

(4.19) $$r_0 = B_2 \max\left\{R^*, M^{1/3}q(1)\right\},$$

where B_2 will be taken sufficiently large depending only on ε_0, c_0. Then in the region

$$r_0 \leqslant r \leqslant 2r_0, \qquad |z| \leqslant r_0,$$

there must be a point $\tilde{x}$ that does not belong to G, and hence $u(\tilde{x}) \leqslant 0$. Taking $x = \tilde{x}$ in the relation $u = V(x) + f(r) + \lambda$, we obtain

(4.20) $$\lambda \leqslant -V(\tilde{x}) - f(\tilde{r}) \qquad [\tilde{r} = r(\tilde{x})].$$

By Lemma 4.6,

$$V(\tilde{x}) \geqslant \frac{C\delta M}{r_0} \, ;$$

here we have taken $B_2 \geqslant A$. Also,

$$-f(\tilde{r}) = \int_{\tilde{r}}^{\infty} \frac{j^2(m(r))}{r^3} \, dr \leqslant \frac{C}{r_0^2} j_1^2 = \frac{C}{r_0^2} M^{4/3} q(1).$$

Putting these inequalities into (4.20), we get

$$\lambda \leqslant -\frac{C\delta M}{r_0} + \frac{C}{r_0^2} M^{4/3} q(1) \leqslant -\frac{C\delta M}{2r_0}$$

by (4.19); this gives (4.18).

Proof of Lemma 4.4. Take a point x^1 on the boundary of G such that $|x^1| = a_1$. Since $u(x^1) = 0$, we get

$$|\lambda| = V(x^1) + f(r_1) \qquad [r_1 = r(x^1)]$$

(4.21) $\leqslant V(x^1).$

By (2.2),

$$V(x^1) \leqslant \frac{CM}{a_1} \log\left(1 + \frac{Ca_1}{M^{1/3}}\right).$$

Substituting this and (4.18) into (4.21), the assertion (4.13) follows.

We next consider the compressible version of Theorem 4.3. In this case it is sufficient to establish the following lemma.

Lemma 4.8. *In the compressible case, assume that (4.5) and the first inequality in (4.6) hold. Then there exists a positive constant B depending only on ε_0, c_0 such that*

(4.22) $$\frac{a_1}{a_*} \leqslant B \log\left(1 + C\frac{a_1}{a_*}\right) \cdot \max\{\hat{Q}, q(1)\},$$

where $\hat{Q}$ is determined by

(4.23) $$C\hat{Q} \log(1 + C\hat{Q}) = A_0 Q,$$

and A_0 (*sufficiently large*) *is a positive constant depending only on* ε_0; C *is a constant independent of* M, j, ε_0, c_0.

The proof of Lemma 4.8 proceeds along the same lines as the proof of Lemma 4.4. Here we let R^* be defined by

$$(4.24) \qquad \frac{CR^*}{a_*} \log\left(1 + \frac{CR^*}{a_*}\right) = A_0 Q.$$

As before, we establish several lemmas.

Lemma 4.9. *Under the assumption of Lemma 4.8,*

$$(4.25) \quad \int_{R^3}\int_{R^3} \frac{\rho(x)\rho(y)}{|x-y|}\, dx\, dy \geqslant \int_{R^3} \frac{j^2(m(r))}{r^2} \rho(x)\, dx + C\int_{R^3} \rho^\gamma(x)\, dx$$

$$+ \frac{CM^2}{R^*} \log\left(1 + \frac{CR^*}{a_*}\right).$$

Proof. Take $\tilde{\rho} = \theta\rho_* I_{\tilde{G}}$, where $\tilde{G}$ is the torus

$$z^2 + (r - R)^2 < s^2, \qquad 2\pi^2\theta\rho_* s^2 R = M, \qquad s < \frac{R}{2}.$$

Then, analogously to Lemma 4.5,

$$\int_{R^3}\int_{R^3} \frac{\tilde{\rho}(x)\tilde{\rho}(y)}{|x-y|}\, dx\, dy \geqslant \frac{CM^2}{R} \log\left(1 + \frac{C\theta^{1/3}R}{a^*}\right).$$

Also,

$$\int_{R^3} \frac{j^2(m(r))}{r^2} \tilde{\rho}(x)\, dx \leqslant \frac{CM}{R^2} j_0^2.$$

Finally,

$$\int_{R^3} \tilde{\rho}^\gamma(x)\, dx \leqslant C\left(\theta\rho_*\right)^{1/\beta} M = C\theta^{1/\beta}\rho_*^{1/3} M^{5/3};$$

here we have used $\gamma = 1 + 1/\beta$ and the definition of a_*. We now conclude that

$$E(\tilde{\rho}) \leqslant -\frac{CM^2}{R} \log\left(1 + \frac{C\theta^{1/3}R}{a_*}\right) + \frac{CM}{R^2} j_0^2 + C\theta^{1/\beta}\rho_*^{1/3} M^{5/3}$$

$$(4.26) \qquad = -\frac{CM^2}{R}\left\{\log\left(1 + \frac{C\theta^{1/3}R}{a_*}\right) - \frac{Ca_*}{R}Q - \frac{CR}{a_*}\theta^{1/\beta}\right\}.$$

If we take θ according to $\theta^{-\alpha} = C_1 R/a_*$ where $\beta < \alpha < 3$ and C_1 is fixed sufficiently large, then the first term on the right-hand side of (4.26) majorizes the third term. Next, we take $R = R^*$ with A_0 sufficiently large to ensure that the condition $s < R/2$ holds and that the first term on the right-hand side of (4.26) majorizes the second. Recalling that $E(\rho) \leqslant E(\tilde{\rho})$, (4.25) follows.

Lemma 4.10. *Assume that the conditions of Lemma 4.8, and set*

$$\mu_A = M^{-1} \int_{\{|x| < AR^*\}} \rho(x)\, dx.$$

Then there exist positive constants A, δ $(0 < \delta \leqslant 1)$ independent of M, j such that

$$(4.27) \qquad\qquad\qquad \mu_A \geqslant \delta.$$

Proof. The proof is similar to the proof of Lemma 4.6. We first need to establish the estimate

$$\iint\limits_{\{r(x),\, r(y) < \eta R^*\}} \frac{\rho(x)\rho(y)}{|x - y|}\, dx\, dy \leqslant \int_{\{r(x) < \eta R^*\}} \frac{j^2(m(r))}{r^2} \rho(x)\, dx.$$

Using a partition as in the proof of Lemma 4.6, we then find that the only point that needs to be verified is that

$$\frac{r_i}{a_*} \log\left(1 + \frac{Cr_i}{a_*}\right) \leqslant c_2 Q$$

for $r_i < \eta R^*$; c_2 is a sufficiently small positive constant. As before, this follows from the definition of R^*.

We next proceed to verify the inequality

$$\frac{CM^2}{R^*} \log\left(1 + \frac{CR^*}{a_*}\right) < \int_{\substack{\{|x| \vee |y| > AR^*\} \\ \{r(x) \vee r(y) > \eta R^*\}}} \frac{\rho(x)\rho(y)}{|x - y|}\, dx\, dy,$$

where A is fixed sufficiently large.

Finally, the proof is completed by applying the second estimate of Lemma 2.1 to the right-hand side of the last inequality.

Lemma 4.11. *Under the assumption of Lemma 4.8,*

$$(4.28) \qquad\qquad |\lambda| \geqslant \frac{M}{B_0} \big[\max\{R^*, a_* q(1)\} \big]^{-1}.$$

where B_0 is a sufficiently large positive constant depending only on ε_0, c_0.

Proof. Let

$$r_0 = B_1 \max \{R^*, a_* q(1)\},$$

where B_1 is a sufficiently large positive constant. In the region

$$r_0 \leqslant r \leqslant 2r_0, \qquad |z| \leqslant r_0$$

there is a point $\tilde{x}$ where $\rho(x)$ attains its minimum. Clearly,

$$\rho(\tilde{x}) r_0^3 \leqslant CM,$$

so, by (1.10),

$$C \left(\frac{M}{r_0^3} \right)^{1/\beta} \geqslant u(\tilde{x}) = V(\tilde{x}) + f(\tilde{r}) + \lambda \qquad [\tilde{r} = r(\tilde{x})].$$

Hence

$$\lambda \leqslant -\frac{C\delta M}{r_0} + \frac{C}{r_0^2} a_* M q(1) + C \left(\frac{M}{r_0^3} \right)^{1/\beta}$$

$$= -\frac{C\delta M}{r_0} \left\{ 1 - \frac{C}{\delta} \frac{a_*}{r_0} q(1) - \frac{C}{\delta} \left(\frac{a_*}{r_0} \right)^{3/\beta - 1} \right\};$$

here we have computed the last term using (2.14), (3.1). Recalling that $R^* > A_1 a_*$ for some positive constant A_1 depending only on ε_0, the assertion (4.28) now follows from the definition of r_0.

The proof of Lemma 4.8 can now be completed in the same way as Lemma 4.4.

From Corollary 4.2 and Theorem 4.3 we have the asymptotic estimate

$$\frac{Q}{C \log Q} \leqslant \frac{a_1}{a_*} \leqslant CQ \log Q \qquad \text{as } Q \to \infty.$$

Naturally, one expects the fluid to be contained in a thin slab about the equatorial plane as the "rapidity of rotation" Q becomes large. The set occupied by the fluid has the form

(4.29) $$G = \{(r, z); |z| < \phi(r), 0 \leqslant r < \infty\}.$$

For any $0 < \mu < 1$, define

$$h_\mu = \max_{r > R_\mu} \phi(r),$$

where R_μ is defined in (4.1). The number h_μ/R_μ is called the *oblateness ratio*.

Theorem 4.12. *If (4.6) holds and Q is sufficiently large, then*

$$(4.30) \qquad \frac{h_\mu}{R_\mu} \leq C_\mu Q^{-p_1} [\log(1 + Q)]^{p_2},$$

where

$$p_1 = \frac{3 - \beta}{2(1 + 2\beta)}, \qquad p_2 = \frac{3(1 + 4\beta)}{2(1 + 2\beta)}$$

in both the compressible case ($0 < \beta < 3$) and the incompressible case ($\beta = 0$); C_μ depends on μ and c_0.

For the incompressible case, (4.30) follows from the following estimate: Set $h(R) = \sup_{r > R} \phi(r)$. If $R > A_0 M^{1/3}$, then

$$(4.31) \qquad \frac{h}{R} \leq C_0 \left(\frac{R}{M^{1/3}} \right)^{-3/2};$$

here A_0 (sufficiently large) and c_0 are absolute constants.

The proof of Theorem 4.12 and (4.31) depend on upper bounds on $|\nabla V|$ and lower bounds on $|V_z|$; for details, see reference 100b.

PROBLEMS

1. Prove that (4.31) implies (4.30) in the incompressible case.

2. Prove that the assertion of Theorem 3.9 remains valid for $M \geq M_0$ provided that $j(m) > 0$ for all $0 < m \leq 1$.
   [*Hint*: Suffices to derive a priori bound on $\rho_0 = \sup \rho : \rho_0 \leq \rho_* = \rho_*(M, j)$. Set $A = R_\mu/a_\mu$, $0 < \mu < 1$ and show for $A > 1$,

   $$\sup V \leq c^* M^{3/2} \rho_0^{1/3} \left\{ \mu^{2/3} + \frac{Ca_0}{R_\mu} \left(\log 1 + \frac{CR_\mu}{u} \right) \right\}.$$

Use $\psi(\rho_0) \leqslant \sup V$ to deduce that

(*) $\qquad \frac{1}{2} c^* M_0^{2/3} \rho_0^{1/3} < c^* M^{2/3} \rho_0^{1/3} \left\{ \mu^{2/3} + \frac{Ca_0}{R_\mu} \log \left(1 + \frac{CR_\mu}{a_0} \right) \right\}$

if $\rho > \bar{\rho}$, where

$$\frac{\psi(t)}{t^{1/3}} \geqslant \tfrac{1}{2} c^* M_0^{2/3} \qquad \forall\, t \geqslant \bar{\rho}.$$

Fix $\mu = M_0/2M$ in (*) and use also (4.2) to deduce

$$\frac{j^2(M_0/3M)}{a_0 M} \leqslant C\hat{A} \log(1 + C\hat{A})$$

for some $\hat{A} > \max(A, 1)$,

$$\left(\frac{M_0}{M} \right)^{2/3} = \frac{C}{\hat{A}} \log(1 + C\hat{A}),$$

or

(**) $\qquad \rho_0^{1/3} \leqslant \frac{CM^{4/3}}{j^2(M_0/3M)} \hat{A} \log(1 + \hat{A}).$

If $A < 1$, then (4.2) gives a stronger inequality.]

3. Set $j_\varepsilon(m) = \varepsilon j(m)$ ($j(m) > 0$ if $m > 0$) and denote by ρ_ε a solution corresponding to $j_\varepsilon(m)$, M (constructed by Problem 1). Let $a_1(\varepsilon) = \sup \{|x| ; \rho_\varepsilon(x) > 0\}$. Prove: If $M \geqslant C_0 M_0$, then

$$\frac{a_1(\varepsilon)}{a_*} \leqslant C_1 \varepsilon^2,$$

where C_0, C_1 are positive constants depending only on Q and $q(1)$. [*Hint*: Follow the proof of Theorem 4.1 in the compressible case.]

4. Establish the estimate: If $R > AM^{1/3}$, then

$$|\nabla V(x)| \leqslant C \left(\frac{M}{R} \right)^{1/2} \qquad [x = (R, 0, Z)]$$

in the incompressible case; A is sufficiently large and C depending on A.

[*Hint:* $|\nabla V| \leq \int_G |x - y|^{-2} \, dy$, $G = G' \cup G''$, $G'' = G \cap \{|R - r(y)| < R/2\}$. Use

$$\int_0^{2\pi} \frac{d\theta}{a - b \cos \theta} = \frac{\pi}{(a^2 - b^2)^{1/2}}$$

to estimate

$$\int_{G'} \frac{dy}{|x - y|^2} \leq C \iint_{\substack{\{|R-r|<R/2\} \\ \{|z| \leq \phi(r)\}}} \frac{dr \, dz}{\left[(z - Z)^2 + (r - R)^2\right]^{1/2}}.$$

Also,

$$\iint_{\substack{\{|R-r|<R/2\} \\ \{|z|<\phi(r)\}}} dr \, dz \leq \frac{CM}{R}.$$

Consider a corresponding maximization problem; see Lemma 2.1.]

5.   THE RINGS OF ROTATING FLUIDS

Set

$$Lu = \frac{\partial^2 u}{\partial r^2} + \frac{1}{r} \frac{\partial u}{\partial r} + \frac{\partial^2 u}{\partial z^2}.$$

If ρ is a solution established in Theorem 3.1, then

(5.1) $u = V + f + \lambda$

is Hölder continuous and, since

(5.2) $\rho = \left(\frac{u^+}{c}\right)^\beta,$

also ρ is Hölder continuous. Applying the Laplacian Δ (in R^3) to (5.1), we obtain

(5.3) $Lu + \gamma(u) = h(r),$

where

(5.4) $\gamma(u) = c_0(u^+)^\beta \qquad (c_0 > 0)$

and

$$(5.5) \qquad h \equiv Lf = \frac{2\pi}{Mr^2}(j^2)'(m(r)) \int \rho(r, z)\, dz - \frac{2}{r^4}j^2(m(r)).$$

We assume in this section that $j^2(m)$ is in $C^{1+\beta}[0, 1]$. Since

$$(5.6) \qquad |m(r) - m(s)| \leqslant C|r - s|, \qquad m(r) \leqslant Cr^2$$

we deduce from (5.5) that h is Hölder continuous (exponent β) for $r > 0$. From (5.3) and elliptic estimates, we then conclude that $u \in C^{2+\beta}$ away from the z-axis.

Recall that u is strictly decreasing in z for $z > 0$: Further, since $u_z \leqslant 0$ if $z > 0$, we obtain from (5.3)

$$Lu_z \geqslant 0 \text{ if } z > 0 \qquad \text{(in the distribution sense)}$$

and the strong maximum principle gives

$$(5.7) \qquad u_z < 0 \qquad \text{if } z > 0.$$

The fluid region G is given by (4.29). Using (5.7), we find that $\phi(r)$ is uniformly continuous in every set $\{0 \leqslant r \leqslant R; \phi(r) > \varepsilon\}$, where $R > 0$, $\varepsilon > 0$. It follows that

$$(5.8) \qquad \phi(r) \text{ is continuous for } 0 \leqslant r < \infty.$$

Definition 5.1. Let $\{r > 0; \phi(r) > 0\} = \cup_i \lambda_i$, where λ_i are disjoint open intervals (a_i, b_i) and $\phi(a_i) = \phi(b_i) = 0$ [If $a_i = 0$ then $\phi(a_i) \geqslant 0$.]. The sets

$$G_{\lambda_i} = \{(r, z); |z| < \phi(r), r \in \lambda_i\}$$

are called the *rings* and λ_i is called the *base* of the ring G_{λ_i}.

Theorem 5.1. *If $h(r)$ (in (5.3)) is analytic in r, $r \in \lambda_i$, then $\phi(r)$ is analytic for $r \in \lambda_i$.*
For the model in Section 2, Problem 2, $h(r)$ is analytic in r if $J(r)$ is analytic in r.

Proof. Take $\lambda = \lambda_i$, $a = a_i$, $b = b_i$. Since

$$u(r, \phi(r)) = 0, \qquad u_z(r, \phi(r)) \neq 0$$

and $u \in C^{2+\beta}$, the implicit function theorem can be applied to deduce that $\phi \in C^{2+\beta}(\lambda)$. To prove analyticity we follow the method of Chapter 2, Section 1, using the hodograph transformation

$$y_1 = x_1, \qquad y_2 = x_2, \qquad y_3 = u(r, z).$$

Since $u_z \neq 0$, we can solve $z = w(y)$. Set $y' = (y_1, y_2)$ and define functions ψ, η by

$$\psi(y) = w(y) \qquad \text{if } y_3 \geq 0,$$

$$\eta(y', y_3) = w(y', -y_3) \qquad \text{if } y_3 \geq 0.$$

Then

$$\psi = \eta, \quad \psi_{y_3} = -\eta_{y_3} \quad \text{on } y_3 = 0.$$

Further ψ and η satisfy, for $y_3 > 0$, the nonlinear elliptic system [see Chapter 2, (1.27)]

$$F(\psi) = h(r) - \gamma(y_3) \qquad \left[r = \left(y_1^2 + y_2^2 \right)^{1/2} \right],$$

$$F(\eta) = -h(r) + \gamma(-y_3),$$

where

$$F(\zeta) = -\frac{\zeta_{33}}{(\zeta_3)^3} + \sum_{i=1}^{2} \left[-\left(\frac{\zeta_i}{\zeta_3} \right)_i + \frac{\zeta_i}{\zeta_3} \left(\frac{\zeta_i}{\zeta_3} \right)_3 \right]$$

and $\zeta_i = \partial \zeta / \partial y_i$. By standard elliptic methods [3b] one can then show that the successive derivatives

$$D_{y'}^m \psi, \qquad D_{y'}^m \eta$$

exist and are continuous; they can be estimated by

$$C_0 C^m (m-2)! \qquad \forall\, m \geq 2.$$

Thus ψ, η are analytic in y'. Since the free boundary $z = \phi(r)$ is given by

$$z = w(y_1, y_2, 0),$$

the analyticity of ϕ follows.

We wish to prove that the number of rings is finite, but first we need two lemmas.

Denote by B_r the ball in R^3 with radius r centered about the origin.

Lemma 5.2. *Let β be a positive noninteger, $\beta \geq \beta_0 > 0$, and let $v(x)$ be a function satisfying:*

(5.9) $$|\Delta v(x)| \leq C_\beta |x|^\beta \text{ in } B_1, \qquad C_\beta \geq 2^\beta.$$

Then

$$(5.10) \qquad v(x) = P(x) + \Gamma(x) \qquad in \ B_1,$$

where $P(x)$ is a harmonic polynomial of degree $[\beta] + 2$ and

$$(5.11) \qquad |\Gamma(x)| \leqslant CC_\beta \frac{\beta}{\langle\beta\rangle} |x|^{\beta+2} \qquad in \ B_1,$$

$$(5.12) \qquad |\nabla\Gamma(x)| \leqslant CC_\beta \frac{\beta^3}{\langle\beta\rangle} |x|^{\beta+1} \qquad in \ B_1,$$

where $\langle\beta\rangle = \min\{\beta - [\beta], 1 + [\beta] - \beta\}$;

C *is a constant depending only on β_0, and on upper bounds on $|v(x)|$, $|\nabla v(x)|$ for $x \in \partial B_1$.*

Proof. Let $e_1 = (1, 0, 0)$. Then (see reference 120, p. 125)

$$\frac{1}{|x - e_1|} = \sum_{n=1}^{\infty} |x|^n P_n(\cos\theta),$$

where $\cos\theta = x_1/|x|$ and $P_n(u)$ is the Legendre polynomial of order n. Notice that

$$\Gamma_n(x) = |x|^n P_n(\cos\theta)$$

is a homogeneous harmonic polynomial of degree n. Since (see reference 120, p. 128)

$$|P_n(u)| \leqslant 1 \qquad if \ |u| \leqslant 1,$$

we have

$$|\Gamma_n(x)| \leqslant |x|^n.$$

More generally, if $|x| < |y|$,

$$\frac{1}{|x - y|} = \frac{1}{|y|} \frac{1}{\left|\frac{x}{|y|} - \frac{y}{|y|}\right|} = \frac{1}{|y|} \sum_{n=1}^{\infty} \Gamma_n\left(M_y^{-1}\left(\frac{x}{|y|}\right)\right),$$

where M_y is an orthogonal transformation that maps e_1 into $y/|y|$. Thus

$$(5.13) \qquad \frac{1}{|x - y|} = \sum_{n=1}^{\infty} \frac{1}{|y|^{n+1}} \Gamma_n^y(x) \qquad (|x| < |y|),$$

where $\Gamma_n^y(x)$ is again a homogeneous harmonic polynomial in x of degree n, and

$$(5.14) \qquad\qquad |\Gamma_n^y(x)| \leqslant |x|^n.$$

Now, by Green's formula,

$$(5.15) \qquad 4\pi v(x) = -\int_{\partial B_1} \left[v(y) \frac{\partial}{\partial \nu} \frac{1}{|x-y|} - \frac{1}{|x-y|} \frac{\partial}{\partial \nu} v(y) \right] dS_y$$

$$- \int_{B_1} \frac{1}{|x-y|} \Delta v(y)\, dy$$

$$\equiv J_1 - J_2.$$

We break up the last integral into two integrals:

$$(5.16) \qquad J_2 = \int_{B_1 \cap B_{(1+1/\beta)|x|}} \frac{1}{|x-y|} \Delta v(y)\, dy$$

$$+ \int_{B_1 \setminus B_{(1+1/\beta)|x|}} \frac{1}{|x-y|} \Delta v(y)\, dy$$

$$\equiv I_1 + I_2.$$

In I_1 we have

$$|\Delta v(y)| \leqslant C_\beta \left(1 + \frac{1}{\beta}\right)^\beta |x|^\beta,$$

so that

$$|I_1| \leqslant CC_\beta |x|^{\beta+2}$$

(we use here the fact that the integral of $1/|x-y|$ taken over a ball with center 0 is smaller than the integral taken over a ball with the same radius but with center y).

For $y \in B_1 \setminus B_{(1+1/\beta)|x|}$, the series

$$\sum \frac{1}{|y|^{n+1}} \Gamma_n^y(x) \Delta v(y)$$

is absolutely and uniformly convergent; we can therefore integrate it term by term. For $n < \beta + 2$ we shall add to the nth term

$$I_2^n = \int_{B_1 \setminus B_{(1+1/\beta)|x|}} \frac{1}{|y|^{n+1}} \Gamma_n^y(x) \Delta v(y)\, dy$$

the integral

$$I_3^n = \int_{B_1 \cap B_{(1+1/\beta)|x|}} \frac{1}{|y|^{n+1}} \Gamma_n^y(x) \Delta v(y) \, dy.$$

The sum

$$Q_n(x) = I_2^n + I_3^n$$

is a homogeneous polynomial of degree n. Also, by (5.9), (5.14),

$$|I_3^n| \leq |x|^n C_\beta \int_{B_1 \cap B_{(1+1/\beta)|x|}} \frac{|y|^\beta}{|y|^{n+1}} \, dy$$

$$\leq C_\beta |x|^n \frac{C}{\beta - [\beta]} \left(1 + \frac{1}{\beta}\right)^{\beta - n + 2} |x|^{\beta - n + 2}$$

where C is a generic constant depending only on β_0. Hence

(5.17) $$\sum_{n < \beta + 2} |I_3^n| \leq C_\beta \frac{C(\beta + 2)}{\beta - [\beta]} |x|^{\beta + 2}.$$

The terms I_2^n with $n > \beta + 2$ are estimated by

$$|I_2^n| \leq |x|^n C \int_{R^3 \setminus B_{(1+1/\beta)|x|}} \frac{1}{|y|^{n+1}} |y|^\beta \, dy$$

$$\leq C_\beta |x|^n \frac{C}{1 + [\beta] - \beta} \left(\left(1 + \frac{1}{\beta}\right)|x|\right)^{\beta - n + 2}$$

$$\leq C_\beta |x|^{\beta + 2} \frac{C}{1 + [\beta] - \beta} \left(1 + \frac{1}{\beta}\right)^{\beta + 2} \left(1 + \frac{1}{\beta}\right)^{-n}.$$

Summing over n, we obtain the bound

$$C_\beta |x|^{\beta + 2} \frac{C}{\beta - [\beta]} (\beta + 1).$$

Collecting the estimates above, we conclude that

(5.18) $$-J_2 = P_0(x) + \Gamma_0(x),$$

where P_0 is a harmonic polynomial of degree $[\beta] + 2$ and Γ_0 is bounded by the right-hand side of (5.11).

From (5.13) and the definition of $\Gamma_n^y(x)$, we obtain an expansion for

$$\frac{\partial}{\partial y_i} \frac{1}{|x - y|}$$

in terms of harmonic polynomials. Since (see reference 120, p. 128)

(5.19) $$|P_n'(u)| \leqslant \frac{n(n + 1)}{2},$$

we find that

(5.20) $$J_1(x) = \sum_{n=0}^{\infty} \sigma_n(x),$$

where $\sigma_n(x)$ is a homogeneous harmonic polynomial of degree n and

(5.21) $$|\sigma_n(x)| \leqslant C\delta^n |x|^n \quad \text{(for any } \delta > 1\text{)};$$

C depends on δ and on upper bounds on $|v|$, $|\nabla v|$ in ∂B_1.
   If $|x| \leqslant \rho$, where $\rho\delta < 1$, then

$$\left| \sum_{n > \beta+2} \sigma_n(x) \right| \leqslant C \sum_{n > \beta+2} \delta^n |x|^n \leqslant C'\delta^\beta |x|^{\beta+2},$$

where C' depends on C, δ, ρ. If $\rho < |x| < 1$, then

$$\left| \sum_{n > \beta+2} \sigma_n(x) \right| = \left| J_1 - \sum_{n < \beta+2} \sigma_n(x) \right| \leqslant C + \sum_{n < \beta+2} |\sigma_n(x)|$$

$$\leqslant C + C \sum_{n < \beta+2} \delta^n \leqslant C'\delta^{\beta+2} \leqslant C'\left(\frac{\delta}{\rho}\right)^{\beta+2} |x|^{\beta+2}$$

with another constant C'.
   Choosing $\delta = \frac{5}{4}$, $\rho = \frac{3}{4}$, we conclude that

(5.22) $$\left| \sum_{n > \beta+2} \sigma_n(x) \right| \leqslant C2^\beta |x|^{\beta+2} \quad \text{if } |x| < 1.$$

Setting

$$P(x) = P_0(x) + \sum_{n < \beta+2} \sigma_n(x),$$

and recalling (5.15), (5.18), (5.20), the assertion (5.11) follows for the function $\Gamma(x)$ defined by (5.10).

From the definition of $\Gamma_n^y(x)$ and (5.19), it follows that

$$|\nabla_x \Gamma_n^y(x)| \leqslant Cn^2 |x|^{n-1}.$$

Using this, we can estimate $\nabla_x \Gamma_0(x)$ in the same way that we have estimated $\Gamma_0(x)$ above, with some minor changes. We find that (5.12) holds for Γ replaced by Γ_0. Since $|P_n''(u)| \leqslant Cn^3$ [which follows by induction, using the relation $P_n'(u) - uP_{n-1}'(u) = nP_{n-1}(u)$; see reference 120, p. 127] we can obtain for $\nabla_x J_1(x)$ an expansion similar to (5.20) with the corresponding $\sigma_n(x)$ still satisfying (5.21). Estimating the remainder term as before, the proof of (5.12) follows.

The next lemma is a nondegeneracy lemma in the spirit of unique continuation. The lemma will be used in Section 17, but a basic part of the proof will be needed in this section.

Lemma 5.3. *Let $v(x)$ be a bounded function in B_1 ($B_1 \subset R^3$) satisfying*

$$(5.23) \qquad |\Delta v(x)| \leqslant C |v(x)|^\alpha \qquad (C > 0, \alpha > 1),$$

$$(5.24) \qquad v(0) = 0,$$

$$(5.25) \qquad v(x) \not\equiv 0.$$

Then there exists a homogeneous harmonic polynomial $H_n(x) \not\equiv 0$ of a precise degree $n \geqslant 1$ such that, in B_1,

$$(5.26) \qquad v(x) = H_n(x) + K(x),$$

where

$$(5.27) \qquad |K(x)| \leqslant \tilde{C} |x|^{n+\delta},$$

$$(5.28) \qquad |\nabla K(x)| \leqslant \tilde{C} |x|^{n+\delta-1} \qquad (\delta > 0, \tilde{C} > 0).$$

Proof. Without loss of generality, we may assume that $1 < \alpha < 2$. By the L^p elliptic estimates, v is in C^1. Therefore, $|v(x)| \leqslant C_0 |x|$. It follows that

$$|\Delta v(x)| \leqslant CC_0^\alpha |x|^\alpha.$$

By Lemma 5.2,

$$v(x) = P_1(x) + \Gamma_1(x) \qquad \text{in } B_1,$$

where $P_1(x)$ is a harmonic polynomial of degree $[\alpha] + 2 = 3$ and

$$|\Gamma_1(x)| \leqslant C_1 |x|^{\alpha+2} \qquad \text{in } B_1.$$

If $P_1(x) \neq 0$, then the assertion of the lemma follows. Otherwise, v coincides with Γ_1, so that

$$|\Delta v(x)| \leqslant CC_1 |x|^{\alpha(\alpha+2)} \qquad \text{in } B_1.$$

Applying Lemma 5.2 once again, we conclude that, provided that $\alpha(\alpha + 2) \neq$ integer,

$$v(x) = P_2(x) + \Gamma_2(x) \qquad \text{in } B_1,$$

where $P_2(x)$ is a harmonic polynomial of degree $\alpha_1 = [\alpha(\alpha + 2)] + 2$ and

$$|\Gamma_2(x)| \leqslant C_2 |x|^{\alpha(\alpha+2)+2};$$

if $\alpha(\alpha + 2)$ is an integer, then the above remains true with $\alpha(\alpha + 2)$ replaced by $\alpha(\alpha + 2) - \varepsilon$ for any $\varepsilon > 0$. If $P_2(x) \neq 0$, then the proof of the lemma follows. Otherwise, we apply Lemma 5.2 once again.

Proceeding in this way step by step, we conclude that if the assertion of the lemma is not true, then

$$(5.29) \qquad |\Delta v(x)| \leqslant C_{\beta_m} |x|^{\beta_m} \qquad \text{in } B_1,$$

$$(5.30) \qquad |v(x)| \leqslant \frac{CB_m}{\langle \beta_m \rangle} C_{\beta_m} |x|^{\beta_m} \qquad \text{in } B_1,$$

where $C_{\beta_m} \geqslant 2^{\beta_m}$ and $\beta_m \uparrow \infty$. Notice that these inequalities hold with β_m such that

$$(5.31) \qquad \beta_m = (\beta_{m-1} + 2)\alpha - \theta_m,$$

where $\theta_m \geqslant 0$ and β_m is not an integer. We can choose the θ_m so that $0 \leqslant \theta_m < 1$ and

$$(5.32) \qquad \langle \beta_m \rangle \geqslant c, \qquad c > 0 \qquad (c \text{ independent of } m)$$

for all m.

We shall estimate the β_m, C_{β_m} by induction. The inductive assumption is that

$$(5.33) \qquad \beta_m \geqslant A\alpha^m \qquad (A > 0),$$

$$(5.34) \qquad \beta_m \leqslant B\alpha^m \qquad (B > 0),$$

$$(5.35) \qquad C_{\beta_m} \geqslant 2^{\beta_m},$$

$$(5.36) \qquad C_{\beta_m} = K_0 K^{\alpha^m D_m} \qquad (K_0 > 1, K > 1, D_m < D)$$

for all $m \leqslant n$, where $D_m = c\sum_{i=1}^{m} i/\alpha^i$, $c > 0$.

From (5.31) with $m = n + 1$, we then get

$$\beta_{n+1} \geq (\beta_n + 2)\alpha - 1 \geq A\alpha^{n+1} + 2\alpha - 1 > A\alpha^{n+1},$$

so that (5.33) holds for $m = n + 1$. Next taking the logarithm in the inequality

$$\beta_{m+1} \leq (\beta_m + 2)\alpha,$$

we get

$$\log \beta_{m+1} \leq \log \alpha + \log \beta_m + \frac{2}{\beta_m}.$$

Summing over m, $1 \leq m \leq n$, and using (5.33),

$$\log \beta_{n+1} \leq n \log \alpha + \log \beta_1 + \sum_{m=1}^{n} \frac{2}{A\alpha^m} \leq n \log \alpha + C.$$

This gives (5.34) for $m = n + 1$.

Since (5.31) holds with $m = n$, $m = n + 1$, and since (5.29) and (5.34) hold with $m = n$, Lemma 5.2 gives

$$|\Delta v(x)| \leq \left(CC_{\beta_n}\beta_n\right)^\alpha |x|^{\beta_{n+1}} \qquad \text{in } B_1.$$

Thus we can take $C_{\beta_{n+1}}$ to be any positive constant satisfying

(5.37) $$C_{\beta_{n+1}} \geq \left(CC_{\beta_n}\beta_n\right)^\alpha.$$

It remains to show that if

$$C_{\beta_{n+1}} = K_0 K^{\alpha^{n+1} D_{n+1}},$$

then both (5.37) and (5.35) (with $m = n + 1$) are satisfied. Now, by (5.34) with $m = n + 1$,

$$2^{\beta_{n+1}} \leq 2^{B\alpha^{n+1}} \leq C_{\beta_{n+1}} \qquad \text{if } K \geq 2, \ D_{n+1} \geq B.$$

As for (5.37), one can easily verify it by using (5.34) with $m = n + 1$, provided that c in the definition of the D_m is sufficiently large, depending on K_0, K, α. From (5.30), (5.32), (5.33), (5.34), (5.36) we obtain the inequality

$$|v(x)| \leq CB\alpha^{n+1}K^{D\alpha^n}|x|^{A\alpha^{n+1}+2}$$

in B_1. If $|x| < (1/K)^{D/A\alpha}$, then

$$|v(x)| \le C\rho^{\alpha^n}, \qquad \text{where } \rho = \rho(x) < 1.$$

Taking $n \to \infty$, we conclude that $v(x) \equiv 0$ if $|x| < (1/K)^{D/A\alpha}$.

Let ρ_0 be the largest number ≤ 1 such that $v(x) = 0$ in B_{ρ_0}. If $\rho_0 < 1$, then we can repeat the previous argument and show that, for some $\rho_1 > 0$, $v \equiv 0$ in any ball with center in B_{ρ_0} and radius ρ_1. Since this contradicts the definition of ρ_0, it follows that $\rho_0 = 1$, that is, $v \equiv 0$ in B_1.

Remark 5.1. From (5.30) we see that v vanishes at $x = 0$ to any order. By well-known results on unique continuation (see references 15, 73) we can then deduce right away that $v \equiv 0$, even if $\alpha = 1$. However, the argument given above will be needed in the sequel.

Definition 5.2. A number $R \ge 0$ is called a *point of accumulation* of rings if there exists a sequence of rings $G_{\lambda_{j'}}$ such that

$$\text{dist}\left(\{R\}, G_{\lambda_{j'}}\right) \to 0 \qquad \text{if } j' \to \infty.$$

Theorem 5.4. *The number of rings is finite in every interval $\{\varepsilon < r < \infty\}$, $\varepsilon > 0$.*

Proof. If the assertion is not true, then there is a number $R > 0$ which is a point of accumulation of rings. Setting

$$d = d(x) = \left((r - R)^2 + z^2\right)^{1/2} \qquad \text{if } x = (r, \theta, z),$$

we shall prove by induction that u satisfies

(5.38) $$\gamma(u(x)) \le C_{\beta_m} d^{\beta_m} \qquad \left(d \le d_0, C_{\beta_m} \ge N^{\beta_m}\right);$$

the constants β_m, C_{β_m} will be determined in the inductive proof, and d_0, N are positive constants independent of m.

Writing

$$\Delta(u - f) = -\gamma(u),$$

we can apply Lemma 5.2 to deduce the expansion

(5.39) $$u - f = P_m + Q_m,$$

where P_m is a polynomial of degree $[\beta_m] + 2$ and Q_m satisfies

(5.40) $$|Q_m(x)| \le CC_{\beta_m}\beta_m d^{\beta_m + 2}, \qquad C \text{ independent of } m.$$

Here β_m is required to be any positive noninteger such that

(5.41) $\beta_m - [\beta_m] \geq c,$ $c > 0,$ c independent of $m.$

Using the relations (5.2), (5.4), and (5.38), we have

$$|m(R) - m(r)| = \left| 2\pi \int_r^R s \int \rho(s, z) \, dz \, ds \right|$$

$$\leq C \left| \int_r^R \gamma(u(s, 0)) \, ds \right|$$

$$\leq C C_{\beta_m} d^{\beta_m + 1}.$$

Writing

$$f(r) = f(R) + \int_R^r j^2(m(R)) \frac{ds}{s^3} + \int_R^r (j^2)'(m(\tilde{R}))(m(s) - m(R)) \frac{ds}{s^3}$$

where $\tilde{R}$ lies in the interval with endpoints $R, r,$ and expanding $1/s^3$ in the first integral in powers of $s - R,$ we find that

(5.42) $$f(r) = \sum_{j=0}^{[\beta_m]+2} a_j(r - R)^j + O(|r - R|^{\beta_m + 2}).$$

Since u and f depend only on $r, z,$ the same must be true of P_m; that is, $P_m = P_m(r, z).$ From (5.39) it follows that also $Q_m = Q_m(r, z).$ Taking $z = 0$ in (5.39) and using (5.42), we get

$$u(r, 0) = \sum_{j=0}^{[\beta_m]+2} b_j(r - R)^j + Q_m(r, 0) + O(|r - R|^{\beta_m + 2}).$$

The function $u(r, 0)$ oscillates an infinite number of times as $r \to R$ from one side. This implies that all the coefficients b_j must vanish. Noting that

$$O(|r - R|^{\beta_m + 2}) \leq C_0 C_{\beta_m} |r - R|^{\beta_m + 2},$$

where C_0 is independent of m and choosing $N > C_0,$ we obtain, after using (5.40),

(5.43) $u(r, 0) \leq C_1 C_{\beta_m} \beta_m |r - R|^{\beta_m + 2},$ C_1 independent of $m,$

provided that $\beta_m \geq 1.$ Therefore,

$$\gamma(u(r, z)) \leq \gamma(u(r, 0)) \leq c_0 (C_1 C_{\beta_m} \beta_m |r - R|^{\beta_m + 2})^\beta.$$

Thus the inductive estimate for $m + 1$ follows with

(5.44)
$$C_{\beta_{m+1}} = c_0 \big(C_1 C_{\beta_m} \beta_m \big)^\beta,$$
$$\beta_{m+1} = \beta(\beta_m + 2) - \theta_m,$$

where $\theta_m \in [0, 1)$ is chosen so that $\beta_{m+1} - [\beta_m] \geqslant c > 0$.

We now proceed similarly to the proof of Lemma 5.3, following (5.30). Indeed, the argument there shows that β_m, C_{β_m} can be chosen as in (5.33)–(5.36) [notice that (5.37), (5.31) are replaced here by (5.44)].

Having established (5.38) with suitable β_m, C_{β_m}, we next show that $\gamma(u) \equiv 0$ in some neighborhood of $(R, 0)$, which is impossible.

Consider now the incompressible case. Here u satisfies

(5.45)
$$Lu + 4\pi I_G = h, \qquad G = \{u > 0\}.$$

G is again given by $\{(r, z); |z| < \phi(r), r \geqslant 0\}$ and we use Definitions 5.1 and 5.2 for this case too. The continuity of $\phi(r)$ is valid as before.

Theorem 5.5. *If $h(r)$ is analytic in r, $r \in \lambda_i$, then $\phi(r)$ is analytic for $r \in \lambda_i$.*

Proof. The proof is similar to the proof of Theorem 5.1. The main difference is that here $u \in C^{1+\alpha}$ for any $0 < \alpha < 1$ (instead of $u \in C^{2+\beta}$). Hence we must write for ψ, η a system of elliptic equations in divergence weak form, namely,

$$H(\psi) = (h(r) - \gamma(y_3))\psi_3,$$

$$H(\eta) = (h(r) - \gamma(-y_3))\eta_3,$$

where

$$H(\zeta) = \sum_{i=1}^{2} \left[-\zeta_{ii} + \left(\frac{1 + \zeta_i^2}{\zeta_3} \right)_3 \right].$$

By $C^{1+\alpha}$ Schauder estimates for such equations [3b] we can deduce (working with finite differences) that ψ, η belong to $C^{2+\varepsilon}$ up to $y_3 = 0$. We can then proceed as in Theorem 5.1.

Theorem 5.6. *If $h(R) \geqslant 0$ for some $R \geqslant 0$, then R is not a point of accumulation of rings.*

The proof involves estimates on the base and height of small rings (see Problems 3 and 4) and on the asymptotic density of $\phi(r)$ as $r \to R$. For details, see reference 58j.

PROBLEMS

1. Let G_λ be a ring for the compressible case with base $\lambda = (a, b)$, and suppose that

$$u_r(b, 0) = 0, \qquad u_{rr}(b, 0) \neq 0.$$

Show that ∂G_λ near $(b, 0)$ is given by

$$z = \pm \mu |r - b| + O(|r - b|^{1 + \beta/2}) \qquad (\mu > 0).$$

2. Consider the model in Section 2, Problem 2. Suppose that $J(r) = 1$ if $r > 2R$, $0 \leqslant J(r) < \delta$ if $0 < r < R$, $\delta \leqslant J(r) \leqslant 1$ if $R < r < 2R$, where $R > 0$, $0 < \delta < 1$. Show that for M sufficiently large any solution $\rho(r, z)$ of Problem (E) satisfies $\rho(r, z) = 0$ if $r < R$.
[*Hint:* If $\int_{r(x) < R} \rho = M_0 > 0$, redistribute the mass in $\{r(x) < R\}$ in $a_1 < r < a_1 + 1$.]

3. Let G_λ be a ring of incompressible fluid, $\lambda = (a, b)$. Set $|\lambda| = b - a$, $h_\lambda = \max_{r \in \lambda} \phi(r)$. Show that

$$h_\lambda \leqslant C |\lambda|.$$

[*Hint:* By the strong maximum principle $u_z(0, z) < 0$; if $h_\lambda = \phi(\bar{r})$, derive $h_\lambda \leqslant C u(\bar{r}, 0)$, and show that $u(\bar{r}, 0) \leqslant C |\lambda|^2$.]

4. If in Problem 3, $h(R) \geqslant 0$ and $\text{dist}(R, \lambda)$ is sufficiently small, then

$$|G_\lambda| \geqslant c |\lambda| \qquad (c > 0).$$

[*Hint:* $u \leqslant C\lambda^2$ in G_λ. If $|G_\lambda| < \varepsilon\lambda^2$ and $S_1 = \{r \in \lambda, \phi(r) > |\lambda|/8\}$, then $|S_1| < 8\varepsilon\lambda$. If $r \in S_2 = \lambda \setminus S_1$, $u < -C|\lambda|^2$ for $|\lambda|/2 < z < |\lambda|$. Show that

$$(*) \qquad \int_{B_{|\lambda|/2}} u^- > -c|\lambda|^4, \qquad \int_{B_{|\lambda|/2}} u^+ < C\varepsilon|\lambda|^4,$$

where B_δ is the disc with center $(r_0, 0)$ $[r_0 = (a + b)/2]$ and radius δ. Represent $u(r_0, 0)$ by Green's function in B_δ, and integrate on δ to obtain

$$u(r_0, 0) = c \int_{B_{|\lambda|/2}} u - \iint_{B_{|\lambda|/2}} \gamma Lu.$$

Use $(*)$ and $h(R) \geqslant 0$ to derive a contradiction.]

5. If $j^2(m) \leqslant Cm^{1+\delta}$ for some $C > 0$, $\delta > 0$, then the number of rings in the compressible case is finite (for $0 \leqslant r < \infty$).

6. If in the compressible model of Section 2, Problem 2, the function $J(r)$ is analytic in r, for $0 \leqslant r < \infty$, then the number of rings is finite.

7. Extend Lemma 5.2 to $n = 2$.

6. VORTEX RINGS

Vortex rings can be produced by ejecting a puff of smoke suddenly from the mouth through rounded lips. (The smoke is just a marking agent which makes the motion visible.) Another example is produced by letting a drop of ink fall approximately 3 centimeters into a glass of water.

In describing the motion of a steady vortex ring in inviscid incompressible fluid, we shall use cylindrical coordinates $x = (r, \theta, z)$ and the associated orthonormal frame $(\mathbf{i}_r, \mathbf{i}_\theta, \mathbf{i}_z)$. The vortex ring is symmetric about the z-axis and propagating with constant speed W in the positive z-direction. With respect to axes fixed in the ring, the velocity vector $\mathbf{v}(x)$ has the form

(6.1) $$\mathbf{v}(x) = v^r(r, z)\mathbf{i}_r + v^z(r, z)\mathbf{i}_z$$

and

(6.2) $$\mathbf{v}(x) = -W\mathbf{i}_z \qquad \text{as } |x| \to \infty.$$

The vorticity field $\omega(x) = \nabla \times \mathbf{v}$ then takes the form

(6.3) $$\omega(x) = \omega(r, z)\mathbf{i}_z, \qquad \omega(r, z) = -v_r^z + v_z^r.$$

Euler's equations are

(6.4) $$\mathbf{v} \cdot \nabla \mathbf{v} = -\nabla p$$

and the equation of conservation of mass is

(6.5) $$\nabla \cdot \mathbf{v} = 0;$$

the unknown pressure p is also to be determined. From (6.5), (6.1) we see that there exists a function $\hat{\psi}$ such that

(6.6) $$v^r = -\frac{1}{r}\hat{\psi}_z, \qquad v^z = \frac{1}{r}\hat{\psi}_r;$$

setting

(6.7) $$\psi = \hat{\psi} + \tfrac{1}{2}Wr^2,$$

the condition (6.2) becomes

(6.8) $$\frac{1}{r}|\nabla\psi| = o(1) \qquad \text{as } |x| \to \infty.$$

The function ψ is the stream function for the associated flow which is at rest at infinity. From (6.3), (6.6), (6.7), we have

(6.9) $$\omega = -\frac{1}{r}L\hat{\psi} = -\frac{1}{r}L\psi,$$

where

(6.10) $$Lv \equiv \frac{\partial^2 v}{\partial r^2} - \frac{1}{r}\frac{\partial v}{\partial r} + \frac{\partial^2 v}{\partial z^2} = r\frac{\partial}{\partial r}\left(\frac{1}{r}\frac{\partial v}{\partial r}\right) + \frac{\partial^2 v}{\partial z^2}.$$

Notice that Euler's equations hold for some p if and only if

(6.11) $$\nabla \times (\mathbf{v} \cdot \nabla \mathbf{v}) = 0.$$

Noting that $\nabla \cdot \mathbf{v} = \nabla \cdot \boldsymbol{\omega} = 0$ and using the identity

$$\mathbf{v} \cdot \nabla \mathbf{v} = \tfrac{1}{2}\nabla |\mathbf{v}|^2 + \boldsymbol{\omega} \times \mathbf{v},$$

we can write

$$\nabla \times (\mathbf{v} \cdot \nabla \mathbf{v}) = \nabla \times (\boldsymbol{\omega} \times \mathbf{v}) = \mathbf{v} \cdot \nabla \boldsymbol{\omega} - \boldsymbol{\omega} \cdot \nabla \mathbf{v}.$$

Also

$$\mathbf{v} \cdot \nabla \boldsymbol{\omega} = (\mathbf{v} \cdot \nabla \omega)\mathbf{i}_\theta,$$

$$\boldsymbol{\omega} \cdot \nabla \mathbf{v} = \omega \frac{v^r}{r}\mathbf{i}_\theta.$$

Thus if we let

(6.12) $$\zeta(r, z) = \frac{1}{r}\omega(r, z),$$

then (6.11) reduces to

(6.13) $$\mathbf{v} \cdot \nabla \zeta = 0,$$

or, in view of (6.6),

(6.14) $$\frac{\partial(\hat{\psi}, \zeta)}{\partial(r, z)} = 0.$$

The last equation in turn is equivalent to the existence of a functional dependence

$$\Phi(\hat{\psi}, \zeta) = 0$$

with $\nabla\Phi \neq 0$. Thus, in particular, if we can find functions ζ, $\hat{\psi}$ such that

(6.15) $\zeta = f(\hat{\psi} - \gamma)$ (γ constant)

and if $\hat{\psi}$, ζ are related by (6.12), (6.9), that is, by

(6.16) $-\dfrac{1}{r^2} L\psi = \zeta,$

then the physical equations (6.4), (6.5) will be satisfied.

Definition 6.1.   The constant γ is called the *flux*, the function $f(t)$ is called the *vorticity function*, and the function

(6.17) $u(r, z) = \psi - \tfrac{1}{2} W r^2 - \gamma$

is called the *adjusted stream function*.

From a physical point of view, the function $f(t)$ is supposed to be nondecreasing on $(-\infty, \infty)$ with $f(t) = 0$ if $t \leq 0$ and $f(t) > 0$ if $t > 0$.

Combining (6.15), (6.16) we obtain

(6.18) $Lu + r^2 f(u) = 0.$

Notice that supp ζ, the *vortex core*, is characterized by

(6.19)   supp $\zeta = \bar{\Omega}$,   where $\Omega = \{x \in R^3; u(x) > 0\}$.

We seek solutions for which $\bar{\Omega}$ is compact.

In what follows we shall need the inverse of $-(1/r^2)L$. This is given in the next lemma.

Lemma 6.1.   *Let*

(6.20) $K(x, x') = \dfrac{rr' \cos(\theta - \theta')}{4\pi |x - x'|},$

where $x = (r, \theta, z)$, $x' = (r', \theta', z')$. *For* $\zeta(r, z) = \zeta(x)$ *any bounded, measurable function with compact support in* R^3, *let*

(6.21) $\psi(r, z) = \psi(x) = \displaystyle\int_{R^3} K(x, x')\zeta(x')\, dx'.$

Then (6.16) holds a.e. and

$$(6.22) \qquad \begin{cases} \dfrac{1}{r}\psi = O(|x|^{-2}) \\ \dfrac{1}{r}|\nabla\psi| = O(|x|^{-3}), \qquad \text{as } |x| \to \infty. \end{cases}$$

Proof. Let $\omega(x) = r\zeta(r,z)\mathbf{i}_\theta$. Then

$$(6.23) \qquad \mathbf{B}(x) = \frac{1}{4\pi}\int_{R^3}\frac{1}{|x-x'|}\omega(x')\,dx'$$

satisfies

$$(6.24) \qquad -\Delta\mathbf{B} = \omega \qquad \text{a.e.}$$

By direct calculation,

$$\int_{-\pi}^{\pi}\frac{1}{|x-x'|}\mathbf{i}_{\theta'}\,d\theta' = \left\{\int_{-\pi}^{\pi}\frac{\cos(\theta-\theta')}{|x-x'|}\,d\theta'\right\}\mathbf{i}_\theta.$$

Thus

$$\mathbf{B}(x) = \left\{\frac{1}{4\pi}\int_{R^3}\frac{\cos(\theta-\theta')}{|x-x'|}r'\zeta(r',z')\,dx'\right\}\mathbf{i}_\theta,$$

and hence, by virtue of (6.20) and (6.21),

$$(6.25) \qquad r\mathbf{B}(x) = \psi(r,z)\mathbf{i}_\theta.$$

We now compute

$$(6.26) \qquad \nabla\times\mathbf{B} = -\frac{1}{r}\psi_z\mathbf{i}_r + \frac{1}{r}\psi_r\mathbf{i}_z.$$

Also, since $\nabla\cdot\mathbf{B}$, we have

$$-\Delta\mathbf{B} = \nabla\times(\nabla\times\mathbf{B}) = -\frac{1}{r}L\psi\,\mathbf{i}_\theta.$$

Combining this with (6.24) yields the desired result (6.16).

To establish the estimates (6.22), we first note that

$$\frac{1}{|x-x'|} = \frac{1}{|x|} + \frac{a(x,x')}{|x|^2},$$

$$\nabla_x\frac{1}{|x-x'|} = \nabla_x\frac{1}{|x|} + \frac{A(x,x')}{|x|^3}$$

for certain functions a, $\mathbf{A} = O(1)$ as $|x| \to \infty$, $x' \in \text{supp } \zeta$. Now using the property

$$\int_{R^3} r'\zeta(r', z') \mathbf{i}_{\theta'} \, dx' = 0,$$

we find that

$$\mathbf{B}(x) = \frac{1}{4\pi} |x|^{-2} \int_{R^3} a(x, x') \omega(x') \, dx'$$

$$\nabla \times \mathbf{B}(x) = \frac{1}{4\pi} |x|^{-3} \int_{R^3} \mathbf{A}(x, x') \times \omega(x') \, dx'.$$

Recalling (6.25) and (6.26), the estimates (6.22) now follow.

We introduce the half plane

$$H = \{(r, z); 0 \leqslant r < \infty, -\infty < z < \infty\}.$$

Green's function of the operator $-(1/r^2)L$ on H with measure $r \, dr \, dz$ is given by

$$G(r, z, r', z') = \int_{-\pi}^{\pi} K(x, x') \, d\theta' \qquad (\theta = 0)$$

(6.27)
$$= \frac{rr'}{4\pi} \int_{-\pi}^{\pi} \frac{\cos \theta' \, d\theta'}{\left[(z - z')^2 + r^2 + r'^2 - 2rr' \cos \theta'\right]^{1/2}},$$

since then (6.21) takes the form

(6.28) $$\psi(r, z) = \iint_H G(r, z, r', z') \zeta(r', z') r' \, dr' \, dz'.$$

Using Lemma 6.1, we have the following expressions for the total kinetic energy of the flow:

$$E = \frac{1}{2} \int_{R^3} \frac{1}{r^2} |\nabla \psi|^2 \, dx = \frac{1}{2} \int_{R^3} \psi \zeta \, dx$$

(6.29)
$$= \frac{1}{2} \int_{R^3} \int_{R^3} K(x, x') \zeta(x) \zeta(x') \, dx \, dx',$$

where the second equality follows from integration by parts.

The total impulse required to generate the flow from rest is defined by

$$(6.30) \qquad \mathbf{P} = \tfrac{1}{2} \int_{R^3} \mathbf{x} \times \boldsymbol{\omega}(x) \, dx.$$

This takes the form

$$(6.31) \qquad \mathbf{P} = P\mathbf{i}_z, \qquad P = \tfrac{1}{2} \int_{R^3} r^2 \zeta(x) \, dx$$

provided that

$$(6.32) \qquad \int_{R^3} rz\zeta(x) \, dx = 0.$$

In the sequel we shall assume that $\zeta(r, z) = \zeta(r, -z)$ holds; hence (6.32) is assured.

In posing the steady vortex ring problem variationally, we shall maximize the energy $E(\zeta)$ over a certain class of functions ζ subject to the constraint that the impulse P be prescribed. We proceed to define the class of admissible functions.

Let $\mathcal{Q}_\lambda$ denote the class of measurable functions $\zeta \geqslant 0$ a.e. on R^3 satisfying the following conditions:

$$(6.33) \qquad \zeta(x) = \zeta(r, z) = \zeta(r, -z),$$

$$(6.34) \qquad \tfrac{1}{2} \int_{R^3} r^2 \zeta(x) \, dx = 1,$$

$$(6.35) \qquad \int_{R^3} \zeta(x) \, dx \leqslant 1,$$

$$(6.36) \qquad \operatorname*{ess\,sup}_{x \in R^3} \zeta(x) \leqslant \lambda, \qquad 0 < \lambda < \infty.$$

Let $\mathcal{Q}_\infty$ denote the (larger) class for which condition (6.36) is removed.

The energy functional $E(\zeta)$ is defined on the class $\mathcal{Q}_\lambda$ $(0 < \lambda < \infty)$ by

$$(6.37) \qquad E(\zeta) = \tfrac{1}{2} \int_{R^3} \int_{R^3} K(x, x') \zeta(x) \zeta(x') \, dx \, dx',$$

where $K(x, x')$ is given by (6.20). We consider the variational problem:

Problem (E_λ). Find a function ζ such that

$$(6.38) \qquad E(\zeta) = \max_{\tilde{\zeta} \in \mathcal{Q}_\lambda} E(\tilde{\zeta}), \qquad \zeta \in \mathcal{Q}_\lambda.$$

Theorem 6.2. *There exists a solution ζ of (6.38). Furthermore, there exist constants $W > 0$, $\gamma \geqslant 0$ such that*

$$(6.39) \qquad\qquad \zeta = \lambda I_\Omega \qquad a.e.,$$

where

$$(6.40) \qquad \Omega = \left\{ x \in R^3; u(x) \equiv \psi(x) - \tfrac{1}{2} W r^2 - \gamma > 0 \right\}$$

with ψ defined by (6.21); Ω is a bounded open subset of R^3, and

$$\Omega = \left\{ x = (r, \theta, z); |z| < Z(r) \right\}$$

for some function $Z(r) \geqslant 0$.

For each value of the free parameter λ the solution obtained clearly represents a steady vortex ring corresponding to the vorticity function

$$(6.41) \qquad\qquad f(t) = \lambda I_{\{t > 0\}}.$$

The constants W, γ arise as Lagrange multipliers for the constraints (6.34), (6.35).

A related family of variational problems is derived from the energy functional

$$(6.42) \qquad E_\beta(\zeta) = E(\zeta) - \beta \lambda \int_{R^3} [\zeta(x)/\lambda]^{1 + 1/\beta} \, dx$$

defined on the class

$$(6.43) \qquad \mathcal{Q}_{\infty, \beta} = \mathcal{Q}_\infty \cap L^{1 + 1/\beta}(R^3) \qquad (0 < \beta < \infty).$$

We consider

Problem $(E_{\lambda, \beta})$. Find a function ζ such that

$$(6.44) \qquad E_\beta(\zeta) = \max_{\tilde{\zeta} \in \mathcal{Q}_{\infty, \beta}} E_\beta(\tilde{\zeta}), \qquad \zeta \in \mathcal{Q}_{\infty, \beta}.$$

Theorem 6.3. *There exists a solution ζ of (6.44) for any prescribed $0 < \lambda < \infty$, provided that $0 < \beta < 5$. Furthermore, there exist constants $W > 0$, $\gamma \geqslant 0$ such that*

$$(6.45) \qquad\qquad \zeta = \lambda \left(\frac{u^+}{1 + \beta} \right)^\beta \qquad a.e.,$$

where

$$(6.46) \qquad\qquad u(x) = \psi(x) - \tfrac{1}{2} W r^2 - \gamma$$

with ψ defined by (6.21); ζ has compact support in R^3, and $\zeta(r, z)$ is nonincreasing as a function of z for $z > 0$.

Clearly, these solutions of the vortex ring problem correspond to a vorticity function of the form

$$(6.47) \qquad f(t) = \begin{cases} \lambda \left(\dfrac{t}{1 + \beta} \right)^{\beta}, & t > 0 \\ 0, & t \leqslant 0. \end{cases}$$

It is evident that the variational equation (6.45) tends, as $\beta \to 0$, to equation (6.39). In fact, Theorem 6.2 is proved by first obtaining the solutions asserted by Theorem 6.3 and then taking the limit of these solutions over some sequence $\beta_j \to 0$; the sequence of solutions converges weakly in $L^p(R^3)$ for every $1 < p < \infty$. One may view problem (E_λ) as a limit of the "penalized" problems $(E_{\lambda, \beta})$, when $\beta \to 0$.

Theorems 6.3 and 6.2 are proved in Section 8. In Section 7 we develop some identities and potential estimates needed for these proofs.

The restriction $0 < \beta < 5$ made in Theorem 6.3 occurs only in establishing Lemma 8.2. This lemma (and consequently also Theorem 6.3) remains valid for any $0 < \beta < \infty$ provided that λ is large enough, depending on β.

There is a similarity between the steady vortex ring problem and the axisymmetric self-gravitating problem studied in Sections 1 to 5. Thus problem (6.38) is analogous to problem (1.19) and problem (6.44) is analogous to problem (1.13). Although the functional to be minimized, for the vortex ring problem, is somewhat simpler looking, the admissible class has more constraints. The presence of (6.36), in particular, among the constraints for problem (6.38) causes technical difficulties; these difficulties will be overcome by obtaining the solution as limit of solutions of the penalized problem (6.44), for which variational conditions can be derived by standard methods.

PROBLEMS

1. Denote by $\mathcal{Q}_{P, \Gamma, \Lambda}$ the class of functions $\zeta \geqslant 0$ satisfying (6.33) and the constraints

$$\frac{1}{2} \int_{R^3} r^2 \zeta(x) \, dx = P \qquad \text{(total impulse)},$$

$$\int_{R^3} \zeta(x) \, dx \leqslant \Gamma \qquad \text{(total circulation)} \times 2\pi,$$

$$\operatorname*{ess\ sup}_{x \in R^3} \zeta(x) \leqslant \Lambda \qquad \text{(vortex strength)},$$

for prescribed positive constants P, Γ, Λ. consider the problem to determine $\zeta \in \mathcal{Q}_{P,\Gamma,\Lambda}$ such that

$$E(\zeta) = \max_{\tilde{\zeta} \in \mathcal{Q}_{P,\Gamma,\Lambda}} E(\tilde{\zeta}).$$

Making the change of variables $x = a\hat{x}$, $\zeta = b\tilde{\zeta}$, and choosing

$$a = P^{1/2}\Gamma^{-1/2}, \qquad b = P^{-3/2}\Gamma^{5/2},$$

show that the problem for $\hat{\zeta}$ coincides with the problem (E_λ), where $\lambda = P^{3/2}\Gamma^{-5/2}\Lambda$.

2. Show that a solution to the vortex ring equations (6.15), (6.16) is given by $f(t) = I_{\{t>0\}}$, $\gamma = 0$,

$$\psi = \begin{cases} \dfrac{1}{10}\lambda R^2(a^2 - R^2)\sin^2\theta & \text{if } R < a \\[2mm] -\dfrac{1}{2}WR^2\left(1 - \dfrac{a^3}{R^3}\right)\sin^2\theta & \text{if } R \geqslant a \end{cases}$$

where $r = R\sin\theta$, $z = R\cos\theta$, and $\lambda a^2/W = \frac{15}{2}$. The vortex core is the ball $\{|x| < a\}$. This solution is called *Hill's vortex* (1894).

7. ENERGY IDENTITIES AND POTENTIAL ESTIMATES

When writing $\int_H h\, dx$ it is understood that $dx = 2\pi r\, dr\, dz$.

We shall often use the integration-by-parts formula

$$(7.1) \qquad \int_H \frac{1}{r^2}(u_r v_r + u_z v_z)\, dx = -\int_H \frac{1}{r^2} u L v\, dx$$

for functions $u(r, z)$, $v(r, z)$ either of compact support or vanishing sufficiently fast as $r^2 + z^2 \to \infty$, with $\int r^{-1} u(r, z) v_r(r, z)\, dz \to 0$ for a sequence $r = r_j \downarrow 0$. In particular, the calculation (6.29) expressing E in terms of ζ is justified using (7.1) and the estimates for the corresponding ψ given in Lemma 6.1, provided that ζ has compact support; see Problem 1.

We now give another formula for $E(\zeta)$.

Lemma 7.1. *Let ζ be any bounded, measurable function on H for which*

$$(7.2)$$

$$E(\zeta) = \frac{1}{2}\int_H \frac{1}{r^2}(\psi_r^2 + \psi_z^2)\, dx < \infty, \qquad |\psi_z(r, z)| \leqslant Cr \text{ if } r < 1 \ (C > 0).$$

Then

(7.3)
$$E(\zeta) = \int_H (r\psi_r + z\psi_z)\zeta\,dx.$$

Proof. Set $\sigma = \frac{1}{2}(r^2 + z^2)$, $\eta = \frac{1}{2}(\psi_r^2 + \psi_z^2)$. Since $L\sigma = 1$, we have

$$E(\zeta) = \int_H \frac{1}{r^2}\eta L\sigma\,dx$$

(7.4)
$$= -\int_H \frac{1}{r^2}(\sigma_r\eta_r + \sigma_z\eta_z)\,dx$$

$$= -\int_H \frac{1}{r^2}(r\eta_r + z\eta_z)\,dx,$$

where (7.1) has been used (the justification will be given below). Expanding the integrand, we find that

$$r\eta_r + z\eta_z = \psi_r(r\psi_r + z\psi_z)_r + \psi_z(r\psi_r + z\psi_z)_z - (\psi_r^2 + \psi_z^2).$$

Putting this into (7.4), we get

$$E(\zeta) = -\int_H \frac{1}{r^2}[\psi_r(r\psi_r + z\psi_z)_r + \psi_z(r\psi_r + z\psi_z)_z]\,dx + 2E(\zeta),$$

so that

(7.5)
$$E(\zeta) = \int_H \frac{1}{r^2}[\psi_r(r\psi_r + z\psi_z)_r + \psi_z(r\psi_r + z\psi_z)_z]\,dx$$

$$= -\int_H \frac{1}{r^2}(r\psi_r + z\psi_z)L\psi\,dx,$$

which yields (7.3).

We must now rigorously justify the use of formula (7.1) in the derivations (7.4) and (7.5) above. It the integration by parts is carried out instead on the finite domain,

(7.6)
$$D_{a,\varepsilon} = \{(r, z) \in H; \varepsilon \leqslant r < a, |z| < a\},$$

for some $\varepsilon > 0$, then we obtain boundary integrals. The boundary integrals on $r = \varepsilon$ are bounded (in both cases) by

$$\int_{-a}^{a} \eta(\varepsilon, z)\,dz + C_a\left\{\int_{-a}^{a} \eta(\varepsilon, z)\,dz\right\}^{1/2}$$

(using $|\psi_z| \leqslant CR$). Since the integral in (7.2) is finite,

$$(7.7) \qquad \int_0^\infty \int_{-\infty}^\infty \frac{\eta(r, z)}{r} \, dz \, dr < \infty,$$

and therefore

$$\int_{-\infty}^\infty \frac{\eta(\varepsilon, z)}{\varepsilon} \, dz = o\left(\frac{1}{\varepsilon}\right)$$

for a sequence $\varepsilon = \varepsilon_n \downarrow 0$. Thus the boundary integrals on $r = \varepsilon_n$ tend to zero. There remains to consider the boundary integrals on $\partial_0 D_{a,0} \equiv \partial D_{a,0} \cap \{r > 0\}$, that is, it remains to prove that

$$\int_{\partial_0 D_{a,0}} \frac{1}{r} \eta \, ds = o\left(\frac{1}{a}\right)$$

for a sequence $a = a_n \to \infty$. But this follows again from (7.7).

For a given vorticity function $f(t)$, let

$$(7.8) \qquad F(t) = \int_0^t f(s) \, ds;$$

then

$$F(t) = \lambda t^+ \qquad\qquad \text{when (6.41) holds,}$$

$$F(t) = \lambda \left(\frac{t^+}{1 + \beta}\right)^{1+\beta} \qquad \text{when (6.47) holds.}$$

Let

$$(7.9) \qquad J(u) = \int_H F(u) \, dx,$$

where $u(r, z)$ is defined in (6.17). Several other expressions for $J(u)$ can be given; namely,

$$(1 + \beta)J(u) = \int_H u\zeta \, dx = -\int_H \frac{1}{r^2} uLu \, dx$$

$$(7.10) \qquad\qquad\qquad = \int_\Omega \frac{1}{r^2} \left(u_r^2 + u_z^2\right) dx,$$

where $\Omega = \{x \in R^3; u(x) > 0\}$ and $\beta > 0$ in case (6.47), $\beta = 0$ in case (6.41). Thus $\frac{1}{2}(1 + \beta)J(u)$ represents the kinetic energy of the steady flow (with stream function u) confined to the vortex core Ω.

Lemma 7.2. *Let ζ be a solution of either* (6.38) *or* (6.44) *with compact support. Then* (*obviously*) (7.2) *holds and*

$$(7.11) \qquad E(\zeta) = -3J(u) + 2W = \tfrac{1}{2}[(1 + \beta)J(u) + W + \gamma].$$

Proof. The assertion that $E(\zeta)$ equals the right-hand side of (7.11) is immediate from

$$\int_H u\zeta \, dx = \int_H \psi\zeta \, dx - \frac{1}{2}W\int_H r^2\zeta \, dx - \gamma\int_H \zeta \, dx$$

$$= 2E - W - \gamma.$$

To prove the first identity in (7.11), take $\psi_r = u_r + Wr$, $\psi_z = u_z$ in (7.3):

$$(7.12) \qquad E(\zeta) = \int_H (ru_r + zu_z)\zeta \, dx + \int_H r(Wr)\zeta \, dx.$$

Integrating by parts, we find that

$$\int_H ru_r\zeta \, dx = \int_H rF(u)_r \, dx = 2\pi \iint_H r^2 F(u)_r \, dr \, dz$$

$$= -4\pi A \iint_H rF(u) \, dr \, dz = -2\int_H F(u) \, dx;$$

and, similarly,

$$\int_H zu_z\zeta \, dx = -\int_H F(u) \, dx.$$

Thus (7.12) becomes

$$E(\zeta) = -3\int_H F(u) \, dx + W\int_H r^2\zeta \, dx = -3J(u) + 2W.$$

Remark 7.1. It is important for later application that we observe that (7.11) remains valid for solutions of (6.44) without the assumption of compact support; it is assumed, however, that ζ satisfies the assumptions of Lemma 7.1. Indeed, if the integration by parts in the proof of Lemma 7.2 is carried out on the finite domain $D_{a,0}$ [defined in (7.6)], then the boundary integrals that arise are of the form

$$\int_{r=a} r^2 F(u) \, dz, \qquad \int_{z=\pm a} rz F(u) \, dr.$$

Noting that $F(u) = \lambda[\zeta(x)/\lambda]^{1+1/\beta}$, we see that there is a sequence $a = a_n \to \infty$ for which these integrals tend to zero since the condition

$$\iint_H r\zeta^{1+1/\beta} \, dr \, dz < \infty$$

implies that

$$\int_{-\infty}^{\infty} r^2 \zeta^{1+1/\beta} \, dz, \qquad \int_0^{\infty} rz\zeta^{1+1/\beta} \, dr = o(1)$$

for $r = a_n$, $|z| = a_n$, respectively.

We now turn our attention to certain estimates for the "potential" ψ, defined in (6.28).

Lemma 7.3. *Let $s = [(r - r')^2 + (z - z')^2]^{1/2}$. There holds*

$$(7.13) \qquad G(r, z, r', z') \le \begin{cases} Cr \log \dfrac{r}{s} & \text{if } s \le \dfrac{r}{2} \\[2mm] \dfrac{Cr^2 r'^2}{s^3} & \text{if } s \ge \dfrac{r}{2}, \end{cases}$$

where C is a (sufficiently large) absolute positive constant.

Proof. Recall that G is defined by (6.27). Letting

$$\xi^2 = \frac{(z - z')^2 + (r - r')^2}{4rr'} \qquad (\xi > 0),$$

we first show that

$$(7.14) \qquad G(r, z, r', z') \le C(rr')^{1/2} \log \frac{C}{\xi} \qquad \text{if } \xi \le 1,$$

$$(7.15) \qquad G(r, z, r', z') \le C(rr')^{1/2} \xi^{-3} \qquad \text{if } \xi \ge 1.$$

Estimate (7.14) is derived from the formula

$$(7.16) \qquad G(r, z, r', z') = \frac{(rr')^{1/2}}{2\pi} \left\{ \left(\frac{2}{k} - k \right) K(k) - \frac{2}{k} E(k) \right\},$$

where K and E are the complete elliptic integrals of the first and second kind [55, formulas 291.00 and 291.01] and

$$(7.17) \qquad \begin{aligned} k^2 &= \frac{4rr'}{(z - z')^2 + (r + r')^2}, \\[3mm] k'^2 &= \frac{(z - z')^2 + (r - r')^2}{(z - z')^2 + (r + r')^2} \qquad [(k^2 + k'^2) = 1]. \end{aligned}$$

Since $\xi \leqslant 1$ implies that $k'^2 \leqslant \frac{1}{2}$, we apply the asymptotic formulas

$$(7.18) \qquad \begin{aligned} K(k) &= \log \frac{4}{k'} + o(1) \qquad (\text{as } k' \to 0) \\ E(k) &= 1 + o(1) \end{aligned}$$

to (7.16); then (7.14) follows directly. Estimate (7.15) follows from the expansion

$$\left[(z - z')^2 + r^2 + r'^2 - 2rr' \cos \theta' \right]^{-1/2} = (4rr')^{-1/2} \left\{ \xi^{-1} + B(\xi, \theta') \right\}^{-3},$$

where $| B(\xi, \theta') | \leqslant C$ for $\xi \geqslant 1$, $-\pi \leqslant \theta' \leqslant \pi$. Applying this to (6.27) and noting the cancellation of the term involving ξ^{-1}, (7.15) follows. The required estimate (7.13) follows directly from (7.14), (7.15).

In the following lemmas we assume that $\zeta \in \mathcal{Q}_{\infty, \beta}$, and we let $N = N(\zeta)$ denote the norm

$$(7.19) \qquad N = \left\{ \int_H \zeta^{1 + 1/\beta} \, dx \right\}^{\beta/(1 + \beta)} \qquad (0 < \beta < \infty).$$

Lemma 7.4. *Assume that $\zeta(r, z)$ is a nonincreasing function of z for $z > 0$. Then for any $0 < \beta \leqslant \beta^*$, $0 < \varepsilon \leqslant 1$, there holds*

$$(7.20) \qquad \psi(r, z) \leqslant \psi(r, 0) \leqslant C(N + 1) \min \{ r, r^{-1 + \varepsilon} \};$$

the constant C depends only on β^*, ε.

Proof. It is clear that $\psi(r, 0) = \max_z \psi(r, z)$. We write $\psi(r, 0) = \psi_1(r, 0) + \psi_2(r, 0)$, where [with s as in Lemma 7.3]

$$\psi_1(r, 0) = \iint_{\{s \leqslant r/2\}} G(r, 0, r', z') \zeta(r', z') r' \, dr' \, dz'$$

$$\leqslant Cr \iint_{\{s \leqslant r/2\}} \log \frac{r}{s} \zeta(r', z') r' \, dr' \, dz',$$

$$\psi_2(r, 0) = \iint_{\{s > r/2\}} G(r, 0, r', z') \zeta(r', z') r' \, dr' \, dz'$$

$$\leqslant Cr^2 \iint_{\{s > r/2\}} \frac{r'^3}{s^3} \zeta(r', z') r' \, dr' \, dz'.$$

To estimate $\psi_2(r, 0)$, note that

$$\frac{r'^2}{s^3} \leq \frac{C}{r} \qquad \text{whenever } s > \frac{r}{2}.$$

Thus, recalling (6.35),

$$\psi_2(r, 0) \leq Cr^2 \iint\limits_{\{s > r/2\}} \frac{C}{r} \zeta(r', z') r' \, dr' \, dz' \leq Cr.$$

Also, recalling (6.34),

$$\psi_2(r, 0) \leq Cr^2 \iint\limits_{\{s > r/2\}} \frac{r'^2}{\left(\frac{1}{2}r\right)^3} \zeta(r', z') r' \, dr' \, dz' \leq Cr^{-1}.$$

Hence

(7.21) $$\psi_2(r, 0) \leq C \min\{r, r^{-1}\}.$$

We proceed to estimate $\psi_1(r, 0)$. For any $0 < \alpha < \infty$, we have, by Hölder's inequality,

(7.22)

$$\iint\limits_{\{s \leq r/2\}} \log \frac{r}{s} \zeta(r', z') r' \, dr' \, dz'$$

$$\leq \left\{ \iint\limits_{\{s \leq r/2\}} \left(\log \frac{r}{s} \right)^{1+\alpha} r' \, dr' \, dz' \right\}^{1/(1+\alpha)} \left\{ \iint\limits_{\{s \leq r/2\}} \zeta^{1+(1/\alpha)} r' \, dr' \, dz' \right\}^{\alpha/(1+\alpha)}.$$

But clearly,

(7.23) $$\iint\limits_{\{s \leq r/2\}} \left(\log \frac{r}{s} \right)^{1+\alpha} r' \, dr' \, dz' \leq Cr \int_0^{r/2} \left(\log \frac{r}{s} \right)^{1+\alpha} s \, ds = C_\alpha r^3.$$

Thus

(7.24) $$\psi_1(r, 0) \leq C_\alpha r^{1+\delta} \|\zeta\|_{L^{1+1/\alpha}(s \leq r/2)} \qquad \left(\delta = \frac{3}{1+\alpha} \right).$$

We now apply the standard interpolation inequality

(7.25) $$\|\zeta\|_{L^{1+(1/\alpha)}} \leq \|\zeta\|_{L^1}^{1-a} \|\zeta\|_{L^{1+(1/\beta)}}^{a} \qquad (\beta \leq \alpha < \infty)$$

where $a = (1 + \beta)/(1 + \alpha)$ $(0 < a \leq 1)$; each L^p norm is taken with respect to the measure $r'dr'dz'$ on the set $\{(r', z') \in H; s \leqslant r/2\}$. Since, by (6.34) and (6.35),

$$\|\zeta\|_{L^1(s \leqslant r/2)} \leqslant C \min\{1, r^{-2}\}$$

we conclude from (7.25) that

$$\|\zeta\|_{L^{1+1/a}(s \leqslant r/2)} \leqslant CN^a \min\{1, r^{-2(-a)}\}.$$

Thus

$$\psi_1(r, 0) \leqslant C_a N^a r^{1+\delta} \min\{1, r^{-2(1-a)}\}.$$

For $r \leqslant 1$ we take simply $\alpha = \beta$, so that

$$\psi_1(r, 0) \leqslant C_{\beta*}(N + 1)r \qquad (0 < \beta \leqslant \beta*).$$

For $r \geqslant 1$ we take α sufficiently large (depending on $\beta*$, ε) so that

$$\psi_1(r, 0) \leqslant C_{\beta*, \varepsilon}(N + 1)r^{-1+\varepsilon} \qquad (0 < \beta \leqslant \beta*);$$

specifically, we choose α so that $(5 + 3\beta*)/(1 + \alpha) < \varepsilon$. Together these estimates yield

(7.26) $$\psi_1(r, 0) \leqslant C_{\beta*, \varepsilon}(N + 1) \min\{r, r^{-1+\varepsilon}\}.$$

Now combining (7.21) and (7.26), we obtain the statement of the lemma.

Lemma 7.5. *Assume that $\zeta(r, z)$ is a nonincreasing function of z for $z > 0$. Then for $0 < \beta \leqslant \beta*$, there holds*

(7.27) $$\psi(r, z) \leqslant C(N + 1)A^{-1+\varepsilon} \min\{r, r^{-1+\varepsilon}\} + Cr^2\left(\frac{A}{z}\right)^3$$

provided that $r/2 < z/A$ and $A > 1$; the constant C depends only on $\beta*$, ε.

Proof. The monotonicity of $\zeta(r, z)$ in z implies that, for any $A > 1$,

$$\iint\limits_{\{|z'-z|<z/A\}} \zeta(r', z')r' \, dr' \, dz' \leqslant \frac{1}{A} \iint\limits_{\{z'>0\}} \zeta(r', z')r' \, dr' \, dz',$$

$$\iint\limits_{\{|z'-z|<z/A\}} r'^2\zeta(r', z')r' \, dr' \, dz' \leqslant \frac{1}{A} \iint\limits_{\{z'>0\}} r'^2\zeta(r', z')r' \, dr' \, dz'.$$

Using these facts, we can modify the proof of the previous lemma for large z.

As in the preceding proof, we write

$$\psi(r, z) = \psi_1(r, z) + \psi_2(r, z).$$

In order to estimate ψ_2 we write

$$\psi_2(r, z) = \psi_2'(r, z) + \psi_2''(r, z),$$

$$\psi_2'(r, z) \leqslant Cr^2 \iint_{\substack{\{s > r/2\} \\ \{|z-z'| < z/A\}}} \frac{r'^2}{s^3} \zeta(r', z') r' \, dr' \, dz'$$

$$\leqslant \frac{C}{A} \min \{r, r^{-1}\}.$$

$$\psi_2''(r, z) \leqslant Cr^2 \iint_{\substack{\{s > r/2\} \\ \{|z-z'| > z/A\}}} \frac{r'^2}{s^3} \zeta(r', z') r' \, dr' \, dz'$$

$$\leqslant Cr^2 \left(\frac{A}{z} \right)^3.$$

Thus

$$\psi_2(r, z) \leqslant \frac{C}{A} \min \{r, r^{-1}\} + Cr^2 \left(\frac{A}{z} \right)^3.$$

To estimate $\psi_1(r, z)$, we follow the reasoning of the previous proof except that now (noting that $s \leqslant r/2$ implies that $|z - z'| < z/A$) we have

$$\|\zeta\|_{L^1(s \leqslant r/2)} \leqslant \frac{C}{A} \min\{1, r^{-2}\}.$$

Thus, as before,

$$\psi_1(r, z) \leqslant C_\alpha N^a A^{-1+\epsilon} r^{1+\delta} \min \{1, r^{-2(-a)}\},$$

with $\delta = 3/(1 + \alpha)$, $a = (1 + \beta)/(1 + \alpha)$, $\beta \leqslant \alpha < \infty$.

Now the proof is completed just as in the preceding proof.

PROBLEM

1. Justify (6.29) for ζ with compact support.

   [*Hint*:

$$\lim_{r \to 0} \int [r^{-1} \psi \psi_r] \, dz = \int \left\{ \int \frac{r' \cos \theta' \zeta(r', z') \, dx'}{4\pi [(z - z')^2 + r'^2]^{1/2}} \right\}^2 dz = 0.]$$

8. EXISTENCE OF VORTEX RINGS

In this section we prove Theorems 6.2 and 6.3. Theorem 6.3 is established by the sequence of Lemmas 8.1 to 8.9, and Theorem 6.2 is obtained in the form of Lemmas 8.10 and 8.11.

Let $\mathcal{C}'_{\infty,\beta}$ denote the class of nonnegative functions $\zeta \in L^{1+1/\beta}(R^3)$ satisfying (6.33), (6.35), and [rather than (6.34)]

$$(8.1) \qquad\qquad \tfrac{1}{2}\int_{R^3} r^2 \zeta(x)\, dx \leq 1;$$

clearly, $\mathcal{C}_{\infty,\beta} \subseteq \mathcal{C}'_{\infty,\beta}$. Consider the problem: Find ζ such that

$$(8.2) \qquad\qquad E_\beta(\zeta) = \max_{\tilde{\zeta} \in \mathcal{C}'_{\infty,\beta}} E_\beta(\tilde{\zeta}), \qquad \zeta \in \mathcal{C}'_{\infty,\beta}.$$

To solve problem (6.44) we first obtain a solution of (8.2); this is necessary because the possible solutions are not known to have bounded support a priori.

Lemma 8.1. *For any prescribed $0 < \lambda < \infty$ there exists a ζ such that (8.2) holds; $\zeta(r, z)$ is a nonincreasing function of z for $z > 0$.*

Proof. In what follows we shall assume that $0 < \beta \leq \beta^*$ for some fixed β^* (and the dependence of any constants on β^* will not be specified). For any $\zeta \in \mathcal{C}'_{\infty,\beta}$, estimate (7.20) implies that

$$(8.3) \qquad\qquad \sup_{x \in R^3} \psi(x) \leq C[N(\zeta) + 1],$$

and hence

$$(8.4) \qquad\qquad E(\zeta) = \tfrac{1}{2}\int_{R^3} \psi\zeta\, dx \leq C[N(\zeta) + 1];$$

here we use the fact that the hypothesis of Lemma 7.4 is satisfied by $\zeta^*(r, z)$, the symmetric rearrangement of $\zeta(r, z)$ in the z variable, and $E(\zeta^*) \geq E(\zeta)$ [by (6.29) and Theorem 7.3 of Chapter 3] while $N(\zeta^*) = N(\zeta)$. Furthermore,

$$(8.5) \qquad \frac{1}{\lambda}N(\zeta) = \|\zeta/\lambda\|_{L^{1+(1/\beta)}(R^3)} \leq C_\varepsilon + \varepsilon\beta\int_{R^3}(\zeta/\lambda)^{1+(1/\beta)}\, dx \qquad (\varepsilon > 0);$$

this follows from the elementary inequality

$$(8.6) \qquad\qquad X \leq C_\varepsilon + \varepsilon\beta X^{1+(1/\beta)} \qquad (\varepsilon > 0)$$

valid for $0 \leq X < \infty$. Applying (8.5) to estimate the energy in (8.4), we get, fixing ε sufficiently small,

$$(8.7) \qquad E(\zeta) \leqslant C_\lambda + \tfrac{1}{2}\beta\lambda \int_{R^3} (\zeta/\lambda)^{1+(1/\beta)} \, dx.$$

We now conclude [recall definition (6.42)] that

$$(8.8) \qquad E_\beta(\zeta) \leqslant C_\lambda \qquad \text{for all } \zeta \in \mathcal{Q}'_{\infty,\beta}.$$

Let $\zeta_j \in \mathcal{Q}'_{\infty,\beta}$ be a maximizing sequence for E_β; that is, $E_\beta(\zeta_j) \leqslant E_\beta(\zeta_{j+1})$ and

$$\lim_{j\to\infty} E_\beta(\zeta_j) = \sup_{\tilde{\zeta} \in \mathcal{Q}'_{\infty,\beta}} E_\beta(\tilde{\zeta}).$$

It is easy to see that (6.42) and (8.7) imply that

$$(8.9) \qquad N(\zeta_j) \leqslant C_\lambda + |E_\beta(\zeta_j)| \qquad (j \geqslant 1).$$

Thus, for a subsequence,

$$(8.10) \qquad \zeta_j \to \zeta \text{ weakly in } L^{1+(1/\beta)}(R^3).$$

The limit ζ is then an element of $\mathcal{Q}'_{\infty,\beta}$ (although not necessarily $\mathcal{Q}_{\infty,\beta}$). Also, by replacing each $\zeta_j(r,z)$ by its symmetrical rearrangement in z [which cannot decrease $E_\beta(\zeta_j)$], we may assume that each $\zeta_j(r,z)$, and hence also $\zeta(r,z)$, is a nonincreasing function of z for $z > 0$.

We claim:

$$(8.11) \qquad \lim_{j\to\infty} E(\zeta_j) = E(\zeta);$$

this will imply that ζ is a solution of (8.2).

To prove (8.11) we first note that for arbitrary $R > 1$, $A > 1$,

$$\int_{\{r>R\}} \psi\zeta \, dx \leqslant C[N(\zeta) + 1] \int_{\{r>R\}} r^{-1+\varepsilon}\zeta \, dx$$

$$(8.12) \qquad\qquad\qquad \leqslant C[N(\zeta) + 1] R^{-3+\varepsilon},$$

$$\int_{\substack{\{0<r\leqslant R\} \\ \{|z|>AR\}}} \psi\zeta \, dx \leqslant C[N(\zeta) + 1] A^{-1+\varepsilon} \int_{R^3} \zeta \, dx + CR^{-3} \int_{R^3} r^2\zeta \, dx$$

$$(8.13) \qquad\qquad\qquad \leqslant C[N(\zeta) + 1] A^{-1+\varepsilon} + CR^{-3},$$

for $0 < \varepsilon < 1$ fixed; (8.12) follows by Lemma 7.4, while (8.13) follows by

Lemma 7.5. Clearly, estimates analogous to (8.12), (8.13) hold with ζ replaced by ζ_j. Thus, as R and A may be taken arbitrarily large, it suffices to show that

$$\int_{\substack{\{0<r\leqslant R\} \\ \{|z|\leqslant AR\}}} \psi_j \zeta_j \, dx \rightarrow \int_{\substack{\{0<r\leqslant R\} \\ \{|z|\leqslant AR\}}} \psi \zeta \, dx \qquad \text{as } j \rightarrow \infty;$$

ψ_j being defined as usual corresponding to ζ_j. Also, recalling Lemma 7.3, for $0 < r \leqslant R, |z| \leqslant AR$,

$$\iint_{\{s>R\}} G(r, z, r', z')\zeta(r', z')r' \, dr' \, dz' \leqslant CR^{-3}r^2,$$

and hence

$$\iint_{\substack{\{0<r\leqslant R\} \\ \{|z|\leqslant AR\}}} \iint_{\{s>R\}} G(r, z, r', z')\zeta(r, z)\zeta(r', z')r \, dr \, dz \, r' \, dr' \, dz' \leqslant CR^{-3};$$

an analogous estimate holds for ζ_j. In view of this, it suffices to show that

$$\iint_D \iint_D G(r, z, r', z')\zeta_j(r, z)\zeta_j(r', z')r \, dr \, dz \, r' \, dr' \, dz'$$

$$\rightarrow \iint_D \iint_D G(r, z, r', z')\zeta(r, z)\zeta(r', z')r \, dr \, dz \, r' \, dr' \, dz',$$

where $D = \{(r, z) \in H: 0 \leqslant r \leqslant 2R, |z| \leqslant (A + 1)R\}$ is a bounded domain. But now the result follows by standard arguments since $G \in L^{1+\beta}(D \times D)$ and $\zeta_j(r, z)\zeta_j(r', z') \rightarrow \zeta(r, z)\zeta(r', z')$ weakly in $L^{1+1/\beta}(D \times D)$ (in the product measure $r \, dr \, dz \, r' \, dr' \, dz'$). This completes the proof of (8.11) and also of the lemma.

Lemma 8.2. *If $0 < \beta < 5\delta$ $(0 < \delta < 1)$, then for all $0 < \lambda < \infty$ there holds*

(8.14) $$E_\beta(\zeta) \geqslant c(\lambda, \delta) > 0.$$

The constant $c(\lambda, \delta)$ is independent of β but degenerates as $\delta \rightarrow 1$ or as $\lambda \rightarrow 0$.

Proof. Let $\zeta_1(x) = \lambda I_{\{|x|<a\}}$ with a determined so that $\frac{1}{2}\int_{R^3} r^2\zeta_1(x) \, dx = 1$. Consider the scaled functions

$$\zeta_\sigma(x) = \sigma^5 \zeta_1(\sigma x) \qquad (0 < \sigma < 1).$$

Then

$$\int_{R^3} \zeta_\sigma \, dx = \sigma^2 \int_{R^3} \zeta_1 \, dx,$$

$$\tfrac{1}{2} \int_{R^3} r^2 \zeta_\sigma \, dx = \tfrac{1}{2} \int_{R^3} r^2 \zeta_1 \, dx,$$

and hence for sufficiently small σ (depending only on λ), we have $\zeta_\sigma \in \mathcal{C}_{\infty,\beta}$. An easy calculation yields

$$E_\beta(\zeta_\sigma) = \sigma^3 E(\zeta_1) - \sigma^{2+(5/\beta)} \beta \lambda \int_{R^3} (\zeta_1/\lambda)^{1+(1/\beta)} \, dx$$

$$= \sigma^3 E(\zeta_1) - \sigma^{2+(5/\beta)} \beta \int_{R^3} \zeta_1 \, dx.$$

Thus $E_\beta(\zeta_\sigma) \geq c(\lambda, \delta)$ for $0 < \beta < 5\delta$ provided that σ is fixed small enough. Since $E_\beta(\zeta) \geq E_\beta(\zeta_\sigma)$, (8.14) follows.

We now deduce the variational conditions for the solution ζ.

Lemma 8.3. *The solution asserted by Lemma 8.1 satisfies (6.45), (6.46) for some constants $W \geq 0$, $\gamma \geq 0$ (uniquely determined by ζ).*

Proof. The positivity of $E_\beta(\zeta)$ given by Lemma 8.2 implies that meas (supp ζ) > 0; therefore, we can find $\delta_0 > 0$ such that meas $\{\zeta \geq \delta_0\} > 0$. We choose bounded, measurable functions h_1, h_2 of the form (6.33) such that

$$\text{supp } h_1, \text{supp } h_2 \subseteq \{\zeta \geq \delta_0\}$$

$$\int_H h_1 \, dx = 1, \qquad \tfrac{1}{2} \int_H r^2 h_1 \, dx = 0,$$

$$\int_H h_2 \, dx = 0, \qquad \tfrac{1}{2} \int_H r^2 h_2 \, dx = 1;$$

Let h be an arbitrary bounded, measurable function [of the form (6.33)] subject to the restriction that

$$h \geq 0 \qquad \text{a.e. on } \{\zeta < \delta\}$$

for some $\delta > 0$. Then $\zeta + \varepsilon \eta \in \mathcal{C}'_{\infty,\beta}$ for

$$\eta = h - \left(\int_H h \, dx \right) h_1 - \left(\tfrac{1}{2} \int_H r^2 h \, dx \right) h_2,$$

provided that $\varepsilon > 0$ is sufficiently small. By the maximality of ζ, we have

$$0 \geqslant \frac{d}{d\varepsilon} E_\beta(\zeta + \varepsilon\eta)\,|_{\varepsilon=0}$$

$$= E_\beta'(\zeta)\eta$$

$$= E_\beta'(\zeta)h - \left(\int_H h\,dx\right) E_\beta'(\zeta)h_1 - \left(\tfrac{1}{2}\int_H r^2 h\,dx\right) E_\beta'(\zeta)h_2$$

with the Frechet differential

$$E_\beta'(\zeta)h = \int_H \psi h\,dx - (1+\beta)\int_H \left(\frac{\zeta}{\lambda}\right)^{1/\beta} h\,dx.$$

Now by the arbitrariness of h (and δ) we obtain the variational conditions

$$(1+\beta)\left(\frac{\zeta}{\lambda}\right)^{1/\beta} = \psi - \gamma - \tfrac{1}{2}Wr^2 \qquad \text{if } \zeta > 0,$$

$$0 \geqslant \psi - \gamma - \tfrac{1}{2}Wr^2 \qquad \text{if } \zeta = 0,$$

with the Lagrange multipliers

$$\gamma = E_\beta'(\zeta)h_1, \qquad W = E_\beta'(\zeta)h_2.$$

These conditions are equivalent to (6.45), (6.46).

Next we show that γ, $W \geqslant 0$. Clearly, there is a sequence of points $(r_n, z_n) \in H$ such that $r_n \to 0$, $z_n \to \infty$ and $\zeta(r_n, z_n) \to 0$. Then the variational conditions imply that

$$\limsup_{n\to\infty} \left\{ \psi(r_n, z_n) - \gamma - \tfrac{1}{2}Wr_n^2 \right\} \leqslant 0.$$

Since $\psi \geqslant 0$ everywhere, we conclude that $\gamma \geqslant 0$. Now take a sequence such that $r_n \to \infty$, $z_n \to 0$, and $\zeta(r_n, z_n) \to 0$. Then since $\psi(r_n, z_n) \to 0$, we conclude, analogously, that $W \geqslant 0$.

To prove uniqueness of the Lagrange multipliers γ, W suppose that γ^*, W^* is another such pair; that is, (6.45), (6.46) hold for γ^*, W^*. This is equivalent to

$$E_\beta'(\zeta)h - \gamma^*\left(\int_H h\,dx\right) - W^*\left(\tfrac{1}{2}\int_H r^2 h\,dx\right) \leqslant 0$$

for any h subject to the restriction

$$h \geqslant 0 \qquad \text{a.e. on } \{\zeta < \delta\}$$

for some $\delta > 0$. In particular, we can take $h = \pm h_1,\ \pm h_2$ (recalling that supp $h_1 \subset \{\zeta \geq \delta_0\}$) and conclude that $\gamma^* = E'_\beta h_1$, $W^* = E'_\beta h_2$.

Lemma 8.4. *For the solution ζ there holds*

$$(8.15) \qquad \frac{1}{r}|\nabla\psi| \leq C(1 + r^{\beta+1}) \qquad (C > 0),$$

$$(8.16) \qquad E(\zeta) = \frac{1}{2}\int_H \frac{1}{r^2}(\psi_r^2 + \psi_z^2)\,dx < \infty.$$

The lemma is not obvious since ζ is not known yet to have compact support.

Proof. By the calculation of Lemma 6.1 [specifically, combining the curl of equation (6.23) with (6.26)],

$$(8.17) \qquad \frac{1}{r}|\nabla\psi(x)| \leq \frac{1}{4\pi}\int_{R^3}\frac{r'\zeta(x')\,dx'}{|x - x'|^2}.$$

Using the fact that

$$(8.18) \quad \int_{R^3}r'\zeta(x')\,dx' \leq \left(\int_{R^3}\zeta(x')\,dx'\right)^{1/2}\left(\int_{R^3}r'^2\zeta(x')\,dx'\right)^{1/2} \leq \sqrt{2},$$

we find that

$$\int_{\{|x-x'|>1\}}\frac{r'\zeta(x')\,dx'}{|x - x'|^2} \leq \sqrt{2}.$$

Since

$$(8.19) \qquad \zeta = C(u^+)^\beta \leq C\psi^\beta \leq Cr^\beta$$

by Lemmas 8.3 and 7.4, we also have

$$\int_{\{|x-x'|<1\}}\frac{r'\zeta(x')\,dx'}{|x - x'|^2} \leq C(r^{\beta+1} + 1),$$

and (8.15) follows.

To prove (8.16) recall that

$$E(\zeta) = -\int_H \frac{1}{r^2}\psi L\psi\,dx.$$

Integrating by parts on the domain

$$D_a = \{(r, z) \in H; 0 \leqslant r < a, |z| < a\},$$

we obtain (using the hint to Problem 1 of Section 7)

$$-\iint_{D_a} \frac{1}{r}\psi L\psi \, dr \, dz = \iint_{D_a} \frac{1}{r}(\psi_r^2 + \psi_z^2) \, dr \, dz - \int_{\partial_0 D_a} \psi \frac{1}{r} \frac{\partial\psi}{\partial\nu} \, ds,$$

$$\partial_0 D_a \equiv \partial D_a \cap \{r > 0\}$$

(ν is the outer unit normal on ∂D_a, and ds is arc length). Therefore, to prove (8.16) it suffices to show that the boundary integral tends to zero as $a \to \infty$.

We claim that for some $\varepsilon_0 > 0$,

$$(8.20) \qquad \int_{\partial_0 D_a} \psi \frac{1}{r} |\nabla\psi| \, ds = O(a^{-\varepsilon_0}) \qquad \text{as } a \to \infty.$$

To prove it we again use (8.17) and write, for $0 < \delta < 1$,

$$V_1(x) = \frac{1}{4\pi} \int_{\{|x-x'|<a^\delta\}} \frac{r'\zeta(x') \, dx'}{|x - x'|^2}, \qquad V_2(x) = \frac{1}{4\pi} \int_{\{|x-x'|\geqslant a^\delta\}} \frac{r'\zeta(x') \, dx'}{|x - x'|^2};$$

we estimate these separately. Using (8.18), it is clear that $V_2(x) \leqslant Ca^{-2\delta}$. Thus, recalling (7.20),

$$\int_{\substack{\{r=a\} \\ \{|z|\leqslant a\}}} \psi V_2 \, dz \leqslant C(1 + a)^{-1+\varepsilon} a^{-2\delta} \int_{-a}^{a} dz \leqslant Ca^{-2\delta+\varepsilon},$$

$$\int_{\substack{\{z=\pm a\} \\ \{0<r\leqslant a\}}} \psi V_2 \, dr \leqslant Ca^{-2\delta} \int_{0}^{a} (1 + r)^{-1+\varepsilon} \, dr \leqslant Ca^{-2\delta+\varepsilon}.$$

Fixing $\varepsilon < 2\delta$, we get

$$(8.21) \qquad \int_{\partial_0 D_a} \psi V_2 \, ds = O(a^{-\varepsilon_0}).$$

We next estimate the corresponding expressions in V_1. It follows from the axisymmetry of $\zeta(x)$ combined with the fact that $\zeta(r, z)$ is a nonincreasing function of z for $z > 0$ that

$$(8.22) \qquad \int_{\{|x-x'|<a^\delta\}} r'\zeta(x') \, dx' \leqslant C_0\left(1 + \frac{r}{a^\delta}\right)^{-1}\left(1 + \frac{|z|}{a^\delta}\right)^{-1}$$

for all $(r, z) \in H$. Indeed, since for $r > 2a^\delta$,

$$\text{meas}\left\{x' \in R^3; |x - x'| < a^\delta\right\}$$

$$\leqslant \frac{Ca^\delta}{r} \text{meas}\left\{x' \in R^3; |r - r'| < a^\delta, |z - z'| < a^\delta\right\},$$

we find [using $\zeta(x') = \zeta(r', z')$] that

$$\int_{\{|x-x'|<a^\delta\}} r'\zeta(x')\, dx' \leqslant \frac{Ca^\delta}{r} \int_{\substack{\{|r-r'|<a^\delta\} \\ \{|z-z'|<a^\delta\}}} r'\zeta(x')\, dx' \leqslant \frac{Ca^\delta}{r} \qquad \text{[by (8.18)]};$$

in turn, for $z > 2a^\delta$ [using $\zeta(r', z')\downarrow$ as $z'\uparrow$, $z' > 0$],

$$\int_{\substack{\{|r-r'|<a^\delta\} \\ \{|z-z'|<a^\delta\}}} r'\zeta(x')\, dx' \leqslant \frac{Ca^\delta}{z} \int_{\{|r-r'|<a^\delta\}} r'\zeta(x')\, dx' \leqslant \frac{Ca^\delta}{z}.$$

Thus (8.22) follows.

Also, by (8.19), (7.27), we have

$$(8.23) \qquad \sup_{|x-x'|<a^\delta} r'\zeta(x') \leqslant C_1 a^{1-\mu} \qquad \text{if } x \in \partial D_a \ (0 < \mu < 1),$$

where μ depends on β. For $r = a$, $|z| \leqslant a$, consider the problem to maximize $V_1(x)$ as a functional of $r'\zeta(x')$ subject to (8.22), (8.23). Clearly, the maximum occurs when

$$r'\zeta(x') = C_1 a^{1-\mu} I_{\{|x-x'|<\rho\}},$$

where ρ is determined so that equality holds in (8.22). Computing $V_1(x)$ in this case, we obtain

$$V_1(x) \leqslant Ca^{(1/3)-(2/3)\mu+(1/3)\delta}\left(1 + \frac{|z|}{a^\delta}\right)^{-1/3}.$$

Thus

$$\int_{\substack{\{r=a\} \\ \{|z|\leqslant a\}}} \psi V_1\, dz \leqslant C(1 + a)^{-1+\varepsilon} a^{(1/3)-(2/3)\mu+(1/3)\delta}\int_{-a}^{a}\left(1 + \frac{|z|}{a^\delta}\right)^{-1/3} dz$$

$$\leqslant Ca^{-(2/3)\mu+(2/3)\delta+\varepsilon}.$$

Similarly, for $z = \pm a$, $0 < r \leqslant a$, we find that

$$V_1(x) \leqslant Ca^{(1/3)-(2/3)\mu+(1/3)\delta}(1 + r/a^\delta)^{-1/3}$$

and thus

$$\int_{\substack{\{z=\pm a\} \\ \{0<r\leqslant a\}}} \psi V_1 \, dr \leqslant Ca^{-(2/3)\mu+(5/3)\delta+(1-\delta)\varepsilon}.$$

Fixing ε and δ sufficiently small, we find that

$$\int_{\partial D_a} \psi V_1 \, ds = O(a^{-\varepsilon_0}).$$

This together with (8.21) proves the claim (8.20) and hence (8.16).

Lemma 8.5. *In the notation of Lemma 8.2, there holds*

$$(8.24) \qquad\qquad W \geqslant \tfrac{1}{2}c(\lambda, \delta) > 0;$$

as a consequence, $\zeta \in \mathcal{C}_{\infty, \beta}$.

In the sequel we shall assume that $0 < \beta < 5\delta$ [δ fixed in $(0, 1)$], and no longer specify the dependence of any constants on δ.

Proof. By Lemma 8.4 (which assures the hypothesis of Lemma 7.1) and the remark following Lemma 7.2, we may apply the identity (7.11) to the solution ζ and obtain

$$(8.25) \qquad\qquad W \geqslant \tfrac{1}{2}E(\zeta)$$

since $J(u) \geqslant 0$. The assertion (8.24) now follows immediately from (8.14) using $E_\beta(\zeta) \leqslant E(\zeta)$.

Since W, the Lagrange multiplier for the constraint (8.1), must equal zero if strict inequality holds in (8.1), we now conclude that the equality (6.34) holds for the solution ζ; that is, $\zeta \in \mathcal{C}_{\infty, \beta}$.

To complete the proof of Theorem 6.3 it remains to show that the solution obtained has compact support. This is established in Lemmas 8.7 and 8.9.

Lemma 8.6. *If $(r, z) \in$ supp ζ, then*

$$(8.26) \qquad\qquad \tfrac{1}{2}r^2|z| \leqslant \frac{E(\zeta)}{\pi W^2}.$$

Proof. Suppose that $z > 0$; then for all $0 < z' < z$, since $(r, z') \in \text{supp } \zeta$, we have

$$\psi(r, z') \geq \tfrac{1}{2} W r^2 + \gamma \geq \tfrac{1}{2} W r^2.$$

Then, since $\psi(0, z') = 0$,

$$\tfrac{1}{2} W r^2 \leq \int_0^r \psi_{r'}(r', z') \, dr'.$$

Integrating in z', we get

$$\tfrac{1}{2} W r^2 z \leq \int_0^r \int_0^z \psi_{r'}(r', z') \, dr' \, dz'$$

$$\leq \left\{ \int_0^r \int_0^z r' \, dr' \, dz' \right\}^{1/2} \left\{ \int_0^r \int_0^z \frac{1}{r'} \psi_{r'}^2(r', z') \, dr' dz' \right\}^{1/2}$$

$$\leq \left(\tfrac{1}{2} r^2 z \right)^{1/2} \left(\frac{E(\zeta)}{\pi} \right)^{1/2};$$

so (8.26) follows.

Lemma 8.7. *There is $r^*(\lambda) < \infty$ such that*

(8.27) $r \leq r^*(\lambda)$ *for all $(r, z) \in \text{supp } \zeta$.*

Proof. By (8.24), we have, for $(r, z) \in \text{supp } \zeta$,

(8.28) $\psi(r, z) \geq \tfrac{1}{2} W r^2 + \gamma \geq c_\lambda r^2$; $c_\lambda > 0$.

Combining this with the estimate (7.20) [recall that $N(\zeta) \leq C_\lambda$], we find

$$c_\lambda r^2 \leq \psi(r, z) \leq C_{\lambda, \varepsilon} r^{-1+\varepsilon} \qquad (0 < \varepsilon < 1).$$

Now fixing ε, we get $r^{3-\varepsilon} \leq \tilde{C}_\lambda$, as required.

Lemma 8.8. *For $0 < r < \rho$, there holds*

(8.29) $\displaystyle\iint\limits_{\substack{\{0 < r' < \rho\} \\ \{-\infty < z' < \infty\}}} G(r, z, r', z') r' \, dr' \, dz' \leq C r^2 \rho^2$;

C is an absolute positive constant.

Proof. Writing $s = [(r - r')^2 + (z - z')^2]^{1/2}$, we consider separately the contributions for $s \leq r/2$ and $s > r/2$ as in the estimates of Section 7.

According to Lemma 7.3,

$$\iint\limits_{\{s \le r/2\}} G(r, z, r', z')r' \, dr' \, dz' \le Cr^2 \iint\limits_{\{s \le r/2\}} \log\frac{r}{s} \, dr' \, dz'$$

$$= Cr^4 \le Cr^2\rho^2.$$

$$\iint\limits_{\substack{\{s > r/2\} \\ \{0 < r' < \rho\}}} G(r, z, r', z')r' \, dr' \, dz' \le Cr^2 \iint\limits_{\substack{\{s > r/2\} \\ \{0 < r' < \rho\}}} \frac{r'^3}{s^3} \, dr' \, dz'.$$

It is easily verified that for $s > r/2$, there holds

$$\frac{r'^3}{s^3} \le \frac{Cr'^3}{\left[(z - z')^2 + (r + r')^2\right]^{3/2}}.$$

Hence it now suffices to estimate

$$\iint\limits_{\{0 < r' < \rho\}} \frac{r'^3}{\left[(z - z')^2 + (r + r')^2\right]^{3/2}} \, dr' \, dz' = 2\int_0^\rho \frac{r'^3}{(r + r')^2} \, dr' \le C\rho^2.$$

Lemma 8.9. *There is* $z^*(\lambda) < \infty$ *such that*

(8.30) $$|z| \le z^*(\lambda) \qquad \text{for all } (r, z) \in \operatorname{supp} \zeta.$$

Proof. By Lemma 8.6 we have $r^2z \le C_\lambda$ (we take $z > 0$); thus to prove (8.30) we may assume that $r < z$, say. Let ρ be defined by

$$\rho^2\frac{z}{2} = C_\lambda.$$

Then

$$\psi(r, z) = \iint\limits_H G(r, z, r', z')\zeta(r', z')r' \, dr' \, dz'$$

$$\le \sup_{0 < r' \le \rho} \zeta(r', z') \iint\limits_{\{0 < r' \le \rho\}} G(r, z, r', z')r' \, dr' \, dz'$$

$$+ \iint\limits_{\{r' > \rho\}} G(r, z, r', z')\zeta(r', z')r' \, dr' \, dz'$$

$$\equiv I_1 + I_2.$$

Using the variational conditions (6.45), (6.46) and noting that $\sup_H \psi \leq C < \infty$, we get $\zeta \leq C$; therefore, upon using Lemma 8.8,

$$I_1 \leq Cr^2\rho^2 \leq C_1 \frac{r^2}{z} \qquad (C_1 > 0).$$

As for I_2, if $(r', z') \in \text{supp} \, \zeta$ and $r' > \rho$, then by Lemma 8.6

$$r'^2|z'| \leq C_\lambda = \rho^2 \frac{z}{2} < r'^2 \frac{z}{2},$$

so that $|z'| < z/2$. Thus in the notation of Lemma 7.3, there holds $s > z/2 > r/2$. Using (7.13), we then get

$$I_2 \leq Cr^2 \iint\limits_{\{r' > \rho\}} \frac{r'^2}{s^3} \zeta(r', z')r' \, dr' \, dz'$$

$$\leq \frac{Cr^2}{z^3} \iint\limits_H r'^2\zeta(r', z')r' \, dr' \, dz' = \frac{Cr^2}{z^3}.$$

Combining the estimates on I_1, I_2 and using (8.28), we obtain

$$C_\lambda r^2 \leq \psi(r, z) \leq \frac{C_1 r^2}{z} + \frac{Cr^2}{z^3},$$

and (8.30) follows.

We have now completed the proof of Theorem 6.3. We proceed to prove Theorem 6.2.

Lemma 8.10. *There exists a solution ζ of (6.38); ζ has compact support and it is nondecreasing in z for $z \geq 0$.*

Proof. Let ζ_β denote the solution obtained in Theorem 6.3 for the penalized problem with a prescribed λ (and $0 < \beta < 1$, say). We know from Lemmas 8.7 and 8.9 that

$$(8.31) \qquad \sup \zeta_\beta \subseteq D \equiv \{(r, z) \in H; 0 \leq r \leq r^*(\lambda), |z| \leq z^*(\lambda)\},$$

independent of β.

Applying (8.7) to ζ_β and recalling that $E_\beta(\zeta_\beta) > 0$, we find that

$$(8.32) \qquad \beta \int_H (\zeta_\beta/\lambda)^{1+(1/\beta)} \, dx \leq C_\lambda.$$

Using the notation in (7.19), this becomes

$$\frac{1}{\lambda} N(\zeta_\beta) \leq \left(\frac{1}{\beta} C_\lambda\right)^{\beta/(1+\beta)},$$

which in turn implies that

(8.33) $$N(\zeta_\beta) \leq \lambda + o(1) \qquad \text{as } \beta \to 0.$$

Now if $\beta < \alpha$, for any fixed $0 < \alpha < 1$, then we can estimate

(8.34) $$\|\zeta_\beta\|_{L^{1+(1/\alpha)}} \leq \|\zeta_\beta\|_{L^1}^{1-a} \|\zeta_\beta\|_{L^{1+(1/\beta)}}^{a} \qquad \left(a = \frac{1+\beta}{1+\alpha}\right)$$

$$\leq [\lambda + o(1)]^a \qquad \text{as } \beta \to 0.$$

Therefore, there exists a sequence $\beta_j \to 0$ such that

(8.35) $$\zeta_{\beta_j} \to \zeta \text{ weakly in } L^{1+(1/\alpha)}(D) \qquad \text{for every } 0 < \alpha < 1.$$

Furthermore, by (8.34),

$$\|\zeta\|_{L^{1+(1/\alpha)}} \leq \liminf_{j \to \infty} \|\zeta_{\beta_j}\|_{L^{1+(1/\alpha)}} \leq \lambda^{1/(1+\alpha)},$$

and taking $\alpha \to 0$, we conclude that

(8.36) $$\text{ess sup } \zeta \leq \lambda.$$

Also, by virtue of (8.31), it follows easily that

(8.37) $$\int_H \zeta \, dx \leq 1, \qquad \frac{1}{2} \int_H r^2 \zeta \, dx = 1.$$

Thus $\zeta \in \mathcal{C}_\lambda$, sup $\zeta \subseteq D$, and $\zeta(r, z)$ is a nonincreasing function of z for $z > 0$.

Finally, we must show that (6.38) holds. In light of (8.31) it is immediate from standard arguments that

$$\lim_{j \to \infty} E(\zeta_{\beta_j}) = E(\zeta).$$

For any $\tilde{\zeta} \in \mathcal{C}_\lambda$ we have

$$E(\zeta) = \lim_{j \to \infty} E(\zeta_{\beta_j}) \geq \lim_{j \to \infty} E_{\beta_j}(\zeta_{\beta_j})$$

$$\geq \lim_{j \to \infty} E_{\beta_j}(\tilde{\zeta}) = E(\tilde{\zeta});$$

the last equality follows since

$$\lim_{\beta \to 0} \beta \lambda \int_H \left(\frac{\tilde{\zeta}}{\lambda} \right)^{1+(1/\beta)} dx = 0.$$

This establishes (6.38).

Lemma 8.11.   *There exist constants $W > 0$, $\gamma \geqslant 0$ such that (6.39), (6.40) hold.*

Proof.   Let W_j, γ_j, ψ_j, u_j denote the quantities in the statement of Theorem 6.3 associated with ζ_{β_j}, $\beta_j \to 0$. Since W_j, $\gamma_j \geqslant 0$, and as a consequence of (7.11), $W_j + \gamma_j \leqslant 2E(\zeta_{\beta_j}) \leqslant C_\lambda$, we may assume (by taking a subsequence) that $W_j \to W$ and $\gamma_j \to \gamma$. Then, $\gamma \geqslant 0$ and, by (8.24), $W \geqslant c_\lambda > 0$.
In view of (8.31), (8.35),

$$\psi_j(x) = \int_{R^3} K(x, x')\zeta_{\beta_j}(x') \, dx'$$

converges pointwise in R^3 to

$$\psi(x) = \int_{R^3} K(x, x')\zeta(x') \, dx'.$$

Hence $u_j(x) \to u(x)$ pointwise on R^3, where $u = \psi - \frac{1}{2}Wr^2 - \gamma$. Also, according to the variational condition (6.45), we have

$$\lim_{j \to \infty} \zeta_{\beta_j}(x) = \begin{cases} \lambda & \text{if } u(x) > 0 \\ 0 & \text{if } u(x) < 0. \end{cases}$$

Furthermore, since $u_z = \psi_z < 0$ for $z > 0$,

$$\text{meas}\{x \in R^3; u(x) = 0\} = 0.$$

Thus the function $\lambda I_{\{u(x)>0\}}$ is the pointwise limit of $\zeta_{\beta_j}(x)$ for a.a. $x \in R^3$; since also $\zeta_{\beta_j} \to \zeta$ weakly, (6.39) follows. Thus the proof of Theorem 6.2 is complete.

Remark 8.1   Theorem 5.1 extends to vortex rings; it yields analyticity of the function $z = z(r)$ representing $\partial\{u > 0\} \cap \{z > 0\}$.

PROBLEMS

1.   Show that

(8.38)   if $\gamma > 0$,   then $\int_{R^3} \zeta(x) \, dx = 1$, where ζ is the solution constructed in Theorem 6.2.

[*Hint*: Otherwise, $\int \zeta_{\beta_j} dx < 1$ and then $\gamma_j = 0$.]

2. Show, with ζ as before, that

(8.39) $$\int \zeta(x)\, dx < 1 \qquad \text{if } \lambda \text{ is sufficiently large.}$$

[*Hint*: If $\Omega_0 = B_R(2R, 0)$, $R = c\lambda^{-1/5}$, then $\zeta_0 = I_{\Omega_0}$ is in $\mathcal{Q}_\lambda$ and $E(\zeta_0) \geqslant c\lambda^{3/5}$ $(c > 0)$. Deduce, using $W \geqslant \frac{1}{2} E(\zeta)$ and Lemma 8.6,

(8.40) $$|z| \leqslant \frac{C}{\lambda^{3/5}} \frac{1}{r^2} \qquad \text{in supp } \zeta.$$

The proof of (8.30) for ζ gives

$$\psi(r, z) \leqslant \left(\frac{C\lambda}{|z|} + \frac{C}{|z|^3} \right) r^2.$$

Using $\psi \geqslant \frac{1}{2} W r^2$ in supp ζ, deduce that

(8.41) $$Z_\lambda \leqslant \frac{C}{\lambda^{1/5}}, \qquad Z_\lambda = 2 + \sup\{|z|, (r, z) \in \text{supp } \zeta\}.$$

Use (8.40), (8.41) to estimate

$$\int_{\{r < \sqrt{2}\}} \zeta\, dx \leqslant C\lambda^{2/5} \log \frac{1}{\varepsilon} \qquad \left(\varepsilon^2 \lambda^{3/5} = C/Z_\lambda \right)$$

$$\leqslant C\lambda^{2/5} \log \frac{1}{\lambda},$$

and recall that $\frac{1}{2} \int r^2 \zeta\, dx = 1$.]

9. A CAPACITY ESTIMATE

We establish a capacity estimate that is useful in studying asymptotic estimates of some free-boundary problems. In Section 10 the estimate is used only in the problems at the end of the section. In Section 13 we shall use this estimate in a more basic way.

Definition 9.1.   Let E, Ω be bounded sets in R^n, E closed, Ω open, $E \subset \Omega$, and set

$$K = \left\{ v \in H_0^1(\Omega), v \geqslant 1 \text{ on } E \right\}.$$

Then the *capacity of E* with respect to Ω is defined by

$$(9.1) \qquad \qquad \text{Cap}_\Omega E = \inf_{v \in K} \int_\Omega | \nabla v |^2 \, dx.$$

Definition 9.2. The solution w of the variational inequality

$$(9.2) \qquad \qquad \int_\Omega \nabla w \cdot \nabla (v - w) \, dx \geq 0 \qquad \forall \, v \in K; \, w \in K$$

is called the *capacitary potential of E* with respect to Ω.
   Clearly,

$$-\Delta w \geq 0 \qquad \text{in } \Omega$$

in the sense of distributions or measures,

$$-\Delta w = 0 \qquad \text{in } \Omega \setminus E,$$

$$w = 1 \qquad \text{a.e. on } E$$

and

$$0 \leq w \leq 1 \qquad \text{a.e. in}$$

(since otherwise $\min \{w^+, 1\}$ [which belongs to K] has a smaller Dirichlet integral than w).
   If ∂E and $\partial \Omega$ are in $C^{1+\alpha}$, then w satisfies:

$$(9.3) \qquad \begin{aligned} \Delta w &= 0 & \text{in } \Omega \setminus E, \\ w &= 1 & \text{in } E, \\ w &= 0 & \text{on } \partial \Omega, \end{aligned}$$

and

$$(9.4) \quad \text{Cap}_\Omega E = \int_{\Omega \setminus E} | \nabla w |^2 = \int_{\partial E} \frac{\partial w}{\partial \nu} \, dS = - \int_{\partial \Omega} \frac{\partial w}{\partial \nu} \, dS = - \int_\Omega \Delta w,$$

where ν is the outward normal to $\Omega \setminus E$; the measure $-\Delta w$ is supported on ∂E.
   From (9.1) one easily infers that

$$(9.5) \qquad \qquad \text{Cap}_\Omega E \uparrow \quad \text{if } E \uparrow,$$
$$(9.6) \qquad \qquad \text{Cap}_\Omega E \downarrow \quad \text{if } \Omega \uparrow.$$

   We shall now restrict ourselves to $n = 2$. One can easily compute that

$$(9.7) \qquad \text{Cap}_{B_R} B_r = \frac{2\pi}{\log(1/r) + O(1)} \qquad \text{as } r \to 0 \qquad (n = 2).$$

We denote the diameter of a set A by $d(A)$. We shall estimate the diameter of a closed domain E from above in terms of the capacity of E with respect to a domain containing E.

Theorem 9.1. *Let E be a closed domain in R^2 such that*

$$E \subset B_{R_1} \subset B_{2R_1} \subset \Omega \subset B_{R_2},$$

where Ω is a domain in R^2. Then

(9.8)
$$\left(1 - \frac{C}{\log(1/d(E))}\right) \frac{2\pi}{\log(2\pi/d(E))} \leqslant \mathrm{Cap}_\Omega E,$$

where C is a positive constant depending only on R_1, R_2.

Proof. Given any small $\varepsilon > 0$, choose points A, B in E such that

$$d(E) - \varepsilon \leqslant |A - B| \leqslant d(E),$$

and connect them by a smooth curve L in E such that

(9.9)
$$L \text{ is linear in a neighborhood of} \\ \text{each of the endpoints } A \text{ and } B.$$

Lemma 9.2. *There exists a positive constant C depending only on R_1, R_2 such that*

(9.10)
$$\mathrm{Cap}_{B_{R_2}} L \geqslant \left(1 - \frac{C}{\log 1/l}\right) \frac{2\pi}{\log(2\pi/l)},$$

where $l = |L| =$ length of L.

Proof. Consider first the case where

(9.11)
$$L \text{ is a line segment}$$

and take for simplicity $R_2 = 1$. Denote by $G(x, y)$ the Green function for $-\Delta$ in B_1 and consider the function

(9.12)
$$f(x) = \int_L G(x, y) \frac{1}{(l/2\pi)\log 2\pi/l} \, ds_y;$$

notice that we are integrating here G against the (roughly) "capacitary distribution" of the ball $B_{l/2\pi}$ of circumference l (equal to the diameter of L); see

(9.7). Since

$$(9.13) \qquad G(x, y) = \frac{1}{2\pi} \log \frac{1}{|x - y|} + O(l) \qquad \text{on } L,$$

$$\max_L f = -\frac{1}{2\pi} \int_L \log r \frac{1}{(l/2\pi) \log 2\pi/l} ds + O\left(\frac{l}{\log l}\right)$$

$$= -\frac{1}{2\pi} 2\left[\int_0^{l/2} \log t \, dt\right] \frac{1}{(l/2\pi) \log 2\pi/l} + O\left(\frac{l}{\log l}\right)$$

$$= -\left(\frac{l}{2} \log \frac{l}{2} - \frac{l}{2}\right) \frac{1}{(l/2) \log 2\pi/l} + O\left(\frac{l}{\log l}\right)$$

$$= \frac{1}{1 - \dfrac{\tilde{C}}{\log 1/l}}, \qquad \tilde{C} \text{ a bounded function.}$$

Hence

$$(9.14) \qquad \left(1 - \frac{C}{\log 1/l}\right) f(x) \leqslant 1 \qquad \text{on } L \quad (\sup|\tilde{C}| = C).$$

Consider the function

$$(9.15) \qquad g(x) = \left(1 - \frac{C}{\log 1/l}\right) f(x)$$

$$= \int_L G(x, y) \left(1 - \frac{C}{\log 1/l}\right) \frac{1}{(l/2\pi) \log 2\pi/l} ds.$$

It satisfies

$$g(x) \leqslant 1 \qquad \text{on } L,$$

$$g = 0 \qquad \text{on } \partial B_1,$$

$$g \text{ harmonic in } B_1 \setminus L.$$

By Problems 2 to 4,

$$(9.16) \qquad \text{Cap}_{B_1} L \geqslant -\int_{\partial B_1} \frac{\partial g}{\partial \nu} = -\left(1 - \frac{C}{\log 1/l}\right) \int_{\partial B_1} \frac{\partial f}{\partial \nu}.$$

On the other hand (using Problems 1 and 2),

$$-\int_{\partial B_1} \frac{\partial f}{\partial \nu} = \int_L \left\{ \text{jump of } \frac{\partial f}{\partial \nu} \text{ across } L \right\} ds = \int_L \frac{ds}{(l/2\pi) \log 2\pi/l} = \frac{2\pi}{\log 2\pi/l}$$

by (9.7). Combining this with (9.16), the assertion (9.10) follows.

We proceed to prove the lemma in case (9.11) is not satisfied. Take a uniform distribution

$$(9.17) \qquad\qquad \frac{1}{(l/2\pi) \log 2\pi/l} \qquad \text{on } \overline{AB}.$$

and redistribute it on L (or on a part of L) in such a way that when we project L orthogonally onto $\overline{AB}$, we get the distribution (9.17). Denote this distribution on L by σ.

Consider the function

$$f(x) = \int_L G(x, y) \sigma \, ds$$

with G Green's function as before. If $x \in L$,

$$(9.18) \qquad f(x) = \frac{1}{2\pi} \int_L \log \frac{1}{|x - y|} \sigma \, ds + O(l) \int_L \sigma \, ds.$$

For each $x \in L$, we perform a substitution in each integral, which corresponds to projecting orthogonally the curve L onto a line segment parallel to $\overline{AB}$ passing through x and having length $|A - B|$; denote it by L_x;

$$(9.19) \qquad\qquad \sigma \, ds \text{ becomes } \frac{d\tilde{s}}{(l/2\pi) \log (2\pi/l)},$$

where $d\tilde{s}$ denotes the length element along L_x. The second integral in (9.18) is then bounded by $C(l/\log l)$. The first integral becomes, at x,

$$\leqslant \frac{1}{2\pi} \int_0^{l/2} \frac{1}{(l/2\pi) \log (2\pi/l)} \log \frac{1}{r} \, dr.$$

Thus we conclude as before that (9.14) holds.

We now proceed to derive

$$\text{Cap}_{B_1} L \geqslant \left(1 - \frac{C}{\log 1/l} \right) \int_L \left\{ \text{jump of } \frac{\partial f}{\partial \nu} \text{ across } L \right\} ds$$

$$= \left(1 - \frac{C}{\log 1/l} \right) \int_L \sigma \, ds = \left(1 - \frac{C}{\log 1/l} \right) \frac{2\pi}{\log 2\pi/l},$$

where (9.19) was used in obtaining the last equality. This establishes Lemma 9.2.

By (9.5), (9.6),

$$\text{Cap}_\Omega E \geqslant \text{Cap}_{B_{R_2}} L.$$

Using Lemma 9.2 and recalling that $d(E) - \varepsilon \leqslant l \leqslant d(E)$ for any $\varepsilon > 0$ [C in (9.10) is independent of ε], the assertion (9.8) follows.

We now define the capacity with respect to the operator L:

$$Lv \equiv r\left[\frac{\partial}{\partial r}\left(\frac{1}{r}\frac{\partial v}{\partial r}\right) + \frac{\partial}{\partial z}\left(\frac{1}{r}\frac{\partial v}{\partial z}\right)\right].$$

Let E be a closed domain and let Ω be a bounded domain, both in the half plane $\{(r, z), r \geqslant 0\}$ and both symmetric with respect to z, and $E \subset \Omega$. Let

$$K = \{v \in H_0^1(\Omega), v \geqslant 1 \text{ on } E, v(r, z) = v(r, -z)\}.$$

Then the *capacity of E* with respect to Ω and L is defined by

$$(9.20) \qquad \text{Cap}_\Omega^L E = \inf_{v \in K}\int_\Omega \frac{|\nabla v|^2}{r}\, dr\, dz.$$

Theorem 9.3. *If $E \subset B_{R_1} \subset B_{2R_1} \subset \Omega \subset B_{R_2}$ and if $(r_0, 0) \in E$, $r_0 \geqslant r_* > 0$, then*

$$\left(1 - \frac{C}{\log\left(1/d(E)\right)}\right)\frac{2\pi}{r_0 \log\left(2\pi/d(E)\right)} \leqslant \text{Cap}_\Omega^L E,$$

where C is a positive constant depending only on R_1, R_2, r_*.

For proof, see Problem 5.

PROBLEMS

1.   If a neighborhood of A intersects L at $\{(x, 0); -a < x < 0\}$ and $A = (0, 0)$, then near A

$$|g_y| \leqslant C, \qquad |g_x| \leqslant C|\log\left(x^2 + y^2\right)|.$$

2.   Show that

$$\int_{B_1 \setminus L}|\nabla g|^2 = \int_{\partial B_1} gg_\nu = -\lim_{\delta \to 0}\int_{\partial L_\delta} gg_\nu = \int_{\partial B_1} g_\nu,$$

where L_δ are suitable neighborhoods of L, decreasing to L as $\delta \to 0$.
[*Hint*: $\partial L_\delta = L_\delta^+ \cup L_\delta^-$, $L_\delta^\pm$ parallel to L at distance $\pm\delta$, except near the endpoints where L_δ^+ connects to L_δ^- at distance ρ from the endpoint, and $\rho = \rho(\delta) \to 0$ in a suitable manner.]

3. Prove that $g \geqslant w$; w as in (9.2).
   [*Hint*: Take $v = w + \max(0, g - w)$ in (9.2) and integrate $\int \nabla g \cdot \nabla \max(0, g - w)$ by parts in $\Omega \setminus L_\delta$, $\delta \to 0$.]

4. Prove that $\int |\nabla g|^2 \leqslant \int |\nabla w|^2$.
   [*Hint*: The difference is equal to

$$-\int |\nabla(g - w)|^2 + 2\int \nabla g \cdot \nabla(g - w).\bigg]$$

5. Prove Theorem 9.3.
   [*Hint*: Green's function $G(X, Y)$ for L in Ω can be written as

$$\frac{1}{2\pi} \log \frac{1}{|X - Y|}(1 + O(|X - Y|));$$

here $X = (r, z)$, $Y = (r', z')$ [a fundamental solution $K(X, Y)$ for $r^{-2}L$ is given by

$$\int_0^{2\pi} K(x, x') \, d\theta, \qquad x = (r, \theta, z), \quad x' = (r', 0, z'),$$

$K(x, x')$ as in (6.20)].]

10. ASYMPTOTIC ESTIMATES FOR VORTEX RINGS

We are interested in the behavior of the vortex ring constructed in Theorem 6.2 as the parameter λ increases to ∞. We shall denote the solution constructed in Theorem 6.2 by ζ_λ and set

(10.1) $$V = V_\lambda = \{(r, z); \zeta_\lambda(r, z) = \lambda\}.$$

The constraint (6.35) gives

(10.2) $$|V_\lambda| \leqslant \frac{1}{2\pi\lambda},$$

where $|A|$ denotes the measure of a set A in H (endowed with measure $r \, dr \, dz$).

Lemma 10.1. *For all λ sufficiently large*

$$(10.3) \qquad\qquad E = E(\zeta_\lambda) \geqslant \frac{1}{8\sqrt{2}\,\pi^2} \log \lambda - C,$$

where C is a constant independent of λ.

  Proof.   Let

$$\zeta_0(r, z) = \lambda I_{B_\varepsilon(\sqrt{2},0)},$$

where $B_\varepsilon(\sqrt{2},0)$ is the disc $(r - \sqrt{2}\,)^2 + z^2 < \varepsilon^2$. Then

$$\int \zeta_0(x)\, dx = 1 \qquad \text{if } 2\pi^2\sqrt{2}\,\varepsilon^2\lambda = 1,$$

that is, if

$$(10.4) \qquad\qquad \varepsilon^2\lambda = \frac{1}{2\sqrt{2}\,\pi^2}.$$

We shall compute $E(\zeta_0)$. For this purpose take first, in $G(r, z, r', z')$, $(r', z') = (\sqrt{2},0)$ and introduce new coordinates about $(\sqrt{2},0)$:

$$r = \sqrt{2} + \varepsilon s, \qquad z = \varepsilon t;$$

set

$$\xi = \sqrt{s^2 + t^2}\,.$$

Then $0 \leqslant \xi \leqslant 1$ as $(r, z) \in \operatorname{supp} \zeta_0$. In terms of the new variables, we find [recall (7.16)–(7.18)] that

$$k'^2 = \frac{\varepsilon^2 t^2 + \varepsilon^2 s^2}{\varepsilon^2 t^2 + \left(2\sqrt{2} + \varepsilon s\right)^2} = \frac{\varepsilon^2 \xi^2}{8}\{1 + O(\varepsilon)\},$$

$$k^2 = 1 + O(\varepsilon),$$

$$\frac{4}{k'} = \frac{8\sqrt{2}}{\varepsilon\xi}\{1 + O(\varepsilon)\}.$$

Hence

$$\left(\frac{2}{k} - k\right)K(k) - \frac{2}{k}E(k) = \log\frac{8\sqrt{2}}{\varepsilon\xi} - 2 + O\left(\varepsilon\log\frac{1}{\varepsilon}\right).$$

Using $(rr')^{1/2} = \sqrt{2} + O(\varepsilon)$, we get

$$G\left(r, z, \sqrt{2}, 0\right) = \frac{\sqrt{2}}{2\pi}\left[\log\frac{8\sqrt{2}}{\varepsilon\xi} - 2\right] + O\left(\varepsilon\log\frac{1}{\varepsilon}\right)$$

$$= \frac{\sqrt{2}}{2\pi}\log\frac{1}{\varepsilon\xi} + O(1) \qquad \text{as } \varepsilon \to 0.$$

Let $\zeta^0 = $ delta function on the circle $r = \sqrt{2}$, $z = 0$, $-\pi < \theta < \pi$, and

$$\psi^0(r, z) = \iint G(r, z, r', z')\zeta^0(r', z')r'\,dr'dz' = \frac{1}{2\pi}G\left(r, z, \sqrt{2}, 0\right).$$

Since $G(r, z, r', z') = G(r, z, \sqrt{2}, 0) + O(1)$ on supp ζ_0, we find that

$$\psi_0(r, z) \equiv \iint G(r, z, r', z')\zeta_0(r', z')r'\,dr'\,dz' = \psi^0(r, z) + O(\varepsilon).$$

Hence

$$E(\zeta_0) = \pi\iint\psi^0(r, z)\zeta_0(r, z)r\,dr\,dz + O(1)$$

$$= \pi\iint\frac{1}{2\pi}\frac{\sqrt{2}}{2\pi}\log\frac{1}{\varepsilon\xi}\zeta(r, z)r\,dr\,dz + O(1)$$

$$= \frac{\sqrt{2}}{4\pi}\iint\log\frac{1}{\varepsilon\xi}\zeta(r, z)r\,dr\,dz + O(1).$$

Since $r = \sqrt{2} + O(\varepsilon)$, $dr\,dz = 2\pi\varepsilon^2\xi\,d\xi$, we get

$$E(\zeta_0) = \frac{\sqrt{2}}{4\pi}\sqrt{2}\,2\pi\varepsilon^2\lambda\int_0^1\xi\log\frac{1}{\varepsilon\xi}d\xi + O(1)$$

$$= \varepsilon^2\lambda\left[\frac{1}{2}\xi^2\log\frac{1}{\varepsilon\xi} + \frac{1}{2}\right]_{\xi=0}^{\xi=1} + O(1)$$

$$= \frac{1}{2}\varepsilon^2\lambda\log\frac{1}{\varepsilon} + O(1)$$

$$= \frac{1}{4\sqrt{2}\,\pi^2}\log\frac{1}{\varepsilon} + O(1)$$

by (10.4). Since

$$\int \zeta_0 \, dx = 1, \qquad \tfrac{1}{2} \int r^2 \zeta_0 = 1 + O(\varepsilon),$$

there is an $O(\varepsilon)$-perturbation ζ_1 of ζ_0 which belongs to $\mathcal{Q}_\lambda$. Hence

$$E(\zeta) \geqslant E(\zeta_1) = E(\zeta_0) + O(1),$$

and (10.3) follows.

Lemma 10.2. *There is a positive constant C independent of λ such that*

(10.5) $\psi(r, 0) \leqslant Cr\big[1 + \log(1 + \lambda r^3)\big] \qquad (0 < r < \infty)$

for any $\lambda > 0$, where ψ is the stream function corresponding to $\zeta = \zeta_\lambda$.

Proof. Write $\psi(r, 0) = \psi_1(r, 0) + \psi_2(r, 0)$ where

$$\psi_1(r, 0) = \int_{\{|x - x'| < r/2\}} K(x, x') \zeta(x') \, dx' \qquad [x = (r, 0, 0)],$$

$$\psi_2(r, 0) = \int_{\{|x - x'| > r/2\}} K(x, x') \zeta(x') \, dx'.$$

To estimate ψ_2, notice that

$$K(x, x') \leqslant \frac{rr'}{|x - x'|} \leqslant 3r \qquad \text{if } |x - x'| > \frac{r}{2}.$$

Thus $\psi_2(r, 0) \leqslant 3r$.

To estimate ψ_1, notice that

$$K(x, x') \leqslant \frac{Cr'}{|x - x'|} \quad \text{if } |x - x'| < \frac{r}{2}$$

and hence

$$\psi_1(r, 0) \leqslant Cr^2 \int_{\substack{\{|r - r'| < r/2\} \\ \{|z'| < r/2\}}} \int_0^{2\pi} \frac{r' \, d\theta}{\big[z'^2 + r^2 + r'^2 - 2rr' \cos\theta\big]^{1/2}} \, dr' dz'.$$

Using (2.7) and proceeding as in the argument following (2.8), we find that

$$\psi_1(r, 0) \leqslant Cr^2 \frac{1}{r} \log \lambda r^3 \qquad \text{provided that } \lambda r^3 \geqslant c$$

for some positive constant c. On the other hand, if $\lambda r^3 \leqslant c$, then we simply use

$$\psi_1(r,0) \leqslant Cr^2\lambda \int_{\{|x-x'|<r/2\}} \frac{dx'}{|x-x'|} \leqslant C\lambda r^4.$$

The assertion (10.5) follows by putting together the estimates on ψ_1, ψ_2.
  Define

$$R_0 = \inf\{r; (r,0) \in \operatorname{supp}\zeta\},$$
$$R_1 = \sup\{r; (r,0) \in \operatorname{supp}\zeta\}.$$

Note that (6.34) implies trivially that

(10.6) $\tfrac{1}{2}R_0^2 \leqslant 1 \leqslant \tfrac{1}{2}R_1^2.$

Lemma 10.3.   *There holds*

(10.7) $R_1 \leqslant C$

where C is a constant independent of λ, $\lambda > 1$.

  Proof.   Since $u = 0$ on ∂V_λ,

$$\tfrac{1}{2}WR_1^2 + \gamma = \psi(R_1,0).$$

Recalling that $\gamma \geqslant 0$ and using Lemma 10.2, we get

(10.8) $\tfrac{1}{2}WR_1^2 \leqslant CR_1 \log\left(2 + \lambda R_1^3\right).$

From Lemmas 7.2 and 10.1, we have

(10.9) $2W = E + 3J(u) \geqslant E \geqslant C \log \lambda$

if λ is sufficiently large. Substituting this estimate into (10.8), we get

$$R_1 \log \lambda \leqslant C \log\left(\lambda R_1^3\right)$$

if λ is sufficiently large (say $\lambda > \lambda_0$), and (10.7) follows. The proof for $1 < \lambda < \lambda_0$ is similar, since $E(\zeta_\lambda) \geqslant c > 0$, c independent of λ.

Lemma 10.4.   *There holds*

(10.10) $J(u) \leqslant C,$

where $J = J(u)$ is defined by (7.9) with $F(t) = \lambda t^+$ and C is a constant independent of λ.

Proof. By (7.10) (with $\beta = 0$),

$$J(u) = 2\pi\lambda \iint_V ur\,dr\,dz = 2\pi \iint_V \frac{1}{r}\left(u_r^2 + u_z^2\right)dr\,dz.$$

Since $R_1 \leqslant C$, we get

(10.11) $$Q \leqslant CJ, \qquad \text{where } Q = \iint_V \left(u_r^2 + u_z^2\right)dr\,dz.$$

We recall the Poincaré inequality,

$$\iint_V u^2\,dr\,dz \leqslant \frac{|V|}{2\pi}\iint_V \left(u_r^2 + u_z^2\right)dr\,dz \qquad \text{if } u\mid_{\partial V} = 0.$$

We use it to derive

$$\frac{J}{2\pi\lambda} \leqslant C\iint_V u\,dr\,dz \leqslant C\,|V|^{1/2}\left(\iint_V u^2\,dr\,dz\right)^{1/2}$$

$$\leqslant C\,|V|\left(\iint_V \left(u_r^2 + u_z^2\right)dr\,dz\right)^{1/2}$$

$$= C\,|V|\,Q^{1/2} \leqslant C\,|V|\,J^{1/2} \qquad [\text{by (10.11)}].$$

Recalling (10.2), the assertion (10.10) now follows.

From Lemmas 7.2, 10.1, and 10.4, we deduce:

Theorem 10.5. *As* $\lambda \to \infty$,

(10.12) $$W = \tfrac{1}{2}E + O(1),$$

(10.13) $$\gamma = \tfrac{3}{2}E + O(1).$$

From (10.12) and Lemma 10.1 it follows that $\gamma > 0$ if λ is sufficiently large. Hence, by Section 8, Problem 1:

Corollary 10.6. *If* λ *is sufficiently large, then*

(10.14) $$|V_\lambda| = \frac{1}{2\pi\lambda}.$$

Lemma 10.7. *If λ is sufficiently large, then*

$$(10.15) \qquad\qquad R_0 \geq c,$$

where c is a positive constant independent of λ.

Proof. We have

$$\gamma \leq \gamma + \tfrac{1}{2}WR_0^2 = \psi(R_0, 0) \leq cR_0\big(1 + \log\big(1 + \lambda R_0^3\big)\big),$$

where Lemma 10.2 was used. Since, by (10.13), (10.3),

$$(10.16) \qquad\qquad \gamma \geq c_0 \log \lambda \qquad (c_0 > 0)$$

if λ is sufficiently large and c_0 is independent of λ, and since $R_0^2 \leq 2$, we obtain

$$c_0 \log \lambda \leq cR_0\big[1 + \log\big(1 + 2^{3/2}\lambda\big)\big];$$

this gives (10.15).

We shall denote various positive constants independent of λ by C. We shall assume in the sequel that for all sufficiently large λ

$$(10.17) \qquad\qquad V_\lambda \text{ is connected.}$$

In reference 33 a formal argument is given suggesting that (10.17) is valid for all $\lambda > 0$.

Lemma 10.8. *If λ is sufficiently large, then*

$$(10.18) \qquad\qquad d(V_\lambda) \leq \frac{C}{\log \lambda}.$$

This implies that

$$(10.19) \qquad\qquad R_1 - R_0 \leq \frac{C}{\log \lambda}.$$

Proof. Consider a family of straight lines l_r, $R_0 \leq r \leq R_1$, each forming an angle $2\pi/3$ with the r axis; l_r cuts the z axis at $z = r$. By (10.17) l_r intersects $\overline{V}$ at a point $(r, 0)$. Denote by (r^*, z^*) the first point of intersection of l_r with $\overline{V}$; thus the segment l_r^* from $(0, r)$ to (r^*, z^*) lies outside V and $(r^*, z^*) \in \partial V$. Then

$$\psi(0, r) - \psi(r^*, z^*) = \int \frac{\partial}{\partial l_r}\psi \, dl$$

and

$$\psi(0, r) = 0, \qquad \psi(r^*, z^*) = \gamma + \tfrac{1}{2} W(r^*)^2 \geq \gamma.$$

Integrating with respect to r, we get

$$\gamma(R_1 - R_0) \leq \int_{R_0}^{R_1} \int \frac{\partial \psi}{\partial l_r} \, dl \, dr \leq C(R_1 - R_0)^{1/2} \left\{ \iint | \nabla \psi |^2 dr \, dz \right\}^{1/2}$$

$$\leq C(R_1 - R_0)^{1/2} \left\{ \iint \frac{1}{r} | \nabla \psi |^2 dr \, dz \right\}^{1/2}$$

since $R_1 \leq C$. The last integral is $\leq CE^{1/2}$; therefore

$$\gamma(R_1 - R_0)^{1/2} \leq CE^{1/2}.$$

Recalling (10.13) and Lemma 10.1, the inequality (10.19) follows.

In order to complete the proof of (10.18), it remains to show that

(10.20) $$Z \leq \frac{C}{\log \lambda} \qquad \text{where } Z = \sup\{z; (r, z) \in V\}.$$

This follow by the same method as before. We begin with

$$-\psi(0, z) + \psi(\bar{r}, z) = -\int_0^{\bar{r}} \frac{\partial}{\partial r} \psi(r, z) dr \qquad (0 < z < Z)$$

where the segment $\{(r, z); 0 < r < \bar{r}\}$ lies outside V and $(\bar{r}, z) \in \partial V$. The left-hand side is equal to

$$\psi(\bar{r}, z) = \tfrac{1}{2} W \bar{r}^2 + \gamma \geq \gamma.$$

Integrating with respect to z and proceeding as before, the assertion (10.20) easily follows.

From (10.6) and (10.19) we obtain:

Corollary 10.9. *If λ is sufficiently large, then*

(10.21) $$| R_1 - \sqrt{2} | \leq \frac{C}{\log \lambda}, \qquad | R_0 - \sqrt{2} | \leq \frac{C}{\log \lambda}.$$

We shall use this result in order to improve Lemma 10.8:

Theorem 10.10. *For all λ sufficiently large,*

(10.22) $$\frac{c}{\lambda^{1/2}} \leq d(V_\lambda) \leq \frac{C}{\lambda^{1/2}} \qquad (0 < c < C < \infty).$$

Proof. Suppose $(r, z) \in V_\lambda$. Then $u(r, z) > 0$ and thus

$$\tfrac{1}{2}Wr^2 + \gamma < \psi(x) = \iint_H G(r, z, r', z')\zeta(r', z')r' \, dr' \, dz'.$$

Set $X = (r, z)$, $X' = (r', z')$. By (7.16),

$$G(r, z, r'z') = \frac{r}{2\pi}\log\frac{1}{|X' - X|} + O(1).$$

Writing $r' = (r' - r) + r$ and noting, by Corollary 10.9, that

(10.23) $$\iint_H \zeta(r', z') \, dr' \, dz' = \frac{1}{2\sqrt{2}\,\pi}\left(1 + O\left(\frac{1}{\log \lambda}\right)\right),$$

we find, after using Theorem 10.5, Lemma 10.1, and the relation $r^2 = 2 + O(1/\log \lambda)$ (which follows from Corollary 10.9) that

$$\frac{\log \lambda}{4\sqrt{2}\,\pi^2} \leqslant \frac{1}{\pi} \iint_H \log\frac{1}{|X' - X|}\zeta(r', z') \, dr' \, dz' + O(1).$$

Choosing $\varepsilon = \lambda^{-1/2}$ and using (10.23), this inequality becomes

(10.24) $$\iint_H \log\frac{\varepsilon}{|X' - X|}\zeta(r', z') \, dr' \, dz' \geqslant -C.$$

For a fixed large A consider the problem of maximizing

$$F \equiv \iint_{\{|X' - X| \leqslant A\varepsilon\}} \log\frac{\varepsilon}{|X' - X|}\zeta(r', z') \, dr' \, dz'$$

among all functions ζ, ζ/λ any characteristic function satisfying (10.23). The maximum is clearly attained for

$$\zeta = \lambda I_{B_\rho(X)}$$

where

$$\pi\rho^2\lambda = \frac{1}{2\sqrt{2}\,\pi}\left(1 + O\left(\frac{1}{\log\lambda}\right)\right).$$

We compute that

$$F \leqslant 2\pi\lambda \int_0^\rho \left(\log\frac{\varepsilon}{r}\right)r \, dr = 2\pi\lambda\varepsilon^2 \int_0^{\rho/\varepsilon}\left(\log\frac{1}{s}\right)s \, ds \leqslant C.$$

Hence (10.24) yields

$$\iint_{\{|X'-X|>A\varepsilon\}} \log\frac{\varepsilon}{|X'-X|}\zeta(r',z')\,dr'\,dz' \geq -C,$$

or

(10.25)
$$\iint_{\{|X'-X|>A\varepsilon\}} \zeta(r',z')r'dr'dz' \leq \frac{C}{\log A} < \frac{1}{4\pi}$$

if A is sufficiently large.

Suppose now that $\tilde{X}$ is another point in V_λ. Then we also have

$$\iint_{\{|X'-\tilde{X}|>A\varepsilon\}} \zeta(r',z')r'\,dr'dz' < \frac{1}{4\pi}.$$

If $|\tilde{X} - X| > A\varepsilon$ then by combining this inequality with (10.25) we get

$$\iint_H \zeta(r',z')r'\,dr'\,dz' < \frac{1}{2\pi},$$

a contradiction. This proves that $d(V_\lambda) \leq A\varepsilon \leq C\lambda^{-1/2}$. Recalling (10.14) we deduce also that $d(V_\lambda) \geq c\lambda^{-1/2}$.

Set $\varepsilon = 1/\sqrt{\lambda}$ and introduce the functions

(10.26)
$$u_\varepsilon(r,z) = u\left(\sqrt{2} + \varepsilon r, \varepsilon z\right),$$

(10.27)
$$\zeta_\varepsilon^0(r,z) = \varepsilon^2\zeta_\lambda\left(\sqrt{2} + \varepsilon r, \varepsilon z\right)$$

and the set

(10.28)
$$V_\lambda^* = \left\{\left(\sqrt{2} + \varepsilon r, \varepsilon z\right); (r,z) \in V_\lambda\right\}.$$

By Theorem 10.10,

(10.29) $\{\rho < \varepsilon_0\} \subset V_\lambda^* \subset \left\{\rho < \dfrac{1}{\varepsilon_0}\right\}$ for some $\varepsilon_0 > 0$,

where (ρ, ϕ) are the polar coordinates about $(\sqrt{2}, 0)$. Notice that

(10.30)
$$2\sqrt{2}\,\pi\int\zeta_\varepsilon^0(r,z)\,dr\,dz = 1 + O(\varepsilon),$$
$$\zeta_\varepsilon^0 = I_{\{u_\varepsilon>0\}},$$

and

$$(10.31) \qquad \zeta_\varepsilon^0(r, z) = \zeta_\varepsilon^0(r, -z), \qquad \int r\zeta_\varepsilon^0(r, z) \, dr \, dz = O(\varepsilon).$$

Let

$$D_0 = \left\{ (r, z); \, r^2 + z^2 < \frac{1}{2\sqrt{2}\,\pi^2} \right\}$$

and denote by $U(R) \, [R = (r^2 + z^2)^{1/2}]$ the $C^{1,1}$ solution of

$$(10.32) \qquad \begin{aligned} \Delta U &= -2I_{D_0} & &\text{in } R^2, \\ U &= 0 & &\text{on } \partial D_0, \\ U &= -A \log \frac{R}{R_0} & &\text{if } R > R_0, \; R_0 = \left(2\sqrt{2}\,\pi^2\right)^{-1/2}. \end{aligned}$$

Notice that A is uniquely determined (and is positive) by requiring that $u \in C^1$ at $R = R_0$.

Theorem 10.11. *As* $\lambda \to \infty$,

$$(10.33) \qquad \begin{aligned} D^i u_\varepsilon(r, z) &\to D^i U \quad \textit{for } i = 0, 1, \textit{ uniformly} \\ &\qquad \textit{in compact sets in } R^2; \end{aligned}$$

further, for all λ *large,* ∂V_λ^* *can be represented in the form* $\rho = h_\lambda(\phi)$, *where*

$$(10.34) \qquad \frac{d^i}{d\phi^i}\left(h_\lambda(\phi) - \frac{1}{\pi\sqrt{2}} \right) \to 0 \qquad \textit{for } i = 0, 1, \textit{ uniformly in } \phi,$$

as $\lambda \to \infty$.

Proof. The u_ε satisfy

$$(10.35) \qquad \Delta u_\varepsilon - \frac{\varepsilon}{\sqrt{2} + \varepsilon r}(u_\varepsilon)_r = -\left(\sqrt{2} + \varepsilon r\right)^2 I_{\{u_\varepsilon > 0\}},$$

$$(10.36) \qquad u_\varepsilon = 0 \text{ on } \partial V_\lambda^*, \qquad u_\varepsilon > 0 \text{ in } V_\lambda^*, \qquad u_\varepsilon \leqslant 0 \text{ outside } V_\lambda^*.$$

We claim that if $r^2 + z^2 < \lambda/C_0$ (for a suitable $C_0 > 0$), then

$$(10.37) \qquad |u_\varepsilon(r, z)| \leqslant C(r^2 + z^2)^{1/2} + \frac{C \log \lambda}{\lambda^{1/2}}(r^2 + z^2)^{1/2} + C.$$

To prove it, we write

$$u_\varepsilon(r, z) = u_\varepsilon(r, z) - u_\varepsilon(\bar{r}, \bar{z}) \qquad \text{for some } (\bar{r}, \bar{z}) \in \partial V_\lambda^*,$$

so that

$$u_\varepsilon(r, z) = \left[\psi\left(\sqrt{2} + \varepsilon r, \varepsilon z\right) - \psi\left(\sqrt{2} + \varepsilon \bar{r}, \varepsilon \bar{z}\right) \right]$$
$$+ \left[\tfrac{1}{2} W\left(\sqrt{2} + \varepsilon r\right)^2 - \tfrac{1}{2} W\left(\sqrt{2} + \varepsilon \bar{r}\right)^2 \right]$$
$$\equiv I_1 + I_2.$$

From Problem 1, we have

$$(10.38) \qquad |\nabla \psi| \leqslant C\lambda^{1/2} = \frac{C}{\varepsilon},$$

where the argument of ψ varies in the interval connecting $(\sqrt{2} + \varepsilon \bar{r}, \varepsilon \bar{z})$ to $(\sqrt{2} + \varepsilon r, \varepsilon z)$. Hence

$$|I_1| \leqslant C(r^2 + z^2)^{1/2} + C.$$

Next,

$$|I_2| \leqslant C(\log \lambda)\varepsilon r + C = \frac{C \log \lambda}{\lambda^{1/2}} r + C;$$

thus (10.37) follows.

From (10.35), (10.37) it follows, by standard elliptic estimates, that for a subsequence $\varepsilon_j \downarrow 0$,

$$u_{\varepsilon_j} \to Z \qquad \text{uniformly in compact sets,}$$
$$\text{together with the first derivatives.}$$

We conclude that

$$(10.39) \qquad \Delta Z + 2I_{\{Z>0\}} = 0 \qquad \text{in } R^2$$

and, by (10.37),

$$(10.40) \qquad |Z(r, z)| \leqslant C(r^2 + z^2)^{1/2} + C.$$

Lemma 10.12. *Z is radially symmetric.*

Proof. Since Z is harmonic and negative in a neighborhood of ∞ [by

(10.29)], and since it has at most a linear growth [by (10.40)], the linear term in its asymptotic expansion near ∞ must vanish. Thus the function $u = -Z$ has the expansion

$$u(X) = \tfrac{1}{2}a \log R^2 + b\frac{A \cdot X}{R^2} + O\left(\frac{1}{R^2}\right)$$

near ∞, where $X = (r, z)$, $R = |X|$. Also,

$$\Delta u - 2I_{\{u<0\}} = 0 \qquad \text{in } R^2.$$

We can now apply a slightly simplified version of Theorem 13.11 (Theorem 13.11 is independent of all the material preceding it). In fact, using the same notation all we need to show is that if $v_\lambda(x, y) \geq 0$ in S_λ^+, then $v_\lambda(x, y) > 0$ in S_λ^+. But since $v_\lambda \geq 0$,

$$\Delta v_\lambda = 2I_{\{u(x, y)<0\}} - 2I_{\{u(-x+2\lambda, y)>0\}} \leq 0 \qquad \text{in } S_\lambda^+,$$

so that $v_\lambda > 0$ in S_λ^+ by the strong maximum principle.

We conclude that u (and thus Z) is radially symmetric with respect to some center (r^0, z^0). Recalling (10.31), it follows that $(r^0, z^0) = (0,0)$.

From Lemma 10.12 it follows that Z is a solution of (10.32) for some disc D_0 with radius $R_0 > 0$. Recalling (10.30), we deduce that $R_0^2 = 1/(2\sqrt{2}\,\pi^2)$. This completes the proof of (10.33).

Since

$$\frac{\partial U(r, z)}{\partial R} \neq 0 \qquad \text{on } R = R_0$$

it follows that

$$c \leq |\nabla u_\varepsilon(r, z)| \leq C \qquad \text{on } \partial V_\lambda^*.$$

Further, ∂V_λ^* can be represented locally in the form $\rho = h_\lambda(\phi)$, where $h_\lambda(\phi) \to R_0$, $h_\lambda'(\phi) \to 0$ uniformly in ϕ as $\lambda \to \infty$.

Corollary 10.13. *As $\lambda \to \infty$,*

(10.41) $$E(\lambda) = \frac{1}{8\sqrt{2}\,\pi^2} \log \lambda + O(1).$$

Indeed, since the boundary of V_λ converges (after scaling) to a circle in the C^1 norm, the computations in Lemma 10.1 can be applied to derive

$$E(\zeta_\lambda) = \frac{1}{8\sqrt{2}\,\pi^2} \log \lambda + O(1)$$

Remark 10.1. The asymptotic results of this section extend to solutions of Theorem 6.3.

PROBLEMS

1. Prove (10.38).

   [*Hint*: Follow the proof of Problem 4 of Section 4.]

2. Let $D = \{(r, z); (r - \sqrt{2})^2 + z^2 < 1\}$. Show that if (10.18) holds then

   $$\inf_{v} \iint_{D \setminus V_\lambda} \frac{|\nabla v|^2}{r} \, dr \, dz \leq \frac{1}{\pi} \frac{E + O(1)}{(\gamma + W)^2},$$

   where v varies over the functions in $H^1(D \setminus V_\lambda)$ with $v = 1$ on ∂V_λ, $v = 0$ on ∂D.

   [*Hint*: $\psi = \frac{1}{2} W r^2 = \gamma + W + g(r)$ on ∂V_λ, where $|g(r)| \leq C$, $|g'(r)| \leq C \log \lambda$. Construct $k(r, z)$ such that $k(r, z) = g(r) h(z)$ in $D_0 = \{\bar{R}_0 < r < \bar{R}_1, |z| < C_0/\log \lambda\}$ ($\bar{R} = R_0 - C_0/\log \lambda$, $\bar{R}_1 = R_1 + C_0/\log \lambda$) and such that $\psi = \gamma + W + k$ on ∂V_λ, $\psi = k$ on ∂D and

   $$\int_{D \setminus V_\lambda} \frac{1}{r} |\nabla k|^2 \, dr \, dz \leq C.$$

   By the Dirichlet principle,

   $$\inf_{w} \int_{D \setminus V_\lambda} \frac{1}{r} |\nabla w|^2 \leq \int_{D \setminus V_\lambda} \frac{1}{r} |\nabla \psi|^2 \leq \frac{E}{\pi},$$

   where $w \in H^1(D \setminus V_\lambda)$, $w = \psi$ on $\partial(D \setminus V_\lambda)$; take $v = (w - k)/(\gamma + W)$.]

3. Use Lemma 9.3 and Problem 2 to deduce that if (10.18) holds then $d(V_\lambda) \leq C/\lambda^{1/2}$. [This is an alternative way of proving Theorem 10.10.]

4. Use the estimate $d(V_\lambda) \geq c/\lambda^{1/2}$ $(c > 0)$ together with Problem 2 and Lemma 9.3 in order to show that

   $$E \leq \frac{1}{8\sqrt{2}\,\pi^2} \log \lambda + C.$$

11. THE PLASMA PROBLEM: EXISTENCE OF SOLUTIONS

The problem of controlled fusion, more specifically, the containment of plasma by a magnetic field, leads to a free-boundary problem. The precise model, which is currently subject to experimentation in the Tokomak machine, involves nonlocal nonlinear operators. However, a simplified model reduces to the following system for an unknown function u:

$$(11.1) \qquad \Delta u - \lambda u_- = 0 \qquad \text{in } \Omega,$$

$$(11.2) \qquad u = c \qquad \text{on } \Gamma = \partial\Omega \qquad (c \text{ constant}),$$

$$(11.3) \qquad \int_\Gamma \frac{\partial u}{\partial \nu} \, dS = I \qquad (\nu \text{ outward normal}),$$

where λ and I are given positive numbers and c is a constant to be determined. The domain Ω is a bounded domain in R^2, but in this section we take Ω to be a bounded domain in R^n, $n \geq 2$.

The sets

$$\Omega_p = \{x \in \Omega; u(x) < 0\},$$

$$\Omega_v = \{x \in \Omega; u(x) > 0\}$$

are called the *plasma set* and the *vacuum set*, respectively, and

$$\Gamma_p = \partial\Omega_p$$

is called the *free boundary*.

We assume that $\Gamma \in C^{2+\alpha}$ for some $0 < \alpha < 1$.

Denote by $\lambda_1, \lambda_2, \ldots$ the increasing sequence of eigenvalues of $-\Delta$ in Ω and by v_1 the positive eigenfunction corresponding to λ_1. We observe that

$$(11.4) \qquad \text{if } \lambda < \lambda_1, \text{ then } c < 0,$$

$$(11.5) \qquad \text{if } \lambda > \lambda_1, \text{ then } c > 0.$$

Indeed, using the formula

$$\int_G (\lambda v_1 u_- + \lambda_1 v_1 u) = \int_G (v_1 \Delta u - u \Delta v_1) = \int_{\partial G} \left(v_1 \frac{\partial u}{\partial \nu} - u \frac{\partial v_1}{\partial \nu} \right)$$

for $\lambda > \lambda_1$, we get

$$0 < -c \int_{\partial\Omega} \frac{\partial v_1}{\partial \nu},$$

and since $\partial v_1/\partial \nu < 0$ it follows that $c > 0$. If $\lambda < \lambda_1$ and $c > 0$, then $\overline{\Omega}_p \subset \Omega$ and taking $G = \Omega_p$ in the formula above, we get

$$0 > \int_{\partial\Omega_p} v_1 \frac{\partial u}{\partial \nu};$$

since $\partial u / \partial \nu \geq 0$, we get a contradiction. Finally, if $c = 0$, then Ω_p must coincide with Ω and u is an eigenfunction with fixed sign in Ω; consequently, $\lambda = \lambda_1$, a contradiction.

Notice, by elliptic regularity, that any solution of (11.1), (11.2) [in $H^1(\Omega)$] must belong to $C^{2+\alpha}(\overline{\Omega}) \cap C^{2+\beta}(\Omega)$ for any $0 < \beta < 1$.

To prove existence of a solution for the plasma problem (11.1)–(11.3), we introduce a variational framework, taking

$$K_0 = \left\{ \rho \in L^2(\Omega), \rho \geq 0 \text{ a.e., } \int_\Omega \rho(x) \, dx = I \right\}$$

as the admissible class, and

$$(11.6) \qquad J_0(\rho) = \frac{1}{2\lambda} \int_\Omega \rho^2(x) \, dx - \frac{1}{2} \int_\Omega \int_\Omega G(x, y) \rho(x) \rho(y) \, dx \, dy$$

as a functional on K_0, where $G(x, y)$ is Green's function for $-\Delta$ in Ω.

Problem (J_0).   Find ρ such that

$$(11.7) \qquad\qquad J_0(\rho) = \min_{\tilde{\rho} \in K_0} J_0(\tilde{\rho}), \qquad \rho \in K_0.$$

This problem is similar to the problem of rotating compressible fluid (in Section 1); it is actually much simpler due to the fact that Ω is a bounded domain.

One can easily compute the Frechet derivative of J_0:

$$(11.8) \qquad J_0'(\rho) = -\int_\Omega G(x, y) \rho(y) \, dy + \frac{1}{\lambda} \rho(x) \qquad \forall \rho \in K_0.$$

Using the inequalities

$$0 \leq G(x, y) \leq \frac{C}{|x - y|^{n-2}} \qquad\qquad \text{if } n \geq 3 \qquad (C > 0),$$

$$0 \leq G(x, y) \leq \frac{1}{2\pi} \log \frac{1}{|x - y|} \qquad \text{if } n = 2$$

and Young's inequality, we find that

$$\left| \int_\Omega G(x, y) \rho(y) \, dy \right|_{L^s(\Omega)} \leq C \left\{ \int_\Omega \rho^2 \right\}^{1/2} \qquad \text{if } s < \frac{2n}{n-4};$$

if $n \leq 4$ then we can take any $s < \infty$.

We can then deduce that, for any $\varepsilon > 0$,

$$\left| \int_\Omega \int_\Omega G(x, y)\rho(x)\rho(y)\, dx\, dy \right| < \varepsilon \int_\Omega \rho^2(x)\, dx + C_\varepsilon.$$

Consequently, J_0 is bounded from below on K_0. Hence, by a standard compactness argument, there exists a solution ρ of problem (J_0) and, as in Section 1, a.e.

$$(11.9) \qquad \begin{aligned} J_0'(\rho) &= -c \quad \text{in } \{\rho > 0\}, \\ &\geq -c \quad \text{in } \{\rho = 0\}, \qquad c \text{ constant.} \end{aligned}$$

Setting

$$(11.10) \qquad u = -\int_\Omega G(x, y)\rho(y) + c$$

and recalling (11.8), the equilibrium conditions (11.9) reduce to

$$(11.11) \qquad \rho = \lambda u_-.$$

Applying Δ to both sides of (11.10), we obtain (11.1). Also, $u = c$ on $\partial\Omega$. Finally,

$$\int_\Gamma \frac{\partial u}{\partial \nu} = \int_\Omega \Delta u = \lambda \int_\Omega u_- = \int_\Omega \rho = I.$$

We have thus proved:

Theorem 11.1. *There exists a solution ρ of problem (J_0), and the function u defined by* (11.10) *is a solution of the plasma problem* (11.1)–(11.3).

Next, we establish uniqueness provided that $\lambda < \lambda_2$.

Theorem 11.2. *If $\lambda < \lambda_2$, then the solution of the plasma problem* (11.1)–(11.3) *is unique.*

Proof. Suppose that u_1, u_2 are two solutions and $u_i = c_i$ on Γ. By (11.4), (11.5) $\operatorname{sgn} c_1 = \operatorname{sgn} c_2$. Suppose for definiteness that $\operatorname{sgn} c_i > 0$ (that is, $\lambda > \lambda_1$) and let $U_i = u_i/c_i$. Setting

$$h(x) = \begin{cases} 0 & \text{if } U_1 = U_2 \\ -\dfrac{(U_1)_- - (U_2)_-}{U_1 - U_2} & \text{if } U_1 \neq U_2, \end{cases}$$

the function $U = U_1 - U_2$ satisfies

$$-\Delta U = \lambda h U \qquad \text{in } \Omega,$$

$$U = 0 \qquad \text{on } \Gamma$$

and $0 \leqslant h \leqslant 1$. We claim:

(11.12)   either $U \geqslant 0$ in Ω or $U \leqslant 0$ in Ω.

Indeed, otherwise let

$$\Omega_1 = \{U > 0\}, \qquad \Omega_2 = \{U < 0\}$$

and choose a positive constant γ so that

$$\gamma \int_{\Omega_1} v_1 U + \int_{\Omega_2} v_1 U = 0.$$

Thus the function

$$\tilde{U} = \begin{cases} \gamma U & \text{on } \Omega_1 \\ U & \text{on } \Omega_2 \end{cases}$$

[which clearly belongs to $H_0^1(\Omega)$] is orthogonal to the principal eigenfunction v_1. From the definition of λ_2,

(11.13) $\displaystyle \lambda_2 = \min_{\int v v_1 = 0} \frac{\int |\nabla v|^2}{\int v^2} \leqslant \frac{\int |\nabla \tilde{U}|^2}{\int \tilde{U}^2} \leqslant \frac{\int |\nabla \tilde{U}|^2}{\int h \tilde{U}^2}.$

But since

$$\lambda \int_{\Omega_1} h \tilde{U}^2 = \gamma^2 \int_{\Omega_1} \lambda h U U = -\gamma^2 \int_{\Omega_1} U \Delta U = \gamma^2 \int_{\Omega_1} |\nabla U|^2$$

$$= \int_{\Omega_1} |\nabla \tilde{U}|^2,$$

and similarly

$$\lambda \int_{\Omega_2} h \tilde{U}^2 = \int_{\Omega_2} |\nabla \tilde{U}|^2,$$

the right-hand side of (11.13) is equal to λ, contradicting the assumption $\lambda < \lambda_2$. We have thus established (11.12). We may then assume that $U_1 \geqslant U_2$.

If $U_1 \not\equiv U_2$, then the domains $D_1 = \{U_1 < 0\}$, $D_2 = \{U_2 < 0\}$ satisfy: $D_1 \subset D_2$, $D_1 \neq D_2$. Since each U_i is the principal eigenfunction of $-\Delta$ in D_i with the same eigenvalue λ, we get a contradiction (see Problem 1). From the identity $U_1 \equiv U_2$ and (11.3), we finally deduce that $c_1 = c_2$, so that $u_1 \equiv u_2$.

Remark 11.1. Uniqueness does not hold, in general, if $\lambda > \lambda_2$. Indeed, for any $\varepsilon > 0$ one can construct domains Ω_ε for which there exist at least two solutions of problem (J_0) with some λ in $(\lambda_2, \lambda_2 + \varepsilon)$. Ω_ε consists approximately of two balls $\{|x| < 1\}$, $\{|x - x_0| < 1\}$ (with $|x_0| > 2$) connected by a thin neck; for details, see reference 158d (see also Section 13, Problem 3).

We shall now introduce another variational principle which yields the same class of solutions of the plasma problem. Here we work directly with u.

The class of admissible functions is

$$(11.14) \quad K = \left\{ v \in H^1(\Omega), \, v = \text{const.} = v(\Gamma) \text{ on } \Gamma, \, \int_\Omega v_- \, dx = \frac{I}{\lambda} \right\},$$

and the functional is

$$(11.15) \qquad J(v) = \tfrac{1}{2} \int_\Omega |\nabla v|^2 \, dx - Iv(\Gamma) - \frac{\lambda}{2} \int_\Omega (v_-)^2 \, dx.$$

Consider the variational problem:

Problem (J). Find u such that

$$(11.16) \qquad\qquad J(u) = \min_{v \in K} J(v), \qquad u \in K.$$

Theorem 11.3. *There exists a solution u of problem (J) and u is a solution of the plasma problem (11.1)–(11.3).*

We first need a lemma.

Lemma 11.4. *There exists a positive constant C such that for any $u \in W^{1,p}(\Omega)$,*

$$(11.17) \quad |u|_{L^q} \leqslant C \left\{ |\nabla u|_{L^p} + \max\left[|u_-|_{L^r}, \frac{(|\nabla u|_{L^p})^{q/(q-r)}}{(|u_-|_{L^2})^{r/(q-r)}} \right] \right\};$$

here $L^s = L^s(\Omega)$, $1/q = (1/p) - (1/n)$ if $p < n$, q is any number in $(1, \infty)$ if $p \geqslant n$, and r is any number $p \leqslant r < q$.

Proof. We may assume that Ω is convex, for otherwise we write Ω as a finite union of domains, each diffeomorphic to a convex domain.

Consider first the case $p < n$. We shall use the Poincaré–Sobolev inequality (see reference 109, Sec. 7.8)

$$(11.18) \qquad\qquad |u - m(u)|_{L^q} \leq C |\nabla u|_{L^p},$$

where $m(u)$ is the average of u in Ω. If $m(u) < 0$, then

$$m(u) \geq -\frac{1}{|\Omega|} \int_\Omega u_-,$$

so that

$$|m(u)| \leq |\Omega|^{-1/r} |u_-|_{L^r}$$

and (11.17) follows from (11.18).

If $m(u) > 0$, then (11.18) gives

$$\int_{\{u \leq 0\}} |u_- + m(u)|^q \leq C(|\nabla u|_{L^p})^q.$$

Using the inequality $(a + b)^q \geq \gamma a^r b^{q-r}$ for $a \geq 0$, $b \geq 0$, with some $\gamma = \gamma(q, r) > 0$, we obtain

$$m(u)^{q-r} (|u_-|_{L^r})^r \leq C(|\nabla u|_{L^p})^q,$$

so that

$$m(u) \leq \frac{C|\nabla u|_{L^p}}{(|u_-|_{L^r})^{r/(q-r)}}.$$

If we use this inequality in (11.18), we again obtain (11.17).

If $p \geq n$, we can take q arbitrarily large in (11.18) and repeat the previous arguments.

Taking $p = r = 2$ in Lemma 11.4, we get

$$|u|_{L^{2n/(n+2)}} \leq C \left\{ |\nabla u|_{L^2} + \max \left[|u_-|_{L^2}, \frac{(|\nabla u|_{L^2})^{n/2}}{(|u_-|_{L^2})^{(n-2)/2}} \right] \right\}$$

if $n > 2$, and

$$|u|_{L^q} \leq C_\varepsilon \left\{ |\nabla u|_{L^2} + \max \left[|u_-|_{L^q}, \frac{(|\nabla u|_{L^2})^{1+\varepsilon}}{(|u_-|_{L^2})^\varepsilon} \right] \right\}$$

if $n = 2$ for any $1 < q < \infty$ and any $\varepsilon > 0$ $[\varepsilon = r/(q - r)]$.

From these inequalities and from the inequality

$$|v(\Gamma)| = \frac{1}{|\Gamma|} \int_\Gamma |v| \leqslant \delta \int_\Omega |\nabla v|^2 + C_\delta \int_\Omega |v|^2$$

for any $\delta > 0$, it is easily seen that the functional $J(v)$ is bounded from below on the set K. Thus, by a standard argument, there exists a solution u of (11.16). We shall show that u satisfies (11.1)–(11.3).

For any $v \in H^1(\Omega)$, $v = \text{const.} = v(\Gamma)$ on Γ, set

$$v_\varepsilon = \frac{u + \varepsilon v}{a_\varepsilon}, \qquad \text{where } a_\varepsilon = \frac{\lambda}{I} \int_\Omega (u + \varepsilon v)_- .$$

By the Lebesgue bounded convergence theorem, we find that if $\varepsilon > 0$, $\varepsilon \downarrow 0$, then

(11.19)
$$\frac{a_\varepsilon - 1}{\varepsilon} \to \frac{\lambda}{I} \int_\Omega (\rho_0 v_- - \rho_1 v),$$

where $\rho_0 = I_{\{u=0\}}$, $\rho_1 = I_{\{u<0\}}$. Since $v_\varepsilon \in K$, we have

$$\lim_{\varepsilon \downarrow 0} \frac{1}{\varepsilon} [J(v_\varepsilon) - J(u)] \geqslant 0.$$

Using (11.19), the last inequality reduces to

(11.20)
$$\langle J'(u), v \rangle \geqslant \gamma \int_\Omega (\rho_0 v_- - \rho_1 v),$$

where

$$\gamma = \frac{\lambda}{I} \left\{ \int |\nabla u|^2 - Iu(\Gamma) - \lambda \int (u_-)^2 \right\}$$

and

(11.21)
$$\langle J'(u), v \rangle = \int_\Omega \nabla u \cdot \nabla v - Iv(\Gamma) - \lambda \int_\Omega u_- v;$$

in deriving (11.20), (11.21) we again use the Lebesgue bounded convergence theorem.

Similarly, taking $\varepsilon < 0$, $\varepsilon \uparrow 0$, we obtain

(11.22)
$$\langle J'(u), v \rangle \leqslant -\gamma \int_\Omega (\rho_0 v_+ + \rho_1 v).$$

Thus

$$\left| \langle J'(u), v \rangle \right| + \gamma \int_\Omega \rho_1 v \leq \gamma \int_\Omega \rho_0 |v| .$$

Since v is arbitrary, we conclude that

$$(11.23) \qquad\qquad -\Delta u + \lambda u_- + \gamma \rho_1 = -\gamma\sigma,$$

where $|\sigma(x)| \leq \rho_0(x)$.

Multiplying (11.23) by $v \in K$ and integrating, we obtain after comparing with (11.20), (11.22),

$$\gamma \int \rho_0 v_- \leq \left(\int_\Gamma \frac{\partial u}{\partial \nu} dS - I \right) v(\Gamma) - \gamma \int_\Omega \sigma v \leq -\gamma \int_\Omega \rho_0 v_+ .$$

If we choose $v(\Gamma) = 1$, $v \geq 0$ with $|v_+|_{L^1}$ arbitrarily small, we then obtain (11.3) and, consequently,

$$\gamma \int_\Omega \rho_0 v_- \leq -\gamma \int_\Omega \sigma v \leq -\gamma \int_\Omega \rho_0 v_+ .$$

It follows that $\gamma \leq 0$. If $\gamma \neq 0$, then $0 \leq \sigma \leq \rho_0$ and integrating (11.23) over Ω gives [after using (11.2), (11.3)]

$$0 > \gamma \int_{\{u<0\}} dx = -\gamma \int_{\{u=0\}} \sigma \geq 0$$

since $\int u_- > 0$. This contradiction shows that $\gamma = 0$ and then (11.23) reduces to (11.1).

Remark 11.2. If u solves problem (J), then $\rho = \lambda u_-$ solves problem (J_0); if ρ solves problem (J_0), then u given by (11.10) with suitable c solves problem (J) and $\inf_K J(v) = \inf_{K_0} J_0(\rho)$. Thus the two problems are equivalent and produce the same class of solutions of (11.1)–(11.3). For proof, see Problem 2.

PROBLEMS

1. If

$$\Delta U_i + \lambda_i U_i = 0 \qquad \text{in } D_i,$$

$$U_i < 0 \qquad \text{in } D_i,$$

$$U_i = 0 \qquad \text{on } \partial D_i,$$

$D_1 \subset D_2$ and there is a C^1 open portion σ of ∂D_1 which is contained in D_2, then $\lambda_1 > \lambda_2$. (Notice that the last condition holds in the proof of Theorem 11.2; indeed, $U_1 > 0$, $\Delta U_1 = 0$ in $\Omega \setminus D_1$ and the inside ball property is satisfied at some points of $\sigma_0 = \partial D_1 \cap D_2$, so that $\partial U_1/\partial \nu \neq 0$ at such points and $\sigma_0 \cap \{U_1 = 0\}$ is then in $C^{2+\alpha}$ near such points.)

[*Hint:* Take a sequence $\varepsilon = \varepsilon_m \downarrow 0$ and use

$$\int_{\partial \{U_1 < -\varepsilon\}} \left(U_1 \frac{\partial U_2}{\partial \nu} - U_2 \frac{\partial U_1}{\partial \nu} \right) = (\lambda_1 - \lambda_2) \int_{\{U_1 < -\varepsilon\}} U_1 U_2.\Big]$$

2. Prove the assertions in Remark 11.2.

   [*Hint:* If $\rho \in K_0$, then the u in K given by (11.10) satisfies $J(u) \leqslant J_0(\rho)$. If $u \in K$, then $\rho = \lambda u_-$ is in K_0 and $J_0(\rho) \leqslant J(u)$.]

3. The Tokomak machine is actually three-dimensional and axially symmetric, and the corresponding plasma problem is obtained by replacing (11.1)–(11.3) by

$$Lu - \lambda u_- = 0 \qquad \text{in } \Omega,$$

$$u = \text{const. on } \Gamma,$$

$$\int_\Gamma \frac{1}{r} \frac{\partial u}{\partial \nu} \, ds = I.$$

where $Lu = ((1/r)u_r)_r + ((1/r)u_z)_z$. Extend Theorems 1.1 to 1.3 to this case, taking

$$J(v) = \frac{1}{2} \int_\Omega \frac{1}{r} \left(v_r^2 + v_z^2 \right) dr \, dz - Iv(\Gamma) - \frac{\lambda}{2} \int_\Omega (v_-)^2 \, dr \, dz.$$

12. THE FREE BOUNDARY FOR THE PLASMA PROBLEM

Theorem 12.1. *Let u be the solution of the plasma problem constructed by Theorem 11.3. Then the plasma set Ω_p and the vacuum set Ω_v are connected sets.*

Proof. If $\lambda \leqslant \lambda_1$, then $\Omega_p = \Omega$ and the theorem is trivial. Suppose that $\lambda > \lambda_1$. Then $u(\Gamma) > 0$ and thus Ω_v contains some Ω-neighborhood of Γ. If G is a component of Ω_v, then ∂G must intersect Γ (otherwise, $u \equiv 0$ in G), and thus Ω_v is connected.

To prove that Ω_p is connected, suppose that G_1 and G_2 are two components of Ω_p. Define

$$\tilde{u} = c_i u \qquad \text{in } G_i \qquad (i = 1, 2)$$

$$= u \qquad \text{in } \Omega \setminus (G_1 \cup G_2),$$

where c_i are positive constants satisfying

(12.1) $$c_1 \int_{G_1} u_- + c_2 \int_{G_2} u_- = \int_{G_1} u_- + \int_{G_2} u_- .$$

Then $\tilde{u} \in K$. Since

$$\int_{G_i} |\nabla u|^2 = -\int_{G_i} u \, \Delta u = \lambda \int_{G_i} (u_-)^2,$$

we have $J(\tilde{u}) = J(u)$, so that $\tilde{u}$ is a minimizer. By the proof of Theorem 11.3 it follows that $\tilde{u}$ solves the plasma problem (11.1)–(11.3) and, in particular, $\tilde{u} \in C^{2+\alpha}(\Omega)$. This is impossible if we choose $c_i \neq 1$, since $\partial u / \partial \nu \not\equiv 0$ on ∂G_i (by the maximum principle).

We shall now restrict n to be equal to 2.

Theorem 12.2. *If $n = 2$ and u is the solution asserted in Theorem 11.3, then the free boundary Γ_p is analytic.*

Proof. First we show that

(12.2) $$\nabla u \neq 0 \qquad \text{on } \Gamma_p.$$

Suppose that the assertion is not true at a point $x_0 \in \Gamma_p$. We shall use complex notation and take $x = (x_1, x_2) = x_1 + ix_2 = z$, $x_0 = 0$. Since $|\Delta u| \leq \lambda |u|$, we can apply Section 5, Problem 7 (that is, Lemma 5.2 for $n = 2$) and Lemma 5.3 for $\alpha = 1$ (see Remark 5.1). It follows that

$$u(z) = \text{Re}\{cz^m\} + O(|z|^{m+\delta})$$

$$= |c| r^m \cos(m\theta + \theta_0) + O(r^{m+\delta}),$$

$$\frac{\partial}{\partial z} u = mcz^{m-1} + O(r^{m-1+\delta}),$$

where $c \neq 0$, $r = |z|$, m is an integer ≥ 2 and $0 < \delta < 1$. Thus the zeros of ∇u are isolated and in some ε-neighborhood of 0 there exist $2m$ smooth curves initiating at 0 which divide this neighborhood into sectors σ_i such that

$$u(z) > 0 \qquad \text{in } \sigma_1 \cup \sigma_3 \cup \cdots \cup \sigma_{2m-1},$$

$$u(z) < 0 \qquad \text{in } \sigma_2 \cup \sigma_4 \cup \cdots \cup \sigma_{2m}.$$

Take points $z_2 \in \sigma_2$, $z_4 \in \sigma_4$. Since Ω_p is connected, we can connect z_2 to z_4 by a curve β lying in Ω_p. Connecting z_2 to 0 and 0 to z_4, we obtain a closed Jordan curve γ contained in $\Omega_p \cup \{0\}$. The interior of γ encloses some σ_j with j odd and the exterior of γ also encloses some σ_i with i odd. Hence Ω_v is not connected, a contradiction.

We have thus proved (12.2). From this follows that $\Gamma_p = \{u = 0\}$ is in $C^{2+\beta}$ for any $0 < \beta < 1$. The proof of analyticity is the same as the proof of Theorem 5.1.

PROBLEMS

1. Extend Theorems 12.1 and 12.2 to the plasma problem given in Section 11, Problem 3.

2. If $\Delta u + c(x)u \geq 0$, $u \leq 0$ in a domain Ω with C^2 boundary, and if $u \not\equiv 0$ and $c(x)$ is a bounded function, then (i) $u < 0$ in Ω, (ii) if $u(x_0) = 0$, $x_0 \in \Omega$, then $(\partial u/\partial \nu)(x_0) \neq 0$.
    [*Hint*: For some $\alpha > 0$, the function $v = e^{-\alpha x_1}u$ satisfies $\Delta v + 2\alpha v_{x_1} \geq 0$.]

3. If $\Delta u = f(u)$ in B_R, $u = 0$ on ∂B_R, $u > 0$ in B_R, and $f(t)$ Lipschitz continuous, then $u = u(|x|)$.
    [*Hint*: Let $T_\lambda = \{x_1 = \lambda\}$, $\Sigma_\lambda = B_R \cap \{x_1 > \lambda\}$, x_λ the reflection of x with respect to T_λ, and consider the set S of λ's in $[0, R]$ such that

    $$u_{x_1} \leq 0 \text{ and } u(x) \leq u(x_\lambda) \qquad \text{if } x \in \Sigma_\lambda.$$

    S is closed. If $\lambda \in S$, then $v(x) = u(x) - u(x_\lambda)$ satisfies $\Delta v + c(x)v = 0$ and Problem 2 gives $v_{x_1} = 2u_{x_1} < 0$ on $x_1 = \lambda$. Also, $u_{x_1} < 0$ on $\partial B_R \cap \{x_1 > 0\}$ since if $f(O) \leq 0$, then $\Delta u \leq f(u) - f(O) = \tilde{c}(x)u$ and if $f(O) > 0$, $u_\nu = 0$, then $u_{rr} > 0$ near $r = R(x_1 > 0)$ and then also $u_r < 0$, $u < 0$, which is impossible. Thus S is open, and $0 \in S$, which imply that $u_{x_1} \leq 0$ on $x_1 = 0$. This is true for any plane through the origin.]

4. Prove that any solution of the plasma problem (11.1)–(11.3) for $\Omega = B_R \subset R^n$ is a function of $|x|$ only, and deduce its uniqueness.

13. ASYMPTOTIC ESTIMATES FOR THE PLASMA PROBLEM

In this section we take $n = 2$, $\lambda > \lambda_1$ and use the notation

$$u = u_\lambda, \qquad \Omega_p = \Omega_{p,\lambda}, \qquad \Omega_v = \Omega_{v,\lambda}, \qquad \Gamma_p = \Gamma_{p,\lambda};$$

u is the solution constructed in Theorem 11.3, that is, u is a solution of Problem (J). We are interested in the behavior of u_λ, $\Omega_{p,\lambda}$, $\Gamma_{p,\lambda}$ as $\lambda \to \infty$. We shall prove that $\Gamma_{p,\lambda}$ is asymptotically a disc and we shall characterize the possible location of $\Gamma_{p,\lambda}$ within Ω. The main results are stated in Theorem 13.13. Theorem 13.11 is of intrinsic interest.

Consider first the case where Ω is the disc $B_R = \{|x| < R\}$ and denote the solution by u_R (see Section 12, Problem 4).

Lemma 13.1. *There holds*

(13.1) $$J(u_R) = -\frac{I^2}{8\pi} \log(\lambda R^2) + \gamma,$$

where γ is a universal constant.

Proof. Set $u = u_R$. If the free boundary is given by $r = \varepsilon$, then

$$u = c\left(1 - \frac{\log r/R}{\log \varepsilon/R}\right) \qquad \text{for } \varepsilon < r < R, c = u(\partial B_R)$$

and (11.3) reduces to

$$c = \left(\frac{1}{2}\pi\right) \log \frac{R}{\varepsilon}.$$

The principal eigenvalue for $-\Delta$ in B_ε is γ_1/ε^2 (γ_1 a universal constant) and the principal eigenfunction is $u_0(r/\varepsilon)$, where $u_0(r)$ is the principal eigenfunction for $-\Delta$ in B_1; we normalize it by

$$u_0 < 0, \qquad \int u_0^2 \, dx = 1.$$

Then

$$u|_{r<\varepsilon} = \gamma_2 u_0\left(\frac{r}{\varepsilon}\right)$$

where γ_2 is chosen so that $\partial u/\partial \nu$ is continuous across $r = \varepsilon$; γ_2 is a universal constant.

One easily computes

$$\frac{1}{2}\int_{\Omega_p} |\nabla u|^2 \, dx = \gamma_3,$$

$$\frac{1}{2}\int_{\Omega_v} |\nabla u|^2 \, dx = \frac{I^2}{4\pi} \log \frac{R}{\varepsilon},$$

$$\frac{\lambda}{2}\int_{\Omega_v} (u_-)^2 \, dx = \gamma_4,$$

$$Iu(\partial B_R) = Ic = \left(\frac{I^2}{2\pi}\right) \log \frac{R}{\varepsilon},$$

where γ_3, γ_4 are universal constants. Combining these results, (13.1) follows.

Lemma 13.2. *If*

$$B_{R_1} \subset \Omega \subset B_{R_2} \qquad (0 < R_1 < R_2 < \infty),$$

then

$$(13.2) \qquad J(u_\lambda) = \inf_{v \in K} J(v) = -\left(\frac{I^2}{8\pi}\right) \log \lambda + O(1),$$

$$(13.3) \qquad u_\lambda(\Gamma) = \left(\frac{I}{4\pi}\right) \log \lambda + O(1),$$

$$(13.4) \qquad \int_{\Omega_{v,\lambda}} |\nabla u_\lambda|^2 \, dx = \left(\frac{I^2}{4\pi}\right) \log \lambda + O(1),$$

where $|O(1)| \leqslant C$, C *a constant depending only on* R_1, R_2.

   Proof. We write $K = K_\Omega$, $J(v) = J_\Omega(v)$ to emphasize the dependence on Ω. Then

$$(13.5) \qquad \Omega_1 \subset \Omega_2 \text{ implies that } \inf_{K_{\Omega_1}} J_{\Omega_1}(v) \geqslant \inf_{K_{\Omega_2}} J_{\Omega_2}(v).$$

Indeed, the minimizer u of J_{Ω_1} can be extended as the positive constant $u(\partial \Omega_1)$ into $\Omega_2 \setminus \Omega_1$ and it then belongs to K_{Ω_2}; since

$$J_{\Omega_1}(u) = J_{\Omega_2}(u),$$

(13.5) follows.
   The assertion (13.2) follows from (13.5) and Lemma 13.1. Multiplying (11.1) by u and integrating over Ω_p, we obtain

$$\int_{\Omega_p} |\nabla u|^2 \, dx = \lambda \int_{\Omega_p} (u_-)^2 \, dx.$$

Similarly,

$$\int_{\Omega_v} |\nabla u|^2 \, dx = u(\Gamma)I.$$

Consequently,

$$(13.6) \qquad J(u_\lambda) = -\frac{I}{2} u_\lambda(\Gamma),$$

and (13.3), (13.4) follow from (13.2).

Lemma 13.3.   *There exists a positive constant C such that*

(13.7)
$$d(\Omega_{p,\lambda}) \le \frac{C}{\log \lambda}$$

for all $\lambda > \lambda_1$.

Proof.   Choose points A, B in $\overline{\Omega}_p$ such that

$$d(\Omega_p) = |B - A|.$$

Consider the family of straight lines γ_x passing through x and orthogonal to $\overline{AB}$ when x varies in $\overline{AB}$. Denote by $\delta_x = y_x z_x$ a segment lying in γ_x such that $y_x \in \Gamma$, $z_x \in \Gamma_p$, and int $\delta_x \subset \Omega_v$. Then

$$u(y_x) - u(z_x) = \int_{\delta_x} \frac{\partial}{\partial \delta_x} u$$

and $u(z_x) = 0$. Using (13.3), we obtain

$$\log \lambda \le C \int_{\delta_x} |Du|.$$

Integrating with respect to x in $\overline{AB}$, we get

$$|B - A| \log \lambda \le C \int_{\Omega_v} |Du| \le C \left(\int_{\Omega_v} |Du|^2 \right)^{1/2} (C|B - A|)^{1/2};$$

recalling (13.4), the assertion (13.7) follows.

Let

(13.8)
$$G(x, y) = \frac{1}{2\pi} \log \frac{1}{|x - y|} - h_x(y)$$

denote Green's function in Ω, that is,

(13.9)
$$\Delta_y h_x(y) = 0 \qquad\qquad \text{if } x, y \in \Omega,$$
$$h_x(y) = \frac{1}{2\pi} \log \frac{1}{|x - y|} \qquad \text{if } x \in \Omega, \ \ y \in \Gamma.$$

Set

(13.10)
$$k(x) = h_x(x).$$

We shall use the notation: $d(x, A) = \text{dist}(x, A)$.

Lemma 13.4. *There holds*

(13.11) $\qquad k(x) \to \infty \qquad$ *uniformly in x, as $d(x, \Gamma) \to 0$.*

Proof. Suppose that $x \to x^0$, $x^0 \in \Gamma$. Then

$$h_x(y) \geqslant -C \qquad\qquad\qquad \text{if } y \in \Gamma, \quad |y - x^0| > \tfrac{1}{2},$$

$$\geqslant C_1 \log \frac{C}{\varepsilon + |y - x^0|} \qquad \text{if } y \in \Gamma, \quad |y - x^0| < \tfrac{1}{2}, \quad |x - x^0| < \varepsilon;$$

here C, C_1 are positive constants. Representing $h_x(x)$ in terms of Green's function, we have

$$h_x(x) = - \int_\Gamma \frac{\partial G(x, y)}{\partial \nu} h_x(y)\, ds_y$$

$$\geqslant \int_{\{\Gamma, |y - x^0| > 1/2\}} C \frac{\partial G(x, y)}{\partial \nu}\, ds_y$$

$$- \int_{\{\Gamma, |y - x^0| < 1/2\}} C_1 \frac{\partial G(x, y)}{\partial \nu} \log \frac{C}{\varepsilon + |y - x^0|}\, ds_y$$

$$= J_1 + J_2;$$

$J_1 \to 0$ if $x \to x^0$, whereas $J_2 \to C_1 \log(C/\varepsilon)$. It follows that

$$\liminf_{x \to x_0} h_x(x) \geqslant C_1 \log \frac{C}{\varepsilon}.$$

Since Γ is smooth, the last inequality can be established uniformly with respect to x^0. Since, finally, ε is arbitrary, the assertion follows.

We shall now find the asymptotic location of the plasma. Let

(13.12) $\qquad\qquad\qquad\qquad k_0 = \min_{x \in \Omega} k(x).$

In view of Lemma 13.4, the minimum of $k(x)$ is indeed attained in Ω, and the set

(13.13) $\qquad\qquad\qquad\qquad S = \{x \in \Omega, k(x) = k_0\}$

is a compact subset of Ω.

Theorem 13.5. *For any $\varepsilon > 0$, there exists a $\lambda_0 = \lambda_0(\varepsilon) > 0$ such that*

(13.14) $\qquad\qquad\qquad d(\Omega_p, S) < \varepsilon \qquad \text{if } \lambda > \lambda_0.$

Proof. Suppose that the assertion is not true. Then there exists an $\varepsilon > 0$ such that

$$(13.15) \qquad d(\Omega_p, S) \geqslant \varepsilon \qquad \text{for arbitrarily large } \lambda \text{'s,}$$

say, $\lambda = \lambda_j$.

Fix a point $y^0 \in \Omega_p$ and let τ be a translation such that $\tau y^0 \in S$. Using Lemma 13.3, we deduce that there exists a positive number $\delta > 0$ such that

$$(13.16) \qquad \delta + h_y(x) < h_{\tau^{-1}y}(\tau^{-1}x) \qquad \text{if } \tau^{-1}x, \tau^{-1}y \in \Omega_p$$

provided that $\lambda = \lambda_j$ is sufficiently large [to ensure that $d(\Omega_p)$ is sufficiently small].

Consider the function

$$(13.17) \qquad u_0(x) = \begin{cases} -u_-(\tau^{-1}x) \text{ if } \tau^{-1}x \in \Omega_p, \text{ harmonic in} \\ \Omega \setminus \tau(\Omega_p), = 0 \text{ on } \partial(\tau(\Omega_p)), = u(\Gamma) \text{ on } \Gamma. \end{cases}$$

Clearly,

$$\int_\Omega (u_0)_- = \frac{I}{\lambda};$$

hence if we show that

$$(13.18) \qquad J(u_0) < J(u),$$

then we derive a contradiction, thus proving the theorem. Since

$$J(u) = \int_{\Omega \setminus \Omega_p} |\nabla u|^2 - Iu(\Gamma), \qquad J(u_0) = \int_{\Omega \setminus \tau(\Omega_p)} |\nabla u_0|^2 - Iu(\Gamma),$$

(13.18) reduces to

$$(13.19) \qquad \int_{\Omega \setminus \tau(\Omega_p)} |\nabla u_0|^2 < \int_{\Omega \setminus \Omega_p} |\nabla u|^2.$$

To prove (13.19), introduce the functions

$$\bar{u} = \frac{u(\Gamma) - u_+}{u(\Gamma)}, \qquad \bar{u}_0 = \frac{u(\Gamma) - (u_0)_+}{u(\Gamma)}.$$

Then (13.19) becomes

$$(13.20) \qquad \int_{\Omega \setminus \tau(\Omega_p)} |\nabla \bar{u}_0|^2 < \int_{\Omega \setminus \Omega_p} |\nabla \bar{u}|^2.$$

Notice that

$$\text{Cap}_\Omega(\Omega_p) = \int_{\Omega\setminus\Omega_p} |\nabla \bar{u}|^2 = \int_{\Gamma_p} \frac{\partial \bar{u}}{\partial \nu} = -\int_{\Gamma} \frac{\partial \bar{u}}{\partial \nu} = -\int_\Omega \Delta \bar{u},$$

$$\text{Cap}_\Omega(\tau(\Omega_p)) = \int_{\Omega\setminus\tau(\Omega_p)} |\nabla \bar{u}_0|^2 = \int_{\partial\tau(\Omega_p)} \frac{\partial \bar{u}_0}{\partial \nu}$$

$$= -\int_\Gamma \frac{\partial \bar{u}_0}{\partial \nu} = -\int_\Omega \Delta \bar{u}_0,$$

where $-\Delta \bar{u}$, $-\Delta \bar{u}_0$ are taken in the sense of measures [with support on Γ_p and $\tau(\Gamma_p)$, respectively].

We can write

$$\bar{u}(x) = -\int_\Omega G(x, y) \Delta \bar{u}(y) \, dy,$$

$$\bar{u}_0(x) = -\int_\Omega G(x, y) \Delta \bar{u}_0(y) \, dy.$$

Consider the function

$$w(x) = -\int_\Omega G(x, y) \Delta \bar{u}(\tau^{-1}) \, dy.$$

Since (13.16) implies that

$$G(x, y) > G(\tau^{-1}x, \tau^{-1}y) \qquad \text{if } x, y \in \overline{\tau(\Omega_p)},$$

we have, for $x \in \Omega_p$,

$$w(\tau x) = -\int G(\tau x, \tau y) \Delta \bar{u}(y) \, dy$$

$$> -\int G(x, y) \Delta \bar{u}(y) \, dy = \bar{u}(x) = 1,$$

so that

$$w(x) > \bar{u}(\tau^{-1}x) = \bar{u}_0(x) \qquad \text{if } x \in \tau(\Omega_p).$$

Since both $w(x)$ and $\bar{u}_0(x)$ are harmonic in $\Omega\setminus\tau(\Omega_p)$ and vanish on Γ, the maximum principle gives

$$w(x) > \bar{u}_0(x) \qquad \text{in } \Omega \setminus \tau(\Omega_p).$$

Hence

$$\frac{\partial w}{\partial \nu} < \frac{\partial \bar{u}_0}{\partial \nu} \qquad \text{on } \Gamma.$$

It follows that

$$\int_{\Omega \setminus \tau(\Omega_p)} |\nabla \bar{u}_0|^2 = -\int_\Gamma \frac{\partial \bar{u}_0}{\partial \nu} < -\int_\Gamma \frac{\partial w}{\partial \nu} = -\int_\Omega \Delta w(y)$$

$$= -\int_\Omega \Delta \bar{u}(\tau^{-1}y) = -\int_\Omega \Delta \bar{u}(y) = \int_\Gamma \frac{\partial \bar{u}}{\partial \nu} = \int_{\Omega \setminus \Omega_p} |\nabla \bar{u}|^2$$

and (13.20) follows.

From Lemma 13.4 and Theorem 13.5, we deduce:

Corollary 13.6. *There exists a $\delta > 0$ and λ_0 positive and sufficiently large such that*

(13.21) $$d(\Omega_{p,\lambda}, \Gamma) > \delta \qquad if \lambda > \lambda_0.$$

Set

$$w = \frac{u(\Gamma) - u}{u(\Gamma)}.$$

Then $w = 0$ on Γ, $w = 1$ on Γ_p and

$$\int_{\Omega_v} |\nabla w|^2 = \frac{I}{u(\Gamma)} = \frac{4\pi}{\log \lambda + O(1)}$$

by Lemma 13.2. Consequently,

(13.22) $$\text{Cap}_\Omega \Omega_{p,\lambda} = \frac{4\pi}{\log \lambda + O(1)}.$$

Lemma 13.7. *There exists a positive constant C such that*

(13.23) $$d(\Omega_{p,\lambda}) \leq \frac{C}{\lambda^{1/2}} \qquad for \ all \ \lambda > \lambda_1.$$

Proof. In view of Corollary 13.6, Theorem 9.1 can be applied to deduce after using (13.22), that

$$\left(1 - \frac{C}{\log 1/l}\right)\frac{1}{\log 1/l} \leq \frac{1}{\log \lambda^{1/2}}, \qquad l = d(\Omega_{p,\lambda}),$$

from which (13.23) easily follows.

We shall now consider scaled solutions. Fix a point y_λ in Ω_p and set

$$\tilde{u}_\lambda(x) = u\left(\frac{x - y_\lambda}{\lambda^{1/2}}\right)$$

$$\tilde{\Omega}_{p,\lambda} = \left\{\frac{x - y_\lambda}{\lambda^{1/2}} ; x \in \Omega_p\right\}, \qquad \tilde{\Omega}_{v,\lambda} = \left\{\frac{x - y_\lambda}{\lambda^{1/2}} ; x \in \Omega_v\right\},$$

$$\tilde{\Omega}_\lambda = \left\{\frac{x - y_\lambda}{\lambda^{1/2}} ; x \in \Omega\right\}, \qquad \tilde{\Gamma}_\lambda = \partial\tilde{\Omega}_\lambda, \qquad \tilde{\Gamma}_{p,\lambda} = \partial\tilde{\Omega}_{p,\lambda}.$$

Then

(13.24)
$$\tilde{u}_\lambda < 0 \text{ in } \tilde{\Omega}_{p,\lambda}, \qquad \tilde{u}_\lambda > 0 \text{ in } \tilde{\Omega}_{v,\lambda}$$

and

(13.25)
$$\Delta\tilde{u}_\lambda + \tilde{u}_\lambda = 0 \qquad \text{in } \tilde{\Omega}_{p,\lambda},$$

(13.26)
$$\Delta\tilde{u}_\lambda = 0 \qquad \text{in } \tilde{\Omega}_{v,\lambda},$$

(13.27)
$$\tilde{u}_\lambda = u(\Gamma) \qquad \text{on } \tilde{\Gamma}_\lambda.$$

Notice that

(13.28)
$$B_{c\lambda^{1/2}} \subset \tilde{\Omega}_\lambda \subset B_{C\lambda^{1/2}} \qquad (c > 0, C > 0).$$

In the sequel we denote by C, C^*, c positive generic constants independent of λ. By Lemma 13.7,

(13.29)
$$\tilde{\Omega}_{p,\lambda} \subset B_{C^*}.$$

Lemma 13.8. *There holds*

(13.30) $\tilde{u}_\lambda(x) \leq C\log(r + 2) \qquad$ *if $x \in \tilde{\Omega}_{v,\lambda}$, $r = |x|$.*

Proof. We first estimate $\tilde{u}_\lambda(x)$ for $|x| = C^* + 2$. Let

$$w = \frac{u(\Gamma) - \tilde{u}_\lambda(\Gamma)}{u(\Gamma)}.$$

Then $w = 1$ on $\tilde{\Gamma}_{p,\lambda}$ and $w = 0$ on $\tilde{\Gamma}_\lambda$. By Green's formula, for $|x| = C^* + 2$,

$$(13.31) \qquad w(x) = \frac{1}{2\pi} \int_{\tilde{\Gamma}_{p,\lambda}} \log \frac{1}{r} \frac{\partial w}{\partial \nu} - \frac{1}{2\pi} \int_{\tilde{\Gamma}_{p,\lambda}} \frac{\partial}{\partial \nu} \log \frac{1}{r}$$

$$+ \frac{1}{2\pi} \int_{\tilde{\Gamma}_\lambda} \log \frac{1}{r} \frac{\partial w}{\partial \nu} - \frac{1}{2\pi} \int_{\tilde{\Gamma}_\lambda} \frac{\partial}{\partial \nu} \frac{1}{r} \cdot w$$

$$\equiv I_1 - I_2 + I_3 - I_4.$$

Clearly, $I_4 = 0$ and $I_2 = 0$. To evaluate I_3, note first that

$$(13.32) \quad \frac{1}{2\pi} \int_{\tilde{\Gamma}_\lambda} \frac{\partial w}{\partial \nu} = \frac{1}{2\pi} \int_{\tilde{\Gamma}_{p,\lambda}} \frac{\partial w}{\partial \nu} = \frac{1}{2\pi} \mathrm{Cap}_{\tilde{\Omega}_\lambda} \tilde{\Omega}_{p,\lambda}$$

$$= \frac{1}{2\pi} \mathrm{Cap}_{\Omega} \Omega_p = \frac{2}{\log \lambda + O(1)} \qquad [\text{by } (13.22)].$$

Since also, by (13.28),

$$\log r = \tfrac{1}{2} \log \lambda + O(1) \qquad \text{on } \tilde{\Gamma}_\lambda,$$

we get

$$I_3 = \frac{1}{2\pi} \left(\frac{1}{2} \log \lambda \right) \int_{\tilde{\Gamma}_\lambda} \frac{\partial w}{\partial \nu} + O(1) \int_{\tilde{\Gamma}_\lambda} \left| \frac{\partial w}{\partial \nu} \right|$$

$$= \frac{\log \lambda}{\log \lambda + O(1)} + O(1) O\left(\frac{1}{\log \lambda} \right) \qquad \left(\text{since } \frac{\partial w}{\partial \nu} \leqslant 0 \right)$$

$$= 1 + \frac{O(1)}{\log \lambda}.$$

Finally, since $|\log r| \leqslant C$ on $\tilde{\Gamma}_{p,\lambda}$,

$$I_1 = O(1) \int_{\tilde{\Gamma}_{p,\lambda}} \left| \frac{\partial w}{\partial \nu} \right| = O(1) \int_{\tilde{\Gamma}_{p,\lambda}} \frac{\partial w}{\partial \nu} \qquad \left(\text{since } \frac{\partial w}{\partial \nu} \geqslant 0 \right)$$

$$= \frac{O(1)}{\log \lambda} \qquad [\text{by } (13.32)].$$

Putting together the estimates for the I_j, we obtain from (13.31),

$$w(x) = 1 + O\left(\frac{1}{\log \lambda} \right) \qquad \text{if } |x| = C^* + 2.$$

Since

$$w(x) = 1 - \frac{\tilde{u}_\lambda}{u(\Gamma)},$$

we get, upon recalling (13.3),

(13.33) $$\tilde{u}_\lambda(x) \leqslant C, \qquad |x| = C^* + 2.$$

By the maximum principle, the same inequality holds for $x \in \tilde{\Omega}_{v,\lambda}$, $|x| < C^* + 2$. Finally, comparing $\tilde{u}_\lambda$ with $C_0 \log(r + 2)$ in $\tilde{\Omega}_\lambda \cap \{|x| > C^* + 2\}$, the assertion of the lemma follows.

Lemma 13.9. *There holds*

(13.34) $$|\tilde{u}_\lambda(x)| \leqslant C \qquad if\ x \in \tilde{\Omega}_{p,\lambda}.$$

Proof. Since $\int u_- = I/\lambda$,

$$\int_{\tilde{\Omega}_{p,\lambda}} |\tilde{u}_\lambda| = I,$$

and (13.25), (13.26) give

$$\int_{\{|x| < A\}} |\Delta\tilde{u}_\lambda| \leqslant I \qquad \text{for any } A > 0.$$

Also, by Lemma 13.8, for any large $A > 0$,

$$\max_{|x| = A} |\tilde{u}_\lambda| \leqslant C \qquad (C \text{ depends on } A).$$

Representing $\tilde{u}_\lambda$ in $|x| < A$ by means of Green's function, we deduce that

$$\int_{\{|x| \leqslant A\}} |\tilde{u}_\lambda|^p \leqslant C \qquad \text{for any } p < \infty.$$

Hence, by (13.25), (13.26),

(13.35) $$\int_{\{|x| \leqslant A\}} |\Delta\tilde{u}_\lambda|^p \leqslant C,$$

and (13.34) follows by applying Sobolev's inequality.

Corollary 13.10. *There exist positive constants C, c such that, for all $\lambda > \lambda_1$,*

(13.36) $$d(\Omega_{p,\lambda}) \geqslant \frac{c}{\lambda^{1/2}},$$

(13.37) $$|\Omega_{p,\lambda}| \geqslant \frac{C}{\lambda}.$$

Indeed, by Lemma 13.9, $(u_\lambda)_- \leqslant C$ and therefore

$$\frac{I}{\lambda} = \int_{\Omega_{p,\lambda}} (u_\lambda)_- \leqslant |\Omega_{p,\lambda}| ;$$

- thus (13.37) follows; (13.36) is a consequence of (13.37).

From Lemma 13.9 and (13.35), we obtain, by elliptic estimates;

(13.38) $|\tilde{u}_\lambda|_{C^{2+\alpha}(B_R)} \leqslant C_R \qquad \forall \text{ ball } B_R$

provided that $\lambda \geqslant \lambda_0(R)$; C_R and $\lambda_0(r)$ depend on R. Hence from any sequence of λ's increasing to infinity, we can extract a subsequence λ_i such that

(13.39) $D^\beta u_{\lambda_i} \to D^\beta U$ in compact subsets of $R^2, 0 \leqslant |\beta| \leqslant 2$,

(13.40) $\Delta U - U_- = 0.$

Furthermore, by Lemma 13.8,

(13.41) $U(x) \leqslant C \log(|x| + 2)$

and, by Lemma 13.7,

(13.42) $\Omega_- \equiv \{x \in R^2; U(x) < 0\} \subset B_{C^*} \qquad (C^* > 0).$

Since also

(13.43) $\int_{R^2} U_- \, dx = I,$

the open set Ω_- is nonempty.

We shall need the following theorem:

Theorem 13.11. *Let u be a solution in R^2 of the equation*

$$\Delta u + f(u) = 0,$$

where f is Lipschitz continuous, and assume that

$$u(x) = \tfrac{1}{2} a \log r^2 + b + \frac{A \cdot x}{r^2} + O\left(\frac{1}{r^2}\right),$$

$$\nabla u(x) = \frac{ax}{r^2} + \nabla\left(\frac{A \cdot x}{r^2}\right) + O\left(\frac{1}{r^3}\right) .$$

as $r \to \infty$, where a is a positive constant. Then u is a radial function, that is, there is a function $g(t)$ $(t \geq 0)$ and a point x^0 such that

$$u(x) = g(|x - x^0|).$$

Proof. By changing variables $x \to x - x^0$ and choosing x^0 in a suitable way, we obtain an expansion for u near infinity in which the term $A \cdot x$ disappears. Without loss of generality we may assume that $a = 1$. Taking also, for simplicity, $x^0 = 0$, and denoting the independent variable by (x, y), we have

$$(13.44) \qquad \nabla u(x, y) = \frac{(x, y)}{r^2} + O\left(\frac{1}{r^3}\right).$$

Consider the rectangles

$$S_\lambda = \{(x, y) : |x - \lambda| < M, |y| < M\},$$

$$S_\lambda^+ = S \cap \{x > \lambda\},$$

where M is any sufficiently large positive number and

$$(13.45) \qquad \frac{K}{M} \leq \lambda \leq 2M.$$

K is a fixed and sufficiently large positive number to be specified later; it is independent of M.

We wish to study the function

$$(13.46) \qquad v_\lambda(x, y) = u(x, y) - u(-x + 2\lambda, y) \qquad \text{in } S_\lambda^+.$$

First we prove that

$$(13.47) \qquad v_\lambda(x, \pm M) > 0 \qquad \text{if } \lambda < x < M + \lambda.$$

Using (13.44), we have

$$v_\lambda(x, \pm M) = \int_{-x+2\lambda}^{x} u_\zeta(\zeta, \pm M) \, d\zeta$$

$$= \int_{-x+2\lambda}^{x} \frac{\zeta}{M^2 + \zeta^2} \, d\zeta + \int_{-x+2\lambda}^{x} O\left(\frac{1}{M^3}\right) d\zeta$$

$$\geq \frac{1}{2} \left[\log\left(M^2 + \zeta^2\right)\right]_{-x+2\lambda}^{x} + O\left(\frac{x - \lambda}{M^3}\right)$$

$$= \frac{1}{2} \left[\log\left(1 + \frac{x^2}{M^2}\right) - \log\left(1 + \frac{(x - 2\lambda)^2}{M^2}\right)\right] + O\left(\frac{x - \lambda}{M^3}\right).$$

Since

$$\frac{(x - 2\lambda)^2}{M^2} \leqslant \frac{x^2}{M^2} < 10$$

and since the function

$$\log(1 + t) - \frac{1}{11}t$$

is monotone increasing in $t, 0 < t < 10$, we get the lower bound

$$\frac{1}{22}\left[\frac{x^2}{M^2} - \frac{(x - 2\lambda)^2}{M^2}\right] + O\left(\frac{x - \lambda}{M^3}\right) \geqslant \frac{(x - \lambda)\lambda}{6M^2} + O\left(\frac{x - \lambda}{M^3}\right).$$

Using (13.45), we obtain

(13.48) $$v_\lambda(x, \pm M) \geqslant \frac{\lambda(x - \lambda)}{7M^2}$$

for a suitably large K, thus establishing (13.47).
    We next show that

(13.49) $$v_\lambda(M + \lambda, y) > 0 \qquad \text{if } |y| \leqslant M.$$

It suffices to consider the case $y \geqslant 0$. We can write

(13.50) $$v_\lambda(M + \lambda, y) = I + J,$$

where

$$I = [u(M + \lambda, y) - u(M + \lambda, M)] - [u(-M + \lambda, y) - u(-M + \lambda, M)],$$
$$J = u(M + \lambda, M) - u(-M + \lambda, M) = v_\lambda(M + \lambda, M).$$

By (13.48),

(13.51) $$J \geqslant \frac{\lambda}{7M}.$$

    To estimate I, write

$$I = -\int_y^M \left[u_\eta(M + \lambda, \eta) - u_\eta(-M + \lambda, \eta)\right] d\eta$$

and, upon using (13.44),

$$I = -\int_y^M \left[\frac{\eta}{\eta^2 + (M+\lambda)^2} - \frac{\eta}{\eta^2 + (M-\lambda)^2} \right] d\eta + O\left(\frac{1}{M^2} \right)$$

$$\geqslant O\left(\frac{1}{M^2} \right)$$

since the integrand is negative (recall that $y > 0$). Combining the last inequality with (13.51), (13.50), and using (13.45), the assertion (13.49) follows.

From the proof of (13.47), it is clear that also

(13.52) $v_\lambda(x, \pm M') > 0$ if $\lambda \leqslant x \leqslant M + \lambda$, $M - 1 \leqslant M' \leqslant M$.

The function v_λ satisfies

$$\Delta v_\lambda + f(u(x, y)) - f(u(-x + 2\lambda, y)) = 0 \quad\quad \text{in } S_\lambda^+,$$

so that

$$\Delta v_\lambda + \tilde{c} v_\lambda = 0 \quad\quad \text{in } S_\lambda^+, \tilde{c} \text{ a bounded function.}$$

From (13.44) it follows that if M is sufficiently large, then $v_\lambda \geqslant 0$ in $\overline{S_\lambda^+}$ if $\lambda = 2M$. We have also proved in (13.47), (13.49) that $v_\lambda > 0$ on $\partial S_\lambda^+ \setminus \{x = \lambda\}$. On $x = \lambda$ we have $v_\lambda = 0$. Thus the strong maximum principle (see Section 12, Problem 2) can be applied to conclude:

$$v_\lambda > 0 \quad\quad \text{in } \overline{S_\lambda^+},$$

and

$$v_{\lambda x}(\lambda, y) > 0 \quad\quad \text{if } |y| < M;$$

hence

$$u_x(\lambda, y) > 0 \quad\quad \text{if } |y| < M.$$

Recalling also (13.52), it follows that if $\lambda' < \lambda$, $\lambda - \lambda'$ sufficiently small, then

$$v_{\lambda'} > 0 \quad\quad \text{in } \overline{S_{\lambda'}^+}.$$

We now apply a "folding" argument (see Section 12, Problem 3). We decrease λ continuously until we reach the smallest possible value of λ, say λ_0, such that either (i) $\lambda_0 = K/M$, $v_{\lambda_0} > 0$ in $\overline{S_{\lambda_0}^+}$, or (ii) $\lambda_0 \geqslant K/M$, $v_{\lambda_0} \geqslant 0$ in $\overline{S_{\lambda_0}^+}$ with equality at some point of $S_{\lambda_0}^+$. The argument given above for v_λ shows that (ii) cannot occur. Thus (i) holds and, in particular,

$$u_x(\lambda_0, y) > 0 \qquad \text{if } |y| < M, \quad \lambda_0 = \frac{K}{M}.$$

Taking $M \to \infty$, we get

$$u_x(0, y) \geq 0 \qquad \text{for all } y.$$

Since the x-axis can be taken in any direction, we get $u_\theta = 0$ where (r, θ) are the polar coordinates with respect to x^0, and the proof of Theorem 13.11 is complete.

Consider now the function U. It satisfies

$$\Delta U + f(U) = 0 \qquad [f(t) = -t_-]$$

and it is harmonic outside a disc [by (13.42)]. It also satisfies (13.41). The relation (13.27) with $u(\Gamma) \sim c_0 \log \lambda$ ($c_0 > 0$) can be used to compare $\tilde{u}_\lambda$ from below with a function of the form $c \log r$ (c positive and sufficiently small). We deduce that

$$U(x) \to \infty \qquad \text{if } |x| \to \infty.$$

Hence all the assumptions of Theorem 13.11 are satisfied. It follows that

$$(13.53) \qquad\qquad U(x) = U_0(|x - x^0|).$$

Let $\rho = |x - x^0|$. The set where $U(x)$ is harmonic must be the exterior of a disc; otherwise, there is a ring $R_1 < \rho < R_2$ in which U is positive and harmonic, and it vanishes on $\rho = R_1$, $\rho = R_2$; this, however, is impossible.
    We have thus shown that

$$(13.54) \qquad\qquad \Omega_- \text{ is a circle } |x - x^0| < R,$$

and

$$(13.55) \qquad U(x) = U_0(\rho) = A \log \frac{\rho}{R} \qquad \text{if } \rho > R; \quad A > 0.$$

Hence, in particular,

$$(13.56) \qquad\qquad \frac{\partial U}{\partial \nu} \neq 0 \text{ on } \partial\Omega_- \qquad (\nu = \text{the normal to } \partial\Omega_-),$$

and from (13.39) we then deduce that

$$(13.57) \qquad\qquad |\operatorname{grad} \tilde{u}_{\lambda_i}| \geq c > 0 \qquad \text{along } \tilde{\Gamma}_{p, \lambda_i}.$$

Thus, every sequence of λ's increasing to ∞ has a subsequence λ_i for which (13.57) holds. This yields:

Lemma 13.12. *There exists a positive constant $c < 1$ such that, for all $\lambda > \lambda_1$,*

$$(13.58) \qquad c \leqslant |\operatorname{grad} \tilde{u}_\lambda| \leqslant \frac{1}{c} \qquad \text{along } \tilde{\Gamma}_{p,\lambda};$$

consequently,

$$(13.59) \qquad c\lambda^{1/2} < |\operatorname{grad} u(x)| \leqslant \frac{1}{c}\lambda^{1/2} \qquad \text{along } \Gamma_p.$$

The second assertion follows from the first one by recalling the definition of $\tilde{u}_\lambda$. Notice that R in (13.54) is determined by the property:

$$(13.60) \qquad \begin{array}{l}\text{the first eigenvalue of } -\Delta \text{ in the}\\ \text{circle } |x| < R \text{ is equal to } 1.\end{array}$$

From (13.39), Theorem 13.11, and Lemma 13.12, we can deduce, using the implicit function theorem:

Theorem 13.13. *For each λ sufficiently large, there exists a point $x_\lambda \in \Omega$ such that the free boundary $\Gamma_{p,\lambda}$ can be represented [in polar coordinates (r, θ) about x_λ] in the form $r = R_\lambda(\theta)$ and*

$$(13.61) \qquad \max_\theta \left| \frac{\partial^i}{\partial \theta^i}[R_\lambda(\theta) - R] \right| \to 0 \qquad \text{if } \lambda \to \infty \quad (i = 0, 1, 2).$$

Further, $\operatorname{dist}(x_\lambda, S) \to 0$ *if* $\lambda \to \infty$.

Thus Γ_p is approximately a disc with radius $R/\lambda^{1/2}$. The last assertion follows from Theorem 13.5.

PROBLEMS

1. The set S is finite.

   [*Hint*: Let $z = g(w)$ be a conformal mapping of $\{|w| < 1\}$ onto Ω. Then

   $$2\pi k(z) = 2\pi h_z(z) = \lim_{w_0 \to w} \left\{ \log|g(w) - g(w_0)| - \log\left|\frac{1 - w\overline{w}_0}{w - w_0}\right| \right\},$$

   so that

   $$2\pi k(g(w)) = -\log|g'(w)| - \log(1 - |w|^2)$$

is subharmonic; in fact, $\Delta k = 4e^{4\pi k}$. If S is not finite, it contains an analytic curve in the interior of which $k \equiv$ const.]

2. If Ω is given by $\{(x, y); |y| \leq \phi(x), a < x < b\}$, then $(\partial/\partial y)k(z) > 0$ if $y > 0$ and, consequently, S lies on the x-axis.
[*Hint*: The formula

$$(13.62) \qquad 2\pi \frac{\partial}{\partial y} k(\zeta) = \int_{\partial\Omega} \left(\frac{\partial G(z, \zeta)}{\partial n_z} \right)^2 \cos(n_z, y) \, ds_z$$

follows using (see references 37 and 38)

$$G_\varepsilon(z, \zeta) - G(z, \zeta) = -\frac{1}{2\pi} \int_{\partial\Omega} \frac{\partial G(z, t)}{\partial n_t} \frac{\partial G(t, \zeta)}{\partial n_t} \delta n_t \, ds_t,$$

where G_ε is Green's function for $\Omega_\varepsilon = \Omega + \varepsilon i$, and

$$\lim_{\varepsilon \to 0} \frac{k(\zeta + i\varepsilon) - k(\zeta)}{\varepsilon} = \lim_{\varepsilon \to 0} \left[G_\varepsilon(z, \zeta) - G(z, \zeta) \right].$$

By the maximum principle,

$$-\frac{\partial G(z, \zeta)}{\partial n_z} + \frac{\partial G(\bar{z}, \zeta)}{\partial n_{\bar{z}}} > 0 \qquad \text{if } \zeta \in \Omega_+ \equiv \Omega \cap \{y > 0\},$$

for any fixed $z \in \partial\Omega_+ \cap \{y > 0\}$.]

3. If S consists of n points, then for all λ sufficiently large there exist at least n solutions of (11.1)–(11.3).
[*Hint*: Let $x_0 \in S$, $K_{G,\Omega} = \{v \in K, v > 0 \text{ in } \Omega \setminus G\}$, G a small neighborhood of x_0, u_0 a minimizer of $J(v)$ over $K_{G,\Omega}$. If $u_0 > 0$ on G, then (11.1)–(11.3) follow as in the proof of Theorem 11.3. Suppose that $\partial\Omega_{p_0} \cap \partial G \neq 0$, where $\Omega_{p_0} = \{u_0 < 0\}$; then $u_0 = 0$ on $\partial G \cap \partial\Omega_{p_0}$ in the trace class sense and if (*) $d(\Omega_{p_0}) \to 0$ as $\lambda \to \infty$, then the proof of Theorem 13.5 gives

$$d(x, x_0) \to 0 \qquad \text{if } x \in \Omega_{p_0}, \quad \lambda \to \infty.$$

To prove (*), use

$$-\frac{I^2}{8\pi} \log \lambda + O(1) \leq \inf_{K_{G,\Omega}} J(v) \leq -\frac{I^2}{8\pi} \log \lambda + O(1),$$

where the right-hand side is obtained by comparing with

$$\inf_{K_{G,G_1}} J(v) \leqslant \inf_{K_{G_1}} J(v),$$

where $G_1 \supset G$ and ∂G_1 is parallel to ∂G at small distance.]

4. Consider the plasma problem described in Section 11, Problem 3. Let $A = \sup\{r; (r, z) \in \Omega\}$. Prove that for any $\varepsilon > 0$,

$$A - \varepsilon \leqslant \frac{\inf J(v)}{(I^2/8\pi)\log \lambda + O(1)} \leqslant A \qquad \text{as } \lambda \to \infty$$

and use it to show that $d(\Omega_{p,\lambda}, \Gamma) \to 0$ if $\lambda \to \infty$.
[*Hint*: Follow the proof of Theorem 13.5.]

14. A VARIATIONAL APPROACH TO CONVEX PLASMAS

In Section 11 we introduced two (equivalent) variational principles for the plasma problem for a general domain Ω in R^n. In this section we restrict Ω to be a convex domain in R^n and introduce a new variational principle. It has the advantage that it yields a solution of the plasma equations (11.1)–(11.3) with a convex plasma.
We assume:

(14.1) Ω is a bounded convex domain with $C^{2+\alpha}$ boundary.

Definition 14.1. A subdomain G of Ω is said to belong to class $\mathcal{B}$ if G is convex and $\bar{G} \subset \Omega$.
For any $\lambda > \lambda_1$ and small $\varepsilon_0 > 0$, define a function $f_{\lambda, \varepsilon_0}$:

(14.2)
$$f_{\lambda, \varepsilon_0}(t) = \begin{cases} \dfrac{t - \lambda}{\varepsilon_0} & \text{if } t > \lambda, \\ \varepsilon_0(t - \lambda) & \text{if } t < \lambda. \end{cases}$$

Consider the functional

$$\Phi_{\lambda, \varepsilon_0}(G) = f_{\lambda, \varepsilon_0}(\lambda_1(G)) + \text{Cap}_\Omega G,$$

where $\lambda_1(G)$ is the principal eigenvalue of G.

Problem ($\mathcal{B}$). Find G such that

(14.3)
$$G \in \mathcal{B}, \qquad \Phi_{\lambda, \varepsilon_0}(G) = \min_{\tilde{G} \in \mathcal{B}} \Phi_{\lambda, \varepsilon_0}(\tilde{G}).$$

To every $G \in \mathcal{B}$ we associate a function u defined by

$$\Delta u = \lambda_1(G)u, \qquad u < 0 \text{ in } G,$$

$$\Delta u = 0 \qquad \text{in } \Omega \setminus \overline{G},$$

(14.4) $$u = 0 \qquad \text{on } \partial G,$$

$$u = \text{const.} = u(\Gamma) > 0 \qquad \text{on } \partial\Omega,$$

$$\int_\Gamma \frac{\partial u}{\partial \nu} \, dS = I.$$

In this section, we prove:

Theorem 14.1. *If ε_0 is sufficiently small, then there exists a solution G of Problem $(\mathcal{B})$ and the corresponding u [defined by (14.4) with suitable $u(\Gamma)$] is a solution of the plasma problem (11.1)–(11.3).*

The last assertion simply means that $u \in C^1$ in a neighborhood of ∂G. We begin with several auxiliary results of intrinsic interest.

Lemma 14.2. *Let w be a $C^2(\Omega)$ solution of an equation*

$$\Delta w = f(x, w, \nabla w)$$

in a bounded convex domain Ω of R^n, and assume that $f(x, w, v_1, \ldots, v_n)$ is a $C^{2+\alpha}$ function nondecreasing in w and convex in (x, w). For any fixed $t \in (0, 1)$, consider the function

$$U(x, y) = w(tx + (1 - t)y) - tw(x) - (1 - t)w(y)$$

in $\Omega \times \Omega$. Then $U(x, y)$ cannot take a negative minimum in $\Omega \times \Omega$.

Proof. Without loss of generality we may assume that f is strictly increasing in w and w is smooth up to the boundary. Indeed, otherwise we first approximate w by w_ε:

$$\Delta w_\varepsilon = f(x, w_\varepsilon, \nabla w_\varepsilon) + \varepsilon w_\varepsilon \qquad \text{in } \Omega_\varepsilon,$$

$$w_\varepsilon = w \qquad \text{on } \partial\Omega_\varepsilon$$

where $\varepsilon > 0$, Ω_ε is convex with smooth boundary and $\Omega_\varepsilon \uparrow \Omega$ if $\varepsilon \downarrow 0$. (The Schauder estimates can be used to prove the existence of a solution w_ε; see Problem 1.)

Suppose that the assertion is false. Then $U(x, y)$ takes negative minimum at a point (x_0, y_0) in $\Omega \times \Omega$. Set $z_0 = tx_0 + (1 - t)y_0$. Since $\text{grad } U = 0$ at

(x_0, y_0),

(14.5) $\nabla w(x_0) = \nabla w(y_0) = \nabla w(z_0)$.

The function $h(x) = U(x_0 + x, y_0 + x)$ takes negative minimum at $x = 0$; hence

(14.6) $\Delta h(0) \geqslant 0$.

Using (14.5) we compute that

$$\Delta h(0) = \Delta w(z_0) - t \Delta w(x_0) - (1 - t) \Delta w(y_0)$$

$$= f(z_0, w(z_0), \nabla w(z_0)) - tf(x_0, w(x_0), \nabla w(z_0))$$

$$- (1 - t)f(y_0, w(y_0), \nabla w(z_0)).$$

Since $U(x_0, y_0) < 0$, we have $w(z_0) < tw(x_0) + (1 - t)w(y_0)$, and thus, by the strict monotonicity of $f(t, w, \nabla w)$ in w,

$$\Delta h(0) < f(z_0, tw(x_0) + (1 - t)w(y_0), \nabla w(z_0))$$

$$- tf(x_0, w(x_0), \nabla w(z_0)) - (1 - t)f(y_0, w(y_0), \nabla w(z_0)) \leqslant 0$$

by the convexity of $f(x, w, \nabla w)$ in (x, w); this contradicts (14.6).

Theorem 14.3. *Let u be a solution of*

$$\Delta u + \lambda q(x)u = 0 \qquad in \ \Omega, \quad \lambda > 0,$$

$$u = 0 \qquad on \ \partial\Omega,$$

$$u > 0 \qquad in \ \Omega,$$

where Ω is a bounded convex domain with C^1 boundary and $q(x) > 0$, $q(x)$ concave in $\overline{\Omega}$. Then $\log u(x)$ is concave.

Proof. We may assume that Ω is strictly convex with C^3 boundary and $q \in C^{2+\alpha}(\overline{\Omega})$, for otherwise we approximate Ω and q by such Ω_ε and q_ε, respectively; if the corresponding principal eigenfunction u_ε is log-concave, then the same is true for $u = \lim u_\varepsilon$.

The function $w = \log u$ satisfies

(14.7) $\Delta w = -\lambda q - |\nabla w|^2 \equiv f(x, \nabla w)$

and

(14.8) $w(x) \to -\infty \qquad$ if dist $(x, \partial\Omega) \to 0$.

Clearly, Lemma 14.2 can be applied. Thus it suffices to show that if

$$(14.9) \qquad \lim_{\substack{x_k \to \bar{x} \\ y_k \to \bar{y}}} U(x_k, y_k) = \inf_{\Omega \times \Omega} U(x, y) < 0,$$

then $\bar{x}$ and $\bar{y}$ both belong to Ω. We shall assume that $\bar{x} \in \partial\Omega$ and derive a contradiction.

Set

$$z_k = tx_k + (1 - t)y_k.$$

Since

$$w(z_k) - tw(x_k) - (1 - t)w(y_k) < 0$$

for k large, it follows from (14.8) that

$$w(z_k) \to -\infty$$

and thus $\text{dist}(z_k, \partial\Omega) \to 0$. Since $\partial\Omega$ is strictly convex, we conclude that

$$y_k \to \bar{x}, \qquad z_k \to \bar{x}.$$

By taking a subsequence we may assume that

$$\lim \frac{x_k - y_k}{|x_k - y_k|} \qquad \text{exists};$$

it determines a direction e. Denote by l the ray from $\bar{x}$ into Ω having direction $+e$ or $-e$.

CASE 1. l nontangential. For x in Ω near $\bar{x}$ and for σ any direction near the direction of l,

$$(14.10) \qquad w_{\sigma\sigma}(x) = \frac{uu_{\sigma\sigma} - u_\sigma^2}{u^2} < 0$$

since $u_\sigma > 0$ at $\bar{x}$ and $u = 0$ at $\bar{x}$. Hence w is strictly concave on the interval $\overline{x_k y_k}$ This implies that $U(x_k, y_k) > 0$, contradicting (14.9).

CASE 2. l is tangential. Suppose that $\partial\Omega$ near $\bar{x}$ has the form $x_n = \phi(x')$, where $x' = (x_1, \ldots, x_{n-1})$, $\bar{x} = (0, \ldots, 0)$, and $\nabla\phi(0) = 0$. Since $(-\nabla\phi, 1)$ is normal to $\partial\Omega$, the vectors $(0, \ldots, 0, 1, 0, \ldots, \phi_{x_i})$ (1 in the ith component) are tangential and

$$T_i = \frac{\partial}{\partial x_i} + \phi_{x_i}\frac{\partial}{\partial x_n} \qquad (1 \leqslant i \leqslant n - 1)$$

is a tangential derivative. Therefore, $T_i^2 u = 0$ along $\partial\Omega$, so that

$$u_{x_i x_i} + \phi_{x_i x_i} u_{x_n} = 0 \qquad \text{at } \bar{x}.$$

Thus $u_{x_i x_i} < 0$ at $\bar{x}$ ($1 \leqslant i \leqslant n - 1$). But, for any x in Ω near $\bar{x}$ and for any direction σ near the direction of l, we have

$$\frac{\partial}{\partial\sigma} = c_n(x)\frac{\partial}{\partial\nu} + \sum_{i=1}^{n-1} c_i T_i,$$

where $|c_n|$ is small and $\sum_{i=1}^{n-1} c_i^2$ is near 1. Consequently,

$$u_{\sigma\sigma} \leqslant -\gamma < 0$$

and, by (14.10),

$$w_{\sigma\sigma} \leqslant \frac{u_{\sigma\sigma}}{u} < 0;$$

a contradiction is now derived as before.

Theorem 14.4. *Let Ω, G be convex domains with C^1 boundary and let u be a solutions of*

$$\Delta u = \gamma(u) \qquad \text{in } \Omega \setminus G,$$

$$u = 1 \qquad \text{on } \partial G,$$

$$u = 0 \qquad \text{on } \partial\Omega,$$

$$0 < u < 1 \qquad \text{in } \Omega \setminus G$$

where $\gamma(t)$ is Lipschitz continuous, nondecreasing, and a nonnegative function of t. Then the level curves of u are convex.

The assertion means that the sets $\{x; u(x) > t\} \cup \overline{G}$ are convex for any $t \in (0, 1)$.

Proof. By approximation we may assume that $\gamma(u)$ is strictly increasing and analytic and that Ω and G are strictly convex (see Problem 2). Extend u by 1 into G. If we can show that

$$(14.11) \qquad M \equiv \sup_{\substack{x, y \in \Omega, u(x) > u(z) \\ z = tx + (1-t)y \text{ for some } t \in (0, 1)}} \{u(y) - u(z)\} \leqslant 0$$

then we get $u(z) \geqslant \min\{u(x), u(y)\}$ whenever $z = tx + (1 - t)y$ for some $t \in (0, 1)$, and the assertion follows.

To prove (14.11) we need the following fact:

$$(14.12) \qquad (x - \tilde{x}) \cdot \nabla u(x) < 0 \text{ in } \Omega \setminus \overline{G} \text{ if } \tilde{x} \in G.$$

To verify it, set $\zeta = (x - \tilde{x}) \cdot \nabla u$. Then

$$\Delta \zeta - \gamma'(u)\zeta = 2\gamma(u) \geqslant 0,$$

and (14.12) follows by the maximum principle.

Suppose now that (14.11) is false. Then there exist points x_k, y_k and $z_k = t_k x_k + (1 - t_k)y_k$ such that

$$u(y_k) - u(z_k) \to M > 0,$$

$$u(x_k) > u(z_k), \qquad u(z_k) = \min \{u(x); x \in \overline{x_k y_k}\},$$

and x_k, y_k belong to Ω,

$$x_k \to x^0, \qquad y_k \to y^0, \qquad z_k \to z^0.$$

Clearly, $z^0 \neq y^0$ since $u(z^0) \neq u(y^0)$.

Since

$$u(x^0) \geqslant u(z^0), \qquad u(y^0) > u(z^0),$$

the interval $\overline{x^0 y^0}$ cannot intersect $\Omega \setminus \overline{G}$, by (14.12). Also, it cannot intersect $\Omega \setminus G$, since otherwise $u = \text{const.}$ along $\overrightarrow{x^0 y^0}$ and, by analyticity, also $u = \text{const.}$ along the entire segment, in $\Omega \setminus G$, which contains $\overline{x^0 y^0}$.

Next, $y^0 \notin \partial \Omega$, since otherwise

$$u(z^0) = u(y^0) - M < 0.$$

The function $u(z)$ along $\overline{x^0 y^0}$ takes minimum at z^0; otherwise, the definition of M is contradicted.

We also have

$$(14.13) \qquad z^0 \notin \partial \Omega.$$

Indeed, if $z^0 \in \partial \Omega$, then the lines through x_k, y_k converge to a line l non-tangential to $\partial \Omega$ at z^0 and thus $\partial u / \partial l \neq 0$ at z^0, from which we deduce that $u(z_k) > u(x_k)$ for k large enough.

Next, we show that

$$(14.14) \qquad z^0 = x^0.$$

Indeed, suppose that $z^0 \neq x^0$. Since $u \not\equiv$ const. along $\overline{x^0 y^0}$, u takes in this interval values larger than $u(z^0)$. Let $x' \in \overline{z^0 x^0}$, $u(x') > u(z^0)$. Then for any x'' near x', $u(x'') > u(z^0)$, and hence

$$u(y'') - u(z^0) \leqslant u(y^0) - u(z^0),$$

where y'' lies in $\overline{x'' z^0}$, y'' near y^0. It follows that u takes local maximum at y^0, which is impossible.

We have proved so far that

(14.15) $$x^0 = z^0 \neq y^0, \qquad \overline{x^0 y^0} \text{ lies in } \Omega \setminus G.$$

Also, since u on $\overline{x_k y_k}$ takes minimum at z_k, and since $z_k \neq x_k$, $z_k \neq y_k$ (if k is large), we have $\partial u(z_k)/\partial l_k = 0$ and, consequently,

(14.16) $$\frac{\partial u(z^0)}{\partial l} = 0;$$

here l_k, l are in the directions of $\overrightarrow{x_k y_k}$ and $\overrightarrow{x^0 y^0}$, respectively.

The next assertion is that

(14.17) $$\nabla u(y^0) \text{ is parallel to } \nabla u(z^0).$$

Indeed, otherwise there is a direction v such that $v \cdot \nabla u(y^0) > 0$, $v \cdot \nabla u(z^0) < 0$. Let $\tilde{y}_0 = y^0 + \alpha v$, $\tilde{z}_0 = z^0 + \beta v$, $\alpha > 0$, $\beta > 0$. If $u(\hat{x}) > u(z^0)$ for some $\hat{x}$ on the ray $\overrightarrow{y^0 z^0}$ beyond z^0, then by choosing $\beta = \alpha$ small and $\tilde{x}_0 = \hat{x} + \alpha v$, we get a triple $\tilde{x}_0$, $\tilde{y}_0$, $\tilde{z}_0$ with

$$u(\tilde{y}_0) - u(\tilde{z}_0) > u(y^0) - u(z^0)$$

$$u(\tilde{x}_0) > u(\tilde{z}_0),$$

contradicting the definition of M. Thus $u(x) \leqslant u(z^0)$ on the ray $\overrightarrow{y^0 z^0}$ beyond z^0. We therefore can choose suitable α and β/α small enough such that $\overrightarrow{\tilde{y}_0 \tilde{z}_0}$ intersects $\overrightarrow{y^0 z^0}$ at a point $\tilde{x}$ beyond z^0 and $u(\tilde{x}) > u(\tilde{z}_0)$ [since $\partial u(z^0)/\partial v < 0$, and, by (14.16), $\partial u(z^0)/\partial l = 0$]. With this triple $\tilde{y}_0$, $\tilde{z}_0$, $\tilde{x}$ we again derive a contradiction to the definition of M.

Having proved (14.17) we set

(14.18) $$\xi = \frac{\nabla u(y^0)}{|\nabla u(y^0)|} = \frac{\nabla u(z^0)}{|\nabla u(z^0)|}$$

and take for simplicity $|y^0 - z^0| = 1$, $|\nabla u(z^0)| = 1$. Set $\lambda = |\nabla u(y^0)|$ and introduce the functions

$$u_1(x) = u(z^0 + x), \qquad u_2(x) = u\left(y^0 + \frac{x}{\lambda}\right)$$

and

$$U(x) = u_2(x) - u_1(x) - \left(u(y^0) - u(z^0)\right),$$

$$w(x) = \nabla u_1(x) \cdot \left(e + \left(1 - \frac{1}{\lambda}\right)x\right), \qquad e = z^0 - y^0.$$

Then

$$\Delta(U + \lambda w) = \frac{1}{\lambda^2}\gamma(u_2) - \gamma(u_1) + \lambda\gamma'(u_1)w$$

$$+ 2\lambda\left(1 - \frac{1}{\lambda}\right)\gamma(u_1).$$

Since $w(0) = 0$ and $\gamma(u(y^0)) - \gamma(u(z^0)) > 0$, we obtain, setting $\varepsilon = \left(\gamma(u(y^0)) - \gamma(u(z^0))\right)/\lambda^2$,

$$\Delta(U + \lambda w)(0) = \varepsilon + \left(\frac{1}{\lambda^2} - 2\lambda - 3\right)\gamma(u(z^0))$$

$$= \varepsilon + \frac{(\lambda - 1)^2(2\lambda + 1)}{\lambda^2}\gamma(u(z^0)) \geq \varepsilon.$$

Thus, for r small enough, the function $U + \lambda w - c|x|^2$ is subharmonic in $B_r(0)$ provided $c < \varepsilon/2n$; it also vanishes at the origin. Using the maximum principle we deduce that

(14.19) $$\max_{\partial B_r(0)} (U + \lambda w) = U(x) + \lambda w(x) \geq cr^2$$

for some $x \in \partial B_r(0)$.

Set

$$y_0^* = y^0 + \frac{x}{\lambda}, \qquad z_0^* = z^0 + x.$$

If $U(x) > 0$ then

$$u(y_0^*) - u(z_0^*) > u(y^0) - u(z^0).$$

If also $w(x) > 0$ then $u(x_0^*) > u(z_0^*)$ where

$$x_0^* = z_0^* + t(z_0^* - y_0^*)$$

for some small $t > 0$. With the triple (x_0^*, y_0^*, z_0^*) we then get a contradiction to the definition of M.

In the general case where $U(x)$, $w(x)$ are not both positive we shall produce a different admissible triple $(\tilde{x}_0, \tilde{y}_0, \tilde{z}_0)$ in the form:

$$\tilde{y}_0 = y^0 + \frac{x}{\lambda} + \frac{h\xi}{\lambda}, \qquad \tilde{z}_0 = z^0 + \frac{x}{\lambda}, \qquad \tilde{x}_0 = \tilde{z}_0 + t(\tilde{z}_0 - \tilde{y}_0)$$

with small $t > 0$, for which

(14.20) $$u(\tilde{y}_0) - u(\tilde{z}_0) > u(y^0) - u(z^0).$$

If $h = -U(x)$ and

$$\tilde{U}(x) = u_2(x + h\xi) - u_1(x) - \left(u(y^0) - u(z^0)\right),$$

$$\tilde{w}(x) = \nabla u_1(x) \cdot \left(e + \left(1 - \frac{1}{\lambda}\right)x - \frac{h}{\lambda}\xi\right),$$

then, since $h = O(r^2)$,

(14.21) $$\tilde{U}(x) = O(r^3)$$

and

$$\tilde{w}(x) = w(x) - \frac{h}{\lambda} + O(r^3).$$

Since, by (14.19), $\lambda w(x) \geq cr^2 + h$, we get

(14.22) $$\tilde{w}(x) \geq \frac{c}{\lambda}r^2 + O(r^3).$$

Noting that $\partial u_2/\partial \xi \neq 0$ in a neighborhood of the origin we deduce from (14.21), (14.22) that by changing the magnitude of h we can achieve simultaneously $\tilde{U}(x) > 0$ and $\tilde{w}(x) > 0$. With this new value of h the triple $(\tilde{x}_0, \tilde{y}_0, \tilde{z}_0)$ defined above is admissible and (14.20) holds; thus the definition of M is contradicted.

We return to Problem ($\mathcal{B}$).

Lemma 14.5. *Problem ($\mathcal{B}$) has a solution.*

Proof. Denote by $r(G)$ the radius of the largest ball contained in G, $G \in \mathcal{B}$. Then (see reference 114)

(14.23) $$\lambda_1(G) \geq \frac{c}{r(G)} \qquad (c > 0)$$

where c is an absolute constant. For any $G \in \mathcal{B}$, denote by V the solution of

$$(14.24) \qquad \Delta V + \lambda_1 V = 0 \text{ in } G, \qquad V \geqslant 0, \qquad \int_G V^2 = 1$$

and by U the solution of

$$(14.25) \qquad \Delta U = 0 \text{ in } \Omega \setminus G, \qquad U = 0 \text{ on } \partial G, \qquad U = 1 \text{ on } \partial \Omega.$$

We claim that

$$(14.26) \qquad\qquad |\nabla V| \leqslant C \qquad \text{in } G,$$

where C is a constant depending only on $\lambda_1(G)$.

Indeed, suppose first that ∂G is smooth and strictly convex. Then (by Problem 4)

$$(14.27) \qquad \begin{array}{c} |\nabla V|^2 + \lambda_1 V^2 \text{ takes its maximum in } \overline{G} \text{ at} \\ \text{an interior point where } |\nabla V| \text{ vanishes.} \end{array}$$

Thus, setting $M = \sup V$, we have

$$(14.28) \qquad\qquad |\nabla V|^2 + \lambda_1 V^2 \leqslant \lambda_1 M^2.$$

By elliptic estimates [109, Theorem 8.25]

$$M^p \leqslant c\lambda_1^p \int_G V^p \qquad \text{if } p > \frac{n}{2}$$

where c is independent of the smoothness of ∂G. Therefore,

$$M^p \leqslant c\lambda_1^p M^{p-2} \int_G V^2 = c\lambda_1^p M^{p-2}$$

so that $M^2 \leqslant c\lambda_1^p$; (14.26) now follows from (14.28).

From (14.26) we obtain

$$(14.29) \qquad\qquad V(x) \leqslant C d(x),$$

where $d(x) = $ distance from x to ∂G.

We now take a minimizing sequence $G_m \in \mathcal{B}$. Since $\lambda_1(G_m)$ remains bounded, (14.23) implies that each G_m contains a fixed ball $B_R(x^0)$ (m is large enough). We claim: There exists a sufficiently small $d_0 > 0$ such that

$$(14.30) \qquad \text{dist}\,(G_m, \partial \Omega) \geqslant d_0 \qquad \text{if } m \text{ is large enough.}$$

Indeed, suppose that $\text{dist}(G_m, \partial\Omega) < d_0$ and let

$$\rho(x) = \text{dist}(x, \partial\Omega),$$

$$\tilde{G}_m = G_m \cap \{\rho(x) > 2d_0\},$$

$$\Sigma_m = G_m \cap \{\rho(x) = 2d_0\}.$$

Denote by $V_m, \lambda_{1,m}$ and $\tilde{V}_m, \tilde{\lambda}_{1,m}$ the V, λ_1 corresponding to $G = G_m$, $G = \tilde{G}_m$, respectively, by (14.24). We may take the G_m to be such that Green's formula applies to $\tilde{G}_m$ (for instance, we can take G_m as polyhedrons).

By Green's formula (suppressing the index m) and (14.26), (14.29),

$$\left(\tilde{\lambda}_1 - \lambda_1\right)\int_{\tilde{G}} V\tilde{V} = \int_{\Sigma} V |\nabla\tilde{V}| \leqslant cd_0 |\Sigma| \leqslant C\,\delta V,$$

where $\delta V = \text{vol}(G \setminus \tilde{G})$. Since

$$\int_{\tilde{G}} V\tilde{V} = 1 + o(1) \qquad \text{as } d_0 \to 0,$$

we get

(14.31)
$$\delta\lambda_1 \leqslant c\,\delta V.$$

We next compute the change in the capacity. Let Z be the harmonic function in $\{0 < \rho(x) < 2d_0\}$ with $Z = 0$ on $\{\rho(x) = 2d_0\}$, $Z = 1$ on $\partial\Omega$. Then

$$|\nabla Z| \geqslant \eta(d_0), \qquad \eta(d_0) \to \infty \text{ if } d_0 \to 0.$$

Also, by the maximum principle,

$$\tilde{U} \geqslant Z,$$

where $\tilde{U}$ is defined by (14.25) with $G = \tilde{G}_m$. The variation δK of the capacity is given by

(14.32)
$$\delta K \equiv \int_\Omega \left(|\nabla\tilde{U}|^2 - |\nabla U|^2\right) = -\int_\Omega |\nabla(\tilde{U} - U)|^2$$
$$+ 2\int_\Omega \nabla\tilde{U} \cdot \nabla(\tilde{U} - U) = -\int_\Omega |\nabla(\tilde{U} - U)|^2,$$

where $U \equiv 0$ in G_m, $\tilde{U} \equiv 0$ in $\tilde{G}_m$ by definition.

For each $p \in \Sigma$, we integrate $\partial \tilde{U} / \partial l$ along the ray $l : \overline{x^0 p}$ from p until we hit $q \in \partial G$:

$$\int | \nabla \tilde{U} | \geqslant \tilde{U}(q) \geqslant Z(q) \geqslant \eta(d_0) |p - q| \; .$$

Integrating with respect to $p \in \Sigma$, we obtain

$$c \eta(d_0) \, \delta V \leqslant \int_{G \setminus \tilde{G}} | \nabla \tilde{U} | \leqslant (\delta V)^{1/2} \left(\int_{G \setminus \tilde{G}} | \nabla \tilde{U} |^2 \right)^{1/2}, \qquad c > 0 .$$

Recalling (14.32), we conclude that

$$\delta K \leqslant - \int_{G \setminus \tilde{G}} | \nabla \tilde{U} |^2 \leqslant -c \big(\eta(d_0) \big)^2 \, \delta V .$$

Combining this with (14.31), we find that

$$\delta \Phi_{\lambda, \varepsilon_0}(G) \leqslant -c \big(\eta(d_0) \big)^2 \delta V + \frac{c}{\varepsilon_0} \delta V \leqslant - \frac{c}{\varepsilon_0} \delta V < - \frac{c}{\varepsilon_0} d_0$$

if d_0 is small enough; here various positive constants are denoted by the same symbol c, and the relation $\delta V \geqslant c d_0$ was used. It follows that $G = G_m$ cannot be a minimizing sequence, and the proof of (14.30) is thereby completed.

Since G_m contains a fixed ball $B_R(x^0)$, one can easily check that $G_m \to G$ in the following sense: Representing ∂G_m, in terms of spherical coordinates about x^0, by $r = r_m(\theta)$, then $r_m(\theta) \to r_0(\theta)$ uniformly in θ, where $r = r_0(\theta)$ represents ∂G. Further, the $r_m(\theta)$ are uniformly Lipschitz continuous (uniformly in m).

It easily follows that $\lambda_1(G_m) \to \lambda_1(G)$, and, using (14.30), $\operatorname{Cap}_\Omega G_m \to \operatorname{Cap}_\Omega G$. This completes the proof of the lemma.

Lemma 14.6. *Let G be a solution of Problem $(\mathcal{B})$. Then, for some positive constants c, C there holds: (i) $V(x) \geqslant c d(x)$; (ii) $U(x) \leqslant c d(x)$; (iii) $| \nabla U(x) | \leqslant C$; (iv) $| \nabla V(x) | \geqslant c$ in a neighborhood of ∂G; (v) $\partial \Omega \in C^1$; (vi) $| \nabla U(x) | \geqslant c$ in a neighborhood ∂G; and finally (vii) $| \nabla U |$, $| \nabla V |$ exist everywhere along ∂G as limits from $\Omega \setminus \overline{G}$ and G, respectively.*

To prove (i), introduce the level surface $\Gamma_t : \{ V = t \}$, which is convex by Theorem 14.3. Let p be a point on Γ_t where $d(x)$ takes its maximum, say $\delta(t)$. Denote by D the smaller part of G cut off by the tangent plane T_p to Γ_t at p, and set $\tilde{G} = G \setminus D$. We can now proceed to estimate $\delta \lambda_1$, δK corresponding to the variation $\tilde{G}$ of G and find (by calculations similar to those in the preceding lemma) that

$$\delta \Phi_{\lambda, \varepsilon_0}(G) \leqslant -c \, \text{area} \left(T_p \cap G \right) \left(\delta(t) - \frac{ct}{\varepsilon_0} \right) .$$

Since $\delta \Phi_{\lambda, \varepsilon_0}(G) \geqslant 0$, we must have $\delta(t) \leqslant ct / \varepsilon_0$, which gives (i).

The proof of (ii) uses the convex level surfaces $U = t$.
For complete details on the proof of Lemma 14.6, see reference 64.

Proof of Theorem 14.1. Let G be a solution of Problem ($\mathcal{B}$) and let u, v be defined by

$$\Delta v + \lambda_1(G) = 0 \quad \text{in } G, v < 0 \text{ in } G,$$

$$v = 0 \quad \text{on } \partial G,$$

(14.33)

$$\Delta u = 0 \quad \text{in } \Omega \setminus G,$$

$$u = 0 \quad \text{on } \partial G,$$

$$u = 1 \quad \text{on } \Gamma.$$

If we can prove that $\lambda_1(G) = \lambda$ and

(14.34)
$$\left| \frac{\partial v}{\partial \nu} \right| = c \left| \frac{\partial u}{\partial \nu} \right| \quad \text{on } \partial G \quad (c \text{ constant}),$$

then extending v into $\Omega \setminus G$ by cu, we obtain a solution v of (11.1)–(11.3), and the proof of Theorem 14.1 is complete.

We make a perturbation of G into the convex domain bounded by $\{u = \varepsilon\}$. This changes the eigenvalue and the capacity by

(14.35)
$$\delta\lambda_1(G) = -\varepsilon \int_{\partial G} \frac{|\nabla v|^2}{|\nabla u|} + o(\varepsilon),$$

(14.36)
$$\delta K = \frac{\varepsilon}{1 - \varepsilon} \, \mathrm{Cap}_\Omega G;$$

the proof of (14.36) is immediate, whereas the proof of (14.35) is based on Lemma 14.6 and it is given in reference 64. If ε_0 is sufficiently small, then we deduce that

$$\delta\Phi_{\lambda, \varepsilon_0}(G) < 0 \quad \text{if } \lambda_1(G) < \lambda.$$

Since G is a minimizer, we conclude that $\lambda_1(G) \geqslant \lambda$. Similarly, one shows that $\lambda_1(G) \leqslant \lambda$; thus $\lambda_1(G) = \lambda$.

In order to prove (14.34) we make further perturbation, moving the surface $\{u = \varepsilon\}$ inward along the normals by a fixed distance s. The total first variations of $\lambda_1(G)$ and $\mathrm{Cap}_\Omega G$ are given by (see reference 64)

$$\delta\lambda_1 = -\varepsilon \int_{\partial G} \frac{|\nabla v|^2}{|\nabla u|} + s \int_{\partial G} |\nabla v|^2 + o(\varepsilon),$$

$$\delta K = \frac{\varepsilon}{1 - \varepsilon} \, \mathrm{Cap}_\Omega G - s \int_{\partial G} |\nabla u|^2 + o(\varepsilon)$$

provided that $s = O(\varepsilon)$. We choose s such that $\delta\lambda_1 = o(\varepsilon)$; that is,

$$s = \varepsilon \int_{\partial G} \frac{|\nabla v|^2}{|\nabla u|} \Big/ \Big(\int_{\partial G} |\nabla v|^2 \Big).$$

The inequality $\delta\Phi_{\lambda,\,\varepsilon_0}(G) \geq 0$ then yields

$$(14.37) \qquad \int_{\partial G} \frac{|\nabla v|^2}{|\nabla u|} \cdot \int_{\partial G} |\nabla u|^2 \leq \int_{\partial G} |\nabla v|^2 \cdot \int_{\partial G} |\nabla u|.$$

Using Hölder's inequality,

$$(14.38) \qquad \Big(\int_{\partial G} |\nabla v| \Big)^2 \leq \int_{\partial G} \frac{|\nabla v|^2}{|\nabla u|} \cdot \int_{\partial G} |\nabla u|,$$

we obtain from (14.37)

$$(14.39) \qquad \Big(\int_{\partial G} |\nabla v| \Big)^2 \cdot \int_{\partial G} |\nabla u|^2 \leq \int_{\partial G} |\nabla v|^2 \cdot \Big(\int_{\partial G} |\nabla u| \Big)^2.$$

Similarly, if we first perturb ∂G into $\{v = -\varepsilon\}$ and then move it outward a distance s, we obtain (14.37) with u, v reversed. Proceeding as before, we also obtain (14.39) with u, v reversed, which means that equality holds in (14.39). But then we must also have equality in (14.38), which gives (14.34).

PROBLEMS

1. If $\Delta w = f(x, w, \nabla w)$ in Ω, $\partial\Omega \in C^{2+\alpha}$, $f \in C^{2+\alpha}$, $w = g$ on $\partial\Omega$, $g \in C^{2+\alpha}(\partial\Omega)$, then there exists a solution of

$$\Delta w_\varepsilon = f(x, w_\varepsilon, \nabla w_\varepsilon) + \varepsilon w_\varepsilon \qquad \text{in } \Omega,$$

$$w_\varepsilon = g \qquad \text{on } \partial\Omega$$

provided that $|\varepsilon|$ is small, and $w_\varepsilon \in C^{2+\alpha}(\overline{\Omega})$.
[*Hint:* $u_\varepsilon = w - w_\varepsilon$ satisfies

$$Lu_\varepsilon = a_{1\varepsilon} u_\varepsilon + a_{2\varepsilon} \cdot \nabla u_\varepsilon \equiv F_\varepsilon(u_\varepsilon);$$

the coefficients of L are independent of u_ε and $|a_{i\varepsilon}|_{1+\alpha} \leq C\varepsilon$ if $|u_\varepsilon|_{2+\alpha} \leq C_0$. Set $Tu_\varepsilon = U_\varepsilon$ if

$$LU_\varepsilon = F_\varepsilon(u_\varepsilon) \text{ in } \Omega, \qquad U_\varepsilon = 0 \text{ on } \partial\Omega.$$

Show that T is a contraction.]

2. Let u be as in Theorem 14.4 and $\gamma_m(u) \in C^1, 0 \leqslant \gamma_m'(u) \leqslant C, \gamma_m(u) \to \gamma(u)$ uniformly in bounded sets, $G_m \uparrow G$, ∂G_m smooth. Denote by u_m the solution u corresponding to γ_m, G_m. Prove that $u_m(x) \to u(x) \ \forall \ x \in G$.

   [*Hint*: Apply the maximum principle to $u_m - u$.]

3. Let f, g be homogeneous harmonic polynomials satisfying: $\{f \geqslant 0\} \subset \{g \geqslant 0\}$. Prove that $g = cf$, where c is a constant.

   [*Hint*: Denote by Δ_S the restriction of the Laplacian to the unit sphere S. Let K_f, K_g be components of $\{f \geqslant 0\}$, $\{g \geqslant 0\}$ on S such that $K_f \subset K_g$. If $\alpha = $ degree f, $\beta = $ degree g, then

$$\Delta_S f + \alpha f = 0, \qquad \Delta_S g + \beta g = 0.$$

Hence

$$(\alpha - \beta) \int_{K_f} fg = \int_{\partial K_f} (fg_\nu - f_\nu g) = -\int_{\partial K_f} f_\nu g \geqslant 0.$$

Deduce that $\alpha \geqslant \beta$ and by working with $-f, -g$, that $\beta \geqslant \alpha$.]

4. Prove (14.27).

   [*Hint*: (reference 166): Show that $\Phi = |\nabla V|^2 + \lambda_1 V^2$ satisfies

$$\Delta \Phi - \frac{W \cdot \nabla \Phi}{2|\nabla V|^2} \geqslant 0 \qquad (W = \nabla \Phi - 4\lambda_1 V \nabla V)$$

and apply the maximum principle.]

15. THE THOMAS–FERMI MODEL

We consider a quantum mechanics problem in which I electrons each of mass m and charge $-e$ are moving about positive charges of magnitude $m_i e$ fixed at positions $a_i \in R^3$ ($1 \leqslant i \leqslant k$). The Thomas–Fermi atomic model views the electrons as "clouds" with density $\rho(x)$, say, so that

$$\rho(x) \geqslant 0, \qquad \int_{R^3} \rho(x) \, dx = I.$$

The repulsive potential energy (the interaction between electrons) is

$$\frac{1}{2} \int_{R^3} \int_{R^3} \frac{\rho(x)\rho(y)}{|x - y|} \, dx \, dy;$$

the attractive potential energy (the interaction between the electrons and the positive nuclei) is

$$\int_{R^3} V(x)\rho(x)\, dx,$$

where

(15.1)
$$V(x) = \sum_{i=1}^{k} \frac{m_i}{|x - a_i|},$$

and the kinetic energy is

$$\int_{R^3} \rho^{5/3}(x)\, dx.$$

Thus the functional

(15.2) $\quad \mathcal{E}(\rho) = \int_{R^3} \rho^{5/3}(x)\, dx - \int_{R^3} V(x)\rho(x)\, dx + \frac{1}{2}\int_{R^3}\int_{R^3} \frac{\rho(x)\rho(y)}{|x - y|}\, dx\, dy$

represents the total energy of the system; it is called the *Thomas–Fermi functional*.

Let

$$K = \left\{ \rho \in L^1(R^3), \rho \geqslant 0, \int_{R^3} \rho(x)\, dx = I \right\},$$

$$K_0 = \left\{ \rho \in L^1(R^3), \rho \geqslant 0, \int_{R^3} \rho(x)\, dx \leqslant I \right\}.$$

The Thomas–Fermi Problem. Find $\bar{\rho}$ such that

(15.3)
$$\mathcal{E}(\bar{\rho}) = \min_{\rho \in K} \mathcal{E}(\rho), \qquad \bar{\rho} \in K.$$

Consider also the problem: Find $\bar{\rho}$ such that

(15.4)
$$\mathcal{E}(\bar{\rho}) = \min_{\rho \in K_0} \mathcal{E}(\rho), \qquad \bar{\rho} \in K_0.$$

It is not difficult to solve problem (15.4). Indeed, observe first that the functional

(15.5)
$$\rho \to \int_{R^3}\int_{R^3} \frac{\rho(x)\rho(y)}{|x - y|}\, dx\, dy \qquad \text{is convex}$$

(since, for instance, $1/|x|$ is the Fourier transform of a positive measure). The function in (15.5) is also continuous on $L^{6/5}(R^3)$. Hence if we take a minimizing sequence ρ_m of problem (15.4) such that $\rho_m \to \bar{\rho}$ weakly in $L^{5/3}(R^3)$, then

$$\liminf \mathscr{E}(\rho) \geq \mathscr{E}(\bar{\rho}).$$

Since clearly, $\bar{\rho} \in K_0$, $\bar{\rho}$ is a solution of (15.4).

As for problem (15.3), if we follow the procedure above we run into the difficulty of verifying that $\bar{\rho} \in K$ or, more specifically, of verifying that $\int \rho(x)\, dx = I$.

We shall now state the main existence results for problem (15.3), setting

$$(15.6) \qquad\qquad M = \sum_1^k m_i.$$

Theorem 15.1. (i) If $0 < I \leq M$, then there exists a unique solution to problem (15.3); (ii) if $I > M$, then there exist no solutions to problem (15.3); (iii) if $0 < I < M$, then the solution $\bar{\rho}$ of problem (15.3) has compact support.

The setting here is somewhat similar to that in the vortex ring. In that problem we had first imposed the (relaxed) constraint

$$\tfrac{1}{2} \int r^2 \zeta(x)\, dx \leq 1 \qquad [\text{see } (8.1)]$$

and eventually established for the minimizer that

$$\tfrac{1}{2} \int r^2 \zeta(x)\, dx = 1.$$

In the same vein the inequality constraint

$$\int \zeta(x)\, dx \leq 1 \qquad [\text{see } (6.35)]$$

turned out to be equality if λ is large enough. Thus it is reasonable to first solve problem (15.4) and then try to show that the minimizer $\bar{\rho}$ is in fact in K. This approach is adopted in reference 139.

Theorem 15.1 will be proved (in a more general form) in the next section. We shall, however, use a different approach. Notice that if $\bar{\rho}$ is a solution, then, by the method of Lemma 1.1, one can derive the equilibrium conditions:

$$(15.7) \qquad \begin{aligned} u - \lambda &= \tfrac{5}{3}\bar{\rho}^{2/3} \qquad \text{if } \bar{\rho} > 0, \\ &\leq 0 \qquad\quad\ \text{if } \bar{\rho} = 0, \end{aligned}$$

where

$$u = V - B\bar{\rho}$$

and

(15.8)
$$B\rho = \int_{R^3} \frac{\rho(y)}{|x - y|}\, dy.$$

Clearly,

$$-\Delta u + 4\pi\bar{\rho} = -\Delta V,$$

and since

$$\bar{\rho} = \left[\tfrac{3}{5}(u - \lambda)^+\right]^{3/2},$$

we obtain

(15.9) $$-\Delta u + c\left[(u - \lambda)^+\right]^{3/2} = -\Delta V \qquad (c > 0).$$

This equation resembles equations (5.3) (for rotating fluids), (6.18) (for vortex rings), and (11.1) (for the plasma problem). Those equations all have the form

$$-\Delta u - \beta(u) = f(x),$$

where $\beta(u)$ is monotone increasing; thus, roughly speaking, $\beta(u)$ appears with the wrong sign for the maximum principle. On the other hand, in (15.9) $\beta(u)$ appears with the good sign; further, $-\Delta + \beta$ is a monotone operator in a bounded domain with zero boundary conditions.

It is because of the "good" sign of β in (15.9) that it appears natural to try solving the Thomas–Fermi problem by first solving the equation (15.9). This, in fact, will be our approach in the next section.

Instead of working with the special case of $r^{5/3}$, we shall work with more general functions $j(r)$ satisfying

(15.10) $j(r)$ is convex in $C^1[0, \infty)$, strictly monotone
                 increasing, and $j(0) = j'(0) = 0$.

We introduce the conjugate convex function

$$j^*(t) = \sup_{s \geqslant 0} \left[ts - j(x)\right].$$

Let V be any function in L^1_{loc} and consider the functional

$$(15.11) \quad \mathcal{E}(\rho) = \int_{R^3} [j(\rho(x)) - V(x)\rho(x)] \, dx + \frac{1}{2} \int_{R^3} \int_{R^3} \frac{\rho(x)\rho(y)}{|x-y|} \, dx \, dy.$$

We introduce a class of admissible functions

(15.12)

$$\mathcal{Q} = \left\{ \rho \in L^1(R^3), \rho \geq 0, j(\rho) - V\rho \in L^1(R^3), \int_{R^3} \int_{R^3} \frac{\rho(x)\rho(y)}{|x-y|} \, dx \, dy < \infty \right\}.$$

Theorem 15.2. *If*

$$(15.13) \qquad \bar{\rho} \in \mathcal{Q} \cap K, \mathcal{E}(\bar{\rho}) \leq \mathcal{E}(\rho) \qquad \forall \rho \in \mathcal{Q} \cap K,$$

then there exists a constant λ such that

(15.14)
$$j'(\bar{\rho}) - V + B\bar{\rho} = -\lambda \qquad a.e. \text{ on } \{\bar{\rho} > 0\},$$
$$j(\bar{\rho}) - V + B\bar{\rho} \geq -\lambda \qquad a.e. \text{ on } \{\bar{\rho} = 0\}.$$

Conversely, if

$$(15.15) \qquad j^*(V(x) + C) \in L^1(R^3) \qquad \text{for some constant } C$$

and $\bar{\rho} \in K$, then (15.14) implies (15.13).

In the Thomas–Fermi case,

$$j(r) = cr^p, \quad j^*(r) = c'(r^+)^{p'} \qquad \left(\frac{1}{p} + \frac{1}{p'} = 1, c > 0, c' > 0 \right)$$

with $p = \frac{5}{3}$. If V is given by (15.1), then (15.15) is satisfied with $C = 1$ and any $p > \frac{3}{2}$.

Proof. The proof that (15.13) implies that (15.14) is similar to the proof of Lemma 1.1 and is therefore omitted. To prove the converse, notice that if $\rho \in L^1$, $\rho \geq 0$, then

$$(15.16) \qquad j(\rho) - j(\bar{\rho}) \geq (V - B\bar{\rho} - \lambda)(\rho - \bar{\rho}).$$

Indeed, by the convexity of j and (15.14), on the set $\{\bar{\rho} > 0\}$

$$j(\rho) - j(\bar{\rho}) \geq j'(\bar{\rho})(\rho - \bar{\rho}) = (V - B\bar{\rho} - \lambda)(\rho - \bar{\rho}),$$

whereas on the set $\{\bar{\rho} = 0\}$ the left-hand side of (15.16) is equal to $j(\rho) \geqslant 0$, and the right-hand side is equal to $(V - B\bar{\rho} - \lambda)\rho \leqslant 0$ by (15.14).

Using (15.16), we can write

$$(15.17) \quad \left[j(\rho) - V\rho + \tfrac{1}{2}\rho B\rho \right] - \left[j(\bar{\rho}) - V\bar{\rho} + \tfrac{1}{2}\bar{\rho}B\bar{\rho} \right]$$

$$\geqslant \left(\tfrac{1}{2}\rho B\rho + \tfrac{1}{2}\bar{\rho}B\bar{\rho} - \rho B\bar{\rho} \right) - \lambda(\rho - \bar{\rho}).$$

For $\rho = 0$, this gives

$$(j(\bar{\rho}) - V\bar{\rho}) + \tfrac{1}{2}\bar{\rho}B\bar{\rho} \leqslant -\lambda\bar{\rho} \in L^1.$$

On the other hand,

$$j(\bar{\rho}) - V\bar{\rho} = j(\bar{\rho}) - (V + C)\bar{\rho} + C\bar{\rho}$$

$$\geqslant -j^*(V + C) + C\bar{\rho} \in L^1.$$

It follows that $j(\bar{\rho}) - V\bar{\rho} \in L^1$ and $\bar{\rho}B\bar{\rho} \in L^1$. Thus $\bar{\rho} \in \mathcal{C}$. Choosing now ρ in $\mathcal{C} \cap K$ in (15.17) and integrating over R^3, we obtain, after using (15.5), $\mathcal{E}(\rho) \geqslant \mathcal{E}(\bar{\rho})$.

16. EXISTENCE OF SOLUTION FOR THE THOMAS–FERMI MODEL

In this section we prove:

Theorem 16.1

(a) *Suppose that*

$$(16.1) \qquad V(x) = \int_{R^3} \frac{h(y)}{|x - y|}, \qquad h \in L^1(R^3),$$

$$(16.2) \qquad V > 0 \qquad \textit{on a set of positive measure}.$$

Then there exists a number $I_0 \in (0, \infty)$ such that:
  (i)   *If $0 < I \leqslant I_0$, then there exists a unique solution $\bar{\rho}, \lambda$ of (15.14) with $\bar{\rho} \in K$.*
  (ii) *If $I > I_0$, then there exist no solutions of (15.14).*
  (iii) *If $I < I_0$ and $V(x) \to 0$ as $|x| \to \infty$, then $\bar{\rho}$ has compact support.*
(b) *Suppose that in (16.1) h is a finite measure in R^3, (16.2) holds, and $j(r) = cr^p$ ($c > 0$) for all r sufficiently large, where $p > \tfrac{4}{3}$. Then:*
  (iv) *The assertions (i) to (iii) remain valid.*
  (v)   *If also V is given by (15.1), then $I_0 = M$.*

It is clear that Theorem 15.1 is a consequence of Theorems 15.2 and 16.1.

The proof of Theorem 16.1 is based on several lemmas. First we need some definitions.

A function f in $L^1_{loc}(R^n)$ is said to belong to the Marcinkiewicz space $M^p(R^n)$ if

$$\|f\|_{M^p} \equiv \sup_{\substack{A \subset R^n \\ 0 < |A| < \infty}} |A|^{-1/p'} \int_A |f(x)| \, dx < \infty,$$

where $(1/p) + (1/p') = 1$. Clearly, $M^p \supset L^p$, $M_p \subset L^q_{loc}$ if $q < p$. It is not difficult to show that

$$(16.3) \qquad |x|^{-\alpha} \in M^{n/\alpha}(R^n) \qquad \text{if } 0 < \alpha < n,$$

and that

$$(16.4) \qquad \|E * f\|_{M^p} \leqslant \|E\|_{M^p} \|f\|_{L^1} \qquad (1 < p < \infty)$$

if $E \in M^p, f \in L^1$. Thus the operator B defined in (15.8) is a bounded operator form $L^1(R^3)$ into $M^3(R^3)$.

Lemma 16.2. *Let $\beta(t)$ be a continuous monotone nondecreasing function, $\beta(0) = 0$. Then for any $f \in L^1(R^3)$ there exists a unique solution $u \in M^3(R^3)$ of*

$$(16.5) \qquad -\Delta u + \beta(u) = f \qquad \text{in } R^3;$$

further, $\beta(u) \in L^1(R^3)$ and

$$(16.6) \qquad \int_{R^3} \beta^+(u) \, dx \leqslant \int_{R^3} f^+(x) \, dx.$$

For proof, see reference 32. The derivation of (16.6) follows by multiplying (16.5) by $H_m(u)$, where $H_m(u) \to 1$ if $u > 0$, $\to 0$ if $u < 0$, and integrating over R^3. Using the fact that

$$(16.7) \qquad \|\nabla u\|_{M^{3/2}} \leqslant C$$

and taking suitable $R = R_m \to \infty$, the inequality (16.6) follows.

Notice that if $\bar{\rho} \in L^1$ and λ satisfy (15.14), then $B\bar{\rho} \in M^3$ and therefore, for any $\delta > 0$,

$$\bar{\rho} < \delta, \qquad |V| < \delta, \qquad |B\bar{\rho}| < \delta$$

except on a set of finite measure. Hence (15.14) yields $-\lambda \leqslant \delta$ for any $\delta > 0$; that is, $\lambda \geqslant 0$.

Setting

$$(16.8) \qquad\qquad u = V - B\bar{\rho},$$

it is clear that ρ, λ satisfy (15.14) if and only if

$$\bar{\rho} = \begin{cases} (j')^{-1}(u - \lambda) & \text{if } u > \lambda \\ 0 & \text{if } u \leqslant \lambda. \end{cases}$$

Since (16.8) is equivalent to

$$-\Delta u + 4\pi\bar{\rho} = -\Delta V,$$

we can express (15.14) in the form

$$(16.9) \qquad -\Delta u + \gamma(u - \lambda) = -\Delta V \qquad \text{in } R^3,$$

where

$$(16.10) \qquad \gamma(t) = \begin{cases} 4\pi(j')^{-1}(t) & \text{if } t > 0 \\ 0 & \text{if } t \leqslant 0. \end{cases}$$

Thus the problem of solving (15.14) with $\bar{\rho} \in K$ reduces to the problem of solving (16.9) with

$$(16.11) \qquad \int \gamma(u - \lambda)\,dx = 4\pi I, \qquad \lambda \geqslant 0.$$

According to Lemma 16.2, for any $\lambda \geqslant 0$ there exists a solution u_λ of

$$u_\lambda \in M^3,$$
$$(16.12) \qquad -\Delta u_\lambda + \gamma(u_\lambda - \lambda) = -\Delta V,$$

provided that $-\Delta V = h \in L^1$, which is ensured by (16.1). Defining

$$I(\lambda) = \int_{R^3} \gamma(u_\lambda - \lambda)\,dx,$$

we are left with the problem of determining λ such that

$$I(\lambda) = 4\pi I.$$

Lemma 16.3. *If (λ) is nonincreasing continuous function with $I(\infty) = 0$, $I(0) > 0$; further, $I(\lambda)$ is strictly decreasing on the set $\{\lambda; I(\lambda) > 0\}$.*

The assertions (i) and (ii) of Theorem 16.1 follow immediately from the lemma.

Proof. Set $\tilde{u}_\lambda = u_\lambda - \lambda$. Then

$$-\Delta\tilde{u}_\lambda + \gamma(\tilde{u}_\lambda) = -\Delta V$$

and

$$\tilde{u}_\lambda(x) \to -\lambda \qquad \text{as } |x| \to \infty$$

in the sense that $\tilde{u}_\lambda + \lambda \in M^3$. Similarly,

$$-\Delta\tilde{u}_\mu + \gamma(\tilde{u}_\mu) = -\Delta V$$

and $\tilde{u}_\mu(x) \to -\mu$ if $|x| \to \infty$. We can now apply a version of the maximum principle to prove that

$$(16.13) \qquad\qquad \tilde{u}_\mu \leq \tilde{u}_\lambda \qquad \text{if } \mu > \lambda$$

(see Problem 1); hence also $\gamma(\tilde{u}_\mu) \leq \gamma(\tilde{u}_\lambda)$, so that $I(\mu) \leq I(\lambda)$.

Since $-\Delta(u_\lambda - V) = -\gamma(u_\lambda - 0) \leq 0$ and since $u_\lambda - V \to 0$ if $|x| \to \infty$ in the sense that $u_\lambda - V \in M^3$, we can deduce by the maximum principle (see the proof of Problem 1) that $u_\lambda - V \leq 0$. It follows that $\gamma(u_\lambda - \lambda) \leq \gamma(V - \lambda)$. Since $\gamma(V - \lambda) \to 0$ if $\lambda \to \infty$, the monotone convergence theorem gives

$$\lim_{\lambda \to \infty} I(\lambda) = 0.$$

We next claim that

$$I(0) > 0.$$

Indeed, if $I(0) = 0$, then $\gamma(u_0) = 0$, so that $u_0 \leq 0$. But since $-\Delta u_0 = -\Delta u_0 + \gamma(u_0) = -\Delta V$, we also have $V = u_0$, contradicting (16.2).

If $\lambda_n \downarrow \lambda$, then by monotonicity of $u_{\lambda_n} - \lambda_n$, it follows that $u_{\lambda_n} \to v$ and v is clearly a solution of the same equation as u_λ, $v \in M^3$. By uniqueness $v = u_\lambda$, and then $I(\lambda_n) \to I(\lambda)$ (by the monotone convergence theorem). Similarly, $\lambda_n \uparrow \lambda$ implies that $I(\lambda_n) \to I(\lambda)$. Thus $\lambda \to I(\lambda)$ is continuous.

It remains to show that $I(\lambda)$ is strictly monotone on the set $\{\lambda; I(\lambda) > 0\}$. Suppose $\lambda < \mu$, $I(\lambda) = I(\mu) > 0$. Since $\gamma(u_\mu - \mu) \leq \gamma(u_\lambda - \lambda)$, we have in fact $\gamma(u_\mu - \mu) = \gamma(u_\lambda - \lambda)$ and thus $\Delta u_\lambda = \Delta u_\mu$, so that $u_\lambda = u_\mu$. The inequality $I(\lambda) > 0$ implies that the set $\{\gamma(u_\lambda - \lambda) > 0\}$ has positive measure. Since $\gamma(t) > 0$ if $t > 0$, also $u_\lambda - \lambda = u_\mu - \mu$ on a set of positive measure; consequently $\lambda = \mu$.

As mentioned above, assertions (i) and (ii) of Theorem 16.1 follow from Lemma 16.3. To prove (iii) note that $0 < I < I_0$ implies $\lambda > 0$. Since $u_\lambda - \lambda \leqslant V - \lambda$, $\bar{\rho} = \gamma(u_\lambda - \lambda) \leqslant \gamma(V - \lambda)$. Using the assumption that $V(x) \to 0$ if $|x| \to \infty$ we find that $\bar{\rho}(x) = 0$ if $|x|$ is sufficiently large.

To prove (iv) we need an extension of Lemma 16.2 to the case where f is a finite (signed) measure.

Lemma 16.4. *Let $\beta(t)$ be a continuous monotone nondecreasing function with $\beta(0) = 0$, and let*

$$(16.14) \qquad \int_{\{|x|<1\}} \beta\left(\pm\frac{1}{|x|}\right) dx < \infty.$$

Then for any finite signed measure f in R^3 there exists a unique solution $u \in M^3(R^3)$ of (16.5); further, $\beta(u) \in L^1(R^3)$ and (16.6) holds.

Proof. Take a sequence f_n such that

$$\|f_n\|_{L^1} \leqslant C, \qquad f_n \to f \text{ in } \mathcal{D}'$$

and let

$$u_n \in M^3,$$

$$-\Delta u_n + \beta(u_n) = f_n.$$

Then, as in the proof of (16.6)

$$\|\beta(u_n)\|_{L^1} \leqslant \|f_n\|_{L^1} \leqslant C$$

and then also

$$\|\Delta u_n\| \leqslant 2C,$$

$$\|u_n\|_{M^3} \leqslant C_1,$$

$$\|\nabla u_n\|_{M^{3/2}} \leqslant C_2 \qquad [\text{see (16.7)}],$$

where C_i are constants.

If we can show that

$$(16.15) \qquad \{\beta(u_n)\} \text{ is relatively compact in } L^1(K)$$

for any compact subset $K \subset R^3$, then, for a subsequence, $u_n \to u$ in L^1_{loc}, $u \in M^3$ and

$$-\Delta u + \beta(u) = f$$

with $\beta(u) \in L^1$; (16.6) clearly also holds. Uniqueness is proved precisely as in Lemma 16.2.

To establish (16.15) it is sufficient to show, by a theorem of Vitali (see, for instance, reference 79, p. 122) that for any $\varepsilon > 0$ there is a $\delta > 0$ such that if $|K| < \delta$, then

$$(16.16) \qquad \int_K |\beta(u_n)| < \varepsilon \qquad \forall n.$$

Setting

$$\bar{\beta}(t) = \beta(t) + |\beta(-t)| \qquad \text{for } t \geqslant 0,$$

we have

$$\int_K |\beta(u_n)| = \int_{K \cap \{|u_n| < R\}} |\beta(u_n)| + \int_{K \cap \{|u_n| > R\}} |\beta(u_n)|$$

$$\leqslant |K| \bar{\beta}(R) - \int_R^\infty \bar{\beta}(\lambda) \, d\alpha_n(\lambda),$$

where

$$\alpha_n(\lambda) = \text{meas}\,[\,|u_n| > \lambda\,] \leqslant \frac{C}{\lambda^3}.$$

By integration by parts,

$$-\int_R^\infty \bar{\beta}(\lambda) \, d\alpha_n(\lambda) = \bar{\beta}(R)\alpha_n(R) + \int_R^\infty \alpha_n(\lambda) \, d\bar{\beta}(\lambda)$$

$$\leqslant \frac{C\bar{\beta}(R)}{R^3} + C\int_R^\infty \frac{1}{\lambda^3} \, d\bar{\beta}(\lambda)$$

$$\leqslant C_1 \int_R^\infty \frac{\bar{\beta}(\lambda)}{\lambda^4} \, d\lambda.$$

Thus

$$\int_K |\beta(u_n)| \leqslant |K| \bar{\beta}(R) + C_1 \int_R^\infty \frac{\bar{\beta}(\lambda)}{\lambda^4}.$$

Since $\bar{\beta}(\lambda)/\lambda^4 \in L^1(1, \infty)$ by (16.14), we can make the second term on the right smaller than $\varepsilon/2$ if R is large enough. We next choose $|K|$ so small that $|K|\bar{\beta}(R) < \varepsilon/2$, and thus derive (16.16).

With Lemma 16.4 at hand, we can now establish the assertion (iv) of Theorem 15.1 by the same arguments as for (i) to (iii). We just have to observe that under the assumptions in (B), the condition (16.14) is satisfied for $\beta = \gamma$.

It remains to prove (v). Notice that $4\pi I_0 = \int \gamma(u_0)$ and

$$-\Delta u_0 + \gamma(u_0) = -\Delta V, \qquad -\Delta V = 4\pi \Sigma m_i \delta_{a_i},$$

where $\delta_a = $ Dirac measure at a. Thus, by (16.6),

$$(16.17) \qquad I_0 \leq \frac{1}{4\pi} \int (-\Delta V)^+ = \sum_{i=1}^{k} m_i = M.$$

To prove the reverse inequality we shall use the well-known formula

$$(16.18) \qquad \frac{1}{4\pi} \int_{|\omega|=1} \frac{d\omega}{|r\omega - y|} = \frac{1}{\max(r, |y|)},$$

where $r\omega$ is identified with a point x in R^3 having spherical coordinates (r, ω). Denote by ρ_0 the $\bar{\rho}$ corresponding to $\lambda = 0$. Then

$$(16.19) \qquad \rho_0 = c(u_0^+)^{1/(p-1)} \qquad (c > 0),$$

$$(16.20) \qquad u_0(x) = V(x) - \int \frac{\rho_0(y)}{|x - y|} dy,$$

and

$$I_0 = \int \rho_0(y) \, dy.$$

We shall consider spherical averages

$$\tilde{v}(r) = \int_{|\omega|=1} v(r, \omega) \, d\omega$$

for $r > \max\{|a_i|\}$.

If we take the spherical averages of both sides of (16.20) and use (16.18), we get

$$\tilde{u}_0(r) = \frac{4\pi M}{r} - 4\pi \int \frac{\rho_0(y)}{\max(r, y)} dy$$

$$\geq \frac{4\pi M}{r} - 4\pi \int \frac{\rho_0(y)}{r} dy = \frac{4\pi(M - I_0)}{r}$$

Hence if $I_0 < M$, then (16.19) gives

$$\tilde{\rho}_0(r) \geq c_0 r^{-1/(p-1)} \qquad (c_0 > 0)$$

and thus

$$\int \rho_0(x)\, dx = 4\pi \int \tilde{\rho}_0(r) r^2\, dr \geq 4\pi c_0 \int r^{2-1/(p-1)}\, dr$$

$$= \infty \qquad (c_0 > 0)$$

since $p > \frac{4}{3}$; a contradiction. Thus $I_0 \geq M$, which complements (16.17).

We conclude this section with a theorem that contains several simple observations. Define p' by

$$\frac{1}{p} + \frac{1}{p'} = 1.$$

Theorem 16.5. *Let $p > \frac{3}{2}$. Then:*

(i) $u = u_\lambda$ *is locally in* $C^{1+[p']+\alpha}$ *away from the set* $\{a_1, \ldots, a_k\}$, *where* $\alpha = p' - [p']$, *provided that* $p' \neq$ *integer, and in* $C^{2+p'+\beta}$ *for any* $0 < \beta < 1$ *if* $p' =$ *integer.*

(ii) *The function* $u_\lambda - V$ *is in* $C^\gamma(R^3)$ *for some* $\gamma > 0$.

(iii) $\bar{\rho}$ *is Hölder continuous in* R^3 *and the open set*

$$\Omega = \{x; \bar{\rho}(x) > 0\}$$

contains a neighborhood of the set $\{a_1, \ldots, a_k\}$.

(iv) *Each component of* Ω *contains at least one point* a_i.

The proof is left to the reader; see Problem 2.

PROBLEMS

1. Prove (16.13).

   [*Hint*: Multiply the equation for $\tilde{u}_\mu - \tilde{u}_\lambda$ by $(\tilde{u}_\mu - \tilde{u}_\lambda)^+$ and integrate over $\{|x| < R\}$, letting $R \to \infty$ and using (16.7).]

2. Prove Theorem 16.5.

   [*Hint*: (i) and (ii) follow from elliptic regularity and Sobolev's inequality. For (iii) and (iv), apply the maximum principle to $\tilde{u}_\lambda$.]

3. Suppose that $0 < I < I_0$ and $|a_i| \leq C$ for $1 \leq i \leq k - 1$. If $|a_k|$ is sufficiently large, then the component of $\{\bar{\rho} > 0\}$ containing a_k does not contain any other a_i.

4. Suppose that $0 < I < I_0$ and $a_1, \ldots, a_{k-1}$ are fixed. If $|a_k - a_{k-1}|$ is sufficiently small, then a_k and a_{k-1} belong to the same component of $\{\bar{\rho} > 0\}$.

17. REGULARITY OF THE FREE BOUNDARY FOR THE THOMAS–FERMI MODEL

In this section we study the regularity of the free boundary

$$\Gamma = \partial\Omega, \qquad \text{where } \Omega = \{\bar{\rho} > 0\}.$$

If we keep $I, a_1, \ldots, a_{k-1}$ fixed and let a_k move from ∞ to a_{k-1}, then (by Problems 3 and 4 of Section 16) there should be an intermediate position of a_k where the boundaries of the components of $\{\bar{\rho} > 0\}$ which contain a_{k-1} and a_k intersect. For this position of a_k we cannot expect the free boundary to be everywhere smooth. Thus our main regularity result (Theorem 17.6) asserts regularity of the free boundary everywhere except for a "thin" subset.

We shall be working with the function $w = u_\lambda - \lambda$, which satisfies

$$(17.1) \qquad \Delta w = cw^q \qquad \left(q = \frac{1}{p-1}, c > 0 \right)$$

for some $1 < q < 2$ (if $\frac{3}{2} < p < 2$). But actually all our results extend to nonzero solutions w of

$$(17.2) \qquad \Delta w = \gamma(w) \qquad (|\gamma(t)| \leqslant c |t|^q, 1 \leqslant q < \infty).$$

We begin by noting that for the solution $w = u_\lambda - \lambda$ of (17.1),

$$(17.3) \qquad \{w = 0\} \text{ has no interior points.}$$

Indeed, if x^0 is an interior point, then Lemma 5.2 implies that $w \equiv 0$ in a δ-neighborhood of x^0, where δ is independent of x^0. It then follows by continuation that $w \equiv 0$ in R^3, which is impossible. Thus:

Lemma 17.1. *If x^0 belongs to the free boundary $\partial\Omega$, then there exists a homogeneous harmonic polynomial $H_n(x)$ of degree $n \geqslant 1$ such that*

$$(17.4) \qquad w(x) = H_n(x - x^0) + \Gamma_n(x),$$

where

$$(17.5) \qquad \begin{aligned} |\Gamma_n(x)| &\leqslant C|x - x^0|^{n+\delta}, \\[6pt] |\nabla\Gamma_n(x)| &\leqslant C|x - x^0|^{n+\delta-1} \qquad (C > 0, \delta > 0) \end{aligned}$$

for all x in some ε_0-neighborhood of x^0.

To study the free boundary more deeply, we develop some notation for harmonic polynomials.

We shall denote points in R^3 by $X = (x, y, z)$.

We denote by Λ_n the space of all polynomials of degree $\leqslant n$, by $\tilde{\Sigma}_n$ the space of all harmonic polynomials of degree $\leqslant n$, and by Σ_n the space of all homogeneous harmonic polynomials of degree n. Since $\tilde{\Sigma}_n$ is a finite-dimensional space, any two norms are equivalent. This is true in particular for the norms

$$L^\infty(B_1), \qquad L^2(B_1)$$

and the norms

$$\|P\|_\Delta = \inf\left\{\|Q\|_{L^\infty(B_1)}; Q \in \Lambda_{n+2}, \Delta Q = P\right\},$$

where B_1 is the unit ball; notice that the last "inf" is actually "min." We shall often drop the symbol B_1 in the sequel.

It is well known that

$$(P_1, P_2)_{L^2(B_1)} = 0 \qquad \text{if } P_1 \in \Sigma_n, \quad P_2 \in \Sigma_m, \quad n \neq m.$$

Writing P in $\tilde{\Sigma}_n$ as a sum

$$P = \sum_{k=0}^n P_k, \qquad P_k \in \Sigma_k,$$

we find that

$$\|P\|_{L^2}^2 = \sum_{k=0}^n \|P_k\|_{L^2}^2 \geqslant \max_k \|P_k\|_{L^2}^2.$$

Hence the inequality

(17.6) $$\|P\| \geqslant C \max_k \|P_k\| \qquad (C > 0)$$

holds in any other norm of $\tilde{\Sigma}_n$.

We now introduce a nonnegative constant $\alpha = \alpha(P_n)$ for $P_n \in \Sigma_n$, which measures the extent to which P_n differs from a polynomial in two variables.

Definition 17.1. Let

$$P_n(X) = \sum_{k=0}^n Q_k(X - X^0), \qquad Q_k(X) \text{ homogeneous polynomials of degree } k,$$

be the expansion of $P_n \in \Sigma_n$ into Taylor's series about a point X^0. Then

(17.7) $$\alpha = \alpha(P_n) = \inf_{\|X^0\|=1} \max_{0 \leqslant k \leqslant n-1} \|Q_k(X)\|_{L^\infty}.$$

Notice that

$$\|P_n\|_{L^\infty(B_1)} = 2^{-n}\|P_n\|_{L^\infty(B_2)} \geqslant 2^{-n}\|P_n(X + X_0)\|_{L^\infty(B_1)}$$

$$= 2^{-n}\left\|\sum_{k=0}^{n} Q_k(X)\right\|_{L^\infty(B_1)}.$$

Utilizing (17.6), we find that

(17.8) $$\alpha(P_n) \leqslant C\|P_n\|_{L^\infty(B_1)}.$$

Lemma 17.2. *Let $P_n \in \Sigma_n$. Then there exist polynomials A, B in Σ_n such that*

(17.9) $$P_n = A + B,$$

(17.10) $$\alpha(A) = 0,$$

(17.11) $$\|B\| \leqslant C_n\alpha(P_n),$$

where C_n is a constant depending only on n.

Proof. Assume that α is attained at $X^0 = (0, 0, 1)$, and write

$$P_n(X) = \sum_{k=0}^{n} a_k(x, y)z^k = \sum_{k=0}^{n} a_k(x, y)((z-1) + 1)^k$$

$$= \sum_{k=0}^{n} a_k(x, y)(z-1)^k + \sum_{k=1}^{n}\sum_{s<k} a_k(x, y)\binom{k}{s}(z-1)^s$$

$$\equiv Q_n(x, y, z-1) + \sum_{l=0}^{n-1} Q_l(x, y, z-1),$$

where

(17.12) $$Q_l = \sum_{n-k+s=l}\binom{k}{s}a_k(x, y)(z-1)^2.$$

Here a_k is a homogeneous polynomial of degree $n - k$ and Q_l is a homogeneous polynomial of degree l [in $(x, y, z - 1)$]. By the definition of α,

(17.13) $$\| Q_l \|_{L^\infty} \leqslant \alpha \qquad \text{for } l < n$$

[where the L^∞ norm is taken for $Q_l(x, y, z)$, (x, y, z) in B_1].

If $l = 0$, then $n - k + s = 0$ in (17.12) implies that $k = n + s \leqslant n$, so that $s = 0$, $k = n$. Thus $a_n(x, y) = Q_0(x, y, z - 1)$. It follows that

(17.14) $$\| a_n \|_{L^\infty} = \| Q_0 \|_{L^\infty} \leqslant \alpha.$$

Similarly, for $l = 1$, $n - k + s = 1$ gives $s = 1$ or $s = 0$. The term corresponding to $s = 1$ is a_n, which is estimated in (17.14). Hence the term corresponding to $s = 0$ can also be estimated, giving

$$\| a_{n-1} \| \leqslant C\alpha.$$

Proceeding step by step, we obtain

(17.15) $$\| a_k \|_{L^\infty} \leqslant C\alpha \qquad \text{for } 1 \leqslant k \leqslant n.$$

From the definition of $\| \cdot \|_\Delta$ it follows that there exists a $\sigma \in \Lambda_n$ such that $\sigma = \sigma(x, y)$,

$$\Delta\sigma(x, y) = \Delta a_0(x, y) = -2a_2(x, y)$$

and

$$\| \sigma \|_{L^\infty} = \| 2a_2 \|_\Delta \leqslant C\alpha.$$

We now take $A = a_0 - \sigma$, $B = P_n - A$. It is clear that $A \in \Sigma_n$, $\| B \| \leqslant C\alpha$. Finally, $\alpha(A) = 0$ with $\alpha(A)$ attained at $X^0 = (0, 0, 1)$.

Remark 17.1. From the proof above we obtain: If $\alpha(P_n) = 0$, then $P_n = P_n(x, y)$ in a suitable system of coordinates.

We shall now suppose that the origin O is a free boundary point and study the behavior of w near O. By Lemma 17.1,

(17.16) $$w(X) = P_n(X) + O\big(| X |^{n+\delta}\big) \qquad (\delta > 0),$$

where $P_n \in \Sigma_n$, $n \geqslant 1$; we may assume that $\delta < 1$.

In the next lemma ε_n is a positive number such that

$$C_n \varepsilon_n < \tfrac{1}{2},$$

where C_n is defined in Lemma 17.2.

Lemma 17.3. *Suppose that*

(17.17) $$\alpha(P_n) < \varepsilon_n \| P_n \|_{L^\infty}.$$

Then, after a suitable rotation of the coordinates, the following is true: If

$$\nabla w(X^1) = 0 \quad \text{for some } X^1 = (x, y, z) \neq 0, |X^1| < 1,$$

then

$$(17.18) \quad (x^2 + y^2)^{(n-1)/2} \leqslant (C\alpha + |X^1|^\delta) |X^1|^{n-1} \qquad [C > 0, \alpha = \alpha(P_n)].$$

Proof. In a suitable coordinate system

$$A(x, y) = C_0 \rho^n \cos n\theta, \qquad \rho = (x^2 + y^2)^{1/2},$$

where C_0 is a real number. In view of (17.11), (17.17),

$$\|A\|_{L^\infty} = |C_0| > (1 - C_n \varepsilon_n) \|P_n\|_{L^\infty} \geqslant \tfrac{1}{2} \|P_n\|_{L^\infty}, \qquad C_0 \neq 0.$$

Since

$$\|\nabla B\|_{L^\infty} \leqslant C_1 \|B\|_{L^\infty} \qquad (C_1 \text{ constant}),$$

we obtain

$$|\nabla P_n(X^1)| \geqslant C_2 \rho^{n-1} - C_3 \|B\| r^{n-1},$$

where C_i are positive constants depending on n (and C_2 depends also on C_0), and $r = |X^1|$. Thus, by Lemma 17.1 [see (17.5)],

$$|\nabla w(X^1)| \geqslant C_2 \rho^{n-1} - (C_2 \|B\| r^{n-1} + O(r^{n-1+\delta})).$$

Since $\nabla w(X^1) = 0$, (17.18) immediately follows.

As usual, we denote by $B_\lambda(X)$ a ball with center X and radius λ.

Lemma 17.4. *Suppose that $\alpha = \alpha(P_n) > 0$. Then there exists a neighborhood $B_{c\alpha^{n/\delta}}(O)$ $(c > 0)$ for which the following is true: If $w(X^0) = 0$ where $X^0 \in B_{c\alpha^{n/\delta}}(O)$, then*

$$(17.19) \qquad w(X) = \tilde{P}_k(X - X^0) + O(|X - X^0|^{k+\delta})$$

for some $k < n$, where $\tilde{P}_k \in \Sigma_k$.

Proof. Let

$$\tilde{X}^0 = \frac{X^0}{r}, \qquad \tilde{X} = \frac{X}{r} \qquad \text{where } r = |X^0|.$$

Writing

$$P_n(X) = \Sigma Q_k(X - X^0),$$

$$P_n(\tilde{X}) = \Sigma \tilde{Q}_k(\tilde{X} - \tilde{X}^0)$$

and comparing the Taylor expansion, we find that

$$\text{coefficients of } Q_k = r^{n-k} \text{ times the coefficients of } \tilde{Q}_k.$$

Hence

$$\|Q_k\|_{L^\infty(B_{Cr^\lambda}(X^0))} \geq c_0 \|\tilde{Q}_k\|_{L^\infty}(r^\lambda)^k r^{n-k} \qquad (\lambda > 1)$$

where $C > 0$, $c_0 > 0$. Choosing k such that

$$\|\tilde{Q}_k\|_{L^\infty} \geq \alpha(P_n)$$

and using (17.6), we get

(17.20) $$\|P_n\|_{L^\infty(B_{r^\lambda}(X^0))} \geq c_0 \alpha r^{\lambda k} r^{n-k} \geq c_0 \alpha r r^{\lambda(n-1)}.$$

Suppose now that the assertion of the lemma is not true. Then

$$w(X) = \tilde{P}_n(X - X^0) + O(|X - X^0|^{n+\delta}).$$

It follows that

(17.21) $$\|w\|_{L^\infty(B_{Cr^\lambda}(X^0))} \leq C_0 r^{\lambda n}.$$

Now, by Lemma 16.1,

(17.22) $$\|P_n\|_{L^\infty(B_{Cr^\lambda}(X^0))} \leq C\|w\|_{L^\infty(B_{Cr^\lambda}(X^0))} + Cr^{n+\delta}.$$

Substituting the estimates from (17.20), (17.21) into (17.22), we obtain

$$\alpha r r^{\lambda(n-1)} \leq Cr^{\lambda n} + Cr^{n+\delta}.$$

Choosing $\lambda = (n + \delta)/n$, we get $\alpha \leq 2Cr^{\delta/n}$, that is, $r > C_1 \alpha^{n/\delta}$ for some $C_1 > 0$. This contradicts the hypothesis that $X^0 \in B_{c\alpha^{n/\delta}}(O)$, provided that $c < C_1$.

Let

(17.23) $$M_n = \{X^0; w(X^0) = 0, \nabla w(X^0) = 0,$$

$$w(X) = P_n(X - X^0) + O(|X - X^0|^{n+\delta})\},$$

where $P_n = P_n^{X^0}$ is a homogeneous harmonic polynomial of degree n (depending on X^0).

Lemma 17.5. *For any $X^0 \in M_n$ there exists a neighborhood $B_{\delta_0}(X^0)$ and a line segment L containing X^0 such that $M_n \cap B_{\delta_0}(X^0)$ is contained in the cusp-like region*

$$\{X; C \mid X - X^0 \mid^{1+\varepsilon} > d(X, L)\},$$

where $\varepsilon = \delta/n(n-1)$, $C > 0$ and $d(X, L) = $ distance from X to L.

Proof. Denote $\alpha(P_n^{X^0})$ by α_{X^0}. If

$$\mid X - X^0 \mid < c(\alpha_{X^0})^{n/\delta}, \qquad X \neq X^0,$$

then, by Lemma 17.4, $X \notin M_n$. Thus it remains to consider the case where

$$\mid X - X^0 \mid \geqslant c(\alpha_{X^0})^{n/\delta}.$$

We shall apply Lemma 17.3 with the origin replaced by X^0 and the axis corresponding to the z-axis denoted by L. Then (17.18) gives, for $X \in M_n \cap B_{\delta_0}(X^0)$,

$$d(X, L)^{n-1} \leqslant C \mid X - X^0 \mid^{n-1} \mid X - X^0 \mid^{\delta/n},$$

that is,

$$d(X, L) \leqslant C \mid X - X^0 \mid^{1+\varepsilon}.$$

We can now state the main result of this section.

Theorem 17.6. *Let $\frac{3}{2} < p < 2$.*

(i) *There exist a finite number of open C^1 curves $\Gamma_1, \ldots, \Gamma_l$ in R^3 such that $\partial\Omega \setminus \cup_{i=1}^l \Gamma_i$ is a $C^{3+\alpha}$ manifold, where $\alpha = 1/(p-1) - 1$.*

(ii) *The positive (negative) limit set of each Γ_i is a single point X_i^+ (X_i^-).*

(iii) *There exists a direction l_i^+ at X_i^+ such that for $X \in \Gamma_i$, X near X_i^+ the angle θ between $\overrightarrow{XX_i^+}$ and l_i^+ satisfies*

$$\theta < C \mid X_i^+ - X \mid^\beta \qquad \textit{for some } C > 0, \beta > 0,$$

that is, X lies in a cusp-like region with axis l_i^+.

Proof of (i). The constants δ_0, C in Lemma 17.5 depend only on a lower bound on $\|P_n\|$. For any $X^1 \in M_n$, denote by $P_n^{X^1}$ the polynomial P_n corresponding to $X^0 = X^1$. From Lemma 17.1 we easily find that

$$\| P_n^{X^1} \| \geqslant C \| P_n \|$$

if X^1 is in a small neighborhood of X^0. It follows that the assertion of Lemma 17.5 is valid for any $X^1 \in M_n \cap B_{\delta_1}(X^0)$, with δ_1, δ_0, and C sufficiently small and independent of X^1. Thus, for any $X^1 \in M_n \cap B_{\delta_1}(X^0)$ ($X^0 \in M_n$), there is a line segment L_{X^1} containing X^1 such that $M_N \cap B_{\delta_1}(X^1)$ lies in the cusp-like region

$$\{ X; C \, | \, X - X^1 |^{1+\varepsilon} \} > d(X, L_{X^1}).$$

From this it is easily seen that

(17.24)   if θ = angle between L_{X^1}, L_{X^2}, then $\theta \leqslant \sigma\big(| X^1 - X^2 |\big)$,

where $\sigma(t) \to 0$ if $t \to 0$.

Now take for simplicity $X^0 = (0, 0, 0)$, $L_{X^0} = z$-axis. Then the plane $z = \zeta$ ($| \zeta |$ small) can intersect M_n in a neighborhood of X^0 in at most one point; otherwise, (17.24) is contradicted. Thus the points of

$$M_n \cap B_{\delta_2}(X^0) \qquad (\delta_2 \text{ sufficiently small})$$

lie on a graph $X = \hat{X}(\lambda)$, where λ is a parameter of L_{X^0}. (We can take, for instance, $\lambda = z$.) If we extend the graph linearly then, in view of (17.24), the extended graph, defined on some λ-interval, is Lipschitz continuous.

We can actually achieve a C^1 extension $X(\lambda)$ of $\hat{X}(\lambda)$. Indeed, take a partition of the λ-interval into m intervals of equal length and choose in each subinterval one point of $\hat{X}(\lambda)$, provided that such a point exists. We connect two adjacent points P_1, P_2 by a C^1 parabolic curve such that the tangents at P_1, P_2 coincide with the directions of L_{P_1}, L_{P_2}, respectively.

Denote this extension by $X_m(\lambda)$. By (17.24) the derivative of $X_m(\lambda)$ has a modulus of continuity in λ, uniform with respect to m. Hence a subsequence of $X_m(\lambda)$ is uniformly convergent to a C^1 curve $X(\lambda)$. Since $X(\lambda) = \hat{X}(\lambda)$ for a dense set of points $\hat{X}(\lambda)$ of $M_n \cap B_{\delta_2}(x^0)$, $X(\lambda)$ is a C^1 extension of $\hat{X}(\lambda)$.

It follows that M_n is contained in a finite union of points or C^1 curves $\Gamma_{n,i}$, the C^1 modulus of continuity depending on a lower bound on the norm $\| P_n \|$.

Next, for every compact region of $\partial\Omega$ there is a finite number n_0 such that $M_n = \varnothing$ if $n > n_0$. It follows that $\cup_n M_n$ is contained in a finite number of C^1 curves Γ_i. Outside these curves, $\nabla w \neq 0$ on $\partial\Omega$. Since $w \in C^{3+\alpha}$, the assertion (i) now follows by the implicit function theorem.

Proof of (ii).   In the construction of a curve $\Gamma_{n,i}$ one can proceed locally as follows: After a segment, say $\tilde{\Gamma}_{n,i}$, is constructed, if an endpoint, say $\tilde{Y}_{n,i}$, belongs to M_n, then we use it as a center of a new coordinate system in order possibly to extend $\tilde{\Gamma}_{n,i}$. The extension takes place if $\tilde{Y}_{n,i}$ is a limit point of points of M_n not lying entirely in $\tilde{\Gamma}_{n,i}$; otherwise, $\Gamma_{n,i}$ terminates at $\tilde{Y}_{n,i}$. When we extend $\Gamma_{n,i}$ beyond $\tilde{Y}_{n,i}$, its new endpoint is again taken to be in M_n.

To prove (ii) we may now parametrize any curve $\Gamma_{n,i}$ by t and assume that it is not C^1 at its (say, right) endpoint; otherwise, there is nothing to prove for this curve. We then have to show that the points $X(t)$ of $\Gamma_{n,i}$ converge to a point $X(\infty)$ as $t \to \infty$. Denote by $\Gamma_{n,i}^+$ the right limit set of $\Gamma_{n,i}$, that is, the limit set of $X(t)$ as $t \to \infty$. Denote by $P_n(t)$ the polynomial P_n^X at $X = X(t)$. We claim that without loss of generality we may assume that

$$(17.25) \qquad\qquad \|P_n(t)\| \to 0 \qquad \text{if } t \to \infty.$$

Indeed, denote by $\tilde{Y}_{n,i,j}$ $(j = 1, 2, \ldots)$ the sequence of points $\tilde{Y}_{n,i}$ used in the construction of $\Gamma_{n,i}$ as $t \to \infty$. Denote by t_j the parameter t corresponding to $\tilde{Y}_{n,i,j}$. We shall first show that we may assume that

$$(17.26) \qquad\qquad \|P_n(t_j)\| \to 0 \qquad \text{if } j \to \infty.$$

The extension of $\Gamma_{n,i}$ beyond $\tilde{Y}_{n,i,j}$ takes place in a ball B_j with center $\tilde{Y}_{n,i,j}$ whose diameter r_j depends on $\|P_n(t_j)\|$. If (17.26) is not true, then $r_j \geq c > 0$ for a subsequence of j's. It follows that two balls B_{j_1}, B_{j_2} $(j_1 < j_2)$ must intersect. From the construction of $\Gamma_{n,i}$ we see that if in the extension of $\Gamma_{n,i}$ we use balls of radii $\leq \delta_2/2$, then the extension of $\Gamma_{n,i}$ from the center point onward can be carried out in such a way as to give a C^1 curve containing all the points of M_n in $B_{j_1} \cup B_{j_2}$. Thus $\Gamma_{n,i}$ will be a closed curve, for which the assertion (ii) is then obvious. Excluding this case, we then must accept the validity of (17.26).

Having established (17.26) and noting that the mapping $X^1 \to \|P_n^{X^1}\|$ is continuous, the assertion (17.25) follows.

From (17.25) it follows that if a curve $\Gamma_{n_0,i}$ is not C^1 up to the (say, right) endpoint, then any point in $\Gamma_{n_0,i}^+$ is in M_{n_0+1}; this is impossible since $M_n = \varnothing$ if $n > n_0$. Thus each $\Gamma_{n_0,i}$ is either a point or a C^1 curve up to the endpoints.

We next consider a curve $\Gamma_{n_0-1,i}$. Since it is connected, also $\Gamma_{n_0-1,i}^+$ is connected and compact. Each point of $\Gamma_{n_0-1,i}^+$ belongs to M_{n_0} [by (17.25)] and therefore it is either contained locally in a C^1 curve whose points lie in M_{n_0} or it is an isolated point. We now assume that

$$(17.27) \qquad\qquad \Gamma_{n_0-1,i}^+ \text{ contains at least two points.}$$

Then, being connected, $\Gamma_{n_0-1,i}^+$ is locally a C^1 curve, and therefore also a global C^1 curve.

By Lemma 17.4, $\alpha(P_{n_0}) = 0$ along $\Gamma_{n_0-1,i}^+$. Therefore, by Lemma 17.3, the set $\cup M_n$ in a neighborhood of each point of $\Gamma_{n_0-1,i}^+$ must lie on a C^1 curve, and this curve must necessarily then be the curve $\Gamma^+_{n_0-1,\,i}$. Since, however,

$$\Gamma_{n_0-1,\,i} \cap \Gamma^+_{n_0-1,\,i} \subset M_{n_0-1} \cap M_{n_0} = \varnothing$$

we cannot have points of $\Gamma_{n_0-1,\,i}$ in a small neighborhood of any given point of $\Gamma^+_{n_0-1,\,i}$; a contradiction. Thus $\Gamma^+_{n_0-1,\,i}$ consists of just one point.

Next we show that for each curve $\Gamma_{n_0-2,\,i}$, if it is a C^1 curve open at one side, say when $t \to \infty$, then its limit set $\Gamma^+_{n_0-2,\,i}$ must consist of a single point. Indeed, since $\Gamma^+_{n_0-2,\,i}$ is a connected, compact set, each of its points is a point of accumulation of the set. Suppose that there exists a point $X \in \Gamma^+_{n_0-2,\,i}$ such that $X \in M_{n_0-1}$, and $\Gamma^+_{n_0-2,\,i}$ consists of more than one point. Then in a neighborhood V of X there cannot be any points of M_{n_0}. By Lemma 17.4, $\alpha(P_{n_0-1}) = 0$ along $\Gamma^+_{n_0-2,\,i}$ (and therefore $\Gamma^+_{n_0-2,\,i} \cap V \subset M_{n_0-1}$), which implies by Lemma 17.3 and the construction of the $\Gamma_{n,\,i}$ that in some neighborhood W of X $\Gamma^+_{n_0-2,\,i} \cap M_{n_0-1}$ is a C^1 curve (containing X) and its points are not in the closure of $(\cup M_n) \setminus \Gamma^+_{n_0-2,\,i}$. Now we get a contradiction as before, because points of $\Gamma_{n_0-2,\,i}$ do in fact converge to any point on that curve.

Consider next the case where no points of $\Gamma^+_{n_0-2,\,i}$ belong to M_{n_0-1}. Then $\Gamma^+_{n_0-2,\,i} \subset M_{n_0}$ and we can proceed as in the case of $\Gamma_{n_0-1,\,i}$.

Proceeding in the foregoing fashion step by step, the proof of (ii) is complete.

Proof of (iii). Let X^0 be an endpoint of some curve Γ_i in M_n. We may assume that $\alpha(P_n^X) = 0$ for a sequence $X = X_j \in \Gamma_i$, $X_j \to X^0$. By Lemma 17.4 it follows that $\alpha(P_n^{X^0}) = 0$. Hence, by Remark 17.1, if we take $X^0 = 0$, then

$$P_n(X) = a(x, y)$$

in a suitable system of coordinates (x, y, z). It follows that

$$|\nabla P_n(X)| \geq Cr^{n-1} \qquad (c > 0),$$

where $r = (x^2 + y^2)^{1/2}$, and Lemma 17.1 gives

$$|\nabla w(X)| > 0$$

if X lies in an opening θ about the z-axis and

$$|z|^{n+\delta-1} < c_0 \theta^{n-1} |z|^{n-1}$$

for some small $c_0 > 0$. The same is true with respect to any other direction along which $\nabla P_n = 0$. Thus the points of $\cup M_i$ that lie in a small neighborhood of X^0 must lie in cusp-like regions with vertex X^0 as described in the assertion (iii).

From the assertions (i) and (ii) of Theorem 17.6, we deduce:

Corollary 17.7. *For any* $\varepsilon > 0$, *the* $(1 + \varepsilon)$-*dimensional Hausdorff measure of the set* $\cup M_n$ *is equal to zero.*

The proof of Theorem 5.1 can be applied to w in a neighborhood of any point where $\nabla w \neq 0$. Thus:

Corollary 17.8. *The portion* $\partial\Omega \setminus \cup_{i=1}^{l} \overline{\Gamma}_i$ *of the free boundary is analytic.*

PROBLEMS

1. Prove Corollary 17.7.

2. Consider the solution u_λ of the Thomas–Fermi problem (with $p = \frac{5}{3}$) corresponding to some $\lambda \geqslant 0$. Prove that

$$| u_\lambda(x) - \lambda | \leqslant C | x |^{-4}.$$

[*Hint*: Consider

$$\psi(x) = CR^4 \big(R^2 - | x - x_0 |^2 \big)^{-4} \quad \text{in } B(x_0, R).$$

Show that $-\Delta\psi + \gamma(\psi) \geqslant 0$ in $B(x_0, R)$, and deduce $\psi \geqslant \pm(u_\lambda - \lambda)$ in $B(x_0, R)$ if $| x_0 | > 2R$, $R > \max | a_i |$.]

18. BIBLIOGRAPHICAL REMARKS

The study of self-gravitating axisymmetric rotating fluids began in the eighteenth century; the initial interest came from astrophysics. A number of papers by Poincaré, for instance, are concerned with the question of the number of rings of a planet. The reader will find in Chandrasekhar's book [69] many explicit formulas (Maclaurin spheroids, Jacobi's ellipsoids, etc.). Existence theorems were first proved by Auchmuty and Beals [18] (the compressible case) and Auchmuty [17] (the incompressible case). The presentation in Section 3 is based on Friedman and Turkington [100d], who also derived the potential estimates of Section 2 in reference 100a. The results of Section 4 are based on reference 100a. P.L. Lions [193a, b] has extended the method of Auchmuty and Beals, developing a general functional analysis argument.

The proof of analyticity of the boundary of the rings (Theorem 5.1) is based on the method of reference 124a. Lemmas 5.2 and 5.3 are due to Caffarelli and Friedman [58e]; their two-dimensional analogs were earlier proved by Hartman and Wintner [113]. For unique continuation results for general second-order elliptic equations, see references 15 and 73. Theorems 5.4 to 5.6 are taken from Caffarelli and Friedman [58j].

Vortex rings with small cross sections were studied by Helmholtz (1858); see references 25 and 92. The first general existence theorem for vortex rings is due to Fraenkel and Berger [92]. They use a variational principle for the stream function; a variant of this approach (using existence of critical points) was developed by Ambrosetti and Mancini [10a, b] and Ni [150], and a functional analysis generalization of this variational procedure is given by Berestycki and P.L. Lions [194]. Benjamin [33] introduced the idea of minimizing $E(\zeta)$; his admissible class consists of all rearrangements of a given ζ_0. The present approach [with admissible class defined by (6.33)–(6.36) in Theorem 6.2, and by (6.33)–(6.35) in Theorem 6.3] is due to Friedman and Turkington [100c]. The results of Sections 6 to 8 and 10 are taken from reference 100c; the present proof of Theorem 10.10 is due to Turkington [175a], the original proof is outlined in Problems 2, 3 of Section 10.

In reference 175b Turkington extended the methods of reference 100c to a wake model. The results of Section 9 are due to Caffarelli and Friedman [58m].

Fraenkel [91a, b] and Norbury [151a, b] have constructed classes of solutions of vortex rings using methods such as nonlinear integral equations; they find solutions bifurcating from the Hill vortex [151a, b] and vortex rings with small cross sections [91a, b]. Norbury [151c] identified known explicit solutions of planar vortex rings as solutions of the variational principle of reference 92. It is not known whether the Hill vortex is a solution of a variational problem.

The derivation of the plasma problem equations can be found, for instance, in reference 170a. The existence of a solution of the plasma problem (11.1)–(11.3) was first established by Temam [170a–c]; he proved Theorem 11.3 and (in reference 170b) Theorem 11.2 for $n = 2$. Theorem 11.2 for $n \geqslant 3$ is due to Puel [155]. Theorem 11.1 is due to Berestycki and Brezis [35a, b]. Damlanian [76b] established Remark 11.1 (see also reference 35b). The results of Section 12 are due to Kinderlehrer and Spruck [125] (see also reference 124 a). The results of Problems 3 and 4 of Section 12 are due to Gidas, Ni, and Nirenberg [105]. The material of Section 13 is taken from Caffarelli and Friedman [58m]. The results of Problems 1 and 2 in Section 13 are due to Baiocchi, Caffarelli, and Friedman [unpublished].

Berger and Fraenkel [36] and Keady [118b] have derived some asymptotic bounds for equations $\Delta u = \lambda g(x, u)$ with $\lambda \to \infty$. Keady and Norbury [119b, c] have established a continuum of solutions (λ, u) with $\lambda \to \infty$.

More complicated (and realistic) plasma models have been studied by several authors; but the existence theory is not completely established as yet. For details, see references 146a, b and 147 and the references given there.

Schaeffer [158d] proved nonuniqueness for the variational problem of the plasma if $\lambda > \lambda_2$. Corresponding bifurcation results (mostly numerical) were obtained by Sermange [160].

The results of Section 14 are due to Caffarelli and Spruck [64]. For $n = 2$, Acker [1d] studied the problem of minimizing $\mathrm{Cap}_\Omega G$, given $\lambda_1(G) = \lambda$, G convex, and established a solution for which the corresponding u [defined by (14.4)] solves the plasma problem.

Theorem 14.4 and the convexity of the level sets of u in Theorem 14.3 follow from the work of Grabriel [101a–c], who also proved the convexity of the level curves for Green's function; see also Lewis [136]. Another proof of Theorem 14.3 was given by Brascamp and Lieb [44]. The present simpler proofs of Theorems 14.3 and 14.4 are due to Caffarelli and Spruck [64]; Lemma 14.2 is due to Korevaar [129a], who independently derived the proof of Theorem 14.3 in reference 129b.

For history and references on the Thomas–Fermi model, see Lieb and Simon [139]. In this paper they prove Theorem 15.1. The results of Section 16 are based on Benilan and Brezis [31, 45c]. They also establish for general V the estimate

$$\int (-\Delta V) \leqslant I_0 \leqslant \int (-\Delta V)^+ .$$

The results of Section 17 are taken from Caffarelli and Friedman [58e]. Asymptotic estimates on $u_\lambda - \lambda$ are derived in Brezis and Lieb [51] and in reference 139 (the result of Problem 2 of Section 17 is taken from reference 51). Another Thomas–Fermi model is studied in reference 51; here the solution ρ does not have compact support. A related model was studied by Benguria, Brezis, and Lieb [29].

5

SOME FREE-BOUNDARY PROBLEMS NOT IN VARIATIONAL FORM

In this chapter we study several types of free-boundary problems which are not formulated as variational principles. In Sections 1 to 5 we deal with gas in porous medium. The equation satisfied by the density function u is

$$\frac{\partial u}{\partial t} = \Delta u^m \qquad (m > 1)$$

and the free boundary is $\partial\{u > 0\}$. It is shown that there exists a unique solution u, and that u and the free boundary are Hölder continuous.

In Sections 6 to 9 we revisit the filtration problem, but here the dam has general shape. Existence and uniqueness of a solution and regularity of the free boundary are established.

Finally, in Section 10 we study the two-phase Stefan problem.

1. THE POROUS-MEDIUM EQUATION: EXISTENCE AND UNIQUENESS

For gas in a porous medium, Darcy's law states that

$$\mathbf{v} = -k\nabla p \qquad (k \text{ positive constant})$$

where $\mathbf{v}$ is the velocity of the gas and p is the pressure. The law of conservation of mass is

$$\nabla \cdot \rho\mathbf{v} + \frac{\partial\rho}{\partial t} = 0$$

where ρ is the density. With the standard isothermic equation of state $p = c\rho^\gamma$ $(c > 0, \gamma > 0)$ it then follows that

$$(1.1) \qquad \frac{\partial u}{\partial t} = \Delta u^m \qquad (m > 1)$$

where $m = 1 + \gamma$ and u is the normalized density. The function u^{m-1} represents the pressure (up to a constant factor). We introduce the initial condition

$$(1.2) \qquad u(x,0) = u_0(x) \qquad (u_0 \geqslant 0),$$

and wish to solve the initial value problem (1.1), (1.2) in some strip

$$Q_T = R^n \times (0, T) \qquad \text{or} \qquad Q = R^n \times (0, \infty).$$

We begin with some preparations.

Definition 1.1. Let X be a real Banach space and let A be an operator from a set $D(A) \subset X$ into X. A is called *accretive* (or *monotone*, when X is a Hilbert space) if

$$(1.3) \qquad \|x_1 - x_2\| \leqslant \|x_1 - x_2 + \lambda(A(x_1) - A(x_2))\|$$

$\forall\, x_1, x_2$ in $D(A)$ and $\forall\, \lambda > 0$. If also

$$(1.4) \qquad \text{range}\,(I + \lambda A) = X \qquad \forall\, \lambda > 0,$$

then A is called *m-accretive* (or *maximal monotone* when X is a Hilbert space).

Notice that (1.3) implies that $I + \lambda A$ is 1-1 and $(I + \lambda A)^{-1}$ is a contraction.

Observe also that if (1.4) holds for just one value $\lambda = \lambda_0 > 0$, then it holds for all $\lambda > 0$. Indeed, for fixed $y \in X$, writing $x + \lambda A x = y$ in the form

$$(1.5) \qquad x = (I + \lambda_0 A)^{-1}\left(\frac{\lambda_0}{\lambda}y + \left(1 - \frac{\lambda_0}{\lambda}\right)x\right) \equiv T_y x,$$

it is clear that $x \to T_y x$ is a contraction with Lipschitz coefficient $|1 - \lambda_0/\lambda| < 1$ if $\lambda > \lambda_0/2$. Thus for any $y \in X$ we can solve (1.5) for x provided that $\lambda > \lambda_0/2$, and (1.4) follows step by step, for all $\lambda > 0$.

One can easily show that if S is a contraction with domain $D(S) = X$, then $I + S$ is *m*-accretive.

Consider the initial value problem

$$(1.6) \qquad \frac{du}{dt} + A(u(t)) = f(t) \qquad (0 < t < T),$$

$$(1.7) \qquad u(0) = x_0.$$

One may try to solve this system by approximating with a finite-difference scheme with mesh size $\lambda_n \downarrow 0$. We solve recursively

$$(1.8) \qquad \frac{x_k^n - x_{k-1}^n}{\lambda_n} + A(x_k^n) = f_k^n \qquad (k = 1, 2, \ldots),$$

$$x_0^n = x_0,$$

where $f_k^n = f(k\lambda_n)$. The approximate solution $u^n(t)$ is defined by $u^n(t) = x_k^n$ if $k\lambda_n \leqslant t < (k + 1)\lambda_n$. Since A is m-accretive, the finite-difference scheme has a unique solution. Set $f^n(t) = f_k^n$ for $k\lambda_n \leqslant t < (k + 1)\lambda_n$.

Theorem 1.1.

(i)   Let A be m-accretive, $x_0 \in \overline{D(A)}$, $f \in L^1(0, T; X)$ and $f^n \to f$ in $L^1(0, T; X)$ as $\lambda_n \to 0$. Then u^n converges uniformly in $[0, T]$ to a function $u(t)$ in $C([0, T]; X)$.

(ii) If also $f(t)$ is Lipschitz continuous and $x_0 \in D(A)$, then $u(t)$ is Lipschitz continuous.

(iii) If, under the assumptions of (i), the problem (1.6), (1.7) has a strong solution $v(t)$ [that is, $v(t)$ is differentiable as X-valued function, $v(t) \in D(A)$ and (1.6) holds a.e., and $\| v(t) - u_0 \| \to 0$ if $t \to 0$], then $u = v$.

(iv) If, under the assumptions of (i), u is differentiable at a point $t = t_0 > 0$ and t_0 is a Lebesgue point of f [that is,

$$\frac{1}{\delta} \int_{t_0 - \delta}^{t_0 + \delta} \| f(t) - f(t_0) \| \, dt \to 0 \qquad \text{if } \delta \to 0 \bigg],$$

then $u(t_0) \in D(A)$ and

$$\frac{du(t_0)}{dt} + Au(t_0) = f(t_0).$$

Theorem 1.1 holds also if $T = \infty$; $[0, T]$ is replaced by $[0, \infty)$. When $f \equiv 0$ we write $u(t) = S(t)x_0$ and call $S(t)$ the (nonlinear) *semigroup generated* by A. Since $(I + \lambda A)^{-1}$ is a contraction, the construction of $S(t)x_0$ shows that $S(t)$ is a weak contraction, that is,

$$\| S(t)x_0 - S(t)x_1 \| \leqslant \| x_0 - x_1 \|.$$

For more details and references on the proof of Theorem 1.1 and on nonlinear semigroups, see references 23 and 85a.

Consider the equation

$$(1.9) \qquad \varepsilon v - \Delta v = g \qquad \text{in } \mathcal{D}'(R^n) \quad (\varepsilon > 0).$$

It is well known (see, for example, references 32 and 168) that for any $g \in L^p(R^n)$ $(1 \leqslant p \leqslant \infty)$ there exists a unique solution $v \in L^p(R^n)$ of (1.9) and, denoting it by $B_\varepsilon g$, there holds

$$(1.10) \qquad \varepsilon \, | \, B_\varepsilon v \, |_{L^p} \leqslant | \, g \, |_{L^p}.$$

One can represent the solution in the form

$$(1.11) \qquad \varepsilon v(x) = \varepsilon^{n/2} \int k\left(\sqrt{\varepsilon}\,(x - y)\right) g(y) \, dy,$$

where the fundamental solution k is given by

$$k(x) = c r^{-(n-2)/2} K_{(n-2)/2}(2\pi r) \qquad (c > 0)$$

and K_m is one of the classical Bessel functions. It is easy to check that

$$(1.12) \qquad \sup_{|x| \geqslant r} k(x) < \infty \qquad \forall \, r > 0,$$

$$(1.13) \qquad \int_{|x| < 1} k(x) \, dx < \infty.$$

Consider next the equation in R^n:

$$(1.14) \qquad v - \Delta \phi(v) = g, \qquad g \in L^1(R^n),$$

where

$$(1.15) \quad \phi(t) \text{ is continuous, strictly monotone increasing, } \phi(0) = 0,$$

and set $\phi(v) = u$, $\beta = \phi^{-1}$. Then the equation (1.14) becomes

$$(1.16) \qquad -\Delta u + \beta(u) = g.$$

Lemma 4.16.2 (for $n = 3$) extends to any n (see reference 32). More precisely, for any $g \in L^1(R^n)$ there exists a unique solution u of (1.16) in $M^{n/(n-2)}(R^n)$ if $n \geqslant 3$, in $\{v \in W^{1,1}_{\text{loc}}(R^2), \; \nabla v \in M^2(R^2)\}$ if $n = 2$, and in $\{v \in W^{1,\infty}(R^2), \; v'' \in L^1(R^1)\}$ if $n = 1$. Further,

$$(1.17) \qquad \int_{R^n} | \, \beta(u) \, |^p \leqslant \int_{R^n} | \, g \, |^p \qquad \forall \, 1 \leqslant p \leqslant \infty$$

[the proof is similar to the proof of (4.16.6)].

Define A by

$$Av = g - v,$$

where $g \in L^1(R^n)$ and v is the unique solution of (1.14) (in a suitable class). Then (1.3) for this A reduces to

$$(1.18) \qquad \| \beta(u_1) - \beta(u_2) - \lambda(\Delta u_1 - \Delta u_2) \| \geq \| \beta(u_1) - \beta(u_2) \|$$

in the $L^1(R^n)$ norm. But, as in the proof of (4.16.6),

$$\int (\beta(u_1) - \beta(u_2) - \lambda \Delta(u_1 - u_2)) h(u_1 - u_2)$$

$$\geq \int (\beta(u_1) - \beta(u_2)) h(u_1 - u_2)$$

if $h(0) = 0$, $0 \leq h'(t) \leq C$. Taking $h(t) \to \operatorname{sgn} t$, (1.18) follows.

Next, (1.4) (with $\lambda = 1$) follows from Lemma 16.2 of Chapter 4. Thus A is m-accretive. Notice also that $\overline{D(A)} = L^1(R^n)$ since $C_0^\infty(R^n) \subset D(A)$.

We can now apply Theorem 1.1 and deduce the existence of a "weak" solution in the sense of that theorem. It is easily seen that

$$(1.19) \qquad u_t - \Delta\phi(u) = 0 \qquad \text{in } \mathcal{D}'(Q).$$

From the finite-difference scheme,

$$\frac{w_{k+1} + A\phi(w_{k+1})}{\lambda_n} = \frac{w_k}{\lambda_n}$$

used to construct the approximation $u^n(t)$ and from (1.17) with $p = \infty$, we get

$$| w_{k+1} |_{L^\infty} \leq | w_k |_{L^\infty},$$

so that

$$(1.20) \qquad | u(t) |_{L^\infty} \leq | u(0) |_{L^\infty}.$$

We summarize:

Theorem 1.2. *If ϕ satisfies (1.15), then for any $u_0 \in L^1(R^n) \cap L^\infty(R^n)$ there exists a solution u of (1.19) in $C([0, \infty); L^1(R^n)) \cap L^\infty(Q)$, and $u(0) = u_0$.*

We now state a uniqueness result:

Theorem 1.3.   *If ϕ satisfies* (1.15) *and if u, $\hat{u}$ are two functions satisfying*:

$$(1.21) \qquad\qquad u, \hat{u} \in L^\infty(Q_T),$$

$$(1.22) \qquad\qquad u_t - \Delta\phi(u) = \hat{u}_t - \Delta\phi(\hat{u}) \qquad in\ \mathcal{D}'(Q_T),$$

$$(1.23) \qquad\qquad u - \hat{u} \in L^1(Q_T),$$

$$(1.24) \quad \int_0^T \int_{R^n} [(u - \hat{u})\psi_t + (\phi(u) - \phi(\hat{u}))\Delta\psi]\, dx\, dt = 0$$

$$\forall\, \psi \in C_0^\infty([0, T) \times R^n),$$

then $u = \hat{u}$ a.e.

Notice that (1.24) is a weak statement on the assumption of the initial values; it is a consequence of (1.22) and

$$(1.25) \qquad\qquad \underset{t \to 0}{\mathrm{ess\,lim}} \int |u(x, t) - \hat{u}(x, t)|\, dx = 0.$$

Theorem 1.3 [with (1.24) replaced by (1.25)] implies that the solution established in Theorem 1.2 is unique.

Proof of Theorem 1.3.   Set $z = u - \hat{u}$, $h = \phi(u) - \phi(\hat{u})$. Since ϕ is continuous and $u, \hat{u} \in L^\infty$, for any $\xi > 0$ there is a $\delta > 0$ such that $|\phi(u(x, t)) - \phi(\hat{u}(x, t))| > \xi$ implies that $|u(x, t) - \hat{u}(x, t)| > \delta$. However, since $u - \hat{u} \in L^1$, we obtain

$$(1.26) \qquad \mathrm{meas}\,\{(x, t) \in Q_T; |h(x, t)| > \xi\} < \infty \qquad \forall\, \xi > 0.$$

Notice next that

$$z_t - \Delta h = 0 \qquad in\ \mathcal{D}'(Q_T),$$

that is,

$$(1.27) \qquad \int_0^T \int_{R^n} (z\psi_t + h\Delta\psi)\, dx\, dt = 0 \qquad \forall\, \psi \in \mathcal{D}(Q_T).$$

We fix $\gamma \in \mathcal{D}(Q_T)$ and set $\psi = B_\varepsilon\gamma$. Then $\psi \in C^\infty$ and $\psi = 0$ near $t = 0$ and near $t = T$. Since $z, h \in L^\infty(Q_T)$, (1.27) continues to hold for ψ in $C^\infty(Q_T) \cap L^1(Q_T)$ with $\psi = 0$ near $t = 0$ and near $t = T$ provided that $\psi_t, \Delta\psi, |\nabla\psi|$ belong to $L^1(Q_T)$, which is the case for our $\psi = B_\varepsilon\gamma$.

Noting that $\Delta B_\varepsilon \gamma = \varepsilon B_\varepsilon \gamma - \gamma$, $(B_\varepsilon \gamma)_t = B_\varepsilon \gamma_t$, we get from (1.27),

$$0 = \int_0^T \int_{R^n} \left[z B_\varepsilon(\gamma_t) + h(\varepsilon B_\varepsilon \gamma - \gamma) \right] dx\, dt$$

$$= \int_0^T \int_{R^n} \left[(B_\varepsilon z)(\gamma_t) + (\varepsilon B_\varepsilon h - h)\gamma \right] dx\, dt$$

(the symmetry of B_ε was used in the second equality). Thus

(1.28) $$(B_\varepsilon z)_t = \varepsilon B_\varepsilon h - h \qquad \text{in } \mathcal{D}'(Q_T).$$

Since z, $B_\varepsilon z$ belong to $L^1 \cap L^\infty$,

$$g_\varepsilon(t) \equiv (B_\varepsilon z(t), z(t))$$

is well defined a.e., where the notation $z(t) = z(\cdot, t)$, $(p, q) = \int_{R^n} p(x) q(x)\, dx$ has been used.

Suppose that we can show:

(1.29) $$\lim_{\varepsilon \downarrow 0} g_\varepsilon(t) = 0 \qquad \text{a.e. in } t.$$

Then it follows that $z(t) = 0$ a.e. Indeed, if $w \in L^2(R^n)$, then $\varepsilon B_\varepsilon w - \Delta B_\varepsilon w = w$, and so

$$(B_\varepsilon w, w) = (B_\varepsilon w, \varepsilon B_\varepsilon w - \Delta B_\varepsilon w) = \varepsilon \| B_\varepsilon w \|_{L^2}^2 + \| \nabla B_\varepsilon w \|_{L^2}^2.$$

Therefore, $(B_\varepsilon w, w) \to 0$ as $\varepsilon \to 0$ implies that $\varepsilon B_\varepsilon w \to 0$ in $L^2(R^n)$ and (since $\nabla B_\varepsilon w \to 0$ in L^2)

$$\Delta B_\varepsilon w = \text{div}\,(\text{grad } B_\varepsilon w) \to 0 \qquad \text{in } \mathcal{D}'(R^n).$$

Consequently, $w = \varepsilon B_\varepsilon w - \Delta B_\varepsilon w \to 0$ in $\mathcal{D}'(R^n)$, and thus $w = 0$. Applying this remark to $w = z(t)$, it follows that (1.29) implies that $z(t) = 0$ a.e.

It remains to verify (1.29). For this we shall prove the following three lemmas.

Lemma 1.4. $g_\varepsilon(t)$ *is absolutely continuous and*

(1.30) $$g_\varepsilon'(t) = 2(\varepsilon B_\varepsilon h(t) - h(t), z(t)) \qquad \text{a.e.}$$

Lemma 1.5. *There holds*

(1.31) $$\operatorname*{ess\,lim}_{t \to 0} g_\varepsilon(t) = 0.$$

Lemma 1.6. *Let $p \in L^\infty(R^n)$, meas $\{x \in R^n, |p(x)| > \xi\} < \infty$ for any $\xi > 0$. Then, for any $q \in L^1(R^n) \cap L^\infty(R^n)$,*

$$\lim_{\varepsilon \to 0} (p, \varepsilon B_\varepsilon q) = 0.$$

Suppose that the lemmas are true. Then we can easily establish (1.29) (thus completing the proof of Theorem 1.3). Indeed, by (1.30), (1.31), and $zh \geqslant 0$,

$$g_\varepsilon(t) \leqslant 2 \int_0^t (\varepsilon B_\varepsilon h(s), z(s)) \, ds.$$

The integrand is bounded by

$$\| \varepsilon B_\varepsilon h \|_{L^\infty} \| z(s) \|_{L^1} \leqslant \| h \|_{L^\infty} \| z(s) \|_{L^1};$$

also, $\| z(s) \|_{L^1} \in L^1(0, T)$. Noting, by Lemma 1.6 [here we use (1.26)], that

$$(\varepsilon B_\varepsilon h(s), z(s)) = (h(s), \varepsilon B_\varepsilon z(s)) \to 0 \qquad \text{a.e. if } \varepsilon \to 0,$$

the Lebesgue bounded convergence theorem gives (1.29).

Proof of Lemma 1.4. For simplicity we take $\varepsilon = 1$ and set $g = g_1$, $Bg = B_1 g$. Let $z_\delta(t)$ be a δ-mollifier of $z(t)$ as a function of t [we defined $z(t) = 0$ outside Q_T]. Thus

$$z_\delta(t, x) = \int_{-\infty}^\infty \rho_\delta(t - s) z(x, s) \, ds = \rho_\delta * z$$

and $\rho_\delta(s)$ is a nonnegative C^∞ function supported on $(-\delta, \delta)$, and $\int \rho_\delta(s) \, ds = 1$, $\rho_\delta(-s) = \rho_\delta(s)$. Then

$$\frac{d}{dt}(Bz_\delta, z_\delta) = 2 \int \left(\frac{\partial}{\partial t} Bz_\delta, z_\delta \right).$$

By (1.28),

$$\frac{\partial}{\partial t} Bz_\delta = \rho_\delta * (Bh - h) \qquad \text{on } (\delta, T - \delta) \times R^n,$$

where we define $h = 0$ outside Q_T; in fact, this follows by applying both sides to any $\gamma \in \mathcal{D}(R^n \times (\delta, T - \delta))$.

From the previous two relations we have, for any $\zeta \in \mathcal{D}(0, T)$ and small enough δ:

$$-\int_0^T (Bz_\delta(s), z_\delta(s)) \zeta'(s) \, ds = 2 \int_0^T (\rho_\delta * (Bh - h)(s), z_\delta(s)) \zeta(s) \, ds.$$

Since $z_\delta \to z$ in $L^1(Q_T)$, $\|z_\delta\|_{L^\infty} \leqslant \|z\|_{L^\infty}$, if follows that

$$-\int_0^T g(s)\zeta'(s)\, ds = 2\int_0^T ((Bh - h)(s), z(s))\zeta(s)\, ds$$

and the lemma follows.

Proof of Lemma 1.5. Let $0 \leqslant a < b < T$, $k(t) = 1$ if $t > b$, $k(t) = 0$ if $t < a$ and $k(t) = (t - a)/(b - a)$ if $a < t < b$. We can approximate k by smooth functions $k_m(t)$ and then taking in (1.27) $k_m(t)\psi(x, t)$ instead of $\psi(x, t)$, we get, as $m \to \infty$:

$$\iint [z(k'\psi + k\psi_t) + hk\Delta\psi] = 0 = \iint (z\psi_t + h\Delta\psi).$$

Taking $a, b \to 0$, we obtain

$$\lim_{\substack{b,\, a \to 0 \\ b > a}} \frac{1}{b - a} \int_a^b \int_{R^n} z(x, t)\psi(x, t)\, dx\, dt = 0 \qquad \forall \psi \in C_0^\infty([0, T); R^n).$$

Since $z \in L^\infty$, we can choose ψ's that approximate any $\psi \in L^1(R^n)$ and thus deduce that

$$(1.32) \qquad \lim_{\substack{b,\, a \to 0 \\ b > a}} \frac{1}{b - a} \int_a^b \int_{R^n} z(x, t)\psi(x)\, dx\, dt = 0 \qquad \forall \psi \in L^1(R^n).$$

From (1.28),

$$B_\varepsilon z(x, t) - B_\varepsilon z(x, s) = \int_s^t [\varepsilon B_\varepsilon h(x, \tau) - h(x, \tau)]\, d\tau.$$

Multiplying both sides by $\psi \in C_0^\infty(R^n)$ and integrating over $x \in R^n$ and $s \in (a, b)$, we get

$$\int_{R^n} B_\varepsilon z(x, t)\psi(x)\, dx - \frac{1}{b - a} \int_a^b \int_{R^n} z(x, s)B_\varepsilon\psi(x)\, dx\, ds$$

$$= \frac{1}{b - a} \int_a^b \int_{R^n} \int_s^t (\varepsilon B_\varepsilon h(x, \tau) - h(x, \tau))\, d\tau\, \psi(x)\, dx\, ds.$$

Taking $a, b \downarrow 0$ and using (1.32), we obtain

$$\int_{R^n} (B_\varepsilon z)(x, t)\psi(x)\, dx = \int_0^t \int_{R^n} (\varepsilon B_\varepsilon h(x, \tau) - h(x, \tau))\psi(x)\, dx\, d\tau.$$

Since $\psi \in C_0(R^n)$ is arbitrary, it follows that

$$B_\varepsilon z(x, t) = \int_0^t (\varepsilon B_\varepsilon h(x, \tau) - h(x, \tau)) \, d\tau,$$

and consequently $\| B_\varepsilon z(t) \|_{L^\infty} \leqslant 2t \| h \|_{L^\infty}$. Hence

$$| g_\varepsilon(t) | \leqslant 2t \| h \|_{L^\infty} \| z(t) \|_{L^1}.$$

Since $\| z(t) \|_{L^1} \in L^1(0, T)$, the assertion ess $\lim_{t \to 0} g_\varepsilon(t) = 0$ follows.

Proof of Lemma 1.6. We can write, for any $\xi > 0$,

$$\left| \int \varepsilon p B_\varepsilon q \, dx \right| \leqslant \int_{\{p > \xi\}} \varepsilon p B_\varepsilon q + \xi \int | \varepsilon B_\varepsilon q |$$

$$\leqslant \text{meas} \{ p > \xi \} \| p \|_{L^\infty} \| \varepsilon B_\varepsilon q \|_{L^\infty} + \xi \| \varepsilon B_\varepsilon q \|_{L^1}$$

$$\equiv I_\varepsilon + J_\varepsilon.$$

Using the representation formula (1.11) we find that, for any $0 < r < \infty$,

$$| \varepsilon B_\varepsilon q(x) | \leqslant \varepsilon^{n/2} C(r) \| q \|_{L^1} + \varepsilon^{n/2} \| q \|_{L^\infty} \int_{\{\sqrt{\varepsilon} |x - y| < r\}} k\left(\sqrt{\varepsilon} (x - y) \right) dy,$$

where $C(r)$ is finite for any $r > 0$ [by (1.12)]. Taking $\varepsilon \to 0$, we obtain

$$\limsup_{\varepsilon \to 0} \| \varepsilon B_\varepsilon q \|_{L^\infty} \leqslant \| q \|_{L^\infty} \int_{\{|y| < r\}} k(y) \, dy.$$

Taking $r \downarrow 0$, we conclude that $I_\varepsilon \to 0$ if $\varepsilon \to 0$. Hence

$$\limsup_{\varepsilon \to 0} \left| \int \varepsilon p B_\varepsilon q \, dx \right| \leqslant \xi \| q \|_{L^1},$$

and the assertion follows by letting $\xi \to 0$.

We shall now specialize to the case of gas in porous medium, that is,

$$(1.33) \qquad\qquad \phi(u) = u^m, \qquad m > 1, \quad u_0(x) \geqslant 0.$$

In this case it is useful to note that the solution can be constructed as a limit of classical positive solutions of the porous medium equation.

In fact, take sequences of balls B_{R_j} and $\varepsilon_j \downarrow 0$ ($j = 1, 2, \ldots$) such that

$$(1.34) \qquad\qquad R_j \to \infty, \qquad \varepsilon_j (R_j)^n \to 0.$$

We assume that $u_0 \in L^1(R^n) \cap L^\infty(R^n)$ and take a sequence $u_{0j} \in C_0^\infty(B_{R_j})$ such that $u_{0j} \geqslant 0$,

$$\|u_{0j}\|_{L^\infty} \leqslant \|u\|_{L^\infty} + \varepsilon_j,$$

$$\|u_{0j} - u_0\|_{L^1(B_{R_j})} \to 0 \qquad \text{if } j \to \infty.$$

Redefine $\phi(t)$ for $t < \varepsilon_j$ such that $\phi'(t) > 0$ for all t and $\phi \in C^2$; denote this function ϕ by ϕ_j. Consider the initial boundary value problem

$$\frac{\partial u}{\partial t} = \Delta\phi_j(u) \qquad \text{in } B_{R_j} \times (0, \infty),$$

(1.35)
$$u = \varepsilon_j \qquad \text{on } \partial B_{R_j} \times (0, \infty),$$

$$u = u_{0j} + \varepsilon_j \qquad \text{on } B_{R_j} \times \{0\}.$$

By standard results for nondegenerate parabolic equations [130], there exists a unique smooth solution $u = u_j$ of (1.35).

By the maximum principle,

(1.36)
$$\varepsilon_j \leqslant u_j \leqslant \|u_0\|_{L^\infty} + 2\varepsilon_j.$$

If we multiply the parabolic equation by pu_j^{p-1} and integrate, we obtain, after noting that $\partial u_j/\partial\nu \leqslant 0$ on $B_{R_j} \times (0, \infty)$ ($\nu =$ outward normal to ∂B_{R_j}),

(1.37)

$$\frac{\partial}{\partial t}\int_{B_{R_j}} u_j^p \, dx + mp(p-1)\int_{B_{R_j}} u_j^{m+p-3}|\nabla u_j|^2 \, dx = 0 \qquad (1 \leqslant p < \infty).$$

It follows that

(1.38) $$\int_{B_{R_j}} u_j^p(x, t) \, dx \leqslant \int_{B_{R_j}} (u_{0j}(x) + \varepsilon_j)^p \, dx \qquad (1 \leqslant p < \infty).$$

Taking $j \to \infty$ and using (1.34), we conclude that for a subsequence, $u_j \to u$ weakly in $L^1(B_R \times (0, T))$ and weak star in $L^\infty(B_R \times (0, T))$ for any $R > 0$, $T > 0$, and $u \in L^1(Q) \cap L^\infty(Q)$. It is clear that u also satisfies (1.1), (1.2) in a weak sense:

(1.39) $$\int_0^\infty \int_{R^n} (u\psi_t + u^m\Delta\psi) \, dx \, dt + \int_{R^n} u_0(x)\psi(x) \, dx = 0$$

for any $\psi \in C_0^\infty(R^n \times [0, \infty))$.

By Theorem 1.3, this solution must coincide with the solution constructed by Theorem 1.2. It follows that the full sequence u_j is convergent a.e. to u and $u \in C([0, \infty); L^1(R^n))$.

Remark 1.1. Replace ε_j, R_j in (1.35) by ε, R and u_0 by a corresponding $u_{0, \varepsilon, R}$, and denote the corresponding solution by $u_{\varepsilon, R}$. Then by the arguments above it also follows that

$$\lim_{R \to \infty} \lim_{\varepsilon \to 0} u_{\varepsilon, R} \quad \left[\text{taken weakly in } L^1_{loc}(R^n \times [0, \infty)) \right]$$

exists and is equal to the solution u.

Theorem 1.7. *In the porous medium case* (1.33), *for any* $u_0 \in L^1(R^n)$ *there exists a unique function u satisfying*

$$u \in C([0, \infty), L^1(R^n)) \cap L^\infty(R^n \times (\varepsilon, \infty)) \quad \forall \varepsilon > 0,$$

(1.40) $$u_t - \Delta\phi(u) = 0 \quad in \; \mathcal{D}'(R^n \times (0, \infty)),$$

$$u(0, \cdot) = u_0.$$

The novelty here is that u_0 is assumed to belong only to $L^1(R^n)$.

Proof. To prove uniqueness we apply Theorem 1.3 to $u(\cdot, t + h)$ and $S(t)u(\cdot, h)$ for any $h > 0$ and conclude that $u(\cdot, t + h) = S(t)u(\cdot, h)$ if $t \geq 0$. Taking $h \to 0$, we get $u(\cdot, t) = S(t)u_0$.

To prove existence let $u = u_{\varepsilon, R}$ ($\varepsilon \geq 0$, $R > 0$) be the solution of (1.35) when $\varepsilon_j = \varepsilon$, $R_j = R$, $u_{0j} = u_{0, \varepsilon, R}$. We shall prove:

Lemma 1.8. *There holds*

(1.41) $$|u(t)|_{L^\infty(B_R)} \leq \frac{C}{t^{n/(2+n(m-1))}} + C' \quad (u = u_{0, R})$$

where C, C' are positive constants depending only on an upper bound on $|u_0|_{L^1(R^n)}$.

Then, for a sequence $R \to \infty$, $u_{0, R} \to u$ weakly in $L^1(B_R \times (0, T))$ for any $R > 0$, $T > 0$, and

(1.42) $$|u(t)|_{L^\infty(R^n)} \leq \frac{C}{t^{n/(2+n(m-1))}} + C'.$$

To complete the proof of existence, it then remains to show that $u \in C([0, \infty); L^1(R^n))$. This can be done as follows:

We first note [see the proof of (1.20)] that $\forall u_0, u_1$ in $L^1(R^n)$,

$$(1.43) \qquad |S(t)u_0 - S(t)u_1|_{L^1(R^n)} \leqslant |u_0 - u_1|_{L^1(R^n)}.$$

The same holds if R^n is replaced by B_R, where $S(t)u_i = 0$ on $\partial B_R \times (0, \infty)$. Let $\tilde{u}_j \in L^1(R^n) \cap L^\infty(R^n)$,

$$(1.44) \qquad |\tilde{u}_j - u_0|_{L^1(R^n)} \to 0 \qquad \text{as } j \to \infty,$$

and denote by $\tilde{u}_{j, R}(t)$ the solutions $u_{0, R}(t)$ corresponding to the initial conditions $\tilde{u}_j$. By Remark 1.1 and Theorem 1.3, $\tilde{u}_j(t) = \lim_{R \to \infty} \tilde{u}_{0, R}(t)$ exists as a weak L^1 limit and

$$(1.45) \qquad \tilde{u}_j(t) \in C([0, \infty); L^1(R^n)).$$

Since (1.43) holds for B_R, it also holds for R^n with $u_1 = \tilde{u}_j$, where $S(t)u_0$ coincides with $u(t)$ and $S(t)\tilde{u}_j$ coincides with $\tilde{u}_j(t)$. From this and from (1.44), (1.45) follows the assertion that $u \in C([0, \infty); L^1(R^n))$.

Proof of Lemma 1.8. Set

$$u_\varepsilon = u_{\varepsilon, R}, \qquad v = u_\varepsilon^{(m+p-1)/2}, \qquad \Omega = B_R, \qquad \varepsilon_0 = \varepsilon^{(m+p-1)/2}.$$

By Sobolev's inequality

$$\int_\Omega |\nabla v|^2 \, dx \geqslant K \left(\int_\Omega |v - \varepsilon_0|^{2^*} \right)^{2/2^*},$$

where K is a universal positive constant, and $2^* = 2n/(n - 2)$ (for simplicity, we take $n \geqslant 3$). Hence

$$\frac{\partial}{\partial t} \int_\Omega u_\varepsilon^p + \frac{4kmp(p-1)}{(m+p-1)^2} \left(\int_\Omega |u^{(m+p-1)/2} - \varepsilon_0|^{2^*} \right)^{2/2^*} \leqslant 0.$$

Taking $\varepsilon \to 0$ we obtain, for $u = u_{0, R}$,

$$(1.46) \qquad \frac{\partial}{\partial t} \int_\Omega u^p \, dx + C_p \left(\int_\Omega u^{ap+b} \, dx \right)^{1/a} \leqslant 0,$$

$$\text{with } a > 1, b > 0.$$

From this inequality one can deduce (1.41); see Problems 1 and 2.

Corollary 1.9. *The solution u of Theorem* 1.7 *satisfies*

$$(1.47) \qquad u(x, t) \leqslant \frac{C}{t^{n/(2+(m-1)n)}} \left(|u_0|_{L^1(R^n)} \right)^{2/(2+(m-1)n)},$$

where C is a constant depending only on m, n.

Proof. From the proof of Lemma 1.8 we obtain the inequality (1.47) provided that

$$\int_{R^n} u_0(x)\, dx = 1 \qquad \text{and} \qquad t \leqslant 1.$$

To prove it in general, introduce the function

$$\tilde{w}(x, t) = \left(\frac{\sigma}{R^2} \right)^{1/(m-1)} u(Rx, \sigma x);$$

this is again a solution for any $\sigma > 0$, $R > 0$, and

$$\int_{R^n} \tilde{w}(x, 0)\, dx = 1$$

if

$$I = R^n \left(\frac{R^2}{\sigma} \right)^{1/(m-1)}, \qquad \text{where } I = \int_{R^n} u_0(x)\, dx.$$

Applying (1.47) to $\tilde{w}(x, t)$ with $t = 1$, we obtain, with $x' = Rx$,

$$u(x', \sigma) \leqslant \frac{C}{\sigma^{n/(2+(m-1)n)}} I^{2/(2+(m-1)n)}.$$

Since $x' \in R^n$, $\sigma \in (0, \infty)$ are arbitrary, the corollary follows.

PROBLEMS

1. Write $u_0 = u_{0, \varepsilon, R}$ when $\varepsilon = 0$. Let $u_0 \in L^{p_0}(\Omega)$ and define $\theta \in (0, 1)$ by

$$\frac{1}{p} = \frac{1 - \theta}{p_0} + \frac{\theta}{ap + b}.$$

Setting $\|v\|_p = \|v\|_{L^p(\Omega)}$, deduce by Hölder's inequality that

$$\|u(t)\|_{ap+b} \geqslant \frac{\|u(t)\|_p^{1/\theta}}{\|u_0\|_{p_0}^{(1-\theta)/\theta}}$$

[using $\|u(t)\|_{p_0} \leqslant \|u_0\|_{p_0}$]. Then use (1.46) to get $\int |u|^p \, dx \leqslant (A/t)^\beta$; compute that

$$\left(\int |u|^p \, dx \right)^{1/(ap+b-p)} \leqslant B \left(\int |u_0|^{p_0} \right)^{1/(ap_0+b-p_0)},$$

$$B = \left[\frac{a(p-p_0)}{C_p t(ap_0 + b - p_0)} \right]^\gamma, \qquad \gamma = \frac{a(p-p_0)}{(ap_0 + b - p_0)(ap + b - p)}.$$

2. Set

$$\phi(p, t) = \log \left(\int |u(x, t)|^p \, dx \right)^{1/(ap+b-p)},$$

$$\psi(p) = \phi \left(p, \lambda \left(\frac{1}{ap_0 + b - p_0} - \frac{1}{ap + b - p} \right) \right), \qquad \lambda > 0.$$

Use Problem 1 to deduce that at any $p = p_0 \geqslant 1$,

$$\psi'(p) \leqslant \frac{a}{(ap + b - p)^2} \log \frac{a(ap + b - p)}{\lambda(a-1)C_p} \equiv h(p).$$

Use this and $C_p \to 4mK$ as $p \to \infty$ to compute that

$$\psi(\infty) \leqslant \phi(p_0, 0) - \frac{a}{a-1} \frac{1}{ap_0 + b - p_0} \log \lambda + C$$

and choose $\lambda = t(ap_0 + b - p_0)$ to deduce that

$$\log \|u(t)\|_{L^\infty} \leqslant (a-1)\phi(p_0, 0) - \frac{a}{ap_0 + b - p_0} \log t + C.$$

3. The function

$$u(x, t) = t^{-k} \left\{ \left(1 - \frac{k(m-1)}{2mn} \frac{|x|^2}{t^{2k/n}} \right)^+ \right\}^{1/(m-1)},$$

where $k = (m - 1 + 2/n)^{-1}$ is a solution of the porous medium equation (with initial data the Dirac measure). It is Hölder continuous with Hölder exponent $\alpha = \min(1, 1/(m-1))$.

4. The function

$$u(x, t) = \left[Ax^2 / (T - t) \right]^{1/(m-1)}$$

with $A = (m - 1)/(2m(m + 1))$ is a solution of the porous medium equation with $n = 1$.

5. If $v = \psi(u) = mu^{m-1}/(m - 1)$ and

$$v(x, t) = C(Ct - (x - \alpha))^+ \qquad (C > 0, \alpha \text{ real}),$$

then $u = \psi^{-1}(v)$ is a solution of the porous medium equation with $n = 1$.

6. If $u_0 \in L^1(R^n) \cap L^p(R^n)$, then the solution in Theorem 1.7 satisfies

$$\int_0^\infty \int_{R^n} |\nabla_x u^q|^2 \, dx \, dt < \infty, \qquad q = \frac{m + p - 1}{2}.$$

2. ESTIMATES ON THE EXPANSION OF GAS

In the next section we establish the Hölder continuity of the solution of the porous medium equation. [Notice that the solution in Problem 3 of Section 1 is Hölder continuous with precise exponent $\min\{1, 1/(m - 1)\}$.] Our result and the various estimates leading to it are all local. For simplicity we assume that

$$(2.1) \qquad u_0 \in L^1(R^n) \cap L^\infty(R^n);$$

however, in view of Theorem 1.7, in asserting regularity for $t > 0$ it suffices to assume that $u_0 \in L^1(R^n)$.

In this section we prepare the tools for proving the Hölder continuity. We establish several lemmas of intrinsic interest regarding the manner in which gas, in porous medium, is expanding.

In Section 1 we have shown how to construct the unique solution u (of Theorem 1.2) as a limit of sequence of positive solutions. We shall need a slightly more refined version of this process.

Let $u_j(x)$ be a sequence of functions satisfying:

$$u_j \in C^\infty(R^n),$$

$$u_j \geqslant \frac{1}{j},$$

$$(2.2) \qquad u_j(x) = \frac{1}{j} \qquad \text{if } |x| > R_j \quad (R_j \to \infty \text{ if } j \to \infty),$$

$$u_j \leqslant |u_0|_{L^\infty} + 1,$$

$$u_j(x)\downarrow \qquad \text{if } j \uparrow,$$

$$\int_{R^n} |u_j(x) - \frac{1}{j} - u_0(x)| \, dx \to 0 \qquad \text{if } j \to \infty.$$

Denote by $u_j(x, t)$ the solution of the porous medium equation corresponding to $u_j(x)$. Using standard regularity results for nonlinear parabolic equations [130] and the maximum principle in a strip [94a], one can then show that

(2.3)

$$D^\alpha\left(u_j(x, t) - \frac{1}{j}\right) \to 0 \qquad \text{if } |x| \to \infty, \text{ uniformly with respect to}$$

$$t \text{ in bounded intervals } 0 \leqslant t \leqslant T \quad (\forall |\alpha| \geqslant 0).$$

The solution u_j can be obtained by solving the porous medium equation in cylinders $B_R \times (0, \infty)$ with data

$$u_j(x) \qquad \text{on } t = 0,$$

$$\frac{1}{j} \qquad \text{on } \partial B_R.$$

The corresponding solution $u_{j, R}$ then satisfies [see (1.38)]

$$\int_{B_R} u_{j, R}(x, t) \, dx \leqslant \int_{B_R} u_j(x) \, dx,$$

so that

$$\int_{B_R}\left(u_{j, R}(x, t) - \frac{1}{j}\right) dx \leqslant \int_{B_R}\left(u_j(x) - \frac{1}{j}\right) dx$$

and $u_{j, R} - 1/j \geqslant 0$ (by the maximum principle). Taking $R \to \infty$, we obtain

(2.4)

$$\int_{R^n}\left(u_j(x, t) - \frac{1}{j}\right) dx \leqslant \int_{R^n}\left(u_j(x) - \frac{1}{j}\right) dx,$$

and $u_j(x, t) \geqslant 1/j$. By comparison [94a, p. 52]

$$u_{j, R}(x, t) \geqslant u_{j+1, R}(x, t),$$

and thus

(2.5) $$u_j(x, t) \downarrow \text{if } j \uparrow.$$

It follows that

(2.6)
$$u(x, t) = \lim_{j \to \infty} u_j(x, t) \text{ exists,}$$

$$u(x, t) \text{ is upper semicontinuous,}$$

and from (2.4) we obtain

$$(2.7) \qquad \int_{R^n} u(x, t)\, dx \leqslant \int_{R^n} u_0(x)\, dx.$$

Thus $u \in L^1$. Since also

$$u_j(x, t) \leqslant 1 + |u_0|_{L^\infty(R^n)},$$

by the maximum principle, the same holds for $u(x, t)$. Thus $u \in L^\infty$. Theorem 1.3 can then be applied. It implies that u coincides with the $C([0, \infty); L^1(R^n))$ solution constructed in Theorem 1.2. Thus u also coincides with the solution of Theorem 1.7.

In order to derive estimates or comparison results for u, we shall often derive them first for u_j and then let $j \to \infty$.

We shall be working also with the pressure function (up to a factor)

$$(2.8) \qquad v = \frac{m}{m-1} u^{m-1}.$$

It satisfies, formally, $\nabla u^m = u \nabla v$ and

$$(2.9) \qquad v_t = (m-1) v \Delta v + |\nabla_x v|^2.$$

Set

$$(2.10) \qquad k = \frac{1}{m - 1 + (2/n)},$$

and introduce the operator

$$(2.11) \qquad Lw \equiv w_t - (m-1) v \Delta w - 2m \nabla_x v \cdot \nabla_x w - \frac{w^2}{k}.$$

If we apply Δ to both sides of (2.9), then we easily deduce that, formally,

$$(2.12) \qquad L(\Delta v) \geqslant 0.$$

Lemma 2.1. *The following inequalities hold*:

$$(2.13) \qquad u_t \geqslant -\frac{k}{t} u,$$

$$(2.14) \qquad v_t \geqslant -\frac{(m-1)k}{t} v,$$

$$(2.15) \qquad \Delta v \geqslant -\frac{k}{t};$$

the derivatives are taken in the distribution sense.

Proof. It suffices to prove these inequalities for each function u_j. Setting

$$v_j = \frac{m}{m-1} u_j^{m-1}, \qquad w_j = \Delta v_j,$$

we have, by (2.12),

$$L w_j \geqslant 0,$$

where the derivatives are taken in the classical sense. Also, $L(-k/t) = 0$ and [by (2.3)]

$$\Delta v_j(x, \varepsilon) > -\frac{k}{\varepsilon} \qquad (x \in R^n)$$

if ε is small enough, depending on ε. By comparison [using (2.3)] we deduce that $\Delta v_j > -k/t$ in $R^n \times (0, \infty)$.

Clearly, (2.13) follows from (2.15), (1.1), and

$$\Delta u^m = u \, \Delta v + \nabla u \cdot \nabla v \geqslant u \Delta v,$$

and (2.14) follows from (2.13).

In the sequel we use the notation

$$B_r(x^0) = \{x; |x - x^0| < r\},$$

$$B_r = B_r(0), \qquad |B_r| = \text{meas}(B_r),$$

$$\fint_{B_r(x^0)} w(x, t)\, dx = \frac{1}{|B_r|} \int_{B_r(x^0)} w(x, t)\, dx.$$

Lemma 2.2. *For any $\eta_0 > 0$ there exist positive constants η, c depending only on m, n, η_0 such that the following is true: Let $x^0 \in R^n$, $t^0 \geqslant \eta_0$, $R > 0$, $0 < \sigma \leqslant \eta$. If*

$$(2.16) \qquad v(x, t^0) = 0 \qquad for\ x \in B_R(x^0)$$

and

$$(2.17) \qquad \fint_{B_R(x^0)} v(x, t^0 + \sigma)\, dx \leqslant \frac{cR^2}{\sigma},$$

then

$$(2.18) \qquad v(x, t^0 + \sigma) = 0 \qquad for\ x \in B_{R/6}(x^0).$$

The physical interpretation of the lemma is that if the gas reached the ball $B_{R/6}(x^0)$ in time $t^0 + \sigma$ and there was no gas in $B_R(x^0)$ at time t^0, then a considerable amount of gas has entered the ball $B_R(x^0)$ in time $t^0 + \sigma$.

Proof. For simplicity we denote $x - x^0$ by x and $t - t^0$ by t. Consider the functions

$$\tilde{v}(x, t) = \frac{\sigma}{R^2} v(Rx, \sigma t),$$

$$\tilde{v}_j(x, t) = \frac{\sigma}{R^2} v_j(Rx, \sigma t).$$

Notice that $\tilde{v}_j$ satisfies the same equation, (2.9), as v_j. By (2.16), (2.17),

(2.19) $\tilde{v}(x, 0) = 0$ in B_1,

(2.20) $\fint_{B_1} \tilde{v}(x, 1)\, dx \leqslant c.$

By Lemma 2.1

(2.21) $\tilde{v}_t \geqslant -\varepsilon \tilde{v},$ where $\varepsilon = \frac{mk}{\eta_0}\eta,$

(2.22) $\Delta \tilde{v} = \sigma \Delta v \geqslant -\varepsilon.$

The last inequality implies that

$$\Delta\left(\tilde{v} + \frac{\varepsilon}{2n}|x|^2\right) \geqslant 0.$$

That is, $\tilde{v} + \varepsilon|x|^2/2n$ is subharmonic in x. The same is true of each $\tilde{v}_j$. Hence for any $x \in B_{1/2}$,

$$\tilde{v}_j(x, 1) + \frac{\varepsilon}{2n}|x|^2 \leqslant \fint_{B_{1/2}(x)}\left(\tilde{v}_j(\xi, 1) + \frac{\varepsilon}{2n}|\xi|^2\right) d\xi.$$

Since $\tilde{v}_j(x, t) \downarrow \tilde{v}(x, t)$ for all (x, t), the dominated convergence theorem gives

$$\tilde{v}(x, 1) + \frac{\varepsilon}{2n}|x|^2 \leqslant \fint_{B_{1/2}(x)}\left(\tilde{v}(\xi, 1) + \frac{\varepsilon}{2n}|\xi|^2\right) d\xi$$

$$\leqslant 2^n \fint_{B_1} v(\xi, 1)\, d\xi + \frac{\varepsilon}{2n}.$$

Using (2.20), we conclude that

(2.23) $\tilde{v}(x, 1) \leqslant 2^n c + \dfrac{\varepsilon}{2n}$ if $x \in B_{1/2}$

From (2.21) it follows that

$$\tilde{v}(x, 1) \geqslant e^{-\varepsilon(1-t)}\tilde{v}(x, t).$$

Combining this with (2.23), we get

(2.24) $\tilde{v}(x, t) \leqslant e^\varepsilon\left(2^n c + \dfrac{\varepsilon}{2n}\right)$ if $x \in B_{1/2}$, $0 < t < 1$.

We introduce a comparison function

(2.25) $V(r, t) = \lambda\left[\alpha^2 t + (r - \tfrac{1}{3})\right]^+$ $(r = |x|, \lambda > 0)$.

It satisfies

(2.26) $V_t \geqslant (m - 1)V\Delta V + |\nabla_x V|^2$ wherever $V > 0$,

provided that

$$1 \geqslant \lambda\left[(m - 1)(n - 1)\frac{\alpha t + (r - 1/3)}{r} + 1\right].$$

We choose $\alpha = \tfrac{1}{6}$. Then $V(r, t) > 0$ if and only if

$$r > \frac{1}{3} - \frac{t}{6}.$$

Thus, for $0 < t < 1$, $0 < r < 1$, (2.26) is valid provided that $\lambda = \lambda(m, n)$ is sufficiently small.
    Next

(2.27) $\tilde{v}(x, 0) = 0 \leqslant V(r, t)$ if $0 \leqslant r \leqslant \tfrac{1}{2}$,

and

(2.28) $\tilde{v}(x, t) < \lambda\alpha\tfrac{1}{12} < V(r, t)$ if $\tfrac{5}{12} < r < \tfrac{1}{2}, 0 \leqslant t \leqslant 1$

by (2.24), provided that ε and c are sufficiently small.
    Setting

$$U = \left(\frac{m-1}{m}V\right)^{1/(m-1)}, \qquad \tilde{u} = \left(\frac{m-1}{m}\tilde{v}\right)^{1/(m-1)}$$

we have

$$U_t \geqslant \Delta U^m, \qquad \tilde{u}_t = \Delta \tilde{u}^m \qquad \text{in } B_{1/2} \times (0, 1),$$

where the last relation is taken in the distribution sense.

We claim that

(2.29) $$\tilde{u} \leqslant U \qquad \text{in } B_{1/2} \times (0, 1).$$

The proof is obtained by employing the sequence of smooth functions

$$\tilde{u}_j = \left(\frac{m-1}{m} \tilde{v}_j\right)^{1/(m-1)}.$$

If we multiply $U_t \geqslant \Delta U^m$ by $(\tilde{u}_j - U)^+$ and $\partial \tilde{u}_j / \partial t = \Delta (\tilde{u}_j)^m$ by $-(\tilde{u}_j - U)^+$, integrate over $B_r \times (0, t)$ and add, we find that

(2.30)

$$\int_{B_r} \left\{ \left[\tilde{u}_j(x, t) - U(r, t)\right]^+ \right\}^2 + \int_{B_r \times (0, t)} \nabla_x (\tilde{u}_j^m - U^m) \cdot \nabla_x (\tilde{u}_j - U)^+$$

$$\leqslant \int_{\partial B_r \times (0, t)} |\nabla_x (\tilde{u}_j^m - U^m)| (\tilde{u}_j - U)^+ + \int_{B_r} \left\{ \left[u_j(x, 0) - U(r, 0)\right]^+ \right\}^2.$$

Taking in (1.37) $p = 1 + m$, we see that

$$\int_0^\infty \int_{R^n} |\nabla_x \tilde{u}_j^m|^2 \, dx \, dt < \infty.$$

Hence if we integrate (2.30) with respect to r, $\frac{5}{12} \leqslant r \leqslant \frac{1}{2}$, we obtain

$$\int_{5/12}^{1/2} dr \int_{B_r} \left\{ \left[\tilde{u}_j(x, t) - U(r, t)\right]^+ \right\}^2$$

$$\leqslant C \int_{(B_{1/2} \setminus B_{5/12}) \times (0, t)} \left\{ (\tilde{u}_j - U)^+ \right\}^2 + \int_{5/12}^{1/2} dr \int_{B_r} \left\{ \left[\tilde{u}_j(x, 0) - U(r, 0)\right]^+ \right\}^2.$$

By (2.27), (2.28), each of the two integrands on the right-hand side converges pointwise and monotonically to zero as $j \to \infty$ [since $\tilde{u}_j(x, t) \downarrow \tilde{u}(x, t)$]. Thus, letting $j \to \infty$ we obtain, by the dominated convergence theorem,

$$\int_{5/12}^{1/2} dr \int_{B_r} \left[\tilde{u}(x, t) - U(r, t)\right]^+ = 0,$$

and (2.29) thereby follows.

From (2.29) we obtain

$$\tilde{v}(x, t) \le V(r, t) = 0 \qquad \text{if } |x| < \tfrac{1}{6},$$

which establishes the assertion (2.18).

Remark 2.1. Lemma 2.2 remains true if η is any positive number, not necessarily small, but then c depends also on η. Indeed, this follows by the same proof applied to

$$\tilde{v}(x, t) = \beta v(x, t^0 + \beta t)$$

with $\beta \eta$ small.

In the following lemma, which is in some sense the converse to Lemma 2.2, ε is the positive number defined in (2.21), γ is a certain positive constant depending only on m, n and λ, ν, C^* are any positive constants satisfying together with ε:

$$(2.31) \quad \varepsilon \le \frac{C^*}{\lambda}, \qquad \varepsilon^\gamma \le C^* \nu |B_1|, \qquad \lambda^m \nu^{m-1} \ge C(C^*) \qquad \left(\varepsilon = \frac{mk}{\eta_0} \eta \right)$$

where $C(C^*)$ is a positive constant depending only on C^*, m, n. These relations are satisfied (i) for any fixed λ, C^* if ε is small enough and ν is sufficiently large, or (ii) for any fixed ν if λ is sufficiently large and ε is sufficiently small.

Lemma 2.3. *Let* $x^0 \in R^n$, $t^0 \ge \eta_0$, $R > 0$, $\sigma > 0$ *and* $\sigma \le \eta$. *If*

$$(2.32) \quad \fint_{B_R(x^0)} u^m(x, t^0) \, dx \ge \nu \left(\frac{R^2}{\sigma} \right)^{m/(m-1)},$$

then

$$(2.33) \quad u^m(x^0, t^0 + \lambda \sigma) \ge c \left(\frac{R^2}{\sigma} \right)^{m/(m-1)},$$

where $c = \nu^{1/m}/\lambda C_0$ *and* C_0 *is a positive constant depending only on* m, n.

The lemma expresses the physical fact that if a large mass of gas was in $B_R(x^0)$ at time t^0, then the gas covers a neighborhood of x^0 at time $t^0 + \sigma$.

Proof. For simplicity we replace $x - x^0$ by x and $t - t^0$ by t. Let

$$\tilde{u}(x, t) = \left(\frac{\sigma}{R^2} \right)^{1/(m-1)} u(Rx, \sigma t).$$

Then

(2.34)
$$\fint_{B_1} \tilde{u}^m(x,0)\, dx \geqslant \nu.$$

We introduce Green's function (for $n \geqslant 3$)

$$G_\rho(r) = r^{2-n} - \rho^{2-n} - \frac{n-2}{2}(\rho^2 - r^2) \qquad (r = |x|)$$

in the ball B_ρ. It satisfies

$$G_\rho(\rho) = G_\rho'(\rho) = 0,\ G_\rho(r) > 0 \qquad \text{if } 0 < r < \rho,$$

and

$$\Delta G_\rho = \alpha_n I_{B\rho} - \gamma_n \delta(x) \qquad [\alpha_n = n(n-2)]$$

in the distribution sense, where γ_n is a positive constant depending only on n, $I_A =$ indicator function of A, and $\delta(x)$ is the Dirac measure.

For $n = 2$ we take

$$G_\rho(r) = \log \frac{\rho}{r} - \frac{1}{2}(\rho^2 - r^2).$$

Then, if $0 < t < 1$, and if $\tilde{u}$ is smooth,

$$\int_{B_1} G_1(r)\tilde{u}(x,t)\, dx \geqslant \int_0^t \int_{B_1} G_1(r)\tilde{u}_s(x,s)\, dx\, ds$$

$$= \int_0^t \int_{B_1} G_1(r)\Delta \tilde{u}^m(x,s)\, dx\, ds$$

$$= -\gamma_n \int_0^t \tilde{u}^m(0,s)\, ds + \alpha_n \int_0^t \int_{B_1} \tilde{u}^m(x,s)\, dx\, ds.$$

It follows that

(2.35)
$$\int_{B_1} G_1(r)\tilde{u}(x,t)\, dx \geqslant -\gamma_n \int_0^t \tilde{u}^m(0,s)\, ds + \alpha_n \int_0^t \int_{B_1} \tilde{u}^m(x,s)\, dx\, ds$$

if u is smooth. This inequality then holds for the solutions u_j and, by approximation, also for the solution u.

Set

$$\phi(t) = \int_{B_1} \tilde{u}^m(x,t)\, dx.$$

and consider first the case where

(2.36) $$n \leqslant 2 \quad \text{or} \quad \frac{n}{n-2} > \frac{m}{m-1}.$$

Then

(2.37) $$\int_{B_1} G_1(r)\tilde{u}(x,t)\, dx \leqslant \left\{\int_{B_1} G_1^q(r)\, dx\right\}^{1/q} \int_{B_1} \left\{(\tilde{u}(x,t)^m)\, dx\right\}^{1/m}$$

where $1/q + 1/m = 1$ and

(2.38) $$\int_{B_1} G_1^q\, dx < \infty.$$

Thus we obtain from (2.35)

(2.39) $$\int_0^t \phi(s)\, ds \leqslant C_1 \int_0^t \tilde{u}^m(0,s)\, ds + C_2(\phi(t))^{1/m}$$

where C_1 are positive constants depending only on m, n.
    Consider next the case where (2.36) is not satisfied, but

(2.40) $$n \leqslant 4 \quad \text{or} \quad \frac{n}{n-4} > \frac{m}{m-1}.$$

Since $G_1(r)$ is bounded, away from $r = 0$,

(2.41) $$\int_{B_1 \setminus B_{1/2}} G_1(r)\tilde{u}(x,t) \leqslant C_3 \int_{B_4 \setminus B_{1/2}} \tilde{u}(x,t)$$

$$\leqslant C_4(\phi(t))^{1/m}.$$

For $x \in B_{1/2}$, we can represent $\tilde{u}^m(x,t)$ by Green's function $G_{1/4}$:

$$\gamma_n \tilde{u}^m(x,t) = \alpha_n \int_{B_{1/4}(x)} \tilde{u}^m(y,t)\, dy - \int_{B_{1/4}(x)} G_{1/4}(|y-x|)\Delta \tilde{u}^m(y,t)\, dy.$$

(Here again we tacitly assume that $\tilde{u}$ is smooth; more precisely, all the calculations are performed for $\tilde{u}_j$, and then in the final inequality we let $j \to \infty$.)
    Recalling that

$$\Delta \tilde{u}^m = \tilde{u}_t \geqslant -\varepsilon\tilde{u},$$

we get

$$\tilde{u}(x, t) = \left[\tilde{u}^m(x, t)\right]^{1/m}$$

$$\leqslant \left[\beta_n \int_{B_{1/4}(x)} \tilde{u}^m(y, t) \, dy + \beta_n \varepsilon \int_{B_{1/4}(x)} G_{1/4}(|y - x|)\tilde{u}(y, t) \, dy\right]^{1/m}$$

$$\leqslant 2\beta_n^{1/m}(\phi(t))^{1/m} + 2\beta_n^{1/m}\varepsilon^{1/m}\left[\int_{B_{1/4}(x)} G_{1/4}(|y - x|)u(y, t) \, dy\right]^{1/m}$$

$$\leqslant 2\beta_n^{1/m}(\phi(t))^{1/m} + C_5\varepsilon^{q/m}$$

$$+ \int_{B_{1/4}(x)} G_{1/4}(|y - x|)\tilde{u}(y, t) \, dy \qquad \left[\beta_n = \max\left(\frac{\alpha_n}{\gamma_n}, \frac{1}{\gamma_n}\right)\right]$$

where $1/q + 1/m = 1$. It follows that

(2.42)

$$\int_{B_{1/2}} G_1(|x|)\tilde{u}(x, t) \, dx \leqslant C_6(\phi(t))^{1/m} + C_5\varepsilon^{q/m}$$

$$+ \int_{B_{1/2}} \int_{B_{1/4}(x)} G_1(|x|)G_{1/4}(|y - x|)\tilde{u}(y, t) \, dy \, dx.$$

The last integral has the form

(2.43)
$$\int_{B_{3/4}} \hat{G}(y)\tilde{u}(y, t) \, dy$$

where, by (2.40),

$$\int_{B_{3/4}} |\hat{G}(y)|^q \, dy < \infty.$$

Evaluating then the last integral in (2.42) by Hölder's inequality and substituting (2.42) and (2.41) into (2.35), we obtain

(2.44) $$\int_0^t \phi(s) \, ds \leqslant C_1 \int_0^t \tilde{u}^m(0, s) \, ds + C_5\varepsilon^{q/m} + C_7(\phi(t))^{1/m}.$$

If (2.40) is not satisfied, but

$$n \leqslant 6 \qquad \text{or} \qquad \frac{n}{n - 7} > \frac{m}{m - 1},$$

then we can proceed as before, but in evaluating the integral (2.43) we first express $\tilde{u}(y, t)$ in terms of Green's function $G_{1/8}$. It is clear that this procedure leads, for any n, to an inequality of the form

$$(2.45) \qquad \int_0^t \phi(s)\, ds \leqslant C_1 \int_0^t \tilde{u}^m(0, s)\, ds + C_8 \varepsilon^\delta + C_9(\phi(t))^{1/m},$$

where C_8, C_9, δ are positive constants depending only on m, n.
By (2.34),

$$\phi(0) \geqslant \nu_0 \qquad (\nu_0 = \nu\,|\,B_1|).$$

By (2.13),

$$(2.46) \qquad (\tilde{u}^m)_t \geqslant -\varepsilon \tilde{u}^m$$

and consequently $\phi'(t) \geqslant -\varepsilon\phi(t)$. Hence

$$(2.47) \qquad \phi(t) \geqslant e^{-\varepsilon\lambda}\nu_0 \qquad \text{for } 0 \leqslant t \leqslant \lambda.$$

Suppose now that (2.33) is not satisfied, that is,

$$\tilde{u}^m(0, \lambda) \leqslant c.$$

Using (2.46), we get

$$(2.48) \qquad \tilde{u}^m(0, t) \leqslant e^{\varepsilon\lambda}c \qquad \text{if } 0 < t < \lambda.$$

Setting

$$(2.49) \qquad \left(C_1 \lambda c e^{\varepsilon\lambda} + C_8 \varepsilon^\delta\right) e^{\varepsilon\lambda/m} / \nu_0^{1/m} \equiv \tilde{C}$$

and using (2.47), we then conclude from (2.45) that

$$(2.50) \qquad \int_0^t \phi(s)\, ds \leqslant \frac{1}{B}(\phi(t))^{1/m} \qquad \left(\frac{1}{B} = C_9 + \tilde{C}\right).$$

Set

$$\psi(t) = \int_0^t \phi(s)\, ds.$$

Then

$$(2.51) \qquad (\psi'(t))^{1/m} \geqslant B\psi(t).$$

From (2.47) it follows that

$$(2.52) \qquad \psi(t) \geqslant At, \qquad A = e^{-\varepsilon\lambda}\nu_0.$$

We compare $\psi(t)$ with the solution,

$$\chi'(t) = (B\chi(t))^m, \qquad t > t_0, \ \chi(t_0) = At_0.$$

In view of (2.51), (2.52),

(2.53) $$\psi(t) \geqslant \chi(t), \qquad t_0 \leqslant t \leqslant \lambda.$$

Now,

$$(m-1)\chi^{m-1}(t) = \frac{1}{C - B^m t},$$

where C is a constant determined by

$$\frac{1}{C - B^m t_0} = (m-1)A^{m-1}t_0^{m-1}.$$

Since $\chi(t) \to \infty$ if $t \to C/B^m$, we also have, by (2.53),

$$\psi(t) \to \infty \text{ if } t \to \frac{C}{B^m}, \qquad \text{provided that } \lambda \geqslant \frac{C}{B^m},$$

which is impossible. Thus (2.33) must be satisfied provided that

(2.54) $$\lambda \geqslant \frac{C}{B^m} = \frac{e^{\varepsilon\lambda(m-1)}}{(m-1)v_0^{m-1}t_0^{m-1}B^m} + t_0,$$

where the definitions of C, A were used.

Choosing $t_0 = \lambda/2$ we easily verify that (2.54) is a consequence of (2.31) and the definitions of B, c.

Lemma 2.4. *Under the same notation as in Lemma 2.3, if*

(2.55) $$\fint_{B_R(x^0)} v(x, t^0) \, dx \geqslant v_0 \frac{R^2}{\sigma} \qquad \left[v_0 = v^{(m-1)/m} \right],$$

then

(2.56) $$v(x^0, t^0 + \lambda\sigma) \geqslant c_0 \frac{R^2}{\sigma} \qquad \left[c_0 = c^{(m-1)/m} \right].$$

Proof. Since

$$\fint_{B_R(x^0)} v(x, t^0) \, dx = \frac{m}{m-1} \fint_{B_R(x^0)} u^{m-1}(x, t^0) \, dx$$

$$\leqslant C_0 \left\{ \int_{B_R(x^0)} u^m(x, t^0) \, dx \right\}^{(m-1)/m}$$

where C_0 is a constant that depends only on m, n, Lemma 2.3 can immediately be applied to deduce (2.56).

PROBLEMS

1. Let u_0, $\tilde{u}_0$ belong to $L^1(R^n)$, $u_0 \geqslant \tilde{u}_0$. If $u(x, t)$, $\tilde{u}(x, t)$ are the solutions of the porous medium equation with the initial data $u_0(x)$ and $\tilde{u}_0(x)$, respectively, as asserted in Theorem 1.7, then $u(x, t) \geqslant \tilde{u}(x, t)$.

2. The solution u asserted in Theorem 1.7 satisfies

$$\int_{R^n} u(x, t)\, dx = \int_{R^n} u_0(x)\, dx.$$

[*Hint*: Use the approximating u_j.]

3. HÖLDER CONTINUITY OF THE SOLUTION

Theorem 3.1. *The solution $u(x, t)$ of the porous medium equation is Hölder continuous in every set $R^n \times (\eta_0, \infty)$, $\eta_0 > 0$.*

Proof. It suffices to prove the Hölder continuity for any smooth positive solution u_j, provided that the Hölder coefficient and exponent are independent of j. For simplicity we denote u_j by u.

We shall first prove a Hölder continuity-type inequality at one point (x^0, t^0) with $t^0 \geqslant 2\eta_0$, $\eta_0 > 0$. Set

(3.1) $\qquad G_k = \big\{ (x, t); |x - x^0| < R_k, t^0 - \sigma(R_k) < t < t^0 \big\},$

(3.2) $\qquad M_k = \max \Big\{ \sup_{G_k} u^m, R_k^\varepsilon \Big\}, \qquad$ for some $\varepsilon > 0$

Let k^* be the positive integer determined by

(3.3) $\qquad \dfrac{1}{c} 2^{-(k^*+1)\varepsilon} < u^m(x^0, t^0) \leqslant \dfrac{1}{c} 2^{-k^*\varepsilon},$

where c is a positive integer depending only on m, n, η_0, to be determined later. We shall choose

(3.4) $\qquad R_k = 2^{-k}, \sigma(R_k) = \overline{C} 2^{-\alpha k} \qquad [\sigma(R_1) < \eta_0]$

for some $1 < \alpha < 2$, $\overline{C} > 1$, and prove by induction that

$$(3.5) \qquad\qquad M_k = R_k^\varepsilon$$

for all $0 \leqslant k < k^*$, where ε is defined by

$$(3.6) \qquad\qquad (\alpha - 2)\frac{m}{m-1} + \varepsilon = 0 \qquad (\alpha < 2, \varepsilon > 0).$$

Notice that for any fixed positive integer k_0 we may assume that (3.5) holds for $k \leqslant k_0$ since, without loss of generality, we may assume that $u \leqslant 2^{-k_0}$. We shall choose k_0 later to depend only on m, n.

To pass from k to $k + 1$, set

$$(3.7) \qquad\qquad A_k = \left(\frac{\sigma(R_k)}{R_k^2}\right)^{m/(m-1)}$$

and introduce the function

$$(3.8) \qquad \begin{aligned} \tilde{u}^m(x, t) &= A_k u^m\big(R_k(x - x^0) + x^0, \sigma(R_k)(t - t_0) + t^0\big) \\ &\quad \text{in } \tilde{G} = \{(x, t): |x - x^0| < 1, t^0 - 1 < t < t^0\}. \end{aligned}$$

Then

$$(3.9) \qquad\qquad A_k M_k = A_k R_k^\varepsilon = C,$$

where

$$(3.10) \qquad\qquad C = \overline{C}^{m/(m-1)}.$$

Therefore,

$$(3.11) \qquad\qquad \sup_{\tilde{G}} \tilde{u}^m = A_k M_k = C.$$

By Lemma 2.3 and (3.3), (3.4), if $k < k^*$ then

$$(3.12) \qquad\qquad \fint_{B_1(x^0)} \tilde{u}^m(x, t^{0^-} - 1)\, dx < \nu$$

provided that $c\nu^{1/m}$ is sufficiently large depending only on m, n, C. Thus we can choose ν such that

$$(3.13) \qquad\qquad \frac{\nu}{C} < \frac{1}{2}$$

and then choose c sufficiently large such that (3.12) holds. From now on, c and v are fixed.

Consider the function

$$w(x, t) = \frac{\tilde{u}^m(x, t + t^0)}{A_k M_k} = \frac{\tilde{u}^m(x, t + t^0)}{C}.$$

By (3.12),

$$(3.14) \qquad \fint_{B_1(x^0)} w(x, -1)\, dx \leqslant \frac{v}{C}$$

and, by (3.11),

$$(3.15) \qquad w \leqslant 1 \qquad \text{in } B(x^0, 1) \times (-1, 0).$$

Also,

$$C\Delta w = \Delta \tilde{u}^m = A_k R_k^2 \Delta u^m \geqslant -A_k R_k^2 C_1 = -C_1 R_k^{2-\varepsilon}$$

(since $\Delta u^m \geqslant -C_1$). Noting that

$$\tilde{u}_t = C^{1/m} \frac{1}{m} w^{1/m-1} w_t,$$

we get

$$(3.16) \qquad C\Delta w - C^{1/m} \frac{1}{m} w^{1/m-1} w_t = 0.$$

If $w_t \leqslant 0$, then

$$C\Delta w - C^{1/m} \frac{1}{m} w_t \geqslant C\Delta w \geqslant -C_1 R_k^{2-\varepsilon},$$

whereas if $w_t > 0$, then

$$C\Delta w - C^{1/m} \frac{1}{m} w_t \geqslant C\Delta w - C^{1/m} \frac{1}{m} w^{1/m-1} w_t = 0$$

by (3.15), (3.16). Thus, in both cases,

$$(3.17) \qquad C\Delta w - C^{1/m} \frac{1}{m} w_t \geqslant C_1 R_k^{2-\varepsilon}.$$

Representing w in $B(x^0, 1) \times (-1, 0)$ in terms of Green's function for the parabolic operator in (3.17), we find that

$$\text{if } |x - x^0| < \frac{1}{2}, \qquad -\frac{1}{2^\alpha} < t < 0,$$

then

$$w(x, t) \leq \theta \frac{\nu}{C} + (1 - \theta) + C^2 R_k^{2-\varepsilon},$$

$$\theta = \theta(x, t) \in (\theta_0, 1 - \theta_0), \quad 0 < \theta_0 < 1,$$

where θ_0 is a fixed number independent of k and C_2 is a constant depending only on C_1, m, n, α; we used here the inequalities (3.14), (3.15), and (3.17). We conclude, upon recalling (3.13), that

(3.18) $w(x, t) \leq 1 - \eta,$ for some $\eta > 0,$

provided $k \geq k_0$, where k_0 is such that

$$C_2 R_{k_0}^{2-\varepsilon} < \tfrac{1}{2};$$

η depends only on m, n, η_0.
    We conclude that

$$\sup_{\tilde{G}} \tilde{u}^m \leq A_k M_k (1 - \eta),$$

that is,

$$\sup_{G_{k+1}} u^m \leq M_k (1 - \eta) = 2^{-k\varepsilon}(1 - \eta).$$

It remains to show that

$$2^{-k\varepsilon}(1 - \eta) \leq 2^{-(k+1)\varepsilon},$$

that is,

$$1 - \eta < 2^{-\varepsilon}.$$

In view of (3.6), this inequality reduces to

$$1 - \eta < 2^{(\alpha - 2)m/(m-1)}.$$

We can clearly satisfy this inequality by choosing $\alpha < 2$, $2 - \alpha$ sufficiently small.
    We have thus proved (3.5) for all $k \leq k^*$. It follows that

$$u^m(x, t) \leq \tilde{C}\big(|x - x^0| + |t - t^0|^\alpha\big)^\varepsilon$$

as long as

$$|x - x^0| + |t - t^0|^\alpha \geq 2^{-k^*}, \qquad t < t^0.$$

On the other hand, if

$$|x - x^0| + |t - t^0|^\alpha < 2^{-k^*}, \qquad t < t^0$$

then, by (3.5) with $k = k^*$,

$$u^m(x, t) \leqslant 2^{-k^* \varepsilon} \leqslant \tilde{C} u^m(x^0, t^0).$$

Here $\tilde{C}$ is a generic constant depending only on m, n, η_0.
   We conclude that

$$(3.19) \qquad u^m(x, t) \leqslant \tilde{C} \max \left\{ \left(|x - x^0| + |t - t^0|^\alpha \right)^\varepsilon, u^m(x^0, t^0) \right\}$$

if $t < t^0$. By continuity (recall that u is one of the smooth functions u_j) (3.19) remains true also if $t = t^0$.
   If $M = \sup u$, $c_* = 1/(mM^{m-1})$, then

$$(3.20) \qquad \Delta u^m - c_* \frac{\partial}{\partial t} u^m \geqslant -\frac{Mk}{\eta_0} \qquad \text{if } t \geqslant \eta_0 \quad (\eta_0 > 0).$$

Indeed, the left-hand side is equal to

$$u_t - c_* m u^{m-1} u_t = \left(1 - c_* m u^{m-1} \right) u_t \geqslant -\left(1 - c_* m u^{m-1} \right) \frac{ku}{t}$$

by (2.13) (since $1 - c_* m u^{m-1} \geqslant 0$), and (3.20) thus follows.
   Let $\tilde{\zeta}$ be the solution of

$$(3.21) \qquad \begin{aligned} \Delta \tilde{\zeta} - c_* \tilde{\zeta}_t &= -\frac{Mk}{\eta_0} \qquad \text{in } R^n \times (t^0, \infty), \\ \tilde{\zeta}(x, t^0) &= \tilde{C} \max \left[|x - x^0|^\varepsilon, u^m(x^0, t^0) \right] \text{ in } R^n. \end{aligned}$$

Then (see Problem 1)

$$(3.22) \qquad \tilde{\zeta}(x, t) \leqslant C^* \max \left\{ \left(|x - x^0| + |t - t^0|^{1/2} \right)^\varepsilon, u^m(x^0, t^0) \right\},$$

where C^* is some constant. Hence, by comparison (using (3.19) with $t = t^0$ and (3.21)), the same estimate holds for u^m.
   We have thus proved:

$$(3.23) \qquad \begin{aligned} u^m(x, t) &\leqslant C \max \left\{ \left(|x - x^0| + |t - t^0|^{1/2} \right)^\varepsilon, u^m(x^0, t^0) \right\} \\ &\quad \text{for any pair } (x, t), (x^0, t^0) \text{ in } R^n \times (\eta_0, \infty); \end{aligned}$$

C and ε depend only on m, n, η_0 and an upper bound on $|u|_{L^1}$.

Lemma 3.2.   *If* $(x^i, t^i) \in R^n \times (\eta_0, \infty)$ *for* $i = 0, 1$ *and if*

$$(3.24) \qquad \left[|x^1 - x^0| + |t^1 - t^0|^{1/2} \right]^\varepsilon \leqslant \frac{1}{C} u^m(x^i, t^i)$$

for $i = 0$ *or* $i = 1$, *then*

$$(3.25) \qquad \frac{1}{C} \leqslant \frac{u^m(x^1, t^1)}{u^m(x^0, t^0)} \leqslant C.$$

Proof. Suppose that (3.24) holds for $i = 0$, that is,

$$(3.26) \qquad \left[|x^1 - x^0| + |t^1 - t^0|^{1/2} \right] \leqslant \frac{1}{C} u^m(x^0, t^0).$$

The inequality $u^m(x^1, t^1) \leqslant C u^m(x^0, t^0)$ is then a consequence of (3.23) and (3.26). To prove the first inequality in (3.25) we take in (3.23) (x^0, t^0) to be (x^1, t^1) and (x, t) to be (x^0, t^0). Recalling (3.26), we then obtain

$$u^m(x^0, t^0) \leqslant C u^m(x^1, t^1).$$

Take now any point (x^0, t^0) with $t^0 > 2\eta_0$ and let k_0 be a positive integer such that

$$(3.27) \qquad M^m(k_0 + 1)^{-\varepsilon} \leqslant u^m(x^0, t^0) < M^m k_0^{-\varepsilon},$$

where $M = \sup u$. Define

$$(3.28) \qquad \Sigma_0 = \left\{ (x, t); |x - x^0| + |t - t^0|^{1/2} \leqslant \frac{c}{k_0}, t > \eta_0 \right\},$$

where c is a positive constant to be determined below (independently of k_0).
If $(x, t) \notin \Sigma_0, t > \eta_0$, then by (3.27), (3.23) and Lemma 3.2,

$$u^m(x, t) \leqslant \tilde{C} \left(|x - x^0| + |t - t^0|^{1/2} \right)^\varepsilon,$$

where $\tilde{C}$ depends only on C, c, M^m. Since the same inequality holds also for $u^m(x^0, t^0)$, we obtain

$$(3.29) \qquad |u^m(x, t) - u^m(x^0, t^0)| \leqslant 2\tilde{C} \left(|x - x^0| + |t - t^0|^{1/2} \right)^\varepsilon$$

$$\text{if } (x, t) \notin \Sigma_0, t > \eta_0.$$

To consider the case where $(x, t) \in \Sigma_0$, notice first that by (3.27) and Lemma 3.2,

$$(3.30) \qquad \frac{1}{C} \leqslant \frac{u^m(\bar{x}, \bar{t})}{u^m(x^0, t^0)} \leqslant C$$

provided that

$$\left(|\bar{x} - x^0| + |\bar{t} - t^0|^{1/2}\right)^\varepsilon \leqslant \frac{M^m}{C}(k_0 + 1)^{-\varepsilon}.$$

Thus, in particular, (3.30) holds provided that

$$(3.31) \qquad |\bar{x} - x^0| < \lambda k_0^{-\varepsilon}, \qquad |\bar{t} - t^0| < \lambda^2 k_0^{-2\varepsilon} \qquad \left(\lambda = \frac{M^m}{C}2^{-\varepsilon}\right).$$

Introduce variables

$$x' = \frac{(x - x^0)k_0^\varepsilon}{\lambda}, \qquad t' = \frac{(t - t^0)k_0^{2\varepsilon}}{\lambda^2}$$

and let $v(x', t') = u(x, t)$. Then

$$(3.32) \qquad\qquad\qquad \frac{\partial v}{\partial t} = \nabla(a(x', t')\nabla v)$$

where $a = mu^{m-1}/\lambda$. If

$$|x'| < 1, \qquad |t'| < 1,$$

then (3.31) holds with $(\bar{x}, \bar{t}) = (x, t)$ and, consequently, (3.30) is satisfied. Thus

$$c_1 < a(x', t') < c_2 \qquad \text{if } |x'| < 1, |t'| < 1,$$

where c_i are positive constants (depending only on C, m, λ). By the Nash-deGiorgi estimate [130] we conclude that for some α, $0 < \alpha < 1$,

$$(3.33) \quad |u(x, t) - u(x^0, t^0)| = |v(x', t') - v(0,0)| \leqslant \overline{C}\left(|x'|^\alpha + |t'|^{\alpha/2}\right)$$

$$= \frac{\overline{C}}{\lambda^\alpha}k_0^{\alpha\varepsilon}\left(|x - x^0|^\alpha + |t - t^0|^{\alpha/2}\right)$$

if $|x'| \leqslant \frac{1}{2}$, $|t'| \leqslant \frac{1}{2}$, where $\overline{C}$ is a constant depending only on c_1, c_2.

If $(x, t) \in \Sigma_0$ and $c = \lambda/2$, then $|x'| \leqslant \frac{1}{2}$, $|t'| \leqslant 1/2$, and thus (3.33) holds. Also,

$$|x - x^0|^{\alpha\varepsilon} + |t - t^0|^{\alpha\varepsilon/2} \leqslant 2\left(ck_0^{-1}\right)^{\alpha\varepsilon} = 2c^{\alpha\varepsilon}k_0^{-\alpha\varepsilon}.$$

Hence (3.33) gives, for $(x, t) \in \Sigma_0$,

$$|u(x, t) - u(x^0, t^0)| \leqslant C^*\left(|x - x^0| + |t - t^0|^{1/2}\right)^{\alpha(1-\varepsilon)}$$

with a constant C^* independent of k_0. Combining this with (3.29), the proof of Theorem 3.1 is complete.

We shall now consider the continuity at $t = 0$.

Theorem 3.3. *If u_0 is Hölder continuous, then $u(x, t)$ is continuous at $t = 0$.*

Proof. Denote by $w(x, t)$ the solution to the porous-medium equation given in Section 1, Problem 3. Then for any $c > 0$, $L > 0$,

$$w_{c, L}(x, t) \equiv cw(Lx, c^{m-1}L^2 t)$$

is again a solution, and so is

$$(3.34) \qquad \tilde{w}(x, t) = w_{c, L}(x - y, t + 1)$$

for fixed $y \in R^n$. Notice that $\tilde{w} \leqslant c$ and

$$\tilde{w}(x, 0) = 0 \quad \text{if } |x - y| > \frac{\beta}{L} \qquad \left(\beta = \left[\frac{2mn}{k(m-1)} \right]^{1/2} \right),$$

$$> 0 \qquad \text{if } |x - y| < \frac{\beta}{L}.$$

We shall now prove the continuity of u at $(y, 0)$. Assume first that $u_0(y) > 0$. Then $u_0(x) > c$ if $|x - y| < \beta/L$, for some $c > 0$, $L > 0$. By comparison (Section 2, Problem 1) it follows that

$$\tilde{w}(x, t) \leqslant u(x, t).$$

Consequently,

$$u(x, t) > 0 \qquad \text{if } |x - y| \leqslant \frac{\beta}{2L}, \quad t \leqslant \delta,$$

for some small $\delta > 0$. Thus u satisfies

$$u_t = \nabla(a \nabla u) \qquad \text{in } B_y(R) \times (0, \delta),$$

where $R = \beta/(2L)$, $c_1 \leqslant a \leqslant c_2$ and c_i are positive constants. Applying the Nash–deGiorgi estimate, we deduce that u is Hölder continuous in $B_y(R/2) \times [0, \delta/2]$.

It remains to prove the continuity of u at any point $(y, 0)$ for which $u_0(y) = 0$. For any $\varepsilon > 0$ there is a $\delta > 0$ such that

$$(3.35) \qquad u_0(x) < \varepsilon \qquad \text{if } |x - y| < \delta.$$

Consider the parabolic system:

$$\frac{\partial z}{\partial t} = \Delta z^m \qquad \text{if } |x - y| < \delta, \quad t > 0,$$

$$z(x, 0) = 2\varepsilon \qquad \text{if } |x - y| < \delta,$$

$$z(x, t) = M + \varepsilon \qquad \text{if } |x - y| = \delta, \quad t > 0.$$

This problem has a classical positive solution. We compare it with u_j with $j > 1/\varepsilon$; in view of (3.35) and the inequality $u_j < M + \varepsilon$ on $|x - y| = \delta$,

$$u_j(x, t) < z(x, t) \qquad \text{if } |x - y| < \delta, \quad t > 0.$$

Taking $j \to \infty$ we obtain $u(x, t) \leqslant z(x, t)$ and, in particular,

$$\limsup_{(x, t) \to (y, 0)} u(x, t) \leqslant \limsup_{(x, t) \to (y, 0)} z(x, t) = 2\varepsilon.$$

Since ε is arbitrary, $u(x, t) \to 0$ if $(x, t) \to (y, 0)$, and the proof is complete.

Definition 3.1. Set

$$\Omega = \{(x, t); t > 0, u(x, t) > 0\},$$

$$\Omega(t) = \{x; u(x, t) > 0\},$$

$$\Gamma = \partial\{u > 0\} \cap \{t > 0\},$$

$$\Gamma(t) = \partial\Omega(t);$$

Γ is called the *free boundary*.

Theorem 3.1 implies that Ω is an open set in $R^n \times (0, \infty)$ and $\Omega(t)$ is an open set in R^n. The inequality (2.13) is equivalent to

$$\frac{\partial}{\partial t}(t^k u) \geqslant 0$$

and it therefore implies that

(3.36) $\Omega(t)$ is increasing with t.

If, in particular,

(3.37) $n = 1, u_0(x)$ continuous, $u_0(x) > 0$ if $a < x < b,$
           $u_0(x) = 0$ if $x < a$ or $x > b \quad (-\infty < a < b < \infty),$

then

(3.38) $\Omega(t) = \{\zeta_1(t) < x < \zeta_2(t)\},$

where $(-1)^i \zeta_i(t)$ is a non-decreasing function of t.

PROBLEMS

1. Prove that (3.21) implies (3.22).
    [*Hint*: Represent $\tilde{\zeta}$ in terms of the fundamental solution.]

2. Assume that (3.37) holds and that $u_0(x) \geqslant (b - x)^{\gamma/(m-1)}$ for some $\gamma < 2$
    and $x \in (b - \eta, \eta)$, $\eta > 0$. Prove that $\zeta_2(t) > \zeta_2(0)$ for all $t > 0$.
    [*Hint*: By comparison we may assume that $v_0(x) = (b - x)^\gamma$ for $b - \delta <$
    $x < b$, where $v_0(x) = v(x, 0)$, $v(x, t)$, as in (2.8). For any small $\varepsilon > 0$ let

    $$\tilde{v}(x, t) = \left[\alpha_\varepsilon^2 t + \alpha_\varepsilon(x - x_\varepsilon)\right]^+ \qquad \text{(see Section 1, Problem 5)}$$

    where $y = \alpha_\varepsilon x + \tilde{x}_\varepsilon$ is tangent to $y = v_0(x)$ at x_ε. Then

    $$v(b - \delta, t) = v(b - \delta, 0) + v_t(b - \delta, \tilde{t})t > \tilde{v}(b - \delta, t)$$

    for $0 < t < T_\delta$ and by comparison [see the proof of (2.29)], $\tilde{v}(x, t) < v(x, t)$
    if $b - \delta < x < b$, $0 < t < T_\delta$. Show that $\tilde{v}(0, t) > 0$ if $t > c\varepsilon^{2-\gamma}$ $(c > 0)$
    and take $\varepsilon \to 0$.]

3. Assume that $u_0(x) > 0$ if $x \in G$, $u_0(x) = 0$ if $x \notin G$, where G is a domain
    in R^n with C^2 boundary. If, further,

    (3.39) $u_0(x) \geqslant \{\text{dist}(x, \partial G)\}^{\gamma/(m-1)}$ $(x \in G)$

    for some $\gamma < 2$, then $\overline{\Omega(0)} \subset \Omega(t)$ for all $t > 0$.
    [*Hint*: Use the comparison function in (3.34).]

4. Assume that (3.37) holds and that $u_0(x) \leqslant C(b - x)^{2/(m-1)}$ for some
    $C > 0$ and $x \in (b - \eta, \eta)$, $\eta > 0$. Prove that $\zeta_2(t) = \zeta_2(0)$ for all t suffi-
    ciently small.
    [*Hint*: Use the comparison function in Section 1, Problem 4.]

5. Let the assumptions of Problem 3 be satisfied with (3.39) replaced by

    $$u_0(x) \leqslant C\{\text{dist}(x, \partial G)\}^{2/(m-1)} \qquad (x \in G)$$

    for some $C > 0$. Prove that if G is convex, then $\Omega(t) = \Omega(0)$ for all t
    sufficiently small.

[*Hint*: Use the comparison function of Section 1, Problem 4 where the space variable is in the normal direction to a given point $x^0 \in \partial G$.]

4.   GROWTH AND HÖLDER CONTINUITY OF THE FREE BOUNDARY

From Lemma 2.2 we immediately get:

Lemma 4.1.   *If in Lemma* 2.1 v *satisfies* (2.16) *and* [*instead of* (2.17)] $(x^0, t^0 + \sigma)$ *belongs to the free boundary, then*

$$(4.1) \qquad \fint_{B_R(x^0)} v(x, t^0 + \sigma)\, dx > \frac{cR^2}{\sigma}.$$

This result and Lemma 2.3 will now be used to study the growth and Hölder continuity of the free boundary.

By comparison with a function of the form (3.34) we find [assuming that $u_0(x) > 0$ in some open subset of R^n] that

$$(4.2) \qquad \inf\{|x|, x \in \Gamma(t)\} \geq ct^{k/n} \qquad (c > 0, t > 1).$$

We shall use the notation

$$\sigma(x^0, t^0) = \{(x^0, t); 0 < t < t^0\}.$$

Theorem 4.2.   *For any point* $(x^*, t^*) \in \Gamma$, *either*:

(i)   *Each point of* $\sigma(x^*, t^*)$ *belongs to* Γ, *or*

(ii)   *No point of* $\sigma(x^*, t^*)$ *belongs to* Γ.

Thus if the free boundary contains a vertical segment, then the extension of this segment down to $t = 0$ also belongs to the free boundary. For $n = 1$ this means that once the free boundary began moving, it never stops moving. Problems 2 to 5 of Section 3 give sufficient conditions as to when the free boundary starts moving immediately from $t = 0$. In view of (4.2), the free boundary must begin to move at some finite time.

Proof.   If the assertion is not true, then there exist points $0 < t^1 < t^2 < t^*$ such that

$$(x^*, t^2) \in \Gamma, \qquad (x^*, t^1) \notin \Gamma.$$

It follows that

$$u(x, t^1) = 0 \qquad \text{if } x \in B_R(x^*)$$

for some $R > 0$. Without loss of generality we may assume that $t^* - t^1$ is sufficiently small. Applying Lemma 4.1 with $\sigma = t^2 - t^1$, we get

$$\fint_{B_R(x^*)} v(x, t^2)\, dx \geqslant \frac{cR^2}{t^2 - t^1}.$$

We can apply Lemma 2.3 with $\lambda\sigma = t^* - t^2$, $\nu = 1$, $\sigma = t^2$ provided that $t^* - t^2 \geqslant C(t^2 - t^1)$ for some sufficiently large constant C (which may certainly be assumed). We then obtain the inequality $v(x^*, t^*) > 0$, which contradicts the fact that $(x^*, t^*) \in \Gamma$.

We can actually use the quantitative nature of Lemmas 2.3 and 4.1 to obtain a Hölder rate of growth of the free boundary.

Theorem 4.3.   *Let $(x^*, t^*) \in \Gamma$ $(t^* \geqslant \eta_0 > 0)$ be such that $\sigma(x^*, t^*)$ does not contain any points of Γ. Then*

$$(4.3) \quad u(x, t) = 0 \quad\quad if\, |x - x^*| \leqslant C(t^* - t) \quad\quad (t^* - h^* < t < t^*)$$

and

$$(4.4) \quad u(x, t) > 0 \quad\quad if\, |x - x^*| \leqslant C(t - t^*) \quad\quad (t^* < t < t^* + h^*)$$

for some positive constants C, γ, h^*.

Proof.   Fix τ such that $0 \leqslant \tau < t^*$ and set $h = t^* - \tau$. Then $u(x, \tau) = 0$ if $x \in B_R(x^*)$, for some $R > 0$. Let

$$t^1 = \tau + (1 - \lambda)h = t^* - \lambda h,$$

where $\lambda \in (0, 1)$ will be determined later on. Suppose that

$$(4.5) \quad\quad\quad\quad\quad\quad \mathrm{dist}\,(x^*, \Gamma(t^1)) < \alpha R,$$

where $\alpha \in (0, 1)$ is to be determined later. Applying Lemma 4.1, we deduce that

$$\fint_{B_{(1-\alpha)R}(x^*)} v(x, t^1)\, dx \geqslant \frac{c(1 - \alpha)^2 R^2}{(1 - \lambda)h} \quad\quad (c > 0),$$

where $x^1 \in \Gamma(t^1)$ is such that $|x^1 - x^*| \leqslant \alpha R$. Hence

$$(4.6) \quad\quad\quad \fint_{B_R(x^*)} v(x, t^1)\, dx \geqslant \frac{c(1 - \alpha)^2 R^2}{(1 - \lambda)h}(1 - \alpha)^n.$$

Given any large C we can choose α, λ so that

$$(4.7) \qquad \frac{c(1-\alpha)^2 R^2}{(1-\lambda)h}(1-\alpha)^n \geq C\frac{R^2}{\lambda h}.$$

But if C is sufficiently large (depending on m, n, η_0), then Lemma 3.3 implies that $u(x^*, t^*) > 0$, which is of course impossible. Thus, if (4.7) holds, then (4.5) cannot be valid, that is,

$$(4.8) \qquad \operatorname{dist}\left(x^*, \Gamma(t^1)\right) \geq \alpha R.$$

Notice that (4.7) is implied by

$$(1-\alpha)^{n+2} \geq C(1-\lambda)$$

with another large constant C, provided that $\lambda > \frac{1}{2}$. Taking $\alpha = \lambda^\gamma$, this reduces to

$$(4.9) \qquad (1-\lambda^\gamma)^{n+2} \geq C(1-\lambda).$$

This inequality is valid for some λ near 1 and γ sufficiently large (for instance, $\lambda = 1 - 1/k$, $\gamma = k$, k sufficiently large).

From (4.8) it follows that

$$u(x, t^1) = 0 \qquad \text{if } x \in B_{\alpha R}(x^*), \quad \alpha = \lambda^\gamma.$$

We can now repeat the previous argument with R replaced by αR, τ replaced by t^1, and h replaced by λh. We deduce that

$$\operatorname{dist}\left(x^*, \Gamma(t^2)\right) \geq \alpha^2 R, \qquad t^2 = t^* - \lambda^2 h.$$

Proceeding step by step, we get

$$(4.10) \qquad \operatorname{dist}\left(x^*, \Gamma(t)\right) \geq \lambda^{\gamma k} R \qquad \text{if } t = t^* - \lambda^k h.$$

By varying h in an interval $\bar{h} < h < h/\lambda$ (the corresponding values of R are bounded from below by a positive number which we again designate by R) and taking $k = 1, 2, \ldots$, we find that (4.10) holds for all t, $t - \bar{h} < t < t^*$; more precisely,

$$\operatorname{dist}\left(x^*, \Gamma(t)\right) \geq c(t^* - t)^\gamma \qquad (c > 0),$$

which gives (4.3).

To prove (4.4), notice that if a point (x, t) of Γ satisfies

$$(4.11) \qquad |x - x^*| < C(t - t^*)^\gamma, \qquad t^* < t < t^* + \hat{h} \quad (\hat{h} \text{ small}),$$

then the segment $\sigma(x, t)$ intersects the interior of the set for which (4.3) holds. Thus $\sigma(x, t)$ contains points not in Γ. By Theorem 4.2 it follows that $\sigma(x, t)$ does not intersect Γ. We can therefore apply the proof of (4.3) with (x^*, t^*) replaced by (x, t) and conclude that $u(x^*, t^*) \neq 0$. Since, however, $u(x^*, t^*) = 0$, we get a contradiction, which proves that the points (x, t) satisfying (4.11) do not belong to Γ. It follows that on the set (4.11) either $u \equiv 0$ or $u > 0$. The first possibility contradicts the assumption that $(x^*, t^*) \in \Gamma$. Thus $u > 0$ on the set (4.11), which completes the proof of (4.4).

Remark 4.1. The constants C, γ in (4.3), (4.4) depend on the choice of τ, R, but are independent of (x^*, t^*).

We shall now assume:

$$G \text{ is a bounded domain in } R^n \text{ with } C^2 \text{ boundary,}$$

(4.12) $u_0(x) \geq c_0 \{ \text{dist}(x, \partial G) \}^\mu$ if $x \in G$, where $c_0 > 0$,

$\mu < 2, u_0(x) = 0$ if $x \notin G$.

By Problem 3 of Section 3 $\Omega(t) \supset \overline{\Omega(0)}$ for all $t > 0$. We can therefore choose τ, R in the proof of Theorem 4.3 to be independent of (x^*, t^*) provided $t^* \geq \eta_0$ for a fixed $\eta_0 > 0$. By Remark 4.1 we conclude that

(4.13) $C \text{ and } \gamma \text{ are independent of } (x^*, t^*).$

Theorems 4.2 and 4.3 imply (when (4.12) holds) that the free boundary is given by a function

$$t = S(x)$$

and, if $S(x^*) \geq \eta_0$,

(4.14) $|S(x) - S(x^*)| \leq C |x - x^*|^{1/\gamma}.$

In view of (4.2), $S(x) < \infty$ for all $x \in R^n$. We summarize:

Theorem 4.4.   *If (4.12) holds, then:*

   (i)   *The free boundary is given by a function $y = S(x)$, where $S(x)$ is finite valued and uniformly Hölder continuous in every set $\{x; S(x) > \eta_0\}$, $\eta_0 > 0$.*

   (ii) *The interior of $\Gamma(t + s)$ contains $(Cs^{1/\gamma})$-neighborhood of the interior of $\Gamma(t)$ provided that $0 < s < 1$, $t > \eta_0$ $(\eta_0 > 0)$, where C, γ are positive constants depending only on η_0 and $|u_0|_{L^1}$.*

We shall complement (ii) by proving:

Theorem 4.5. $\Gamma(t + s)$ *is contained in $(Cs^{1/2})$-neighborhood of $\Gamma(t)$, where $0 < s < 1$, $t \geq \eta_0$ $(\eta_0 > 0)$, and C is a constant depending only on η_0 and $|u_0|_{L^1}$.*

Proof. Let $u(x^0, t^0) = 0$, $\text{dist}(x^0, \Gamma(t^0)) = a$ and consider the function

$$U(r, t) = \left\{ \lambda \left[\alpha^2(t - t^0) + \alpha(r - b) \right]^+ \right\}^{1/(m-1)} \qquad (b > 0, \alpha > 0)$$

for $t^0 < t < t^0 + h$, where

$$r = |x - x^0|, \qquad h < t^0 + \frac{b}{a}.$$

$U(r, t) > 0$ if and only if $r > b - \alpha(t - t^0)$. As in the proof of Lemma 2.2,

$$\Delta U^m \leqslant U_t$$

provided that

$$(4.15) \qquad \lambda \left[(m - 1)(n - 1) \frac{\alpha(t - t^0) + r - b}{r} + 1 \right] \leqslant 1$$

We choose $b < a$ and

$$(4.16) \qquad \alpha(a - b) = \frac{N_1}{\lambda} \equiv N, \qquad \text{where } N_1 \geqslant u^{m-1} + 1.$$

Then on $t = t^0$, $r \leqslant a$,

$$u = 0 \leqslant U$$

and on $t^0 < t < t^0 + h$, $r = a$,

$$U \geqslant \left[\lambda \alpha(a - b) \right]^{1/(m-1)} = N_1^{1/(m-1)} > u.$$

By comparison [see the proof of (2.29)], we then get $u \leqslant U$ in $B_a(x^0) \times (t_0, t_0 + h)$; in particular,

$$(4.17) \qquad u(x, t^0 + h) \leqslant U(x, t^0 + h).$$

We now define b by

$$(4.18) \qquad a - b = \sqrt{Nh}.$$

Then α, defined by (4.16), is given by

$$(4.19) \qquad \alpha = \frac{\sqrt{N}}{\sqrt{h}}.$$

The free boundary of U is given by $r = b - a(t - t^0)$, and at $t = t^0 + h$,

$$r = b - \alpha h = a - \sqrt{Nh} - \alpha h = a - 2\sqrt{Nh} .$$

Choosing

$$(4.20) \qquad\qquad h = \frac{a^2}{16N}$$

so that

$$(4.21) \qquad\qquad 2\sqrt{Nh} = \frac{a}{2},$$

we conclude, upon recalling (4.19), that

$$u(x, t^0 + h) = 0 \qquad \text{if } |x - x^0| < b - \alpha h = \frac{a}{2}.$$

This establishes the assertion of the theorem, provided that λ can be chosen (independently of a) to satisfy (4.15). It suffices to choose λ such that $\lambda \leqslant \frac{1}{2}$ and

$$(4.22) \qquad\qquad \lambda \frac{\alpha h + (a - b)}{r} \leqslant \frac{1}{2(m - 1)(n - 1)} = \gamma.$$

Since $r \geqslant b - \alpha h = a/2$, (4.22) is a consequence of

$$(4.23) \qquad\qquad \frac{2}{a} \lambda 2\sqrt{Nh} \leqslant \gamma;$$

here we have used (4.18), (4.19). In view of (4.21), (4.23) is equivalent to $\lambda \leqslant \gamma$. Thus we choose $\lambda = \min(\gamma, \frac{1}{2})$.

Remark 4.2.   If (4.12) holds, then Theorem 4.5 implies that

$$(4.24) \qquad\qquad |S(x^1) - S(x^2)| \geqslant c |x^1 - x^2|^2 \qquad (c > 0)$$

if $|x^1 - x^2|$ is sufficiently small and x^1, x^2 do not belong to $\overline{G}$.

We conclude this section with a theorem on the asymptotic behavior of $u(x, t)$ as $t \to \infty$. The special solution in Section 1, Problem 3 will play a basic role. We scale it by setting, for any $L > 0$,

$$(4.25) \qquad V_L(r, t) = L^{1/(m-1)} \frac{1}{(Lt)^k} G\left(\frac{r}{(Lt)^{k/n}}\right), \qquad r = |x|,$$

where

$$G(s) = \left\{ \left[\beta^2 - c^2 s^2 \right]^+ \right\}^{1/(m-1)}, \qquad c^2 = \frac{k(m-1)}{2mn}$$

and β is a positive constant such that $\int_{R^n} G(|x|) \, dx = 1$. Notice that

(4.26)
$$\int_{R^n} V_L(r, t) \, dx = L^{1/(m-1)}$$

and that the support of the function $r \to V_L(r, t)$ is given by

$$\left\{ r \le \frac{(Lt)^{1/n}}{c/\beta} \right\}.$$

Also, for any $x^0 \in R^n$, τ real,

(4.27)
$$t^k \left[V_L(|x - x^0|, t + \tau) - V_L(|x|, t) \right] \to 0$$

if $t \to \infty$, uniformly with respect to x, $x \in R^n$.

Theorem 4.6. *Let u be the solution asserted in Theorem 1.7, and set $I = \int_{R^n} u_0(x) \, dx$. Then*

(4.28)
$$t^k |u(x, t) - V_{L_0}(r, t)| \to 0 \qquad \text{as } t \to \infty,$$

uniformly with respect to x in any set $\{|x| < Ct^{k/n}\}$, $C > 0$, where $L_0 = I^{m-1}$.

Thus for large times the gas behaves as if it were initially concentrated at the origin.

The proof is outlined in the following problems.

PROBLEMS

1. For fixed $L > 0$, $t > 0$, if ε is positive and sufficiently small, then

$$V_L(r, t - \varepsilon) > V_L(r, t) \qquad \text{for } 0 < r < \theta \frac{(Lt)^{k/n}}{c/\beta},$$

$$V_L(r, t - \varepsilon) < V_L(r, t) \qquad \text{for } \theta \frac{(Lt)^{k/n}}{c/\beta} < r < \frac{(Lt)^{k/n}}{c/\beta},$$

where $\theta = \theta_\varepsilon \to \theta_0$ as $\varepsilon \to 0$, $\theta_0 = \sqrt{(m-1)k} \in (0, 1)$.

2. In Problems 2 to 10 it is assumed that $u_0 \in C_0(R^n)$ and $u_0(0) > 0$. Show that

(4.29) $$V_{L_1}(r, t + \tau_1) < u(x, t) < V_{L_2}(r, t + \tau_2)$$

for some $L_i > 0$, $\tau_i > 0$.

3. In proving Theorem 4.6, we may assume that for any $L > 0$, τ real, $t > 0$, $t + \tau > 0$,

$$u(x, t) \not\equiv V_L(r, t + \tau) \qquad (x \in R^n).$$

4. For fixed $t > 0$, denote by Σ_t the set of points (L, τ) such that $L > 0$, $\tau \geqslant 0$ and $u(x, t) \geqslant V_L(r, t + \tau)$, and let

$$L(t) = \sup \{L; (L, \tau) \in \Sigma_t\}.$$

Prove that there exist $(L^*, \tau^*) \in \Sigma_t$ such that $L(t) = L^*$ and $\tau^* \leqslant C(t + 1)$, where C is a constant independent of t.
[*Hint*: By (4.29), $L_1 \leqslant L(t) \leqslant I^{m-1}$.]

5. $L(t)$ is monotone nondecreasing and there exists a sequence $t_j \uparrow \infty$ such that

(4.30) $$L(t_j) < L(t_{j+1}).$$

[*Hint*: If $L = L(t_0)$, there exist $\xi > 0$ such that $u(x, t_0 + \xi) \not\equiv V_L(r, t_0 + \tau + \xi)$ for

$$r < \frac{L^{k/n}(t_0 + \tau + \xi)^{k/n}}{c/\beta} \equiv r_0$$

(using Problem 3); hence also if $r < \tilde{\theta} r_0$, where $0 < \tilde{\theta} < 1$, $1 - \tilde{\theta}$ sufficiently small. Deduce that

$$u(x, t_0 + \eta) > V_L(r, t_0 + \tau + \eta) \qquad \text{if } r < \theta_1 r_0$$

for any $\eta > \xi$, $\eta - \xi$ small, $\theta_0 \in (0, 1)$. By Problem 1,

$$u(x, t_0 + \eta) > V_L(r, t_0 + \tau + \eta - \varepsilon) \qquad (\varepsilon > 0)$$

in the support of $r \to V_L(r, t_0 + \tau + \eta - \varepsilon)$ if ε is small enough, and thus

$$u(x, t_0 + \eta) > V_{L^1}(r, t_0 + \tau + \eta - \varepsilon) \qquad \text{for some } L^1 > L.]$$

6. If $u_\lambda(x, t) = \lambda^n u(\lambda x, \lambda^{n/k} t)$, $\lambda > 0$, then for any sequence $\lambda_j^* \uparrow \infty$ there is a subsequence $\lambda_j \uparrow \infty$ such that

$$u_{\lambda_j}(x, t) \to w(x, t)$$

uniformly in compact sets of $R^n \times (0, \infty)$.

7. Define $L_0 = \lim_{t \to \infty} L(t)$. Then, for any $t > 0$,

$$(4.31) \qquad w(x, t) = V_{L_0}(r, t + \tilde{\tau}_t), \qquad \tilde{\tau}_t \leqslant Ct.$$

[*Hint*: $u(x, t) \geqslant V_{L(t)}(r, t + \tau_t)$ implies that

$$u_\lambda(x, t) \geqslant \lambda^n V_{L(\lambda^{n/k} t)}(\lambda r, \lambda^{n/k} t + \tau_{\lambda, t})$$

with

$$\tau_{\lambda, t} \leqslant \frac{1}{\lambda^{n/k}} \tau_{\lambda^{n/k} t} \leqslant Ct,$$

giving " $\geqslant$ " in (4.31). If " $=$ " is not here, then (see problem 5)

$$w(x, t + \eta) > V_{\tilde{L}}(r, t + \tilde{\tau}_t + \eta - \varepsilon)$$

on the support of $r \to V_{\tilde{L}}$, for some $\tilde{L} > L_0$, and by Problem 6,

$$L\left(\lambda_i^{n/k}(t + \eta)\right) > \tilde{L}$$

if λ_i is large enough.]

8. Show that $\tilde{\tau}_t = \tilde{\tau}_0$.

9. Show that $\tilde{\tau}_0 = 0$.
[*Hint*: For any small $\delta > 0$, $u_\lambda(0, \delta) \geqslant C/\delta^k$ by (4.29) ($C > 0$). Also,

$$u_{\lambda_i}(0, \delta) \to w(0, \delta) = V_{L_0}(0, \delta + \tilde{\tau}_0).]$$

10. $L_0 = I^{m-1}$ and $u_\lambda(x, t) \to V_{L_0}(r, t)$ as $\lambda \uparrow \infty$, uniformly for (x, t) in compact subsets of $R^n \times (0, \infty)$.

11. Complete the proof of theorem 4.6 when support u_0 is bounded.
[*Hint*: Work with $u(x - x^0, t + \varepsilon)$, using (4.27).]

12. Prove Theorem 4.6 for general u_0 in $L^1(R^n)$.
[*Hint*: Let $u_0^N(x) = u_0(x)$ if $|x| < N$, $u_0^N(x) = 0$ if $|x| > N$, and denote by $u^N(x, t)$ the solution with initial data u_0^N. Set $u_\lambda^N(x, t) =$

$\lambda^n u^N(\lambda x, \lambda^{n/k} t)$. By Corollary 1.9,

$$u_\lambda(x, t) \leqslant \frac{C}{t^k} I^{2k/n}$$

and, by Theorem 3.1, for any sequence $\lambda_j^* \uparrow \infty$ there is a subsequence $\lambda_j \uparrow \infty$ such that $u_{\lambda_j}(x, t) \to w(x, t)$ uniformly in compact subsets of $R^n \times (0, \infty)$. By Problem 11, $u_k^N(x, t) \to V_{L_N}(r, t)$ as $t \to \infty$, where $L_N \uparrow L_0 = I^{m-1}$. Also, $u \geqslant u^N$ implies that $w \geqslant V_{L_N}$ and thus $w \geqslant V_{L_0}$. Use the relation

$$\int w(x, t)\, dx \leqslant \lim\inf \int u_{\lambda_i}(x, t)\, dx = I. \Big]$$

5.   THE DIFFERENTIAL EQUATION ON THE FREE BOUNDARY

In this section we shall prove that the rate of change (in space) of the pressure at the free boundary is equal to the rate of growth (in time) of the free boundary.

We fix a point $y \notin \overline{\Omega(\tau)}$ ($\tau > 0$) and set

$$(5.1) \qquad h(t) = \max\{R;\, v(x, t) \equiv 0 \text{ in } B_R(y)\};$$

that is, $h(t)$ is the distance from y to $\Omega(t)$.

By Theorem 4.5,

$$(5.2) \qquad h(t) > 0 \qquad \text{in some interval } 0 < t < \tau + \delta_0 (\delta_0 > 0).$$

We introduce the function

$$(5.3) \qquad \gamma_t(\varepsilon) = \max_{x \in B_{h(t)+\varepsilon}(y)} \frac{v(x, t)}{\varepsilon} \qquad (\varepsilon > 0, 0 < t < \tau + \delta_0)$$

and set

$$(5.4) \qquad \gamma_t(t) = \lim_{\varepsilon \to 0} \gamma_t(\varepsilon)$$

if the limit exists.

We shall prove in this section the following theorem.

Theorem 5.1.   (*i*) *The limit in* (5.4) *exists,* (*ii*) $h'(t + 0)$ *exists, and* (*iii*) *the following relation holds*:

$$(5.5) \qquad \gamma_t(0) = -h'(t + 0) \qquad (0 < t < \tau + \delta_0).$$

We shall divide the proof into several lemmas. It is clearly sufficient to establish the assertions of Theorem 5.1 at the point $t = \tau$. For simplicity we take $y = 0$. We also set

$$R = h(\tau), \qquad \gamma(\varepsilon) = \gamma_\tau(\varepsilon), \qquad \gamma = \gamma_\tau(0).$$

Thus

(5.6)
$$\gamma(\varepsilon) = \max_{x \in B_{R+\varepsilon}} \frac{v(x, \tau)}{\varepsilon},$$

(5.7)
$$\gamma = \lim_{\varepsilon \to 0} \gamma(\varepsilon) \qquad \text{if the limit exists.}$$

Lemma 5.2. *The limit in (5.7) exists.*

Proof. Suppose that $n \geqslant 3$. Consider the functions

(5.8) $\quad M(r) = \dfrac{R^{n-1}}{n-2}(R^{2-n} - r^{2-n}), \qquad r = |x|,$

$$w(r) = \varepsilon\gamma(\varepsilon)\frac{M(r)}{M(R+\varepsilon)} + C(R + \varepsilon - r)(r - R) \qquad (C > 0)$$

for $R < r < R + \varepsilon$. Then

$$\Delta w < -C \qquad \text{if } \varepsilon \text{ is sufficiently small, say if } \varepsilon < \frac{R}{n}.$$

Also,

$$w(r) = 0, \qquad v(x, \tau) = 0 \text{ on } \partial B_R,$$

$$w = \varepsilon\gamma(\varepsilon) \geqslant v(x, \tau) \text{ on } B_{R+\varepsilon} \qquad [\text{by } (5.6)].$$

In view of (2.15),

$$\Delta w < \Delta v \qquad \text{in } R < r < R + \varepsilon$$

provided that the constant C is suitably chosen (namely, $C > 2k/\tau$). Applying the maximum principle, we conclude that

(5.9)
$$v(x, \tau) \leqslant w(r) \qquad \text{if } R < r < R + \varepsilon.$$

Taking the maximum when $x \in B_{R+\delta}$, where $0 < \delta < \varepsilon$, we get

$$\delta\gamma(\delta) \leqslant \varepsilon\gamma(\varepsilon)\frac{M(R+\delta)}{M(R+\varepsilon)} + C(\varepsilon - \delta)\delta.$$

Thus

$$(5.10) \qquad \frac{\delta\gamma(\delta)}{M(R+\delta)} \leq \frac{\varepsilon\gamma(\varepsilon)}{M(R+\varepsilon)} + \frac{C(\varepsilon-\delta)\delta}{M(R+\delta)}.$$

Since $M(R) = 0$, $M'(R) = 1$,

$$\frac{C(\varepsilon-\delta)\delta}{M(R+\delta)} < 2C(\varepsilon-\delta)$$

and, consequently, (5.10) implies that the function

$$(5.11) \qquad g(\varepsilon) = \frac{\varepsilon\gamma(\varepsilon)}{M(R+\varepsilon)} + 2C\varepsilon \text{ is monotone increasing.}$$

Consequently, $\lim_{\varepsilon \to 0} g(\varepsilon)$ exists, and thus also the limit in (5.7) exists.

For $n = 2$ the proof is the same, except that $M(r)$ is now defined as

$$M(r) = R\log\frac{r}{R}.$$

Take a sequence $\varepsilon_m \downarrow 0$ and points x^m in $B_{R+\varepsilon_m}$ such that

$$(5.12) \qquad \frac{v(x^m, \tau)}{\varepsilon_m} = \gamma(\varepsilon_m), \qquad x^m \to x^0 \in \partial B_R.$$

For any $0 < \mu < 1$ introduce the cone

$$K_\mu = \{x; (x - x^0) \cdot x^0 \geq \mu |x - x^0|\}.$$

K_μ is a cone with vertex x^0; it lies in the exterior of B_R (except for its vertex), and K_μ increases, as $\mu \downarrow 0$, to a half space whose boundary is tangent to ∂B_R at x_0.

Lemma 5.3. *For any $\mu > 0$ there exists a function $g_\mu(t)$ satisfying*

$$(5.13) \qquad \frac{g_\mu(t)}{t} \to 0 \qquad \text{if } t \downarrow 0,$$

such that

$$(5.14) \qquad v(x, \tau) - (r - R)\gamma = g_\mu(r - R) \qquad \text{if } x \in K_\mu.$$

Proof. Since

$$\frac{v(x, \tau)}{r - R} \leq \gamma(r) \leq \gamma + o(1),$$

we get

(5.15) $$v(x, \tau) - (r - R)\gamma \leqslant o(1)(r - R).$$

Thus it remains to establish the lower bound

(5.16) $$v(x, \tau) - (r - R)\gamma \geqslant \eta_\mu(r - R)$$

where

$$\frac{\eta_\mu(t)}{t} \to 0 \qquad \text{if } t \downarrow 0.$$

Fix a point $z \in K_\mu$ and set

$$|z| = \rho_1, \qquad \delta = \rho_1 - R.$$

We shall prove that an inequality of the form

(5.17) $$\frac{v(z, \tau)}{\delta} \leqslant \gamma - c$$

cannot hold if c is a fixed positive constant and δ is arbitrarily small; this will clearly establish (5.16).

We shall work with the scaled solutions

$$v_\delta(x, t) = \frac{1}{\delta} v\left(x^0 + \delta(x - x^0), \tau + \delta t\right) \qquad (0 < \delta < 1)$$

of (2.9). By (5.15),

(5.18) $$v(x, \tau) \leqslant C_0 \operatorname{dist}(x, B_R).$$

We compare $v(x, t)$ for $\tau < t < \tau + \sigma_1$ $(\sigma_1 < 1)$, $|x| < R_1$ $(R < R_1 < R + 1)$ with the radial supersolution [see (2.25)]

(5.19) $$V(r, t) = \lambda\left[C_1^2(t - \tau) + C_1(r - R)\right]^+ \qquad (r = |x|),$$

where $C_1\lambda = 2C_0$, $C_1\sigma_1 < \min(1, R/2)$ and $\lambda = \lambda(m, n)$ is sufficiently small. In view of (5.18),

$$V(r, \tau) \geqslant v(x, \tau) \qquad \text{if } |x| < R_1$$

with strict inequality on $|x| = R_1$, so that

$$V(r, t) > v(x, t) \qquad \text{if } |x| = R_1, \quad \tau < t < \tau + \sigma_1$$

for a suitably chosen small σ_1. Hence, by comparison,

$$v(x, t) \leq \lambda \left[C_1^2(t - \tau) + C_1(|x| - R) \right]^+ .$$

It follows that

$$v_\delta(x, t) \leq \lambda \left(C_1^2 t + C_1 |x - x^0| \right)^+ .$$

Recalling (2.14) we can extend this inequality to $t < 0$ (with a different constant C_1). Thus, for any $A > 0$,

(5.20) $v_\delta(x, t) \leq C$ when $|x| \leq A, \quad |t| \leq A,$

C is a constant independent of δ.

Notice also that

(5.21) $\Delta v_\delta \geq -C\delta.$

With these two inequalities at hand, we can now apply the proof of Theorem 3.1 and deduce that

(5.22) $v_\delta(x, t)$ is uniformly Hölder continuous in
bounded sets $|x| \leq A, |t| \leq A$, with coefficient
and exponent independent of δ.

Set

$$0^* = x^0 - \frac{x^0}{\delta}, \qquad z^* = x^0 + \frac{z - x^0}{\delta}$$

and denote by Σ', Σ'' the spheres with center 0^* and radii R/δ and $(R/\delta) + 1$, respectively. Then $x^0 \in \Sigma'$ and $z^* \in \Sigma''$. In general, when x varies in Σ' (Σ''), $x^0 + \delta(x - x^0)$ varies in ∂B_R ($\partial B_{R+\delta}$). When $x^0 + \delta(x - x^0)$ varies in the cone K_μ, x varies in the same cone.

Denote by G the domain bounded by Σ', Σ''. Denote by Q_0 a cylinder whose axis coincides with the axis of K_μ and which contains $K_\mu \cap G$. We can take the diameter of Q_0 to be a positive constant A_μ depending only on μ ($A_\mu \to \infty$ if $\mu \to 0$).

Set $Q = Q_0 \cap G$.

For any point x and a set B, write

$$d(x, B) = \text{dist}(x, B).$$

Then (5.15) gives, for $x \in Q$,

(5.23) $v_\delta(x, 0) \leq \gamma d(x, \Sigma') + \eta_0(\delta) d(x, \Sigma'),$
$$0 \leq \eta_0(\delta) \to 0 \quad \text{if } \delta \to 0.$$

Consider the harmonic function (we take, for definiteness, $n \geq 3$)

$$k(x) = \frac{(\gamma + \eta(\delta))\Lambda^{n-1}}{n-2}(\Lambda^{2-n} - \rho^{2-n}) \qquad \left(\Lambda = \frac{R}{\delta}\right),$$

where

$$\eta(\delta) = \eta_0(\delta) + C_0\delta, \qquad \rho(x) = |x - 0^*|, \qquad C_0 > 0.$$

It satisfies, on Σ',

$$k = 0, \qquad \frac{\partial k}{\partial \rho} = \gamma + \eta(\delta), \qquad \frac{\partial^2 k}{\partial \rho^2} \leq C\delta.$$

Recalling (5.23), we conclude that, provided C_0 is large enough,

(5.24) $\qquad k(x) \geq \max\{v_\delta(x,0), \gamma d(x, \Sigma')\} \qquad$ in Q.

Consider the function

(5.25) $\qquad \tilde{k}(x) = k(x) + C_0\delta(\Lambda + 1 - \rho)(\rho - \Lambda) \qquad (C_0 > 0).$

If δ is small enough, then $\Delta\tilde{k} < -C_0\delta$. Choosing C_0 sufficiently large and recalling (5.21), we get

(5.26) $\qquad \Delta\tilde{k}(x) < \Delta v_\delta(x,0) \qquad$ in Q.

Clearly also,

(5.27) $\qquad \tilde{k}(x) \geq k(x) \geq \max\{v_\delta(x,0), \gamma d(x, \Sigma')\} \qquad$ in Q,

by (5.24).

Suppose now that (5.17) holds; we shall derive a contradiction. From (5.17) we have

$$v_\delta(z^*, 0) < \gamma - c$$

and, by (5.22), if $|x - z^*| < c_0$, then

$$v_\delta(x,0) < \gamma - c - C_1 c_0^\alpha \qquad (C_1 > 0, \alpha > 0).$$

Since $|x - z^*| < c_0$ implies that $d(x, \Sigma') > 1 - c_0$, we then obtain, using (5.27),

$$\tilde{k}(x) > \gamma(1 - c_0) = \gamma - \gamma c_0 > v_\delta(x,0) = \frac{c}{2} \qquad \text{if } |x - x^*| < c_0$$

provided that c_0 is small enough.

Recalling (5.26), (5.27) we can apply the maximum principle to the function $w(x) = \tilde{k}(x) - v_\delta(x, 0)$ in Q and deduce that

$$\frac{\partial w}{\partial \nu} \geq C_2 \qquad \text{on } \partial Q \cap \Sigma' \quad (C_2 > 0)$$

where ν is the inward normal; more precisely,

(5.28) $$w(x) \geq C_2(\rho - \Lambda) \qquad \text{in } Q.$$

It follows that

$$v_\delta(x, 0) \leq k + C_0\delta(\Lambda + 1 - \rho)(\rho - \Lambda) - w$$

$$\leq (\gamma + \eta(\delta) + C_0\delta - C_2)(\rho - \Lambda) \qquad \text{in } Q.$$

Taking in particular $x = x^0 + \delta(x^m - x^0)$, we get

$$v(x^m, \tau) \leq (\gamma + \eta(\delta) + C_0\delta - C_2)d(x^m, \partial B_R)$$

for all m sufficiently large. Thus, by (5.12),

$$\gamma(\varepsilon_m) < \gamma + \eta(\delta) + C_0\delta - C_2,$$

which is impossible for m sufficiently large. This completes the proof of the lemma.

Lemma 5.4. *The following inequality holds*:

(5.29) $$\liminf_{t \downarrow \tau} \frac{|h(t) - h(\tau)|}{t - \tau} \geq \gamma.$$

Proof. Take the same $v_\delta(x, t)$ as before, where δ is any small positive number. In view of (5.22), any sequence $\delta_m \downarrow 0$ then has a subsequence $\delta_{m'}$ such that

(5.30) $$v_\delta \to v_0 \text{ uniformly in } (x, t), \quad |x| \leq A, \quad |t| \leq A$$
$$\text{for any } A > 0 \quad (\delta = \delta_{m'} \downarrow 0),$$

and $v_0(x, t)$ is a solution of (2.9). By Lemma 5.3,

$$v_\delta(x, 0) = \gamma\frac{\delta(\tilde{\rho} - R)}{\delta} + \frac{g_\mu(\delta(\tilde{\rho} - R))}{\delta},$$

where $\tilde{\rho} = r(x^0 + \delta(x - x^0))$; the last term tends to zero as $\delta \to 0$, whereas $\tilde{\rho} \to -x_n$ if the axis of the cones K_μ is taken to be the negative x_n-axis. It follows that

$$v_0(x, 0) = -\gamma x_n.$$

Let B be a ball with radius R_0 lying in $x_n < 0$ with ∂B tangent to $x_n = 0$ at the origin. We shall compare v_0 with a suitable solution $\tilde{w}$ as in (3.34); we denote the corresponding pressure function by W. We choose $\tilde{w}$ and R_0 such that the support of $x \to \tilde{w}(x, 0)$ coincides with B and

$$\frac{\partial W}{\partial x_n} = \frac{\partial v_0}{\partial x_n} \qquad \text{at the origin.}$$

Since $W(x, 0)$ is concave, whereas $v_0(x, 0)$ is linear,

$$W \leq v_0 \qquad \text{on } t = 0.$$

We can now compare both solutions in a cylinder $|x| < A$, $t > 0$ with $A > R_0$. We conclude that $W \leq v_0$. Hence, if we denote by $h_W(t)$, $h_0(t)$ the functions $h(t)$ corresponding to W and v_0, respectively, with respect to any point y_0 lying on the positive x_n-axis, then

$$(5.31) \qquad h_0(s) \leq h_W(s).$$

One can check that

$$(5.32) \qquad h_W(s) = h_W(0) - \gamma s + o(s^2) \qquad (s \to 0).$$

From (5.30) it follows that

$$h_\delta(s) \leq h_0(s) + \sigma(\delta), \qquad \sigma(\delta) \to 0 \text{ if } \delta = \delta_{m'} \to 0$$

provided that $h_\delta(s)$ (defined with respect to v_δ) and $h_0(s)$ are computed with respect to the same center y_0; $\sigma(\delta)$ converges to zero independently of the choice of y_0, if, say, $|y_0| > y_* > 0$. From (5.31), (5.32) we therefore get

$$(5.33) \qquad h_\delta(s) \leq h_\delta(0) - \gamma s + \sigma(\delta) + Cs^2.$$

We have thus proved that for any sequence δ_m there is a subsequence $\delta_{m'}$ such that (5.33) holds for all $\delta = \delta_{m'}$; C is independent of the sequence. It follows that (5.33) holds for all δ sufficiently small.

From the definition of v_δ we see that

$$(5.34) \qquad h_\delta(t) = \frac{1}{\delta} h(\tau + \delta t)$$

provided that y_0 (above) is suitably chosen. Using (5.33), we obtain

$$h(\tau + \delta s) \leq h(\tau) - \gamma \delta s + \delta \sigma(\delta) + Cs^2 \delta.$$

For any small s, we set $\tau + \delta s = t$ and deduce, as $\delta \to 0$, that

$$\liminf_{t\downarrow\tau} \frac{|h(t) - h(\tau)|}{t - \tau} \geqslant \liminf_{t\downarrow\tau} \left(\gamma - \frac{\sigma(\delta)}{s} - Cs \right) = \gamma - Cs.$$

Since s is arbitrary, (5.29) follows.

Lemma 5.5. *The following inequality holds*:

$$(5.35) \qquad\qquad \limsup_{t\downarrow\tau} \frac{|h(t) - h(\tau)|}{t - \tau} \leqslant \gamma.$$

Proof. From Lemma 5.2 we deduce that for any $\varepsilon > 0$,

$$v(x, \tau) \leqslant \left(\gamma + \frac{\varepsilon}{2} \right)(|x| - R) \qquad \text{if } |x| < R + d_0,$$

where d_0 is a positive number depending on ε.

Let $V(r, t)$ $(r = |x|)$ be the radial supersolution (5.19) with $C_1 = (\gamma + \varepsilon)/\lambda$. Then

$$V(r, \tau) > v(x, \tau) \qquad \text{if } R < r < R + d_1$$

provided that d_1 is sufficiently small. By continuity, also

$$V(r, t) > v(x, t) \qquad \text{if } r = R + d_1, \quad 0 \leqslant t - \tau \leqslant \sigma,$$

provided that σ is sufficiently small. We can now use comparison to conclude that $V \geqslant v$ if $0 \leqslant r \leqslant R + d_1$, $0 \leqslant t - \tau \leqslant \sigma$. In particular,

$$v(x, t) = 0 \qquad \text{if } |x| - R = -(\gamma + \varepsilon)(t - \tau),$$

which gives

$$\frac{|h(t) - h(\tau)|}{t - \tau} \leqslant \gamma + \varepsilon \qquad \text{if } \tau \leqslant t \leqslant \tau + \sigma.$$

Taking $t \downarrow \tau$, (5.35) follows.

Proof of Theorem 5.1. The assertion (i) follows from Lemma 5.2. From Lemmas 5.4 and 5.5 we deduce that $h'(\tau + 0)$ exists and it is equal to $-\gamma$; thus the assertions (ii) and (iii) follow.

We shall give an application of Theorem 5.1, assuming that

$$(5.36) \qquad\qquad h(t) \geqslant h_0 > 0 \qquad \text{if } \tau \leqslant t < \tau + \sigma_0$$

for some positive constants h_0, σ_0. This condition means that the gas does not "close in" on the point y as long as $t \leqslant \tau + \sigma_0$.

Theorem 5.6. *There exists a positive constant K depending on h_0 (but not on σ_0) such that*

$$(5.37) \qquad |d(y, \Omega(t_2)) - d(y, \Omega(t_1))| \leqslant K(t_2 - t_1)$$

provided that $\tau < t_1 < t_2 < \tau + \sigma_0$.

Thus the free boundary grows at most linearly, as long as it does not "close in" on any point.

Proof. Let $g(x)$ be the solution of

$$\Delta g = -C \qquad \text{if } h(t) < |x - y| < h(t) + 1,$$

$$g = 0 \qquad \text{if } |x - y| = h(t) + 1,$$

$$g = N \qquad \text{if } |x - y| = R^*,$$

where $N \geqslant \sup v$. Then $g(x)$ majorizes $v(x, t)$ and, consequently,

$$|\nabla_x v(x, t)| \leqslant |\nabla_x g(x)| \qquad \text{on } |x - y| = h(t).$$

In view of (5.36),

$$(5.38) \qquad |\nabla g(x)| \leqslant K,$$

where K may depend on h_0. It follows that

$$\gamma_t(0) \leqslant K$$

and, by Theorem 5.1,

$$(5.39) \qquad |h'(t + 0)| \leqslant K.$$

Since $h(t)$ is monotone decreasing and right continuous (by Theorem 4.5), by a standard theorem in real variables,

$$h(t_2) - h(t_1) \geqslant \int_{t_1}^{t_2} h'(t)\, dt = \int_{t_1}^{t_2} h'(t + 0)\, dt$$

$$\geqslant -K(t_2 - t_1) \qquad (t_1 < t_2)$$

[by (5.39)], which implies (5.37).

Remark 5.1.   If we do not make the assumption (5.36), we only get

$$(5.40) \qquad\qquad |(h^2)'(t+0)| \leqslant K.$$

Theorem 5.1, in the special case $n = 1$, gives the following:

Corollary 5.7.   *Suppose that (3.37) holds. Then, for any $t > 0$, $v_x(\zeta_2(t) - 0, t)$ and $\zeta_2'(t+0)$ exist and*

$$(5.41) \qquad\qquad v_x(\zeta_2(t), t) = -\zeta_2'(t+0).$$

A similar conclusion is valid for ζ_1.

The assertion that $v_x(\zeta_2(t) - 0, t)$ exists is a consequence of $v_{xx} \geqslant -C$ [established in (2.15)] since then $v_x + Cx$ is monotone increasing in x.
   We conclude this section with a stronger regularity result on $\zeta_i(t)$.

Theorem 5.8.   *Suppose (3.37) holds. Then $\zeta_i(t)$ is continuously differentiable for all $t > t_i^*$, where $t_i^* = \sup \{t > 0; \zeta_i(t) = \zeta_i(0)\}$. More precisely,*

$$(5.42) \qquad\qquad (-1)^i\zeta_i(t) = \xi_i(t) + \eta_i(t) \qquad \text{for } t > t_i^*$$

where $\xi_i(t)$ is a C^1 convex function and $\eta_i(t)$ is a $C^{1,1}$ function.
   The proof is outlined in Problems 7 to 13.

PROBLEMS

1. v_x is bounded in any strip $t \geqslant \eta_0$ ($\eta_0 > 0$).
     [*Hint:* $v_x + C_x$ is increasing in $x \in \Omega(t)$ and bounded on $\Gamma(t)$ since by (5.11), (5.7),

$$|v_x(\zeta_2(\tau), \tau)| \leqslant \frac{\gamma_\tau(1)}{M(R+1)} + 2C.\Bigg]$$

2.   Set $\zeta(t) = \zeta_2(t)$ and take for simplicity $t^* = 0$. Let $\tilde{w}$ be a solution of the form (3.34) whose free boundary $x = \tilde{\zeta}(t)$ is tangent to $x = \zeta(t)$ at $t = t_0$ ($t_0 \geqslant \eta_0 > 0$) and $w_{xx}(x, t_0) = -2P$ whenever $w > 0$; here $v_{xx} \geqslant -2P$, P nonnegative constant (by (2.15)). Then

$$\tilde{\zeta}(t) \leqslant \zeta(t) \qquad (t > t_0)$$

$$\tilde{\zeta}(t) - \tilde{\zeta}(t_0) - (t - t_0)\tilde{\zeta}'(t_0) \leqslant \zeta(t) - \zeta(t_0) - (t - \cdot t_0)\zeta'(t_0).$$

Compute

$$\tilde{\zeta}''(t_0) = -\gamma P \tilde{\zeta}'(t_0) \qquad (\gamma \text{ positive constant depending only on } m),$$

$$\tilde{\zeta}''(t) = \tilde{\zeta}''(t_0) + O(t - t_0),$$

and deduce that

$$\Phi_h(t) \equiv \frac{\zeta(t + h) - \zeta(t) - h\zeta'(t + 0)}{h^2/2} \qquad (h > 0)$$

satisfies

$$\Phi_h(t) + \gamma P \zeta'(t + 0) \geq O(h) \qquad \text{at any point } t_0.$$

3. Prove that for some finite positive measure μ,

    (5.43) $$\zeta'' + \gamma P \zeta' = \mu$$

    in the sense of distributions.
    [*Hint*: $\Phi_h \geq -C$, $\int_\delta^T \Phi_h(t)\, dt \leq C$. Take a sequence $\Phi_{h_n} \to \mu_0$, weakly, where μ_0 is a signed measure.]

4. Prove (5.42) with ξ_i convex and η_i in $C^{1,1}$.
    [*Hint*: Take $\xi(t) = \int_\delta^t \int_\delta^\tau d\mu(x)\, d\tau$,

    $$\eta(t) = \zeta(\delta) + \zeta'(\delta + 0)(t - \delta) - \gamma P \int_\delta^t (\zeta(s) - \zeta(\delta))\, ds. \Big]$$

5. Deduce from (5.43) that

    $$\zeta'(t_2 - 0)e^{\gamma P t_2} \geq \zeta'(t_1 + 0)e^{\gamma P t_1} \qquad (t_2 > t_1)$$

    and consequently, $\zeta'(t \pm 0) \geq c > 0$ if $t \geq \eta_0 > 0$.

6. Let $d(\bar{x}, \bar{t})$ denote the distance from $(\bar{x}, \bar{t})$ to the free boundary $x = \zeta(t)$; $(\bar{x}, \bar{t})$ belongs to N_δ, a δ-neighborhood of $(\zeta(t_0), t_0)$ intersected with $\{v > 0\}$. Then

    $$C_1 \leq \frac{v(x, t)}{d(x, t)} \leq C_2 \qquad (C_i > 0).$$

    [*Hint*: $c_1 d(x, t) \leq |x - \zeta(t)| \leq c_2 d(x, t)$ and

    $$v(x, t) \geq |v_x(\zeta(t), t)| |x - \zeta(t)| - P(x - \zeta(t))^2. \Big]$$

7. Show that

$$|v_t| \leq C, \qquad |v_{tt}| \leq \frac{C}{d(x,t)} \quad \text{in } N_\delta.$$

[*Hint*: Let S be a square with center $(\bar{x}, \bar{t})$ and side $\gamma = d(\bar{x}, \bar{t})/2$. Consider

$$\tilde{v}(x,t) = \frac{1}{\gamma} v(\gamma x + \bar{x}, \gamma t + \bar{t}) \qquad \text{in } |x| < 1, \quad |t| < 1.$$

It satisfies

$$\tilde{v}_t = (m-1)\tilde{v}\tilde{v}_{xx} + \tilde{v}_x^2, \qquad C_1 < \tilde{v} < C_2.$$

Apply the Nash–deGiorgi estimate and then Schauder's estimates.]

8. Set $\beta = \zeta'(t_0 + 0)$, $t_0 > 0$. Then

$$-v_x(x,t) \leq \beta + \sigma(\varepsilon) \qquad \text{in } N_\varepsilon,$$

where $\sigma(\varepsilon) \to 0$ if $\varepsilon \to 0$.
[*Hint*: If $-v_x(x_n, t_n) > \beta + \delta$, $(x_n, t_n) \to (x_0, t_0)$, then

$$v(x, t_n) \geq v(x_n, t_n) - (\beta + \delta)(x - x_n) - P(x - x_n)^2 \qquad (x < x_0),$$

$$v(x, t_0) \geq -(\beta + \delta)(x - x_0) - P(x - x_0)^2.$$

But $-v_x(x, t_0) \to \beta$ if $x \to x_0$.]

9. $v_t(x, t_0) \leq \beta^2 + \sigma(\varepsilon)$ if $(x, t_0) \in N_\varepsilon$, where $\sigma(\varepsilon) \to 0$ if $\varepsilon \to 0$.
[*Hint*: If $v_t(x_n, t_0) \geq \beta^2 + \delta$, $\varepsilon_n = x_0 - x_n$, then

$$v(x_n, t_0 + \alpha\varepsilon_n) = v(x_n, t_0) + v_t(x_n, t_0)\alpha\varepsilon_n + \tfrac{1}{2}\alpha^2\varepsilon_n^2 v_{tt}$$

$$\geq \beta\varepsilon_n - P\varepsilon_n^2 + (\beta^2 + \delta)\alpha\varepsilon_n - C\alpha^2\varepsilon_n.$$

Also,

$$v(x_n, t_0 + \alpha\varepsilon_n) = \int v_x \leq -(\beta + \sigma_n)(x_n - \zeta(t_0 + \alpha\varepsilon_n))$$

$$= -(\beta + \sigma_n)(x_n - x_0 - \alpha\varepsilon_n\beta + \varepsilon_n\tilde{\sigma}_n) \qquad (\sigma_n, \tilde{\sigma}_n \to 0).$$

Choose $\alpha < C/\delta$ to derive a contradiction.]

10. Take for simplicity $(\zeta(t_0), t_0) = (0, 0)$ and let

$$v_\gamma(x, t) = \frac{1}{\gamma} v(\gamma x, \gamma t).$$

Then, for a sequence $\gamma \to 0$, $v_\gamma(x, t) \to V(x, t)$ uniformly in compact sets, where V is again a solution with free boundary $x = \zeta_0(t)$,

$$\zeta_0(t) = \beta \qquad \text{if } t > 0$$

$$= \beta_0 \qquad \text{if } t < 0$$

and $\beta = \zeta'(t_0 + 0)$, $\beta_0 = \zeta'(t_0 - 0)$. Let $w(x, t) = \beta(\beta t - x)^+$. To complete the proof of Theorem 5.8 [that is, that $\zeta_i(t) \in C^1$], it suffices to show that $V \equiv w$.

11. Show that $V(x, 0) = w(x, 0)$.
    [*Hint*: Otherwise, since $V_{xx} \geq 0$, there exists a solution $w_0(x, t) = \bar{\beta}(\bar{\beta} t - x - \delta)^+$ with $\bar{\beta} > \beta$ and $\delta > 0$ such that $V(x, 0) \geq w_0(x, 0)$ and, by comparison, $V \geq w_0$ if $t \geq 0$. But then $\zeta_0(t) > \beta t$ if $t > \delta/(\bar{\beta} - \beta)$.]

12. Prove that $V(x, -1) \geq w(x, -1)$.
    [*Hint*: Apply Problems 7 to 9 to v_γ ($\gamma \to 0$) to deduce that

    $$\liminf_{M \to \infty} \{V(-M, -1) - w(-M, -1)\} \geq 0,$$

    $$\liminf_{M \to \infty} \{V_x(-M, -1) - w_x(-M, -1)\} \geq 0$$

    and note that $V_{xx} \geq 0$.]

13. Prove that $V \equiv w$.
    [*Hint*: By Problem 12 and comparison, $V \geq w$ if $t > -1$; now use Problem 11.]

14. Prove that v_x is continuous up to the free boundary $x = \zeta_i(t)$, $t > t_i^*$.
    [*Hint*: Use Theorem 5.8 and $v_{xx} \geq -2P$ to estimate $-v_x(x, t) \geq -v_x(\zeta(t_0), t_0) - \sigma(\varepsilon)$ in N_ε, and recall Problem 8.]

15. Prove that v_t is continuous up to the free boundary $x = \zeta_i(t)$, $t > t_i^*$ and $v_t(\zeta_i(t), t) = (\zeta_i'(t))^2$.
    [*Hint*: $v_{xx} \geq -2P$ and Problem 14 imply that $\underline{\lim} \, v_t \geq (\zeta')^2$. The method of Problem 9 gives $\overline{\lim} \, v_t \leq (\zeta')^2$.]

16. The assertion $|v_x| \leq C$ in Problem 1 is valid for any $u_0 \in L^1(R^n)$ [that is, the assumption (3.37) is not needed].
    [*Hint*: If $v \leq N$, let $\psi(r) = \frac{1}{3}Nr(4 - r)$ $(0 < r < 1)$, $v = \psi(w)$, $p = w_x$. Write the differential equation for w, differentiate in x, and multiply by p to derive a differential equation $Lp = 0$. Let $\zeta \in C_0^2(R^n \times (0, \infty))$, $0 \leq \zeta \leq 1$, $z = \zeta^2 p^2$. If z takes maximum at (x_0, t_0), then $z_x = 0$, $\psi z_{xx} - \psi_t \leq 0$ at (x_0, t_0). Use these conditions in $Lp = 0$ to deduce that $\zeta^2 p^4 \leq C_1 p^2 + \zeta C_2 |p|^3$ at (x_0, t_0). Hence max $z \leq C$ and then also $|v_x|$ is bounded.]

6. THE GENERAL TWO-DIMENSIONAL FILTRATION PROBLEM: EXISTENCE

In the filtration problem studied in Chapter 1, Section 5 and Chapter 2, Section 6 the dam was rectangular. We shall now consider a general two-dimensional dam Ω and establish existence and uniqueness, and regularity of the free boundary.

We denote a point in the plane by $X = (x, y)$. We assume that $\partial\Omega$ consists of three parts: S_1, the impervious part; S_2, the part in contact with the air; and S_3, the part in contact with the water reservoirs; see Figure 5.1. We assume that there is only a finite number of disjoint reservoirs R_j ($1 \leqslant j \leqslant k$) in which the water at a height $y = h_j$, and we set $S_{3,j} = \partial R_j \cap \{y < h_j\}$. Thus $S_3 = \cup_{j=1}^{k} S_{3,j}$.

Denote by A the wet portion of Ω, that is, the portion filled with water. The boundary of A consists of four parts:

$$
\begin{aligned}
&\Gamma_1 \subset S_1 &&\text{(the impervious part)},\\
&\Gamma_2 \subset \Omega &&\text{(the free boundary)},\\
&\Gamma_3 = S_3 &&\text{(the part in contact with the reservoirs)},\\
&\Gamma_4 \subset S_2 &&\text{(the wet part of the dam, in contact with}\\
& && \text{the air; it is called the \textit{seepage line}).}
\end{aligned}
$$

Γ_3 is given but all the other Γ_i are not a priori known.

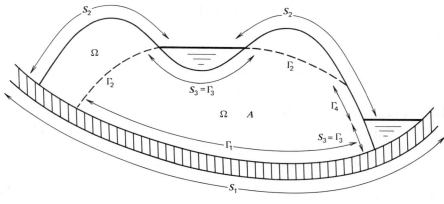

FIGURE 5.1

The hydrostatic pressure on S_3 is given by $h_i - y$. We extend this function into S_2 by setting

$$u^0(x, y) = \begin{cases} h_i - y & \text{on } S_{3,i} \\ 0 & \text{on } S_2. \end{cases}$$

Denote by ν the exterior normal to A. If we denote the pressure function by u, then as in Chapter 1, Section 5,

$$\Delta u = 0 \qquad \text{in } A,$$

$$\frac{\partial}{\partial \nu}(u + y) = 0 \qquad \text{on } \Gamma_1,$$

(6.1) $$u = 0 \text{ and } \frac{\partial}{\partial \nu}(u + y) = 0 \qquad \text{on } \Gamma_2,$$

$$u = u^0 \qquad \text{on } \Gamma_3,$$

$$u = u^0 \text{ and } \frac{\partial}{\partial \nu}(u + y) \leq 0 \qquad \text{on } \Gamma_4.$$

We shall henceforth assume:

(6.2) S_1 and $S_2 \cup S_3$ are continuous and piecewise $C^{1+\alpha}$ curves; both are graphs in the direction y, and $S_2 \cup S_3$ lies above S_1.

Thus there is an interval $\sigma_* \leq x \leq \tau_*$ such that any line $x = x_0$ with $x_0 \in [\sigma_*, \tau_*]$ intersects S_1 (and $S_2 \cup S_3$) in either one point or one closed interval, and the functions

$$S^-(x) = \sup\{x; (x, y) \in S_1\},$$

$$S^+(x) = \inf\{x; (x, y) \in S_2 \cup S_3\}$$

satisfy

$$S^-(x) < S^+(x) \qquad (\sigma_* < x < \tau_*).$$

For $x_0 < \sigma_*$ or $x_0 > \tau_*$ the line $x = x_0$ does not intersect $S_1 \cup S_2 \cup S_3$. We further assume that

(6.3) $S^-(x), S^+(x)$ are piecewise continuous.

Notice that

$$\Omega = \{(x, y); S^-(x) < y < S^+(x), \sigma_* < x < \tau_*\}.$$

We denote by $\pi(E)$ the projection of a set E on the x-axis, and index the $S_{3,i}$ such that

(6.4) $$\pi(S_{3,i}) = \{x; \sigma_i \leqslant x \leqslant \tau_i\}, \qquad \tau_i < \sigma_{i+1},$$

where σ_i may coincide with τ_i. Thus, in the case of the rectangular dam in Chapter 1, Section 5, $\sigma_1 = \tau_1 = \sigma_* = 0$ and $\sigma_2 = \tau_2 = \tau_* = a$.

Observe that

$$\frac{\partial u}{\partial \nu} = -\frac{\partial y}{\partial \nu} > 0 \qquad \text{on } \Gamma_1.$$

Since also $u = u^0 \geqslant 0$ on Γ_3, the maximum principle gives $u > 0$ in A.

If $\zeta \in C^1(\overline{\Omega})$, $\zeta = 0$ on Γ_3, $\zeta \geqslant 0$ on Γ_4 then by integration by parts,

$$\int_A \nabla u \cdot \nabla \zeta + \int_A \zeta_y = \int_{\Gamma_1 \cup \Gamma_2 \cup \Gamma_4} \left(\frac{\partial u}{\partial \nu} + \nu \cdot e \right) \zeta = \int_{\Gamma_4} \frac{\partial}{\partial \nu}(u + y)\zeta \leqslant 0,$$

where (6.1) and the relation $\partial y / \partial \nu = \nu \cdot e$ have been used; here e is the unit vector in vertical direction:

$$e = (0, 1).$$

Extending u by zero into $\Omega \setminus A$ and introducing the Heaviside function

$$H(t) = 1 \text{ if } t > 0, \qquad = 0 \text{ if } t \leqslant 0,$$

we get

(6.5) $$\int_\Omega \nabla \zeta \cdot (\nabla u + H(u)e) = \int_\Omega \nabla \zeta \cdot \nabla u + \int_A \zeta_y \leqslant 0,$$

We shall now give a weak formulation of the problem. Since Γ_4 is not a priori known, we shall take the test functions ζ to satisfy $\zeta \geqslant 0$ on all of S_2.

Problem (A). Find a pair (u, γ) where $u \in H^1(\Omega)$, $\gamma \in L^\infty(\Omega)$ such that $0 \leqslant \gamma \leqslant 1$, $\gamma = 1$ on $\{u > 0\}$, $u \geqslant 0$, $u = u^0$ on $S_2 \cup S_3$, such that

(6.6) $$\int_\Omega \nabla \zeta \cdot (\nabla u + \gamma e) \leqslant 0 \qquad \forall \zeta \in H^1(\Omega), \quad \zeta \geqslant 0 \text{ on } S_2, \quad \zeta = 0 \text{ on } S_3.$$

If we can show (as will be done later) that $\gamma = I_{\{u>0\}}$, then (6.6) coincides with (6.5).

Theorem 6.1. *Problem (A) has a solution (u, γ).*

Proof. Suppose first that $\partial\Omega$ is in $C^{1+\alpha}$. Introduce the Lipschitz continuous functions

$$H_\varepsilon(t) = \begin{cases} 0 & \text{if } t \leqslant 0 \\ \dfrac{t}{\varepsilon} & \text{if } 0 \leqslant t \leqslant \varepsilon \qquad (\varepsilon > 0) \\ 1 & \text{if } t > \varepsilon \end{cases}$$

and the classes of functions

$$K = \{v \in H^1(\Omega); v = u^0 \text{ on } S_2 \cup S_3\},$$

$$K_0 = \{v \in H^1(\Omega); v = 0 \text{ on } S_2 \cup S_3\}.$$

Consider the penalized problem: Find u_ε such that

(6.7)
$$u_\varepsilon \in K,$$

$$\int_\Omega \nabla\zeta \cdot (\nabla u_\varepsilon + H_\varepsilon(u_\varepsilon)e) = 0 \qquad \forall\,\zeta \in K_0.$$

To solve this problem consider, for any $v \in L^2(\Omega)$ the problem: Find $w \in K$ satisfying

(6.8)

$$a(w, \eta - w) \equiv \int_\Omega \nabla w \cdot \nabla(\eta - w) = -\int_\Omega H_\varepsilon(v)\frac{\partial}{\partial y}(\eta - w) \equiv F(\eta - w)$$

for all $\eta \in K_0$. Notice that

$$|F(\eta)| \leqslant C_0 |\eta|_{H^1(\Omega)} \qquad (C_0 \text{ constant})$$

and $a(w, \eta)$ is a coercive bilinear form on K. Thus there exists a unique solution of (6.8). Further, we easily find that

(6.9)
$$|w|_{H^1(\Omega)} \leqslant C,$$

where C is a constant independent of ε. It follows that the mapping $w = Tv$ maps the set $\{v \in L^2(\Omega), |v|_{L^2} \leqslant C\}$ into a compact subset and, by Schauder's fixed-point theorem, there exists a solution w of $w = Tw$. This w is precisely a solution u_ε of (6.7).

Since $u_\varepsilon = u^0 \geqslant 0$ on $S_2 \cup S_3$, we can take $\zeta = u^-$ as a test function in (6.7) and obtain

$$-\int |\nabla u_\varepsilon^-|^2 = \int H_\varepsilon(u_\varepsilon)(u_\varepsilon^-)_y = 0.$$

It follows that $u_\varepsilon^- = 0$, that is,

(6.10) $u_\varepsilon \geq 0.$

Since also $u_\varepsilon = 0$ on S_3, we get

(6.11) $\dfrac{\partial u_\varepsilon}{\partial \nu} \leq 0 \quad$ on S_3.

Next,

$$-\Delta u_\varepsilon = \frac{\partial}{\partial y} H_\varepsilon(u_\varepsilon)$$

and $0 \leq H_\varepsilon \leq 1$. Also, by (6.9),

(6.12) $|u_\varepsilon|_{H^1(\Omega)} \leq C.$

But then standard elliptic estimates give

(6.13) $|u_\varepsilon|_{W_{\text{loc}}^{1,p}(\Omega)} \leq \overline{C} \quad \forall\, p \in (1, \infty),$

where $\overline{C}$ is a constant independent of ε.

We now choose a sequence $\varepsilon = \varepsilon_m \downarrow 0$ such that

$$u_{\varepsilon_m} \to u \quad \text{uniformly in } C_{\text{loc}}^\alpha(\Omega) \qquad (0 < \alpha < 1),$$

$$u_{\varepsilon_m} \to u \quad \text{weakly in } H^1(\Omega),$$

$$H_{\varepsilon_m}(u_{\varepsilon_m}) \to \gamma \quad \text{weakly star in } L^\infty(\Omega).$$

Clearly, $0 \leq \gamma \leq 1$, and $\gamma = 1$ on the set $\{u > 0\}$. Also, $u \geq 0$ and $u = u^0$ on $S_2 \cup S_3$.

For any ζ as in (6.6),

$$\int_\Omega \nabla \zeta \cdot (\nabla u_\varepsilon + H_\varepsilon(u_\varepsilon)e) = \int_{S_2} \frac{\partial u_\varepsilon}{\partial \nu} \zeta \leq 0$$

by (6.11). Taking $\varepsilon = \varepsilon_m \downarrow 0$ we obtain the inequality in (6.6). Thus (u, γ) is a solution of problem (A).

We have assumed above that $\partial\Omega \in C^{1+\alpha}$. In the general case of piecewise $C^{1+\alpha}$ boundary, we approximate Ω by smooth $C^{1+\alpha}$ domains by smoothing the corners of $\partial\Omega$. It is easy to see that the corresponding solutions converge to a solution for Ω.

Definition 6.1.   A function ζ as in (6.6) is called a *test function* [for (A)].

From the preceding proof we have actually obtained a solution in $C_{\text{loc}}^{1+\alpha}(\Omega)$ for any $0 < \alpha < 1$ [recall (6.13)]. We shall now prove:

Theorem 6.2.   *For any solution (u, γ) of problem (A), $u \in C_{\text{loc}}^{0,1}(\Omega)$.*

First we establish a lemma.

Lemma 6.3.   *If $B_r(X) \subset \Omega \cap \{u > 0\}$ and $\partial B_r(X) \cap \partial \{u > 0\} \neq \varnothing$, then $u(X) \leq Cr$, where C is a universal constant.*

Proof.   For any $\delta > 0$ let v be the harmonic function in the ring

$$D = B_{r+\delta}(X) \setminus B_{r/2}(X)$$

taking the boundary values

$$v = \lambda \equiv \inf_{\partial B_{r/2}(X)} u \qquad \text{on } \partial B_{r/2}(X),$$

$$v = 0 \qquad \text{on } \partial B_{r+\delta}(X).$$

The functions $\zeta = \pm \max(v - u, 0)$ when extended by zero into $\Omega \setminus D$ are test functions. Hence, by (6.6),

(6.14) $$\int_D \nabla \max(v - u, 0) \cdot (\nabla u + \gamma e) = 0.$$

We can write

$$\int_D \nabla \max(v - u, 0) \cdot \nabla u$$

$$= \int_D \nabla \max(v - u, 0) \cdot \nabla(u - v)$$

$$+ \int_D \nabla \max(v - u, 0) \cdot \nabla(v + y) - \int_D \nabla \max(v - u, 0) \cdot e.$$

The first integral on the right-hand side is equal to

$$-\int_{D \cap \{u > 0\}} |\nabla \max(v - u, 0)|^2 - \int_{D \cap \{u = 0\}} |\nabla v|^2.$$

The second integral vanishes since $v + y$ is harmonic, whereas $\max(v - u, 0)$

vanishes on ∂D. Using this information in (6.14), we obtain

$$\int_{D \cap \{u>0\}} |\nabla(v-u)^+|^2 = -\int_{D \cap \{u=0\}} |\nabla v|^2 + \int_D \nabla \max(v-u,0)(\gamma-1)e.$$

In the last integral, if $u > 0$, then $\gamma = 1$ and the integrand vanishes. Thus the integral is bounded above by

$$\int_{D \cap \{u=0\}} |\nabla v|,$$

and we obtain

$$(6.15) \qquad \int_{D \cap \{u>0\}} |\nabla(v-u)^+|^2 \leqslant \int_{D \cap \{u=0\}} |\nabla v|(1 - |\nabla v|).$$

We claim that

$$(6.16) \qquad \int_{D \cap \{u>0\}} |\nabla(v-u)^+|^2 > 0.$$

Indeed, otherwise $v \leqslant u$ in $\{u > 0\}$ and, since D contains free-boundary points, v will vanish at these points, contradicting the maximum principle.

Notice next that $|\nabla v|$ is a strictly monotone-decreasing function of the distance from X. Hence, from (6.15), (6.16),

$$|\nabla v| < 1 \qquad \text{on } B_{r+\delta}(X).$$

On the other hand,

$$|\nabla v| \geqslant c \frac{\lambda}{r+\delta} \qquad \text{on } \partial B_{r+\delta}(x),$$

where c is a universal positive constant. Thus $\lambda \leqslant r/c$. We have thus proved that

$$\inf_{\partial B_{r/2}(X)} u \leqslant \frac{r}{c}$$

and by Harnack's inequality it follows that $u(X) \leqslant Cr$.

Proof of Theorem 6.2.   Using Lemma 6.3, the proof is similar to the proof of Theorem 3.2 of Chapter 3.

We denote by I_K the characteristic function of a set K.
In the following sections we shall use the following lemma.

Lemma 6.4. *Let* $u \in H^1(\Omega) \cap C(\Omega)$ *and denote by K a component of* $\{u > 0\}$. *Then $I_K u$ is in $H^1(\Omega)$ and*

$$\nabla(I_K u) = I_K \nabla u.$$

For proof, see Problem 2.

PROBLEMS

1. Prove that the solution of problem (6.7) is unique.
    [*Hint*: If u_1, u_2 are two solutions and $w = u_1 - u_2$, then

    $$\left| \int \nabla w \cdot \nabla \zeta \right| \leqslant \int |w| |\zeta_y|.$$

    Take $\delta > 0$, $\zeta = (w - \delta)^+ / w$ and show that

    $$I_\delta \equiv \int_{[w>\delta]} \frac{|\nabla w|^2}{w^2} \leqslant C_0 \int_{[w>\delta]} \left| \frac{w_y}{w} \right|, \qquad I_\delta \leqslant C,$$

    $$\int \left| \nabla \log \left(1 + \frac{(w - \delta)^+}{w} \right) \right|^2 \leqslant I_\delta.$$

    Use Poincaré's inequality and take $\delta \to 0$.]

2. Prove Lemma 6.4.
    [*Hint*: If $u \in C(\overline{\Omega})$, Ω bounded, let $\varepsilon > 0$, $K_\varepsilon = \{x \in K, u(x) \geqslant \varepsilon\}$, $\zeta_\varepsilon \in C_0^\infty(R^n \setminus K)$, $0 \leqslant \zeta_\varepsilon \leqslant 1$, $\zeta_\varepsilon = 1$ on K_ε. Then

    $$\nabla \left[I_{K_\varepsilon}(u - \varepsilon) \right] = \nabla \left[\zeta_\varepsilon (u - \varepsilon)^+ \right] = \zeta_\varepsilon \nabla (u - \varepsilon)^+;$$

    let $\varepsilon \downarrow 0$. In general, take $\overline{\Omega}_m \subset \Omega_{m+1} \uparrow \Omega$.]

7.   REGULARITY OF THE FREE BOUNDARY

In this section we denote by (u, γ) any solution of problem (A).

Theorem 7.1.   *The following relations hold in Ω in the distribution sense:*

(7.1) $$\Delta u + \gamma_y = 0,$$

(7.2) $$\Delta u \geqslant 0,$$

(7.3) $$\gamma_y \leqslant 0.$$

Proof. Taking $\zeta \in C_0^\infty(\Omega)$ in (6.6), (7.1) follows. To prove (7.2), let $\zeta \in C_0^\infty(\Omega)$, $\zeta \geq 0$, $\varepsilon > 0$. Since the functions $\zeta = \pm\min(u, \varepsilon\zeta)$ are test functions,

$$\int_\Omega \nabla u \cdot \nabla \min(u, \varepsilon\zeta) + \int_\Omega \gamma[\min(u, \varepsilon\zeta)]_y = 0.$$

Since $\gamma = 1$ on $\{u > 0\}$, the second integral is equal to

$$\int_\Omega [\min(u, \varepsilon\zeta)]_y = \int_{\partial\Omega} \min(u, \varepsilon\zeta)(y \cdot \nu) = 0.$$

Hence

$$0 = \int_\Omega \nabla u \cdot \nabla \min(u, \varepsilon\zeta)$$

$$= \varepsilon \int_{\{u > \varepsilon\zeta\}} \nabla u \cdot \nabla\zeta + \int_{\{u \leq \varepsilon\zeta\}} |\nabla u|^2,$$

and consequently,

$$\int_\Omega I_{\{u > \varepsilon\zeta\}} \nabla u \cdot \nabla\zeta \leq 0.$$

Taking $\varepsilon \to 0$, we obtain $\int_\Omega \nabla u \cdot \nabla\zeta \leq 0$. Since ζ is arbitrary nonnegative function in $C_0^\infty(\Omega)$, (7.2) follows. Finally, (7.3) is a consequence of (7.1), (7.2).

Lemma 7.2. *If* $\{x_0\} \times [y_0 - \varepsilon, y_1 + \varepsilon] \subset \Omega \cap \{u = 0\}$, $\varepsilon > 0$, *then*

(7.4) $$u(x, y) = o(|x - x_0|) \qquad \text{as } x \to x_0,$$

uniformly with respect to y *in* $[y_0, y_1]$.

Proof. Let $\sigma = \{x_0\} \times [y_0, y_1]$ and set

$$\lambda = \limsup_{(x, y) \to \sigma} \frac{u(x, y)}{|x - x_0|}.$$

Notice that λ is finite since $u \in C^{0,1}$. Take a sequence $X_r = (x_r, y_r) \to \sigma$ such that $r = |x_r - x_0|$ and

$$\frac{1}{r} u(X_r) \to \lambda$$

and consider the blowup sequence (u_r, γ_r) with respect to balls $B_r(x_0, y_r)$; that is,

$$u_r(x, y) = \frac{1}{r} u(x_0 + rx, y_r + ry),$$

$$\gamma_r(x, y) = \gamma(x_0 + rx, y_r + ry).$$

Then $|\nabla u_r| \leqslant C$ in every bounded set and thus, for a subsequence,

$$\nabla u_r \to \nabla \tilde{u} \quad \text{weakly star in } L^\infty_{\text{loc}}(R^2),$$

$$u_r \to \tilde{u} \quad \text{in } C^\alpha_{\text{loc}}(R^2) \quad \forall \alpha < 1,$$

$$\gamma_r \to \tilde{\gamma} \quad \text{weakly star in } L^\infty_{\text{loc}}(R^2).$$

It is easy to see that $(\tilde{u}, \tilde{\gamma})$ is a solution of problem (A) in R^2; that is,

(7.5) $\quad \int \nabla \zeta \cdot (\nabla \tilde{u} + \tilde{\gamma} e) = 0 \quad \forall \zeta \in H^1(R^2)$ with compact support.

Arguing as in Theorem 7.1, we deduce that

$$\Delta \tilde{u} \geqslant 0 \quad \text{in } R^2.$$

We also have

$$\tilde{u}(x, y) \leqslant \lambda |x|$$

and, assuming for definiteness that the $x_r - x_0$ are positive,

$$\tilde{u}(1, 0) = \lambda.$$

Applying the maximum principle to the subharmonic function $\tilde{u} - \lambda x$ in $\{x > 0\}$, we conclude that $\tilde{u} \equiv \lambda x$ in $\{x > 0\}$. If we show that $\lambda = 0$, then (7.4) follows.

For any $\delta > 0$, let

$$d_\delta(x) = \max\left\{\min\left(1 + \frac{x}{\delta}, 1\right), 0\right\}.$$

Observe that $d_\delta(x) = 1$ if $x > 0$. Since also $\tilde{u} = \lambda x$ if $x > 0$, we can write for any $\zeta \in C^1_0(R^2)$,

(7.6) $\quad \lambda \int_{\{x=0\}} \zeta \, dy = -\int_{\{x>0\}} \nabla \zeta \cdot (\nabla \tilde{u} + e)$

$$= -\int_{\{x>0\}} \nabla(d_\delta \zeta) \cdot (\nabla \tilde{u} + \tilde{\gamma} e).$$

Since $d_\delta \zeta$ is a test function for (7.5), the right-hand side is equal to

$$\int_{\{-\delta<x<0\}} d_\delta \nabla \zeta \cdot (\nabla \tilde{u} + \tilde{\gamma} e) + \frac{1}{\delta} \int_{\{-\delta<x<0\}} \zeta \frac{\partial \tilde{u}}{\partial x}.$$

The first integral tends to zero as $\delta \to 0$; the second integral is equal to

$$-\frac{1}{\delta} \int_{\{-\delta<x<0\}} \tilde{u} \zeta_x - \frac{1}{\delta} \int_{\{x=-\delta\}} \tilde{u} \zeta \, dy.$$

Here the first term tends to zero as $\delta \to 0$ since the integral is $O(\delta^2)$ (recalling that $\tilde{u} \in C^{0,1}$). Thus we obtain from (7.6),

$$\lambda \int_{\{x=0\}} \zeta \, dy = -\lim_{\delta \to 0} \frac{1}{\delta} \int_{\{x=-\delta\}} \tilde{u} \zeta \, dy.$$

Choosing $\zeta \geq 0$, we conclude that $\lambda \leq 0$.

Theorem 7.3. *The set* $\Omega \cap \{u > 0\}$ *has the form*

$$\{(x, y) \in \Omega; S^-(x) < y < \phi(x)\},$$

where $\phi(x)$ *is lower semicontinuous, and*

$$(7.7) \qquad\qquad \phi(x) = S^+(x) \qquad if \, x \in \bigcup_{i=1}^{k} [\sigma_i, \tau_i].$$

The last assertion means that the dam is wet below S_3.

Proof. Suppose that $u(x_0, y_0) > 0$. Then $u > 0$ in some disc $B_\delta(X_0)$ where $X_0 = (x_0, y_0)$. It follows that $\gamma = 0$ in $B_\delta(X_0)$. Since $0 \leq \gamma \leq 1$ and $\gamma_y \leq 0$, we must also have that $\gamma = 0$ in

$$K_\delta = \{(x, y) \in \Omega; |x - x_0| < \delta, y < y_0\} \cup B_\delta(X_0).$$

But then (7.1) implies that $\Delta u = 0$ in K_δ. Observing that $u(x_0, y_0) > 0$, $u \geq 0$ in K_δ, the maximum principle gives $u > 0$ in K_δ. We have thus proved that if

$$\phi(x_0) = \sup \{y; u(x_0, y) > 0\},$$

then

$$u(x_0, y) > 0 \qquad \text{for } y < \phi(x_0),$$
$$= 0 \qquad \text{for } y \geq \phi(x_0).$$

The preceding proof also shows that $\phi(x)$ is lower semicontinuous.

Since $u > 0$ on $S_3 \cap \cup_{i=1}^k (\sigma_i, \tau_i)$, it follows that $u > 0$ if $x \in (\sigma_i, \tau_i)$, $(x, y) \in \Omega$.

Suppose finally that $(x_0, y_0) \in \Omega$, $x_0 = \sigma_i$ or $x_0 = \tau_i$. We claim that $u(x_0, y_0) \neq 0$. Indeed, otherwise $u(x_0, y) = 0$ for $y > y_0$ and the maximum principle

$$\frac{\partial}{\partial x} u(x_0, y) \neq 0.$$

However, this contradicts Lemma 7.2.

Lemma 7.4. *Suppose that* $u(x_1, y) = u(x_2, y) = 0$ *for all* $(x_i, y) \in \Omega$, $y > h$, $i = 1, 2$, *and* (x_1, x_2) *lies in* $\pi(S_2)$ *(the projection of S_2 on the x-axis). Let*

$$Z_h = \{(x, y); x_1 < x < x_2, y > h\} \cap \Omega.$$

Then

$$(7.8) \qquad \int_{Z_h} (\gamma u_y + \gamma^2) \leq 0.$$

Proof. Let $\zeta \in H^1(Z_h) \cap C(\overline{Z}_h)$, $\zeta \geq 0$, $\zeta(x, h) = 0$. We claim that

$$(7.9) \qquad \int_{Z_h} \nabla \zeta \cdot (\nabla u + I_{\{u>0\}} e) \leq \int_{x_1}^{x_2} \zeta(x, \phi(x)) \, dx.$$

Indeed, for any $\varepsilon > 0$, $\min(u, \varepsilon \zeta)$ is a test function (when extended by zero outside Z_h). Therefore,

$$\int_{Z_h \cap \{u \leq \varepsilon \zeta\}} |\nabla u|^2 + \varepsilon \int_{Z_h \cap \{u > \varepsilon \zeta\}} \nabla u \cdot \nabla \zeta + \int_{Z_h} \gamma [\min(u, \varepsilon \zeta)]_y \leq 0.$$

But the first integral is nonnegative and the last integrand vanishes on $\{u = 0\}$ and equals $[\min(u, \varepsilon \zeta)]_y$ on $\{u > 0\}$. Consequently,

$$\int_{Z_h} \left[I_{\{u > \varepsilon \zeta\}} \nabla u \cdot \nabla \zeta + I_{\{u > 0\}} \left[\min\left(\frac{u}{\varepsilon}, \zeta \right) \right]_y \right] \leq 0,$$

or

$$\int_{Z_h} I_{\{u>\varepsilon\zeta\}} \nabla u \cdot \nabla \zeta + I_{\{u>0\}}\zeta_y \leq \int_{Z_h} I_{\{u>0\}}\left(\zeta - \frac{u}{\varepsilon}\right)_y^+$$

$$= \int_{x_1}^{x_2} dx \left[\lim_{\delta \downarrow 0} \int_h^{\phi(x)-\delta}\left(\zeta - \frac{u}{\varepsilon}\right)_y^+ dy\right]$$

$$\leq \int_{x_1}^{x_2}\left(\zeta - \frac{u}{\varepsilon}\right)^+ (x, \phi(x)) \, dx \leq \int_{x_1}^{x_2}\zeta(x, \phi(x)) \, dx.$$

Taking $\varepsilon \to 0$, (7.9) follows.

Now let $\eta(x)$ be a function in $C_0^\infty(x_1, x_2)$ such that $0 \leq \eta \leq 1$, $\eta = 1$ in $(x_1 + \varepsilon, x_2 - \varepsilon)$. Then

$$\int_{Z_h}(u_y + \gamma) = \int_{Z_h}\left[\nabla u \cdot \nabla(y - h) + \gamma(y - h)_y\right]$$

$$= \int_{Z_h}\{\nabla u \cdot \nabla[\eta(y - h)] + \gamma[\eta(y - h)]_y\}$$

$$+ \int_{Z_h}\{\nabla u \cdot \nabla[(1 - \eta)(y - h)] + \gamma[(1 - \eta)(y - h)]_y\}.$$

The first integral on the right-hand side is nonpositive since $\eta(y - h)$ is a test function (when extended by 0 outside Z_h). The second integral, by (7.9), is smaller than

$$\int_{x_1}^{x_2}(1 - \eta)(\phi(x) - h) \, dx + \int_{Z_h}(1 - \eta)(\gamma - I_{\{u>0\}}).$$

Letting $\varepsilon \to 0$, we get $\int_{Z_h}(u_y + \gamma) \leq 0$. Since finally $u_y\gamma = u_y$, $\gamma^2 \leq \gamma$ a.e., (7.8) follows.

Corollary 7.5. *If $u = 0$ in $B_r(X_0) \subset \Omega$, then $\gamma = 0$ a.e. in $B_r(X_0)$.*

Indeed, we can take Z_h with base $(x_0 - r, x_0 + r)$ and by applying Lemma 7.4 obtain

$$\int_{Z_h}\gamma^2 = \int_{Z_h}\left(\gamma u_y + \gamma^2\right) \leq 0,$$

which yields the assertion.

Lemma 7.6. *Let the assumptions of Lemma 7.4 hold and suppose that $u(x, h) \leqslant M$ if $x_1 \leqslant x \leqslant x_2$. Then*

$$u(x, y) = 0 \qquad if \ x_1 \leqslant x \leqslant x_2, \quad y \geqslant h + M.$$

Proof. The proof involves a comparison with a function v:

$$v(y) = \max [C - (y - h), 0], \qquad C > M.$$

For any $\delta > 0$, $\varepsilon > 0$, let

$$\psi_\delta(s) = \max \left(0, \min \left(\frac{s}{\delta}, 1 \right) \right)$$

$$d_\varepsilon(X) = \min \left\{ \frac{1}{\varepsilon} \, \text{dist} \left(X, Z \cap (v > 0) \right), 1 \right\},$$

where $Z = Z_h$.
  Then

$$-d_\varepsilon(\psi_\delta(u) - 1) + \psi_\delta(u - v)$$

is a test function when extended by zero outside Z_h. Hence

$$\int_Z \nabla [d_\varepsilon(\psi_\delta(u) - 1) - \psi_\delta(u - v)] \cdot (\nabla u + \gamma e) \geqslant 0.$$

Since $\nabla v = (0, -1)$, also

$$\int_Z \nabla \psi_\delta(u - v) \cdot (\nabla v + I_{\{v > 0\}} e) = 0.$$

Adding, we obtain

$$0 \leqslant \int_Z \left\{ \nabla \psi_\delta(u - v) \cdot (\nabla(v - u) - (\gamma - I_{\{v > 0\}}) e \right\}$$

$$- \int_Z \nabla [d_\varepsilon(1 - \psi_\delta(u))] \cdot (\nabla u + \gamma e).$$

Therefore, by adding and subtracting

$$\int_{Z \cap \{v = 0\}} \nabla \psi_\delta(u) \cdot \gamma e,$$

we can write

$$(7.10) \quad \int_{Z \cap \{v > 0\}} \psi'_\delta(u - v) |\nabla(u - v)|^2 \leqslant$$

$$- \int_{Z \cap \{v = 0\}} \nabla \psi_\delta(u) \cdot (\nabla u + \gamma e)$$

$$- \int_Z \nabla \psi_\delta(u - v) \cdot \left[(\gamma - I_{\{v > 0\}}) e - \gamma I_{\{v = 0\}} e \right]$$

$$+ \int_Z d_\varepsilon \nabla \psi_\delta(u) \cdot (\nabla u + \gamma e)$$

$$- \int_Z (1 - \psi_\delta(u)) \nabla d_\varepsilon \cdot (\nabla u + \gamma e)$$

$$= J_1 + J_2 + J_3 + J_4.$$

Notice that in J_2, if $u = 0$, then $\nabla \psi_\delta(u - v) = 0$, whereas if $u > 0$, then $\gamma = 1$ and the second factor in the integrand vanishes. Thus

$$(7.11) \qquad\qquad\qquad\qquad J_2 = 0.$$

Next, in J_3 the integral can be taken just over $\{v = 0\}$ since $d_\varepsilon = 0$ if $v > 0$. Thus

$$(7.12)$$

$$J_1 + J_3 = - \int_{Z \cap \{v = 0\}} (1 - d_\varepsilon) \nabla \psi_\delta(u) \cdot (\nabla u + \gamma e) \to 0 \qquad \text{if } \varepsilon \to 0, \forall \delta.$$

The part of J_4 where $u = 0$ is negative since the integrand is $\nabla d_\varepsilon \cdot \gamma e \geqslant 0$. Thus

$$J_4 \leqslant C \int_{B_\varepsilon(\partial(v > 0) \cap \Omega) \cap \{0 < u < \delta\}} |\nabla d_\varepsilon|,$$

where $B_\varepsilon(E) = \varepsilon$-neighborhood of E; here we have used the fact that $u \in C^{0,1}$. Denoting by H^1 the one-dimensional Hausdorff measure, we then obtain (using again the Lipschitz nature of u)

$$J_4 \leqslant C H^1 \big[\partial(v > 0) \cap \Omega \cap \{0 < u < \delta + C\varepsilon\} \big].$$

But as $\delta + \varepsilon \to 0$, the set in brackets shrinks to the empty set. Hence

$$\limsup_{\delta + \varepsilon \to 0} J_4 \leqslant 0.$$

Recalling also (7.11), (7.12), we deduce from (7.10) that

$$(7.13) \quad \int_{Z \cap \{v>0\} \cap \{0<u-v<\delta\}} \psi'_\delta(u-v) \, |\nabla(u-v)|^2 \to 0 \qquad \text{if } \delta \to 0.$$

Now we can easily complete the proof of the lemma. Let $w = \max(u - v, 0)$. Then for any $\zeta \in C_0^\infty(B)$ (B any ball),

$$\int_{Z \cap B} \psi_\delta(w) \nabla \zeta \cdot \nabla w = \int_{Z \cap B} \nabla(\psi_\delta(w)\zeta) \cdot \nabla(u-v) - \int \zeta \nabla \psi_\delta(w) \cdot \nabla w.$$

The first integral on the right-hand side is equal to zero since $u - v$ is harmonic wherever $u - v > 0$ and $\psi_\delta(w)\zeta = 0$ on the boundary of $Z \cap B$. The last integral is bounded in absolute value by

$$\frac{C}{\delta} \int_{Z \cap B \cap \{0<w<\delta\}} |\nabla w|^2 \to 0$$

by (7.13), provided that $B \supset \{v > 0\}$. Taking $\delta \to 0$ we obtain

$$\int_{Z \cap B} \nabla \zeta \cdot \nabla w = 0 \qquad \text{provided that } \{v > 0\} \subset B.$$

Thus w is harmonic in $Z \cap \{v > 0\}$. Since $w = 0$ and $v > 0$ in a Z-neighborhood of $\partial Z \cap \{y = h\}$, we deduce by analytic continuation that $w = 0$ in $Z \cap \{v > 0\}$; that is, $u \le C - (y - h)$ if $h < y \le C + h$. Taking $C \downarrow M$, the assertion of the lemma follows.

Theorem 7.7. *If $S^-(x_0) < \phi(x_0) < S^+(x_0)$, then $\phi(x)$ is continuous in some δ-neighborhood of x_0 and $\gamma = 0$ if $y > \phi(x)$, $|x - x_0| < \delta$.*

Proof. Set $y_0 = \phi(x_0)$. Then $u(x_0, y) = 0$ for $y_0 < y < S^+(x_0)$. Observe next that u cannot be positive in any $\{x > x_0\}$-neighborhood of any point of the interval $\{(x_0, y); y_0 < y < S^+(x_0)\}$. Indeed, otherwise we shall have $u_x \ne 0$ at that point (by the maximum principle), which contradicts Lemma 7.2. It follows that for any $h > y_0$ there is a sequence $(\bar{x}_i, \bar{y}_i)$ such that

$$\bar{x}_i \downarrow x_0, \qquad \bar{y}_i \in (y_0, h), \qquad u(\bar{x}_i, \bar{y}_i) = 0.$$

Similarly, there is a sequence (x_i, y_i) such that

$$x_i \uparrow x_0, \qquad y_i \in (y_0, h), \qquad u(x_i, y_i) = 0.$$

Setting

$$Z_h = \Omega \cap \{(x, y); x_i < x < \bar{x}_i, y > h\},$$

we are now in a situation where Lemma 7.6 can be applied. Since $u \in C^{0,1}$,

$$u(x, h) \leq C\delta_i, \qquad \text{where } \delta_i = \bar{x}_i - x_i.$$

Consequently,

$$(7.14) \qquad u(x, y) = 0 \qquad \text{if } x_i < x < \bar{x}_i, \, y > h + C\delta_i.$$

Since h can be arbitrarily close to y_0, (7.14) implies that $\phi(x)$ is upper semicontinuous at $x = x_0$. Since ϕ is also lower semicontinuous, we conclude that $\phi(x)$ is continuous at $x = x_0$. Thus $\phi(x)$ is continuous as long as $(x, \phi(x))$ remains in Ω.

The last assertion of the theorem follows from Corollary 7.5 and the continuity of $\phi(x)$.

Remark 7.1. From the proofs of lower and upper semicontinuity of $\phi(x)$ we also see that if $\phi(x_0) = S^{\pm}(x_0)$ is continuous at $x = x_0$, then $\phi(x)$ is continuous at $x = x_0$.

Corollary 7.8. $\gamma = I_{\{u>0\}}$ a.e.

We finally prove:

Theorem 7.9. If $(x_0, \phi(x_0)) \in \Omega$, then $\phi(x)$ is analytic in a neighborhood of $x = x_0$.

Proof. Take a small interval $x_1 < x < x_2$ containing x_0 and let h, H be such that

$$S^{-}(x) < h < \phi(x) < H < S^{+}(x) \qquad \text{for } x_1 \leq x \leq x_2.$$

Introduce

$$Z = \{(x, y); \, x_1 < x < x_2, \, h < y < \infty\} \cap \Omega,$$

$$Z_H = \{(x, y); \, x_1 < x < x_2, \, h < y < H\}$$

and define a function w by

$$w(x, y) = \int_y^{\infty} u(x, s) \, ds = \int_y^{H} u(x, s) \, ds;$$

we conveniently define $u = 0$ in $Z \setminus Z_H$.

For any $\zeta \in C_0^\infty(Z_H)$, define a function $\eta \in C^\infty(Z)$ such that

$$\eta(x, y) = \int_h^y \zeta(x, s)\, ds \qquad \text{if } h < y < H,$$

$$= 0 \qquad \text{near } \partial Z \cup S_2.$$

Then $\pm\eta$ are test functions for (u, γ) (since they vanish on $\partial Z \cup S_2$) and therefore

$$(7.15) \qquad \int_Z \nabla\eta \cdot (\nabla u + I_{\{u>0\}}e) = 0.$$

By integration by parts,

$$\int_Z \eta_y u_y = \int_Z \zeta u_y = -\int_Z \zeta_y u = \int_Z \zeta_y w_y$$

and by change of order of integration

$$\int_Z \eta_x u_x = \int_Z \zeta_x w_x.$$

Noting also that $I_{\{u>0\}} = I_{\{w>0\}}$ we obtain from (7.15)

$$(7.16) \qquad \int_Z (\nabla\zeta \cdot \nabla w + \zeta I_{\{w>0\}}) = 0 \qquad \forall \zeta \in C_0^\infty(Z_H);$$

that is, $-\Delta w + I_{\{w>0\}} = 0$. Thus w is a solution in Z_H of the variational inequality

$$w \geq 0, \qquad -\Delta w \geq -1, \qquad w(\Delta w + 1) = 0.$$

Since also $w_y < 0$, we can apply Chapter 2, Theorems 6.1 and 6.2 to deduce the analyticity of $\phi(x)$.

PROBLEM

1. Let $Z = \{(x, y); x_1 < x < x_2, h < y < \infty\} \cap \Omega$ and suppose that $u \geq M - (y - h)$ on ∂Z for some $M > 0$. Prove that $u \geq M - (y - h)$ in Z.

[*Hint:* $\int_Z \nabla \max(v - u, 0) \cdot (\nabla u + \gamma e) = 0$ implies that

$$\int_{Z \cap \{u > 0\}} |\nabla \max(v - u, 0)|^2 = \int_Z \nabla \max(v - u, 0) \cdot (\nabla v + I_{\{v > 0\}} e)$$

$$- \int_{Z \cap \{u = 0\}} \nabla v \cdot (\nabla v + (1 - \gamma) e).\Big]$$

8. UNIQUENESS FOR THE FILTRATION PROBLEM

Definition 8.1. A solution (u_0, γ_0) is called S_3-*connected* if

$$\bar{C}_i \cap S_3 \neq \emptyset \qquad \text{for any component } C_i \text{ of } \{u_0 > 0\} \cap \Omega.$$

Theorem 8.1. *The solution of problem* (A) *is unique up to groundwater re-servoirs; that is, there is a unique S_3-connected solution u_0 such that any other solution u has the form*

(8.1) $$u(x, y) = u_0(x, y) + \Sigma (H_i - y)^+ I_{D_i},$$

where H_i are some real numbers and D_i are connected components of $\Omega \cap \{y < H_i\}$.

The terms $(H_i - y)^+ I_{D_i}$ represent groundwater reservoirs, or wells.

Clearly, if S_1 is given by either a monotone curve $y = k(x)$ or by a curve $y = k(x)$ having one local maximum but no local minimum, then no ground-water reservoirs can exist and the solution is therefore unique.

Proof. We first show that if D is a connected component of $\Omega \cap \{u > 0\}$ [u is a solution of problem (A)] such that

(8.2) $$\bar{D} \cap S_3 = \emptyset,$$

then

(8.3) $$u(x, y) = (H - y)^+ \qquad \text{in } D.$$

In view of Theorem 7.3, (8.2) implies that $\pi(D) \subset \pi(S_2)$. Let $\pi(D) = (x_0, x_1)$ and

$$Z = \{(x, y) \in \Omega, x_1 < x < x_2\}.$$

Then $\pm I_Z u = \pm I_D u$ is in $H^1(\Omega)$, by Lemma 6.4, and it is therefore a test function. Thus

$$(8.4) \qquad \int_Z \left(|\nabla u|^2 + \gamma u_y \right) = 0.$$

Next, by Lemma 7.4,

$$\int_Z \left(\gamma u_y + \gamma^2 \right) \le 0.$$

Combining this with (8.4), we obtain

$$\int_Z \left(u_x^2 + (u_y + \gamma)^2 \right) \le 0.$$

It follows that $u_x = 0$, $u_y = -\gamma$ in Z. Since $\gamma = I_D$ we easily deduce that $u = (H - y)^+$ in Z, for some real number H.

Now let $(\tilde{u}, \tilde{\gamma})$ be any solution of problem (A) and denote by D_i the components of $\{\tilde{u} > 0\}$ such that $D_i \cap S_3 = \varnothing$. Setting

$$(8.5) \qquad u_0 = \tilde{u} - \Sigma I_{D_i} \tilde{u}, \qquad \gamma_0 = \tilde{\gamma} - \Sigma I_{D_i}$$

we have, for any test function ζ,

$$\int_\Omega \nabla \zeta \cdot (\nabla u_0 + \gamma_0 e) = \int_\Omega \nabla \zeta \cdot (\nabla \tilde{u} + \tilde{\gamma} e) - \Sigma \int_{D_i} (-\zeta_y + \zeta_y) = 0.$$

Thus (u_0, γ_0) is S_3-connected solution and from the previous proof we obtain the decomposition (8.1).

To complete the proof of Theorem 8.1, it remains to establish:

Lemma 8.2. *There exists at most one S_3-connected solution of problem (A).*

Proof. Suppose that (u_1, γ_1), (u_2, γ_2) are two S_3-connected solutions and denote their free boundaries by $x = \phi_1(x)$, $x = \phi_2(x)$, respectively. Set

$$u_0 = u_1 \wedge u_2, \qquad \gamma_0 = \gamma_1 \wedge \gamma_2, \qquad \phi_0 = \phi_1 \wedge \phi_2,$$

$$A = \{u_0 > 0\}.$$

Then, for any $\zeta \in H^1(\Omega) \cap C(\overline{\Omega})$, $\zeta \ge 0$,

$$(8.6) \qquad \int_\Omega \nabla \zeta \cdot [\nabla(u_i - u_0) + (\gamma_i - \gamma_0)] \le \int_{L_i} \zeta(x, \phi_i(x)) \, dx,$$

where

$$L_i = \left\{ x \in [\sigma_*, \tau_*], \phi_0(x) < \phi_i(x) \right\}.$$

Indeed, if $\xi = \min[u_i - u_0, \varepsilon \zeta]$, $\varepsilon > 0$, then $\xi = 0$ on $S_2 \cup S_3$ and thus $\pm \xi$ is a test function. It follows that for $i \neq j$,

$$\int_\Omega \nabla \xi \cdot \left[\nabla(u_i - u_j) + (\gamma_i - \gamma_j)e \right] = 0,$$

or

$$\int_\Omega \nabla \xi \cdot \nabla(u_i - u_0) + \int_\Omega (\gamma_i - \gamma_0)\xi_y = 0.$$

Since the first integral is equal to

$$\int_{\Omega \cap [u_i - u_0 \leq \varepsilon \zeta]} | \nabla(u_i - u_0) |^2 + \varepsilon \int_{\Omega \cap [u_i - u_0 > \varepsilon \zeta]} \nabla \zeta \cdot \nabla(u_i - u_0),$$

we obtain

$$(8.7) \quad \int_{\Omega \cap [u_i - u_0 > \varepsilon \zeta]} \nabla \zeta \cdot \nabla(u_i - u_0) + \int_\Omega (\gamma_i - \gamma_0)\zeta_y$$

$$\leq \int_\Omega (\gamma_i - \gamma_0)\left[\zeta - \min \left(\zeta, \frac{u_i - u_0}{\varepsilon} \right) \right]_y.$$

Observing that on $\{u_0 > 0\}$, we have $u_i > 0$ and $\gamma_i = \gamma_0 = 1$, the last integral becomes (using Corollary 7.8)

$$\int_{\Omega \cap \{u_0 = 0, \zeta > u_i/\varepsilon\}} (\gamma_i - \gamma_0)\left(\zeta - \frac{u_i}{\varepsilon} \right)_y^+ \leq \liminf_{\delta \downarrow 0} \int_{\{\phi_0(x) < y < \phi_i(x)\}} \left(\zeta - \frac{u_i}{\varepsilon} \right)_y^+$$

$$= \liminf_{\delta \downarrow 0} \int_{L_i} \left(\zeta - \frac{u_i}{\varepsilon} \right)^+ (x, \phi_i(x) - \delta) \, dx$$

$$\leq \int_{L_i} \zeta^+ (x, \phi_i(x)).$$

If we substitute this into (8.7), we obtain

$$\int_{\Omega \cap [u_i - u_0 > \varepsilon \zeta]} \nabla \zeta \cdot \nabla(u_i - u_0) + \int_\Omega (\gamma_i - \gamma_0)\zeta_y \leq \int_{L_i} \zeta^+ (x, \phi_i(x)) \, dx.$$

We now let $\varepsilon \to 0$. Since $u_i \geq u_0$ and since $\nabla(u_i - u_0) = 0$ on the set $\{u_i - u_0 = 0\}$, (8.6) follows.

Let B_r be a disc with center at one of the components $S_{3,j}$ of S_3 and with radius r sufficiently small so that $\pi(B_r) \subset \pi(S_{3,j})$, and set $\Gamma_0 = \partial B_r \cap \Omega$. Let W be a solution of

$$\Delta W = 0 \qquad \text{in } B_r,$$

$$W = 1 \qquad \text{on } \Gamma_0,$$

$$0 \leq W \leq 1 \qquad \text{on } \partial B_r \setminus \Gamma_0, W \not\equiv 1,$$

$$W = 1 \qquad \text{in } \Omega \setminus B_r, W \in H^1(\Omega).$$

Then, by the maximum principle

$$(8.8) \qquad\qquad\qquad\qquad 0 \leq W \leq 1,$$

$$(8.9) \qquad \frac{\partial W}{\partial \nu} > 0 \text{ on } \Gamma_0 \qquad (\nu = \text{outward normal to } \partial B_r).$$

By Green's formula,

$$\int_{\Gamma_0} (u_i - u_0) \frac{\partial W}{\partial \nu} = \int_{\Omega \cap B_r} \nabla(u_i - u_0) \cdot \nabla W;$$

hence

$$(8.10) \qquad \int_{\Gamma_0} (u_i - u_0) \frac{\partial W}{\partial \nu} = \int_{\Omega} \left[\nabla(u_i - u_0) \cdot \Delta W + (\gamma_i - \gamma_0) W_y \right].$$

For any small $\varepsilon > 0$, let $\alpha_\varepsilon \in C_0^\infty$, $\alpha_\varepsilon = 1$ on A_0, $\alpha_\varepsilon = 0$ if $\text{dist}(X, A_0) > \varepsilon$, $0 \leq \alpha_\varepsilon \leq 1$. In view of (8.8), $(1 - \alpha_\varepsilon)W$ is a nonnegative function; it vanishes on S_3 provided that ε is small enough. Thus it is a test function. We therefore have

$$\int_{\Omega} \left\{ \nabla u_i \cdot \nabla((1 - \alpha_\varepsilon)W) + \gamma_i [(1 - \alpha_\varepsilon)W]_y \right\} \leq 0.$$

But since $(1 - \alpha_\varepsilon)W = 0$ on A_0 and $u_0 = \gamma_0 = 0$ outside A_0 (by Corollary 7.8),

$$\int_{\Omega} \left\{ \nabla u_0 \cdot \nabla((1 - \alpha_\varepsilon)W) + \gamma_0 [(1 - \alpha_\varepsilon)W]_y \right\} = 0.$$

Subtracting this from the preceding inequality, we obtain

$$(8.11) \quad \int_{\Omega} \left[\nabla(u_i - u_0) \cdot \nabla W + (\gamma_i - \gamma_0) W_y \right]$$

$$\leqslant \int_{\Omega} \left[\nabla(u_i - u_0) \cdot \nabla(\alpha_\varepsilon W) + (\gamma_i - \gamma_0)(\alpha_\varepsilon W)_y \right].$$

Adding both sides of (8.10), (8.11) and evaluating the right-hand side of (8.11) by (8.6), we get

$$\int_{\Gamma_0} (u_i - u_0) \frac{\partial W}{\partial \nu} \leqslant \int_{L_i} (\alpha_\varepsilon W)(x, \phi_i(x)) \, dx.$$

As $\varepsilon \downarrow 0$, $\alpha_\varepsilon(x, \phi_i(x)) \to 0$ for any $x \in L_i$. Therefore, we obtain

$$\int_{\Gamma_0} (u_i - u_0) \frac{\partial W}{\partial \nu} \leqslant 0.$$

Recalling (8.9), we conclude that $u_i = u_0$ on Γ_0 for $i = 1, 2$; that is, $u_1 = u_2$ on Γ_0.

If we denote by C_j^i the component of $\{u_i > 0\}$ that contains $\Omega \cap B_r$, then the previous proof (with $B_{r'}$, $0 < r' < r$) gives $u_1 = u_2$ in $\Omega \cap B_r$. By analytic continuation, we then also have $u_1 = u_2$ in $C_j^1 \cap C_j^2$ and therefore also $C_j^1 = C_j^2$; this completes the proof.

PROBLEMS

1. Suppose that S_3 is connected with $\pi(S_3) = [\sigma_1, \tau_1]$ and S_1 is a curve $y = k(x)$ such that for some $x_1 > \tau_1$: (i) $k(x)$ is monotone if $x > x_1$; (ii) if $y_0 = \max_{\tau_1 \leqslant x \leqslant x_1} k(x)$, then the line segment $\{\tau_1 \leqslant x \leqslant x_1, y = y_0\}$ lies below S_2. Show that any solution u is positive on $\{(x, k(x)), x > \tau_1\}$.
    [*Hint*: If $u(x, y) = 0$ for $x > x^*$, $x^* \in (\tau_1, x_1)$, let $Q_\varepsilon = \{(x, y) \in \Omega, x < x^*, y < y_0 + \varepsilon\}$ and take for small enough ε the test function $-(u - (y_0 + \varepsilon - y))^- I_{Q_\varepsilon}$; show that $\int_{Q_\varepsilon} |\nabla[u - (y_0 + \varepsilon - y)^-]|^2 \leqslant 0$.]

2. Re-index the reservoirs so that $h_1 \geqslant h_2 \geqslant \cdots \geqslant h_k$. Prove that the S_3-connected solution u satisfies $u \leqslant (h_1 - y)^+$. Similarly, if C is a component of a solution u such that $\overline{C} \cap S_{3,i} = \varnothing$ for $1 \leqslant i \leqslant j$, then $u \leqslant (h_{j+1} - y)^+$ in C.
    [*Hint*: $\zeta = (u - (h_1 - y)^+)^+$ vanishes on $S_2 \cup S_3$, and ζ is then a test

function. Deduce

$$\int_{\Omega \cap [y \leqslant h_1]} \left| \nabla \left(u - (h_1 - y)^+ \right)^+ \right|^2 + \int_{\Omega \cap [y > h_1]} |\nabla u|^2 + \int_{\Omega \cap [y > h_1]} \gamma u_y \leqslant 0$$

and use Lemma 7.4 ($\Omega \cap [y > h_1]$ is the union of sets Z_h) to deduce that

$$\left[\int_{\Omega \cap [y \leqslant h_1]} \left| \nabla \left(u - (h_1 - y)^+ \right)^+ \right|^2 + \int_{\Omega \cap [y > h_1]} \left[u_x^2 + (u_y + \gamma)^2 \right] \leqslant 0. \right]$$

3. Take again $h_1 \geqslant h_2 \geqslant \cdots \geqslant h_k$ and set $h_{k+1} = \inf S^-$. Let u be the S_3-connected solution and let $h \in [h_{j+1}, h_j)$, $U_h = \{(x, y) \in \Omega; u(x, y) > (h - y)^+\}$ (U_h is empty if $h > h_1$; see Problem 2). Denote by $C_{h,i}$ the connected component of U_h satisfying $\overline{C_{h,i}} \supset S_{3,i}$. Show that $U_h = C_{h,1} \cup C_{h,2} \cup \cdots \cup C_{h,j}$ (some of the $C_{h,i}$ may coincide).
   [*Hint*: Assume for simplicity that $C_{h,i}$ are all disjoint and let $C_h' = \Omega \setminus \cup_{i=1}^j C_{h,i}$. Then $\zeta = I_{C_h'}(u - (h - y)^+)^+$ is a test function. Derive

$$\int_{C_h' \cap [y \leqslant h]} \left| \nabla \left(u - (h - y)^+ \right) \right|^2 + \int_{C_h' \cap [y > h]} \left(|\nabla u|^2 + \gamma u_y \right) \leqslant 0.$$

By Lemma 7.4,

$$\int_{C_h' \cap [y > h]} \left(\gamma u_y + \gamma^2 \right) \leqslant 0,$$

so that

$$\int_{C_h' \cap [y \leqslant h]} \left| \nabla \left(u - (h - y)^+ \right)^{\mathsf{r}} \right|^2 + \int_{C_h' \cap [y > h]} \left[u_x^2 + (u_y + \gamma)^2 \right] \leqslant 0.$$

Deduce that $u \geqslant 0$ in $C_h' \cap [y > h]$ and $u \leqslant (h - y)^+$ in C_h'.]

4. Let u be the S_3-connected solution and let $(x, h) \in \Omega \ \forall x_1 \leqslant x \leqslant x_2$. If (x_1, h), (x_2, h) belong to the same connected component $C_{h,i}$ of U_h, show that $u(x, y) > 0$ if $(x, y) \in \Omega \cap \{[x_1, x_2] \times (-\infty, h)\}$.
   [*Hint*: Suppose that $u(x_0, y_0) = 0$, $(x_0, y_0) \in \Omega$, $x_1 \leqslant x_0 \leqslant x_2$, $y_0 < h$ and connect (x_1, h) to (x_2, h) in $C_{h,i}$ by a curve Γ; Γ intersects $\{x = x_0\}$ below y_0. We may take Γ to lie below $\{y = h\}$ and denote by D the domain bounded by Γ and $\{y = h\}$. Then $-I_D(u - (h - y))^-$ is a test function, giving

$$\int_{D \cap \{u > 0\}} \left| \nabla (u - (h - y))^- \right|^2 \leqslant 0,$$

i.e., $(u - (h - y))^- = C$ in $D \cap \{u > 0\}$,

where C is a constant. Derive a contradiction.]

5. Suppose that there is only one reservoir, and set $\pi(S_3) = [\sigma_1, \tau_1]$. Let S_2 be given by $y = k(x)$, where $k(x) \wedge h_1$ is a monotone increasing function for $x < \sigma_1$ and monotone decreasing for $x > \tau_1$. Then for the S_3-connected solution there holds: $\phi(x)$ is monotone increasing for $x < \sigma_1$ and monotone decreasing for $x > \tau_1$.

   [*Hint:* Suppose that the assertion is false for $x > \tau_1$. Then there exist $\tau_1 < x_0 < x_1$, $h < \phi(x_1) < h_1$, $y_0 = \phi(x_0) < h$. Connect $(x_1, \phi(x_0))$ to a point (α_1, β_1) near S_3 by a curve Γ in $C_{h,1}$ and derive a contradiction to the result of Problem 4.]

6. Suppose that there are two reservoirs, $h_1 \geqslant h_2$, $\pi(S_i) = [\sigma_i, \tau_i]$, $\sigma_* < \sigma_1 < \tau_1 < \sigma_2 < \tau_2 = \tau_*$ and S_2 is given by $y = k(x)$, where $k(x) \wedge h_1$ is monotone increasing for $x \in (\sigma_*, \sigma_1)$ and monotone decreasing for $x \in (\tau_1, \sigma_2)$. Show that (i) the free boundary $\phi(x)$ is increasing for $x \in (\sigma_*, \sigma_1)$; (ii) $\phi(x)$ is decreasing in $x \in (\tau_1, \sigma_2)$ in any subinterval $(\tau_1, \tilde{x})$, where $\phi(x) \geqslant h_2$; (iii) if $\min_{\tau_1 \leqslant x \leqslant \sigma_2} \phi(x) = \phi(x^*) < h_2$, then $\phi(x)$ is increasing for $x \in (x^*, \sigma_2)$.

   [*Hint:* Use the method of Problem 5.]

7. Consider the case of two reservoirs with $\pi(S_{3,1}) = [\sigma_1, \tau_1]$, $\pi(S_{3,2}) = [\sigma_2, \tau_2]$, $\sigma_1 = \sigma_*$, $\tau_2 = \tau_*$, $h_1 > h_2$. Let S_2 be given by $y = k(x)$ with $k(x) \wedge h_1$ monotone decreasing and suppose that S_1 lies below $y = h_2$. Prove that there is a unique solution and that its free boundary $y = \phi(x)$ is a monotone decreasing function (for $\tau_1 < x < \sigma_2$).

   [*Hint:* Use $\zeta = \pm (u - (h_2 - y)^+)^-$ as a test function to obtain

   $$-\int_{[y < h_2]} |\nabla(u - (h_2 - y))^-|^2 + \int_{[y < h_2]} (\gamma - 1)(u - (h_2 - y))_y^- = 0.$$

   The second integral equals $-\text{meas}\,\{[u = 0] \cap [y < h_2]\}$. Thus $u > 0$ in $[y < h_2]$ and in particular $\phi(x) \geqslant h_2$. It follows that u is S_3-connected and thus uniquely determined. The monotonicity of ϕ follows as in Problem 6(ii).]

9. THE FILTRATION PROBLEM IN N DIMENSIONS

In this section we generalize the results of Section 6 to 8 to n dimensions. Wherever possible we shall use the same notation as before. Thus the dam is an n-dimensional domain Ω bounded by surfaces S_1 (the impervious part), S_2 (the part in contact with air), and S_3 (the part in contact with the water reservoir).

We assume [see (6.2)]:

(9.1)

S_1 and $S_2 \cup S_3$ are piecewise $C^{1,1}$ surfaces, both being graphs in the y-direction, and $S_2 \cup S_3$ lies above S_1; further, S_2 is open and is given by $y = f(x)$ ($x \in B_0$, B_0 open) where $f \in C^{1,1}(\overline{B}_0)$.

Here by "S piecewise $C^{1,1}$" we mean that there is a finite triangulation $\{\Sigma_i\}$ of S where Σ_i are uniformly in $C^{1,1}$ and $\partial \Sigma_i$ are piecewise $C^{1,1}$ ($n-2$)-dimensional manifolds.

We denote by $X = (x, y)$ points in R^n, where x is a variable point in R^{n-1} and y is the height parameter.

We shall also assume that

(9.2)

S_3 consists of a finite number of disjoint components $S_{3,i}$; if $(x_0, y_0) \in \Omega$ and $x_0 \in \pi(\partial S_{3,i_0})$, then there exists a ball K in R^{n-1} such that $K \subset \pi(S_{3,i_0})$ and $x_0 \in \partial K$.

The formulation of problem (A) remains unchanged. Theorems 6.1 and 6.2 remain valid with the same proof. The proof that $u \in W_{loc}^{1,p}(\Omega)$ actually extends up to the boundary, near any $C^{1,1}$ portion of the boundary. In particular,

$$u \in C^{\beta}(\overline{\Omega}_0) \qquad \text{for any } 0 < \beta < 1,$$

where Ω_0 is any subdomain of Ω with $\overline{\Omega}_0 \subset \Omega \cup S_2$.

We shall now extend the results of Section 7. First, Theorem 7.1 clearly remains valid without change in the proof. Next:

Lemma 9.1. *The set $\Omega \cap \{u > 0\}$ has the form*

$$\{(x, y); S^-(x) < y < \phi(x)\},$$

where $\phi(x)$ is lower semicontinuous, and

$$\phi(x) = S^+(x) \qquad \text{if } x \in \text{int}\,(\pi(S_3)).$$

The proof is the same as for Theorem 7.3.

The proof of Lemma 7.2 is strictly two-dimensional. We shall now extend this result to n dimensions. This is, in fact, the main new step needed for establishing the smoothness of the free boundary in n dimensions.

Fix points

$$X_1 = (x_0, y_1) \in S_2,$$

$$X_0 = (x_0, y_0) \in \Omega, \qquad y_0 < y_1.$$

Lemma 9.2. *If $u(x_0, y_0) = 0$, then for any $0 < \beta < 1$, $\varepsilon > 0$,*

$$(9.3) \qquad u(x, y) = O\big(|x - x_0|^{1+\beta}\big) \qquad \forall \, y_0 + \varepsilon < y < y_1.$$

The proof depends on several lemmas. We choose small positive numbers δ and δ_0 such that

$$B_\delta(x_0) \subset \pi(S_2),$$

$$B_\delta(x_0) \times \{h\} \subset \Omega, \qquad \text{where } h = y_0 - \delta_0.$$

Let

$$Z = B_\delta(x_0) \times (h, H),$$

where H is sufficiently large, say

$$H > S^+(x) \qquad \forall \, x \in \overline{\Omega},$$

and extend u by 0 into $Z \setminus \Omega$.

Lemma 9.3. *There exists a bounded nonpositive measurable function $\theta(x)$, $x \in B_\delta(x_0)$, such that*

$$(9.4) \qquad \int_Z \nabla \zeta \cdot (\nabla u + I_{\{u>0\}} e) = \int_{B_\delta(x_0)} \theta(x) \zeta(x, f(x)) \, dx$$

for any $\zeta \in C_0^1(Z)$.

Proof. Let u_Ω be the solution of

$$(9.5)$$

$$\int_\Omega \nabla \zeta \cdot (\nabla u_\Omega + e) = 0 \qquad \forall \, \zeta \in H^1(\Omega), \quad \zeta \geq 0 \text{ on } S_2, \quad \zeta = 0 \text{ on } S_3,$$

$$u_\Omega \in H^1(\Omega), \quad u_\Omega = u^0 \text{ on } S_2 \cup S_3.$$

By comparison (see Problem 1) we have $u \leq u_\Omega$. Since also $u = u_\Omega = 0$ on S_2, we obtain, formally,

$$\left| \frac{\partial u}{\partial \nu} \right| \leq \left| \frac{\partial u_\Omega}{\partial \nu} \right| \leq C \qquad \text{on } S_2 \cap Z,$$

where C is a constant, and (9.4) easily follows with

$$\theta = -\left|\frac{\partial u}{\partial \nu}\right|\frac{dS}{dx}, \qquad dS = \text{surface element on } S_2.$$

However, since $\partial u/\partial \nu$ is not well defined, we proceed somewhat differently. Set $Z_0 = Z \cap \Omega$. The function $v \equiv u - u_\Omega$ satisfies

$$\Delta v = -\nabla\left(I_{\{u>0\}}e\right) \equiv \mu_1 \qquad \text{in } \mathcal{D}'(Z_0),$$

where μ_1 is positive measure in Z_0; indeed, if $\zeta \in C_0^1(Z_0)$, then [since $\Delta u_\Omega = 0$ in Z_0 and $\pm\zeta$ is a test function for problem (A)] the distribution Δv applied to ζ gives

$$\langle \Delta v, \zeta \rangle = -\int \nabla v \cdot \nabla\zeta = -\int \nabla u \cdot \nabla\zeta$$

$$= \int I_{u>0}e \cdot \nabla\zeta = \int_{\{u>0\}} \zeta_y = \int \zeta(x, \phi(x))\, dx$$

$$\equiv \langle \mu_1, \zeta \rangle.$$

We extend the function v by 0 into $Z \setminus Z_0$. Since $v \leqslant 0$ in Z_0 and $v = 0$ on $\partial Z_0 \cap S_2$, the extended function v satisfies (see Problem 2)

(9.6) $$\Delta v \leqslant \mu_1 \qquad \text{in } \mathcal{D}'(Z);$$

that is, for any $\zeta \in C_0^1(Z)$, $\zeta \geqslant 0$,

$$\int v\,\Delta\zeta \leqslant \langle \mu_1, \zeta \rangle \equiv \int I_{\{u>0\}}e \cdot \nabla\zeta.$$

Since $v \in C^0(\overline{Z})$, the left-hand side is equal to

$$-\int \nabla v \cdot \nabla\zeta = -\int \nabla u \cdot \nabla\zeta + \int \nabla u_\Omega \cdot \nabla\zeta.$$

We thus obtain

$$0 \geqslant \int (\nabla u + I_{\{u>0\}}e) \cdot \nabla\zeta \geqslant \int \nabla u_\Omega \cdot \nabla\zeta \geqslant \int_{S_2 \cap Z} \left|\frac{\partial}{\partial \nu} u_\Omega\right|\zeta,$$

from which (9.4) easily follows. Indeed, the functional

$$F(\zeta) = \int (\nabla u + I_{\{u>0\}}e) \cdot \nabla\zeta$$

is a bounded functional on $C_0^0(S_2 \cap Z)$ and

$$0 \leqslant -F(\zeta) \leqslant C \int_{S_2 \cap Z} \zeta \qquad \text{if } \zeta \geqslant 0.$$

By the Radon–Nikodym theorem it follows that

$$F(\zeta) = \int_{B_\delta(x_0)} \theta(x)\zeta(x, f(x)) \, dx, \qquad 0 \leqslant -\theta(x) \leqslant C.$$

Finally, since any $\zeta \in H_0^1(Z)$ can be written as $\zeta^+ - \zeta^-$, (9.4) follows for any $\zeta \in C_0^1(Z)$.

Lemma 9.4. *The function*

(9.7) $$w(x, y) = \int_y^H u(x, s) \, ds$$

belongs to $C^{1+\beta}(\overline{Z}_0)$, *for any* $0 < \beta < 1$.

Proof. Taking

$$\zeta(x, y) = \int_h^y \xi(x, s) \, ds, \qquad \xi \geqslant 0, \quad \xi = 0 \text{ on } \partial Z \cap \{y > h\},$$

we can transform (9.4) into [see the derivation of (7.16)]

$$\int_Z (\nabla \xi \cdot \nabla w + \xi I_{\{w>0\}}) = \int_{B_\delta(x_0)} \theta(x) \left(\int_{f(x)}^\infty \xi(x, s) \, ds \right) dx.$$

It follows that

(9.8)
$$\Delta w = I_{\{w>0\}} + \mu_1,$$
$$\mu_1 \text{ bounded positive measure in } Z,$$

and the assertion follows from elliptic estimates.

Lemma 9.5. *There holds*

$$u_y \leqslant Cu \qquad \text{in } Z_0,$$

where C *is a positive constant.*

Proof. We first show that

(9.9) $$u_y \leqslant c \qquad \text{in } Z_0, \quad c \text{ constant.}$$

Consider finite differences

$$\Delta^\gamma_y u(x, y) = \frac{1}{\gamma}(u(x, y) - u(x, y - \gamma)), \qquad \gamma > 0$$

in Z_0. Recall that u (and thus $\Delta^\gamma_y u$) is continuous in $\overline{Z}_0$. Also, $\Delta^\gamma_y u$ is harmonic in $Z_0 \cap \{u > 0\}$. Let v^γ be the solution of

$$\Delta v^\gamma = 0 \qquad\qquad \text{on } Z_0,$$

$$v^\gamma = \max(\Delta^\gamma_y u, 0) \qquad \text{on } \partial Z_0.$$

Since $v^\gamma \geq 0$ and $\Delta^\gamma_y u \leq 0$ on $Z_0 \cap \partial\{u > 0\}$, the maximum principle gives

$$v^\gamma \geq \Delta^\gamma_y u \qquad \text{in } Z_0 \cap \{u > 0\}.$$

Since S_2 is in $C^{1,1}$ we can represent v^γ in the form

$$v^\gamma(X) = -\int_{\partial Z_0} \frac{\partial G}{\partial \nu}(X, Y) \max(\Delta^\gamma_y u, 0)\, dS,$$

where G is Green's function in Z_0. It follows that

$$\Delta^\gamma_y u(X) \leq v^\gamma(X) \leq C\int_{\partial_0 Z_0} |\Delta^\gamma_y u|\, dS$$

provided that $\text{dist}(X, \partial_0 Z_0) \geq \varepsilon_0 > 0$, where

$$\partial_0 Z_0 = [B_\delta(x_0) \times \{h\}] \cup [(\partial B_\delta(x_0) \times (h, H)) \cap \Omega];$$

C depends on ε_0. Integrating with respect to δ and h, in some small intervals, we obtain

$$\Delta^\gamma_y u(X) \leq C\int |\Delta^\gamma_y u|\, dx\, dy,$$

where the integration is over some n-dimensional neighborhood of $\partial_0 Z_0$. Recalling that $u_y \in L^2(\Omega)$, we obtain

$$\Delta^\gamma_y u(X) \leq C\int_\Omega |u_y|,$$

and (9.9) (with slightly smaller δ and h in the definition of Z_0) follows upon taking $\gamma \to 0$.

Consider the function

$$z = \varepsilon u_y - Mu + w \qquad \text{in } Z_0,$$

where $\varepsilon > 0$ is small and M is large.

If $\text{dist}(X, \partial\{u > 0\}) > \varepsilon$, then $u \geqslant c_0(\varepsilon) > 0$ and by choosing $M = M(\varepsilon)$ sufficiently large, we obtain $z(X) < 0$. If $\text{dist}(X, \partial\{u > 0\}) < \varepsilon$, then $w < C\varepsilon$ (by Lemma 9.4) and $|\varepsilon u_y| \leqslant c\varepsilon$ (by (9.9)); thus $z \leqslant C_0\varepsilon$ (C_0 constant independent of ε).

Recalling that, by (9.8),

$$\Delta z = \Delta w \geqslant 1 \qquad \text{in } Z_0 \cap \{u > 0\},$$

we can apply the argument used in the proof of Chapter 2, Theorem 6.3 (with w replaced by $-z$) to deduce that $z < 0$ in $Z_0' \equiv [B_{\delta'}(x_0) \times (h, H)] \cap \Omega$, for any $\delta' < \delta$, provided that ε is sufficiently small. Thus

$$u_y \leqslant Cu \qquad \text{in } Z_0';$$

by starting with a slightly larger δ, we can then achieve this estimate in Z_0.

Proof of Lemma 9.2. Take $h < h' < y_0 < y_1$. For any $h < y' < h' < y < y_1$, $x \in B_\delta(x_0)$,

$$u(x, y) = u(x, y') + \int_{y'}^{y} u_s(x, s)\, ds$$

$$\leqslant u(x, y') + C\int_{y'}^{y} u(x, s)\, ds \qquad \text{by Lemma 9.5,}$$

$$\leqslant u(x, y') + Cw(x, y').$$

Integrating over y', $h < y' < h'$, yields

$$(9.10) \qquad u(x, y) \leqslant C_1 w(x, h), \qquad C_1 = \frac{1}{h' - h} + C.$$

Since $w(x_0, y) = 0$ if $y > y_0$, $w \geqslant 0$, $w \in C^{1+\beta}$ (by Lemma 9.4), we have $\nabla w(x_0, y) = 0$ and

$$w(x, y) \leqslant C|x - x_0|^{1+\beta}.$$

Combining this with (9.10), the assertion of Lemma 9.2 follows.

Corollary 9.6. *If $x_0 \in \pi(\partial S_3)$, then $u(x_0, y) > 0$ for all $(x_0, y) \in \Omega$.*

Indeed, if $u(x_0, y_0) = 0$ for some $(x_0, y_0) \in \Omega$, then by assumption (9.2) and the strong maximum principle,

$$\frac{\partial u}{\partial \nu} \neq 0 \qquad \text{at } (x_0, y) \in \Omega, \ y > y_0,$$

where the derivative is taken as a limit of finite differences in a suitable direction perpendicular to the *y*-axis. This contradicts Lemma 9.2.

Remark 9.1. The ball property assumed in (9.2) is used only in the proof of Corollary 9.6. Using Problem 7.6 of Chapter 2, it is sufficient to replace the ball *K* by a cone with vertex at x_0.

We next prove:

Theorem 9.7. *Suppose that* $X_0 = (x_0, y_0)$ *belongs to* $\Omega \cap \partial\{u > 0\}$. *Then, for some* $\delta > 0$, *the free boundary in* $[B_\delta(x_0) \times (y_0 - \delta, \infty)] \cap \Omega$ *is given by*

$$y = \phi(x), \qquad \phi(x) \text{ analytic;}$$

further, $\gamma = 0$ *if* $y > \phi(x)$.

Proof. Lemma 7.4 extends to *n* dimensions if we assume in this lemma that

$$u(x, y) = 0 \qquad \text{on } \partial B_\delta(x) \times (h, H).$$

Thus also Corollary 7.5 remains valid.

We shall next use Lemma 9.2 and the argument of Lemma 7.6 to establish the upper semicontinuity of $\phi(x)$.

Consider a domain *D* as in Figure 5.2; *D* is bounded below by $y = h$, laterally by $\partial B_\rho(x_0) \times (h, \infty)$, and above by a smooth surface $y = N(x)$; *D* is contained in the cylinder $Z = B_\rho(x_0) \times (h, \infty)$. We choose $N(x)$ such that

(9.11) $$e \cdot \nu \geqslant \text{const} > 0 \qquad \text{on } \{y = N(x)\} \cap \overline{\Omega},$$

where ν is the outward normal.

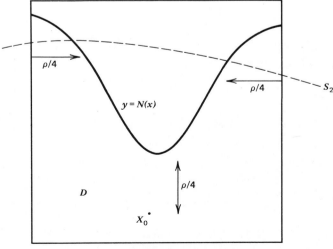

FIGURE 5.2

Let v_0 be a function satisfying

$$\Delta v_0 = 0 \qquad \text{in } D,$$

$$v_0 = 0 \qquad \text{on } y = N(x),$$

$$v_0 = \rho \text{ on the remaining parts of } \partial D.$$

Extend v_0 by 0 into $Z_0 \equiv Z \cap \Omega$. Then $v_0 \in H^1(Z_0)$. We claim that if ε and ρ are suitably small, then

$$(9.12) \qquad\qquad \varepsilon v_0 > u \qquad \text{on } \partial Z_0 \cap \Omega,$$

$$(9.13) \qquad \nabla(\varepsilon v_0 + e) \cdot \nu \geq 0 \qquad \text{on } \{y = N(x)\} \cap Z_0.$$

Indeed, in view of Lemma 9.2, (9.12) is satisfied if $\varepsilon \geq c\rho^{1+\beta}$. Next, (9.13) holds if

$$\varepsilon \nabla v_0 \cdot \nu + e \cdot \nu > 0,$$

that is, by (9.11), if

$$\varepsilon |\nabla v_0| \leq c_0, \qquad c_0 \text{ some positive constant.}$$

By scaling we find that $|\nabla v_0| \leq C$, so that if $\varepsilon \leq C_1$ (for a suitable small $C_1 > 0$), then (9.13) holds. Thus (9.12) and (9.13) hold if we choose $\varepsilon = C_1$, $c\rho^{1+\beta} \leq C_1$.

Consider the function $v = \varepsilon v_0$. In view of (9.12),

$$(9.14) \qquad\qquad \begin{aligned} v &\geq u \qquad \text{on } \tilde{S}_2, \\ v &> u \qquad \text{on } \partial Z_0 \cap \Omega, \end{aligned}$$

where

$$\tilde{S}_2 = (\partial Z_0 \cap \Omega) \cup (\partial Z_0 \cap S_2).$$

From (9.13) we also infer that

$$(9.15)$$

$$\int_{Z_0} \nabla \zeta \cdot (\nabla v + I_{\{v>0\}} e) \geq 0 \qquad \forall \zeta \in H^1(Z_0), \quad \zeta \geq 0, \quad \zeta = 0 \text{ on } \tilde{S}_2.$$

The functional v is thus a "supersolution" in Z_0 in the same sense as the function v used in the proof of Lemma 7.6. In fact, the same proof shows that

$$u \leq v \qquad \text{in } Z_0.$$

Recalling the shape of D in Figure 5.2, we deduce that $y = \phi(x)$ is indeed upper semicontinuous at x_0. It follows that $\phi(x)$ is continuous at $x = x_0$.

From the continuity of $\phi(x)$ and Corollary 7.5, we deduce that $\gamma = 0$ if $y > \phi(x)$. Finally, the analyticity of $\phi(x)$ follows as in Theorem 7.9.

We define the concept of S_3-connected solution exactly as in the case $n = 2$.

Theorem 9.8. *Theorem* 8.1 *extends to the case of n dimensions.*

Since $\gamma = I_{\{u > 0\}}$, the proof proceeds precisely as in the case $n = 2$.

PROBLEMS

1. Prove that $u \leq u_\Omega$.

   [*Hint*: Note that $u_\Omega > 0$ in Ω and take $\zeta = \pm \min(0, u_\Omega - u)$ as test functions.]

2. Prove (9.6).

   [*Hint*: Let $\tilde{d}$ = distance function to ∂Z_0, $d_\varepsilon = \tilde{d}/\varepsilon$ if $\tilde{d} < \varepsilon$, $d_\varepsilon = 1$ if $\tilde{d} > \varepsilon$. Let $\zeta \in C_0^1(Z)$, $\zeta \geq 0$. Then

   $$-\int \nabla v \, \nabla(\zeta d_\varepsilon) = \int I_{\{u > 0\}} e \cdot \nabla(\zeta d_\varepsilon)$$

   and

   $$\int \zeta I_{\{u > 0\}} e \cdot \nabla d_\varepsilon < 0 \qquad \text{if } \varepsilon \text{ is small.}$$

   Thus

   $$-\int \nabla v \cdot \nabla \zeta + I \leq \int I_{u > 0} e \cdot \nabla \zeta,$$

   where

   $$I = - \lim_{\varepsilon \to 0} \int \zeta \nabla v \cdot \nabla d_\varepsilon.$$

   Introduce tangential parameters $(s_1, \ldots, s_{n-1}) = s$ and normal parameter ρ near $S_2 \cap \partial Z_0$ and write

   $$-\nabla d_\varepsilon = \frac{1}{\varepsilon} \nu + \frac{1}{\varepsilon} O(\rho^\theta), \qquad \nu \text{ outward normal, } 0 < \theta < 1,$$

   $$\zeta = \zeta(s) + O(\rho).$$

   Then

   $$I = \overline{\lim} \left\{ \int_{Z_0 \cap \{\text{dist}(X, S_2) < \varepsilon\}} \zeta(s) \frac{\partial v}{\partial \rho} \right\}$$

   $$= \overline{\lim} \int_{Z_0 \cap \{\text{dist}(X, S_2) = \varepsilon\}} [-\zeta(s) v(s, \rho)] \geq 0. \Bigg]$$

10. THE TWO-PHASE STEFAN PROBLEM

The one-phase Stefan problem was introduced in Chapter 1, Section 9. In the two-phase problem the temperature of the ice is not identically zero. To give a classical formulation of this problem, we introduce bounded domains G_i in R^n,

$$G_0 \subset \bar{G}_0 \subset G_1 \subset \bar{G}_1 \subset G_2$$

and let

$$D_1 = G_1 \setminus \bar{G}_0, \qquad D_2 = G_2 \setminus \bar{G}_1.$$

Initially, D_1 is occupied by water and D_2 is occupied by ice. Set

$$\Gamma_1 = \partial G_0, \qquad \Gamma_2 = \partial G_2, \qquad \Gamma_0 = \partial D_1 \cap \partial D_2.$$

We denote the temperature in $\bar{D}_1 \cup \bar{D}_2$ by θ and seek a surface $\Gamma : \Phi(x, t) = 0$ and a function θ satisfying the following conditions:

$$\theta(x, 0) = h_i(x) \qquad \text{on } D_i,$$

$$\theta(x, t) = g_i(x, t) \qquad \text{on} \Gamma_i \times (0, \infty) \quad (i = 1, 2),$$

(10.1)

$$\theta = 0, \; k_1 \nabla_x \theta_1 \cdot \nabla_x \Phi - k_2 \nabla_x \theta_2 \cdot \nabla_x \Phi = \alpha \Phi_t \qquad \text{on } \Gamma,$$

$$k_i \Delta \theta_i - \frac{\partial \theta_i}{\partial t} = 0 \qquad \text{in } Q_i \equiv \bigcup_{t > 0} D_i(t),$$

where $D_i(s)$ is the domain in $\{t = s\}$ bounded by $\Gamma_i \times \{s\}$ and $\Gamma(s) : \Phi(x, s) = 0$, and $\theta_i = \theta$ in $\bar{Q}_i$. Here k_i, α are given positive numbers, h_i and g_i are given functions, and

(10.2) $$(-1)^{i-1} h_i \geqslant 0, \qquad (-1)^{i-1} g_i \geqslant 0.$$

Finally, $\Gamma(0)$ is supposed to coincide with Γ_0.

Γ is called the *free boundary*.

The formulation above is too restrictive. We therefore proceed to give a weak formulation. Set

$$\alpha_i = \frac{1}{k},$$

$$g = \frac{g_i}{\alpha_i} \qquad \text{on } \Gamma_i \times (0, \infty),$$

$$h = \frac{h_i}{\alpha_i} \qquad \text{on } D_i \times \{0\}$$

and introduce the function

$$a(u) = \begin{cases} \alpha_1 u & \text{if } u > 0 \\ \alpha_2 u - \alpha & \text{if } u \leqslant 0. \end{cases}$$

We shall denote by ν the outward normal to

$$G \equiv G_2 \setminus \overline{G}_0$$

and set

$$u = \frac{\theta_i}{\alpha_i} \qquad \text{in } Q_i.$$

Finally, we define

$$Q_T = G \times (0, T), \qquad G(t) = G \times \{t\}.$$

Definition 10.1. A pair (u, γ) is called a *weak solution* of the Stefan problem in Q_T if $u \in L^\infty(Q_T)$, $\gamma \in L^\infty(Q_T)$, $-\alpha \leqslant \gamma \leqslant 0$ a.e. on the set $\{u = 0\}$ and $\gamma = a(u)$ a.e. on $\{u \neq 0\}$; finally,

$$(10.3) \qquad \iint_{Q_T} (u\Delta\phi + \gamma\phi_t) \, dx \, dt = \int_0^T \int_{\partial G} g \frac{\partial\phi}{\partial\nu} \, dS \, dt - \int_G \gamma(x, 0)\phi(x, 0) \, dx$$

for any function ϕ satisfying:

$$D_x\phi, \quad D_x^2\phi, \quad D_t\phi \in C(\overline{Q}_T),$$

$$\phi = 0 \qquad \text{on } G(T) \text{ and on } \partial G \times (0, T).$$

ϕ is called a *test function*.

One can show that:

$$(10.4) \qquad \begin{array}{l} \text{if } u \text{ is a classical solution of } (10.1), \\ \text{then it is also a weak solution.} \end{array}$$

We shall assume:

$$(10.5) \qquad \begin{array}{l} \partial G \in C^{2+\beta} \quad (0 < \beta < 1); \ g_1 \geqslant \gamma_1 > 0, \quad g_2 \geqslant \gamma_2 > 0; \quad h_1 > 0 \\ \text{in } D_1, \quad h_2 < 0 \text{ in } D_2, \quad h \in C(\overline{G}) \cap H^{1,2}(G); \quad g \text{ is the restriction} \\ \text{of a function } \Psi \text{ with } D_x\Psi, \ D_x^2\Psi, \ D_t\Psi \text{ in } C(\overline{Q}_T). \end{array}$$

Theorem 10.1. *There exists a unique weak solution* (u, γ) *of the Stefan problem, and*

$$(10.6) \qquad \iint_{Q_T} |u_t|^2 < \infty,$$

$$(10.7) \qquad \operatorname*{ess\,sup}_{0 < t < T} \int_{G(t)} |\nabla_x u|^2 < \infty.$$

By uniqueness we mean that if (u, γ), $(\hat{u}, \tilde{\gamma})$ are two solutions, then $u = \hat{u}$ a.e.

Proof.   Let $a_m(u)$ be C^∞ functions which approximate the graph $a(u)$:

$$0 < \alpha_0 \leqslant a'_m(u), \qquad a_m(u) = a(u) \text{ if } |u| > \frac{1}{m}.$$

Consider the parabolic problem:

$$a'_m(v)v_t - \Delta v = 0 \qquad \text{in } Q_T,$$

(10.8)
$$v = g \qquad \text{on } \partial G \times (0, T),$$

$$v = h \qquad \text{on } G(0).$$

By reference 130 there exists a unique solution $v = v_m$ of (10.8). One can further show (see Problem 2) that there exists a δ_0-neighborhood W_i of Γ_i $(i = 1, 2)$ and an $\varepsilon_0 > 0$ such that

(10.9)
$$(-1)^{i-1}v_m \geqslant \varepsilon_0 \qquad \text{in } (W_i \cap D_i) \times (0, T).$$

By the maximum principle we also have

(10.10)
$$|v_m| \leqslant C,$$

where C is independent of m.

From (10.9) it follows that

$$\alpha_i \frac{\partial v_m}{\partial t} = \Delta v_m \qquad \text{in } (W_i \cap D_i) \times (0, T).$$

Using also (10.10) and standard parabolic estimates, we get

(10.11)
$$|\nabla v_m| \leqslant C \qquad \text{on } \partial G \times (0, T).$$

We now multiply both sides of the differential equation in (10.8) by $\partial v_m/\partial t$ and integrate over Q_σ. We easily find that

$$\iint_{Q_\sigma} \left(\frac{\partial v_m}{\partial t} \right)^2 + \int_{G(\sigma)} |\nabla v_m|^2 \leqslant C.$$

The existence proof can now be completed by taking a subsequence such that

$$v_m \to u \qquad \text{weakly in } H^{1,2}(Q_T) \text{ and a.e.},$$
$$a_m(v_m) \to \gamma \qquad \text{weak star in } L^\infty(Q_T).$$

The details are left to the reader.

To prove uniqueness, suppose that (u, γ) and $(\hat{u}, \hat{\gamma})$ are two solutions. Then

$$(10.12) \qquad \iint_{Q_T} (\gamma - \hat{\gamma})(\phi_t + e\Delta\phi) = 0$$

for any test function ϕ, where

$$e(x, t) = \begin{cases} \dfrac{u(x, t) - \hat{u}(x, t)}{\gamma - \hat{\gamma}} & \text{if } u \neq \hat{u} \\ 0 & \text{if } u = \hat{u}. \end{cases}$$

Notice that $0 \leq e \leq c$, c constant.

Choose a sequence $\tilde{e}_m \in C^\infty(Q_T)$, $\tilde{e}_m \geq 0$ such that

$$|\tilde{e}_m - e|_{L^2(Q_T)} \leq \frac{1}{m}, \qquad \tilde{e}_m \leq 1 + c$$

and take $e_m = \tilde{e}_m + 1/m$. Then

$$e_m \leq 2 + c,$$

$$(10.13) \qquad |e_m - e|_{L^2(Q_T)} \to 0,$$

$$\left| \frac{e}{e_m} \right|_{L^2(Q_T)} \leq C.$$

For any function $f \in C_0^\infty(Q_T)$, let ϕ_m be the solution of

$$(10.14) \qquad \frac{\partial \phi_m}{\partial t} + e_m\Delta\phi_m = f \qquad \text{in } Q_T,$$

$$(10.15) \qquad \phi_m = 0 \qquad \text{on } G(T) \text{ and on } \partial G \times (0, T).$$

Multiplying both sides of (10.14) by $\Delta\phi_m$ and integrating over Q_σ, we easily deduce that

$$(10.16) \qquad \int_{G(\sigma)} |\nabla\phi_m|^2 + \iint_{Q_\sigma} e_m(\Delta\phi_m)^2 \leq C \iint_{Q_\sigma} \frac{f^2}{e_m}.$$

We shall now estimate

$$J \equiv \iint_{Q_T} (\gamma - \hat{\gamma})f.$$

If we substitute the test function $\phi = \phi_m$ into (10.12), we obtain

$$(10.17) \quad \iint_{Q_T} (\hat{\gamma} - \gamma)\left(\frac{\partial \phi_m}{\partial t} + e_m\Delta\phi_m \right) = \iint_{Q_T} (\hat{\gamma} - \gamma)(e_m - e)\Delta\phi_m \equiv J_m$$

and

$$|J_m| \leqslant C \iint_{Q_T} |e_m - e| |\Delta\phi_m| \leqslant C \iint_{Q_T} \sqrt{e_m} |\sqrt{e_m} - \sqrt{e}| |\Delta\phi_m|$$

$$+ C \iint_{Q_T} \sqrt{e} |\sqrt{e_m} - \sqrt{e}| |\Delta\phi_m| \equiv J_{m,1} + J_{m,2}.$$

For any small $\eta > 0$, let

$$E_\eta = Q_T \cap \left\{ |\sqrt{e_m} - \sqrt{e}| > \eta \right\}.$$

Since $e_m \to e$ in L^2, for any small $\lambda > 0$ there is an $m_1 = m(\eta, \lambda)$ such that $\text{meas}(E_\eta) > \lambda$ if $m > m_1$. Setting $F_\eta = Q_T \setminus E_\eta$, we have

$$J_{m,1} \leqslant C\eta \iint_{F_\eta} \sqrt{e_m} |\Delta\phi_m| + C \iint_{E_\eta} \sqrt{e_m} |\Delta\phi_m|.$$

Thus

$$J_{m,1}^2 \leqslant C\eta^2 \iint_{Q_T} e_m |\Delta\phi_m|^2 + C\lambda \int_{E_\eta} e_m |\Delta\phi_m|^2$$

$$\leqslant C(\eta^2 + \lambda) \iint_{Q_T} \frac{f^2}{e_m} \qquad [\text{by } (10.16)].$$

Similarly,

$$J_{m,2}^2 \leqslant \left[C\eta^2 \iint_{Q_T} \frac{e}{e_m} + C \iint_{E_\eta} \frac{e}{e_m} \right] \iint_{Q_T} \frac{f^2}{e_m},$$

and by (10.13),

$$J_{m,2}^2 \leqslant \left(C\eta^2 + C\sqrt{\lambda} \right) \iint \frac{f^2}{e_m}.$$

We have thus proved that

(10.18) $$J_m^2 \leqslant \left(C\eta^2 + C\sqrt{\lambda} \right) \iint_{Q_T} \frac{f^2}{e_m}.$$

Choosing a sequence $f = f_j \in C_0^\infty(Q_T)$ converging to $\hat{u} - u$ in $L^2(Q_T)$, we then obtain (10.18) for $f = \hat{u} - u$. Upon substituting the result into (10.17), we find that

$$(10.19) \qquad \iint_{Q_T} (\hat{\gamma} - \gamma)(\hat{u} - u) \leqslant C\left(\eta^2 + \sqrt{\lambda}\right) \iint_{Q_T} \frac{(\hat{u} - u)^2}{e_m}$$

if m is sufficiently large, depending on η, λ.

Lemma 10.2. *If $f = \hat{u} - u$ and $f^2/e = 0$ by definition whenever $\hat{u} = u$, then*

$$\lim_{m \to \infty} \iint_{Q_T} \frac{f^2}{e_m} = \iint_{Q_T} \frac{f^2}{e}.$$

Suppose that the lemma is true. Since

$$\iint \frac{f^2}{e} = \int (\hat{\gamma} - \gamma)(\hat{u} - u),$$

we obtain from (10.19)

$$\iint_{Q_T} (\hat{\gamma} - \gamma)(\hat{u} - u) = 0,$$

which gives the assertion $\hat{u} = u$.

Proof of Lemma 10.2. For any small $\delta > 0$, set

$$\Omega_\delta = \{(x, t) \in Q_T; |\hat{u}(x, t) - u(x, t)| < \delta\},$$

$$\Sigma_\delta = Q_T \setminus \Omega_\delta,$$

$$\tilde{\Omega}_\delta = \{(x, t) \in Q_T; 0 < |\hat{u}(x, t) - u(x, t)| < \delta\}.$$

Then

$$(10.20) \qquad \iint_{Q_T} \frac{f^2}{e_m} = \iint_{\Omega_\delta} \frac{f^2}{e_m} + \iint_{\Sigma_\delta} \frac{f^2}{e_m}$$

and

$$(10.21) \qquad \iint_{\Omega_\delta} \frac{f^2}{e_m} = \iint_{\tilde{\Omega}_\delta} \frac{(\hat{u} - u)^2}{e_m} = \iint_{\tilde{\Omega}_\delta} \frac{\hat{u} - u}{a(\hat{u}) - a(u)} \frac{a(\hat{u}) - a(u)}{e_m}(\hat{u} - u)$$

$$\leqslant C \iint_{\tilde{\Omega}_\delta} \frac{e}{e_m} |\hat{u} - u| \leqslant C\delta \iint_{\tilde{\Omega}_\delta} \frac{e}{e_m} \leqslant C\delta,$$

where (10.13) was used. Similarly,

$$(10.22) \qquad \iint_{\Omega_\delta} \frac{f^2}{e} = \iint_{\tilde{\Omega}_\delta} (a(\hat{u}) - a(u))(\hat{u} - u) \leqslant C \iint_{\tilde{\Omega}_\delta} |\hat{u} - u| \leqslant C\delta.$$

Since Σ_δ is measurable, for a.a. $(x_0, t_0) \in \Sigma_\delta$,

$$\lim_{\varepsilon \to 0} \frac{|\Sigma_\delta \cap B_\varepsilon|}{|B_\varepsilon|} = 1,$$

where B_ε is a ball of radius ε centered at (x_0, t_0) and $|A|$ denotes the Lebesgue measure of A. If we set

$$E_i = \left\{ (x, t) \in \Sigma_\delta, |\Sigma_\delta \cap B_\varepsilon| > \tfrac{1}{2}|B_\varepsilon| \ \forall \ 0 < \varepsilon < \frac{1}{i} \right\},$$

thus $E_i \subset E_{i+1}$, and $\Sigma_\delta \setminus (\cup_i E_i)$ has measure zero. It follows that for any $\varepsilon > 0$ there is a $j = j(\varepsilon)$ such that meas $(\Sigma_\delta \setminus E_j) < \varepsilon$. Hence [see (10.21)]

$$(10.23) \qquad \iint_{\Sigma_\delta \setminus E_j} \frac{f^2}{e_m} \leqslant C \iint_{\Sigma_\delta \setminus E_j} \frac{e}{e_m} \leqslant C\sqrt{\varepsilon}.$$

Similarly,

$$(10.24) \qquad \iint_{\Sigma_\delta \setminus E_j} \frac{f^2}{e} \leqslant C\sqrt{\varepsilon}.$$

On the set $\Sigma_\delta \ e(x, t) \geqslant \lambda(\delta)$, where $\lambda(\delta) > 0$. We can choose the $\tilde{e}_m$ to be mollifiers of e_m and this certainly ensures that $e_m \geqslant \tilde{e}_m \geqslant \lambda(\delta)/2$ on E_j provided that m is large enough. Thus $0 \leqslant f^2/e_m \leqslant C$ on E_j. Since $e_m \to e$ a.e., the Lebesgue bounded convergence theorem gives

$$(10.25) \qquad \lim_{m \to \infty} \iint_{E_j} \frac{f^2}{e_m} = \iint_{E_j} \frac{f^2}{e}.$$

Combining (10.21)–(10.25), we obtain from (10.20)

$$\limsup_{m \to \infty} \left| \iint_{Q_T} \frac{f^2}{e_m} - \iint_{Q_T} \frac{f^2}{e} \right| \leqslant C(\delta + \sqrt{\varepsilon}).$$

Since δ and ε are arbitrary positive number, the lemma follows.

The method of proof of Theorem 10.1 has several by-products, which we briefly mention. First, a comparison result:

(10.26)   If (u, γ) and $(\hat{u}, \hat{\gamma})$ are solutions corresponding to
$(h, g), (\hat{h}, \hat{g})$, and if $h \geqslant \hat{h}, g \geqslant \hat{h}$, then $u \geqslant \hat{u}$.

The next result is concerned with stability. We suppose that (u, γ) and $(\hat{u}, \hat{\gamma})$ are solutions corresponding to (h, g) and $(\hat{h}, \hat{g})$, respectively, and that $\hat{g}$ can be extended into a function $\hat{\Psi}$ in Q_T as in (10.5). Then there holds:

(10.27)

$$\iint_{Q_T} \alpha(\hat{u} - u)^2 \leqslant \iint_{Q_T} (\hat{u} - u)(\hat{\gamma} - \gamma)$$

$$\leqslant C \iint_{G(0)} |\hat{h} - h|^2 + C \iint_{Q_T} \left[|\hat{\Psi} - \Psi|^2 + |\nabla_x(\hat{\Psi} - \Psi)|^2 \right],$$

where C is a constant independent of T.

Suppose in particular that for some function $g_\infty \in C^{2+\beta}(\partial G)$,

(10.28) $$\int_1^\infty \int_{\partial G} \left[|g - g_\infty|^2 + |\nabla_x(g - g_\infty)|^2 \right] dS_x \, dt < \infty$$

and let w be the solution of

(10.29)
$$\Delta w = 0 \qquad \text{in } G,$$
$$w = g_\infty \qquad \text{on } \partial G.$$

Then we can infer that

(10.30) $$\int_0^\infty \int_G |u(x, t) - w(x, t)|^2 \, dx \, dt < \infty.$$

Using (10.6), which is valid also for $T = \infty$, one can deduce from (10.30) that

(10.31) $$\int_G |u(x, t) - w(x)|^2 \, dx \to 0 \qquad \text{if } t \to \infty.$$

The two-phase Stefan problem can also be reformulated as follows. Find a solution u of

(10.32) $$\frac{\partial}{\partial t} a(u) = \Delta u \qquad \text{in } \mathcal{D}'(Q_T)$$

satisfying the appropriate initial boundary conditions in some weak sense; see Chapter 1, Section 9, Problem 3.

We introduce the inverse of $a(u)$:

$$(10.33) \qquad \phi(w) = \begin{cases} \dfrac{\alpha}{\alpha_2} + \dfrac{w}{\alpha_2} & \text{if } w < -\alpha \\[2mm] 0 & \text{if } -\alpha < w < 0 \\[2mm] \dfrac{w}{\alpha_1} & \text{if } w > 0. \end{cases}$$

Then (10.32) can be written in the form

$$(10.34) \qquad w_t - \Delta\phi(w) = 0 \qquad \text{in } \mathcal{D}'(Q_T).$$

Notice that $\phi(w)$ is Lipschitz continuous and $\phi'(w) = 0$ in some interval. The equation of gas in a porous medium has the form (10.34).

Theorem 10.3. *Let ϕ be given by* (10.33). *Then any solution of* (10.34) *is continuous in Q_T.*

Thus the temperature for the two-phase Stefan problem is a continuous function.

Theorem 10.3 is due to Caffarelli and Evans [57], who also extended it to a large class of functions $\phi(w)$, including $\phi(w) = |w|^m \operatorname{sgn} w$, $m > 1$. Alternative proofs were subsequently given by DiBenedetto [77a, b] and Ziemer [183], who further replaced Δ by a general elliptic operator (nonlinear, of divergence type). The proof of reference 77a and b gives continuity up to the boundary.

The proofs of references 57 and 77a and b establish also some modulus of continuity for the solution; however, this is substantially weaker then the moduli of continuity obtained in the special cases of $\phi(w)$ given by Theorems 9.1 of Chapter 2 and 3.1 of Chapter 5.

PROBLEMS

1. Prove (10.4).

2. Show that (10.9) is valid.
   [*Hint*: Compare with $\Delta w = 0$ in G, $w = \varepsilon$ on ∂G_1, $w = K$ on ∂G_2.]

3. Prove the comparison result (10.26).
   [*Hint*: Compare v_m with the corresponding $\hat{v}_m$.]

4. Establish (10.27) with C independent of T.

5. Prove that if (10.28) holds, then (i) (10.30) holds, (ii) (10.31) holds, and (iii) $\int_G |u(x, t) - w(x)|^p dx \to 0$ if $t \to \infty$ for any $p < 2n/(n - 2)$ if $n \geq 2$, and $u(x, t) \to w(x)$ uniformly in x if $n = 1$.

11. BIBLIOGRAPHICAL REMARKS

The porous-medium equation for one space dimension ($n = 1$) was studied by Oleinik, Kalashnikov, and Yui-Lin [152], who proved existence and uniqueness. Properties of the free-boundary curves were studied by Aronson [11a–c], Caffarelli and Friedman [58h] and Knerr [127]. For $n \geq 1$, existence and uniqueness were first proved by Sabinina [157]. More general uniqueness theorems were proved by Brezis and Crandall [47] and Vol'pert and Hudjaev [178]. The uniqueness proof given in Section 1 is based on reference 47. Corollary 1.9 is due to Benilan [30] and Veron [177]. The special solution in Problem 3 of Section 1 was discovered by Barenblatt [24].

Lemma 2.1 is due to Aronson and Benilan [12]. Problems 2 and 4 of Section 3 are taken from Knerr [127]. Theorem 4.6 was established by Friedman and Kamin [96]. All the other material of Sections 2 to 5 up to Theorem 5.6 is based on Caffarelli and Friedman [58g, k]. Corollary 5.7 was proved by Knerr [127]; [see also reference 11a–c]. Theorem 5.8 is due to Caffarelli and Friedman [58h]; the proof outlined in the problems is in part a simplification due to Aronson, Caffarelli, and Kamin [13]. In reference 13 it is shown that if $v_0(x) \leq \beta(b - x)^2$ for $x < 0$, $v_0(x) = 0$ for $x > 0$, $v_0(x)(b - x)^{-2} \to \beta$ as $x \uparrow b$, then $t_2^* = \beta/(2(m + 1))$. Problems 14 and 15 of Section 5 are due to reference 58h and Problem 16 is based on reference 11a.

Uniqueness of solutions for initial data which are measures was established by Pierre [153]; thus, in particular, the Barenblatt solution is the unique solution with Dirac initial data. Related results characterizing the class of initial measures for which solutions exist were more recently obtained by Aronson and Caffarelli [185] and by Benilan, Crandall and Pierre [186]. Asymptotic behavior (as $t \to \infty$) for solutions of the porous-medium equation in a bounded domain was studied by Aronson and Peletier [14] and Donnelly [78]. For other results on asymptotic behavior when $n = 1$, see Alikakos and Rostamian [4] and Vazquez [176].

For the equation $u_t = \Delta u^m$ with $m < 1$, see Berryman and Holland [39] and the references given there. In this case, if $m > (n - 2)/n$, then for any $T > 0$ there exists a unique solution u which is strictly positive everywhere (provided that $u_0 \geq 0$, $u_0 \not\equiv 0$). If $0 < m \leq (n - 2)/n$, then u is strictly positive in some finite interval $(0, T)$, and $u(x, T) \equiv 0$.

Theorem 6.1 is due to Alt [5c] and to Brezis, Kinderlehrer, and Stampacchia [50]. In reference 5c the penalization is more general and in itself represents a physical problem; capillarity is included. Theorem 6.1 extends to variable permeability $k(x, y)$ [replacing $\nabla^2 u$ by $\nabla(k \nabla u)$], but the assertion $\gamma_y \leq 0$ in

Theorem 7.1 requires, in general, the assumption that $k_y \leq 0$. Theorem 6.2 is due to Alt [5c].

The results of Section 7 are based primarily on Alt and Gilardi [8]; Lemma 7.4 and the uniqueness proof in Section 8 are taken from Carrillo and Chipot [67a]. Another proof of uniqueness, which applies to more general dams, was given by Alt and Gilardi [8] (see also reference 63d). In reference 8 the authors also study the behavior of the free-boundary curve as it hits the fixed boundary; Caffarelli and Gilardi [61] earlier studied the behavior of the free boundary as it hits S_2.

Problems 1 to 6 of Section 8 are based on Carrillo and Chipot [67b]. The result of Problem 7 in that section was established in reference 61, but the present proof is different.

The regularity of the free boundary in n dimensions (established in Section 9) is due to Alt [5b]; for an inhomogeneous dam it was established by Alt and Caffarelli [6b].

Theorem 10.1 and its consequences are base on Friedman [94b]. The uniqueness part was originally established by Kamenomostskaja (Kamin) [117], who also obtained an existence theorem by means of finite differences; see also Brezis [45(d)]. Damlanian [76(a)], and Alt and Luckhaus [9] for other methods. Very little is known about the free boundary for the two-phase Stefan problem if $n > 1$; for $n = 1$, see the references mentioned in Chapter 1, Section 12. Recently Marimanov [143] proved, for $n > 1$, the existence of a classical solution for small times.

REFERENCES

1. (a) A. Acker, A free boundary optimization problem, *SIAM J. Math. Anal.* **9**, 1179–1191 (1978).
   (b) A free boundary optimization problem II, *SIAM J. Math. Anal.* **11**, 201–209 (1980).
   (c) Interior free boundary problems for the Laplace equation, *Arch. Ration. Mech. Anal.* **75**, 157–168 (1981).
   (d) On the convexity of equilibrium plasma configurations, *Math. Methods Appl. Sci.* **3**, 435–443 (1981).

2. S. Agmon, *Lectures on Elliptic Boundary Value Problems*, Van Nostrand, Princeton, NJ, 1965.

3. (a) S. Agmon, A. Douglis, and L. Nirenberg, Estimates near the boundary for solutions of elliptic equations satisfying general boundary conditions I, *Commun. Pure Appl. Math.* **12**, 623–727 (1959).
   (b) Estimates near the boundary of elliptic partial differential equations satisfying general boundary conditions II, *Commun. Pure Appl. Math.* **17**, 35–92 (1964).

4. N. Alikakos and R. Rostamian, Large time behavior of solutions of Neumann boundary value problem for the porous medium equation, *Indiana Univ. Math. J.*, **30**, 749–785 (1981).

5. (a) H. W. Alt, A free boundary problem associated with the flow of ground water, *Arch. Ration. Mech. Anal.* **64**, 111–126 (1977).
   (b) The fluid flow through porous media. Regularity at the free surface, *Manuscr. Math.* **21**, 255–272 (1977).
   (c) Strömungen durch inhomogene poröse Medien mit freiem Rand, *J. Reine Angew. Math.* **305**, 89–115 (1979).

6 (a) H. W. Alt and L. A. Caffarelli, Existence and regularity for a minimum problem with free boundary, *J. Reine Angew. Math.* **105**, 105–144 (1981).
   (b) Fluid flow through inhomogeneous porous media. Regularity of the free boundary, to appear.

7 (a) H. W. Alt, L. A. Caffarelli, and A. Friedman, Axially symmetric jet flows, *Arch. Ration. Mech. Anal.*, to appear.
   (b) Asymmetric jet flows, *Commun. Pure Appl. Math.* **35**, 29–68 (1982).
   (c) Jet flows with gravity, *J. Reine Angew. Math.*, to appear.
   (d) Jets with two fluids I: One free boundary, to appear.
   (e) Jets with two fluids II: Two free boundaries, to appear.

8. H. W. Alt and G. Gilardi, Fluid flow through porous media. Behavior of the free boundary, *Ann. Scu. Norm. Sup. Pisa.*, to appear.

9. H. W. Alt and S. Luckhaus, Quasi-linear elliptic–parabolic differential equations, to appear.

10. (a) Ambrosetti and G. Mancini, On some free boundary problems, in *Recent Contributions to Nonlinear Partial Differential Equations*, H. Berestycki and H. Brezis, Eds., Pitman, London, 1981, pp. 24–35.

     (b) A free boundary problem and a related semilinear equation, *Nonlinear Anal. Theory, Methods Appl.* **4**, 909–915 (1980).

11. (a) D. G. Aronson, Regularity properties of flows through porous media, *SIAM J. Appl. Math.* **17**, 461–467 (1969).

     (b) Regularity properties of flows through porous media: the interface, *Arch. Ration. Mech. Anal.* **37**, 1–10 (1970).

     (c) Regularity properties of flows through porous media: a counter example, *SIAM J. Appl. Math.* **19**, 200–307 (1970).

12. D. G. Aronson and Ph. Benilan, Régularité des solutions de l'équation de millieux dans R^n, *C. R. Acad. Sci. Paris* **288**, 103–105 (1979).

13. D. G. Aronson, L. A. Caffarelli, and S. Kamin, How an initially stationary interface begins to move in porous medium flow, *SIAM J. Math. Analysis*, to appear.

14. D. G. Aronson and L. A. Peletier, Large time behavior of solutions of the porous medium equation in bounded domains, *J. Differ. Eq.* **39**, 378–412 (1981).

15. N. Aronszajn, A unique continuation theorem for solutions of elliptic partial differential equations or inequalities of second order, *J. Math. Pures Appl.* **36**, 235–249 (1957).

16. I. Athanasopoulos, Stability of the coincidence set for the Signorini problem, *Indiana Univ. Math. J.* **30**, 235–247 (1981).

17. J. F. G. Auchmuty, Existence of axisymmetric equilibrium figures, *Arch. Ration. Mech. Anal.* **65**, 249–261 (1977).

18. J. F. Auchmuty and R. Beals, Variational solutions of some nonlinear free boundary problems, *Arch. Ration. Mech. Anal.* **43**, 255–271 (1971).

19. (a) C. Baiocchi, Su un problema a frontiera libera connesso a questioni di idraulica, *Ann. Mat. Pura Appl.* **92**, 107–127 (1972).

     (b) Studio di un problema quasi-variazionale connesso a problemi di frontiera libera, *Boll. Unione Math. Ital.* **11**(4), 589–613 (1975).

20. C. Baiocchi and A. Capelo, Disequazioni variazionali e quasi variazionali, applicazioni a problemi di frontiera libera, I, II, *Quaderni dell' U.M.I*, Pitagora, Bologna, 1978.

21. C. Baiocchi, V.Comincioli, E. Magenes, and G. A. Pozzi, Free boundary problems in the theory of fluid flow through porous media: existence and uniqueness theorems, *Ann. Mat. Pura Appl.* **96**(4), 1–82 (1973).

22. C. Baiocchi and A. Friedman, A filtration problem in a porous medium with variable permeability, *Ann. Mat. Pura Appl.* **114**, 377–394 (1977).

23. V. Barbu, *Nonlinear Semigroups and Differential Equations in Banach Spaces*, Nordhoff, Leyden, 1976.

24. G. I. Barenblatt, On one class of exact solutions of the plane one-dimensional problem of unsteady filtration of a gas in a porous medium, *Akad. Nauk SSSR Prikl. Mat. Meh.*, **17**, 739–742 (1953).

25. G. K. Batchelor, *An Introduction to Fluid Dynamics*, Cambridge University Press, New York, 1974.

26. J. Bear, *Dynamics of Fluids in Porous Media*, Elsevier, New York, 1972.

27. H. Behnke and F. Sommer, *Theorie der analytischen Funktionen einer komplexen Veränderlichen*, Springer-Verlag, New York, 1976.

28. V. Benci, On a filtration problem through a porous medium, *Ann. Mat. Pura Appl.* **100**, 191–209 (1974).

29. R. Benguria, H. Brezis, and E. Lieb, Thomas–Fermi–Von Weizsacker Theory of Atoms and Molecules, *Commun. Math. Phys.* 79, 167–180 (1981).

30. Ph. Benilan, Opérateurs accrétifs et semi-groupes dans espaces L^p ($1 \leqslant p \leqslant \infty$), in *Functional Analysis and Numerical Analysis*, H. Fujita, Ed., Japan Society for the Promotion of Science, Tokyo, 1978, pp. 15–53.

31. Ph. Benilan and H. Brezis, Nonlinear problems related to the Thomas–Fermi equation, in preparation.

32. Ph. Benilan, H. Brezis, and M. Crandall, A semilinear equation in L^1, *Ann. Scu. Norm. Sup. Pisa* **2**, 523–555 (1975).

33. T. B. Benjamin, The alliance of practical and analytic insights into the nonlinear problems of fluid mechanics, in *Applications and Methods of Functional Analysis to Problems of Mechanics*, Lecture Notes in Mathematics No. 503, Springer-Verlag, New York, 1976, pp. 8–29.

34. A. Bensoussan and J. L. Lions, *Applications des inégalités variationnelles en controle stochastique*, Dunod, Paris, 1978.

35. (a) H. Berestycki and H. Brezis, Sur certains problèmes de frontière libre, *C. R. Acad. Sci. Paris* **283**, 1091–1094 (1976).
    (b) On a free boundary problem arising plasma physics, *Nonlinear Anal. Theory Methods Appl.* **4**, 415–436 (1980).

36. M. S. Berger and L. F. Fraenkel, Nonlinear desingularization in certain free boundary problems, *Commun. Math. Phys.* **77**, 149–172 (1980).

37. S. Bergman, *The Kernel Function and Conformal Mapping*, Math. Surveys 5, American Mathematical Society, Providence, RI, 1950.

38. S. Bergman and M. Schiffer, *Kernel Functions and Elliptic Differential Equations in Mathematical Physics*, Academic Press, New York, 1953.

39. J. G. Berryman and C. J. Holland, Stability of the separable solution for fast diffusion, *Arch. Ration. Mech. Anal.* **74**, 379–388 (1980).

40. L. Bers, F. John, and M. Schecter, *Partial Differential Equations*, Interscience, New York, 1964.

41. G. Birkhoff and E. H. Zarantonello, *Jets, Wakes, and Cavities*, Academic Press, New York, 1957.

42. P. Boieri and F. Gastaldi, Convexity of the free boundary in a filtration problem, *J. Differ. Eq.*, to appear.

43. J. -M. Bony, Principe du maximum dans les espaces de Sobolev, *C. R. Acad. Sci. Paris* **265**, 333–336 (1967).

44. H. J. Brascamp and E. H. Lieb, On extension of the Brunn–Minkowski and Prekoja–Leindler theorems, including inequalities for log concave functions, and with an application to the diffusion equation, *J. Funct. Anal.* **22**, 366–389 (1976).

45. (a) H. Brezis, Problèmes unilatéraux, *J. Math. Pures Appl.* **51**, 1–168 (1972).
    (b) Multiplicateur de Lagrange en torsion "elasto-plastique," *Arch. Ration. Mech. Anal.* **49**, 32–40 (1972).
    (c) Some variational problems of the Thomas–Fermi type, in *Variational Inequalities and Complimentarity Conditions*, R. W. Cottle, F. Gianessi, and J. L. Lions, Eds., Wiley-Interscience, New York, 1980, pp. 53–73.
    (d) On some degenerate nonlinear parabolic equations, *Nonlinear Functional Analysis*, *Symposia in Pure Mathematics*, Vol. 18, American Mathematics Society, Providence, R.I., 1970, pp. 28–38.

46. H. Brezis, L. A. Caffarelli, and A. Friedman, Reinforcement problems for elliptic equations and variational inequalities, *Ann. Math. Pura Appl.* **123**, 219–246 (1980).

47. H. Brezis and M. G. Crandall, Uniqueness of solutions of the initial-value problem for $u_t - \Delta\phi(u) = 0$, *J. Math. Pure Appl.* **58**, 153–163 (1979).

48. H. Brezis and A. Friedman, Estimates on the support of solutions of parabolic variational inequalities, *Ill. J. Math.* **20**, 82–97 (1976).

49. H. Brezis and D. Kinderlehrer, The smoothness of solutions to nonlinear variational inequalities, *Indiana Univ. Math. J.* **23**, 831–844 (1974).

50. H. Brezis, K. Kinderlehrer, and G. Stampacchia, Sur une nouvelle formulation du problème de l'écoulement à travers une digue, *C. R. Acad. Sci. Paris* **287**, 711–714 (1978).

51. H. Brezis and E. H. Lieb, Long range atomic potentials in Thomas–Fermi theory, *Commun. Math. Phys.* **65**, 231–246 (1979).

52. H. Brezis and Sibony, Équivalence de deux inéquations variationelles et applications, *Arch. Ration. Mech. Anal.* **41**, 254–265 (1971).

53. (a) H. Brezis and G. Stampacchia, Sur la régularité de la solution d'inéquations elliptiques, *Bull. Soc. Math. Fr.* **96**, 153–180 (1968).
    (b) Une nouvelle méthode pour l'étude d'écoulements stationnaires, *C. R. Acad. Sci. Paris* **276**, 129–132 (1973).
    (c) The hodograph method in fluid dynamics in the light of variational inequalities, *Arch. Ration. Mech. Anal.* **61**, 1–18 (1976).
    (d) Remarks on some fourth order variational inequalities, *Ann. Scu. Norm. Sup. Pisa* **4**(4), 363–371 (1977).

54. P. J. Budden and J. Norbury, Sluice-gate problems with gravity, *Math. Proc. Camb. Philos. Soc.* **81**, 157–175 (1977).

55. R. F. Byrd and M. D. Friedman, *Handbook of Elliptic Integrals for Engineers and Scientists*, Springer-Verlag, New York, 1971.

56. (a) L. A. Caffarelli, The smoothness of the free surface in a filtration problem, *Arch. Ration. Mech. Anal.* **63**, 77–86 (1976).
    (b) The regularity of free boundaries in higher dimension, *Acta Math.* **139**, 155–184 (1977).
    (c) Some aspects of the one-phase Stefan problem, *Indiana Univ. Math. J.* **27**, 73–77 (1978).
    (d) Further regularity for the Signorini problem, *Commun. P. D. E.* **4**, 1067–1076 (1979).
    (e) Compactness methods in free boundary problems, *Commun. P. D. E.*, **5**, 427–448 (1980).
    (f) A remark on the Hausdorff measure of a free boundary and the convergence of coincidence sets, *Boll. Unione Mat. Ital.* **18**, 1297–1299 (1981).

57. L. A. Caffarelli and L. C. Evans, Continuity of the temperature for the two phase Stefan problem, *Arch. Ration. Mech. Anal*, to appear.

58. (a) L. A. Caffarelli and A. Friedman, Asymptotic estimates for the dam problem with several layers, *Indiana Univ. Math. J.* **27**, 551–580 (1978).
    (b) The dam problem with two layers, *Arch. Ration. Mech. Anal.* **68**, 125–154 (1978).
    (c) Continuity of the temperature in the Stefan problem, *Indiana Univ. Math. J.* **28**, 53–70 (1979).
    (d) The obstacle problem for the biharmonic operator, *Anu. Scu. Norm. Sup. Pisa* **6**(4), 151–184 (1979).
    (e) The free boundary in the Thomas–Fermi atomic model, *J. Differ. Eq.* **32**, 335–356 (1979).
    (f) The free boundary for elastic–plastic problems, *Trans. Amer. Math. Soc.* **252**, 65–97 (1979).
    (g) Continuity of the density of a gas flow in a porous medium, *Trans. Amer. Math. Soc.* **252**, 99–113 (1979).
    (h) Regularity of the free boundary for the one dimensional flow of gas in a porous medium, *Amer. J. Math.* 1193–1218 (1979).
    (i) Regularity of the solution of the quasi-variational inequality for the impulse control problem, *Commun. P. D. E.* **4**, 279–291 (1979).
    (j) The shape of axisymmetric rotating fluids, *J. Funct. Anal.* **35**, 109–142 (1980).
    (k) Regularity of the free boundary of a gas flow in an *n*-dimensional proous medium, *Indiana Univ. Math. J.* **29**, 361–369 (1980).
    (l) A free boundary problem associated with a semilinear parabolic equation, *Commun. P. D. E.* **5**, 969–981 (1980).
    (m) Asymptotic estimates for the plasma problem, *Duke Math. J.* **47**, 705–742 (1980).

(n) Reinforcement problems in elasto-plasticity, *Rocky Mt. J. Math.* **19**, 155–184 (1980).

(o) Sequential analysis of several simple hypotheses for a diffusion process and the corresponding free boundary problem, *Pac. J. Math.* **93**, 49–94 (1981).

(p) Axially symmetric infinite cavities, *Indiana Univ. Math. J.*, **30**, 135–160 (1982).

(q) Unloading in the elastic–plastic torsion problem, *J. Differ. Eq.*, **41**, 186–217 (1981).

59. L. A. Caffarelli, A. Friedman, and G. A. Pozzi, Reflection methods in elastic–plastic torsion problems, *Indiana Univ. Math. J.* **29**, 205–228 (1980).

60. (a) L. A. Caffarelli, A. Friedman, and A. Torelli, The free boundary for a fourth order elliptic operator, *Ill. J. Math.* **25**, 402–422 (1981).

(b) The two-obstacle problem for the biharmonic operator, *Pacific J. Math* to appear.

61. L. A. Caffarelli and G. Gilardi, Monotonicity of the free boundary in the two-dimensional dam problem, *Ann. Scu. Norm. Sup. Pisa* **7**(4), 523–537 (1980).

62. L. A. Caffarelli and D. Kinderlehrer, Potential methods in variational inequalities, *J. Anal. Math.* **37**, 285–295 (1980).

63. (a) L. A. Caffarelli and N. M. Riviere, Smoothness and analyticity of free boundaries in variational inequalities, *Ann. Scu. Norm. Sup. Pisa* **3**(4), 289–310 (1976).

(b) Asymptotic behavior of free boundaries at their regular points, *Ann. Math.* **106**, 309–317 (1977).

(c) The smoothness of the elastic–plastic free boundary of a twisted bar, *Proc. Amer. Math. Soc.* **63**, 56–58 (1977).

(d) Existence and uniqueness for the problem of filtration through a porous medium, *Not. Amer. Math. Soc.* **24**, 576 (1977).

(e) On the Lipschitz character of the stress tensor when twisting an elastic plastic bar, *Arch. Ration. Mech. Anal.* **69**, 31–36 (1979).

64. L. A. Caffarelli and J. Spruck, Convexity properties of solutions of some classical variational problems, *Commun. in P.D.E.*, to appear.

65. G. Capriz and G. Cimatti, On some singular perturbation problems in the theory of lubrication, *Appl. Math. Optim.* **4**, 285–297 (1978).

66. L. Carelson, *Selected Topics on Exceptional Sets*, Van Nostrand, Princeton, NJ, 1967.

67. (a) J. Carrillo Menendez and M. Chipot, Sur l'unicité de la solution du problème de l'écoulement à travers une digue, *C. R. Acad. Sci. Paris* **292**, 191–194 (1981).

(b) On the dam problem, *J. Differ. Eq.*, to appear.

68. (a) D. S. Carter, Existence of a class of steady plane gravity flows, *Pac. J. Math.* **11**, 803–819 (1961).

(b) Uniqueness of a class of steady plane gravity flows, *Pac. J. Math.* **14**, 1173–1185 (1964).

69. S. Chandrasekhar, *Ellipsoidal Figures of Equilibrium*, Yale University Press, New Haven, CT, 1963.

70. M. Chicco, Principio di massimo per soluzioni di equazioni ellitiche del secondo ordine di tipo Cordes, *Ann. Mat. Pura Appl.*, **100**, 239–258 (1974).

71. (a) M. Chipot, Sur la régularité de la solution d'inéquations variationnelles elliptiques, *C. R. Acad. Sci. Paris* **288**, 543–546 (1979).

(b) Some remarks about an elastic–plastic torsion problem, *Nonlinear Anal. Theory Methods Appl* **3**, 261–269 (1979).

72. G. Cimatti, On a problem of the theory of lubrication governed by a variational inequality, *Appl. Math. Optim.* **3**, 227–242 (1977).

73. H. O. Cordes, Über die Bestimmtheit der Lösungen elliptischer Differentialgleichungen durch Anfangsvorgaben, *Nachr. Akad. Wiss. Goettingen Math.-Phys. K1. IIa*, pp. 239–258 (1956).

74. P. Courant and D. Hilbert, *Methods of Mathematical Physics*, Vol. 2: *Partial Differential Equations*, Interscience, New York, 1962.

75. C. W. Cryer, A proof of the convexity of the free boundary for porous flow through a rectangular dam using the maximum principle, *J. Inst. Math. Appl.* **25**, 111–121 (1980).

76. (a) A. Damlanian, Some results on the multiple-phase Stefan problem, *Commun. P. D. E.* **2**, 1017–1044 (1977).
    (b) Application de la dualité non convexe à un problème non linéaire à frontière libre (équilibre d'un plasma confiné), *C. R. Acad. Sci. Paris* **286**, 153–155 (1978).

77. (a) E. DiBenedetto, Continuity of weak solutions to certain singular parabolic equations, *Ann. Mat. Pura Appl.*, to appear.
    (b) Continuity of weak solutions to general porous media equations, *Indiana Univ. Math. J.*, to appear.

78. H. Donnely, Asymptotic expansions for the solution of certain nonlinear parabolic problems, I, II, *Math. Ann.* **254**, 189–200 (1980); II, *Commun. P. D. E.* **5**, 1251–1271 (1980).

79. N. Dunford and J. T. Schwartz, *Linear Operators*, Part I: *General Theory*, Interscience, New York, 1967.

80. G. Duvaut, Résolution d'un problème de Stefan (fusion d'un bloc de glace à zéro degré), *C. R. Acad. Sci. Paris* **276**, 1461–1463 (1973).

81. G. Duvaut and J. L. Lions, *Les inéquations en méchaniques et en physique*, Dunod, Paris, 1972.

82. A. Dvoretzky and P. Erdös, Some problems on random walk in space, *Proc. Second Berkeley Symp. Math. Stat. Prob.*, *1950*, University of California Press, Berkeley, 1951, 353–357.

83. C. M. Elliott and V. Janovsky, A variational inequality approach to Hele–Shaw flow with a moving boundary, *R. Soc. Edin.* **88A**, 93–107 (1981).

84. A. Erdélyi et al., *Higher Transcendental Functions, Bateman Manuscript*, Vol. 1, McGraw-Hill, New York, 1953.

85. (a) L. C. Evans, Application of nonlinear semigroup theory to certain partial differential equations, in *Nonlinear Evolution Equations*, M. Crandall, Ed., Academic Press, New York, 1978, pp. 163–188.
    (b) A second order elliptic equation with gradient constraint, *Commun. P. D. E.* **4**, 555–572 (1979).

86. L. C. Evans and R. F. Gariepy, Wiener's criterion for the heat equation, *Arch. Ration. Mech. Anal.*, to appear.

87. (a) L. C. Evans and B. F. Knerr, Instantaneous shrinking of the support of nonnegative solutions to certain nonlinear parabolic equations and variational inequalities, *Ill. J. Math.* **23**, 153–166 (1979).
    (b) Elastic–plastic plane stress problems, *Appl. Math. Optim.* **5**, 331–348 (1979).

88. E. F. Fabes and N. M. Riviere, L_p-estimates near the boundary for solutions of the Dirichlet problem, *Ann. Scu. Norm. Sup. Pisa* **24**, 491–553 (1970).

89. G. Fichera, Problemi elastostatici con vincoli unilaterali: il problema di Signorini con ambigue condizioni al contorno, *Atti Accad. Naz. Lincei Mem. Cl. Sci. Fis. Mat. Nat. Sez. Ia*, **7**(8), 91–140 (1963–1964).

90. R. Finn, Some theorems on discontinuity of plane fluid motion, *J. Anal. Math.* **4**, 246–291 (1954–1956).

91. (a) L. E. Fraenkel, On steady vortex rings of small cross-section in an ideal fluid, *Proc. R. Soc. Lond. A* **316**, 29–62 (1970).
    (b) Examples of steady vortex rings of small cross-section in an ideal fluid, *J. Fluid Mech.* **51**, 119–135 (1972).

92. L. E. Fraenkel and M. S. Berger, A global theory of steady vortex rings in an ideal fluid, *Acta Math.* **132**, 13–51 (1974).

93. (a) J. Frehse, On the regularity of the solution of a second order variational inequality, *Boll. Unione Mat. Ital.* **6**(4), 312–315 (1972).

   (b)   Zum Differenzierbarkeitsproblem bei Variationsungleichungen höherer Ordnung, *Hamburg Univ. Math. Sem., Abhand.* **36**, 140–149 (1971).

   (c)   On the regularity of the solution of the biharmonic variational inequality, *Manuscr. Math.* **9**, 91–103 (1973).

   (d)   On the Signorini problem and variational problems with thin obstacles, *Ann. Scu. Norm. Sup. Pisa* **4**(4), 343–362 (1977).

   (e)   On the smoothness of variational inequalities with obstacle, to appear in Proceedings of the Semist. Partial Differential Equation, Banach Center, Warsaw, 1978 (Bonn Technical Report 418, 1978).

94. (a)   A. Friedman, *Partial Differential Equations of Parabolic Type*, Prentice-Hall, Englewood Cliffs, NJ, 1964.

   (b)   The Stefan problem in several space variables, *Trans. Amer. Math. Soc.* **133**, 51–87 (1968).

   (c)   *Partial Differential Equations*, Holt, Reinhart and Winston, New York, 1969; reprinted by Krieger, New York, 1976.

   (d)   Stochastic games and variational inequalities, *Arch. Ration. Mech. Anal.* **51**, 321–326 (1973).

   (e)   Regularity theorems for variational inequalities in unbounded domains and applications to stopping time problems, *Arch. Ration. Mech. Anal.* **52**, 134–160 (1973).

   (f)   Parabolic variational inequalities in one space dimension and smoothness of the free boundary, *J. Funct. Anal.* **18**, 151–176 (1975).

   (g)   *Stochastic Differential Equations and Applications*, Vol. 2, Academic Press, New York, 1976.

   (h)   Analyticity of the free boundary for the Stefan problem, *Arch. Ration. Mech. Anal.* **61**, 97–125 (1976).

   (i)   A problem in hydraulics with non-monotone free boundary, *Indiana Univ. Math. J.* **25**, 577–592 (1976).

   (j)   Variational inequalities in sequential analysis, *SIAM J. Math. Anal.* **12**, 385–397 (1981).

   (k)   Asymptotic behavior for the free boundary of parabolic variational inequalities and applications to sequential analysis, *Ill. J. Math.* **26** (1982), to appear.

95. (a)   A. Friedman and R. Jensen, Elliptic quasi-variational inequalities and application to a non-stationary problem in hydraulics, *Ann. Scu. Norm. Sup. Pisa* **3**(4), 47–88 (1976).

   (b)   Convexity of the free boundary in the Stefan problem and in the dam problem, *Arch. Ration. Mech. Anal.* **67**, 1–24 (1978).

   (c)   A non-steady flow of liquid in a porous pipe with variable permeability, *J. Differ. Eq.* **34**, 1–24 (1979).

96.   A. Friedman and S. Kamin, The asymptotic behavior of gas in an *n*-dimensional porous medium, *Trans. Amer. Math. Soc.* **262**, 551–563 (1980).

97.   A. Friedman and D. Kinderlehrer, A one-phase Stefan problem, *Indiana Univ. Math. J.* **24**, 1005–1035 (1975).

98.   A. Friedman and G. A. Pozzi, The free boundary for elastic–plastic torsion problems, *Trans. Amer. Math. Soc.* **257**, 411–425 (1980).

99.   A. Friedman and A. Torelli, A free boundary problem connected with non-steady filtration in a porous media, *Nonlinear Analy. Theory Methods Appl.* **1**, 503–545 (1977); correction **2**, 513–518 (1978).

100. (a)   A. Friedman and B. Turkington, Asymptotic estimates for axisymmetric rotating fluid, *J. Funct. Anal.* **37**, 136–163 (1980).

   (b)   The oblateness of an axisymmetric rotating fluid, *Indiana Univ. Math. J.* **29**, 777–792 (1980).

   (c)   Existence and asymptotic estimates for vortex rings, *Trans. Amer. Math. Soc.* **268**, 1–37 (1981).

   (d)   Existence and dimensions of rotating white dwarfs, *J. Differ. Eq.* **42**, 414–437 (1981).

702

101. (a) R. M. Gabriel, An extended principle of the maximum for harmonic functions in 3-dimensions, *Lond. Math. Soc. J.* **30**, 388–501 (1955).
    (b) A result concerning convex level-surfaces of 3-dimensional harmonic functions, *Lond. Math. Soc. J.* **32**, 286–294 (1957).
    (c) Further results concerning the level surfaces of Green's function for a 3-dimensional convex domain (I), (II), *Lond. Math. Soc. J.* **32**, 295–302, 303–306 (1957).

102. P. R. Garabedian, H. Lewy, and M. Schiffer, Axially symmetric cavitational flow, *Ann. Math.* **56**, 560–602 (1952).

103. P. R. Garabedian and D. C. Spencer, Extremal methods in cavitational flows, *J. Ration. Mech. Anal.* **1**, 359–409 (1952).

104. (a) C. Gerhardt, Regularity of solutions on a nonlinear variational inequalities, *Arch. Ration. Mech. Anal.* **52**, 389–393 (1973).
    (b) Regularity of solutions of nonlinear variational inequalities with a gradient bound as constraint, *Arch. Ration. Mech. Anal.* **58**, 309–315 (1975).

105. B. Gidas, W. M. Ni, and L. Nirenberg, Symmetry and related topics via the maximum principle, *Commun. Math. Phys.* **68**, 209–243 (1979).

106. (a) G. Gilardi, Studio di una disequazione quasivariazionale relativa ad un problem di filtrazione in tre dimensione, *Ann. Math. Pura Appl.* **113**(4), 1–17 (1977).
    (b) A new approach to the evolution free boundary problem, *Commun. P.D.E.* **4**, 1099–1122 (1979).

107. (a) D. Gilbarg, Uniqueness of axially symmetric flows with free boundaries, *J. Ration. Mech. Anal.* **1**, 309–320 (1952).
    (b) Jets and cavities, in *Handbuch der Physik*, Vol. 9, Springer-Verlag, New York, 1960.

108. D. Gilbarg and L. Hörmander, Intermediate Schauder estimates, *Arch. Ration. Mech. Anal.* **74**, 297–318 (1980).

109. D. Gilbarg and N. S. Trudinger, *Elliptic Partial Differential Equations of Second Order*, Springer-Verlag, New York, 1977.

110. M. I. Gurevich, *Theory of Jets in Ideal Fluids*, Academic Press, New York, 1965.

111. Ei-Ichi Hanzawa, Classical solution of the Stefan problem, *Tohoku Math. J.* **33**, 297–335 (1981).

112. G. H. Hardy, J. E. Littlewood, and G. Polya, *Inequalities*, Cambridge University Press, Cambridge, 1934.

113. P. Hartman and A. Wintner, On the local behavior or non-parabolic partial differential equations, *Amer. J. Math.* **85**, 449–476 (1953).

114. W. K. Hayman, Some bounds for principal frequency, *Appl. Anal.* **7**, 247–254 (1978).

115. R. A. Hummel, The hodograph method for convex profiles, *Ann. Scu. Norm. Sup. Pisa*, to appear.

116. (a) R. Jensen, Structure of the non-monotone free boundaries in a filtration problem, *Indiana Univ. Math. J.* **26**, 1121–1135 (1977).
    (b) The smoothness of the free boundary in the Stefan problem with supercooled water, *Ill. J. Math.* **22**, 623–629 (1978).
    (c) Boundary regularity for variational inequalities, *Indiana Univ. Math. J.* **29**, 495–504 (1980).
    (d) Regularity for elasto-plastic type variational inequalities, to appear.

117. S. L. Kamenomostskaja, On Stefan's problem, *Mat. Sb.* **53**(95), 485–514 (1965).

118. (a) G. Keady, The jet from a horizontal slot at large Froude number, *Proc. Camb. Philos. Soc.* **73**, 515–529 (1973).
    (b) An elliptic boundary value problem with a discontinuous nonlinearity, *Proc. R. Soc. Edinb. A*, to appear.

119. (a) G. Keady and J. Norbury, The jet from a horizontal slot under gravity, *Proc. R. Soc. Lond. A* **344**, 471–478 (1975).
    (b) A semilinear elliptic eigenvalue problem, I, *Proc. R. Soc. Edinb. A*, **87**, 65–82 (1980).
    (c) A semilinear elliptic eigenvalue problem, II: The plasma problem, *Proc. R. Soc. Edinb. A*, **87**, 83–109 (1980).

120. O. D. Kellogg, *Foundations of Potential Theory*, Dover, New York, 1953.

121. J. T. Kemper, Temperatures in several variables: kernel functions, representations, and parabolic boundaries, *Trans. Amer. Math. Soc.* **167**, 243–262 (1972).

122. (a) D. Kinderlehrer, How a minimal surface leaves an obstacle, *Acta Math.* **130**, 221–242 (1973).
    (b) The free boundary determined by the solution to a differential equation, *Indiana Univ. Math. J.* **25**, 195–208 (1976).
    (c) Variational inequalities and free boundary problems, *Bull. Amer. Math. Soc.* **84**, 7–26 (1978).
    (d) The smoothness of the solution of the boundary solution obstacle problem, *J. Math. Pure Appl.* **60**, 193–212 (1981).
    (e) Remarks about Signorini's problem in linear elasticity, *Ann. Scu. Norm. Sup. Pisa*, to appear.
    (f) Estimates for the solution and its stability in Signorini's problem, *Appl. Math. Optimization*, to appear.

123. (a) D. Kinderlehrer and L. Nirenberg, Regularity in free boundary value problems, *Ann. Scu. Norm. Sup. Pisa* **4**(4), 373–391 (1977).
    (b) The smoothness of the free boundary in the one phase Stefan problem, *Commun. Pure Appl. Math.* **31**, 257–282 (1978).
    (c) Analyticity at the boundary of solutions of nonlinear second-order parabolic equations, *Commun. Pure Appl. Math.* **31**, 282–338 (1978).

124. (a) D. Kinderlehrer, N. Nirenberg, and J. Spruck, Regularity in elliptic free boundary problems, I, *J. Anal. Math.* **34**, 86–119 (1978).
    (b) Regularity in elliptic free boundary problems, II: Equations of higher order, *Ann. Scu. Norm. Sup. Pisa* **6**(4), 637–683 (1979).

125. D. Kinderlehrer and J. Spruck, Regularity in free boundary problems, *Ann. Scu. Norm. Sup. Pisa* **5**(4), 131–148 (1978).

126. (a) D. Kinderlehrer and G. Stampacchia, A free boundary problem in potential theory, *Ann. Inst. Fourier (Grenoble)*, **25**, 323–344 (1975).
    (b) *An Introduction to Variational Inequalities and Their Applications*, Academic Press, New York, 1980.

127. B. F. Knerr, The porous medium equation in one dimension, *Trans. Amer. Math. Soc.* **234**, 381–415 (1977).

128. V. A. Kondratev, Boundary problems for elliptic equations in a domain with conical or angular points, *Tr. Mosk. Math. Obshchest.* **16**, 209–292 (1967). [*Trans. Moscow Math. Soc.* **16**, 227–341 (1968) (American Mathematical Society, Providence, RI).]

129. (a) N. Korevaar, Capillary surface convexity above convex domains, *Indiana Univ. Math. J.*, to appear.
    (b) Convex solutions to nonlinear elliptic and parabolic boundary value problems, *Indiana Univ. Math. J.*, to appear.

130. O. A. Ladyženskaja, V. A. Solonnikov, and N. N. Ural'ceva, *Linear and Quasilinear Equations of Parabolic Type*, Amer. Math. Soc. Transl., American Mathematical Society, Providence, RI, 1968.

131. H. Lanchon, Torsion élastoplastique d'une barre cylindrique de section simplement ou multiplement connexe, *J. Mech.* **13**, 267–320 (1974).

132. N. S. Landkoff, *Foundations of Modern Potential Theory*, Springer-Verlag, New York, 1972.

133. B. E. Larock and R. L. Street, A nonlinear theory for a full cavitating hydrofoil in a transverse gravity field, *J. Fluid Mech.* **29**, Part 2, 317–336 (1967).

134. M. Lavrentieff, On certain properties of univalent functions and their application to wake theory, *Math. Sb.* **46**, 331–458 (1938).

135. J. Leray, Les problèmes de représentation conforme de Helmholtz I, II, *Commun. Math. Helv.* **8**, 149–180; 250–263 (1935).

136 J. L. Lewis, Capacity functions in convex rings, *Arch. Ration. Mech. Anal.* **66**, 201–224 (1979).

137 (a) H. Lewy, On a variational problem with inequalities on the boundary, *J. Math. Mech.* **17**, 861–884 (1968).

 (b) On a refinement of Evans law in potential theory, *Atti Accad. Naz. Lincei* **48**(8), 1–9 (1970).

 (c) On the coincidence set in variational inequalities, *J. Differ. Geom.* **6**, 497–501 (1972).

138. (a) H. Lewy and G. Stampacchia, On the regularity of the solution of a variational inequality, *Commun. Pure. Appl. Math.* **22**, 153–188 (1969).

 (b) On the smoothness of superharmonics which solve a minimum problem, *J. Anal. Math.* **23**, 224–236 (1970).

 (c) On existence and smoothness of solutions of some noncoercive variational inequalities, *Arch. Ration. Mech. Anal.* **41**, 241–253 (1971).

139. E. Lieb and B. Simon, The Thomas–Fermi theory of atoms, molecules and solids, *Adv. Math.* **23**, 22–116 (1977).

140. J. L. Lions, *Quelques méthodes de résolution de problèmes aux limites non linéaires*, Dunod, Paris, 1969.

141. J. L. Lions and G. Stampacchia, Variational inequalities, *Commun. Pure Appl. Math* **20**, 493–519 (1967).

142. W. Littman, A strong maximum principle for weakly *L*-subharmonic functions, *J. Math. Mech.* **8**, 761–770 (1959).

143. A. M. Marimanov, On a classical solution of the multidimensional Stefan problem for quasilinear parabolic equations, *Mat. Sb.* **112**, 170–192 (1980).

144. (a) P. van Moerbeke, An optimal stopping problem for linear reward, *Acta Math.* **132**, 1–41 (1974).

 (b) Optimal stopping and free boundary problems, *Arch. Ration. Mech. Anal.* **60**, 101–148 (1975/76).

145. C. B. Morrey, *Multiple Integrals in the Calculus of Variations*, Springer-Verlag, New York, 1966.

146. (a) J. Mossino, Some nonlinear problems involving a free boundary in plasma physics, *J. Differ. Eq.* **34**, 114–138 (1979).

 (b) A priori estimates for a model of Grad Mercier type in plasma confinement, *Appl. Anal.*, to appear.

147. J. Mossino and R. Temam, Directional derivative of the increasing rearrangement mapping and application to a queer differential equation in plasma physics, *Duke Math. J.*, **43**, 475–496 (1981).

148. B. Muchenhoupt and E. M. Stein, Classical expansions and their relation to conjugate harmonic functions, *Trans. Amer. Math. Soc.* **118**, 17–92 (1965).

149. M. Murthy and G. Stampacchia, A variational inequality with mixed boundary conditions, *Isr. J. Math.* **13**, 188–224 (1972).

150. W. M. Ni, On the existence of global vortex rings, *J. Anal. Math.* **37**, 208–247 (1980).

151. (a) J. Norbury, A steady vortex ring close to Hill's spherical vortex, *Proc. Camb. Philos. Soc.* **72**, 253–284 (1972).

 (b) A family of vortex rings, *J. Fluid Mech.* **57**, 417–431 (1973).

(c) Steady planar vortex pairs in an ideal fluid, *Commun. Pure Appl. Math.* **28**, 679–700 (1975).

152. O. A. Oleinik, A. S. Kalashnikov, and C. Yui-Lin, The Cauchy problem and boundary problems for equations of the type of nonstationary filtration, *Izv. Akad. Nauk SSSR, Ser. Mat.* **22**, 667–704 (1958).

153. M. Pierre, Uniqueness of the solutions of $u_t - \Delta\phi(u) = 0$ with initial data on a measure, *Nonlinear Analy. Theory Methods Appl.* **6**, 175–187 (1982).

154. G. Polya and G. Szegö, *Isoperimetric Inequalities in Mathematical Physics*, Princeton University Press, Princeton, NJ, 1951.

155. J. P. Puel, Sur un problème de valeur propre non linéaire et de frontière libre, *C. R. Acad. Sci. Paris* **284**, 861–863 (1977).

156. L. I. Rubinstein, *The Stefan Problem*, Am. Math. Soc., Transl., Vol. 27, American Mathematical Society, Providence, RI, 1971.

157 E. S. Sabinina, On the Cauchy problem for the equation of nonstationary gas filtration in several space variables, *Dokl. Akad. Nauk SSSR* **136**, 1034–1037 (1961).

158 (a) David G. Schaeffer, A stability theory for the obstacle problem, *Adv. Math.* **17**, 34–47 (1975).

(b) A new proof of the infinite differentiability of the free boundary in the Stefan problem, *J. Differ. Eq.* **20**, 266–269 (1976).

(c) Some examples of singularities in a free boundary, *Ann. Scu. Norm. Sup. Pisa* **4**(4), 131–144 (1977).

(d) Non-uniqueness in the equilibrium shape of a confined plasma, *Commun. P.D.E.* **2**, 587–600 (1977).

159. L. Schwartz, *Théorie des distributions*, Vol. 1, Hermann, Paris, 1957.

160. M. Sermange, Une méthode numérique en bifurcations—application à un problème à frontière libre de la physique des plasmas, *Appl. Math. Optim.* **5**, 125–151 (1979).

161. (a) J. Serrin, Existence theorems for some hydrodynamical free boundary problems, *J. Ration. Mech. Anal.* **1**, 1–48 (1952).

(b) Uniqueness theorems for two free boundary problems, *Amer. J. Math.* **74**, 492–506 (1952).

(c) Two hydrodynamic comparison theorems, *J. Ration. Mech. Anal.* **1**, 563–572 (1952).

(d) On plane and axially symmetric free boundary problems, *J. Ration. Mech. Anal.* **2**, 563–575 (1953).

(e) On the Harnack inequality for linear elliptic equations, *J. Anal. Math.* **4**, 292–308 (1955–1956).

(f) The problem of Dirichlet for quasilinear elliptic differential equations with many independent variables, *Philos. Trans. R. Soc. Lond. Ser. A*, **264**, 413–496 (1969).

162. (a) E. Shimborsky, Variational methods applied to the study of symmetric flows in laval nozzles, *Commun. P.D.E.* **4**, 41–77 (1979).

(b) Variational inequalities arising in the theory of two dimensional potential flows, *Nonlinear Anal. Theory Methods Appl.* **5**, 433–444 (1981).

163. S. L. Sobolev, On a theorem of functional analysis, *Amer. Math. Soc. Transl.*, Ser. 2, **34**, 39–68 (1963).

164. V. A. Solonnikov, A priori estimates for equations of second order parabolic type, *Tr. Mat. Inst. Steklov* **10**, 133–142 (1964). [*Amer. Math. Soc. Transl.*, **65**, 51–137 (1967).]

165. F. Spitzer, Some theorems concerning two dimensional Brownian motion, *Trans. Amer. Math. Soc.* **87**, 187–197 (1958).

166. I. Stakgold and L. E. Payne, Nonlinear problems in nuclear reactor analysis, in *Nonlinear Problems in the Physical Sciences and Biology*, Lecture Notes in Mathematics No. 322, Springer-Verlag, New York, 1973, pp. 298–307.

167. (a) G. Stampacchia, Variational inequalities, theory and applications of monotone opera-tors, *Proc. NATO Adv. Study Inst.*, Oderisi-Gubbio, 1959.
     (b) On the filtration of liquid through a porous medium, *Usp. Math. Nauk SSSR* **29**(4), 89–101 (1974).

168. E. M. Stein and G. Weiss, *Introduction to Fourier Analysis on Euclidean Spaces*, Princeton University Press, Princeton, NJ, 1971.

169. G. Talenti, Best constant in Sobolev inequality, *Ann. Mat. Pura Appl.* **110**(4), 353–372 (1976).

170. (a) R. Temam, A non-linear eigenvalue problem: equilibrium shape of a confined plasma, *Arch. Ration. Mech. Anal.* **60**, 51–73 (1975).
     (b) Applications de l'analyse convexe au calcul des variations, in *Nonlinear Operators and the Calculus of Variations, Bruxelles, 1975*, Lecture Notes in Math. No. 543, Springer-Verlag, New York, 1976, pp. 208–237.
     (c) Remarks on a free boundary problem arising in plasma physics, *Commun. P.D.E.* **2**, 563–585 (1977).

171. (a) David F. Tepper, Free-boundary problem, *SIAM J. Math. Anal.* **5**, 841–846 (1974).
     (b) On a free-boundary problem, the starlike case, *SIAM J. Math. Anal.* **6**, 503–505 (1975).

172. (a) T. W. Ting, Elastic–plastic torsion of simply connected cylindrical bars, *Indiana Univ. Math. J.* **20**, 1047–1076 (1971).
     (b) Elastic–plastic torsion problem over multiply connected domains, *Ann. Scu. Norm. Sup. Pisa* **4**(4), 291–312 (1977).
     (c) The unloading problem for severely twisted bars, *Pac. J. Math.* **64**, 559–582 (1976).
     (d) The repeated loading–unloading process of elastic–plastic torsion of solid bars, *Ann. Mat. Pura Appl.* **119**, 333–378 (1979).

173. F. Tomarelli, Un problème de fluidodynamique avec les inéquations variationnelles, *C. R. Acad. Sci. Paris* **286**, 999–1002 (1978).

174. (a) A. Torelli, Su un problema non lineare con una condizione di evoluzione sulla frontiera, *Ann. Mat. Pura Appl.* **112**(4), 91–106 (1977).
     (b) Existence and uniqueness of the solution of a non steady free boundary problem, *Boll. Unione Mat. Ital.* **14-B**(5), 423–466 (1977).
     (c) On a free boundary value problem connected with a non steady filtration phenomenon, *Ann. Scu. Norm. Sup. Pisa* **4**(4), 33–59 (1977).

175. (a) B. Turkington, Steady vortex flows in two dimensions, I, *Commun. P.D.E.*, to appear.
     (b) On steady vortex flow in two dimensions, II, *Commun. P.D.E.*, to appear.

176. J. L. Vasquez, Asymptotic behavior and propagation properties of the one-dimensional flow of gas in a porous medium, to appear.

177. L. Veron, Effets régularisants de semi-groupes non linéaires dans des espaces de Banach, *Ann. Fac. Sci. Tolouse*, to appear.

178. A. I. Vol'pert and S. Hudjaev, Cauchy's problem for degenerate second order quasi-linear parabolic equations, *Mat. Sb.* **78**(120), 374–396 (1969). [*Math. USSR Sb.* **7**, 365–387 (1969).]

179. A. Weinstein, Zur Theorie der Flüssigkeitsstrahlen, *Math. Zeit.* **31**, 424–433 (1929).

180. K. O. Widman, Inequalities for the Green function and boundary continuity of the gradient of solutions of elliptic differential equations, *Math. Scand.* **21**, 17–37 (1967).

181. M. Wiegner, The $C^{1,1}$ character of solutions of second-order elliptic equations with gradient constraint, *Commun. P.D.E* **6**, 361–371 (1981).

182. T. Y. Wu, Cavity and wakes flows, in *Annual Review of Fluid Mechanics*, Vol. 4, George Banta Co., Palo Alto, CA, 1972.

REFERENCES

183. W. P. Ziemer, Interior and boundary continuity of weak solutions of degenerate equations, *Trans. Amer. Math. Soc.*, to appear.

184. J. Frehse and U. Mosco, Irregular obstacles and quasi-variational inequalities of stochastic impulse control, *Ann. Scu. Norm. Sup. Pisa*, to appear.

185. D. G. Aronson and L. A. Caffarelli, The initial trace of a solution of the porous medium equation, to appear.

186. Ph. Benilan, M. G. Crandall and M. Pierre, Solutions of the porous medium equation in R^N under optimal conditions of initial values, to appear.

187. (a) D. Phillips, A minimization problem and the regularity of solutions in the presence of a free boundary, *Indiana Univ. Math. J.*, to appear.
    (b) Hausdorff measure estimates of a free boundary for a minimum problem, *Commun. in P.D.E.*, to appear.

188. B. Schild, A regularity result for polyharmonic variational inequalities with thin obstacles, to appear.

189. (a) V. M. Isakov, Inverse theorems concerning the smoothness of potentials, *Differential Equations* **11**, 50–56 (1974) [translated from Russian].
    (b) Analyticity of solutions of nonlinear transmission problems, *Differential Equations* **12**, 41–47 (1976) [translated from Russian].

190. H. Brezis and G. Duvaut, Écoulements avec sillages autour d'un profil symétrique sans incidence, *C. R. Acad. Sci. Paris*, **276**, 875–878 (1973).

191. H. Ishii and S. Koike, Boundary regularity and uniqueness for an elliptic equation with gradient constraint, Commun. in *P. D. E.*, to appear.

192. C. M. Elliott and J. R. Ockendon, *Weak and Variational Methods for Moving Boundary Problems*, Pitman, London, 1982.

193. (a) P. L. Lions, Minimization problems in $L^1(\mathbb{R}^N)$, *J. Funct. Anal.* **41**, 236–275 (1981).
    (b) Principe de concentration-compacité en calcul de variations, *C. R. Acad. Sci. Paris* (1982), to appear.

194. H. Berestycki and P. L. Lions, A direct variational approach to the problem of vortex rings in an ideal fluid, in preparation.

INDEX